SENSORS AND ENVIRONMENTAL APPLICATIONS OF REMOTE SENSING

EARSeL

PROCEEDINGS OF THE 14TH EARSeL SYMPOSIUM
GÖTEBORG/SWEDEN/6-8 JUNE 1994

Sensors and Environmental Applications of Remote Sensing

Edited by
JAN ASKNE
Chalmers University of Technology, Göteborg, Sweden

A.A. BALKEMA/ROTTERDAM/BROOKFIELD/1995

ACKNOWLEDGEMENT

For making the conference and this publication possible many persons have contributed and been very helpful, too many to be named. The help of Mrs M.Godefroy, EARSeL Secretariat, is particularly acknowledged.

Cover picture: SPOT image over Göteborg, Sweden

The texts of the various papers in this volume were set individually by typists under the supervision of each of the authors concerned.

Published by
A.A.Balkema, P.O.Box 1675, 3000 BR Rotterdam, Netherlands (Fax: +31.10.413.5947)
A.A.Balkema Publishers, Old Post Road, Brookfield, VT 05036, USA (Fax: 802.276.3837)

ISBN 90 5410 524 0

Printed in the Netherlands

Sensors and Environmental Applications of Remote Sensing, Askne (ed.) © 1995 Balkema, Rotterdam, ISBN 90 5410 524 0

Table of contents

Preface XI

Welcome address

ESA earth observation programmes and plans for the future 3
L. Emiliani

Plenary session

The activities of the Institute for Remote Sensing Applications, Joint Research Centre, Ispra 9
J. Mégier

R&D-needs in a rapidly changing marketplace 15
C.-G. Borg

Land applications

CORINE Land Cover – A pilot project in Sweden 19
E. Ahlcrona

Effect on agricultural area estimates of using sample units shared by more than one CORINE GIS Stratum 23
F. J. Igualada

Image processing for CORINE Land Cover – Hungary 31
G. Büttner & G. Maucha

Classification of agricultural crops in Denmark using ERS-1 SAR and SPOT imagery 37
I. Sandholt, B. Fog, J. N. Poulsen, M. Stjernholm & H. Skriver

The use of ERS-1 data in JRC's MARS project 45
H.-G. Kohl, E. Nezry, M. Mróz & H. de Groof

Towards a satellite-observatory for mapping and monitoring the degradation of Mediterranean ecosystems 53
J. Hill, S. Sommer, W. Mehl & J. Mégier

Assessment of the potential of optical and microwave satellite remote sensing techniques for land surface monitoring of the impact of land reform 63
P.Winkler, P.Somogyi & Gy.Zelei

Land transformation analysis in Kassala Province, Eastern Sudan 69
H.Larsson

A strategy for spectral time-profiling in coniferous forest using Landsat TM 75
P.Syrén

Terrestrial net primary production of the Baltic drainage basin, based on global vegetation index, spatial stratification and local growing season estimates 79
K.B.-E.Olsson

Drought monitoring in Spain using satellite remote sensing 87
F.González Alonso, J.M.Cuevas, J.L.Casanova, A.Calle & P.Illera

Application of airborne polarimetric SAR data for forest classification 91
B.Koch, B.Förster & G.Hunner

Retrieval of forest biomass using SAR 97
H.Israelsson & J.Askne

Estimation of soil moisture from the MAESTRO-1 SAR data of Flevoland 103
Jiankang Ji, H.Skriver & P.Gudmandsen

Use of ERS-1 SAR data for soil moisture assessment 111
K.Dabrowska-Zielinska, M.Gruszczynska, M.Janowska & K.Stankiewicz

The use of ERS-1 SAR imagery for groundwater assessment 117
D.Wladis & M.S.Rosenbaum

Change detection in Swedish land cover from a year of ERS-1 SAR data 123
D.B.Michelson & J.W.Seaquist

The Eurasian large area ecosystem mapping experiment 131
S.Folving & J.Mégier

EC initiatives on European forest ecosystems mapping and forest statistics 135
S.Folving, P.Kennedy & N.McCormick

Improved forest mapping by combining corrections of atmospheric and topographic effects in Landsat TM imagery 143
J.Hill, W.Mehl & V.Radeloff

Cartographic aspects

Maps and remote sensing in mountainous areas of Third World countries: Example of the upper Mustang district in Nepal 155
R.Kostka & W.Krämer

Multisource data integration with an integrated and unified geometric modelling 163
T.Toutin

Oceans

Mesoscale and submesoscale atmospheric phenomena studied by the synthetic aperture radar aboard the ERS-1 satellite 177
W. Alpers

An operational, satellite-based, European oil slicks monitoring system 183
D. R. Sloggett & I. S. Jory

Gravity profiles from satellite altimetry compared with ice gravity profiles in the Gulf of Bothnia 189
J. Noréus, K. Hayling & M. Nyborg

Sea ice

Sea ice – A remote sensing challenge 197
P. Gudmandsen

Arctic sea ice climate system parameters as observed by synthetic aperture radar 203
J. Askne

Classification of ice properties in the Arctic during freeze-up using ERS-1 SAR and SSM/I 211
M. I. Pettersson, J. Askne & D. J. Cavalieri

On the accuracy of estimation of ice type concentration from ERS-1 SAR images 219
H. Skriver

Active microwave signatures of sea ice in the Greenland Sea during summer 1993 227
B. B. Thomsen, L. T. Pedersen & H. Skriver

Influence of ice and snow properties on ice ridge mapping by SAR 235
A. Carlström, L. M. H. Ulander & W. Dierking

An analysis of daily ice motion over the Yermak Plateau 243
Y. Sun & J. Askne

Cryosphere

Real time runoff forecasts for two hydroelectric stations based on satellite snow cover monitoring 253
K. Seidel, W. Brüsch, C. Steinmeier, J. Martinec & J. Wiedemeier

Snow cover monitoring in a high altitude mountain using SPOT and ERS data 259
F. Cauneau, I. Pons, L. Castagnas & M. Albuisson

The discrimination of glacier facies using multitemporal ERS-1 SAR data 263
G. J. Marshall, W. G. Rees & J. A. Dowdeswell

Marine and inland waters

Remote sensing and water quality in the Netherlands 273
R. Allewijn, J. M. M. Kokke, A. G. Dekker & R. J. Vos

Estimation of water quality by passive optical remote measurements 281
T. Kutser, H. Arst & S. Mäekivi

Operational uses of ocean colour data – Perspectives for the OCTOPUS Programme 289
D. R. Sloggett, M. Srokosz, J. Aiken & S. Boxall

Use of fluorescence based estimates of phytoplankton production to validate primary production estimates based on remote sensing techniques 305
M. Olaizola, D. W. van der Linde & P. M. Schlittenhardt

Bathymetry measurement in shallow water areas using airborne spectrometers 311
J. Bijaoui, S. Jullien & F. Cauneau

Interannual variations in the extent of surface cyanobacterial accumulations in the Baltic Sea detected by satellites 315
O. Rud & M. Kahru

Thermal plume monitoring from ATM data 319
R. A. Vaughan, Chan-Chuan Wen & W. Mansell

Atmosphere

Determination of true surface temperature using TOVS and AVHRR data 329
M. Arbelo, F. Herrera, F. J. Expósito & V. Caselles

Comparison between the aerosols optical depth data obtained by the NOAA satellites and the Optronic OL752 spectroradiometer 333
F. J. Expósito, M. Arbelo, F. Herrera, J. P. Díaz & A. Díaz

An algorithm for the reconstruction of columnar aerosol size distribution from spaceborne spectrometer data 337
A. Bartoloni, G. Milillo & M. Mochi

Analysis of cloud type and distribution using NOAA-AVHRR-APT data and ground observations in Central Europe 343
U. Heinzmann

Synergy of Meteosat and NOAA/AVHRR data for the assessment of the role of clouds upon the air-sea sensible heat flux in the Northeastern Tropical Atlantic Ocean 349
O. Bauer & F. Cauneau

Hazards

Volcanic gas emission studies using optical remote sensing techniques 357
H. Edner, P. Ragnarson, S. Svanberg, E. Wallinder, R. Ferrara, R. Cioni, B. Raco & G. Taddeucci

The Near Ultraviolet Spectrometer (NUVS): A new instrument for volcanic sulphur dioxide measurements 363
S. Mattei, S. Panachia, V. Schena & G. Hoover

Estimating forest fire danger from AVHRR data 371
M. P. Martín, L. Domínguez & E. Chuvieco

Fire monitoring in Spain using NOAA thermal data and NDVI 377
P. Illera, A. Fernández & J.A. Delgado

Emergency planning outside major hazard installations through territorial monitoring by means of satellite remote sensing 385
M. Ippoliti, L. Lepore, U. Poli, M. Battista, G. Morrone & N. Napolitano

Sensors – Optical technique

Reduction of high resolution spectra: Application to characterization of salinity status of soils 393
L. Pásztor & F. Csillag

An improved method of estimating measurement error in remotely sensed images 399
P. M. Atkinson & A. R. Harrison

Specially cultivated fields as test targets for satellite imagery 407
R. Kuittinen & J. Laaksonen

Sensors – Microwave-technique

The ENVISAT Advanced Synthetic Aperture Radar (ASAR) 417
S. Karnevi

The calibration of the ERS-1 synthetic aperture radar using UK-PAF imagery 423
P. J. Meadows

ERS-1 Scatterometer data processing over the Mediterranean sea: Algorithm development and validation activity at I-PAF 431
A. Bartoloni, C. D'Amelio, G. Milillo & F. Nirchio

Use of C-band active radar calibrators in spaceborne and airborne SAR missions 439
G. De Carolis, S. Ponte, F. Posa & V. D. Schena

CARABAS VHF synthetic aperture radar: Image quality and radiometric calibration 447
L. M. H. Ulander & H. Hellsten

Remote sensing methodology

Neural networks in remote sensing multisensor data processing 457
V. Cappellini, A. Chiuderi & S. Fini

Comparison of neural network and maximum likelihood classifications in an urban area 463
L. Halounová

Scaling in the fractal spatial structure of environment and implementation for imagery interpretation 469
L. Vasiliev

The temporal observation of natural phenomenon – Preliminary report 475
B. Bullen, R. Vaughan, G. Wright & D. Miller

Training in remote sensing of the environment, from research to application: The TELSAT project 481
M. Ippoliti, L. Lepore, U. Poli, P. Bragatto, N.A. Minciotti & S. Mugnoli

Author index 485

Colour plates 489

Sensors and Environmental Applications of Remote Sensing, Askne (ed.) © 1995 Balkema, Rotterdam, ISBN 90 5410 524 0

Preface

The 14th Symposium of the European Association of Remote Sensing Laboratories, EARSeL, was held in Göteborg, Sweden 6-8 June 1994 with the theme of 'Sensors and Environmental Applications'. This was the first time the symposium was held in Sweden. The symposium included roughly one hundred participants and most of the papers presented are included in this volume.

The cover of this book is a composition of a SPOT satellite image over Göteborg made by the Swedish Space Corporation. In the context of sensors and environmental applications of remote sensing it may symbolise the transition from local to global aspects of remote sensing and the need to have sensors covering those different scales. These aspects are particularly important for environmental studies.

A main theme of the development of remote sensing for a long time has been the environmental aspects and the importance has increased tremendously during recent years. On the other hand all efforts using remote sensing are based on the ability to develop sensors which are sensitive to the properties of interest and to signal processing algorithms to extract the information. We see relatively few examples when a single sensor measures something from space unambiguously and exactly. Instead in most cases we must rely on a combination of measurements from different sensors and with various types of extra information such as topography, climatology etc. using information systems. This is why the development of sensors and environmental applications must go hand in hand.

M.von Glehn mentioned in her welcome address how a strong competence in remote sensing has been built up in Sweden with the support of the Swedish National Space Board, and Sweden has thus developed a capacity to contribute to global environmental programmes. The plans for establishing a 'Centre for Environmental Data from Satellites' in Kiruna are to be seen in this context. ODIN with a planned launch in 1997 is a Swedish initiative for a spectroscopic research satellite which will be the first to perform limb scanning measurements in the submillimetre wavelength region to investigate ozone chemistry relations. It will be unique in the sense that it is designed to conduct both atmospheric and astronomical studies using essentially the same instrumentation. The satellite project will be carried out in co-operation with Canada, France, and Finland.

Papers are grouped in this book according to broad application areas, such as land applications and cartography, oceans, sea ice, marine and inland waters, atmosphere, and hazard detection. On the sensor side properties on developments of optical and microwave sensors are presented as well as remote sensing methodology.

A workshop on Topography from Space followed the Symposium and the workshop papers will, after review, be published in the EARSeL journal 'Advances of Remote Sensing'. Aspects of the workshop theme were covered in a common part of the symposium and workshop and these are included in the present volume.

The introduction of the symposium included welcome addresses and overviews given by a number of people, J. Kahn from the Swedish Ministry of Environment and Natural Resources, M. von Glehn, Swedish National Space Board, G. Konecny, Chairman of EARSeL. G. Duchossois presented an overview of the ERS-1/2 programme and its environmental applications. Three more overviews were given and these are included more or less verbatim in the present volume, by L. Emiliani, who is the new director of the ESA Directorate for Observation of the Earth and its Environment regarding ongoing programmes and future plans of the agency, J. Mégier on behalf of the new director of the Institute of Remote Sensing Applications, Joint Research Centre presented the activities of the institute, and by C.-G. Borg, from the Swedish Space Corporation who presented a personal view as seen from the remote sensing marketplace.

Scientific keynote papers were presented by P. Gudmandsen on Sea ice – A remote sensing challenge, and by S. Svanberg on Laser radar techniques for monitoring of atmospheric pollutants and vegetation status* (not included in the proceedings). Remote sensing of water quality in the Netherlands was presented by R. Allewijn, Mesoscale atmospheric phenomena by W. Alpers, and the ENVISAT ASAR by S. Karnevi. This volume includes a coverage of remote sensing activities in most countries of Europe including many of the eastern countries.

The environmental issues are a challenge for our generation as we may live at a unique turning point of our world and just as Marianne von Glehn finished her welcome address it may be appropriate to finish with citations from 'Aniara', written by the Swedish 1974 Nobel Prize laureate, Harry Martinsson:

'We come from Earth
from Dorisland
the jewel
in our solar system
the only orb
where Life obtained
a land of milk
and honey.
Describe the landscapes
found back there,
the days those dawns
could breed.'

Jan Askne
Editor

*S. Svanberg, Differential absorption lidar (dial), Chapter 3 in Air Monitoring, Spectroscopic Techniques, Editor M. Sigrist, Wiley, New York, 1994.

Welcome address

Sensors and Environmental Applications of Remote Sensing, Askne (ed.) © 1995 Balkema, Rotterdam, ISBN 90 5410 524 0

ESA earth observation programmes and plans for the future

Lanfranco Emiliani
Observation of the Earth and its Environment, European Space Agency, Paris, France

ABSTRACT: A review is given on what has been happening in the ESA Earth Observation directorate over the past year by reviewing ongoing programmes and outline plans for the immediate future.

It gives me great pleasure to address you on this occasion. This is my first opportunity, since I took office, to address the EARSeL Symposium and what I should like to do is to review what has been happening in the ESA Earth Observation directorate over the past year. I shall begin by reviewing our ongoing programmes and then outline our plans for the immediate future.

As you know, ERS-1 was launched about three years ago and has been operating nominally since then. This mission is unique in many aspects due to its instrumentation which allows an all-weather as well as day-and-night observing capability. Its success can be measured by the widespread interest it has aroused in the user community, both in terms of science and applications. The following figures are relevant in this respect :

- 150 contributions to the first ERS-1 Symposium (1992) in Cannes (450 participants)
- 270 contributions to the second ERS-1 Symposium (1993) in Hamburg (700 participants)
- About 150 pilot projects covering many applications (e.g. agriculture, forestry, offshore applications)
- Over 400 proposals coming from all over the world in response to a second AO (covering also ERS-2).

Continuity of ERS-1 data will be ensured by ERS-2, which is scheduled for launch at the end of this year or early next year.

Tandem operations of ERS-1 and ERS-2 would bring numerous benefits amongst which the most significant are:

- improved conditions for SAR interferometry
- increased space/time sampling

This is a matter which is being addressed to our Programme Board in order to get their approval, together with the relevant funding authority.

ERS-2 will be followed by the ENVISAT-1 mission, currently scheduled for launch at the end of 1998, again to ensure continuity of data.

As a prime objective ENVISAT-1 will focus on making a significant contribution to environmental studies, notably in the areas of atmospheric chemistry and ocean/ice studies, including marine biology of the ocean surface. Additional objectives include enhancement of the capability to monitor and manage the Earth's resources and a contribution to a better understanding of land processes. Furthermore, ENVISAT-1 should capitalise on the experience gained from the ERS missions, extending and consolidating the range of parameters observed.

To realise these objectives, it is necessary to observe a wide range of parameters which can be loosely grouped according to whether the variable related to the atmosphere, land, ocean, or snow and ice.

Eleven instruments will be flown on ENVISAT-1, which together provide the means to realise the mission objectives.

A Meteosat Second Generation (MSG) programme has very recently been initiated. It is planned to launch the first of a new series in mid-2000, thus ensuring continuity of service with the first generation satellites. The MSG programme is an ESA-EUMETSAT co-operative programme conducted under the terms of an agreement between the two Agencies which entered into force in February 1994.

The MSC system objectives may be broadly defined as follows:

- making a contribution to the Global Observing System (GOS) of the World Weather Watch by providing observations from a geostationary orbit as part of the global system of geostationary satellites;

- addressing specific European needs for data from geostationary orbit for nowcasting, numerical weather prediction and climate/environmental monitoring.

Compared to the "classic" Meteosat, the MSG system offers enhanced performances, in line with the current state of the art, in terms of higher spatial and spectral resolution imaging (visible and infrared), radiometric accuracy, more frequent imaging, data collection and dissemination.

In polar orbit meteorology, preparatory activities for the METOP-1 mission are currently under way. Actually, this mission, for which the launch of the first satellite is scheduled late in the year 2000, is not solely dedicated to meteorology in that it has two broad objectives. The first is to fulfil the needs of the operational meteorological community, represented by EUMETSAT. The second is to fulfil requirements related to the global study and monitoring of the Earth's climate system expressed by international co-operative programmes such as the Global Climate Observing System, the International Geosphere and Biosphere Programme and the World Climate Research Programme. The second objective is complementary to the climate monitoring measurements of ENVISAT. The mission is again planned to be an ESA/EUMETSAT co-operative programme.

In order to fulfil the mission objectives, the METOP satellites are planned to carry two instrument packages:

- a core operational meteorology payload mainly procured and/or financed by EUMETSAT;
- a climate monitoring payload partly procured and/or developed by ESA.

The final payload selection is currently being defined in consultation with EUMETSAT, aiming at a series of missions which is "affordable" by the two organisations yet which fulfils the mission objectives.

It should perhaps be added that, as far as the operational meteorological objectives are concerned, the METOP satellites, together with the NOAA polar satellites, were originally intended to form part of an integrated meteorological system providing data coverage approximately every six hours; EUMETSAT would be responsible for the "morning" orbit service and NOAA for the "afternoon" orbit service. The recent decision by the US Government to implement a "converged" system which integrates NOAA and DoD systems introduces some changes to the original plans. However, it has confirmed METOP to be one of the elements of the converged satellite constellation to be put in place in the first years of the next decade. As the convergence will be fully achieved in the USA in 2004, its full effects are something to be considered in relation to METOP-3 and beyond.

As you will have gathered, our on-going programmes will take us into the early part of the next century. Nevertheless, forward planning of the programmes to follow should not be delayed, as the time it will take for a set of approved programmes to emerge may well be measured in years. And we need to ensure continuity in the data flow to users. This calls for a longer-term European Earth Observation strategy, work on which has started; it will be part of the ESA space plan to be submitted to the Ministerial Council scheduled for the second half of 1995.

Formulation of this strategy involves many interrelated aspects.

The first consists of a thorough review of the foreseeable future requirements of user communities in the scientific, applications and commercial domains. For this reason we have recently requested the support of a group of well-known European experts, in which EARSeL is represented, and we also plan to organise a large user consultation workshop next October, to tell the widest possible audience about our initial ideas and establish a dialogue with them, which we hope will be constructive and fruitful.

Confirmation of existing user requirements and formulation of new ones is only the first logical step. The ultimate definition of a mission scenario for the next century will require the interplay of numerous variables. These include affordable budgetary requirements, continuity with current on-going programmes, co-operation and co-ordination with other entities dealing with Earth Observation (European Union, EUMETSAT, NASA, etc.). What we have in mind is to propose two categories of missions: one category which is essentially research oriented, funded by a mandatory budget, and a second category which is applications oriented, to be initially financed by an ESA optional budget and later by operational entities to be eventually put in place to oversee land/ocean processes.

The next aspect of our Earth observation strategy will concern the ground segment, a very significant element of the whole system. Clearly Earth Observation data will need to be processed and distributed to users in such a way as to satisfy their requirements in terms of timeliness, validity and facility (including cost-effectiveness) of access. This must address a number of policy questions such as:

- identification of data products (in terms of processing levels and numbers) to be made available through ESA funding as opposed to further, higher-level products to be generated via separate funding sources (e.g. the European Union) in order to satisfy the widest spectrum of user communities;
- formulation of a common ESA Earth Observation data policy which identifies standard principles to be applied to all ESA programmes although their specific implementation may differ from mission to mission (or even from instrument to instrument).
- definition of an overall ESA ground segment concept and architecture which, building on the facilities serving ERS and ENVISAT, as well as those for the acquisition of data from foreign satellites, would be capable of optimally and continuously supporting the needs of the user communities.

These various processes cannot be finalised without close interaction with the Space Organisations of our Member States, with the appropriate European Union authorities, with EUMETSAT and our partners in the International Co-ordination Working Group (including the Space Organisations of the US, Canada and Japan) and with the Committee on Earth Observation Satellites (CEOS) as all of these entities have complementary and fundamental activities in the various fields which concern us.

We are therefore looking forward to achieving maximum synergy between our activities and those of the organisations I have just mentioned, particularly in the areas where these activities complement one another. Our ultimate aim is thus to develop a strong partnership within Europe in order to meet the needs for data continuity, long-term archiving and data access, as well as transfer of technology and responsibility to operational entities. I would like to mention that the extensive world-wide network of ground stations for the reception of data from ESA satellites is already an aspect of the international co-operation brought forward by needs of Earth observation from space, which addresses both developed and developing countries. The relevant activities, initiated in the context of ERS-1, will be continued to serve the needs of further missions. In addition we are pursuing close co-operation with the USA (NASA and NOAA), Canada and Japan, through reciprocal provision of instruments and agreements for the exchange of data originating from different programmes.

I hope you have found this short update of our activities in Earth Observation useful and that you will be able to continue to provide advice to ESA and to influence your governmental leaders to ensure that Earth Observation is properly financed and given the priority it deserves. I should like to conclude by saying once again what a pleasure it is to be here and how grateful I am to have this opportunity of bringing you up to date with what is going on in ESA Earth Observation and what are our plans at this moment in time.

Plenary session

Sensors and Environmental Applications of Remote Sensing, Askne (ed.) © 1995 Balkema, Rotterdam, ISBN 90 5410 524 0

The activities of the Institute for Remote Sensing Applications, Joint Research Centre, Ispra

J. Mégier
Joint Research Centre, Ispra Establishment, Italy

THE INSTITUTE OBJECTIVES: The global environment is constantly in a state of flux; indeed there is considerable concern that, as a result of either man's intervention or natural occurrences, changes to the earth's environment will have a long term effect on the global climate. In addition, regional, national and international governments continue to seek to maximise the use and benefit of their renewable and non-renewable resources. In order to fully address these needs there is a requirement for timely, accurate information, often from regional to continental scales; data from earth observing satellites can provide an important source of such information.

In order to address this requirement, the Institute for Remote Sensing Applications (I RSA) has been set a number of objectives:

a) to evaluate and demonstrate possible applications of remote sensing in support of the European Commission's sectorial policies, in areas such as:

- Common Agricultural Policy (CAP) (agricultural statistics, land use),
- Environmental Policy (land and sea protection),
- Development (food resources and environmental protection in developing countries),
- Fisheries (resources evaluation and conservation),
- Regional Aid.

b) to undertake research on advanced methods for the interpretation and utilisation of satellite data including their integration with geographical data.

c) to help to stimulate the scientific community in the use of earth observing satellites such as ERS-1, SPOT and the Polar Orbiting Platforms; this objective is complementary to that of the European Space Agency (ESA).

The Commission's scientific, regulatory and administrative bodies are IRSA's principal "customers" in various forms, the main one of which is the Specific Programme, incorporating a research body that is defined and approved by the EU Member States via the JCR's Governing Board.

The second, and increasingly more important form is the work that IRSA undertakes as Scientific and Technical Support to the Commission's Sectorial Policies. IRSA undertakes research on behalf of the Commission's Directorates General (DG's), the objectives of the research work being given by the DG's. Currently IRSA undertakes such work for DG I (External Relations), DG VI (Agriculture), DG VIII (Development Aid) and DG XI (Environment).

The third element of IRSA's programme is entitled Exploratory Research. This is a modest part of the Specific Programme that allows IRSA scientists to test new ideas that are not currently contained within the Specific Programme, but do look promising for future applications. This research is selected by the JRC's Governing Board on the basis of proposals made by JRC scientists, and funded as a part of the Specific Programme.

In addition, IRSA's programme provides the opportunity for national governments and private corporations to utilise IRSA's resources to carry out contract research. On this basis IRSA has set out to address a number of specific issues of importance to Europe, using the full range of satellite data now available from earth observing satellites.

THE INSTITUTE STRUCTURE AND PROGRAMME

The Institute is made up of five Units, with 92 permanent staff and an additional number of doctorate and post-doctorate students, visiting scientists and detached national experts. The Units' main activities are summarised below:

AGRICULTURE INFORMATION SYSTEMS

Some 127 million hectares of Europe are classified as agricultural. Decision and policy makers in the EC require accurate and timely data on these areas in order to execute the Common Agricultural Policy (CAP).
This is not always possible with conventional data gathering techniques. IRSA is therefore conducting a study on behalf of Directorate General VI (Agriculture) and the European Statistical Office (EUROSTAT) to define and demonstrate how remote sensing can be used operationally to supplement, interpret and standardise data provided by conventional techniques.

Remote sensing has several advantages for agricultural monitoring. The technique is objective and centralised, but, at the same time, the methodology is applicable in any part of Europe. Additionally, a range of scales, from single fields up to the entire continent, can be addressed an short-term changes can be monitored within days of data acquisition.

A number of specific actions are currently being undertaken within this context. These include an annual inventory of areas under crop in a number of European regions in Portugal, France, Spain, Greece and Italy. Each inventory region covers between 2 and 8 million hectares with in the most agriculturally productive parts of the country concerned. This study has helped a number of regional and national governments to integrate remote sensing into the national system of agricultural statistics.

In addition earth observation data have been used to monitor the conditions and development of vegetation at a continental scale. Status bulletins are provided to Directorate General VI (Agriculture) at 10-day intervals giving up-to-date information on the state of vegetation over the whole of Europe. These data are collected at low spatial resolution (one kilometre at nadir) and can be used as an indicator of crop condition, allowing the monitoring and mapping of the extent and severity of drought or frost.

Obtaining yield (and hence production) figures from remotely sensed data is not easy. A far better source of information on yield is an agrometeorological model of plant growth. These models use observed environmental conditions to simulate crop growth and to derive yield. Their results can be used to complement the information from satellites or conventional sources. IRSA is therefore conducting a study to provide this essential information by means of agrometeorological models at a continental scale. These models function with a grid-cell size o the order of 50 kilometres.

In order to satisfy Directorate General VI's need for frequent updates on the probable areas under various crops as the season progresses, IRSA has undertaken a study to provide rapid estimates of changes in acreage and potential yield using high-resolution satellite data at sample sites scattered over Europe. A team of interpreters examines each new image as it arrives to identify each crop as accurately as possible. Early images tend to contain more general information (spring or winter crops, for example), and information gets more precise as the tear goes by. Agricultural status bulletins are published by fax every 2 weeks between April and October.

Finally, in order to control agricultural surfaces within the new CAP regulations, and to produce registers of a number of agricultural crops, remote sensing techniques (mainly satellite imagery but also aerial photography when needed are used to establish and maintain registers on vineyards, olive groves and citrus orchards. Earth observation satellites are also use to help in controlling crop declarations. Their main contribution is in reducing the ground work required and increasing the objectivity of the controls.

ENVIRONMENTAL MAPPING AND MODELLING

The main purpose of the research concerns monitoring major components of the European landscape and developing methods for applying remotely sensed data together with other geocoded information to the management and protection of the European environment.

These activities are conducted along two main lines: a) forest grassland and natural vegetation, and b) Mediterranean land degradation. Furthermore activities in image understanding are aimed at continuing the development of methodologies or automatic classification, mapping and map generalisation from remote sensing data. These developments are necessary in order to make the maximum use of the potential of remote sensing in the stated fields of application.

Action on Mediterranean land degradation mapping and monitoring stems from two technical opportunities:

a) quantitative methods are now becoming available (and IRSA is actively contributing to their development, together with a few other laboratories world-wide) to map complex "natural" landscapes with operational satellites, where the main components of the landscape are present in varying quantitatives inside most of the pixels even with high resolution data, and

b) reasonable time series of suitable data are now becoming usable (Landsat TM from 1984 SPOT from 1986) in order to study vegetation and soil evolution.

The research started on a reduced scale in 1991 and continued from 1992 to date by using high resolution imaging spectrometer airborne data (GER II and AVIRIS) then TM data. Various abundances of green vegetation, soil development states and rock outcrops (parent rocks) were mapped in the South of the Département Ardèche (F). The investigation was then extended to test sites in the Northern Peloppenese area to map vegetation degradation levels over a five year period The investigation will now also consider Spanish and Italian areas in order to enlarge the test on the capability of the method to analyse various European landscapes.

The second thematic line corresponds to the effort to investigate the strength of spatial remote sensing to analyse the diversity and to monitor the behaviour of the large permanent vegetation communities of Europe: forest, grassland and natural vegetation.

A first study on the mapping of forest species, including broad-leaved and coniferous varieties, was conducted in Portugal with convincing results. Forest mapping in the entire Grand Duchy of Luxembourg was also undertaken and is now carried on by taking into account the effect of slope and exposure on the pixel radiometry. Forest and grassland mapping in the Italian Alpine Val Malenco watershed was the preliminary step of a run-off simulation study also to be continued.

Two large projects were prepared and launched concerning mapping and monitoring forest ecosystems. The first one deals with forest and land degradation in six Eastern European countries within the frame of the PHARE project (DG I), the second one called FIRS (Forest Information from Remote Sensing) has the purpose to design and develop a forest information system for all of Europe, with equal emphasis on sectorial (production) and environmental information.

The participation in a network of Chinese and Russian research organisations, together with the IIASA Institute (Vienna, A) for the study of the Arctic, sub-Arctic and temperate Eurasian ecosystems has also been initiated because of its great importance for global change issues.

MARINE ENVIRONMENT

Understanding the earth as a system is not possible without a good knowledge of the role of the oceans. The oceans cover 70% of the globe and have a crucial influence on all meteorological and climatic phenomena. The different processes in the ocean at local and regional scales are elements of the global system and must therefore be understood and observed for the modelling and understanding of global climate change, and equally for local investigations and applications.

Within this context remote sensing can be a key element in the observation and understanding of the ocean as a part of the global climate system, as well as playing an important role in providing data for environmental protection at a local scale.

On this basis IRSA is developing, demonstrating and validating methodologies for the use of data from space and airborne observations in both, operational applications and scientific investigations related to the marine environment.

Basically this involves an investigation of two main parameters that can be derived from earth observation sensors, and their role in different oceanic processes.

The first is ocean colour, which is an expression of the substances in the near surface water layer and the biological activity in the euphotic zone. The second is 5 the sea surface temperature, which is an indicator of the heat energy budget at the sea / air interface and is an element of the general hydrodynamic processes.

Within this context three main activities are being undertaken by IRSA. Firstly ocean colour data, as collected by the Coastal Zone Colour Scanner (CZCS) from 1978 to 1986, represent an important source of information for understanding bio-geochemical and physical processes in the sea. These data over marine regions of European concern are being pre- processed to a standard format and archived. From these data geophysical parameters are derived in order to create a data base of bio-optical information on the marine environment.

Secondly, remote sensing techniques are being applied in the northwest African upwelling area to help to evaluate the important oceanic phenomenon where cold nutrient rich water flows to the surface resulting in high biological activity. The water masses in these areas are characterised by low temperature and high phytoplankton concentrations, two parameters which can be observed by remote sensing techniques. The remotely sensed data describe the temporal and spatial distribution of coastal upwelling and related processes.

Finally, the role of the ocean in the global system is being investigated using remote sensing techniques. The test site off the North West African coast has been selected for particular investigation where ocean dynamics, including upwelling phenomena and air-sea exchange, and the primary productivity of the marine biota are under analysis.

MONITORING TROPICAL VEGETATION

Due to the immense scales involved, the monitoring of vegetation in tropical and sub-tropical areas is an area where remote sensing can play an important role in supplying accurate and timely data.

Monitoring changes in the forests of the tropical belt, for example, represents a formidable challenge for remote sensing. Indeed, these changes are taking place over large areas and are sometimes accelerating in pace. A recent FAO evaluation indicates that the rates of deforestation in the 1981-90 period (168.000 km^2/year) may be nearly double that of the 1976-80 period (92.000 km^2 /year).

In order to address this problem, research has concentrated upon the use of low spatial resolution data for assessing changes in the tropical forest canopy. This is seen as the first step in preparing a global forest inventory and monitoring exercise.

In addition, fire is a natural feature of many ecosystems; the African savannahs for example are often identified as fire climax ecosystems. Indeed fire has major impacts on local, regional and global ecosystems. Burning almost instantaneously changes the vegetation characteristics of the burned area and repeated burning can effectively alter the spatial boundaries of ecosystems forcing transition from one to another.

If the spatial distribution and timing of fire in an ecosystem remains constant these effects are usually in a state of equilibrium. However pressures of population growth, economic demands and demographic shifts are likely to change burning patterns. This can alter the balance and make fire an agent of environmental destruction. Burning patterns are therefore useful indicators of environmental change over large areas.

In order to map and monitor these effects at a regional scale, techniques have been developed to identify active fires using 1 km spatial resolution data over West Africa. These techniques result in the ability to extract fire statistics and to describe the areal and thermal characteristics of the fires.

In order to extend this work to a continental-scale study of fire for the whole of Africa, it is possible to utilise the 4 km spatial resolution data together with the 1 km data, where the latter can act as a 'control' to quantitatively evaluate the quality of the 4 km data product.

Remote sensing-based methods for assessing rainfed foodcrop production in Sahelian Countries is also being undertaken. This includes the derivation of acreage and yield estimates from satellite measurements combined with field observation at scales ranging from the administrative district to the subcontinent.

In addition, the problem of modelling and monitoring global climate change using earth observation and ancillary data is also addressed. This includes investigations related to climate and human impact upon terrestrial ecosystems with particular emphasis being given to mathematical modelling of the physics of radiation transfer leading to the development of vegetation indices. Furthermore, methods of numerical inversion for the extraction of surface parameters relevant to climatic processes are investigated.

ADVANCED TECHNIQUES

Besides the development of applications of proven space borne sensors in the visible and infra-red range, there is a need for the evaluation and promotion of more advanced remote sensing techniques, involving basic, systematic and long-term research.

This is particularly true for microwave remote sensing. Spaceborne microwave systems will play an important role in remote sensing in the immediate future as is clear from the launch of the E RS-1 and JERS-1 satellites, the coming launch of the E RS-2 and Radarsat systems, plus the Shuttle based systems SIR-C and X-SAR. These systems are a first important step towards the introduction of microwave remote sensing to a larger user community throughout the next decade.

Currently, however, there is neither the technology nor the fundamental knowledge of the information content of the backscattered signal to either fulfil the requirements of the user community or to specify the requirements for advanced microwave systems in the future.

To start with, IRSA has launched a co-ordinated research programme within Europe. This programme is based upon a number of elements including the implementation of a European microwave signature laboratory (EMSL), an equivalent facility operating at optical wavelength, the Gonimeter, the implementation of a European Airborne Remote Sensing Capability (EARSEC) the promotion of application-oriented airborne experiments in collaboration with ESA and national institutions, the development of application-oriented radar data interpretation algorithms and the promotion of application based pilot projects.

Additionally, new techniques for earth observation are being evaluated, such as imaging spectrometry. This technique, which splits visible and infra-red light into high spectral resolution channels, is seen as extremely important for the future development of earth observation.

Lastly, several years ago IRSA embarked on the development of an advanced airborne sensor to analyse the spectral and temporal response of remote targets excited by laser light; the time resolved laser fluorosensor (TRLF). This sensor is now functioning, with the current application being the characterisation and, possibly, the thickness of oil spills on the sea surface. In addition, it is also intended to examine other potential applications such as the measurement of water quality parameters, including chlorophyll, yellow substance and various chemicals.

Sensors and Environmental Applications of Remote Sensing, Askne (ed.) © 1995 Balkema, Rotterdam, ISBN 90 5410 524 0

R&D-needs in a rapidly changing marketplace

Claes-Göran Borg
Strategy Division, Swedish Space Corporation, Solna, Sweden

ABSTRACT: A personal view on the R&D needs of remote sensing based on experience of the remote sensing market is given.

During the last 20 years I have seen remote sensing from various perspectives: scientific, technical, political, operational and even commercial. I hope I can draw some conclusions.

It is not just me that is up for a challenge. We are all up for a big challenge during the rest of this century.

Up till now it has been an easy ride! From the early elated 70's when everything was new and fantastic. Over the rather dull 80's when not very much seemed to happen contradicting our hopes. To the present situation where the promised land is still just behind the horizon.

True - much has happened - the technique and the methodology has matured. Computers and GIS are here to help us. The Space Agencies' plans for the coming years are truly impressing.

The Space Agencies have done their job. We have fantastic plans for marvellous satellites giving us incomparable data. Maybe this week we will even decide on the Ground Segment for ENVISAT.

However, some sacrifices had to be done on the way:

* all satellites had to be "painted green" (even those already up in the sky),
* the satellites had to be expensive enough to satisfy the space industry,
* all big nations - and some small like Sweden - must have their piece of the pie of the ground segment. Maybe not an operational optimum - but a political.
* finally, big promises were given about applications and application needs that will be satisfied. But there was regretfully no time to really ask the users.

But why dwell on these pettinesses.

Back to the challenge. The Space Agencies have spent all their money. They have left us with tremendous amounts of data. We must show that there is meaningful information in these data. If not - there will be no more satellites. The back-lash will be painful and we will all be out of jobs. And we do not have too much time.

We must develop routine applications of significant importance. And the truth is there are terrifying few such applications today. The next generation of remote sensing satellites will have to be paid by the users - like the meteorological satellites are today.

The users may be the scientific community (e.g. Global Change), the operational public benefit users (e.g. environmental monitoring, sea-ice mapping) or perhaps even a commercial user (and I hope there will be more of those).

To develop routine applications we must understand that remote sensing research has very limited value if it does not pave the way for real use in any of these fields.

Therefore what we must do - and this is not very original but deserves saying over and over again - is to go to the users, develop applications that are well anchored in user defined needs, see to it that our methods are tailored to their systems and requirements. You can lead a horse to the water but you can not force it to drink. Make sure the user is thirsty - not hungry - when we arrive at the golden river. It is just as important for the remote sensing.

A year ago at a similar symposium someone gave a paper where he found it to be a big disadvantage that there were no remote sensing experts in the town bureaucracy. He said we have to train them. I say no, we have to train ourselves to see things their way.

Let me give you an example in sea-ice mapping in the Baltic. The captain on the icebreakers love getting SAR imagery for navigation on the bridge. Asked if he was prepared to pay substantial amounts for this information he was however rather reluctant. "I have my helicopters that fly with people anyhow. I talk to my fellow captains on the other ships. Within a few hours we know the situation" the captain usually answers. From his perspective the situation is very different.

This tells me there are no remote sensing users, just people who needs relevant information.

We must develop systems, not just methods. End-to-end systems that solves the whole problem at affordable cost, that fits into established structures, and that makes use of available techniques and other scientific fields as well.

But things are happening that will change the situation for the R&D-world. New structures are evolving. Some years ago it was like building the house in the 19th century. You had to know everything. Not very many houses were build.

This is now changing. Today we buy prefabricated doors and windows at stores for building material. A similar development will come in remote sensing. New entities are evolving and soon we can look in the catalogue for the right prefabricated building blocks. And the final analysis is left with the end user side. Because he is the one with the knowledge in the application field.

We also see more formal efforts internationally. CEOS is creating a dialogue with the research users and exploring ways to communicate with the operational users. EU starts the CEO initiative building a structure that in a network connects the providers and the users and there is a lively Data Policy dialogue.

We must influence these new initiatives and make use of the new structures. The data providers must be sensitive to your requirements - otherwise we do not use their data because now there will be competition.

Remember - its the buyers market! So, specify your requirements:
* the providers will love it,
* it is not easy,
* and be careful - overspecifying will cost money.

In Sweden we have a project to help bridge the gap between providers and users in the environmental field. The Centre for Environmental Data from Satellites is put up in Kiruna. It will be a Swedish Node in CEO. It is set up with the Environmental Protection Agency and other users.

The Centre will:
* prepare data bases to user requirements,
* make it easier to search for data and order data that are quality assured,
* archive datasets of historical value,
* help develop new methodology,
* have some environmental monitoring tasks.

We hope that this datacentre will be a focal point not just in Sweden and that it will be a good contribution to make the access to Earth Observation data much easier and thus foster applications in the environmental field.

I would like to summarise my message in the following brief points:

* There is a big challenge in front of us.
* The Space Agencies have done their job.
* Now we must see to it that the data is turned into valuable information - if not there will be no more satellites.
* Turn to the users and develop the applications on their terms.
* Adjust to the new structures evolving and put pressure on the providers - they will love it.

I have been working with remote sensing during all my professional life. I love it. But we have to put it into perspective. When ERS-1 was brand new, all the images were put on exhibition to the PB-EO delegates at ESA. They were truly impressed. One ESA executive came up to a colleague of mine and said: "Are they not wonderful" - and believe me, each of these SAR images can be the subject of a doctors theses. This scares me more than anything else.

Remote sensing is bridging the gap between technology, science and application, and culture. Recently an exhibition - "Erdsicht - The Earth - Global Change" - was opened at the Museum of Modern Art in Stockholm. Let this be a sign that we are leaving the laboratories and are moving out into society to help solve the many information needs out there.

Land applications

Sensors and Environmental Applications of Remote Sensing, Askne (ed.) © 1995 Balkema, Rotterdam, ISBN 90 5410 524 0

CORINE Land Cover – A pilot project in Sweden

Eva Ahlcrona
Swedish Space Corporation, Solna, Sweden

ABSTRACT: The overall aim is to propose a method which will allow a country-wide CORINE Land Cover mapping to be made in Sweden. The recommended mapping method is based on a combination of computer-based classification of Landsat TM data combined with digital map data, computerized generalization, and visual interpretation or editing of generalized result with the use of Landsat TM data and ancillary information. Two land cover products will be generated. One product will meet the requirements of the CEC and one product will meet the national requirements.

1. INTRODUCTION

The CORINE programme was initiated by the CEC in 1985 with the purpose to give information on the status and the changes of the environment, ensuring compatibility between countries. Within the programme a number of digital environmental databases are being created. Information on land cover is one of the databases created within the programme (CEC, 1993).

The final product of CORINE Land Cover is a digital land cover database made up of 44 classes. The smallest mapping unit covers 25 hectares. The method adopted by most of the countries that have conducted CORINE Land Cover projects is visual interpretation of satellite images (Landsat TM) at a scale of 1:100 000, in combination with ancillary information. It is planned to periodically update the land cover database.

At the request of the Swedish Environmental Protection Agency, Swedish Space Corporation has conducted a pilot project with the overall aim to propose a method which will allow a country-wide CORINE Land Cover mapping to be made (SSC, 1994). The purpose of the study is to lay the ground for Swedish participation in CORINE Land Cover.

2. METHODOLOGY

In the initial phase of the pilot project it was decided that two CORINE Land Cover products should be created. One product should correspond to the requirements of the CEC and one product should meet national requirements. To meet national requirements, national subclasses have been added, and the smallest mapping unit has been reduced to 5 hectares. The working hypothesis has been that the mapping should be organized in such a way that information corresponding to national requirements is generated first. The information is then adapted (simplified) to comply with CEC requirements.

The land cover information should in the first place to be collected from existing data sets. If suitable current data sets do not exist, the information should be derived from satellite imagery.

The pilot project may be divided into the following main parts:

- Establishment of national subclasses which meet national requirements, and which at the same time support cost-effective mapping. The proposal of national subclasses has been prepared together with a swedish working group and a nordic technical CORINE Land Cover group.

- Compilation of existing national information (maps, statistics), and assessment of its applicability in a country-wide CORINE Land Cover mapping.

- Investigation of methods for land cover mapping based on satellite imagery. Tests of visual interpretation and computer-based classification of Landsat TM data were carried out separately in an area (50 x 50 km) in the county of Örebro. Around 30 land cover classes (including national subclasses) are represented in the study area. The visual interpretation was based on Landsat TM imagery (1:50 000) in combination with ancillary information. The computer-based classification (supervised M-L) was based on Landsat TM data in combination with digital map data and other ancillary information. The result was subjected to checks for accuracy based on more than 700 control points obtained from field survey and interpretation of infrared aerial photographs.

- Generalization. To ensure that the final result from the computer-based classification corresponded to the criteria of the CEC, and to those established at the national level, a computerized generalization method was developed. The method generalizes from the classified 25 m by 25 m pixels which resulted from the computer-based classification procedure, to areas of at least 5 hectares, and then to areas of at least 25 hectares. In converting from national (5 hectares) to European standards (25 hectares), subclasses at national level were aggregated into classes according to the European standard. The computerized generalization method thus developed is based on the GRID module in ARC/INFO. The processing is divided into eight steps: pre-filtering, combining national subclasses into European standard classes, merging small areas with neighbouring areas of the same class, merging small areas with adjacent areas of a similar class, delimiting areas to be assigned to heterogeneous classes, division of small areas which have no neighbouring areas of a similar class, smoothing of area delimitations and vectorization.

3. RESULT

The proposal to national subclasses resulted in that around 15 subclasses were added (see table 1).

Existing information at present judged to be most useful are: the vegetation map, the vegetation map of the Swedish mountain regions, the topographic map, and the economic or general land use map (including orthophotographs). The use of the vegetation map of the Swedish mountain regions and of the ordinary Swedish vegetation map is of major interest since these maps contain very useful information and cover around 1/3 of the total area of the country.

The test of satellite-based mapping methods resulted in a good total classification accuracy (90 %) for visual interpretation of Landsat TM images combined with ancillary information. The classification accuracy for almost all individual classes was satisfying as well. The classification accuracy was 93% when aggregating the national subclasses to standard European classes.

The total classification accuracy for computer-based classification combined with digital map data was 79%. As to the identification of certain individual classes, the computer-based classification showed a non-acceptable low accuracy. The classification accuracy was 89% when aggregating the national subclasses to standard European classes.

The approach chosen for the generalization methodology proved to be effective. The most serious practical difficulty was the large amount of computer processing time required. The result as such, of the processing, met the formulated requirements to all intents and purposes. A decisive advantage of the proposed methodology is that processing is carried out with well-defined criteria, which allows for easy adaptation in collaboration with clients and users, as well as with experts within fields such as biology and geography.

4. COUNTRY-WIDE CORINE LAND COVER MAPPING

Visual interpretation of Landsat TM images in combination with ancillary information is not recommended as a country-wide mapping method. One reason is that the mapping would be very expensive to perform due to the high amount of working-hours required. Other reasons are that visual interpretation is based on subjective judgements, which may cause problems in for example the updating, and that the possibilities to create other products are restricted.

Neither is computer-based classification of Landsat TM data in combination with digital map data and subsequent computerized generalization recommended

Table 1. CORINE Land Cover classes in Sweden and proposal to national subclasses (italics)

Continuous fabric	Mixed forest
Discontinuous fabric	*Older mixed forest, not on mire*
Industrial or commercial units	*Younger mixed forest, not on mire*
Roads and rail networks and associated land	Mixed forest on mire
Port areas	Natural grassland
Air ports	Moors and heathland
Mineral extraction sites	Transitional woodland/shrub
Dump sites	*Thickets*
Construction sites	*Open cleared-felled or regeneration area*
Green urban areas	*Regrowing cleared-felled or regeneration area*
Sport and leisure facilities	Beaches, dunes and sand plains
Non-irrigated arable land	Bare rocks
Fruit trees and berry plantation	Sparsely vegetated areas
Pastures	*Sparsely vegetated areas with scant bottom layer*
Complex cultivation patterns	*Sparsely vegetated areas with rich bottom layer*
Land principally occupied by agriculture with significant presence of natural vegetation	Burnt areas
Broad-leaved forest	Glaciers and permanent snowfields
Older broad-leaved forest, not on mire	Inland marshes
Younger broad-leaved forest, not on mire	Peatbogs
Broad-leaved forest on mire	*Mires with dwarf shrubs*
Coniferous forest	*Mires not with dwarf shrubs*
Dense older coniferous forest, not on mire	*Peat extraction sites*
Sparse older spruce-dominated coniferous forest, not on mire	Salt-marshes
Sparse older pine-dominated coniferous forest, not on mire	Water courses
Sparse older pine-dominated coniferous forest, not on outcrop, not on mire	Water bodies
Younger coniferous forest, not on mire	Estuaries
Coniferous forest on mire	Sea and ocean

as a country-wide mapping method. The main reason is that certain classes cannot be mapped with satisfying accuracy.

As a country-wide mapping method we recommend a combination of both methods. Computer-based classification is used to very fast map large areas and a subsequent visual interpretation or editing of classified and generalized result is performed to map certain classes or areas with satisfying result. The country-wide mapping proposal recommended can briefly be described as:

thematic information in digital maps are translated into land cover classes or groups of land cover classes in a GIS environment. Mapping of changes, and of those classes which are not represented in the maps, is done through computer-based classification of Landsat TM data, in combination with digital maps. Generalization into 5 hectare areas is done through computerized methods. The subsequent editing (correction) of certain classes in the generalized result is carried out through visual interpretation at the computer screen with support from ancillary information. Generalization to the CEC standard is then performed using computerbased methods and subsequent visual editing.

This mapping proposal is the most cost-effective mapping method and also judged to be the most accurate one. Other advantages of the mapping method is that it is mainly based on objective estimates, which is an advantage when updating the database, and that it can be used for other applications since the original classification has a resolution of 25 m by 25 m pixel size.

A country-wide land cover mapping performed according to the mapping method recommended can be completed within 4½ years time by a project group of about 10 persons.

5. CONCLUSIONS

A country-wide CORINE Land Cover mapping of Sweden is recommended to be conducted through a combination of computer-based classification of Landsat TM data combined with digital map data, computerized generalization, and visual interpretation or editing of generalized result with the use of Landsat TM data and ancillary information. The mapping can be completed within 4½ years time by a project group of 10 persons.

Two land cover databases will be generated. One product will meet the requirements of the CEC and one product will meet the national requirements.

Prior to the initiation of a full-scale CORINE Land Cover mapping of the entire country it is recommended that further development work and a full-scale test in e g a whole county is undertaken.

6. REFERENCES

SSC, 1994: CORINE Land Cover - a pilot project in Sweden. Main report (in Swedish). Swedish Space Corporation, Solna.

CEC, 1993: CORINE Land Cover - Guide technique. Commission des Communautés Européennes. EUR 12585.

Sensors and Environmental Applications of Remote Sensing, Askne (ed.) © 1995 Balkema, Rotterdam, ISBN 90 5410 524 0

Effect on agricultural area estimates of using sample units shared by more than one CORINE Gis Stratum

Francisco José Igualada
Western European Union Satellite Centre, Torrejón, Spain & Cranfield University, Silsoe College, Bedford, UK

ABSTRACT: This study examines the potential usefulness of several statistical estimation methods through a Geographical Information System and a Remote Sensing Operational chain.
Sampling units shared by strata and therefore irregular in shape are superior in that they can manage detailed stratifications obtained by photo-interpretation in each stratum. Throughout the study, a number of computational tables were compiled, which were used to compute Ratio and Expansion statistical values.
An increase in the stratification efficiency and a reduction of the variance have been obtained for the Ratio method. However, by comparing it with the method using a weighting expansion, it appears that the precision of the estimator with this latter method is generally a little better. On the basis of the statistical results, a theoretical justification has been included in order to recommend a new Area Frame strategy. Accuracy assessments for both approaches have been carried out.
Two satellite image processing and classification strategies have been used to obtain information on pixels classified as a certain crop by including the Ratio as an auxiliary variable, using a Multiple Regression correction.

1 INTRODUCTION

Crop area estimates is an important economical factor to manage the European agricultural sector. The way in which this information is obtained needs to be methodologically reliable and cost-effective. This research was stimulated by the need to improve crop area estimations.
Current methods use ground survey measurements from a random sample of sites to obtain crop areas. With this technique,the error tends to be unacceptably high because of the sampling. Attempts have been made to reduce the sampling error by integrating the ground survey measurements with remotely sensed data and by improving the ground survey design. One of the ways to improve the efficiency of the survey is by stratification which involves dividing the total area of the crop survey into subregions,each having different but more internally uniform characteristics. Results from studies in the European Community (CEC J.R.C.) suggest that the current stratification is not sufficiently efficient. This work seeks to explore an alternative and more efficient way of stratifying, by examining the potential usefulness of several statistical estimation methods through a Geographical Information System and a Remote Sensing operational chain. The important feature of this procedure is the integration of both a CORINE Land Cover stratification (automatic) at 1:100.000 scale compared with the one outlined by photointerpretation (manual) at 1:5.000 scale together with and an existing Area Frame Sampling survey (AFS). The combination of these datasets and sampling units (quadrats) allowed the definition of new segments for estimation resulting from the intersection of quadrats with the strata boundaries, implementing in this way, four approaches based on the Ratio method of computation.

2 OVERVIEW OF THE METHODOLOGY

The method was supported by the availability of a complete and recent Area Frame Survey (AFS) carried out by the Statistical Section of the Spanish Ministry of Agriculture and Fisheries (M.A.P.A.) together with ten digital map sheets from the CORINE project (CEC) at 1:100.000 scale. This determined the choice of the region (804000 has) of

Palencia province (Spain) for the current feasibility study. In addition, the 55% predominance of agricultural land was also considered.
The overall project is divided into three main Phases.

2.1 Preliminary phase

The purpose of CORINE is to provide consistent and comparable thematic mapping of land cover classes of the whole Community, taking into account the variations in types of natural spaces and providint a general summary of those easily comparable from the different Member States. The main task consisted of grouping the digital GIS land use classes into six strata for estimation, (fig.1) followed by the preparation of the final stratification map with four strata.

Percentage	Description	CORINE code (class description)	
31.50	Non-agricultural	3.1.1	Broad leaved forest
		3.1.2	Coniferous forest
		3.1.3	Mixed forest
		3.2.1	Natural grasslands
		3.2.2	Moors & heathland
		3.2.3	Sclerpohylluss vegetation
		3.2.4	Transitional woodland-
		2.3.1	scrub
10.15	Irrigated land	2.1.2	Pasture
54.60	Rainfed agriculture	2.1.1	Permanently irrigated land
0.01	Permanent crops	2.2.1	Non-irrigated arable land
		2.2.2	Vineyards
		2.2.3	Fruit trees
1.74	Mixed annual and permanent	2.4.1	Olive groves
2.00	Mixed agriculture and natural areas	2.4.3	Annual crops ass. with perm.
		2.4.4	Land principally occupied by agricultures with significant areas of natural vegetation
			Agroforestry (Dehesas)

Fig. 1 Reclassification of CORINE Land Use

Production of the Frame table (Fig.2) giving the area for each strata, or the overall distribution of the population quadrats (700x700m) in the region. It is computed on the digitised CORINE stratification using a grid not coinciding with the original AFS grid.
Followed, a subsampling of the ground data was carried out to reduce the amount of data to be handled. Only one replication was used for each (10x10 Km) UTM block in the agricultural strata and half a replication from the remaining non-agricultural areas. This half replication is formed by the quadrats of the first replication in the even blocks. Each quadrat is associated for the sampling with the strata in which it has a larger area included. Classically, the strata are defined so that no quadrat straddles the border between two strata.

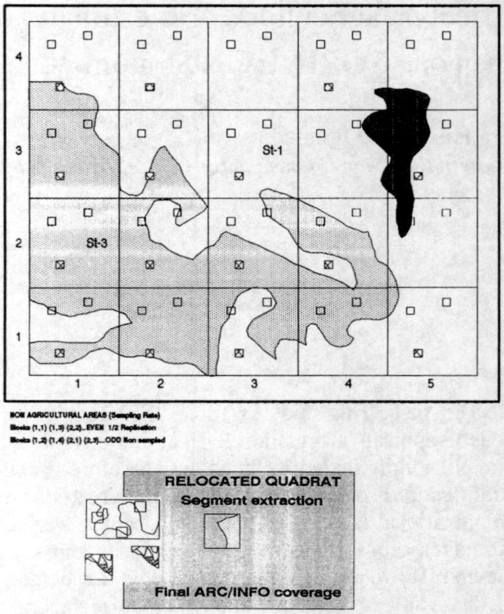

Fig. 2 SUBSAMPLING USING A 10x10 Km BLOCK FRAME
selection of the first replicate

To refine and improve the definition of the segments prior to the splitting, a relocation of the quadrats using recognisable features for shifting through the satellite image was performed. For the majority of the quadrats, it was not difficult to relate the polygon boundaries to the image data and in some cases the shifts required were rather important. There was not any defined trend in the distributions of x and y shifts. The cluster of shifted points showed a very low correlation (0.178) centred in a mean value for xdiff = -27.05 m and for ydiff = 7.82 m. One every two quadrats are shifted at least 1 Landsat-TM pixel, with respect or away from the origin (x) and 1 out of ten quadrats is relocated towards the origin of the same amount. This accounted for errors of parallax, geom. correction, digitising and scale of interpretation.

2.2 Geographic Information Systems

Each quadrat is subdivided into segments accordingly using the CORINE 1:100000 scale photo products and class boundaries from Landsat-TM (*Approach-A, Automatic splitting with CORINE*). At the same time, the sample units are split using the original 1:5.000 scale aerial photographs which were

reinterpreted according to defined strata classes. (*Approach B, Reinterpretation of quadrats*).
Two methods of division were evaluated in relation to their degree of accuracy as well as the statistical results yielded for both in terms of coeffiecient of variation for each crop type. Thus the reliability of CORINE and Photointerpretation approaches were assessed in two supplementary ways: by using proximity surfaces (buffers) along the interpreted boundaries and by comparing the interpreted area for each quadrat.

2.2.1 Area accuracy assessment

For every kind of land use strata within the quadrats, the area corresponding to the CORINE and photo strata were compared, obtaining the difference between both types of stratification for each quadrat area of 49 ha.

Stratum 3. Example of area accuracy corresponding to Dryland Agriculture

Row	Quadrat	St3C	St3P	3C3P	3Aac%
1	3863	22.11	48.56	-26.45	-53.958
2	3864	45.61	48.41	- 2.8	- 5.172
3	3964	23.11	48.32	-25.21	-51.428
4	4063	29.94	48.56	-18.62	-37.985
5	4064	48.48	39.45	9.03	18.421
6	4164	48.93	29.53	19.4	39.576
7	3763	24.58	32.93	- 8.35	-17.034
8	3563	48.93	34.01	14.92	30.437
9	3764	26.76	48.81	- 22.05	-44.982
10	3664	37.78	45.55	- 7.77	-15.851
11	3564	46.92	44.14	2.78	5.6712
12	3865	48.89	48.64	0.25	0.51

Shows an example for each quadrat in Stratum 3, where St3C corresponds to CORINE interpretation, St3P is photo interpreted, 3C3P means the difference in area and 3Aac% is a percentage of area difference or accuracy referred to 49 has (100%).

Overall statistics of area segments per strata

Column	Sum	Mean	Median	SDE
St1C	506	10.33	3.49	14.24
St1P	486.9	9.937	4.13	12.76
St2C	339.6	21.22	6.72	23.47
St2P	208.4	13.02	2.18	17.25
St3C	2029	34.39	43.6	17.87
St3P	2142	36.31	41.1	14.48
St5C	24.34	6.085	0	12.17
St5P	16.41	4.102	3.29	4.288
St6C	37.03	4.114	0	6.893
St6P	64.81	7.201	4.45	10.75

Shows the overall area per strata using both methods CORINE (St1C) and photo (St1P).

The increase in mean value form the area marked by the CORINE strata means that it has identified more area with the split quadrats by CORINE than it would really existed as a mean value for photo-interpretation (reference).

2.2.2 Line accuracy differences

Four buffer corridors were selected in order to compare the scale errors and the way in which the interpreted boundaries differ from the CORINE ones.

The percentages of overlap with respect to the total area of the quadrat were computed for each quadrat which had any common split segment (lines). The larger the percentage for a certain buffer range, the closer the lines were interpreted.

Relation between buffer ranges and scale errors

Ranges (m)	CORINE (error)	Photo (error)
(1) 0-30	0.3 mm	6 mm
(2) 0-60	0.6 mm	12 mm
(3) 0-120	1.2 mm	24 mm
(4) 0-180	1.8 mm	36 mm

Mean relationship between buffers

	Buffers	Mean	Sdev	RE	CV.	
1	0-30 m	7.14%	6.00		84	
2	0-60 m	21.63%	13.61	8.3	63	R :q (1 wih 2) 0.88
3	0-120 m	54.48%	24.07	3.3	44	R :q (1 wih 3) 0.70
4	0-180 m	80.69%	20.73	1.5	25	R :q (1 with 4) 0.34
TOTAL		40.99%	15.18			

As the correlations between the first buffer and the rest becomes smaller, the line overlap (averaged) gets larger. This line overlap is considered as an indication of the line matching accuracy for that specific range.

2.3 Preparations for statistical calculations

The use of GIS for data extraction was the main task, leading to the production of two tables for estimation in each approach. This enable to test a modified prototype software for computing direct expansion or Ratio values.
The creation of the SST (Sample Segment Table) and FdT (Field Table) for the CORINE extracts (Fig.3), involves the selection of different quadrats (squared sample units of 700x700 m) and manually assigning fields to them. In case a portion of the field parcel falls outside the CORINE segment in question, then a query is made to the database to establish the percentage of field which falls into the adjacent CORINE segment. If this percentage is 25% or less, then it is added to the larger field portion and considered to be one field.
In the case of Approach-B, to visualize the locational inaccuracies, CORINE strata boundaries were overlaid into each Arc/Info quadrat coverage. Such limits do not usually relate to any field parcel boundary identified on the photograph. Therefore, new limits were inferred helped with the use of aerial photographs in conjunction with quadrat's mosaic extracted form the satellite image . In

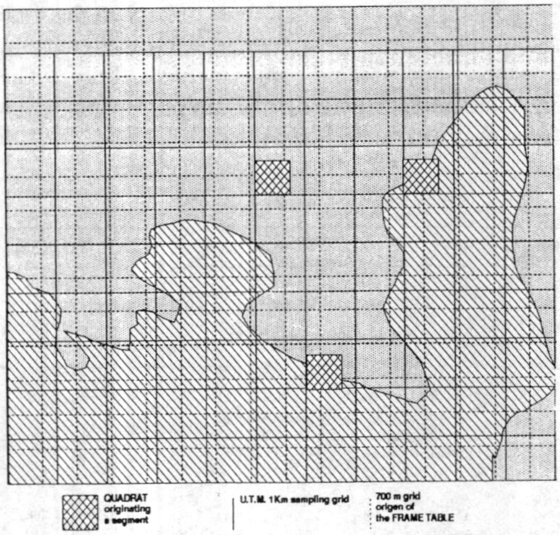

Fig. 3 PRODUCTION OF Frame Table AND Sample Frame Table

addition to this any reference derived from any field survey forms which proved to be highly useful for redrawing new strata photo boundaries. The use of accurate photo limits to obtain a new set of tables was important in order to avoid any effect from the misinterpretation of Landsat-TM, which would have substantially affected the estimation results.

2.4 Implication of the CORINE-based stratification on statistical calculations

The estimation based on proportions was introduced due to the limitations presented by the traditional Direct expansion method used for MARS project (J.R.C.). This later yielded relatively better results for certain crops provided proportionality conditions. The Ratio method (Gallego & Rueda 1991) improves the variance using the information on the segment size di in relation to the area of the estimated crop and the overall surface of the strata. The method is just a modification of the previous due to the introduction of a detailed stratification without the need of approximating it to an existing sampling grid, normally used for selecting the sampling units.

$$Zh = \hat{Y}hDh = \frac{Zh}{dh}$$

$$\text{Variance} \quad Var(Zn) = \frac{Nh^2\,(1-fn)}{nh(nh-1)}\,\Sigma\,(zi - \hat{Y}hdi)^2$$

$$\text{with } fn = \frac{nh}{Nh}$$

The Ratio can be generalised to include an auxiliary variable X to get a multiple regression estimator (Cárdenas 1978 and Konijn 1974).

A comparison of results between the following estimation methods was made

1 CORINE (Ratio Splitting)
2 CORINE (Quadrats by stratum)
3 PHOTO-INT. (Ratio Splitting)
4 PHOTO-INT. (Quadrats by stratum)
5 DIRECT EXPANSION (using 3 replicates and MARS stratification)

Methods 1 and 3 are the usual procedures to divide a quadrat into one or more estimation units. Whereas in the Methods 2 and 4, the largest segment of each quadrat is kept (approximating the stratification to the sampling grid) and given the same weight. This estimator is not very strict, in the sense that it does not comply with a number of requirements from a statistical point of view.

Finally and as a reference it was computed the usual Direct expansion, using the physiographic stratification of MARS project.

2.5 Remote sensing phase

Satellite image pre-classification processing was used in order to prepare the Landsat-TM dataset for further processing. This involved geometric correction to locate ground observations and register the various data levels for strata masking to enable a better classification performance and easy handling.

Extraction of satellite data for the sample units producing a composite of tiles for further image training area selection. The signature selection was an interactive four-step process of refinement.

It follows the Maximum Likelihood classification of the satellite data. After the class assignment is completed, there was still a need for a 'post facto' exercise to determine the accuracy and frequency of the error.

The most important step was the extraction of the classification results for sample units and the compilation of the computational tables, named the Classified Segment table and the Classified Strata totals. These are used to explore the potential usefulness of a multiple regression estimator to correct the values obtained by the Ratio method.

Two satellite image processing and classification strategies were used to obtain information about the auxiliary variables or number of pixels classified as a certain crop. This includes the Ratio as an auxiliary variable, using a Multiple regression correction. After the computational tables were produced, software problems did not allow a final statistical test of this supplementary method.

3 CONCLUSIONS AND RESULTS

Throughout this research two types of conclusions are considered relevant: *particular* or related to various tasks performed and *general* which corresponds to the overall study results and are summarised below.

CORINE			SPLITTING CORINE			
			Estimate	Std. Dev.	EFF.	C.V.
(AG)	0010	Water	425.0	435.3	0.51	102.3
(MM)	0063	Wood	44626.5	20901.2	0.77	46.8
(BA)	0071	Fallow land	60245.9	12813.6	0.85	21.2
ML	0072	Sclero-phy	61260.6	20517.9	0.87	33.4
ER	0074	Open with	22500.4	9068.2	0.75	40.3
NA	0107	Non-agric.	23133.4	15499.3	0.67	66.9
TR	1100	Wheat	82611.1	11581.5	1.14	14.0
CN	1150	Rye	42317.8	19598.7	0.78	46.3
CB	1160	Barley	278508.2	23971.1	1.69	8.6
AV	1180	Oats	17233.4	5292.4	1.07	30.7
MA	1200	Grain-maize	4687.5	4322.9	0.55	92.2
TT	1212	Triticale	433.4	468.5	1.08	108.0
GA	1311	Peas	42.7	45.9	1.09	107.1
GS	1320	Fodder-pea	915.1	708.4	0.44	77.4
JS	1330	Beans	1430.3	1298.9	1.07	90.7
LE	1341	Lentils	1471.0	918.2	1.12	62.4
VE	1342	Vetches	9597.4	4856.1	1.11	50.5
HS	1349	Other-dried	2463.7	2863.8	0.57	116.2
PT	1360	Potatoes	1159.6	615.4	0.76	53.0
RM	1370	Sugar beet	11652.4	4162.4	1.38	35.7
GI	1450	Sunflower	12995.5	6303.5	0.81	48.5
HO	1600	Fresh veg.	113.8	71.9	0.91	63.1
VI	2410	Vineyards	3337.0	1418.1	1.19	42.4
AF	2612	Other annual	19750.0	5465.7	1.23	27.6
VF	2680	Temporary	12180.2	4572.3	0.80	37.5
PR	2721	Permanent	9280.9	4525.6	0.83	48.7
PS	2722	Rough grazing	66577.1	22388.8	0.88	33.6
MD	3230	Agro-forestry	1006.1	1056.0	0.71	104.9
JM	3888	Non-productive	1330.5	1417.4	0.99	106.5
AT	3999	Lupins	87.5	94.3	1.09	107.7

PHOTO-INTERPRETATION						
			Estimate	Std. Dev.	EFF.	C.V.
(AG)	0010	Water	390.8	292.8	1.11	74.9
(MM)	0063	Wood	32103.5	18144.9	0.77	56.5
(BA)	0071	Fallow land	57905.1	10765.0	1.14	18.5
ML	0072	Sclero-phy	66069.4	21163.0	1.03	32.0
ER	0074	Open with	24417.4	8296.7	0.92	33.9
NA	0107	Non-agric.	25732.6	17092.9	0.64	66.4
TR	1100	Wheat	80124.1	11639.0	1.07	14.5
CN	1150	Rye	44756.0	20029.6	0.84	44.7
CB	1160	Barley	258671.6	17938.5	2.66	6.9
AV	1180	Oats	16728.6	4736.2	1.31	28.3
MA	1200	Grain-maize	5315.9	3007.7	1.10	56.5
TT	1212	Triticale	408.9	404.4	1.40	99.0
GA	1311	Peas	40.6	39.9	1.41	98.2
GS	1320	Fodder-pea	594.0	487.2	0.92	81.9
JS	1330	Beans	2494.3	1801.0	0.54	72.2
LE	1341	Lentils	1394.8	795.3	1.45	57.0
VE	1342	Vetches	8789.5	4146.3	1.47	47.1
HS	1349	Other-dried	2136.4	3202.2	0.30	149.1
PT	1360	Potatoes	1376.4	579.0	0.92	42.0
RM	1370	Sugar beet	16786.7	3219.6	2.23	19.1
GI	1450	Sunflower	17349.1	4922.2	1.28	28.3
HO	1600	Fresh veg.	127.2	53.0	1.61	41.7
VI	2410	Vineyards	3829.9	1259.2	1.59	32.8
AF	2612	Other annual	24700.2	4811.4	1.49	19.4
VF	2680	Temporary	12599.4	4747.1	0.84	37.6
PR	2721	Permanent	11239.1	3792.6	1.19	33.7
PS	2722	Rough grazing	73624.8	21375.4	1.23	29.0
MD	3230	Agro-forestry	1150.5	1141.2	0.71	99.2
JM	3888	Non-productive	2434.0	2408.7	0.47	98.9
AT	3999	Lupins	83.4	82.2	1.40	98.5

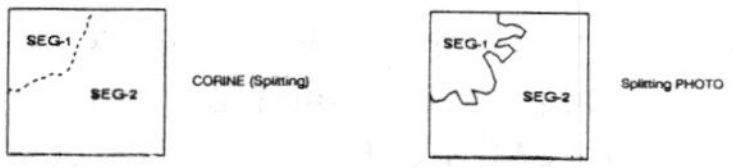

Fig. 4

3.1 Photo Ratio compared with CORINE Ratio

For the main crops of interest, wheat, barley, rye, oats, fallow and sugar beet, the stratification efficiency is usually larger using the photo approach, with the exception of wheat where there is a difference of 0.07 in favour of the CORINE method. In the case of barley and sugar beet, this difference es rather pronounced, being about 1/3 larger than the efficiency of CORINE. This effect on barley may be related to important errors in the CORINE stratifications, as barley as well as zones where wheat and barley have almost the same proportion. This point can be regarded as an important justification for stratification improvement.

3.2 Relation between Photo Ratio and Photo Quadrat by stratum

As far as the stratification efficiency of the photointerpretation is concerned, small improvements are obtained in relation to the photo (Ratio splitting), with the important exception on barley which increased 0.4 in the photo quadrat by stratum method reaching 2.66 nearly doubling the splitting by CORINE. This remarkable improvement of the photo method might be considered as a very likely reason for any further implementation. Sugar beet also slightly improved with the Photo splitting since small sample sizes are highly dependent upon exact stratification boundaries to obtain reliable results. This can also be linked with the reduction of the Coefficient of Variation (C.V.) for the photo splitting.

3.3 General considerations on the results

-. The Ratio method can only be applied successfully under specific conditions of proportionality between the two variables involved, which represent a linear regression through the origin between the segment size and the crop area. These conditions should be studied before attempting to implement it and compute estimates. In stratum-1, the variability increases drastically (CVy>2CVx), especially regarding the different crop areas (Fig.5) in each segment. With respect to other crops such as sugar beet, the largest variability is due to crop area not to size and the same thing happens to occur with woodland considering all the segments in the strata together. The Ratio is more efficient than the expansion except in the isolated case, where the ρxy

CORINE			QUADRATS BY STRATUM			
			Estimate	Std. Dev.	EFF.	C.V.
(AG)	0010	Water	447.9	325.2	0.91	12.5
(MM)	0063	Wood	46226.8	21516.5	0.73	46.5
(BA)	0071	Fallow land	59123.6	10641.9	1.23	17.9
ML	0072	Sclero-phy	62740.1	20758.4	0.85	33.0
ER	0074	Open with	23460.3	8809.4	0.79	37.5
NA	0107	Non-agric.	24409.0	15130.4	0.70	61.9
TR	1100	Wheat	80100.1	10070.7	1.51	12.5
CN	1150	Rye	43355.6	19551.0	0.79	45.0
CB	1160	Barley	278178.9	22742.6	1.88	8.1
AV	1180	Oats	19154.4	4563.9	1.44	23.8
MA	1200	Grain-maize	4837.1	2982.1	1.16	61.6
TT	1212	Triticale	413.7	412.7	1.39	99.7
GA	1311	Peas	40.8	40.7	1.39	99.7
GS	1320	Fodder-pea	588.4	428.9	1.19	72.7
JS	1330	Beans	1388.4	1150.0	1.36	82.8
LE	1341	Lentils	1411.9	821.8	1.40	58.1
VE	1342	Vetches	8936.8	4290.7	1.42	48.0
HS	1349	Other-dried	2760.2	2663.8	0.66	96.4
PT	1360	Potatoes	1190.6	579.4	0.86	48.6
RM	1370	Sugar beet	12028.3	3123.8	2.45	25.9
GI	1450	Sunflower	13192.7	4673.1	1.47	35.4
HO	1600	Fresh veg.	112.3	60.0	1.31	53.5
VI	2410	Vineyards	4067.6	1176.2	1.74	28.9
AF	2612	Other annual	19824.6	4431.7	1.88	22.3
VF	2680	Temporary	12818.7	4379.2	0.87	34.1
PR	2721	Permanent	9423.5	3903.9	1.11	41.4
PS	2722	Rough grazing	68534.2	22528.9	0.87	32.8
MD	3230	Agro-forestry	1099.8	1098.3	0.66	99.8
JM	3888	Non-productive	1201.1	1198.3	1.39	99.7
AT	3999	Lupins	83.8	83.6	1.38	99.7

PHOTO-INTERPRETATION			QUADRATS BY STRATUM			
			Estimate	Std. Dev.	EFF.	C.V.
(AG)	0010	Water	682.5	522.8	0.35	16.5
(MM)	0063	Wood	36752.8	20489.5	0.60	55.7
(BA)	0071	Fallow land	60834.3	10352.0	1.23	17.0
ML	0072	Sclero-phy	66531.3	24311.9	0.78	36.5
ER	0074	Open with	25541.3	8301.9	0.92	32.5
NA	0107	Non-agric.	27487.3	17549.3	0.61	63.8
TR	1100	Wheat	78788.3	10443.4	1.33	13.2
CN	1150	Rye	43442.5	21748.5	0.71	50.0
CB	1160	Barley	265934.2	26033.5	1.26	9.7
AV	1180	Oats	16093.2	4377.5	1.54	27.1
MA	1200	Grain-maize	5253.5	3404.6	0.86	64.8
TT	1212	Triticale	370.4	369.5	1.67	99.7
GA	1311	Peas	36.8	36.7	1.66	99.7
GS	1320	Fodder-pea	591.9	420.8	1.23	70.9
JS	1330	Beans	1276.6	1058.4	1.55	82.9
LE	1341	Lentils	1272.0	740.2	1.67	58.1
VE	1342	Vetches	8014.9	3862.9	1.69	48.1
HS	1349	Other-dried	2485.5	2379.5	0.55	97.7
PT	1360	Potatoes	1211.4	620.1	0.80	51.1
RM	1370	Sugar beet	18183.7	3323.4	2.10	18.2
GI	1450	Sunflower	19566.6	6307.5	0.78	32.2
HO	1600	Fresh veg.	108.4	55.9	1.45	51.7
VI	2410	Vineyards	2839.5	1280.8	1.54	45.0
AF	2612	Other annual	23023.3	5699.0	1.06	24.7
VF	2680	Temporary	12531.8	4923.7	0.78	39.2
PR	2721	Permanent	10265.9	4398.9	0.89	42.8
PS	2722	Rough grazing	69377.9	22165.2	1.15	31.9
MD	3230	Agro-forestry	1283.0	1281.3	0.56	99.8
JM	3888	Non-productive	1292.8	1280.0	0.85	99.7
AT	3999	Lupins	77.2		1.60	99.6

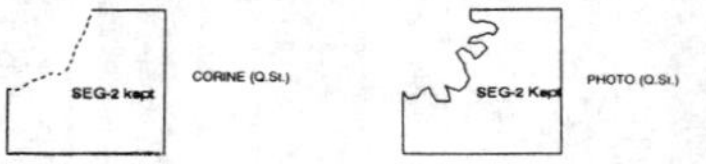

= 0 (correlation) where they are equally efficient. This relation holds true for barley and wheat, although in the case of these two cereals from stratum-3, it is very near the "limit allowed". As far as woodland, fallow land and sugar beet are concerned, the relation does not apply succesfully. The closer to the value of 2, the better the Ratio will perform.

-. An improvement can be made on the final estimates by eliminating small pieces belonging to a single strata, which in the Ratio act as a sort of outliers altering the proportionality. If this elimination is carried out a weighting expansion method needs to be applied instead.

-. By comparing the Ratio method with the one using the largest piece of quadrat that falls into an

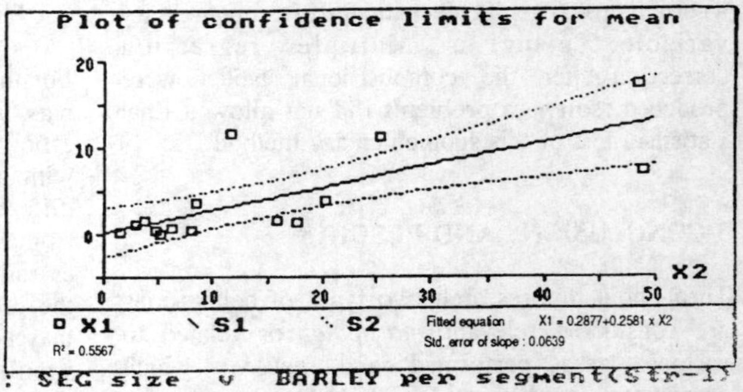

Fig. 5 Ratio Conditions for barley
x=segment size; y=area of barley per segment

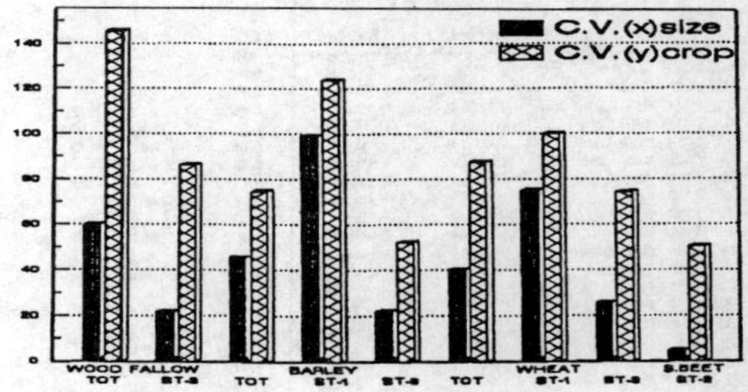

Fig. 6 Relation between coefficients of variation and justification of the efficiency of the Ratio for the major crops

strata or weighting expansion, it appears that the precision of the estimator with the latter method is generally a bit better than the one obtained with the Ratio (all pieces of the sample). From this situation, it might be concluded that using more information and thus increasing the size of the samples, a marked contradiction was obtained, as the results turned out to be worse. It might be that the Ratio estimator would not be the most suitable under these particular conditions.

-. The relatively important differences in Stratification Efficiency obtained for major crops like barley, maize and sugar beet through the methods under study, may indicate the need for a further stratification improvement using GIS procedures, before planning any AFS of these characteristics.

-. The approach using photointerpretation with the Ratio estimate is considered as the most appropriate in the light of the results obtained, as the case of barley. Even though the method lacks an interactive graphic procedure.

-. Accuracy assessments for both approaches have been carried out and used as a validation measure, leading to a remarkably good boundary coincidence regardless of the scale used.

-. Although the Ratio method has some important restrictions, it might,however be considered as the only one acceptable in a case when a complex stratification is used for attempting an integration with remote sensing data.

EXPANSION ESTIMATION			Total Area Estimated	Standard Error	C.V.	Rel. Eff.
(AG)	0010	Water	1512	505	33.4	1.11
(MM)	0063	Wood	104569	16070	15.1	0.70
(BA)	0071	Fallow land	60000	8000	13	1.06
ML	0072	Sclero-phy	41212	9661	23.4	0.57
NA	0107	Non-agric.	41000	10000	23	0.57
TR	1100	Wheat	74046	6582	8.9	1.23
CN	1150	Rye	21162	5226	24.7	1.06
CB	1160	Barley	257251	11382	4.4	2.19
AV	1180	Oats	13252	2652	20.0	1.07
MA	1200	Grain-maize	2550	903	35.4	1.31
TT	1212	Triticale	176	143	81.3	1.15
GA	1311	Peas	285	175	61.4	1.13
GS	1320	Fodder-pea	1000	1000	42	1.18
JS	1330	Beans	609	397	65.2	1.16
LE	1341	Lentils	269	237	88.1	1.14
VE	1342	Vetches	6978	2312	33.1	1.15
HS	1349	Other-dried	0	0		
PT	1360	Potatoes	980	357	36.4	1.05
RM	1370	Sugar beet	9370	1680	17.9	1.31
GI	1450	Sunflower	14176	2396	16.9	1.23
HO	1600	Fresh veg.	59	51	86.4	1.14
VI	2410	Vineyards	1757	518	29.5	1.16
AF	2612	Other annual	29000	4000	12	1.27
VF	2680	Temporary				
PR	2721	Permanent	10000	3000	30	0.88
PS	2722	Rough grazing	41385	11371	27.5	0.84
MD	3230	Agro-forestry	6229	2403	38.6	0.53
JM	3888	Non-productive	5121	2584	50.5	0.87
AT	3999	Lupins	31	31	100	1.02

-. The use of a reduced sample size might have been another possible limitation on the present study. The use of only 1 replication led to an amalgamation into four strata for computation.

-. In order to select a project in terms of cost-efficiency,two important technical features should be met:having a low C.V. and high Eff.Str.Using this reasoning, an "approximate cost function" proved that apart from the Expansion method using MARS strata, the most cost-effective solutions are: CORINE (quadrat by stratum) and Photo (Ratio) with a very small difference between them. This fact can only be taken as an indication, and any possible extrapolation should be cautiously regarded.

REFERENCES

Cárdenas M., G.A. Hanuschak 1978. Multiple Regression Estimation using Classified Landsat data. Report NASS-USDA. Washington.

Cochran W. 1977. *Sampling Techniques*. New York: John Wiley & Sons

Gallego F.J., and Rueda C., 1991, Estimates with sampling units divided by strata. J.R.C. Int. Report.

Gallego, F.J. & J. Delincé 1990. Crop Area Estimations through Area Frame Sampling and Remote Sensing. JRC Publication

González, F., S. López, J.M. Cuevas 1990. Influencia de la información contenida en Imágenes de Satélite sobre las precisión de la superficie cultivada de cereal en Navarra. INIA Monografías (MAPA)

Igualada, F.J. and V. del Barrio 1991. Estimación de Superficies de cultivo de verano utilizando imágenes de satélite. TECNO-Ambiente nº9 Julio

Konjin H.S. 1974. *Statistical Theory of sample survey design and Analysis*. Elsevier - North Holland

Sensors and Environmental Applications of Remote Sensing, Askne (ed.) © 1995 Balkema, Rotterdam, ISBN 90 5410 524 0

Image processing for CORINE Land Cover – Hungary

G. Büttner & G. Maucha
FÖMI Remote Sensing Centre, Budapest, Hungary

ABSTRACT: The extensions of the main CORINE inventories have started for Central and Eastern-Europe in the frames of the PHARE Programme. The Land Cover database, as one of the most important component of CORINE, is created by visual interpretation of precisely rectified satellite images. The paper describes the image processing methodology which has resulted a full Landsat TM image map (1:100.000) coverage of Hungary within four months.

1 INTRODUCTION

The CORINE (Coordination of Information on the Environment) programme of the Commission of the European Communities started in 1985 as an experimental project for gathering, coordinating and ensuring the consistency of information on the state of the environment and natural resources in the Community. The Land Cover project is part of the CORINE programme. Its aim is to provide up to date information on land cover of Europe at scale 1:100.000. The nomenclature is organised into a three level hierarchy comprising 44 categories on the third level. Classification is performed by computer assisted visual interpretation using precisely rectified satellite image maps (European Communities 1993).

In 1991 it has been decided to extend the CORINE programme for Central and Eastern-Europe within the PHARE Regional Environmental Programme. In 1993 a tender was launched for satellite image processing covering six countries of Central and Eastern Europe. As a result, Hungary is the only country that has produced Landsat TM image maps for her own area.

The goal was to rectify ten full Landsat TM scenes, image enhancement, and production of hardcopy photomaps at scale 1:100.000 for the entire area of Hungary (93.000 km^2, 84 map sheets). This paper gives an overall description of the work performed between 15 November, 1993 and end of March, 1994.

This was the first project in Hungary for the applications of high resolution satellite imagery that covered the whole country.

2 LANDSAT TM DATA

System corrected Landsat TM computer compatible tapes (CCT) were supplied by Eurimage as part of the CORINE Land Cover - Hungary project (Table 1). Only the easternmost three scenes were made during the same pass of the satellite (path 186). The images are cloud free and relatively up to date. All the images were acquired in summer and early autumn, which is considered optimal for land cover interpretation in Central Europe, because of the maximum contrast between agriculture (crops, pasture) and natural vegetation (forest, grassland, wetland).

Table 1. Landsat TM data

scene	date
186-26	July 30, 1992
186-27	July 30, 1992
186-28	July 30, 1992
187-26	Oct 7, 1991
187-27	July 21, 1992
187-28	Oct 7, 1991
188-27	July 23, 1990
188-28	Aug 29, 1992
189-27	Oct 5, 1991
189-28	Aug 20, 1992

3 DATA PROCESSING

Main parameters of the processing are summarised in Tables 2 and 3, as defined by the Central CORINE Technical Team working with the European Environment Agency (EEA). Figure 1 shows the flow chart of data processing.

A moderately priced high end PC system (a 486 CPU, Imagraph 1024*1024*32 board image display board, 1.7 Gbyte Winchester, 1 Mbyte removable MO diskdrive, Ethernet connection to a microVAX) with ERDAS 7.5 and pcARC/INFO 3.4D sofware was used for geometric correction and enhancement.

Table 2. Parameters of Landsat TM image rectification

number of GCPs	≥40 / proportional to the actual image size
no. of verification points	≥20 / proportional to the actual image size
control RMSE	≤25 m
verification RMSE	≤25 m
max. error verific. points	≤50 m
pixel size	25 m
resampling	cubic convolution
processing	all 7 bands

Table 3. Main parameters of photomap production

sheet size	48 km * 32 km plus 2-2 km margin
colour composite	TM4=R, TM5=G, TM3=B
scale of master film	1:250.000
scale of prints	1:100.000
type of paper	soft, rc coated, semi-matt

3.1 *Cosmetic operations*

All the processed scenes contained more or less partial missing scan lines. Sometimes it was very difficult to detect these artifacts by eye, especially when the missing segment was short. A computer programme has been written to detect these type of glitches automatically. Suppression has been made by using ERDAS 7.5 software.

An unique error, not seen on any other scenes processed within this project has been detected on scene 189-27. At the bottom of the image there were two blocks (consisting of 80 and 96 lines) shifted by six pixels along the direction of scanning. These could easily be detected as a break on the shore line of lake Balaton. The errors have been corrected by shifting the misplaced blocks backward.

A general problem with the quality of processed images was that linear elements (roads, rivers, etc.) appeared broken if their direction was nearly perpendicular to the scanline. This is the consequence of nearest neighbour resampling applied by Eurimage to correct for the panoramic distorion of scanning. This error could be avoided by using raw satellite data.

3.2 *Geometric correction*

The ERDAS system was applied for identification

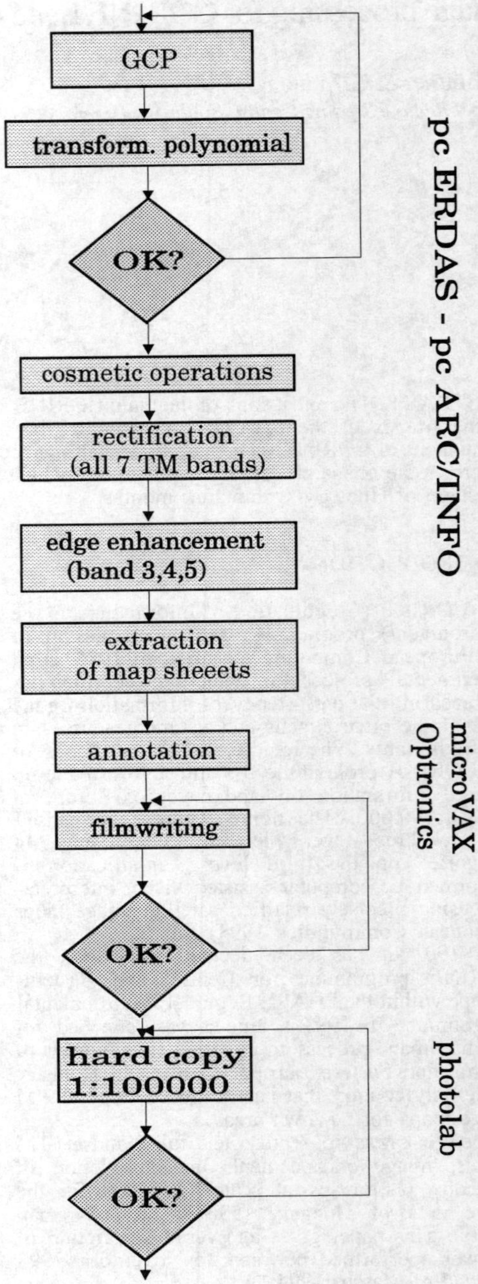

Figure 1. Flowchart of image processing

ground control points (GCP) on screen and geometric correction. Sheets of the Hungarian Civil Map System (EOTR) at scale 1:100.000 were used as cartographic basis. Map coordinates were measured using ARC/INFO software. For all scenes 2nd order polynomial was derived for subsequent cubic convolution resampling (Mather 1987). An independent set of verification GCPs was also selected to check the performance of the transformation polynomial. Table 4 lists data for each scene describing the parameters of geometric correction. Usually, twice the required number of GCPs had to be measured (either for control or for verification purpose) to achieve the necessary accuracy.

Scenes 186-26, 186-27 and 186-28 were acquired in the course of the same overpass of the satellite and were artificially cut to scenes by EURIMAGE with some overlap to the successive scene. In order to be able to produce full map sheets also in the overlap zone, scenes 186-26 and 186-27 were extended suitably, using identical anchor points, which were identified within the overlap area.

To facilitate later updating of the Land Cover data base, a GCP chip library has also been constructed, storing the 21*21 pixels neighbourhood of accepted GCPs. These imagettes can be used in automatic GCP location on a new satellite image covering the same area (Benny 1983)

Table 4 gives technical data for the rectification of each scene.

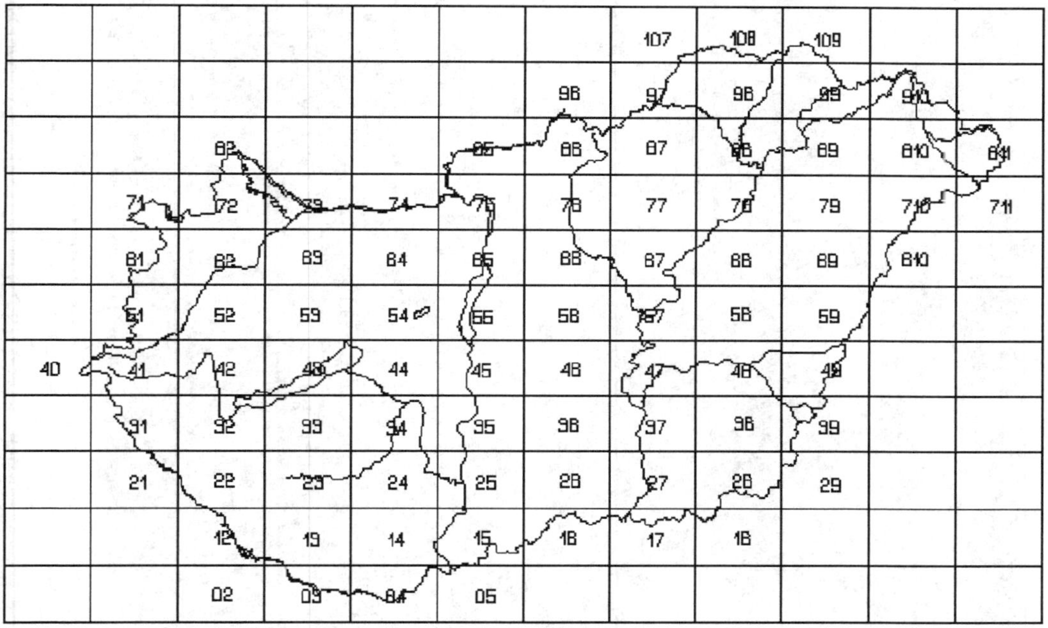

Figure 2. 1:100.000 scale grid for Hungary

Table 4. Results of rectification

Scene	Date	Effective area km^2	No. of sheets	No.of GCPs	Control RMSE m	No. verif. GCPs	Verif. RMSE m	Max.verif. error m
186-26	07/30/92	6200	6	30	23.69	15	24.23	36.95
186-27	07/30/92	10200	9	45	24.44	20	24.82	34.19
186-28	07/30/92	3600	5	28	23.98	14	24.24	32.37
187-26	10/07/91	4600	5	28	24.17	14	24.42	35.27
187-27	07/21/92	24400	14	40	24.36	20	24.71	38.54
187-28	10/07/91	11700	9	34	24.74	17	24.82	36.59
188-27	07/23/90	21300	10	40	23.46	20	24.16	38.64
188-28	08/29/92	12500	9	32	21.83	16	23.56	34.84
189-27	10/05/91	16000	13	40	24.47	20	24.60	34.18
189-28	08/20/92	5000	7	32	23.79	16	24.33	31.33

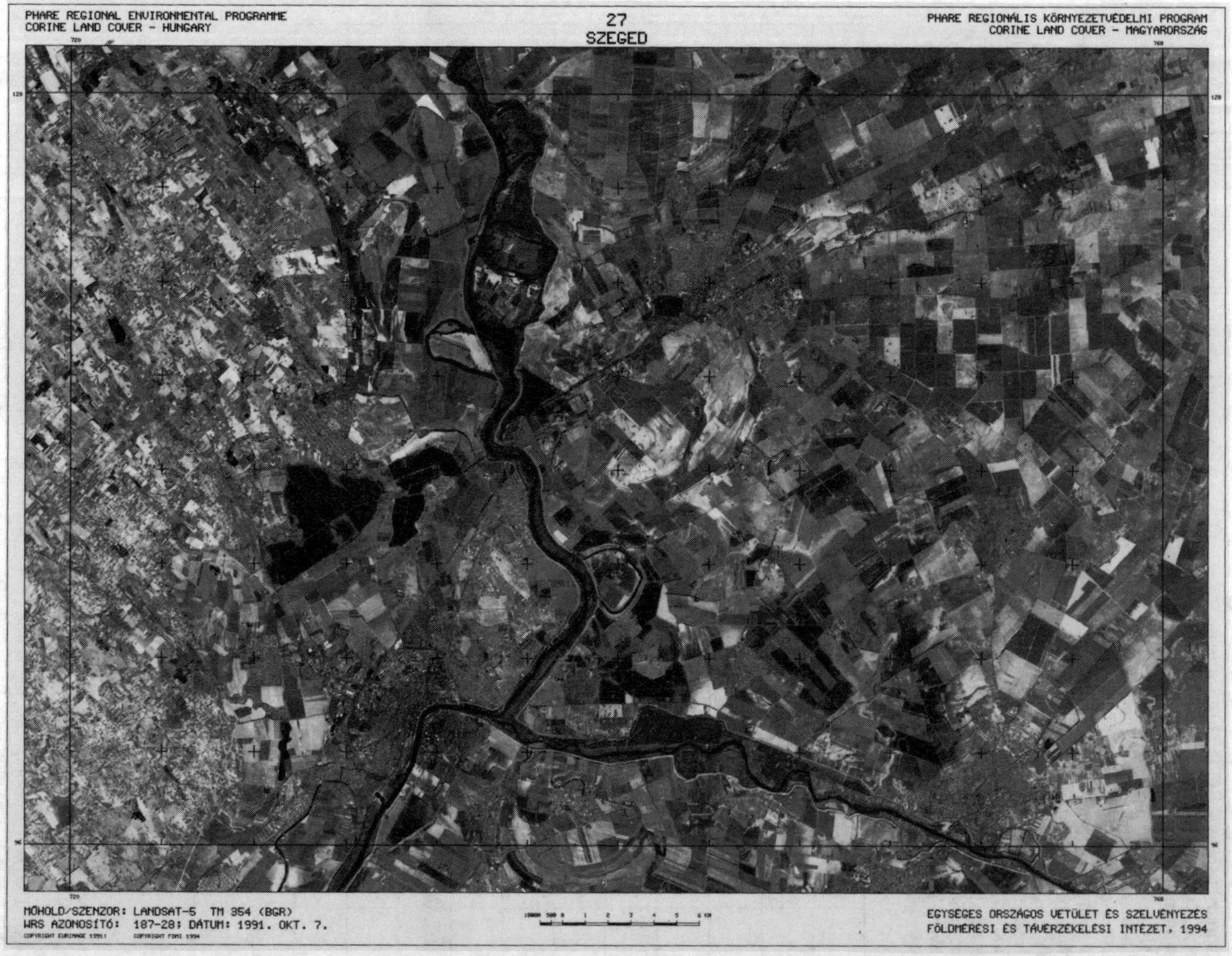

Plate 1. Example of photomap sheets produced by FÖMI RSC for CORINE Land Cover Hungary

Colour plate see page 489

3.3 *Image enhancement*

To improve the interpretability of images, high spatial frequency enhancement has been applied using a suitable 3*3 convolution filter. Filters have been specified for each scene and each of the TM3, TM4 and TM5 bands (with a special care to avoid over-filtration) carefully.

3.4 *Extraction of map sheets*

There exist 84 EOTR map sheets for the country at scale 1:100.000 (Figure 2.). One map sheet covers a rectangular area of 48 km * 32 km. Together with the 2 - 2 km margin each image map sheet covers a 52 km * 36 km area.

Most of the map sheets could be extracted using a single satellite scene thanks to the large overlap between neighbouring paths. In some cases however, because of the small along track overlap between scenes, two scenes should be used to cover a sheet. Two different solutions were applied, depending on the size of missing area:

a. For seven sheets a small area was missing only from a complete coverage. In these cases the almost complete sheet has been supplemented by data, derived from an other scene. The two incomplete sheets have been stitched together along a straight line. (Real mosaicing was not allowed not to confuse photointerpreters.) Histogram matching has been applied for the overlap area to minimize colour differences arising from the different image acquisition dates. In all cases the excellent geometrical matching is an indirect evidence of the good quality geometric correction, because the two scenes have been independently rectified.

b. For three sheets a larger area was missing from a complete coverage. In these cases two different color prints have been produced for the same sheet resulting a region without information (black) on both photomaps. The overlap between the two photomaps (representing usually different years and/or seasons) may have significant benefit for the photointerpretation.

3.5 *Graphical layer*

The following annotation has been applied for all sheets, using an in-house developed software running on microVAX:
- data for image acquisition (satellite, sensor, bands, path, row, date),
- map-sheet identification (number, title),
- reference to the PHARE CORINE Land Cover programme (in English and Hungarian),
- reference to the projection system,
- corner coordinates,
- scale-bar,
- map-sheet border lines,
- copyright information.

A regular network of small crosses (4 * 4 km grid) has been written onto the sheets to facilitate checking geometric quality and thematic validation of the final land cover map.

3.6 *Photographic restitution*

An Optronics C-4500 filmwriter has been used to producing a provide negative films at scale 1:250.000. Linear contrast stretching has been applied simultaneously to the filmwriting, using 1% upper and lower cutoff frequencies computed from the histogram.

A densitometric wedge has been exposed onto each film to monitor the procedure (status of the light source of the filmwriter, colour balance, photographic development). The wedge has provided a feedback for the calibration of the filmwriter, which was fundamental concerning the quality of the images.

An example photomap sheet (contact print) is presented on Plate 1 .

3.7 *Paper print production*

A Durst CLS 2000 enlarger has been used with Agfa semi-matt RC photographic paper.

The aim was to produce the same colours on all sheets resulted from a given scene. Sometimes however, we had to deviate from this principle, because of local atmospheric effect on the image or large differences on the terrain. Therefore small colour differences within the overlap zones can be observed.

The enlarger system has been kept fixed during the period of hard copy production. According to quality control there may be ±0.5 mm differences relative to the nominal sheet size. This can be explained by slight differences in the film size, mounting errors, and drift of the photographic paper.

4 ARCHIVING

Many different data archiving steps have been taken during the execution of the project:
- The ten original Landsat TM CCTs have been written into five compact-disks (CD-ROM) for a long term archive at FÖMI RSC.
- Original and rectified Landsat TM data are stored on magneto-optical (MO) cartridges as ERDAS 7.5 .LAN files.
- Rectified Landsat TM data have been delivered to EEA on two DAT tapes.
- Processing characteristics and GCP chips for each scene have been archived at FÖMI RSC and delivered also to EEA.
- Annotated photomap sheets in digital format have been archived to facilitate a later color print generation with a more advanced technology.
- 87 master films and contact hardcopies of

photomap sheets have been archived to facilitate service to regular clients at FÖMI RSC.

- Each of the 87 photomap hardcopies have been delivered to EEA and transferred to the Hungarian CORINE Land Cover interpreters' team.

5 CONCLUSIONS

Satellite image maps at scale 1:100.000 have been produced within a short time (4.5 month) for the entire area of Hungary (93.000 km^2, 84 map sheets) using ten full Landsat TM scenes. This was the first country-wide project for the applications of high resolution satellite imagery in Hungary.

The RMS error of the rectification is less than 25 meter. Data were enhanced to allow for high quality (both colour and geometry) hardcopy images suitable for visual photo-interpretation within the CORINE Land Cover project. All the seven rectified Landsat TM channels with 25 meter pixel size are available also in digital form for further computerised analysis.

It is not only the CORINE programme which benefits from the results. It is planned to apply the photomaps in updating 1:100.000 topographic maps, and there is a considerable interest within the clients of FÖMI RSC to use the image maps in various thematic applications.

A spin-off of the processing is a GCP chip library for the whole country which makes Landsat TM image rectification a more automated procedure later.

REFERENCES

Benny, A.H. 1983. Automatic relocation of ground control points is Landsat imagery. *International Journal of Remote Sensing*. 2:335-342.

European Communities 1993. CORINE Land Cover, Guide Technique. EUR12585, Bruxelles, Luxembourg.

Mather, P.M. 1987. *Computer processing of remotely sensed images.* John Wiley & Sons.

ACKNOWLEDGEMENT: This project has been supported by PHARE as part of the Regional Environmental Programme under contract ZZ9211/05.01/B003.

Advices of Mr. Chris Steenmans, member of the Central CORINE Technical Team for the improvement of the quality of image products are greatly acknowledged.

The following colleagues also contributed to the successful implementation of the project: Mr. P. Winkler project preparation, Ms. M. Bíró and dr. É. Csató GCP selection, Ms. G. Nádor image processing, Mr. L. Kollár filmwriting, Mr. L. Pálinkás and Mr. Gy. Rehorovics photography.

Preparation for the implementation of the project has been supported by the National Office for Technological Developments, Hungary (MEC-920017/033).

Sensors and Environmental Applications of Remote Sensing, Askne (ed.) © 1995 Balkema, Rotterdam, ISBN 90 5410 524 0

Classification of agricultural crops in Denmark using ERS-1 SAR and SPOT imagery

Inge Sandholt
Dina KVL, Royal Veterinary and Agricultural University, Copenhagen, Denmark

Bjarne Fog
Institute of Geography, University of Copenhagen, Denmark

Jan Nyholm Poulsen & Michael Stjernholm
Department of Land Data, Danish Institute of Plant and Soil Science, Vejle, Denmark

Henning Skriver
Electromagnetics Institute, Technical University of Denmark, Denmark

ABSTRACT: The combined use of SPOT and ERS-1 SAR data in classification of agricultural fields is addressed. It is shown that classification accuracies increases when SAR data is included as compared to SPOT data alone. A method for geometrical rectification of ERS-1 SAR data including a digital elevation model is presented. Data has been filtered and transformed to make a Gaussian assumption valid as required by the Maximum Likelihood Classification Algorithm used in this study. A simple 3x3 mean filter combined with a logarithmic transformation is suggested.

1 INTRODUCTION

Today optical satellite data are applied operationally in land use and crop monitoring in Denmark. The data sources are SPOT and Landsat TM data, and the methodology used is multispectral and multitemporal classification, also using information stored in a GIS. Optimal identification of crops requires satellite data from three periods of the growing season. For Danish climatic conditions this is not always possible because of cloud cover. This has lead to the initiation of a research project aiming at investigating the extent to which data from space-borne synthetic aperture radar (SAR) may be applied in combination with optical satellite data for crop monitoring. The advantage of SAR-data is that they may be acquired independently of clouds and time of day. On the other hand, the presently available space-based SAR, the ERS-1 SAR, operates at a single frequency (5.3 GHz) and with one polarization (VV), which means that only a single channel is produced. This single-channel image is affected by a number of factors, notably surface roughness and soil moisture which in the single band case can not be separated. Also the effective spatial resolution, taking the radar speckle into account, is lower than those of SPOT and Landsat TM. Thus there is a scope for combining optical data with SAR data. A preliminary study was carried out in 1993 in order to study these potentials (Fog et al., 1993). The project started out with a study of crop classification in a study area of 10·10 km, where information about the crops in 1992 on training and test fields were obtained by interviewing the farmers in the spring 1993. The ground truth data were dominated by rather large fields compared to normal Danish conditions, (20 ha), and the number of different crops was limited to five. The results indicated that combined use of SPOT multispectral and ERS-1 multitemporal data carries a potential for agricultural land use mapping and monitoring.

This study presents land use classification results achieved by the combined use of SPOT and ERS-1 satellite images from the growing season in 1993 tested against collected reference data for an extended number of fields and crop types. Section 2 describes the study area and the collection of reference data; in Section 3 the available satellite data are described. Preprocessing of SAR images are described in Section 4, and in Section 5 methods and results from the classification are given and discussed.

2 THE STUDY AREA - REFERENCE DATA COLLECTION

The study area is situated in the southern part of Fyn (Figure 1). The elevation varies between sea level and 112 m above sea level. During the period of reference data collection, the weather conditions were very stable all over Denmark. Only local showers were measured and the day temperatures were within the range of 15 to 25 degrees Celsius.

Figure 1. The location of the study area in the southern part of Fyn, Denmark.

Total cloud cover appeared a couple of times in the period; else clouds only appeared as scattered clouds. Due to these facts, sudden changes in soil moisture, which is an important parameter influencing the microwave data, were reduced to a minimum.

The basis of this study is a collection of reference data. In Denmark common field sizes are small compared to satellite scale, even SPOT with resolution 20 m sometimes has difficulties mapping the smaller fields. When SAR-images are included it may be necessary to use even larger fields. Therefore 42 fields of different size ranging from less than 1 ha (normal for Danish conditions) to fields of approx. 35 ha have been chosen in the area. The fields belong to farms or estates in the southern Fyn area. All reference data were collected *in situ*. For each field the crop was identified and parameters such as crop height (soil surface to top of plant), and phenological state were measured. The fields were visited twice, on May 27, 4 days after the second ERS-1 acquisition, and on June 11, the day of the third ERS-1 acquisition. The main crop types are: spring barley (sb), winter barley (wb), winter wheat (ww), winter oil-seed rape (wr), sugar beets (sb), field peas (fp), temporary grass (gr) and grain maize (co). The distribution of crop types is shown in Table 1. As test material, we have used 69 fields, where ground truth data were collected in connection with another study taking place at the same time in the same area.

Table 1. The distribution of crop types in training and test fields.

Crop type	No. of training fields	No. of test fields
Spring barley	6	9
Winter barley	4	5
Winter wheat	12	13
Winter oilseed rape	5	13
Field peas	5	7
Temporary grass	4	11
Sugar beets	3	3
Grain maize	3	8

3 SATELLITE DATA

The SAR on ERS-1 is an active microwave remote sensing instrument, that produces images independently of clouds and sunlight with a spatial resolution of approximately 30 m · 30 m in a 100 km swath. It operates at C-band (wavelength = 5,7 cm) and with VV-polarization. The data used are PRI--data (Precision Image, pixel spacing 12.5 m · 12.5 m), from the UK and the German PAF. The data values have been transformed into backscatter coefficient values using the method described in Laur (1992).

In the analysis time period the satellite was configured in a 35-days repeat cycle. At the Danish latitudes, however, there is an approximate 17-days repeat cycle, where the overlap between the images is approximately 60%. Three acquisitions from the early part of the growing season (May 7, May 23 and June 11) are used in the study, together with one multispectral SPOT acquisition from May 17.

4 PREPROCESSING OF THE SATELLITE IMAGES.

In this project satellite data are acquired at different dates and with different sensors, which means that the satellite data have to be treated quite differently according to sensor type. The only preprocessing of the SPOT image is a geometrical registration to UTM including correction for distortions due to elevation differences.

ERS-1 SAR amplitude data are supplied in 16-bit integer format. The image processing system used for geometrical correction and classification (CHIPS, Rasmussen et al., 1992) can only handle 8-bit integers, therefore the data have to be compressed. Also, the large variation due to speckle noise has to be reduced for the classification.

The ERS-1 SAR data are processed to 3-looks, and

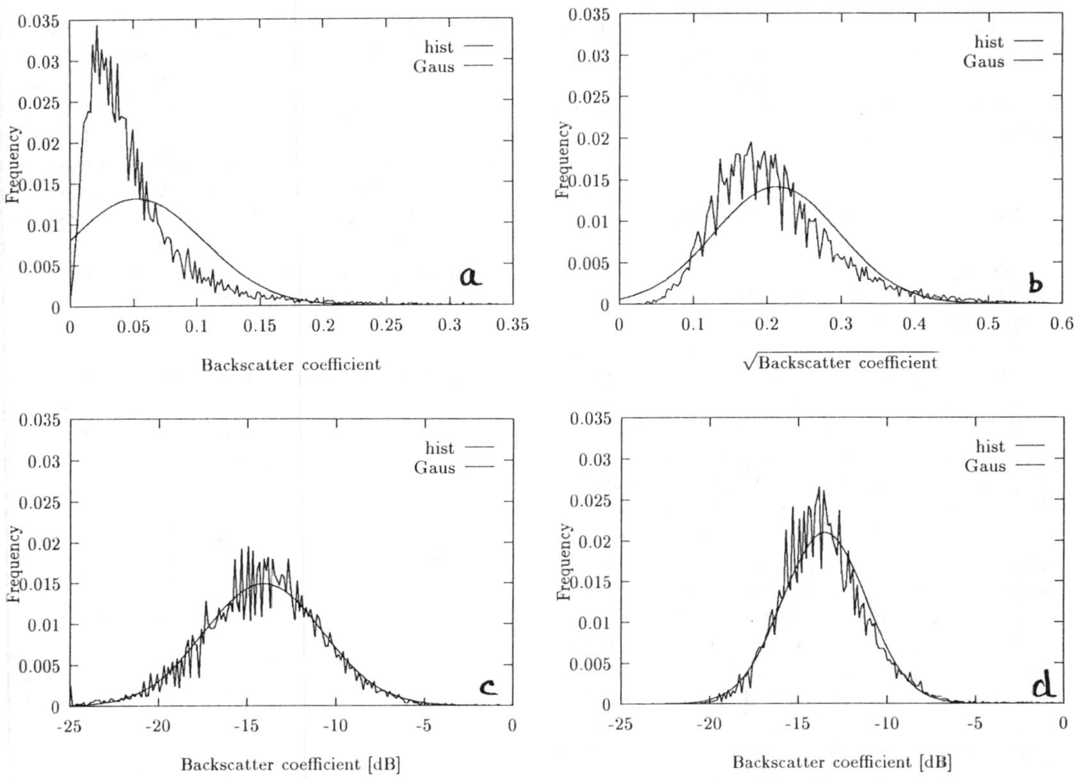

Figure 2. Histograms of the backscatter coefficients for winter wheat training fields together with a Gaussian distribution, 2a) intensity SAR data, 2b) amplitude SAR data linear compressed from 16-bit to 8-bit, 2c) logarithm transformed amplitude SAR data, 2d) 3 x 3 averaged and logarithm transformed amplitude SAR data.

due to overlapping looks the equivalent number of looks (ENL) is approximately 2.7. The probability density function for intensity data (backscatter coefficient) for a homogeneous area is the Gamma-distribution, and with an ENL of 2.7 this distribution is far from being a Gaussian distribution as required by standard classification algorithms. This is illustrated in Figure 2a, where the histogram of the backscatter coefficient for the training fields for winter wheat is shown together with a Gaussian distribution. We get closer to the Gaussian distribution by using the amplitude data instead of the intensity data, as seen from Figure 2b, where the histogram of the amplitude data for winter wheat is shown together with a Gaussian distribution.

A possible compression algorithm from 16-bit to 8-bit would be linear scaling of the amplitude data.

Another compression algorithm that has been considered is the logarithm transformation.

The logarithm transformation transforms the above-mentioned Gamma-distribution for a homogeneous area into a distribution with shape like a slightly skewed Gaussian with constant speckle variance independent of the mean (Horgan, 1994), i.e. the transformed data are well suited as input to standard classification algorithms. Figure 2c shows the histogram for logarithm transformed data (in dB) for the winter wheat training fields together with a Gaussian distribution. It is clearly seen that the Gaussian distribution fits the histogram very well. However, the speckle noise is still present, and a simple mean filter has been used for speckle reduction. As a trade-off between reduction of speckle and preservation of edge information a filter size of 3 x 3 is found well suited. The averaging is performed on the intensity data before logarithm transformation,because averaging in the logarithm transformed image corresponds to multiplication of intensity data, which is not proportional to the backscatter coefficient. The histogram of 3 x 3 averaged and logarithm transformed data is shown in Figure 2d, and a reduction of the variation compared with Figure 2c is observed. The latter algorithm is hereafter used to produce 8-bit data for classification.

4.1 Geometric coregistration of images

The co-registration process has several elements, involving the identification of relation between image and map projection, identification of relation between the individual images, and finally a correction for the different influence of the terrain for the different sensor systems. An earlier study (Fog et al., 1993) showed the difficulties of identifying corresponding control points in the topographical map and the ERS-1 SAR image. A SPOT image was therefore geometrically rectified and resampled into a UTM-projection. Corresponding control points between the SPOT image and the ERS-1 images could now be obtained, using features like field boundaries normally not available from the topographical maps. Since the SAR images contain speckle and in some periods show relative low contrast, the SAR image with the highest contrast and the most distinct features was selected as MASTER image. The transformation between the other SAR images and the MASTER were determined allowing the use of more fuzzy features as ground control points (GCPs), using the cursor cross hairs to interpolate e.g. a field corner. Finally the transformation between the MASTER image and the SPOT image was determined.
The correction of the images for terrain distortions is carried out using the following steps: 1) The observed sample coordinates from the image are projected to the position they would have had, given that the points were at sea level. The resampling transformation is calculated as a second order polynomial using the transformed GCP coordinates. 2) A relatively dense resampling grid is calculated from the resampling polynomium. 3) The sampling positions of the resampling grid are transformed according to the sensor system and the elevation at the point.

The different image formation of SAR images compared to optical images has the impact that given a view angle of 23°, the displacement in ERS-1 image is 2.35 times the elevation difference, 1/tan(v), while the displacement in SPOT is only 0.42 times the elevation difference, tan(v).

A digital elevation model with a gridspacing of 50 m and an estimated accuracy of 1 m was used for the terrain correction process.

The SAR images to MASTER image transformation had a rms of approx. 1.3 pixel, while the MASTER to SPOT (UTM) transformation had a rms of 1.1 pixel. The resampling of the SAR images were done using nearest neighbor resampling. The vectors identifying training and test fields were digitized from the UTM rectified SPOT image from May 17. Using the determined transformations, the vector files could be transformed to the original SAR image geometry. This allowed for the calculation of statistics for the training and test fields based on the 16 bit SAR images.

5 CLASSIFICATION AND SEPARABILITY.

The possibility to distinguish between crop classes using ERS-1 data, is visualized in Figure 3, where

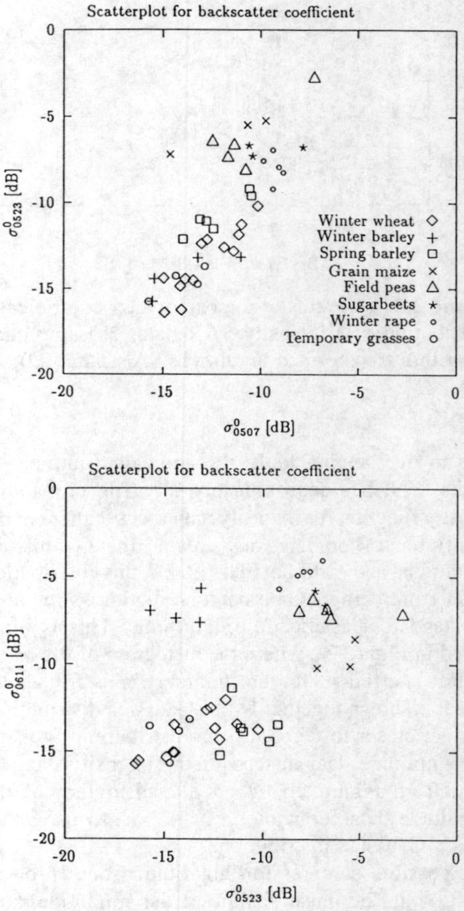

Figur 3. Scatterplots of the mean backscatter coefficients for field segments of two combinations of ERS-1 acquisitions.

scatterplots of the mean backscatter coefficient for the crops are drawn for two combinations of ERS-1 acquisitions. When the two May images are used, it is not possible to separate any classes, but when the June image is used, in combination with either of the two May images, more can be distinguished. The best two band combination is May 23 and June 11. This is due to the fact, that the two early images are showing almost the same crop evolution stage, whereas, at least for winter barley, there has been an evolution in the phenological stage of the crops between the May and June images. This has been verified by visits to the fields, where we found that the ears of winter barley have been bent down between the late May image and the early June image. In Figure 4, we have drawn the mean backscatter coefficient for each crop in the training set in the three SAR images as a function of time. It is seen that the temporal evolution in backscatter coefficient is slightly different for the different crops. This can be exploited in the classification of the crops.

To select the optimal combination of images in the classification, the Jeffries-Matusita (JM) distance separability index (or Matusita Index) (McLachlan, 1992) is calculated for several image combinations, three are shown in Table 2a, b and c respectively. Using SAR alone, is not very promising. The JM-distance for fully separable classes is 1.41. and the highest separability for SAR is 1.34 for grass and beets, overall separability is 1.03. For the three SPOT bands the separability is significantly higher, 1.28, but when SPOT and SAR are combined, the overall separability raises to 1.33. Corn and beets are confused, as well as winter wheat and grass.

Table 2a. The Jeffries-Matusita distance based on three ERS-1 images.

no.	1	2	3	4	5	6	7	8
1	0.00				see legend			
2	1.23	0.00						
3	1.22	0.69	0.00					
4	1.12	1.36	1.29	0.00				
5	0.98	1.25	1.14	0.74	0.00			
6	1.15	0.26	0.80	1.37	1.26	0.00		
7	1.24	1.18	0.98	1.02	0.60	1.26	0.00	
8	1.15	1.30	1.17	0.55	0.42	1.34	0.71	0.00

Overall class separability : 1.03

Legend:
crop 1: winter barley (wb)
crop 2: winter wheat (ww)
crop 3: spring barley (sb)
crop 4: field peas (fp)
crop 5: grass (gr)
crop 6: oil-seed rape (wr)
crop 7: grain maize (gm)
crop 8: sugar beets (sb)

Table 2b. The Jeffries-Matusita distance based on three SPOT bands

	1	2	3	4	5	6	7	8
1	0.00				see legend			
2	1.03	0.00						
3	1.28	1.12	0.00					
4	1.40	1.41	1.41	0.00				
5	1.41	1.41	1.29	1.41	0.00			
6	1.00	0.93	1.38	1.41	1.41	0.00		
7	1.41	1.41	1.39	1.41	0.82	1.41	0.00	
8	1.41	1.41	1.39	1.41	0.98	1.41	0.66	0.0

Overall class separability : 1.28

Table 2c. The Jeffries-Matusita distance based on three SPOT bands, and three SAR images.

	1	2	3	4	5	6	7	8
1	0.00				see legend			
2	1.31	0.00						
3	1.39	1.21	0.00					
4	1.40	1.41	1.41	0.00				
5	1.41	1.41	1.37	1.41	0.00			
6	1.31	0.97	1.40	1.41	1.41	0.00		
7	1.41	1.41	1.41	1.41	1.00	1.41	0.00	
8	1.41	1.41	1.41	1.41	1.04	1.41	0.92	0.00

Overall class separability : 1.33

5.1 Results

The classification of pixels have been performed using a standard method, based on a Bayesian maximum likelihood per pixel algorithm. The solution has been obtained using classical quadratic discrimination (se e.g. McLachlan 1992). We have assumed that data follows a Gaussian distribution, which is addressed in Section 4 for the log-transformed SAR images. The feature vectors consist of pixel values for various image and band combinations, as summarized in Table 3.

Table 3. Image combinations used in the classifications and overall classification accuracy.

Combination	SAR images 07/05	23/05	11/06	SPOT image 17/05	overall accuracy
1	x	x	x		50.6%
2			x		74.5%
3	x	x	x	x	76.8%
4		x	x	x	78.5%

x: image included in classification

Confusion matrices for the classification results are shown in Table 4. The poorest performance is found for the three SAR images alone, whereas the

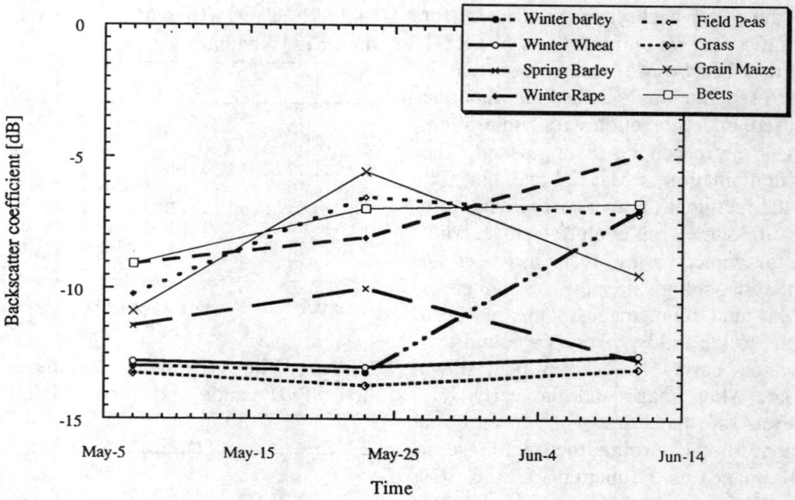

Figure 4. Temporal evolution in backscatter coefficient for each crop type.

overall accuracy for SPOT and SPOT in combination with the SAR images are higher and of the same magnitude. But if the confusion matrixes are studied more closely, it is seen, that the classification accuracy for the cereal types is raised significantly, for instance winter barley from 69.4 for SPOT alone to 89.4 for the SPOT/SAR combination. Corn and sugar beets are confused in either case. The overall accuracy for the four combinations is summarized in Table 3.

Table 4a. Classification based on SAR data (three images). (combination 1)

PERCENTAGE OF TEST PIXELS CLASSIFIED AS:

		wb	ww	sb	wr	fp	gr	co	be	SUM /
T	wb	78.2	3.7	5.0	2.5	3.3	7.1	0.0	0.2	100.0
E	ww	0.6	42.7	16.2	0.0	0.2	40.2	0.1	0.0	100.0
S	sb	0.3	9.2	82.4	0.0	0.3	3.7	3.4	0.6	100.0
T	wr	3.9	0.0	0.6	73.1	9.6	0.1	0.5	12.2	100.0
	fp	2.6	0.0	3.5	5.3	52.0	0.1	21.5	15.1	100.0
	gr	13.5	24.9	6.2	0.1	8.1	44.7	2.1	0.4	100.0
	co	0.6	1.4	25.8	3.3	20.3	1.7	20.3	26.7	100.0
	be	1.4	0.5	11.4	7.6	35.8	1.4	21.4	20.4	100.0

AVERAGE ACCURACY : 51.7 % (Average of diagonal)
OVERALL ACCURACY : 50.6 % (Sum of correctly classified test / sum of test pixels)

Table 4b. Classification based on SPOT image (three bands). (combination 2)

PERCENTAGE OF TEST PIXELS CLASSIFIED AS:

		wb	ww	sb	wr	fp	gr	co	be	SUM
T	wb	69.4	10.6	0.7	0.8	0.0	18.4	0.0	0.0	100.0
E	ww	1.8	87.4	1.3	0.0	0.0	9.5	0.0	0.0	100.0
S	sb	1.9	4.7	86.5	0.0	7.0	0.0	0.0	0.0	100.0
T	wr	0.8	0.2	0.2	98.2	0.0	0.6	0.0	0.0	100.0
	fp	0.0	0.0	2.1	0.0	83.0	0.0	14.6	0.3	100.0
	gr	4.8	21.6	26.0	0.0	0.1	47.5	0.0	0.0	100.0
	co	0.0	0.6	1.7	0.0	9.7	0.0	43.9	44.2	100.0
	be	0.7	0.5	0.8	0.9	8.6	0.9	50.6	37.0	100.0

AVERAGE ACCURACY : 69.1 % (Average of diagonal)
OVERALL ACCURACY : 74.5 % (Sum of correctly classified test / sum of test pixels)

Table 4c. Classification based on three SAR images, SPOT (three bands). (combination 3)

PERCENTAGE OF TEST PIXELS CLASSIFIED AS:

		wb	ww	sb	wr	fp	gr	co	be	SUM
T	wb	89.4	7.0	0.0	0.2	0.0	3.5	0.0	0.0	100.0
E	ww	0.2	87.3	0.9	0.0	0.0	11.7	0.0	0.0	100.0
S	sb	0.2	8.7	90.0	0.0	1.0	0.0	0.0	0.0	100.0
T	wr	1.1	0.0	0.2	98.4	0.0	0.3	0.0	0.0	100.0
	fp	0.0	0.0	0.9	0.0	86.7	0.0	8.9	3.5	100.0
	gr	1.9	18.3	24.4	0.0	0.4	55.0	0.0	0.0	100.0
	co	0.0	0.8	4.7	0.0	27.8	0.0	28.3	38.3	100.0
	be	0.0	0.7	0.9	0.9	14.2	1.3	45.2	36.8	100.0

AVERAGE ACCURACY : 71.5 % (Average of diagonal)
OVERALL ACCURACY : 76.8 % (Sum of correctly classified test / sum of test pixels)

Table 4d. Classification based on two SAR images, SPOT bands 1,2 and 3. (combination 4)

PERCENTAGE OF TEST PIXELS CLASSIFIED AS:

		wb	ww	sb	wr	fp	gr	co	be	SUM
T	wb	89.7	6.6	0.0	0.2	0.0	3.5	0.0	0.0	100.0
E	ww	0.3	87.4	1.5	0.0	0.0	10.8	0.0	0.0	100.0
S	sb	0.1	8.5	90.0	0.0	1.4	0.0	0.0	0.0	100.0
T	wr	1.3	0.0	0.1	98.3	0.0	0.3	0.0	0.0	100.0
	fp	0.0	0.0	0.7	0.0	92.6	0.0	5.8	0.9	100.0
	gr	1.7	19.6	24.9	0.0	0.6	53.2	0.0	0.0	100.0
	co	0.0	0.6	3.6	0.0	10.6	0.0	62.5	22.8	100.0
	be	0.1	0.5	0.9	0.9	8.9	1.4	44.8	42.6	100.0

AVERAGE ACCURACY : 77.0 % (Average of diagonal)
OVERALL ACCURACY : 78.5 % (Sum of correctly classified test / sum of test pixels)

When the early SAR image is excluded, the overall accuracy is raised further. The reason is that early in the growing season, field peas, corn and beets are confused, because these crops are only slightly developed in this period. When the crops become larger, they are more easily distinguished.

6 CONCLUSION AND DISCUSSION.

The combination of SPOT and ERS-1 SAR data showed that an improvement in the classification compared with the SPOT data alone is possible. Still some classes have a rather low classification accuracy and the results are expected to improve if a Landsat image was included, taking advantage of the mid infrared band. As we have stated before, the accuracy for individual crop types like barley, which is one of the most common crops in Denmark, is improved significantly, when SAR from ERS-1 is included, compared to using a single SPOT acquisition. It is found that the preprocessing of the images is important, and if one wishes to use a maximum likelihood classifier, the preprocessing is essential. Logarithmic transformation of the data is found to give good results, as the pixels thereby may be treated as belonging to a normal distribution. A simple mean filter has shown to be adequate for reducing the noise within well defined fields when logarithmic transformation has been performed. To perform a satisfactory classification of crops, it is of great importance to select time of acquisition with respect to the phenological evolution of the individual crops. Similar accuracies are found in a previous study (Fog et al., 1993). The data in that study were not transformed logarithmical, but the accuracies are comparable. This is due to the fact, that the greatest influence on the classification result is the mean value of each class, and their relative positions are not affected by the transformation. The covariance matrix on the other hand, is affected, and if standard tests have to be performed assuming Gaussian distributions, transformations of data are highly recommended.

7 ACKNOWLEDGEMENTS.

This study was financed by the Danish Space Board. Thanks to Mats Rudemo for invaluable comments. We would like to thank all the farmers involved in this study for their cooperation.

8 REFERENCES.

Fog, B., Poulsen, J., Sandholt, I., Stjernholm ,M., Skriver, H., 1993, "Monitoring land cover and crop types in Denmark using ERS-1 SAR and optical satellite images". In Proc. Int. Symp. From Optics to Radar, SPOT and ERS applications, Paris, CNES series no 312, pp 71-80.

Horgan, G., 1994, "Choosing weight functions for filtering SAR" Int. J. Remote Sensing, vol. 15, no. 5.

Laur, H., 1992, "Derivation of Backscatter Coefficient in ERS-1.SAR.PRI Products", Issue 7 Rev 0, ESA Document.

McLachlan, G. J., 1992, " Discriminant Analysis and Statistical Pattern Recognition", Wiley Series in Probability and Mathematical Statistics, New York.

Andersen, H.S., Fog, B., Grundtmann, J., Rasmussen, K., 1993: CHIPS version 3.0, User Guide, Routine Reference and Technical Reference.

Sensors and Environmental Applications of Remote Sensing, Askne (ed.) © 1995 Balkema, Rotterdam, ISBN 90 5410 524 0

The use of ERS-1 data in JRC's MARS project

Hans-Günter Kohl, Edmond Nezry & Hugo de Groof
Joint Research Centre of the European Communities (JRC), Institute for Remote Sensing Applications (IRSA), Agriculture Information System (AIS), Ispra (Va), Italy

Marek Mróz
Akademia Rolniczo-Techniczna, Olsztyn, Poland

ABSTRACT

In previous studies conducted at AIS [1], it was shown that ERS-1 imagery improves significantly classification accuracy of agricultural crops, if compared to a monotemporal SPOT scene. Recently, the work has been directed towards the analysis of high dimensional, multitemporal ERS-1 SAR PRI data sets. 9 images of the agricultural season 1992-1993 over the Seville (Spain) site of the MARS project and 2 images of the Olsztyn (Poland) site were analyzed, in both cases in combination with SPOT data. The methods used were based on an improved structure retaining Gamma-Gamma MAP speckle filter [2] taking now into account also second order statistics (Sum average or autocorrelation functions) [3].
First results for crop areas estimates obtained so far did allow the rice and cotton surface estimation in the Guadalquivir valley, South of Seville. The identification of other crops proves to be still difficult, but the inclusion of historical land-use information stored on a GIS (occurrence masks) seems to be a promising approach. A short evaluation of the SPOT/ERS-1 synergism for agricultural land-use discrimination has also be conducted on a Polish test-site (Olsztyn).

1. INTRODUCTION

The Pilot Project for Remote Sensing applied to Agricultural Statistics includes a specific action aimed at crop acreage estimation (Action 4). For 53 European sites, optical imagery from SPOT and TM is nominally acquired four times during the growing season. These data are then successively analyzed by computer aided photo-interpretation techniques. Within 10 days after data reception an updated forecast bulletin for the main crops is published. Due to unfavourable weather conditions, in northern Europe, it proved to be difficult to obtain a sufficient coverage from optical sensors. This time and weather constraints, as well as the potentialities demonstrated by ERS-1, emphasize the interest to introduce ERS SAR data into Action 4.
The main objective of the research is to develop strategies for automated detection and interpretation of land use/cover/changes with regard to the integration of ERS SAR data into an Advanced Agriculture Information System (AIS) developed within the framework of the European MARS Project. These strategies include the use of existing change detection techniques; we will also develop new approaches for ERS SAR time series processing and information extraction. Research is conducted, with emphasis to an assessment of the following points:

- Level and enhancement of pre-processing methods needed,
- Information content of ERS SAR data for agriculture monitoring,
- Cost-effectiveness in the Agriculture Information System.

This research will contribute to get more insight into these aspects of the problem. It is conducted with the support of a limited number of contractors in Europe.
Emphasis is also put on the assessment of synergism between ERS SAR and SPOT data and its temporal and/or spectral complementary. A first important aspect of complementarity of ERS SAR data with regard to SPOT data is, for instance, the expected possibility to produce some early area

estimation of agricultural crops in Europe using ERS SAR data.
The experiments conducted at the moment at JRC/-IRSA/AIS in this domain (Test-sites of Seville, Great Driffield, and Olzstyn) are the start point for these studies.
The final longer-term goal is to integrate ERS SAR data into an operation oriented system taking also advantage of backscattering and agro-meteorological models for a better understanding of the phenomena and of their spatial behaviour.

2. OBJECTIVES OF CURRENT INVESTIGATION

The overall objective of this investigation is to explore the potential integration of agricultural management practices, existing agricultural land cover, environmental conditions, and SPOT/TM and ERS SAR data in a knowledge based system, in order to obtain :

- Construction of "occurrence masks", allowing to indicate a level of likelihood for the crops to appear at a given place. This approach is seen as a first step towards an expert system designed to reduce interpretation ambiguities (identification of crop type) using either SAR, or preferably combined SAR/optical data.

- Already early in the year (bare soil conditions), the areas covered by a number of economically important agricultural crops on a regional basis, with ERS SAR data,

- Better monitoring of crop area estimates during the growing season, taking profit of the all weather capabilities of both ERS-1 and ERS-2 SAR sensors,

- Better crop discrimination during the growing season, taking profit of the comple-mentarity of information carried by ERS SAR and SPOT/TM data.

Previous ERS-1 experience of the investigators result in:

- Analysis and assessment of the information content of ERS-1 SAR single scenes and time series [1],

- Combination of SPOT and ERS-1 SAR data gave encouraging results for synergetic use of optical and SAR time-series for land-use and agriculture monitoring,

- Correlation between ERS-1 SAR backscatter and environmental parameters was also established [1,4].

More generally, the main conclusions of the studies conducted up to now, either by the investigators, or by other research groups in Europe, are that:

- Similar crops show similar distinctive temporal backscatter profiles (but not similar backscatter level) whatever the soil type. Thus, the use of temporal backscatter profiles matching techniques seems a promising method for crops identification,

- The backscatter of crops is highly variable at early growth stages,

- Effects of meteorological parameters need to be more deeply investigated, but could be considered as discriminant factors for statistical analysis of ERS time series, and identification of economically important crops.

- Some level of ERS SAR data pre-processing is needed, in order to retrieve ERS SAR information (speckle filtering, discriminant analysis of time series, mutual information between acquisitions, mutual information between ERS SAR and SPOT/TM data, etc...).

According to the current status of research, some topics of interest for the introduction of ERS SAR data into an operational remote sensing system, as the Advanced Agriculture Information System in development within the framework of the European MARS Project, have been identified.

2.1. Enhancement and operational use of ERS SAR preprocessing methods

In the case of SAR data, the first step in low-level processing, includes image adaptive speckle filtering, in order to reduce problem complexity by enhancing all desirable image properties as, for instance, signal to noise ratio, backscattering coefficient, edges and structural features, textural prop-

erties. These methods have been enhanced and transformed into operational tools, optimized for both ERS SAR images higher level processing or photo-interpretation.
The next research on ERS SAR image preprocessing will be the development of land cover/use detectors, especially designed for ERS SAR time series. Although preliminary work was made for this purpose at JRC/IRSA, the results achieved up to now are not yet fully satisfactory for the purpose of fully operational use.
Therefore, as first steps, we recently improved:

- The restoration of ERS SAR reflectivity and texture, by enhancing the performances of structure retaining adaptive speckle filters. These filters, already successful for radar reflectivity restoration, have been optimized for better preservation of ERS PRI images texture in presence of correlated speckle [3], making texture readily available as an additional discriminator.

- Field boundaries can preferably be extracted from SPOT or Russian KFA imagery, thus allowing to perform further classifications or high-level processing on a per field basis. Nevertheless, due to the limitations in SPOT/TM data availability, direct ERS SAR image segmentation methods are under development [5].

- Change detectors will be improved in terms of spatial accuracy by improving the image processing schemes, making available quantitative measurement of the changes as well as error analysis on these measurements.

2.2. Early area estimation of agricultural crops in Europe using ERS SAR

Conventional agricultural inventory and monitoring systems using optical spaceborne data concentrate on imagery acquisition and analysis during the growing season of the crops under investigation. Traditionally, data are collected during several time-windows from March until October within a calendar year. Rapid estimates of crop surfaces early in the season are often hampered by the lack of suitable spaceborne remote sensing data within this time-window due to cloud cover. Optical remote sensing imagery of late fall and winter periods is excluded from the analysis not only for the problem of cloud cover, but as well for the lack of an agricultural vegetation cover which would result in characteristic reflectance values.
Although the vegetative cycle comes to a halt during the winter period, the land preparation and agricultural practices during this period modify often dramatically the characteristics of an agricultural environment. Various crops require different field preparation during the fall and the winter. The preparation of the fields is guided by both environmental conditions such as temperature, rainfall, soil composition, and common management practices such as crop rotation schemes.
The sensitivity of optical remote sensing sensors to these type of land surface alterations is generally low.
An alternative exists in the use of a spaceborne SAR system such as the AMI instrument on board the European satellites ERS-1 and ERS-2. This system is sensitive to the geometrical characteristics of the soil surface layer, commonly referred to as the soil "roughness", and the soil humidity.
Another important motivation to incorporate ERS SAR data in an early estimation system is the advanced stage of microwave backscatter modelling theory for bare surfaces. Microwave backscatter models for vegetation are still in an experimental stage due to the complexity of interactions within the vegetation layer and the multiple reflections between surface and vegetation [6]. Backscattering of bare soil surfaces is far better understood and its implementation in a mathematical computer model rather straightforward. Feeding these models with both environmental parameters and surface characteristics would allow for the development of a database of characteristic backscatter values for various types of land preparation.
Given that we know the typical land preparation for a given crop type and the geographical parameters such as type of soil, temperature, rainfall, and soil humidity, the relation between the backscatter values observed by the SAR system and those resulting from the backscatter models can be established. This would allow an identification of the type of crop to be cultivated in the following growing season without having to rely on expensive ground observations.
Within the framework of this exploratory research, the collection of ground observations during the fall and winter periods needs to be carried out. Here as well, advanced techniques of data collection and distribution could be investigated. Common ground observation practices in operational and semi-operational European projects do not provide timely information to the remote sensing

data analysts. The observations are only used at the end of a calendar year to evaluate the accuracy of the estimates obtained during the year. Therefore, this research will actively evaluate methods for the gathering and dissemination of ground observations within very short timeframes. The availability of these ground observations in the regional knowledge base will be evaluated in terms of precision of the estimates on crop type coverage, and the ease with which more advanced data classification procedures will be implemented afterwards.

2.3. Analysis of ERS time series for the monitoring of land use changes in agriculture

An expert workshop [7] took place last February in JRC/IRSA/AIS Ispra, in order to collect ideas and expertise, especially in the fields of general SAR image understanding, modelling, segmentation, and interpretation. The recommendations lead to the following orientations of the research work.
In the high-level processing, meaningful descriptions of the physical objects observed by the ERS SAR sensors include the geographical identification of changing areas (since agriculture involves seasonal changes), as well as the type of change. This implies the use of multiple source data, i.e. the following approaches are already under investigation:

- The use of physical backscattering models can help to reach some level of backscatter prediction. This needs more investigation of the relationships between:
  i) C-band, 23 degrees ERS backscatter, ii) soil moisture content, iii) soil composition, iv) soil roughness, and v) crops evolution (type, phenology, height). As an example, correlation between ERS backscatter and rainfall (responsible for soil moisture) has been already experimentally investigated. This research imposes to use ERS fully calibrated PRI images.

- On the other hand, one can consider soil moisture, dielectric properties of soils, texture, etc... as instantaneous parameters. For instance, soil moisture (i.e. dielectric constant), a predominant parameter influencing the mean backscatter of agricultural surfaces during the growing season, is often considered as a measurement disturbance. In this case, a fully statistical approach involving analysis of mutual information between ERS acquisitions through maximum entropy and/or standardized principal component analysis is expected to provide good crops classification. This point does not exclude the previous one (backscatter models), since both approaches could provide complementary results.

- Change detection techniques will also be investigated and improved in order to provide a dynamic image of the agricultural landscape. This research will also use ERS PRI images.

- Integration of the change detection and classification results (ERS SAR time series) into the GIS used by the Agriculture Information System, which will provide the ancillary data (soil maps, topographic maps, data from meteorological networks, previous years classifications, etc...) needed for their final interpretation, in terms of physical and thematic meaning.

- We will try to establish an interactive and comprehensive link between the parameters introduced in the predictive agro-meteorological models used by the Agriculture Information System and the ERS SAR change detection results, to make easier change interpretation and adjustment of the model's parameters possible.

2.4. ERS SAR and SPOT data complementarity and data fusion

This is a field in which the Agriculture Unit (AIS) has already a good experience. Additional efforts to relate changes occurring in ERS time series to optical data are undertaken to establish further strategies for the complementary use of both sensors, with the scope to improve the efficiency of remote sensing applied to agriculture:

- Analysis of the crop information content of ERS SAR data in comparison to SPOT/TM imagery will concentrate on crop types for which SPOT/TM data does not provide clear separability.

- On the other hand, when based SPOT/TM

based crop separability is satisfactory, ERS SAR data together with agrometeorological and backscatter models, can be used to give quantitative results on crop growth, even if neither radiometric nor textural separability between these crops is achieved by the radar itself.

3. EXPERIMENTAL PLAN

Ground truth collection on the test sites will be ensured, either by JRC/IRSA/AIS, or through contracts with the University of Stuttgart (Institute for Navigation), and the Instituto Nacional de Investigaciones Agrarias, Madrid. The later will provide us with ground information on 250 segments, sampled 7 times a year by the Junta de Andalucía. In order to get a more complete knowledge on land cultivation practices, crop rotation, etc..., ground truth has been collected, from 1988 up to now.

A Pilot Project proposal has been submitted to ESA. The proposal intends to evaluate standard ERS SAR image (PRI) products for their integration in the context of AIS.
Experimental studies already began on the Action 4 test sites of Seville (Andalusia, Spain) and Great Driffield (UK), as well as on an additional test site in Poland (Olsztyn).

4. FIRST RESULTS

4.1. Seville test site

The spatial extent of available information (ground truth, DTM, ERS-1 frame 2853 coverage) made it possible to extend our Seville test site [1] from 40x40 km to 62.5x75 km. 4 SPOT-XS, 9 ERS-1 and the ground segments from 1991/1992 have been georeferenced to a panchromatic SPOT orthoimage. Raster and vector information have been integrated in a database.
A pre-processing chain for ERS-1 PRI images has been implemented, consisting in full image calibration, image resampling to 20x20 meters (linear weighted intensity method), and adaptive stucture retaining filter [2, 3].
In order to address image segmentation, a KFA-1000 (5m resolution) frame of the Seville test-site has been also evaluated. This evaluation is still under way, and seems to be promising.

Table 1 Classification results, percent of total area

	RICE	COTTON	WATER	OTHER
SPOT (AIS)	31.8	16.1	3.5	48.6
SPOT (S)	29.8	17.9	7.3	45.0
ERS-1	24.8	14.6	5.5	55.1

SPOT (AIS) : 4-date, performed by AIS
SPOT (S) : 4-date, performed by contractor
ERS-1 : 4-date, performed by AIS

Based on available crop calendars [8], appropriate dates of SPOT and ERS-1 acquisitions have been selected. Figure 1 shows the results of 4-date SPOT classification, and of a 4-date ERS-1 classification, directed towards the identification of rice and cotton fields in an area South of Seville. In addition the figure shows also the classification results obtained in the operational framework of the Rapid Estimates Project. These three classifications are consistent, although they show minor differences. This is confirmed by the acreage estimates presented in Table 1.

4.2. Olsztyn test site

In the perspective of the extension of the EC-MARS Project to Eastern Europe, a pathfinder study has been conducted at JRC/IRSA/AIS in the framework of the GDTA training program (CETEL). The study was oriented towards the joint use of optical and SAR data for land-use mapping of the Olsztyn (Mazuria, Poland) test site. Two ERS-1 PRI (April & May 1992) and one SPOT-XS (April 1992) were combined. The main results obtained on this test-site [9] with regard to ERS-1 can be qualitatively summarized as follows :

- ERS-1 data allow for accurate discrimination (better than SPOT) between built-up and natural (forest, agriculture) vegetated surfaces,

- Discrimination between the principal land-use components, bare soil/grasslands versus agricultural crops and forested areas can also be sucessfully achieved by ERS-1.

The Figure 2 shows a subset of a filtered ERS-1 image of April, 17th. Contour lines have been digitized from a topographic map at the scale 1:25000 and overlaid on the image. This illustra-

Figure 1 Classification results
Colour plate see page 490

tive example demonstrates how the radar backscatter is influenced by the local relief variations. In addition (Fig. 3), the radar signal has been combined with SPOT-XS using an IHS transform. This kind of product has demonstrated its usefulness for the mapping of areas with high water or soil moisture content.

4.3. Great Driffield test site

8 ERS-1 dates (16 PRI frames, year 1992) of the Great Driffield (UK) test site are now available at JRC/IRSA/AIS. They will be used soon in agriculture land-use study of this MARS Project / Action 4 test site.

5. PERSPECTIVES AND CONCLUSION

The added value of integration of the ERS SAR PRI data to the MARS project through the Advanced Agriculture Information System would consist in :

- Spectral complementarity to SPOT/TM

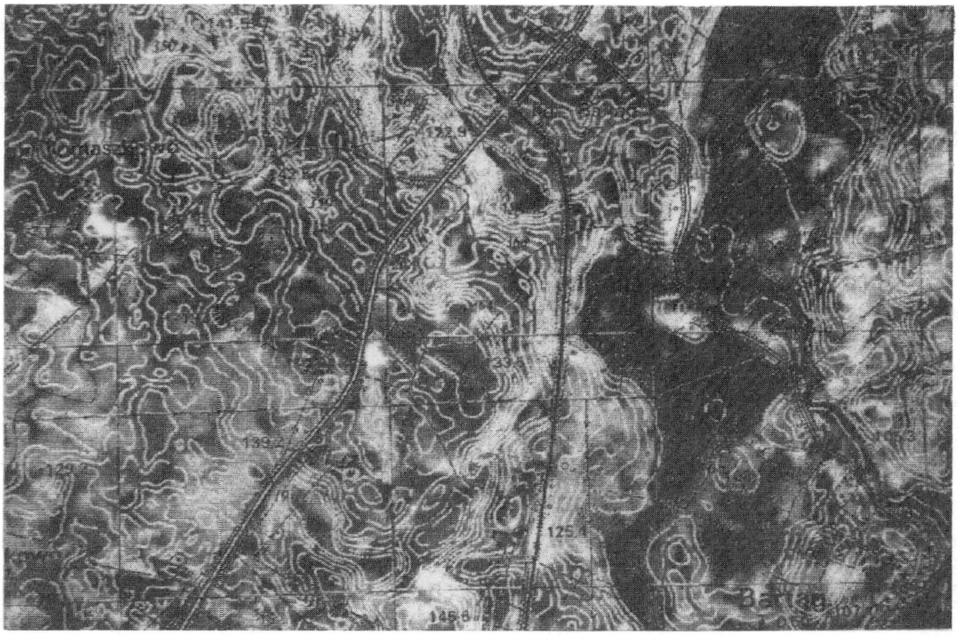

Figure 2 Olsztyn (Poland), filtered PRI image with topographic map overlaid

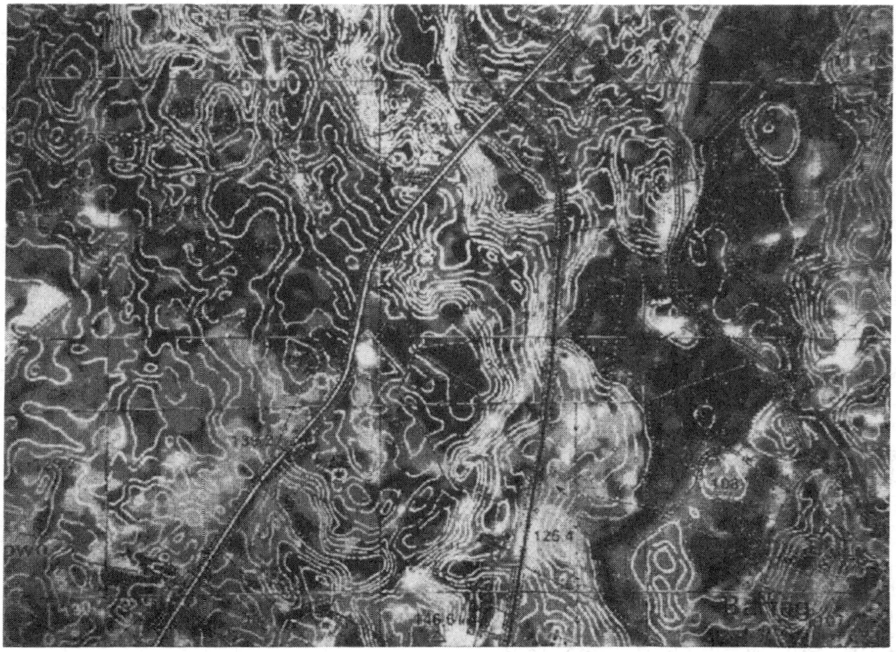

Figure 3 Olsztyn (Poland), IHS transform of filtered PRI image, SPOT-XS and topographic map. Dark tones indicate areas with potential water accumulation.

optical data : the expected advantages are better discrimination of target crops using ERS/optical data and derived higher level products.

- Temporal complementarity to SPOT/TM optical data: early detection/classification of crops appears possible using ERS SAR PRI data, and will therefore be assessed,

- Eventual substitution of ERS SAR PRI data to optical remote sensing data in the case of SPOT/TM missing sites,

- Pricing and rapid delivery by ESA should prove valuable arguments for operational use of ERS SAR PRI data. Reduction of the overall agriculture monitoring costs can also be reasonably foreseen at longer term.

ACKNOWLEDGMENTS

The authors would like to thank the European Space Agency, and especially Dr. G. Kohlhammer and Dr. H. Laur from ESA/ESRIN, for the rapid data delivery and related general assistance. A debt of gratitude is due to the participants to the Expert Workshop held at JRC/IRSA/AIS on 16-17 February 1994 [7], whose recommendations have been considered in the present paper.

REFERENCES

[1] H.G. Kohl, C. King, H. De Groof, 1993: "Agricultural Statistics : Comparison of ERS-1 and SPOT for the Crop Acreage Estimation of the MARS Project", Proceedings of the 2nd ERS-1 Symposium, ESA SP-361, Vol.1, pp.87-92, Hamburg (Germany), 11-14 October 1993.

[2] A. Lopes, E. Nezry, R. Touzi, H. Laur, 1993: "Structure detection and statistical adaptive speckle filtering in SAR images", Int. J. Remote Sensing, Vol.14, n°9, pp.1735-1758.

[3] E. Nezry, H.G. Kohl, H. De Groof, 1994: "Restoration of textural properties in SAR images using second order statistics", to be published in the Proceedings of IGARSS'94, Pasadena (CA), 8-12 August 1994.

[4] G. Kattenborn, E. Nezry, G. De Grandi, A. Sieber, 1994: "high resolution detection and monitoring of changes using ERS-1 time series", Proceedings of the 2nd ERS-1 Symposium, ESA SP-361, Vol.1, pp.635-642, Hamburg (Germany), 11-14 October 1993.

[5] R.P.H.M. Schoenmakers, H.J.C. Van Leeuwen, G.G. Lemoine, E. Nezry, 1994: "Segmentation of combined high resolution optical and radar imagery for the determination of field inhomogenities", to be published in the Proceedings of IGARSS'94, Pasadena (CA), 8-12 August 1994.

[6] M.A. Karam, A.K. Fung, R.H. Lang, N.S. Chauhan, 1992: "A microwave scattering model for layered vegetation", IEEE Transactions of GRS, Vol.30, nº4, pp.767-784, July 1992.

[7] "Integration of ERS-1/2 SAR's in the Advanced Agriculture Information System", Expert Workshop, held at JRC/IRSA/AIS, Ispra (Italy), 16-17 February 1994.

[8] J.M. Moreira Madueño, 1991: "Capacidad de uso y erosion de suelos. Una aproximación a la evaluación de tierras in Andalucía", Junta de Andalucía, Consejería de Cultura y Medio Ambiente, Agencia de Medio Ambiente.

[9] M. Mróz, 1994: "L'utilisation des images ERS-1 (SAR) et SPOT XS pour l'étude d'occupation des sols dans le nord-est de la Pologne", Rapport de stage CETEL 1993/ 1994, GDTA Toulouse.

Sensors and Environmental Applications of Remote Sensing, Askne (ed.) © 1995 Balkema, Rotterdam, ISBN 90 5410 524 0

Towards a satellite-observatory for mapping and monitoring the degradation of Mediterranean ecosystems

Joachim Hill, Stefan Sommer, Wolfgang Mehl & Jacques Mégier
Institute for Remote Sensing Applications, Environmental Mapping and Modelling (EMAP), Commission of the European Communities, Joint Research Centre, Ispra (Va), Italy

ABSTRACT. The degradation of the permanent semi-natural vegetation and the resulting acceleration of soil degradation and erosion processes constitute major elements of land degradation in the Mediterranean basin. Given the European Commission's interest to map and control desertification phenomena in the Mediterranean member states, the "Environmental Mapping and Modelling Unit" (EMAP) of the Joint Research Centre is investigating possibilities to use operational earth observation satellites for mapping and repeated monitoring of vegetation and soil characteristics. In the context of previous experiments, approaches have been developed which can already be routinely applied to large regions. Here, we discuss problems related to the retrieval of remotely sensed primary parameters (such as reflectance), concepts for the thematic interpretation of reflectance data, and the definition of satellite-derived degradation indices. We further present requirements for the design of an operational satellite observatory for Mediterranean land degradation monitoring.

1 INTRODUCTION

Land degradation processes which imply a reduction of the potential productivity of the land (e.g., soil degradation and accelerated erosion, reduction of the quantity and diversity of natural vegetation) are widely spread in the Mediterranean basin. In continuation of a long history of human pressure upon land resources, the main environmental impact results from interactions between climatic characteristics and ecologically unbalanced human interventions which, in the sense of recent definitions of the United Nations Environmental Programme*, are often summarised as 'desertification processes'. An overview of the ecological, physical, social, economic and cultural issues which are collectively contributing to the increasing risk of further degradation of Mediterranean lands has recently been presented by Perez-Trejo (1994).

2 MAIN SOURCES AND PROCESSES OF MEDITERRANEAN ECOSYSTEM DEGRADATION

The criteria which have been associated with desertification processes, are in some ways inappropriate descriptions of Mediterranean landscapes but biased towards African examples. Perez-Trejo (1994) therefore concludes that a reconceptualisation of desertification - one more appropriate for the European situation - is needed in which the role of urban-industrial expansion, tourism and agriculture in relation to the allocation of water resources are seen as significant contributors to the problem. Inadequate land use practises (e.g., excessive grazing, fuelwood collection, uncontrolled fires) further contribute to the acceleration of degradation processes which result primarily from complex interactions of plant growth and erosion processes. It is now widely agreed that accelerated water erosion is one of the most important sources of soil degradation which, together with the destruction of vegetation cover and structure, contributes to the potential increase of desertification in the Mediterranean Europe. Both are often connected, since the degree of soil degradation is in many ways a reflection of the state of vegetation that covers and conditions it (Perez-Trejo, 1994). We conclude that the most important *physical indicators* of land degradation in the Mediterranean basin are therefore related to the destruction of soil and vegetation resources.

3 EXTENSION AND DYNAMICS OF DEGRADATION PROCESSES

The degradation processes exhibit an enormous spatial variability within the Mediterranean basin: degraded areas are

* UNEP (1991) has redefined desertification in the sense of considering it as "the land degradation in arid, semiarid and dry-semihumid areas resulting mainly from adverse human impact" where the term 'land' includes land and local water resources, the land surface and its natural vegetation.

found in the direct vicinity of apparently stable ecosystems, and both, degrading or recovering systems, may occur under a large variety of climatic and physiographic conditions. In other words: to understand whether and where 'desertification' in the European Mediterranean is primarily driven by (changing?) climatic conditions or adverse human impact first of all requires a thorough understanding of degradation processes, but also a good knowledge about the spatial extension of stable and endangered ecosystems at regional level. However, although ecosystem processes with their implications for the future progress of desertification are intensely studied at numerous field sites in the Mediterranean region (e.g., MEDALUS, 1993) it is not yet clear how findings from field studies at patch-scale can be extrapolated and upscaled to relatively large areas. It is believed that remote sensing systems can significantly contribute to solve this problem. Satellite remote sensing also provides the means for a cartographic inventory of degraded (i.e., environmentally sensitive) areas, and it is virtually the only data source which permits a repeated monitoring of land degradation dynamics. Both, mapping and monitoring, represent important prerequisites for drafting and implementing development plans and policy decisions about a sustained use of Mediterranean land resources.

3.1 *Monitoring land degradation with remote sensing systems*

Operational earth observation satellites with multi-spectral sensor systems (e.g., Landsat-MSS and -TM, SPOT-HRV) are characterised by high spatial (30 or 20 m) and intermediate spectral resolution (6 or 3 bandpasses in the reflective part of the solar spectrum), but have rather large standard revisit intervals (16 - 26 days). The optical systems onboard these satellites register that part of the downwelling solar radiation which has been reflected by the earth's surface and transmitted through the atmosphere. One of the most important issues is to identify indicators for land degradation processes which have some general applicability to the Mediterranean as a whole, and which can be observed with operational (i.e., spaceborne) remote sensing systems. However, although it is agreed that remote sensing provides a convenient source of information, the problem is that the data collected by these instruments do not directly correspond to the information we need. We must therefore interpret the signal which has interacted with remote objects in terms of the properties of these remote objects (Verstraete, 1994).

3.2 *Remotely sensed primary parameters, thematic concepts and derived indices*

Engineering data about the detector sensitivity (i.e., calibration coefficients) allow us to reconvert encoded image grey values into measured radiances, and radiative transfer calculations can be used to correct for atmospheric effects, such that the surface-reflected radiance is restored from the satellite-measured signal. Dividing this term by the downwelling solar irradiance provides us with an important *primary parameter* which is termed 'bidirectional reflectance' ρ (figure 1).

Although of relevance to the global radiation budget, albedo and reflectance changes *per se* are not direct indicators of land degradation processes, in particular when we consider spatially complex areas like the European Mediterranean. It should be recalled that a simple increase in albedo/reflectance (surely characteristic for the desert encroachment in the African Sahel) might here be due to changes in land surface characteristics (i.e., increase of greenhouse areas, maturing cereals, non-photosynthetic vegetation, clearing of woodlands, etc.) which do not necessarily imply negative effects. It is therefore not sufficient to simply map albedo or its changes over time, but we need to infer the environmental impact of reflectance/albedo changes by characterising their *physical nature* in terms of land surface conditions. In other words, albedo and reflectance changes must be analysed in comparison to the multi-spectral characteristics of a range of known surface materials. We thus need an appropriate *scene model* which can be used to convert multi-spectral reflectance into thematic information (figure 1). Here the question is not so much to identify a particular surface type, but to characterise it (Verstraete, 1994). A variety of methods have been proposed which range from empirical spectral indices to the design and inversion of physically-based models. While the applicability of the various approaches depends on the nature and accuracy of the desired information and the availability of resources (i.e., sensor characteristics), an important prerequisite for their operational use is that they must satisfy specific requirements in terms of standardisation and portability.

Both the development of suitable indices and their interpretation in the thematic context of land degradation monitoring requires a *conceptual framework* which allows to draw concise conclusions about the land surface conditions. Though these underlying concepts might vary as a function of regional ecosystem characteristics (i.e., depending on physiographic conditions such as parent material, aridity etc.), we request that the results from different regions can be

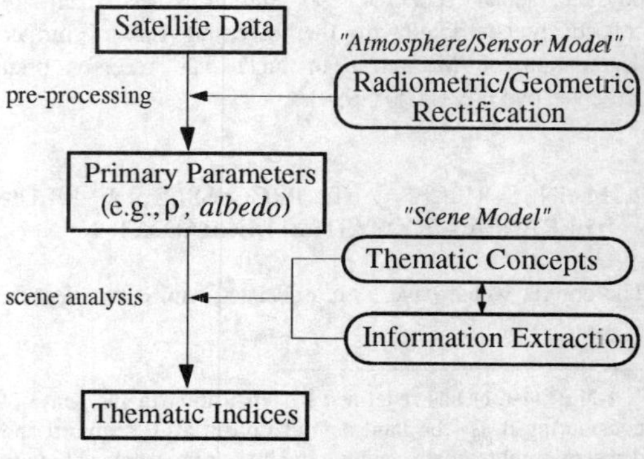

Figure 1. The conversion of satellite raw data into standardised thematic information layers.

consistently evaluated on a higher hierarchical level, such that the system's susceptibility to further degradation can be assessed by using image-derived, but also ancillary information layers. Important conclusions will nevertheless depend on the capability to analyse multi-annual time series, and it is for this reason that retrospective studies are so important with respect to developing approaches for a continuous monitoring of environmental changes (Graetz, 1994).

4 TOWARDS STANDARDISED APPROACHES

The most important issue in satellite remote sensing is to conceptualise and streamline the data analysis in a way that consistent indicators for land degradation can be obtained. This implies the development of standardised processing schemes which can be efficiently applied to a variety of landscape types in the Mediterranean basin.

4.1 *Radiometric pre-processing issues*

Today, we are in the favourable situation that several radiative transfer codes are available which, if constrained with the same atmospheric parameters, provide consistent results (e.g., Conel et al., 1988). Reflectance factor retrievals from the Landsat TM bands have been achieved with an accuracy of ± 0.005 to 0.02 (Holm, 1989; Moran et al., 1992; Wrigley et al., 1992; Markham et al., 1992), and it is important to note that this level of precision can also be accomplished when the atmospheric parameters are estimated from scene data itself (Hill and Sturm, 1991; Hill and Aifadopoulou, 1990). The major problem in retrieving reflectance factors from satellite data lies in the sensitivity to the absolute radiance calibration of the sensors. However, there is an increasing awareness of the scientific user community towards this problem. In-flight calibrations for Landsat-TM and SPOT data are conducted at various high-reflectance sites which is useful for determining at least an absolute radiometric calibration gain and monitoring the sensor degradation with time (e.g., Slater et al., 1987; Thome et al., 1993).

4.2 *Conceptual framework for the analysis of vegetation and soil conditions*

Thematic concepts provide the rational for translating remote measurements of primary parameters (e.g., ρ) into relevant thematic information about vegetation and soil conditions. Ideally, such concepts should be valid for any location in large ecosystems, but in reality some adaptations might be necessary in order to account for regional variations.

4.2.1 Mapping soil conditions. Eroded soils are often recognised through typical soil colour changes which are due to the removed topsoil. It is nevertheless difficult to define a universally applicable concept which accounts for a variability of soil types and the corresponding sequence of pedogenetic horizons. Our approach refers to basic concepts which consider soil development to be either *progressive* or *regressive* with time (Birkeland, 1990). Under progressive development, soils become better differentiated by horizons, and horizon contrasts become stronger. In contrast, regressive pedogenesis refers to the addition of material to the surface at a rate that suppresses soil formation (i.e. eolian dunes, glacial moraines, distal fans, etc.), or the suppression of pedogenesis and the truncation of soil horizons by surface erosion. Both, progressive and regressive pedogenesis cause alterations of the soil surface which, due to corresponding colour changes, are detectable through the wavelength-dependant variations of ρ (e.g., Baumgardner et al., 1985; Escadafal, 1994). The intensity of brunification and rubification, and the organic matter content of the topsoil material thus provide important diagnostic features for the spectral identification of a majority of undisturbed Mediterranean soils (e.g., cambisols, fluvisols, luvisols, vertisols, rendzinas). Compared to that, soil erosion produces truncated soil profiles which are characterised by decreasing amounts of iron oxides and organic carbon, while the proportion of parent material increases (e.g., lithosols, regosols). Most parent materials differ spectrally from developed soil substrates, in particular due to specific spectral absorption features and increased albedo levels.

The resulting concept, which is based on the spectral contrast between developed substrates and parent materials, seems to provide a widely applicable framework for relating spectrally detectable surface phenomena to Mediterranean soil conditions, thereby satisfying an important requirement for the successful application of remote sensing techniques (Hill, 1993; Hill et al., 1994a). However, the validity of such concepts has to be carefully analysed in the context of the specific physiographic conditions under which they should be applied. Modifications might be required, for example, in cases of extreme aridity where soil forming processes do not permit the accumulation of noticeable amounts of organic components (e.g., Escadafal, 1994).

4.2.2 Mapping vegetation abundance. Vegetation attributes are usually described by structure, dynamics and taxonomic composition, of which taxonomy is the least important of the three. The classification which is most compatible with remote sensing relates to the projected foliage cover (PFC, or cover) and the life form of the tallest vegetation stratum (Graetz, 1989). However, about 75 % of the earth's surface is covered by sparse vegetation which transmits the colour of the soil beneath, i.e. the projected foliage cover (PFC) is below 1. In particular semi-arid ecosystems, such as the Mediterranean, are dominated by such vegetation communities. For these, the soil surface itself should be as much an object of attention as is the vegetation (see Graetz, 1989, and above), and the key issue is therefore to provide accurate estimates of green vegetation abundance which are not biased by the spectral contribution of background components (i.e., soils and rock outcrops). Attention should also be given to the spectral characteristics of non-green components of plant canopies and associated plant litters, which largely contribute to the reflectance of terrestrial surfaces in semi-arid ecosystems (Elvidge, 1990). Although we know that the spectral resolution of earth observation satellite systems is not adequate to consistently differentiate between dry plant components and soils, efforts have to be made to resolve ambiguities from the image context (Smith et al., 1990).

4.3 *Advanced data interpretation - the spectral mixing paradigm*

It results from our conceptual considerations that we require information extraction methods which provide largely unbiased estimates for green vegetation cover, permit the identification of soil related spectral information, and allow sufficient standardisation for multi-temporal monitoring. It has already been argued that traditional multi-spectral classification approaches as well as most vegetation indices are not ideally suited to fulfill these requirements (Hill et al., 1994b). Since the inversion of physically-based bidirectional reflectance models against satellite data is not feasible with currently available data sets, we wish to draw the attention on suitable semi-empirical models.

One of the most promising approaches has become known as "Spectral Mixture Analysis" (e.g., Adams et al., 1989; Smith et al., 1990). Spectral mixture analysis (SMA) assumes that most of the spectral variation in multi-spectral images is caused by mixtures of a limited number of surface materials, and it attempts to model the multispectral reflectance ρ as a mixture of representative 'prototype' spectra, the so-called 'spectral endmembers' (i.e., vegetation, soil and bedrock components, 'shade' as an illumination component, etc.). Linear mixing within the footprint of a multichannel spectrometer or imaging sensor further assumes that the surface components are large and/or opaque enough to allow photons to interact with only one component, i.e. the radiative transfer processes are additive. Spectra can then be 'unmixed' by inverting the linear mixing equation

$$R_i = \sum_{j=1}^{n} F_j \cdot RE_{ij} + \varepsilon_i \quad \text{and} \quad \sum_{j=1}^{n} F_j = 1$$

through a least squares regression, while constraining the sum of fractions (i.e., the proportional weights) to one; R_i denotes the reflectance of the mixed spectrum in band i, RE_{ij} the reflectance of the endmember spectrum j in band i; F_j is the fraction of endmember j, and ε_i the residual error in band i. A unique solution is possible as long as the number of spectral components (endmembers) does not exceed the number of bands plus one.

The 'mixing paradigm' provides a type of image enhancement which is not only intense but also physically meaningful (Craig, 1994). Its objective is of course to isolate the spectral contributions of important surface materials ("endmember abundance") before these are edited and recombined to produce thematic maps (Adams et al., 1989). We have successfully adopted this approach to analyse the spectral information content related to the erosional state of

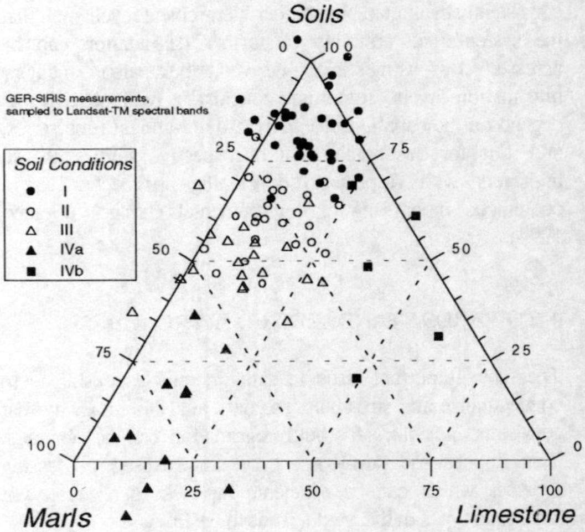

Figure 2. Ternary plot of about 100 field- and lab-measured soil spectra at Landsat-TM spectral resolution, where the mixing volume is defined by one soil (vertic cambisol) and two bedrock spectra (marls and limestone). Soil conditions range from undisturbed soils (I) to substrates that are increasingly affected by erosion (II-IV) (Hill et al., 1994b).

soils (figure 2), but also to derive precise maps of soil conditions and improved estimates of green vegetation abundance from various types of multispectral images (e.g., Hill, 1993; Hill et al., 1994a). An essential part of SMA consists in identifying optimised sets of spectral endmembers which are representative for the major variations of regional soil and vegetation conditions. Though endmembers can be retrieved from the image itself, we prefer to choose them from collections of spectroradiometric field- and/or laboratory measurements, the so-called spectral libraries. The spectral library concept may be questioned with regard to the infinite number of spectral individuals encountered in the field. However, there is reason to believe that we can represent the spectral variability in which we are interested by a relatively small number of base categories (i.e., developed soils, fresh or weathered rock outcrops, green vegetation, non-photosynthetic vegetation). The compilation of such 'libraries' has already been initiated (e.g., Altherr et al., 1991), and is being continued in the frame of additional field campaigns in Greece, Spain, France and Italy.

4.4 *Susceptibility analysis through the evaluation of satellite-derived information layers*

It is considered very important to further evaluate temporal changes in soil and vegetation conditions in the context of a more ecologically-based framework. Again, repeated monitoring requires standardisation, mainly because any improvements in the analysis need to be easily incorporated into existing monitoring schemes by simply reprocessing already available image-derived information layers (e.g., soil

Figure 3. Peloponnese, Greece: Comparison of land surface properties based upon a synoptic evaluation of Landsat-derived information layers on soil conditions and vegetation abundance. The numbers refer to "desertification risk levels" in the sense of varying susceptibility to further degradation, ranking from "stable" (1: vegetation cover more than 50% and soil condition index I) to "irreversible degradation" (9: vegetation cover less than 20 % and soil condition index IV).

Colour plate see page 491

condition index, estimated green vegetation cover). As a first attempt, we have proposed a simplified formalism to combine soil- and vegetation-related information layers (Hill, 1993; Hill et al., 1994a). It is based on the fact that, in a more synoptic perspective, an "environmental ranking" of specific soil conditions in Mediterranean ecosystems can be defined as a function of green vegetation cover (because of its protective role), thereby distinguishing for example between regrowth/succession on eroded soils and sparse vegetation cover on well-preserved soil resources (figure 3).

The proposed susceptibility index only represents a first attempt to combine image-derived information layers of soil and vegetation conditions through a strictly formalised procedure, and we are aware of the fact that it should be revised and further developed in close cooperation with ecologists and specialists from various geosciences. New concepts or findings from detailed field studies are easily integrated by updating the rule base module such that they can be readily applied to new satellite images and archived data products. The incorporation of terrain parameters from digital elevation data (e.g., slope, exposition) appears particularly attractive for obtaining an improved evaluation of ecosystem characteristics, such as water availability or erosion risks.

5 A SATELLITE OBSERVATORY FOR MEDITERRANEAN LAND DEGRADATION MONITORING

Finally, we wish to discuss basic elements of a future operational environmental monitoring system for the Mediterranean basin. This requires a concise definition of the region to be observed, but also includes strategies for implementing efficient processing schemes as a function of existing data types, archives, present and future sensor availability, and overpass cycles. Although this proposal is focused on the use of operational earth observation satellite systems (i.e., Landsat and SPOT), it is believed that existing and future low resolution satellites with more frequent coverage (e.g., the NOAA AVHRRs, the planned VEGETATION instrument onboard SPOT-4, MERIS, etc.) and active microwave systems (e.g., ERS-1, ERS-2, RADARSAT) should be incorporated at a later stage.

5.1 Operational land degradation mapping and monitoring

It seems obvious that the identification of degraded areas in the sense of an environmental inventory provides the fundamental basis for such approaches. In order to better understand the *processes* of land degradation from their spatial context, it is essential to monitor soil conditions and the disturbance regime of plant communities over time, including their successional recovery. With regard to the

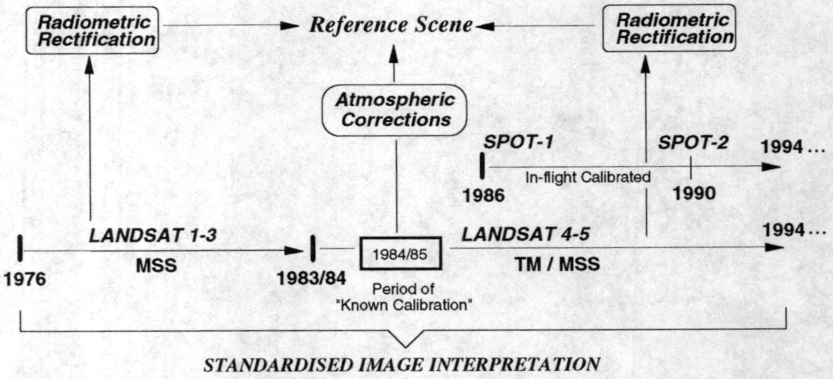

Figure 4. Radiometric rectification pathways for the analysis of Landsat and SPOT time series

available data archives* we wish to particularly emphasise the importance of retrospective studies which may provide the key for understanding the present situation, but also for optimising our approaches for regular monitoring.

Since the retrospective analysis of archived data and regular monitoring are sensitive to the absolute radiance calibration of the sensors, we require practical approaches to minimise the uncertainty about these calibration coefficients. We have already mentioned (section 4.1) that in-flight calibration coefficients for earth observation satellites are available from specifically designed experiments (e.g. Slater et al., 1987; Thome et al., 1993). Though the coefficients are not continuously updated, they are at least valid for specific periods of the sensor's lifetime. This allows us to follow alternative pathways which do not require that each individual scene is radiometrically corrected through radiative transfer calculations (figure 4). It would be sufficient to apply atmospheric corrections only to reference scenes from so-called 'periods of known calibration', and to routinely adjust the radiometry of earlier or later image acquisitions with reference to pseudo-invariant surfaces in the scenes (Schott et al., 1988). This approach, to which we refer as 'radiometric rectification', has already been successfully applied to a time series of Landsat TM data (Hill et al., 1994a).

5.2 Stratified sampling for environmental monitoring

The criteria for defining the regional extension of Mediterranean-type ecosystems have been a scientific issue in itself (e.g., Di Castri and Mooney, 1973; Di Castri et al., 1981). In principal, we have to consider the Iberian Peninsula (except its humid parts in the north), the French *Midi*, the Apennine Peninsula, the Croatian-Albanian coast, Greece, western and southern Turkey, western Syria, Lebanon, Israel, parts of Lybia, northern and central Tunisia, and Algeria and Morocco north of the Atlas massifs. Even when we limit ourselves to the Mediterranean member states of the European Union, we should accept that a complete coverage with data from earth observation satellites is not feasible. In particular the repeated monitoring for change detection would largely exceed our financial and data analysis capacities.
Instead, we propose to define a sampling frame of representative sites which accounts for the major physiographic and agro-economic variations within the Mediterranean basin. Such a sampling scheme must be based on the path/row reference system of the Landsat and SPOT orbiting patterns, but is to be complemented by stratifying the region into zones with homogeneous climatic and edaphic characteristics (ecoregions). This is important because any comparative analysis of land degradation dynamics must account for the physiographic variability within the Mediterranean basin. We expect that these ecoregions can be derived largely from the evaluation of already available cartographic documents and topographic data (e.g., Commission of the European Communities, 1985; Council of Europe and Commission of the European Communities, 1987). Each image frame then represents a measurement for analysing how changes in climate, drought, land use and fire regimes generate regional changes in ecosystem processes and patterns which can influence the future progress of land degradation. As such, they may be integrated in a nested hierarchy of aerial units with similar response to desertification ("Desertification Response Units") which has been recently proposed by Imeson et al. (1994).

5.3 Standardised thematic interpretation.

The proposed scheme consists of several standardised modules (characterised by the ellipsoids in figure 5). Radiometrically rectified satellite images provide the primary parameter ρ for a given control site, and we then use spectral mixture analysis (SMA) to convert the multispectral surface reflectance into soil and vegetation related information layers. Although a unique set of standard endmembers has been

* Landsat-MSS data for European areas are available from 1976, Landsat-TM coverage started in 1983.

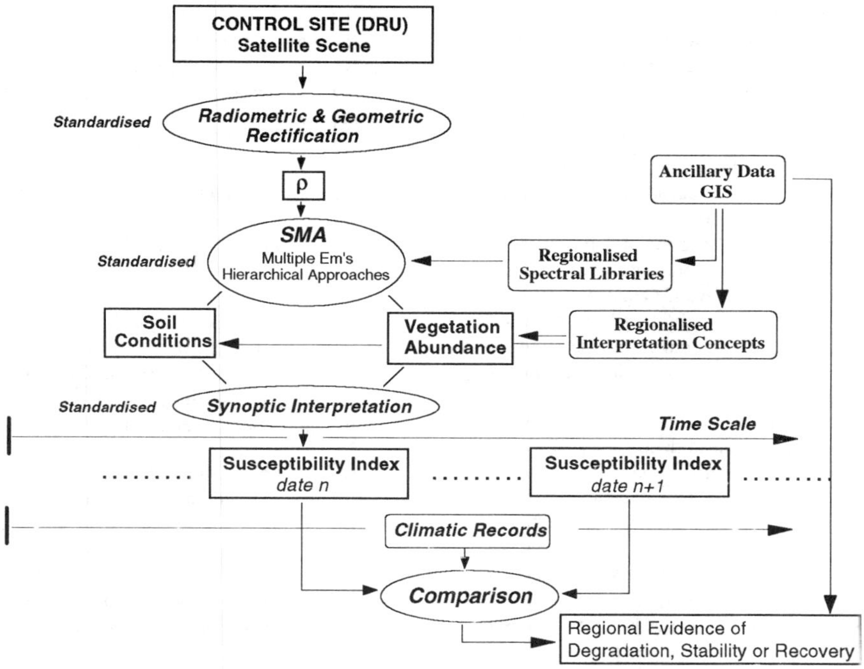

Figure 5. Processing scheme for deriving standardised land degradation indices from earth observation satellite data

identified which can be used for different areas of primarily carbonatic rocks, it is believed that the spectral unmixing can sometimes be optimised by accounting for regional endmember characteristics (figure 5). The conceptual framework may require modifications in view of particular regional conditions, too. But it should also be mentioned that the concept which we presented in section 4.2, looses validity only in areas of extreme aridity.

The 'synoptic interpretation' module uses the intermediate information layers (e.g., soil condition index, projected foliage cover) for computing an index of degradation and susceptibility to land degradation processes (figure 5), and the comparison of susceptibility indices from different years will provide evidence of either degradation, stability or recovery on a regional scale. However, an important objective of the thematic interpretation is to separate the rhythmic phenological changes of growth and senescence from episodic alterations introduced by climate or human induced disturbance. We must therefore also incorporate the climatic records in order to understand whether the meteorological situation in both years was at all comparable (figure 4). Since the use of abnormally humid or dry years should be avoided, we can impossibly define a fixed recurrence rate. Instead, we propose a flexible monitoring scheme which guaranties a minimum of one scene per successive five-year-interval. It should also be avoided to directly compare images from different seasons in order to exclude artefacts due to phenological effects, and to minimise radiometric distortions which result from bi-directional effects and illumination differences.

6 CONCLUSIONS

We have summarised important sources and processes of Mediterranean ecosystem degradation in order to discuss options for monitoring land degradation with remote sensing systems. Unlike meteorological approaches which concentrate on the exchange of matter, momentum and radiation between the earth's surface and the atmosphere, we have emphasised the importance of thematic interpretation pathways for the characterisation of actual land surface conditions

We have further illustrated how remotely sensed primary parameters, such as the spectral surface reflectance ρ, can be converted into a standardised characterisation of soil conditions and vegetation abundance. In this context, the term of 'thematic concepts' has been introduced, by which we understand the conceptual background for identifying functional links between surface reflectance and vegetation and soil characteristics. Such concepts are primarily based on research in geosciences and ecology, and we require that they are not only valid for local experiments, but can be applied to extended regions with specific physiographic conditions (e.g., bioclimate, lithology, soil forming processes, etc.). As long as they can be related to regional scale ecosystems, we even foresee to use multiple concepts in response to the physiographic conditions of particular regions.

The actual data analysis employs linear spectral mixture analysis as core element. SMA provides an intense, but also physically meaningful image enhancement which ideally complements the conceptual framework we have introduced. Selective editing of the resulting fraction images permits an efficient separation of vegetation and soil related spectral information. This is an important advantage for obtaining more objective estimates of green vegetation abundance, and for mapping substrate-related spectral soil properties more independently from the disturbing influence of sparse vegetation cover or soil moisture and illumination differences. It further holds the potential to be largely standardised in terms of required processing parameters. The inputs from individual analysts can be minimised, which is an advantage for designing operational processing schemes.

Although some processing modules are already available which appear suited for a standardised data analysis within operational schemes, additional efforts are required, in particular for the definition of suitable sampling schemes (i.e. the selection of monitoring sites which are representative for large ecoregions), for extending the thematic concepts for an interpretation of remotely sensed primary parameters as a function of ecosystem characteristics, for the incorporation of ancillary information (climatic records, lithology, topographic information), and for improving various processing modules with respect to the physiographic variability within the Mediterranean basin.

Finally, we wish to emphasise the importance of a retrospective analysis of earth observation satellite images which are available for almost 20 years. The integrated interpretation of the satellite-derived information layers, available climatic records and results from detailed field studies may provide a new perspective to understand land degradation processes in the European Mediterranean.

7 REFERENCES

Adams, J.B., M.O. Smith and A.R Gillespie, 1989, Simple models for complex natural surfaces: a strategy for the hyperspectral era of remote sensing, Proc. of the IGARSS '89 Symposium, July 10-14, Vancouver, vol. 1, 16-21.

Altherr, M., J. Hill, M. Mehl, and P. Pollicini, 1991, *NASA multi-sensor airborne campaign: reflectance spectroscopy Ardèche, June-July 1991*, Technical Note no. I.91.120, (JRC, Inst. for Remote Sensing Applications: Ispra, Italy).

Birkeland, P.W., 1990, Soil-geomorphic research - a selective overview, *Geomorphology*, 3, 207-224.

Craig, M.D., 1994, Minimum-volume transforms for remotely sensed data, *IEEE Transactions on Geoscience and Remote Sensing*, vol. 32, no. 3, 542-552.

Conel, J.E., R.O. Green, R.E. Alley, C.J. Bruegge, V. Carrere, J.S. Margolis, G. Vane, T.G. Chrien, P.N. Slater, S.F. Biggar, P.M. Teillet, R.D. Jackson, and M.S. Moran, 1988, In-flight radiometric calibration of the Airborne Visible/Near-Infrared Spectrometer (AVIRIS), *in* P.N. Slater, ed., *Recent advances in sensors, radiometry, and data processing for remote sensing*, Proc. SPIE, vol. 924, 179-195.

Commission of the European Communities, 1985, Soil map of the European Communities, scale 1:1,000,000, (Office for Official Publications of the European Communities: Luxembourg).

Council of Europe and Commission of the European Communities, 1987, Map of the natural vegetation of the member countries of the European Community and the Council of Europe, (Office for Official Publications of the European Communities: Luxembourg).

Di Castri, F. and H. A. Mooney, eds., 1973, *Mediterranean type ecosystems, origin and structure*, (Springer Verlag: Berlin ...).

Di Castri, F., D.W. Goodall, and R.L. Specht, eds., 1981, *Mediterranean-type shrublands*, Ecosystems of the World, 11, (Elsevier Scientific Publishing Company: Amsterdam ...)

Elvidge, C.D., 1990, Visible and near infrared reflectance characteristics of dry plant materials, *Int. J. Remote Sensing*, vol. 11, 1775-1795.

Graetz, R.D., 1990, Remote sensing of terrestrial ecosystem structure: an ecologist's pragmatic view, in R.J. Hobbs and H.A. Mooney, eds., *Remote sensing of biosphere functioning*, (Springer Verlag: New York ...), 5-30.

Graetz, R.D., 1994, Empirical and practical approaches to land surface characterisation and change detection, *in* J. Hill and D. Peter, eds., *Remote sensing for land degradation and desertification monitoring in the Mediterranean basin*, Proc. of a Workshop held in Valencia, Spain, 13-14 June 1994, in preparation.

Grenon, M., and M. Batisse, eds., 1989, Futures for the Mediterranean basin, the Blue Plan, (Oxford University Press: Oxford).

Hill, J. and D. Aifadopoulou, 1990, Comparative analysis of Landsat-5 TM and SPOT HRV-1 data for use in multiple sensor approaches, *Remote Sensing of Environment*, 34, 55-70.

Hill, J. and B. Sturm, 1991, Radiometric correction of multitemporal Thematic Mapper data for use in agricultural land cover classification and vegetation monitoring, *Int. J. Remote Sensing*, 12, 1471-1491.

Hill, J., 1993, Monitoring land degradation and soil erosion in Mediterranean environments, ITC Journal, 4, 323-331.

Hill, J., W. Mehl, M.O. Smith, and J. Mégier, 1994a, Mediterranean ecosystem monitoring with earth observation satellites, Proc. 13th EARSeL Symposium on "Remote Sensing - From Research to Operational Applications in the New Europe", Dundee, Scotland, UK, 28 June - 1 July, (Springer Verlag: Budapest ...), 131-141.

Hill, J., J. Mégier, and W. Mehl, 1994b, Land degradation, soil erosion and desertification monitoring in Mediterranean ecosystems, *Remote Sensing Reviews*, in press.

Holm, R.G., M.S. Moran, R.D. Jackson, P.N. Slater, B. Yuan, and S.F. Biggar, 1989, Surface reflectance factor retrieval from Thematic Mapper data, *Remote Sensing of Environment*, 27, 47-57.

Imeson, A.C., L.H. Cammeraat, and F. Perez-Trejo, 1994. The response of landscape-units to desertification, *in* J. Thornes and J. Brandt, eds., *Mediterranean desertification and land use*, (John Wiley: New York ...), in press

Le Houérou, H.N., 1990, Ecological guidelines to control land degradation in European Mediterranean countries, *in* J.L. Rubio and R.J. Rickson, eds., *Strategies to combat desertification in Mediterranean Europe*, EUR 11175 EN/ES, (Office for Official Publications of the European Communities: Luxembourg), 331-360.

Markham, B.L., R.N. Halthore and S.J. Goetz, 1992, Surface reflectance retrieval from satellite and aircraft sensors: results of sensor and algorithm comparisons during FIFE, *J. of Geophys. Res.*, 97, 18,785-18,795.

MEDALUS, 1993, Mediterranean desertification and land use. Executive summary, abstracted from the first phase of the MEDALUS Project, (Bristol, UK).

Moran, M.S., R.D. Jackson, P.N. Slater, and P.M. Teillet, 1992, Evaluation of simplified procedures for retrieval of land surface reflectance factors from satellite sensor output, *Remote Sensing of Environment*, 41, 169-184.

Perez-Trejo, F., 1994, *Desertification and land degradation in the European Mediterranean*, EUR 14850 EN, (Office for Official Publications of the European Communities: Luxembourg).

Schott, J.R., C. Salvaggio, and W.J. Volchok, 1988, Radiometric scene normalisation using pseudo-invariant features, *Remote Sensing of Environment*, 26, 1-16.

Slater, P.N., S.F. Biggar, R.G. Holm, R.D. Jackson, Y. Mao, M.S. Moran, M. Palmer, and B. Yuan, 1987, Reflectance- and radiance-based methods for the in-flight absolute calibration of multispectral sensors, *Remote Sensing of Environment*, 22, 11-37.

Smith, M.O., J.B. Adams, and D.E. Sabol, 1994, Mapping sparse vegetation canopies, *in* J. Hill and J. Mégier, eds., *Imaging spectrometry - a tool for environmental observations*, (Kluwer Academic Publishers: Dordrecht ...), 221-235.

Smith, M.O., S.L. Ustin, J.B. Adams, and A.R. Gillespie, 1990, Vegetation in deserts: I. A regional measure of abundance from multispectral images, *Remote Sensing of Environment*, 31, 1-26.

Thome, K.J., D.I. Gellmann, R.J. Parada, S.F. Biggar, P.N. Slater, and M.S. Moran, 1993, In-flight radiometric calibration of Landsat-5 Thematic Mapper from 1984 to present, *Proc. of Photo-Optical Instrum. Eng. Symp.*, 12-16 April, Orlando, Fla., 126-130.

UNEP, 1991, *Status of desertification and implementation of the United Nations plan of action to combat desertification*, (UNEP: Nairobi).

Verstraete, M.M., 1994, Scientific issues and instrumental opportunities in remote sensing and high resolution spectrometry, *in* J. Hill and J. Mégier, eds., *Imaging spectrometry - a tool for environmental observations*, (Kluwer Academic Publishers: Dordrecht ...), 25-38.

Wrigley, R.C., M.A. Spanner, R.E. Slye, R.F. Pueschel, and H.R. Aggarwal, 1992, Atmospheric correction of remotely sensed image data by a simplified model, *J. of Geophys. Res*, 97, 18,797-18,814.

Sensors and Environmental Applications of Remote Sensing, Askne (ed.) © 1995 Balkema, Rotterdam, ISBN 90 5410 524 0

Assessment of the potential of optical and microwave satellite remote sensing techniques for land surface monitoring of the impact of land reform

P. Winkler, P. Somogyi & Gy. Zelei
FÖMI Remote Sensing Centre, Budapest, Hungary

ABSTRACT: The study assesses the potential of ERS-1 SAR date for monitoring changes in land sizes and for the differentiation of land cover categories on a test field in HUNGARY. A colour composite of three ERS-1 images is used in the study with dates of 8th July, 12 August and 16th September. The five main agricultural plants on the area are wheat, maize, sugarbeet, sunflower and alfalfa. On a 128x128 pixels size area the effect of the Lee sigma Filter is evaluated and the optimal window-size and other filter parameters is applied to the three images to smooth speckle and enhance the visual interpretability of the original data. We report some numerical results for the studied plant categories that include the calculation of the mean amplitude values on homogeneous areas, and we try to recognise and offer reasons for the colours that are characteristic for the different plants.

1 INTRODUCTION

In Hungary, after the collapse of the so called "socialist system" in 1990, the freely elected new parliament and government had decided to change not only the political, but also the economical structure of the country. This means - among other things - that the previous "socialist properties" should be privatized, e.g. many citizen, who had before the II. world war any property, can get back his ownership under certain circumstances determined by the new law. This privatization would effect the structure of the Hungarian agriculture land use as well. Most of the large agricultural cooperatives will be privatized. During the previous regime, the average size of agricultural fields was about 50-80 ha-s. After the privatization this will be very similar to the Western-European systems, where the average size of agricultural fields is 3 - 5 ha-s.

During the last 14 years of activities of the FÖMI Remote Sensing Centre we worked out technologies to use remote sensing techniques for land use mapping, vegetation status assessment, monitoring of cultivated area, change detection, environmental monitoring etc. based on Landsat MSS, TM, SPOT and airborne data.

In the frame of this project we would like to get answer, how can we use our alredy existing technologies and the ERS-1 data for the detection of landscape changes due to privatization.

2 DATA BASIS

In the following we summarize how we chose our test area and what preliminary work has been done.

2.1 *Selection of the test area*

The test area had been chosen with the active participation of our potential users, experts from the Ministry of Agriculture and the Ministry for Environmental Protection and Regional Planning, taking into consideration the following factors:

- the advancement of the privatization
- the area should be a characteristical agricultural region
- environmental considerations
- availability of different types of data from earlier investigations, surveys
- the knowledge to be gained could be connected to other national and international projects (CORINE, MARS, etc)
- the test area should be easily reached (because of limitid budget)

On the base of the above considerations we had chosen a test area, which is located in Pest county, about 50 km in the south of Budapest and has the following corner coordinates (geographical coordinates): N47.6/E19.0 N47.6/E20.0 N46.6/E19.0 N46.6/E20.0 This area is corresponding to the 1:100000 scale topographic map sheet of number 55 in the Hungarian national unified map projection system.

In the frame of ESA/FAO cooperation we received three ERS-1 SAR images of this area, acquired on the 8th July, 12th August and 16th September, respectively.

2.2 *Reference materials*

The following materials were available for the test area:

- a "COSMOS" satellite image that contains the eastern part of the test area (it is a panchromatic image made in 1979 and has a resolution of 5m)
- a LANDSAT TM composite (band 2, 3, 4) made in 04. 04. 1985.
- a LANDSAT TM composite (band 3, 4, 5) made in 09. 14. 1986.
- a LANDSAT TM composite (band 2, 3, 4) made in 08. 18. 1988.
- a LANDSAT TM composite (band 2, 3, 4) made in 03. 17. 1990.
- black and white air photos in the scale 1:10 000, made for soil mapping
- soil maps in the scale 1:10 000, made by the Soil Mapping Group of the Remote Sensing Department of FÖMI RSC
- colour infrared air photos in the scale of 1:22 000 from July 1992, covering the whole Pest county
- result of the interpretation of an environmental study in the scale of 1:100 000 with 19 classes, covering the whole Pest county (made by the Hungarian-German program-office and sponsored by the Ministry of Environmental Protection in Germany)
- 6 m resolution airborne SAR images (X-band), covering part of the test area, aquired by INTERA (Canada) in the frame of the "Open Sky"
- 1:100 000 scale topographic map
- 1:100 000 scale agrotopographic map
- 1:10 000 scale topographic map
- agrometeorological data (including measurements of rainfall, wind and temperature in six different places of the test area from the time of the three overpass of the satellite)

2.3 *Field data collection*

In the possession of the available data we went to the test area five times to collect field data. On the field we tried to localize those parts of the test area where the privatization had already started. We had chosen such fields that could be used as training sets in the course of computer processing. We recorded the boundaries of regions with different land surface coverage, and on the agricultural territories we marked the boundaries of the fields with different plants.

As the result of the field data collection we can conclude the following:

- Agriculture is the dominant activity in the region of the test area.
- The most important plants are: winter cereals, different kinds of maize, alfalfa and sunflower.
- On the fields already privatized we found quite mixed cultivation. The new owners grow vegetables (for example cabbage, grapes, fodder-plants as well as such industrial plants as the sunflower and potato. The size of the privatized fields does not exceed the 2-5 hectares.
- The test area consists of part of the National Park of Kiskunság which is a natural grass land.
- Forests can also be found on smaller parts of the test area.
- The important surface waters of the area are the Danube and the different mine lakes.
- On the test area the density of the roads and railways corresponds to the characteristic hungarian circumstances. We can find a highway, roads, smaller roads with different surface materials in the region as well as railways and railway junctions.

3 IMAGE PROCESSING

For image processing - because of the lack of a complete radar image processing software package - we had to use a system, that was far from optimal for SAR processing. The system includes a MicroVAX II. with softwares including the in-house written Remote Sensing Package for image processing, an IBM PC with the ERDAS 7.5 image processing software package, and a Pericolor 2000 image processing computer for displaying the images. The processing steps were as follows:

1) on the MicroVAX:
- reading the tapes containing the images

2) on the PC:
- cutting out a subimage that contains the test area
- because of the reversed position of the subimage reflecting it around its vertical axis

3) on the MicroVAX and the Pericolor:
- reduction from 16 bits to 8 bits
- filtering
- image to image registration
- geometric correction
- interpretation

We will describe in details only the filtering method, the image to image registration and the geometric correction of the images.

3.1 *Filtering*

Filtering is a very essential step in processing radar images, because SAR images contain speckle noise even after the multilook processings. There has been introduced many filters so far which take into account the specific properties and statistical characteristics of radar images and speckle noise - the best known among them are the the Lee Sigma and Lee Local Statistics filter (Lee, 1986), the Frost (Frost et al., 1982) and the MAP (Lopes et al., 1993) filter.

We implemented one of the simplest of the different speckle specific filters, the Lee Sigma Filter (Lee, 1986). Its main features can be summarized as follows:

a) It uses the multiplicative model for the speckle noise, on the base of which the following equation

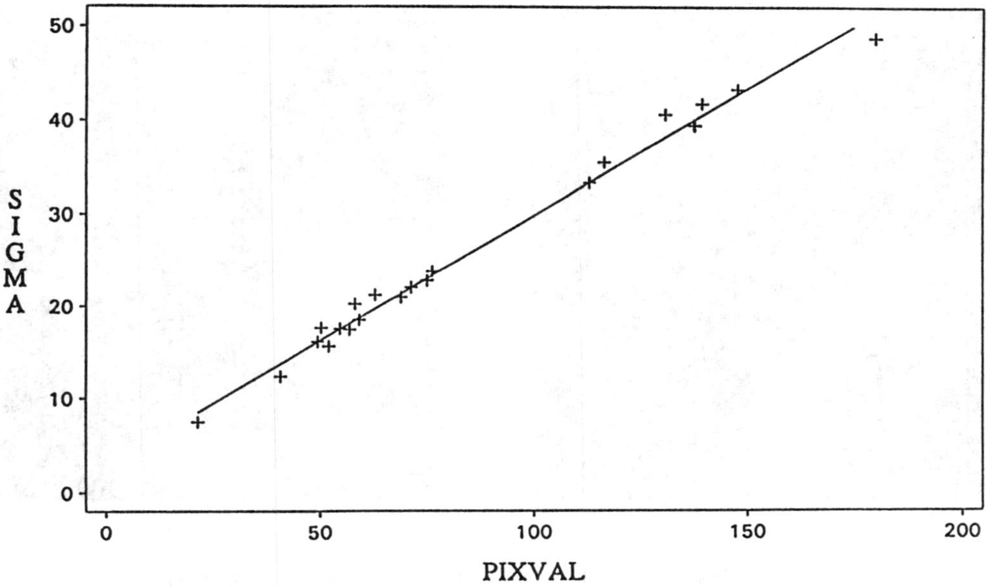

Figure 1. ERS-1 speckle noise characteristics

can be derived (valid for homogeneous, flat areas only):

$$\sigma_v = \frac{\sqrt{var(z)}}{\bar{z}}, \qquad (3.1)$$

where σ_v is the standard deviation of the noise, var(z) and z are the variance and the mean of the observed values on different homogeneous areas, respectively.

b) The Lee filter is a spatial filter in the sense that it uses a moving window with variable size for filtering the image, and replaces the central pixel with the average of certain pixels in the window. The selection of the pixels included into the averaging is based on the sigma probability of the Gaussian distribution, which selection can be summarized with the following inequality:

$$(1-C\sigma_v)z_{cp} < z_{i,j} < (1+C\sigma_v)z_{cp} \qquad (3.2)$$

where z_{cp} is the grey value of the central pixel, $z_{i,j}$ is a pixel in the window, σ_v is the standard deviation of the noise, and C is a user-defined constant (1< =C< =2). In this inequality we have already made use of the simple equation in a), which tells us that the standard deviation of the pixel values in the window, σ_z can be expressed by the product of the mean, z and the standard deviation of the noise, σ_v.

For determining what filter parameters should be used, we chose a subimage of 256x256 pixels for which area we applied the Sigma filter with different parameters. We had to determine the window size, the value of σ_v and the value of the C constant. For determining the value of σ_v, we calculated the mean and standard deviation of pixel values on homogeneous areas, and delineated them in a coordinate system. The value of σ_v is equal to the tangent of the line we fitted onto the discrete points (Figure 1. shows the result for the August 12 image).

After the calculation of the value of σ_v, we determined the other two parameters (C, window-size) by running the filter to the subimage several times with different parameters and evaluated the results both numerically and visually. Table 1. shows how the window-size influences the mean and standard deviation of the pixel values considering both the whole subimage and different fields within the subimage.

We found that considering both the effectiveness of the filters with different parameters and the computational time, we can reach the best result by filtering the images in an iterative method, first applying a window 7x7 with C=1.5, then using a window size 3x3 with C=1.

This way we can preserve the mean value quite well, while reducing the standard deviation significantly. In Figure 2. we show a subimage of 128x128 pixel size, first the unfiltered image, then the result of the iterative filtering.

a)

b)

Figure 2. The result of the applied filtering method on the subimage

3.2 *Geometric correction and image to image registrations*

To make a multitemporal image from the three SAR images acquired on different dates we had to register the images. We found that there was no need for the usual calculations using second or higher order polynomials to make the registration, a single shift between the images was enough to bring them into a total correspondence.

Table 1. Numerical results of the filtering for the whole subimage and within-fields using the Sigma filter with different window size

	Filters	Mean	Stand.Dev
Subimage	original	81.0	36.9
	5x5	78.8	28.7
	7x7	78.8	27.6
	9x9	78.5	27.1
	11x11	78.3	26.8
	7x7 and 3x3	78.9	26.9
Sunflower	original	127.3	44.3
	5x5	124.4	28.4
	7x7	124.4	25.9
	9x9	123.8	24.8
	11x11	123.3	24.2
	7x7 and 3x3	124.5	24.3
Wheat	original	60.3	18.7
	5x5	58.9	10.2
	7x7	58.9	8.8
	9x9	58.8	8.6
	11x11	58.8	8.6
	7x7 and 3x3	59.0	7.8

We also transformed the three radar images into the Hungarian National Projection System (Hungarian abbreviation is EOV) using GCP-s (ground control points). It was an essential step because our GIS database that we would like to apply in the future to help the interpretation is in the same map system. The transformation was accomplished using polinoms of the first order. The scale of the maps from which we took the reference points was 1:10000 to ensure the appropriate accuracy. The RMS error of the transformation was smaller then 1 pixel and we used at least 15 appropriately spreaded GCP-s for accomplishing the transformation in the case of each images. The composite image we made from the three transformed images was restored on our Colormation filmwriter (Figure 3 - multitemporal ERS-1 SAR image). In the image the July image is in blue, the August image is in red, and the September one is in green.

By resampling the radar images to have a pixel size characteristic to the Landsat TM imagery, we also registered the three images to a TM acquired in 1990 with the aim that we could observe the changing field sizes due to privatization and that we could find out what additional information can be retrieved from the radar images comparing them to a Landsat TM.

4 INTERPRETATION AND CONCLUDING REMARKS

The purpose of the interpretation we have done so far was twofold:

1) we tried to assess the possibility of using the ERS-1 and Landsat TM images to monitor the changes in the size of the agricultural fields due to privatization and

2) we investigated in what extent we are able to distinguish between the different types of crops.

4.1 *Monitoring changes in the field sizes*

As the TM image was made in 1990 when privatization was still a theoretical thing and the SAR images are from 1993 we expected that comparing the TM to the SAR images as well as the SAR images to each other we would be able to detect the changes in the field sizes. But, as the distribution of the fields to the new owners hasn't begun but in a few places our aim to follow the process of the privatization wasn't possible to reach. However, we succeeded to reach some results concerning the size of the fields. By visual interpretation we can conclude that we can differentiate small parcels in some places depending on their cultivation and vegetation cover. The smallest field size we was still able to detect based on our field visits was 50m by 200m, so about 1 ha. This is a promising result for the future taking into consideration that the recent field sizes are about 80 ha-s.

4.2 *Separation of different crops*

We tried to recognize using the multitemporal SAR image the agricultural plants that are grown on the fields. There are five main plants in this area: wheat, sunflower, alfalfa, maize and sugar-beet. In addition to these plants, we also observed orchards, vineyards, potato fields, pastures and on small parcels vegetables like cabbage and cauliflower. Table 2 shows some results concerning the mean of the pixels' amplitude on homogeneous fields from the different periods.

On the base of these numerical results and visual interpretation we can conclude the following:

a) Water bodies can be recognized on the image by their characteristic dark violet colour. This can be explained that in July they gave high backscatter due to the strong wind observed at the time of the overpass.

b) Wheat hasn't got a characteristic trend in the change of the mean values from the different dates, and can be observed with different colours on the multitemporal image. This can be explained partly with the different date of the harvesting of the wheat (in a lot of places wheat has already been harvested by the beginning of July, so from these fields we get low backscatter) - partly with the difference in the date of starting the autumn plough on the fields. The extremely high values from September can be due to the effect of the plough that was parallel with the orbit of the satellite.

c) Sunflower has a characteristic change in the mean values from the three different dates and can be easily recognized on the image with a bluish-pink-violet colour. The high backscattering in July can be explained that sunflower canopy is a very rough surface for the incoming radar waves and thus gives high backscattering. For August it has already dried out partly that's why it gives lower backscattering.

d) Sugar beet also has a characteristic change in the mean values and therefore it can be recognized on the image with a greenish colour. It gives the highest backscattering in August.

e)Alfalfa gives law backscattering in all of the three dates and therefore has a dark, almost black colour. It can be confused with pasture that has the same law backscattering.

f) Maize, as wheat, doesn't have a characteristic colour. This can be explained with that there are different types of maize and they are harvested at different dates.

Considering these results we can claim that for recognizing the different plants these three dates are not good enough, and we would need at least two, but rather three image from the growing season. The images even after filtering were still too noisy for making an automatic classification, so that's why we interpreted the images only visually. We can also claim that for our aim a single-date, black and white radar image is not enough, we have to use the multitemporal image for visual interpretation.

As for our future plans, we would like to continue our investigations concerning both the assessment of

Table 2. Characteristic mean amplitude values for the different crops from the 3 different dates.

Names	July	August	Sept.
Sunflower	131.9	111.5	137.6
	145.9	118.1	135.8
	129.5	93.6	86.2
	140.2	121.5	119.3
	143.4	125.9	124.3
Maize	123.3	103.2	96.3
	118.1	137.0	112.0
	86.5	84.7	68.5
	97.2	104.1	146.6
Sugarbeet	119.6	143.8	99.3
	110.4	135.6	108.6
	105.2	120.3	104.0
	104.4	115.8	104.4
Wheat	59.1	71.6	74.1
	74.3	73.4	101.2
	124.7	112.8	221.3
	70.0	78.1	95.1
	94.9	78.1	95.1
	53.7	83.1	77.5
	60.7	68.5	67.5
Alfalfa	43.2	44.9	54.7
	54.3	54.4	68.0
	40.2	40.0	47.0
	64.7	64.1	63.4

using the ERS-1 images to monitor the changes due to privatization and the separability of the different crops on the fields. We have to find out which dates are optimal for crop separability (doing this we will make use of Table 3, in which we describe the growing and cultivating phases of the main crops), and by using a better filter we would like to try out the possibility for making an automatic classification. We can state that - considering both our results and the fact that according to our statistics 40-45% of the LANDSAT and SPOT images are partly cloud covered - the ERS-1 SAR images should be good complementary data to the optical data.

REFERENCES:

Lee, J.S. 1986. Speckle suppression and analysis for synthetic aperture radar images. *Optical Engineering. Vol.25.No.5*: 636-643.

Frost, V. S., Stiles, J. A., Shanmugan, K. S., Holtzmann, J. C. A model for radar images and its application to adaptive digital filtering of multiplicative noise. *IEEE. Transactions on Pattern Analyses and Machine Intelligence. PAMI-4*: 157-166.

Lopes, A. Nezry, E. Touzi, R. Laur, H. Structure detection and statistical adaptive speckle filtering in SAR images. *International Journal of Remote Sensing, 1993, Vol.14.No.9*: 1735-1758.

Sensors and Environmental Applications of Remote Sensing, Askne (ed.) © 1995 Balkema, Rotterdam, ISBN 90 5410 524 0

Land transformation analysis in Kassala Province, Eastern Sudan

H. Larsson
Remote Sensing and GIS Laboratory, Lund University, Sweden

ABSTRACT: Regression models for canopy cover estimation in Acacia woodlands, Eastern Sudan, using Landsat TM, MSS and SPOT XS data and an impact analysis of rainfall on agriculture in Eastern Sudan were carried out. The study resulted in three models and correlation coefficients and trend diagrams on yield and mean annual rainfall. The trend diagrams indicates lower rainfall and lower yields.

1 INTRODUCTION

The research group at the Remote Sensing and GIS laboratory, Lund university, Sweden, has been engaged in the application of remote sensing and GIS research for desertification studies in African drylands since mid-seventies (Rapp 1974, Helldén 1984, Olsson L. 1985, Olsson K. 1985 and Ahlcrona 1988)

The project "Land transformation analysis in the Kassala Province, Eastern Sudan" started in January 1990 and is financed by Swedish National SpaceBoard. The project aims to analyse and explain the land transformation in the Kassala Province, Eastern Sudan. This will be accomplished by carrying out the following project components: 1) establish linear regressions for canopy cover estimation in acacia woodlands using Landsat TM, MSS and SPOT XS data, 2) analyse rainfall and agricultural statistics from the last three decades, 3) analyse historical data in the form of maps and journey reports from the last one hundred years and 4) analyse satellite data from 1972, 1979 and 1990. This paper deals with the first two components.

2 REGRESSION MODELS FOR CANOPY COVER ESTIMATION IN ACACIA WOODLANDS, EASTERN SUDAN, USING LANDSAT TM, MSS AND SPOT XS DATA

The aim of this study is to establish models for estimation of canopy cover of Acacia woodlands. The models were based on NDVIs and established using Landsat TM data, Landsat MSS data and SPOT Multi Spectral (XS) data. The Acacia woodlands in the Kassala Province, eastern Sudan were used in establishing the models. Extensive studies in the Sahelian zone of Sudan and Ethiopia (Helldén 1987, Olsson K. 1985) have demonstrated a relation between the woody dry weight of trees and crown diameter. Several studies investigated the shadowing effect of a tree canopy (Prince 1988, Franklin *et al.* 1991).

2.1 *Methodology and data*

Canopy cover was measured in 30 plots of 50x50 meters. The positions of the field-measured plots were determined using a Magellan NAV 1000 Global Positioning System receiver (GPS). Within the plots, the crown diameter of each tree was measured, the area of each tree crown was calculated and the percentage canopy coverage in each field-measured plot was calculated. The field data were collected in October, 1990.

Altogether 19 field-measured plots (merged or singled) were used. Field-measured plots situated next to each other were merged into one. The field-measured plots were collected in the Rawashda forest reserve in Kassala Province, The Sudan, located at 14°15'N and 35°40'E. The average annual rainfall is 400 mm, the relief is flat, the ground cover is low and the soils are black cotton soils. The dominating tree species are *Acacia seyal, Acacia senegal* and *Acacia mellifera*.

The satellite data used in the study were; SPOT

XS data registered on 26 February, 1990 and Landsat TM and MSS data registered on 29 January, 1990. Only 15 of the 19 calibration plots are covered by the SPOT scene.

Relative and at-satellite radiometric calibrations of the Landsat TM and MSS data were calculated according to Markham and Barker (1986). The relative and at-satellite radiometric calibrations of SPOT XS data were calculated according to Spotimage (1986), Markham and Barker (1986) and Hill and Aifadopoulou (1990).

The SPOT XS data was geometrically corrected to four ground control points positioned with the GPS. A nearest neighbour transformation were used. Three transformations were made with pixel sizes 20x20 m, 30x30 m and 50x50 m. The Landsat TM data was geometrically corrected to the SPOT XS data with a pixel size of 30x30 m and the Landsat MSS was corrected to the SPOT XS data with a pixel size of 50x50 m. Nearest neighbour transformation were used in both cases.

The NDVI for Landsat TM, MSS and SPOT XS data were used as the variable explaining the canopy cover. The NDVI were defined as:

$$NDVI_{TM} = \frac{TM4 - TM3}{TM4 + TM3} \qquad (1.1)$$

$$NDVI_{MSS} = \frac{MSS4 - MSS2}{MSS4 + MSS2} \qquad (1.2)$$

$$NDVI_{SPOT} = \frac{XS3 - XS2}{XS3 + XS2} \qquad (1.3)$$

Single VISRed/NIR channels were also correlated to field measured canopy cover. The correlation coefficients were around 0.5 and thereby lower than the correlations between NDVI and field-measured canopy cover.

The 95 per cent confidence limits for the correlation coefficients between NDVI and field-measured canopy cover were calculated (Williams 1984). The RMA method minimises the sum of the cross products $(\hat{y} - y_i)(\hat{x} - x_i)$ of the differences on both axes. The line $y = \hat{a} + \hat{b}x$ is fitted from (Curran & Hay 1986):

$$\hat{\beta} = \frac{\sigma_y}{\sigma_x} \qquad (1.4)$$

$$\hat{a} = \bar{y} - \hat{\beta}\bar{x} \qquad (1.5)$$

The RMA method was applied on the data, where y was the percentage canopy cover and NDVI was the x variable. The RMA method was chosen instead of the conventional regression model, since it regards errors in both x and y variables.

Ninety-five per cent confidence interval for the true mean of predicted canopy cover were calculated.

2.2 *Results*

The correlation coefficients between field-measured canopy cover and the satellite data (NDVI) are presented in table 1.

r_{TM} is significant at the 0.02 significance level, r_{MSS} is significant at the 0.001 significance level and r_{SPOT} is significant at the 0.01 significance level.

The 95 per cent confidence limits for the correlation coefficients are presented in Table 1.

Table 1. Correlation coefficients between field-measured canopy cover and the satellite data (NDVI) and the 95 per cent confidence limits for the correlation coefficients (Larsson a1993).

	Correlation between satellite data and field-measured data	95 per cent confidence limits for the correlation coefficient
r_{TM}	0.552	$0.131 < r < 0.875$
r_{MSS}	0.698	$0.357 < r < 0.875$
r_{SPOT}	0.718	$0.325 < r < 0.899$

The linear regression models for the Landsat TM, MSS and SPOT XS data are presented in Table 2.

Table 2. The linear regression models for the Landsat TM, MSS and SPOT XS data (Larsson a1993).

Models
$y_{TM} = -82.5 + 512x_{TM\_ndvi}$
$y_{MSS} = -29.9 + 345x_{MSS\_ndvi}$
$y_{SPOT} = -47.1 + 486x_{SPOT\_ndvi}$

The linear regression models were established with RMA (see section 2.1). Figures 1, 2 and 3 illustrate the field-measured canopy cover regressed on NDVI, the established model and a 95 per cent confidence interval for the true mean of y, predicted canopy cover, according to the fitted line.

The regression line y_{TM}= -82.5 + 512x_{TM_ndvi} fitted according to *RMA* (Figure 1). Ninety-five per cent confidence interval for the true mean

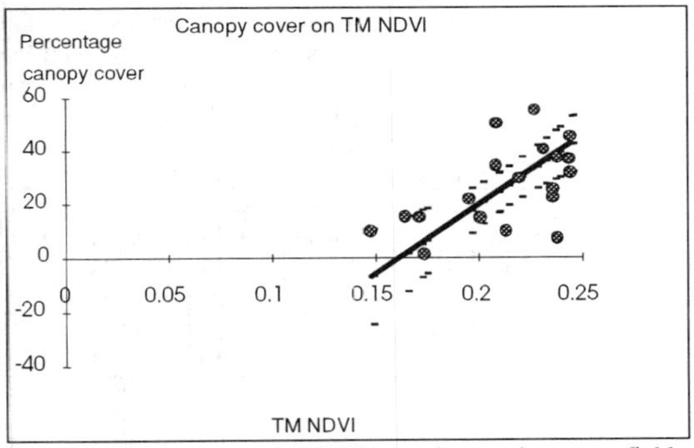

Figure 1. Landsat TM NDVI on the x-axis versus field-measured canopy cover on the y-axis (Larsson a1993).

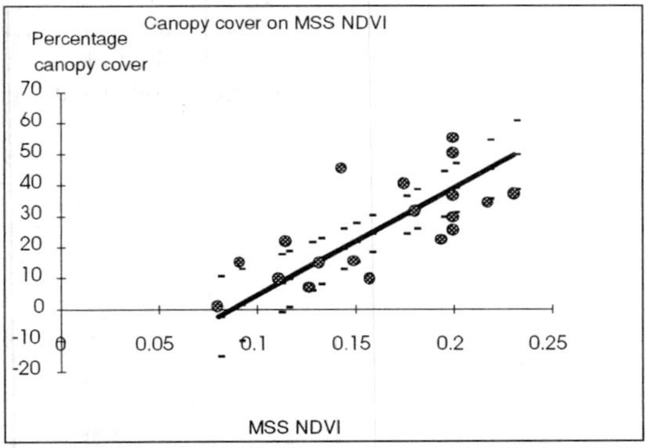

Figure 2. Landsat MSS NDVI on the x-axis versus field-measured canopy cover on the y-axis (Larsson a1993).

of predicted canopy cover. n = 19. r = 0.552 (Larsson a1993).

The regression line $y_{MSS} = -29.9 + 345x_{MSS\_ndvi}$ fitted according to *RMA* (Figure 2). Ninety-five per cent confidence interval for the true mean of predicted canopy cover. n = 19. r = 0.698 (Larsson a1993).

The regression line $y_{SPOT} = -47.1 + 486x_{SPOT\_ndvi}$ fitted according to *RMA* (Figure 3). Ninety-five per cent confidence interval for the true mean of predicted canopy cover. n = 15. r = 0.718 (Larsson a1993).

2.3 *Discussion and Conclusion*

The wide 95 per cent confidence interval for the correlation coefficients indicates that a larger number of calibration plots may have made the models safer. That the correlation coefficients differ, from 0.55 to 0.72, between the different data sets is probably because of the small number of calibration plots.

The models are established on satellite data from Eastern Sudan in the dry season. The models are valid on satellite data from the dry season when most of the ground cover is gone. Dark soils should prevail and the dominating specie should be *Acacia seyal.* When applying the models on different data sets the satellite data should be calibrated according to the data in this study (see section 2.1). Larsson (a1993) presents further information about the statistical methods used in the establishment of the models.

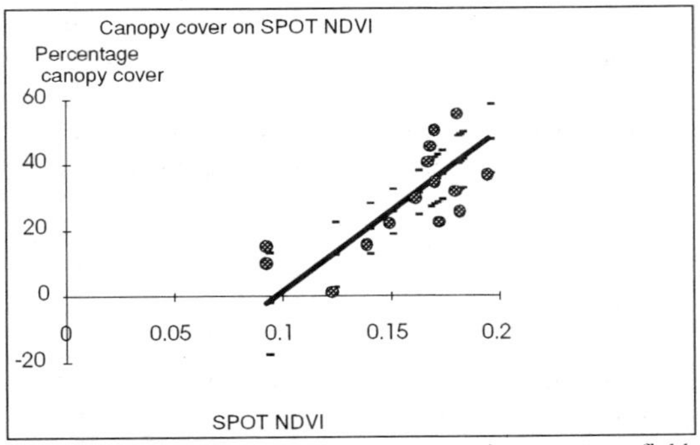

Figure 3. SPOT NDVI on the x-axis versus field measured canopy cover on the y-axis (Larsson a1993).

3 IMPACT OF RAINFALL ON AGRICULTURE IN KASSALA PROVINCE, EASTERN SUDAN

Years of droughts, crop failure and famine has been frequent in the Sahelian zone. Several studies on the impact of rainfall on the agriculture in the Sahelian zone has been carried out among others by Olsson L. (1985, b1993), Ahlcrona (1988) and Sivakumar (1992). Olsson L. (1985, b1993) found that the correlation between millet yield and amount of rainfall in western Sudan was generally lower than the correlation between millet yield and number of days with at least 1 mm rainfall, was generally very high and that the decrease in crop yield experienced since early 1960s is due to mainly natural factors beyond the control of man.

This study aims to correlate the amount of rainfall to production and yield of sorghum, millet and sesame in the Kassala Province.

The study area is the Kassala Province in the Eastern Sudan. The Province is 125 120 km^2 and it is located between latitudes 12°40' and 17°57' N and longitudes 33°34' and 37°01' E (Van der Kevie & Buraymah 1976). It includes climatic zones from desert in the north to semi-humid savannahs and forests in the south. The savannah region is cultivated and there are large mechanized rainfed and irrigated schemes.

3.1 *Methodology*

The rainfall statistics used in the analysis are from the stations in Table 3.

Table 3. The stations and the area used in the study (Larsson b1993).

Station	Area
Halfa El Gadida	New Halfa
Ungwatiri Gedaref Ghadambaliya Showak Wad El Huri	Gedaref
Hawata Mafaza	Rahad
Aroma Kassala	Gash
All stations	Kassala Province

In the agriculture statistics the areas are in Feddans which is approximately 0.42 hectare, the production are in M. Tons (thousand tons) and the yields are in Kg/Feddan for Millet, Sesame and Sorghum. The agriculture statistics are from the agriculture schemes New Halfa agriculture scheme which is irrigated from the Khashm El Girba dam, Gash agriculture scheme which is flooded from the Gash river, Rahad agriculture scheme which is irrigated from the Rahad river and Gedaref Mechanized Farming scheme which is rainfed.

The correlation coefficients were calculated between mean rainfall for different months (see Table 4) in the Kassala Province and production and yield for sorghum, millet and sesame. In the correlation analysis the years from 1960 and 1961 are used, including years with zero registration.

3.2 *Results*

The highest correlation coefficients were 0.65 and 0.67 were found between yield of sorghum and mean rainfall of July, August and September and between yield of sorghum and mean annual rainfall (Table 4). For the crops millet and sesame the correlations are < 0.5 (Table 4).

Table 4. Mean of July, August and September rainfall separately, July-August-September rainfall together and mean annual rainfall correlated towards sorghum, millet and sesame, production and yield (Larsson b1993).

		Jul_ mean	Aug_ mean	Sep_ mean	JuAu Se	Year
Sorgh	*Prod*	-0.01	0.06	0.34	0.18	0.15
	Yield	0.48	0.48	0.28	0.65	0.67
Millet	*Prod*	-0.20	0.44	-0.04	0.17	0.21
	Yield	-0.24	0.21	-0.21	-0.10	-0.07
Sesame	*Prod*	0.16	-0.40	0.20	-0.10	-0.01
	Yield	0.39	-0.14	0.10	0.16	0.20

Sorghum, millet and sesame which are rainfed crops are plotted against mean annual rainfall in Figure 4-6. In Figure 4 the sorghum yield and the mean annual rainfall are plotted against each other, the correlation coefficient is 0.67. Figure 5 illustrates the millet yield and mean annual rainfall in the Kassala Province plotted against each other with the

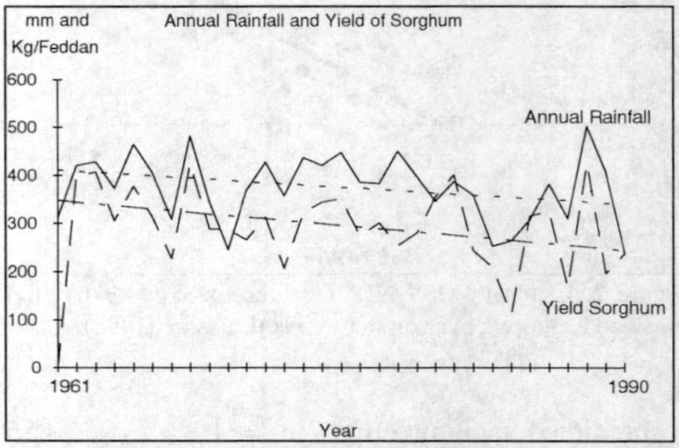

Figure 4. Mean annual rainfall in Kassala Province plotted against the yield of sorghum, with data from 1961 to 1990. The annual rainfall is expressed in mm/year and the yield is expressed in Kg/Feddan. r = 0.67 (Larsson b1993).

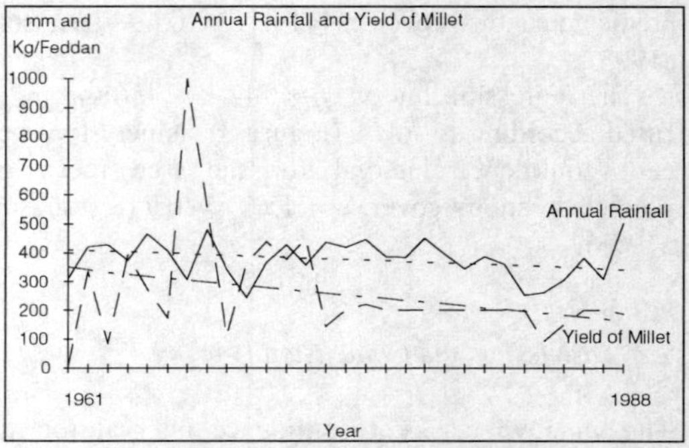

Figure 5. Mean annual rainfall in Kassala Province plotted against the yield of millet, with data from 1961 to 1988. The annual rainfall is expressed in mm/year and the yield is expressed in Kg/Feddan. r = -0.07 (Larsson b1993).

correlation coefficient -0.07. Sesame yield is plotted against mean annual rainfall in the Kassala Province in Figure 6. The correlation coefficient between the graphs is 0.20.

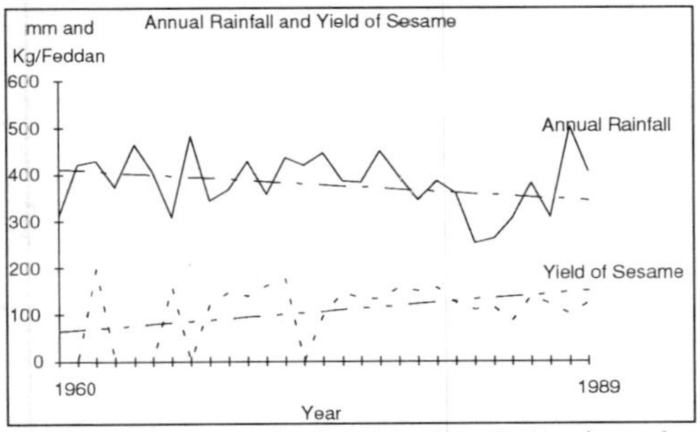

Figure 6. Mean annual rainfall in Kassala Province plotted against the yield of sesame, with data from 1960 to 1989. The annual rainfall is expressed in mm/year and the yield is expressed in Kg/Feddan. r = 0.20 (Larsson b1993).

3.3 *Discussion and Conclusion*

The highest correlation coefficients were between sorghum and mean annual and mean July-Aug-Sep rainfall, 0.6-0.7. There is no evidence of soil mining but the amount of rainfall is decreasing and there is high correlation between rainfall and sorghum yield, which indicates that the annual yield is most dependent on climate.

Sorghum is grown on the heavy clayey soils, the black cotton soils, and is not as sensitive for rainfall fluctuations as millet which is grown on sandy soils. Olsson (a1993) shows that yield of millet and rainfall on the sandy soils in the Kordofan, Western Sudan, has a correlation coefficient of 0.8 while millet grown on the black cotton soil in the Kassala Province has a correlation coefficient of 0.2 (Table 4). The precipitation has probably not as big influence on millet in the Kassala Province as in the Kordofan.

Larsson (b1993) presents a deeper analysis between amount and intensity of rainfall and agriculture. Larsson (b1993) also presents a soil nutrient analysis in the Kassala Province.

ACKNOWLEDGEMENT

This study was financed by the Swedish National SpaceBoard, Dno. 9/93. I am grateful to my colleagues for their comments on the results. Many thanks to Dr. Abbas S. Musa, Adam A. Yassin at the Geography Department, University of Khartoum.

REFERENCES

Ahlcrona, E. 1988. The Impact of Climate and Man on Land Transformation in Central Sudan. Applications of Remote Sensing. Lund University Press, ISBN 91-7966-028-2. Sweden: 140.

Curran, P. and Hay, A.M. 1986. The Importance of Measurement Error for Certain Procedures in Remote Sensing at Optical Wavelengths. *Photogrammetric Engineering and Remote Sensing*, Vol. 52, No. 2: 229-241.

Franklin, J., Prince, S.D., Strahler, A.H., Hanan, N.P. and Simonett, D.S. 1991. Reflectance and transmisson properties of West African savanna trees from ground radiometer measurements. *International Journal Remote Sensing*, 12: 1369-1385.

Helldén, U. 1984. Drought impact monitoring. A Remote Sensoing Study of Desertification in Kordofan, Sudan. Lunds universitets naturgeografiska institution. Rapporter och notiser 61. ISSN 0348-3339: 61.

Helldén, U. 1987. An Assessment of Woody Biomass, Community forests, Land use and Soil erosion in Ethiopia. - A feasibility study on the use of remote sensing and GIS-analysis for planning purposes in developing countries. Lund Studies in Geography. Ser. C. General, mathematical and Regional Geography. No. 14. ISBN 0-86238-158-4: 73.

Hill, H. and Aifadopoulou, D. 1990. Comparative analysis of Landsat-5 TM and SPOT HRV-1 Data for Use in Multiple Sensor approaches. *Remote Sensing of Environment*, Vol. 34, No. 1: 55-70.

Larsson, H. a1993. Linear regressions for canopy cover estimation in Acacia woodlands using Landsat-TM, -MSS and SPOT HRV XS data. *Int. J. Remote Sensing*, 1993, VOL. 14, NO. 11: 2129-2136.

Larsson, H. b1993. Analysis of rainfall, agriculture and soils in the Kassala Province, Eastern Sudan: 17. (unpublished material)

Markham, B.L. and Barker, J.L. 1986. Landsat MSS and TM Post-Calibration Dynamic Ranges. Exoatmospheric Reflectances and At-satellite Temperatures. EOSAT. *Landsat Technical Notes*, No. 1: 229-241.

Olsson, K. 1985. Remote Sensing for fuelwood resources and land degradation studies in Kordofan, The Sudan. Meddelanden från Lunds universitets Geografi ska institution, avhandlingar C. ISBN 91-7222-988-8: 182.

Olsson, L. 1985. An Integrated Study of

Desertification. Department of Physical Geography. University of Lund. ISBN 91-7222-975-6. Sweden: 170.

Olsson, L. a1993. On the Causes of Famine - Drought, Desertification and Market Failure in the Sudan. Ambio 1993: 17.

Olsson, L. b1993. Dept. of Physical Geography, University of Lund. Solvegatan 13, 223 62 Lund, Sweden.

Prince, S.D. 1988. Measurement of canopy interception of solar radiation by stands of trees in sparsely wooded savanna. *International Journal of Remote Sensing,* 8: 1747-1766.

Rapp, A. 1974. A review of desertization in Africa. Water, vegetation and man. SIES Report No1. NFR, Stockholm: 77.

Sivakumar, M.V.K. 1992. Climate Change and Implications for Agriculture in Niger. Climate Change 20: 297-312.

Spotimage 1986. Document: SI/AT/85.0113, Version: A.A, Doc: V01.94.

Van der Kevie, W. & Buraymah, I.M. 1976. Exploratory Soil Survey of Kassala Province. A study of Physiography, Soils and Agricultural Potential. Soil Survey Reports No. 13: 158.

Williams, R.B.G. 1984. Introduction to statistics for geographers and earth scientists. MacMillian Publishers Ltd., Hong Kong. ISBN: 0-333-35275-0: 349.

Sensors and Environmental Applications of Remote Sensing, Askne (ed.) © 1995 Balkema, Rotterdam, ISBN 90 5410 524 0

A strategy for spectral time-profiling in coniferous forest using Landsat TM

P. Syrén
Department of Physical Geography, Stockholm, Sweden

ABSTRACT: Change detection in coniferous forest, with the use of satellite imagery, can be divided into two broad categories. That which deals with spatial difference of specific object classes from one time to another, and that where the spectral development over time is investigated. The first case is easily studied with image enhancement techniques, preferably integrated with a geographical information system. The latter category is much more complex to investigate, since the spectral response to a change in health status or biomass may be obscured by a number of factors; natural spectral variation and development of the object, differences in atmospheric conditions and illumination and viewing geometry between registrations, non-Lambertian behaviour of the object and instability in the sensor system. In order to arrive at reliable estimates of spectral change between two or more satellite registrations it is necessary to systematically handle these sources of error in a correction scheme. In this study, such a scheme is applied to coniferous forests of Norway spruce *(Picea abies)* and Scots pine *(Pinus Sylvestris)* in central Sweden. The proposed correction scheme consists of four steps. 1/ Reflectance calibration of satellite imagery through simultaneous low altitude reflectance measurements. 2/ Correction for slope effects, using a digital elevation model and empirically developed Minnaert constants. 3/ A sub standwise approach for estimate of reflectance statistics, i.e., each stand is segmented according to slope-aspect intervals. 4/ And finally, the object dependent reflectance correction, using empirically derived relationships between stand characteristics and solar angle for standardization of reflectance to a standard solar angle. The data used in accomplishing the correction scheme consists of two Landsat TM scenes (September 1986 and July 1992) and a digital elevation model with 50 m resolution. The spectral changes obtained (between 1986 and 1992) in visible, near-, and middle-infrared wavelengths provide the possibility to explain the cause for spectral change within these stands.

1 INTRODUCTION

Change detection in coniferous forest, with the use of satellite imagery, can be divided into two broad categories. That which deals with spatial difference of specific object classes from one time to another, and that where the spectral development over time is investigated. The first case is easily studied with image enhancement techniques, preferably integrated with a geographical information system. The latter category is much more complex to investigate, since the spectral response to a change in health status or biomass may be obscured by a number of factors; natural spectral variation and development of the object, differences in atmospheric conditions and illumination and viewing geometry between registrations, non-Lambertian behaviour of the object and instability in the sensor system. In order to make reliable estimates of spectral change between two or more satellite registrations it is necessary to systematically handle these sources of error in a correction scheme. In figure 1 a principle scheme of spectral dynamics of coniferous canopy reflectance is outlined.

2 METHOD

In this study a radiometric correction scheme was applied to coniferous forests of Norway spruce *(Picea abies)* and Scots pine *(Pinus Sylvestris)* in the Stockholm area, Sweden. Two Landsat TM-scenes, acquisitions dates 1986-09-17 and 1992-07-24 (09.21 and 09.17 GMT), were analysed in combination with a digital terrain model. Both images and the terrain model were resampled at 25 m resolution and georeferenced.

Prior to the resampling process the 1992 image was classified with a maximum likelihood classifier. The areas classified as mature stands of spruce or

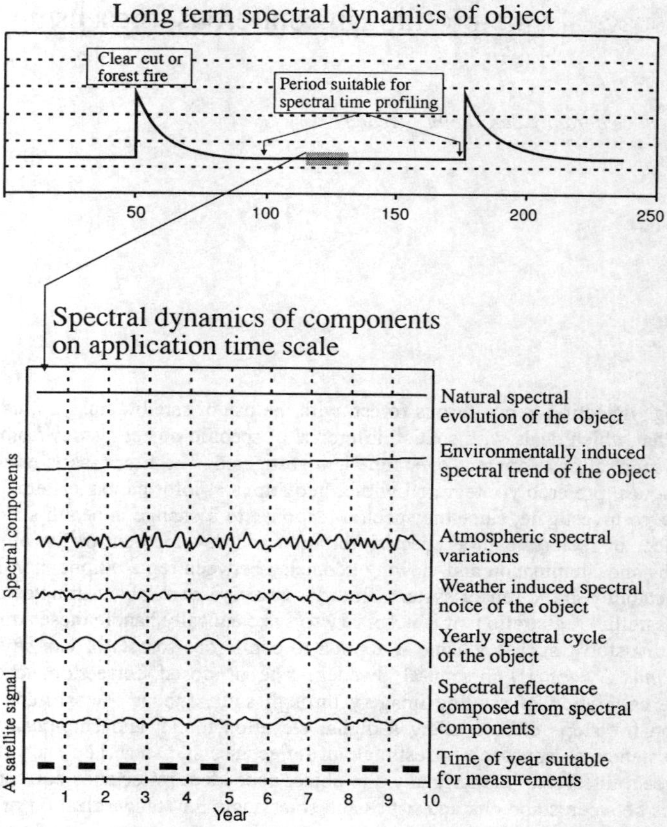

Figure 1. Schematic outline of long and short term spectral dynamics of coniferous canopy reflectance. To make reliable estimates of a spectral change between two or more satellite registrations (in order to find the environmentally induced spectral trend), accurate estimates of atmospheric conditions are crucial. Also, the influence from natural spectral evolution, weather induced spectral noise and the yearly spectral cycle of the object has to be minimized prior interpretation of the composed reflectance at satellite.

pine were used as feature images to extract digital numbers from the 1986 and 1992 image, as well as slope and aspect values from the terrain model.

2.1 Processing of the feature image

Spectral comparisons of mean object reflectance over time, should be carried out with a standwise approach. Depending on available information of the object in question, e.g. inventory data or forest maps, the rules for defining homogenous stands will differ somewhat . When using a traditional computer based classifier, the classified image must be processed to achieve standwise homogenous objects. This post-processing includes steps to:

- Shrink the classified areas with a "mixed-pixel" border zone. Depending on resampling errors this step was found relevant in order to exclude a border zone around each classified area to assure comparison of reflectance means from corresponding objects. Also, this procedure will reduce adjacency effects from neighbouring objects with high spectral difference.
- Clear up the image of objects of a specified minimum size, and to assign an index to each remaining stand.
- Create a standwise segmentation depending on slope- and aspect intervals. For stands that extend over several slope-aspect classes it was necessary to compute standwise reflectance means weighted according to the segment area in each slope and aspect class.

2.2 Reflectance calibration of Landsat TM data

Image normalisation is achieved through simultaneous low-altitude reflectance measurements from helicopter and satellite overpass (Vogelman and Rock, 1988; Kleman and Alm, 1989). This empirical method depends on accurate reflectance measurements from objects of a wide spectral range that can be localised in the satellite image. Within the test area, spectral observations with the sensor in nadir position were collected from low-, high- and medium reflectance objects, in wavelength bands corresponding to the Landsat TM-bands 1, 3, 4 and 5, visible blue and red (485 and 654 nm), near-infrared and middle-infrared (NIR, 841 and MIR, 1676 nm). Spectral irradiance estimates from a fully illuminated and horizontal barium sulfate panel, corrected for deviation from the Lambertian assumption Kimes and Kirshner (1982), made before and after the flight were used to interpolate the expected value at the time of object measurement. The reflectance factor per band was calculated as the ratio of the object radiance and corrected irradiance estimate. In order to identify the location in the satellite images, colour prints were taken of each object using a fisheye-lens. For each object at least 5 pixels were chosen to give an average digital number of the corresponding measured reflectance. The regressions were used to recalculate the image data to reflectance factors. At this point irradiance-, atmospheric- and sensor system differences between the images were removed. The remaining discrepancies are composed of topographic effects, true spectral change of the object and differences in sun-target-sensor geometry between registrations.

2.3 Correction for topography

The influence of topography on forest canopy reflectance has been described in several papers e.g. Lepriuer *et al.* (1988), Teillet *et al.* (1982) and Ekstrand (1993). Terrain location indices, defined as the effective incidence angle (Smith *et al.*, 1980; Holben *et al.*, 1980), were used to calculate Minnaert constants for radiometric correction of forest reflectance in each spectral band Teillet *et. al.* (1982).

2.4 Reflectance standardisation

The relationship between solar zenith angle (SZA), forest characteristics and canopy reflectance can be described by linear functions in different wavelength bands (Syrén, in press). The functions for mature spruce and pine canopies were used to calculate the object dependent reflectance correction factor, as the ratio of the regression estimate at the acquisition solar zenith angle (SZA_a) and the regression estimate at the standard solar zenith angle (SZA_S). The factor obtained corrects the reflectance factor at SZA_a to that expected at the standard solar zenith angle. The difference in SZA between the 1986 and 1992 image was 16° (59.3° and 43.3°), and the SZA_s was set to 45°.

2.5 Database contents

The processed feature images of mature spruce and pine stands were used to extract mean values of DN, incidence angle, slope and aspect for each feature segment in the 1986 and 1992 images and entered to a database. From the database, standwise mean values of terrain corrected DN, reflectance, and standardised reflectance with the corresponding standard errors were calculated.

3 RESULTS

In figure 2 the spectral difference between 1992 and 1986 (1992 minus 1986) are plotted for each stand, using the slope-aspect weighted mean values with confidence limits at the 0.05 level, together with the respective difference in class mean with confidence limits.

4 CONCLUSIONS

To accomplish reliable estimates of a spectral change between satellite registrations, a far reaching radiometric correction scheme is required. *Absolute* or *relative* calibration are the standard procedures to accomplish comparable image data. In both cases the accuracy dependent on several factors as; 1/ a stable sensor system 2/ well known calibration constants 3/ accurate estimates of atmospheric conditions if the influence from atmosphere is modelled using a radiative transfer code, and 4/ spectrally stable reference targets if used in relative calibration. The empirical method applied in this study "simultaneously obtained reflectances that are regressed on satellite pixel values from corresponding sites" must be considered as a strong alternative since no assumptions on the sensor system or atmospheric conditions are made. This method will also allow statements on spectral deviation in absolute values. The anisotropic reflectance response of coniferous forest with varying solar elevation must also be controlled. if there is a difference in illumination geometry between satellite registrations. This correction was accomplished

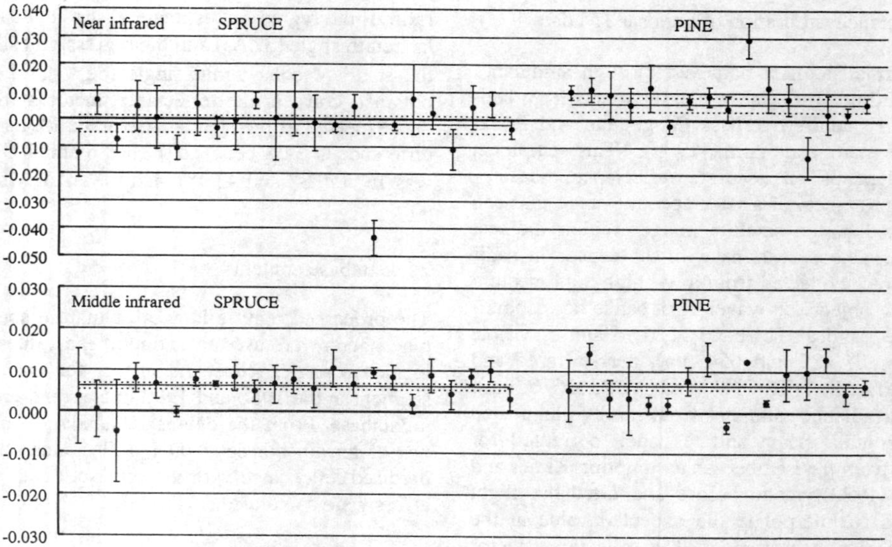

Figure 2. Reflectance difference (1992 minus 1986) of mature spruce and pine stands, with confidence limits at the 0.05 level.

by empirically established relationships between solar angle and canopy reflectance of the investigated forest types. The topographic influence on the radiometric response was of less importance due to the low relief of the study area.

REFERENCES

Ekstrand, S., (1993): Assessment of forest damage using Landsat TM, elevation models and digital forest maps. Doctoral dissertation. Royal Institute of Technology, Stockholm, Sweden.

Holben, B. N. and Justice, C. O., (1980): The topographic effect on Spectral Response from Nadir-pointing sensors.. *Photogramm. Eng. Remote Sens.* 46:1191-1200.

Kimes, D. S., Kirchner, J. A., (1982): Irradiance measurement errors due to the assumption of a Lambertian reference panel, *Remote Sens. Environ.* 12:141-149.

Kleman, J. and Alm, G., (1989): Estimation of the atmospheric influence in a Landsat TM-scene using simultaneous low-altitude radiometer data. Research report 75, Department of Physical Geography, Stockholm University, pp. 28.

Lepriuer, C. E., Durand, J. M., and Peyron, J. L., (1988): Influence of topography on forest reflectance using Landsat Thematic Mapper and digital terrain data, *Photogramm. Eng. Remote Sens.* 54:491-496.

Smith, J. A, Lin, T. L. and Ranson, K. J., (1980): The Lambertian assumption and Landsat Data, *Photogramm. Eng. Remote Sens.* 46:1183-1189.

Syrén, P., (in press): Reflectance Anisotropy for Nadir Observations of Coniferous Forest Canopies, *Remote Sens. Environ.*

Teillet, P. M., Guindon, B. and Goodenough, D. G., (1982): On the slope-aspect correction of multispectral scanner data, *Can. J. Remote Sens.* 8:84-106.

Vogelmann, J. E. and Rock, B. N., (1988): Assessing forest damage in high-elevation coniferous forest in Vermont and New Hampshire using Thematic Mapper data. *Remote Sens. Environ.* 24:227-246.

Sensors and Environmental Applications of Remote Sensing, Askne (ed.) © 1995 Balkema, Rotterdam, ISBN 90 5410 524 0

Terrestrial net primary production of the Baltic drainage basin, based on global vegetation index, spatial stratification and local growing season estimates

Kjell B.-E. Olsson
Department of Physical Geography, Stockholm University, Sweden

ABSTRACT: Net primary production, NPP, of the terrestrial biosphere is important in the context of sustainable development. At global level, good correlation between NPP and integrated NDVI has been found. High latitude areas, complex landscape patterns and sparse vegetation in arid areas have been identified as complicated targets. This paper gives an overview of a project investigating the applicability of the integrated NDVI approach over Northern Europe, an area including the three difficult issues. Improvements are expected from using spatial stratification, and from using local growing season as the integration period. Preliminary qualitative results are in accordance with known features of the area.

1 INTRODUCTION

Net primary production, NPP, is the source of energy to the biosphere (Prince et al 1994). It provides the principal resources of food, fuel, construction material etc. for the inhabitants of a region. Consequently it is of interest trying to estimate the NPP in the context of sustainable development of a region. One of the objectives of the Baltic Drainage Basin Project, BDBP, an EU project under DG XII (Folke et al, 1993), is to model the so called "appropriated carrying capacity" of the Baltic Sea drainage basin. An estimate of NPP on a more local level was desired as input, preferably on at least a river drainage basin level.

For larger regions and global studies models based on climatic factors or remote sensing have been used. Box et al (1989) found good global correlation between integrated Normalised Difference Vegetation Index, NDVI, and NPP. This relationship has also been shown based on physical reasoning by Sellers (1985, 1987, 1989) and proved valid also on stand level (Sellers et al, 1992). However, Box et al (1989) found difficulties, and low correlation, over complex landscape patterns, at high latitudes and over sparse vegetation in arid areas.

2 OBJECTIVES

The objectives of the present project have been to try to satisfy the needs of the BDBP using the integrated NDVI approach and evaluate the methodology with special reference to the three more difficult issues addressed by Box et al (1989). To improve the estimates a spatial stratification into three-dimensional, i e woody vegetation, and two dimensional, i e non-woody vegetation, is made. Further improvement is expected from using the actual local growing season as the integration period.

3 NET PRIMARY PRODUCTION

Net primary production, NPP, is the surplus of the gross primary production, GPP, after the maintenance activities of the plant have used up their part. GPP is the total photosynthetic energy gathered from the irradiating sunlight. The major part of maintenance is the respiration of the plant. In areas with high respiration plants may have high GPP, but low NPP. The equation of NPP is often given as

$$NPP = GPP - respiration \quad (1)$$

Several, mainly stress-related, factors influence the NPP. Agricultural crops often experience less stress and may therefore have higher NPP compared to natural vegetation.

NPP includes all tissue that is produced during a specific time period, including all leaves, fruits, roots and litter, also parts and whole plants that may die during the period. It should not be confused with Net Ecosystem Production, NEP, that is the increment of plant mass over the time period (Whittaker & Marks, 1975). A community with high losses may have high NPP, but low NEP.

NPP includes both above and below ground biomass production. The below ground production is often not very well known and more difficult to observe. The consequence of this is often that merely the above-ground production is studied, particularly if remote sensing is utilised. Conversion factors are then used to assess the total NPP, including below-ground production. A similar procedure is used to convert from measured yield to whole plant.

NPP is a measure of photosynthesis and consequently related to the amount of absorbed photosynthetically active radiation, APAR. APAR is about 50 % of the irradiating solar spectrum, 400-700 nm. APAR, being absorbed radiation, is correlated to the reflected radiation, which can be detected using satellite remote sensing, and consequently also to NDVI. An approach in line with that of Kumar & Monteith (1981) is generally utilised. The product of NDVI, irradiating PAR and a radiation utilisation efficiency coefficient, ε, is integrated over a period of time, normally a year (e.g. Goward et al, 1991, Prince, 1991b, Ruimy et al, 1994).

$$NPP = \int \varepsilon(APAR)S\,dt = \int \varepsilon(NDVI)S\,dt \qquad (2)$$

where

ε = radiation utilisation efficiency coeff. i e how efficiently light is turned into matter

APAR = Absorbed Photosynthetically Active Radiation

NDVI = Normalised Difference Vegetation Index, (NIR - R) / (NIR + R), NIR = Near Infrared, R = Red

S = irradiating PAR

Various values of ε are given in the literature, not quite consistent, but often within a limited range (Ruimy et al, 1994). Ruimy et al (1994) has made a compilation of ε-values in the global context. ε is influenced by stress and a different approach, incorporating this is being developed by the SAGES group, lead by I.C. Prentice, in Lund, Sweden (Alex Haxeltine, personal communication, April 1994). Lo Seen Chong et al (1993) has compiled NPP values for Africa, which can be used for ε assessments.

4 GLOBAL VEGETATION INDEX

On a larger regional or global level the most appropriate available NDVI product is the Global Vegetation Index, GVI, produced on a regular basis and distributed by National Oceanic and Atmospheric Administration, NOAA, USA (Kidwell, 1990). The lineage of this product includes certain sampling procedures that have impacts on the final product and the use of it. Some of these effects are particularly significant at higher latitudes. The spatial resolution is ca 15-20 km (Belward 1991), depending on projection. GVI is based on Global Area Coverage, GAC, data from the Advanced Very High Resolution Radiometer, AVHRR, (Hastings & Emery, 1992) on the TIROS satellites. The original Local Area Coverage, LAC, pixels, about 1 x 1 km at nadir, are resampled onboard the satellite to create GAC, ca 1 x 4 km representing 3 x 5 km. GAC pixels are later mapped to a geo-referenced grid, using the last pixel per day per grid cell (Kidwell, 1991). The difference between GVI and GAC pixel resolution is roughly the same as between GAC and LAC, about 1:15. The daily GVI images are then combined using Maximum Value Compositing techniques, MVC. The objectives with the MVC is to reduce degrading effects related to clouds, view angle, illumination etc. (e g Holben, 1986). By averaging, sampling and temporal composition the size of a global data set is reduced to about 0.1 % of its original (LAC) volume (Tarpley, 1991). Some processed versions of the GVI exist and in this study mainly the Monthly Generalised Global Vegetation Index, MGV, by Kineman & Hastings, included on the Environmental Protection Agency's, EPA, Global Ecosystems Database 1.0 CD-ROM (EPA, 1992), is being utilised. It is mainly data from the NOAA-9 satellite. It is less influenced by outliers than other comparable, available products due to a median selection procedure.

5 CORRECTION FOR THE TERMINATOR EFFECT

GVI based on raw data contain a particular effect at high latitudes, called the terminator effect by Holben (1986). It is located to the twilight zone and gives a "foggy" appearance to the north and south edges of the data set. It is probably primarily related to atmospheric effects at low sun elevation and is most pronounced during the winter season.

Empirical correction factors have been established using best fit polynomials, based on strips of open ocean. Ocean water is supposed to have about zero NDVI. A method described by Davis (1986) has been used, with a slight modification. It uses an F-test to check if a higher order of polynomial will result in a significant improvement of fit. If this is not the case, the lower order is kept. It has been limited to maximum third order polynomials. This approach avoids matrix singularity and reduces polynomial complexity. The correction factors also compensate for the sensor drift of NOAA-9. Preliminary analysis indicate that they are applicable on hemispherical level. They have not yet been applied to the data.

6 IMPROVEMENT BY USING LOCAL GROWING SEASON INTEGRATION PERIOD

Normally NDVI is integrated over the whole year to estimate annual NPP, according to the integration formula above (2)

In most locations this will include a period during which no photosynthetic activity takes place, in Africa normally due to draught, but at high latitudes often caused by low temperatures. The NDVI values are nevertheless often significantly above zero, see fig. 1. Including these values in the integration will overestimate the NPP. They are not invariant and consequently they introduce an unsystematic error. Fig. 1 shows some annual sample curves from various test-sites within the study area.

In West Africa fixed growing season, based on the movement of the Inter Tropical Convergence Zone, ITCZ, has been introduced as the integration period (Prince, 1991a, Wylie et al, 1991). The actual growing season is however not the same all over a region, but influenced by local factors, like topography, latitude, distance to the coast etc. Kammerud (1993) studied the improvement achieved by using local growing season integration period over Mali, West Africa, compared to fixed season or whole year integration. He found significantly better production estimates using the actual growing season.

Preliminary tests have been made over the study area using fixed and actual growing season integration.

The fixed season was first based on visual analysis of the temporal NDVI curves from the test sites, like in fig. 1. Later the International Institute for Applied Systems Analysis, IIASA, Database for Mean Monthly Values of Temperature, Precipitation, and Cloudiness on a Global Terrestrial Grid (Leemans & Cramer, 1991), has been used to find the growing season. The season was adjusted to two months shorter based on this climatological data, April - October. A monthly mean temperature of 5°C is presently being used as the limit, mainly based on Swedish conditions (e g Johnsson, 1965). Above about 50°N the temperature is the dominating limiting factor for growth (Kurt Perttu, personal communication, Nov. 1993). The actual local growing season is based on the same criteria, but evaluated on pixel level. No quantitative comparison has been made, because evaluation data has not yet been compiled.

Alternatively the growing season can be determined directly from analysing temporal NDVI profiles (Loudjani et al 1993). This procedure probably needs some adjustments to be useful at high latitudes.

7 IMPROVEMENT BY DIMENSIONAL STRATIFICATION

Different plants have different production capabilities. The carbon structure varies between plants, the major cathegories are named C3 plants and C4 plants. The effect of these differences is a more or less efficient utilisation of the incoming light. For example Richter (1988) gives some details on these differences. C3 plants are the most common within the study area.

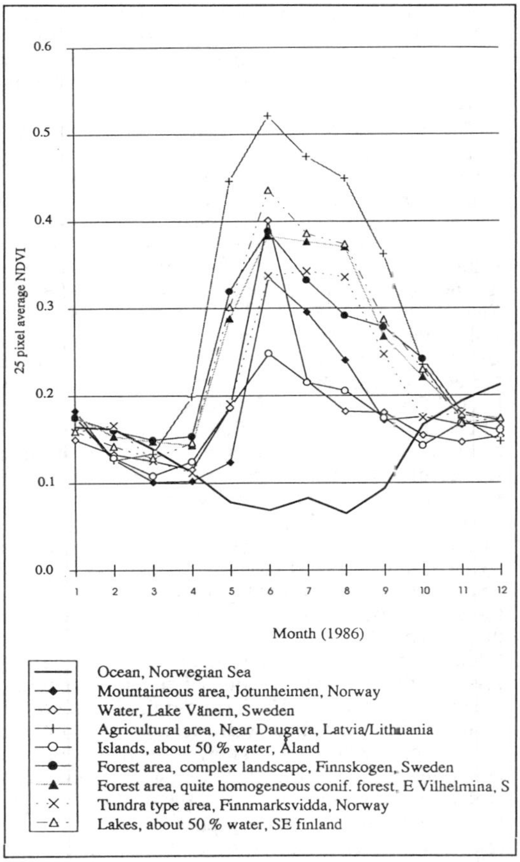

Fig. 1 Examples of annual NDVI profiles for different types of surfaces. The values are based on uncorrected original GVI, averaged over 25 pixels, for selected test-sites within the study area. The peak on the white diamond curve is not real, but caused by a data error in the GVI MVC.

Another mode of separation is by leaf form and permanence, like Running et al (1994) have proposed, also partly related to production pattern. The most obvious way to divide plants is probably, however, according to dimensionality, i e two or three dimensional vegetation. This concept is at least suitable for the temperate latitudes, including the study area. Three dimensional vegetation is generally forest, while two dimensional vegetation is crops, grass etc. The division is also useful in the context of global change modelling (Wolfgang Cramer, personal communication, Aug. 1993). The two categories differ in amount of woody, green and senescent components, water utilisation and needs, rooting depths, stress reactions etc.

In this study the dimensionality stratification is based on the European Space Agency's Remote Sensing Forest Map, ESA RSFM (ESA 1992, GAF

1992). This product is based on NOAA AVHRR LAC and has four classes: forest, open land, water and data gap. The forest definition is more or less where there are enough trees to be detected in the image.

The ESA RSFM has a resolution corresponding to about $1'^2$ over the study area. Each part (N Europe, Central Europe and CIS) was rasterised separately with 1' resolution. The raster versions did not match completely, when merging them to one data set. Pixels along the joining edges ended up without value, as data gaps. It was also discovered that lakes did not always match their real world counterpart. They did not have anything like the correct shape and some that did have the correct shape was labelled with terrestrial classes. A separate correction layer has been created, based on the Digital Chart of the World, DCW.

The frequency distribution for the three major classes, forest, open land and water, 1' resolution, was studied for the extent of the study area and a surprisingly even distribution was found.

For each 10' pixel of the study area, corresponding to the GVI resolution, the fractions of each land cover (from ESA RSFM 1' raster version) was computed. These fraction layers have then been used for various purposes. Two examples are illustrated in this paper.

Classifications have been made using the feature space made up of the fraction of forest vs. annually integrated, unscaled NDVI, see fig. 2. The objective of this is to study how well various classification results correspond to known features of the environment within the area. In fig. 2 a systematic classification scheme is illustrated. Seven equal areas of interest have been used for the classification, covering most of the feature space. The three first classes roughly have a forest cover of about 50 - 99 % forest. From left to right they represent low, medium and high values of integrated NDVI, corresponding to low, medium and high production. The four remaining classification areas have a forest cover of about 1 - 50 %, i e it is more or less open land, including mountainous areas. The integrated NDVI values, or production, increase from left to right, class 4-7. 0 % forest is excluded as it mainly represents water. 100 % forest is excluded due to reasons of symmetry.

Another application of the land cover fractions has been to compute normalised production curves vs. latitude. They are made summing up the product of a certain land cover category fraction and annually integrated NDVI for the pixels of each 10' latitude line. This value is then divided by the linesum of the fractions, for normalisation. This makes the values valid for the terrestrial part of the 10' line.

$$\left(\sum_{4°E}^{40°E} f\left(\int_{Jan.}^{Dec.} NDVI\,dt\right)\right) \Bigg/ \left(\sum_{4°E}^{40°E} f\right) \qquad (3)$$

This method is also intended to be used on areal basis in the evaluation process.

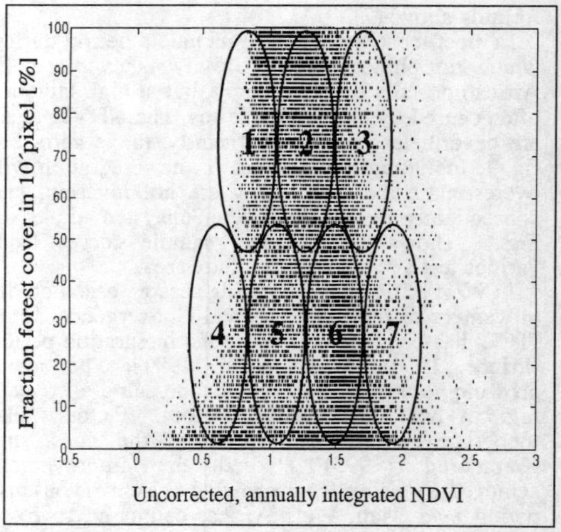

Fig. 2. The feature space of fractional forest cover vs. uncorrected, annually integrated NDVI. The seven systematic classification ellipses are those used to achieve the classification result shown in Fig. 3. Class 1-3 represent "forest" with increasing productivity from 1 to 3. Class 4-7 represent "open land" with increasing productivity from 4 to 7. The apparent striping is partly an artificial effect from the printing. Forest fraction is however integer values, while the integrated NDVI is decimal values.

8 RESULTS

The result of the classification is seen in fig. 3. The pattern corresponds well with a priori knowledge about the area. Altitudinal effects are rather obvious. The "holes" in the Scandinavian mountain ridge correspond to areas with low fractional forest cover and low productivity. Productivity generally decreases towards the higher altitudes. The darkest forest class (3, high productivity) include areas for example in Sweden with more deciduous forest. The production of the open land is highest in W central Europe and decreases with higher latitude. In central Europe, Sweden and Finland, open land corresponds to agricultural land, but in Norway much of it is treeless mountain areas and in Northern Fennoscandia and on the Kola peninsula it is more tundra type areas. One interesting feature is the area of high values open land, more or less corresponding to Kaliningrad. This has not yet been explained. Another interesting area is the combination of high productive forest next to high productive open land (lower centre of image). The "low production open land" along the coasts can be either real or caused by mixels of land and water.

The result of the normalised weighted production linesums vs. latitude is illustrated in fig. 4. All curves are actually rather coincident. For visibility the three forest curves have been replaced 20 units to the right. The curves are very similar in shape, but some inter-annual variations can be seen. The

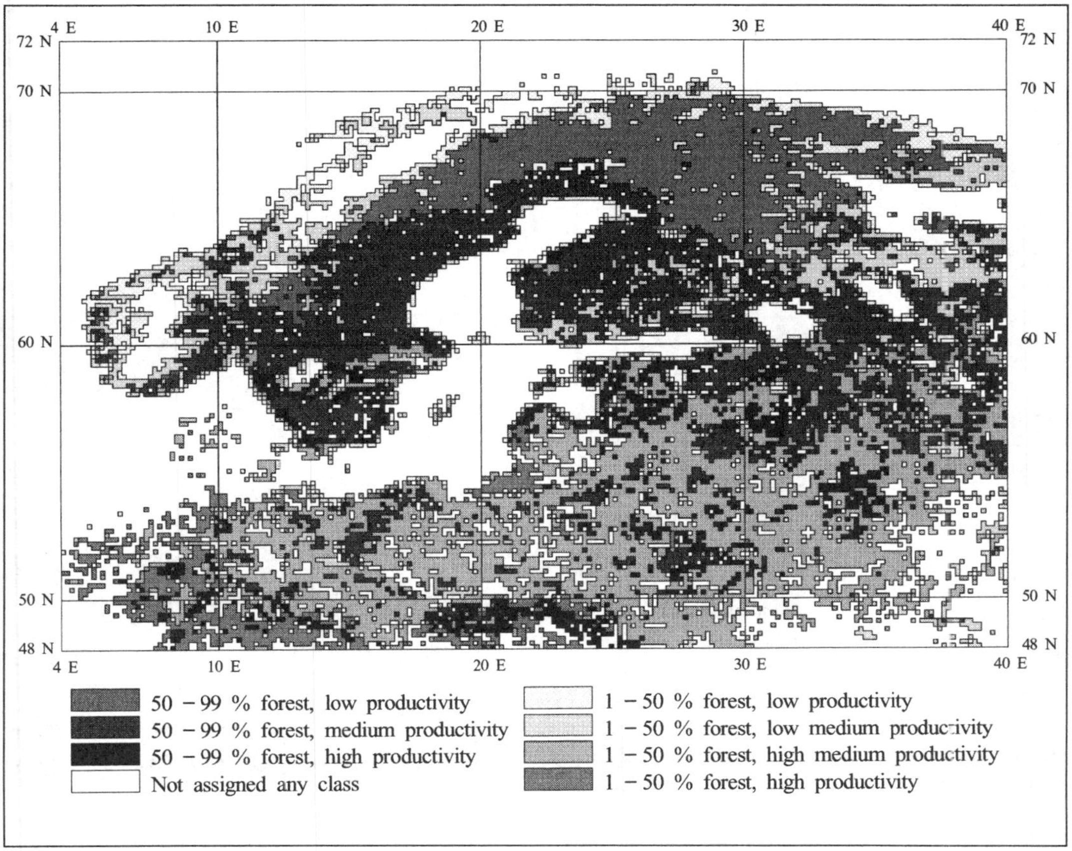

Fig. 3. Systematic classification from the feature space of annually integrated average (1986-1988) NDVI vs forest fraction. Class limits are illustrated in figure 2. Forest fraction from ESA Remote Sensing Forest Map. NDVI from Monthly Generalised Global Vegetation Index.

general trend is decreasing production wit higher latitude, as could be expected. It is clear that the curves are more equal, and show rather invariant production, S of about 58°N. The consistent feature at about 63°N is very interesting. It is clearly more pronounced for the open land, compared to the forest. A similar feature, but inverted can be seen at about 68°N. The cause of these anomalies is yet to be found.

9 CONCLUSIONS AND FUTURE PERSPECTIVES

The results so far are encouraging. They indicate that the method is functional also at higher latitudes, i e that NPP is correlated to integrated NDVI also here. The results of the analysis are realistic, despite the fact that they have been based on uncorrected data until now. Introducing the correction for the terminator effect, including the correction for sensor drift, will make the latitudinal trend even more significant. Excluding the period without photosynthetic activity will also act in this direction. The spatial, dimensional stratification is believed to improve the reliability of the estimates. It also gives some correction for the mixel effects. There is still a lot of difficulties, but there clearly seems to be a potential for NPP estimates over the Baltic Drainage Basin area using this technique.

ACKNOWLEDGEMENTS

This research was carried out within the framework of the European Union DG XII-D (Contract EV5V-CT92-0183) and funded by the Swedish and Norwegian Research Councils.

UNEP/GRID -Arendal supplied the GVI data and the climate data. The latter as part of the set of cut-outs from global data sets, having the same geographical extent as this study.

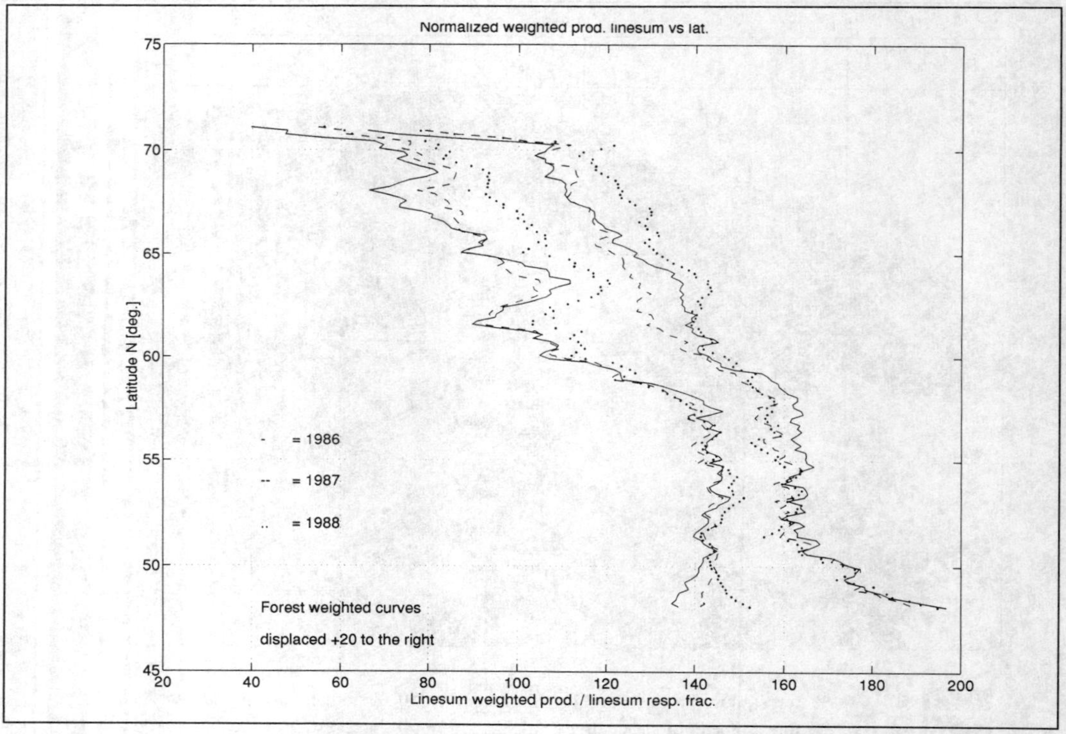

Fig. 4. Normalised fraction weighted production summed over 10' latitude lines plotted vs latitude. Forest curves are displaced by 20 units for better separability. Solid line = 1986. Dashed line = 1987. Dotted line = 1988.

ACRONYM LIST

APAR	Absorbed Photosynthetically Active Radiation
AVHRR	Advanced Very High Resolution Radiometer
BDBP	Baltic Drainage Basin Project
CD-ROM	Compact Disc Read Only Memory
CIS	Collaborating Independent States (?)
DCW	Digital Chart of the World
DG	Directorate General
EPA	Environmental Protection Agency (USA)
ESA	European Space Agency
ESTEC	European Space Technology Centre
EU	European Union
EUMETSAT	European Meteorological Satellites (?)
GAC	Global Area coverage
GAF	Gesellschaft für Angewandte Fernerkundung
GPP	Gross Primary Production
GRID	Global Resources Information Database
GVI	Global Vegetation Index
IGBP_DIS	International Geosphere-Biosphere Program Data and Information System
IIASA	International Institute for Applied Systems Analysis
ISY	International Space Year
ITCZ	Inter Tropical Convergence Zone
LAC	Local Area Coverage
MGV	Monthly Generalised global Vegetation index
MVC	Maximum Value Composite
NCDC	National Climatic Data Centre
NDVI	Normalised Difference Vegetation Index
NEP	Net Ecosystem Production
NESDIS	National Environmental Satellite Data and Information Service
NGDC	National Geophysical Data Centre
NOAA	National Oceanic and Atmospheric Administration
NPP	Net Primary Production
PAR	Photosynthetically Active Radiation
RSFM	Remote Sensing Forest Map
SAGES	Skåne Association for Global Environmental Studies
SDSD	Satellite Data Services Division
TIROS	Television InfraRed Observation Satellite
UNEP	United Nations Environment Program
USA	United States of America

REFERENCES

Box, E. O., B. N. Holben, V. Kalb 1989. Studies of the AVHRR vegetation index as a predictor of biomass, primary productivity and net CO_2 flux. *Vegetatio* 80:71-89.

Belward, A. S. 1991. Remote sensing for vegetation monitoring on regional and global scales. In A. S. Belward

& C. R. Valenzuela (eds.), *Remote Sensing and Geographical Information Systems for Resource Management in Developing Countries*: 169-188. Dordrecht: Kluver Academic Publ.

Cramer, W. 1993. Potsdam Institute for Climate Impact Research, Germany. Personal communication.

Davis, J. C. 1986. *Statistics and data analysis in geology*. 2:nd ed. New York: John Wiley & Sons.

EPA 1992. *Global ecosystems database. Version 1.0 (on CD-ROM). Documentation manual Disc-A*, EPA/600/R-92/194b, NGDC key to Geophysical Records Doc No 27 : A01-2 - A01-13.

ESA 1992. *Remote sensing forest map of Europe*. ESA/ESTEC, Nordwijk, the Netherlands.

Folke, C., S. Langaas, J. Larsson, J. Sweitzer 1993. *The Baltic drainage basin project. Sub-project 1: Land use and ecological carrying capacity of the Baltic drainage basin*. Project report distributed at the meeting 13 Nov. 1993. Royal Academy of Sciences, Stockholm.

GAF 1992. *Feasibility study of regional forest inventory maps of the ISY - "World Forest Watch" project*. Final report prepared for ESA/ESTEC. GAF, Munich, Germany.

Goward, S. N., B. Markham, D. G. Dye, W. Dulaney, J. Yang 1991. Normalized Difference Vegetation Index measurements from the Advanced Very High Resolution Radiometer. *Remote Sensing of Environment* 35:257-277.

Hastings, D. A., W. J. Emery 1992. The Advanced Very High Resolution Radiometer (AVHRR): A brief reference guide. *Photogrammetric Engineering and Remote Sensing* 58:1183-1188.

Haxeltine, A. 1994. SAGES, Department of Plant Ecology, University of Lund, Sweden. Personal communication.

Holben, B. N. 1986. Characteristics of maximum-value composite images from temporal AVHRR data. *International Journal of Remote Sensing* 7:1417-1434.

Johnson, B. 1965. *Medeltemperaturens och vegetationsperiodens nutida förändringar i Sverige*. Thesis, Department of Geography, University of Stockholm.

Kammerud, T. A. 1993. *The influence of soil on satellite monitoring of Sahelian grasslands*. MSc thesis, Department of Geography, University of Oslo.

Kidwell, K. B. 1990. *Global vegetation index user's guide*. NOAA/NESDIS/NCDC/SDSD, Washington D.C.

Kumar, M., J. L. Monteith 1981. Remote sensing of plant growth. In H. Smith (ed.), *Plants and the Daylight Spectrum*: 133-144. London: Academic Press.

Leemans, R., W. P. Cramer 1991. *The IIASA database for mean monthly values of temperature, precipitation and cloudiness on a global terrestrial grid*. IIASA report RR-91-18.

Lo Seen Chong, D., E. Mougin, J. P. Gastellu-Etchegorry 1993. Relating the global vegetation index to net primary productivity and actual evapotranspiration over Africa. *International Journal of Remote Sensing* 14:1517-1546.

Loudjani, P., B. Saugier, G. Saint, J. C. Menaut 1993. Estimation of regional net primary production using coarse resolution satellite data. A case study: West Africa. *Proc. 6th AVHRR data users' meeting*, 29 June - 2 July 1993, Belgirate, Italy, EUMETSAT EUM P12: 125-134.

Perttu, K. 1993. Swedish University of Agricultural Sciences. Personal communication.

Prince, S. D. 1991a. Satellite remote sensing of primary production: comparison of results for Sahelian grasslands 1981-1988. In S. D. Prince & C. O. Justice (eds.), *Coarse Resolution Remote Sensing of the Sahelian Environment, International Journal of Remote Sensing* 12:1301-1312.

Prince, S. D. 1991b. A model of regional primary production for use with coarse resolution satellite data. In S. D. Prince & C. O. Justice (eds.), *Coarse Resolution Remote Sensing of the Sahelian Environment, International Journal of Remote Sensing* 12:1313-1330.

Prince, S. D., C. O. Justice, B. Moore III (eds.) 1994. *Monitoring and modelling of terrestrial net and gross primary production*. IGBP_DIS Working Paper # 8.

Richter, G. 1988. *Stoffwechselphysiologie der Pflanzen. Physiologie und Biochemie des Primär- und Sekundärstoffwechsels*. 5:th ed. Stuttgart: Georg Thieme Verlag.

Ruimy, A., B. Saugier, G. Dedieu 1994. Methodology for the estimation of terrestrial net primary production from remote sensing data. *Journal of Geophysical Research* 99(D3):5263-5283.

Running, S. W., T. R. Loveland, L. L. Pierce 1994. A vegetation classification logic based on remote sensing for use in global biogeochemical models. *Ambio* 23:77-81.

Sellers, P. J. 1985. Canopy reflectance, photosynthesis and transpiration. *International Journal of Remote Sensing* 6:1335-1372.

Sellers, P. J. 1987. Canopy reflectance, photosynthesis and transpiration. II. The role of biophysics in the linearity of their interdependence. *International Journal of Remote Sensing* 21:143-183.

Sellers, P. J. 1989. Vegetation-canopy spectral reflectance and biophysical processes. In G Asrar (ed.), *Theory and Applications of Optical Remote Sensing*. New York: John Wiley & Sons.

Sellers, P. J., J. A. Berry, G. J. Collatz, C. B. Field, F. G. Hall 1992. Canopy reflectance, photosynthesis and transpiration. III. A reanalysis using improved leaf models and a new canopy integration scheme. *Remote Sensing of Environment* 42:187-216.

Tarpley, J. D. 1991. The NOAA global vegetation index product - A review. *Paleogeography, Paleoclimatology, Paleoecology (Global and Planetary Change Section)* 90:189-194.

Whittaker, R. H., P. L. Marks 1975. Methods of assessing terrestrial productivity. In H. Lieth & R. H. Whittaker (eds.), *Primary productivity of the Biosphere*: 55-118. New York: Springer Verlag.

Wylie, B. K., J. A. Harrington Jr, S. D. Prince, I. Denda 1991. Satellite and ground-based pasture production assessment in Niger: 1986-1988. In S. D. Prince & C. O. Justice (eds.), *Coarse Resolution Remote Sensing of the Sahelian Environment, International Journal of Remote Sensing* 12:1281-1300.

Sensors and Environmental Applications of Remote Sensing, Askne (ed.) © 1995 Balkema, Rotterdam, ISBN 90 5410 524 0

Drought monitoring in Spain using satellite remote sensing

F. González Alonso & J. M. Cuevas
Laboratorio de Teledetección, Instituto Nacional de Investigación y Tecnología Agraria y Alimentaria, Carretera de La Coruña, Madrid, Spain

J. L. Casanova, A. Calle & P. Illera
Laboratorio de Teledetección, Departamento de Física Aplicada I, Universidad de Valladolid, Spain

ABSTRACT: In 1993, large areas of Spain were affected by drought as a result of the restricted rainfall during the first months of the agricultural year. The main regions affected were Andalusia, Extremadura, Aragon and Castile La Mancha. The NOAA-11 pictures appear to offer a useful alternative in the monitoring of these types of situations.

1 INTRODUCTION

Drought is a recurring and extremely serious phenomenon in those countries with arid and semi-arid climates. These account for approximately 30% of the earths surface.

Spain, as a country, is particularly affected by the drought phenomenon given that in the period from 1880 to 1980 more than 50% of the years were designated as dry.

In Table 1 we show those years described as dry or very dry for the period 1940-1992. It can be seen that during the period 1980-1992, eight of the years were deemed very dry and two as dry. That is to say, ten of the twelve years have registered a water deficit and have been described as dry or very dry.

Table 1. Drought in Spain: 1940-1992

Period	Dry years	Very dry years
1940-1949	3	3
1950-1959	2	3
1960-1969	1	2
1970-1979	1	3
1980-1989	1	6
1990-1992	1	2

From a climatological point of view, drought is a deficit of relative humidity over a period of time and the consequence of an imbalance between the rainfall and the potential evapo-transpiration.

The drought phenomenon can be analyzed from different points of view: thus we may speak of meteorological drought (divergence of the precipitation with respect to normality), agricultural drought (insufficient humidity present in the earth for adequate crop growth), or hydrological drought (based on parameters such as river flow or the amount of water stored in reservoirs). Drought is also a phenomenon which can be characterized by its duration, intensity and the surface area affected.

The definitions of a drought can vary greatly but, generally speaking, they refer to a significant rainfall decrease which involves important reductions in the growth and development of the plants, whether we speak of cultivated crops or natural vegetation.

Palmer (1965) defines a drought as the time interval, generally in months or years, in which the actual availability of humidity in a specific place decreases considerably with respect to the humidity typical of the climate of that region.

Tucker and Choudhury (1987) understand a drought as a period of time in which plant growth is reduced, in comparison with the historical average, as a consequence of a decrease in the amount of rainfall.

The sensors installed in the earth-observation satellites are able to quantify the capacity which the vegetation cover present in the earth´s surface possesses for initiating the process of photosynthesis. When a drought exists, due to a notable reduction of the rainfall, the capacity to carry out the chlorophyllian function on the part of the vegetation is notably reduced. This occurrence is demonstrated by the spectral response which the affected vegetation covers provide.

The chlorophyllous pigments produce two maximum absorption radiation zones in the green plants, one in the blue region of the spectrum (0·43 micrometers) and another in the red region (0·66 micrometers). On the other hand, the mesophyll of the leaves - provided with irregularly shaped cells which constitute a surface with large inter-cellular spaces - is very reflective of the radiation incident in the neighbouring infrared. (0·75 - 1·1 micrometers). Thus, the response of the green vegetation (in a good physiological and healthy state) is characterized by a substantial absorption in the red region and a large reflection in the infrared region near the electromagnetic spectrum.

On the other hand, it has been observed that the vegetation which is unhealthy, ageing or subject to conditions of stress (such as could be the case with vegetation affected by drought), increases its reflectance in the red region of the spectrum while it decreases in the nearby infrared.

2 DROUGHT MONITORING AND THE VEGETATION INDICES

Numerous researchers have developed techniques in order to analyze, qualitatively and quantatively, the condition of the vegetation cover using spectral measurements.

The denominated "vegetation indices" have been defined with the object of reducing the number of specific dimensions of the spectral measurements to a single dimension. These vegetation indices basically use the information of the red and infrared regions near the spectrum in order to reduce the particular characteristics of the vegetation such as the type of cover, foliar surface, stress and phytomass (Tucker, 1979; Perry and Lautenschlager, 1984).

One of the said vegetation indices is the so-called NDVI (Normalized Difference Vegetation Index), (Tucker, 1979). The NDVI is defined as the difference between the infrared radiation and the red radiation reflected by a surface, normalized by the sum of both. That is to say:

$$NDVI = (IR-R)/(IR+R)$$

The range of variation of the NDVI fluctuates between -1 and +1. In general terms the water, snow and clouds have negative values. The rocks and the bare ground have values near to zero and only the vegetation has positive values: these latter being higher where the health of the green vegetation is greater.

Hayes (1985) indicates that the NDVI is highly correlated with the foliar surface index, the primary net productivity and the biomass.

The NDVI vegetation index can be determined using the information supplied by the AVHRR sensor (Advanced Very High Resolution Radiometer) installed on board the North American meteorological satellites of the NOAA type. (Kidwell, 1988).

The NOAA satellites (initially called TIROS-N) have been in orbit since the year 1978. The complete configuration of the NOAA system consists of two satellites in, more or less, polar orbit. They are complementary, one of them crossing the Ecuator at 7·30 and 19·30 and the other at 02·30 and 14·30 hours. The even numbered satellites cover the orbit in the morning and the odd numbered ones cover the orbit in the evening. At the moment, the satellites NOAA-11 and NOAA-12 are in operational order.

As a result of their orbital characteristics and given that there exist two satellites in operation, all the points on the Ecuator are scanned twice a day and twice at night. In the middle latitudes, as is the case with Spain, this degree of coverage is approximately double, which makes the satellites especially appropriate for carrying out the monitoring of large regions, countries or even entire continents.

The AVHRR sensor possesses a channel in the visible band of the spectrum, another in the neighbouring infrared band and three in the thermal infrared band. The spatial resolution is of the order of 1·1 km. in the nadir in the LAC (Local Area Coverage) complete resolution mode.

The possibility of using the NOAA pictures in the detection and monitoring of the drought phenomenon has begun to be explored as a consequence of the important droughts which have occurred in the Sahel countries in the 1980´s.

The experiments have mainly been carried out concerning droughts which have occurred in Africa (Hielkema and Van Herwaarden, 1993; Tucker and Choudhury, 1987), India (Thiruvengaddchari and Gopalkrishna, 1993), Australia (Bryceson, 1993) and the United States (Teng, 1990). The results arrived at in the studies mentioned may be considered quite satisfactory and promising given the fact that they are based on the pictures provided by the most operational Earth observation satellite of those in existence at the moment.

The existence of AVHRR picture libraries since the year 1981 and the missions programmmed for the present decade, ensure a continuity in the flow of information and allows us to refer back to previous years for which there exists a complete breakdown of the prevalent climatic and agronomic conditions.

3 METHODOLOGY AND RESULTS

In the first six months of the year 1993 NOAA-11 pictures (in HRPT format) were obtained by the receiver antenna located in the Remote Sensing Laboratory of the University of Valladolid. These pictures were accordingly processed in order to carry out their radiometric and geometric correction and the NDVI was subsequently calculated.

The cartography of the zones affected by drought was carried out by means of a non-supervised multi-temporal process of automatic classification.

The pictures, with the NDVI corresponding to the pictures most free of clouds and prior to the harvesting of the cereals, have been categorised by means of an automatic algorithm classification in 15 groups. The average value of the NDVI on the different dates for the different groups have been interpreted and those groups in which the NDVI has successively shown very low values are those which were considered to correspond to zones affected by drought. Specifically, the cereal regions of dry farming most affected by the 1993 drought are represented in yellow in Photograph 1 and were the provinces of Seville, Jaen, Badajoz, Toledo, Ciudad Real and Saragossa.

In Table 2 we present, in the provinces indicated, the ratio between the production of cereals corresponding to 1993 with respect to 1991. This was considered by the authorities of the Ministry of Agriculture as a "normal" year from the point of view of agricultural production.

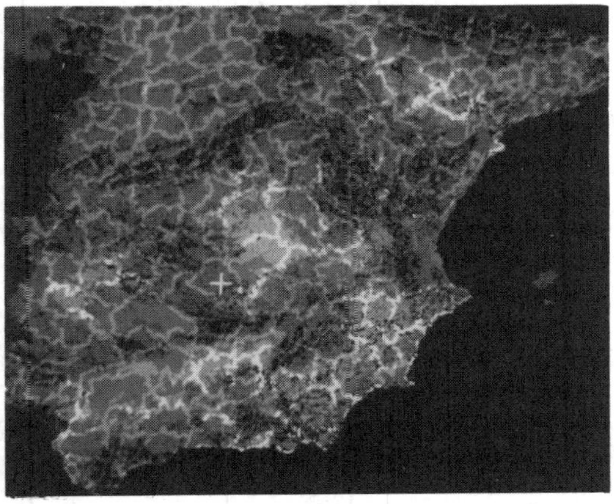

Photograph 1. 1993 drought zones (in yellow)
Colour plate see page 491

Table 2. Cereal production: 1993 vs.1991

Region	1993/1991 %
Seville	18·94
Jaen	53·70
Badajoz	3·89
Toledo	70·52
Ciudad Real	100·64
Saragossa	59·82

From the observation of the previously mentioned Table we can deduce that, with the exception of the province of Ciudad Real, in the rest of the provinces very important reductions have been registered in the production of cereals with respect to 1991. In the specific case of Badajoz this reduction was of the order of 96%.

4 CONCLUSIONS

The high definition pictures obtained by the NOAA-AVHRR contain interesting advantages for the study of the drought phenomenon in the Spanish context. These include: the possibility of obtaining cloud-free pictures, the considerable radiometric precision of the AVHRR sensor in 10 bits, the spatial resolution adapted to the scale of work, the reasonable cost of acquiring the pictures and the coverage of all the national peninsular territory in a single picture.

The results obtained in the present study are encouraging although it appears necessary to study in greater depth the relationship existing between the indices of drought estimation and the remote sensing data as well as that between the remote sensing data (in terms of the NDVI) and the production estimate.

5 REFERENCES

Bryceson, K.P., 1993. Locust survey and drought monitoring in Australian government agencies. International Symposium Operationalization of Remote Sensing, ITC, Enschede, pp. 237-248.

Hayes, L., 1985. The current use of of TIROS-N series of meteorological satellites for land cover studies. International Journal of Remote Sensing, 6: 35-35.

Hielkema, J.V. and Van Herwaarden, G.J., 1993. FAO uses real-time environmental data to secure crop health. Earth Observation Magazine, 9: 32-34.

Kidwell, K.B., 1988. NOAA Polar Orbiter Data Users Guide, NESDIS, NOAA, Washington, D.C.

Lopez Bermudez, F., 1993. Desertificación: factores y procesos. En "La Teledetección en el seguimiento de los fenómenos naturales. Climatología y Desertificación". Eds. S. Gandía y J. Meliá. Universidad de Valencia. 308 págs.

Palmer, W.C., 1965. Meteorological drought. Res. Pap. No 45, US Weather Bureau, Washington, DC, pp. 1-58.

Perry, C.R. and Lautenschlager, L.C., 1984. Functional equivalence of spectral vegetation indices. Remote Sensing of Environment, 14: 169-182.

Puigdefabregas, J., 1994. Desertificación: una perspectiva sobre la cuenca mediterránea. Revista FRONTERAS, CSIC, 3: 15-19.

Teng, W.L., 1990. AVHRR monitoring of U.S. crops during the 1988 drought. Photogrammetric Engineering and Remote Sensing, 56: 1143-1146.

Thiruvengadachari, H.R. and Gopalkrishna, H.R., 1993. An integrated PC environment for assessment of drought. International Journal of Remote Sensing, 14: 3201-3208.

Tucker, C.J., 1979. Red and photographic infrared linear combinations for monitoring vegetation. Remote Sensing of Environment, 8: 127-150.

Tucker, C.J. and Choudhury, B.J., 1987. Satellite remote sensing of drought conditions. Remote Sensing of Environment, 23: 243-251

Sensors and Environmental Applications of Remote Sensing, Askne (ed.) © 1995 Balkema, Rotterdam, ISBN 90 5410 524 0

Application of airborne polarimetric SAR data for forest classification

Barbara Koch
Department of Photogrammetry and Remote Sensing, University Freiburg, Germany

Bernhard Förster & Gerhard Hunner
Institute of Landuse Planning and Nature Conservation, University of Munich, Germany

ABSTRACT: Investigations on the use of multifrequent (C-, L- and P-band), polarimetric airborne radar image data for forest mapping were carried out based on a data set assessed during the E-Mac campaign in 1991 over an forest area south west of Munich. The results proved a good separability between forest and non forest land in P- and L- band independent from polarisation status. Within the forest, deciduous stands, old spruce stands, young spruce stands, pine stands and mixed stands were separable in L- and P-band. Best polarisation combination seems to be HV followed by HH.

1 INTRODUCTION

During the E-Mac campaign in 1991 the use of multipolarimetric and multifrequent radar image data for the separation of forest stands has been tested. The data take has been carried out over a forest site located south-west of Munich at a latitude of 48° 04‘ and longitude of 11° 11‘. The test area is characterized by pure and mixed spruce (picea abies), pine (pinus sylvestris) and beech (fagus sylvatica) stands of different natural age classes. The terrain is fairly flat with some smooth rolling hills at the edges of the test site. The overflight was carried out on July the 12th 1991 at 1 15 pm GMT.

The image parameters are characterized as follows:

pixel projection	slant range
pixel resolution	6.6 m slant range
pixel resolution	12.5 m azimuth
pixel size	12.5 m x 12.5 m
polarisation	Quad pol mode
frequencies	C-Band (5.6 cm) L-Band (24 cm) P-Band (68 cm)
local incidence angle	45° runway of the German Space Agency (DLR)
flight direction	180°
flight altitude	6000 m
image size	1099 x 1024 az x rg

2 DATA PROCESSING AND ANALYZATION

The data have been calibrated with the software modul POLCAL and rectified with the RSG software based on 42 control points and a 50m x 50m DGM with 2 m resolution. The quality control for the rectification showed a RMS smaller than 1 Pixel. The good results were also confirmed by visuel checks. With the SYNTHESIZE software the stoke format data were decompressed and transformed into digital numbers within a 16 bit range.

The first data check showed a decrease of the grey values in slant range direction. The strongest influence has been observed for the L-Band data

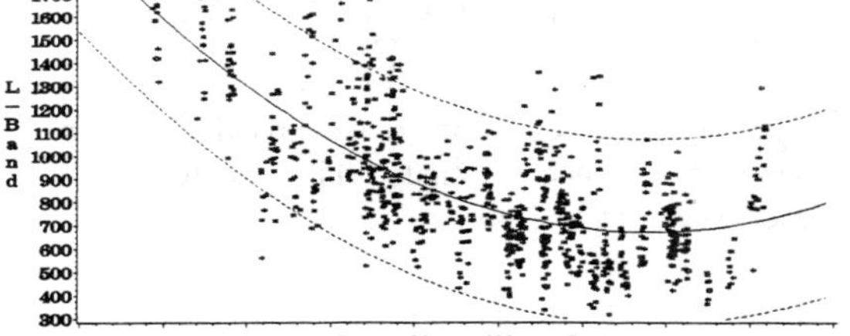

Fig. 1: Grey value decrease in L-Band in slant range direction

(Fig. 1).This influence has been corrected by a quadratic regression function model (Fig. 2)

For the extraction of the training plots the rectified image was overlayed by the polygons of the stand borders drawed from a digital forest map. 228 training plots were picked representing 20 different forest classes. In order to analyze the full polarimetric information the backscatter characteristic were visulized in 3 D graphs for all training plots. The 3 D graph provided the backscatter pattern in each band with 5° polarisation resolution. (Fig. 3)

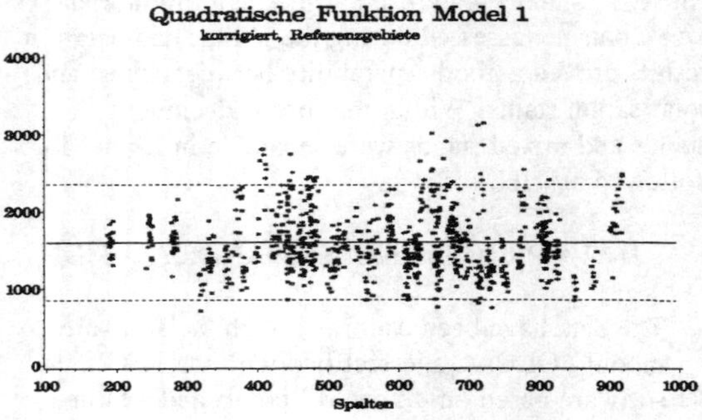

Fig 2: Corrected L-Band grey values in slant range direction

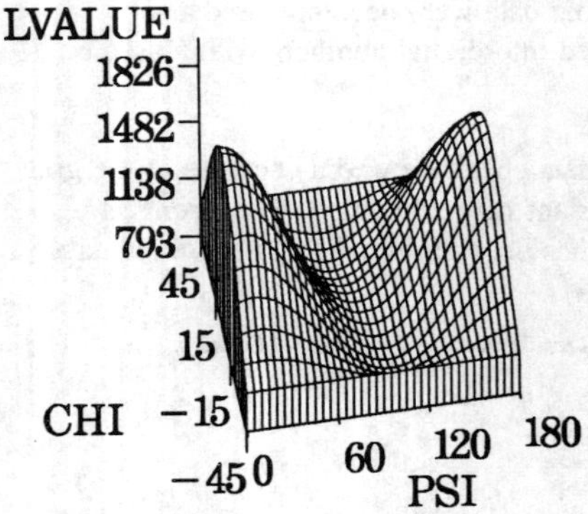

Fig 3: 3-D graph for co-polarized L-band data over orientation (psi) and ellipticity (chi) angle for spruce timber in 5° steps.

Based on the 3-D graphs the most informative polarisation combinations were defined. In each band 50 polarisation states including the main polarisations have been selected as input data for statistical discriminant analysis. With the statistical software package SAS a stepwise discriminant analysis has been performed. Each stand type was tested against each other and the discriminant power, expressed by the F-statistic and the discriminator wilk's lambda, was registered.

Based on the aforementioned statististical tests final classes were selected for a two level classification. In the first level the landuse classes
- forest
- housing areas
- agriculture
- water
- golf course,

in the second level the forest classes
- deciduous stands
- spruce timber
- mixed conifer-deciduous stands
- spruce pole stands
- pine pole stands
- mixed pine-deciduous pole stands

were determined.

In addition for the radar image data also a 4 level classification was carried out

1. level of separation	- forest - housing areas - agriculture land - water bodies - golf course
2. level of separation	- deciduous stands - spruce timber - pole spruce stands - mixed stands
3. level of separation	- mixed stands - pole stands
4. level of separation	- mixed stands - pole pine deciduous stands

Upon the same training plots as defined for radar data also the discrimination power of Landsat-TM data and SPOT-XS data was calculated for the aforementioned classes.

For the classificaction the maximum likelihood method was applied to the radar image data as well as for the optical data. A prior to classification the Lee Filter was applied to the data for speckle reduction and after classification a 3 x 3 median filter. As input channels for the radar image data classification the set of the 16 best channels selected according to the stepwise discrimination analysis and Jeffries Matusita divergency were provided. Only channels which did show an approximative normal shaped distribution of the data were accepted for classification.

3 RESULTS

3.1 Analysation of the 3 D - Graphs

The visuell analysation on the interdependence between polarisation status and backscatter intensities proved:

1. for cross-polarized data for all forest classes and all 3 frequencies the same signal response pattern over the orientation angle (0° - 180°)
2. for co-polarized data specific differences in backscatter pattern over the orientation angle.(0° - 180°) for landuse and forest classes.The differences in backscatter pattern were mainly observed in L- and P-band between spruce and deciduous stands, while pine stands show a pattern characteristic in between spruce and broad leaf trees.

These differences in backscatter pattern along the orientation angle were considered for the selection of the set of input channels for the discrimination analysis.

3.2 Results of the discrimination analysis and classification of the land use classes

The discrimination analysis provided in each band a ranking of the 50 input channels according to their discrimination power. The statistical analysation proved that the discrimination between the land use classes
- water
- forest
- golf course
- agriculture and
- housing areas

was, at this vegetation stadium, in general more effective on the basis of radar data than for optical data. According to the statistical evaluations, classification results for the radar image data were superior between agriculture and forest as well as agriculture and golf course compared to the optical data.

Tab. 1: Jeffries Matusita Distance for the training areas of landuse classes

based on optical data sets and radar image data

sensor type	radar	Landsat-TM	Spot-XS
bands	P-band chi 0/0 psi 0/90 C-band chi 25/25 psi 0/0	band 2 band 5	band 2 band 3
land use classes			
golf-agri	1201	1154	904
golf-forest	1414	1414	1374
golf-water	1414	1414	1414
agri-forest	1409	1195	1286
agri-water	1406	1413	1414
forest-water	1414	1414	1405
Ave	1386	1334	1299
Min	1201	1194	904

*Jeffries-Matusita-Distance best separability is 1414

Only the separability between housing areas and forests was higher for the optical data as the classification proved, even so this is not expressed for the Jeffries Matusita distances clearly, due to the fact that the distances refer to training pixels picked in dense building areas, and not in areas with single distributed houses in gardens, as it is more typical for the test area. Nevertheless due to the texture in the radar image a visual separation between open housing areas and forests was possible.

After masking the housing areas a differentiation between forest and nonforest land close to 100% in P-band data and only slightly less in L-band was achieved. (Fig. 4)

In comparison the ERS-1 frequency and polarization was applied for land use classification. The classified image proved that the separation of forest and agriculture land on the basis of a C-VV channel is highly incorrect based on a single data set from that specific vegetation period.

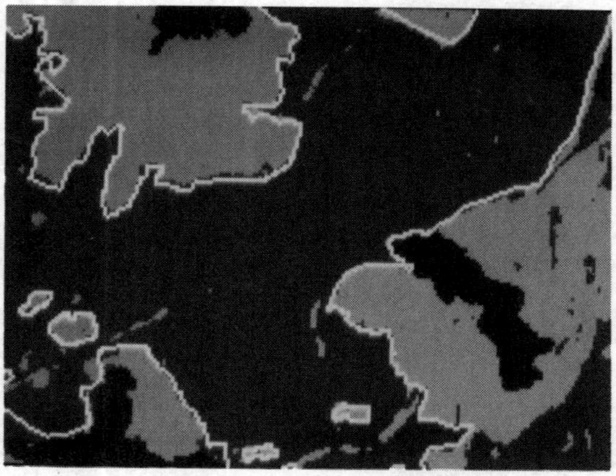

Fig. 4: Classified radar image overlaped by the border of the forest land out of a forest map.
forest =dark
agricultural land =light
housing areas =black

Colour plate see page 491

3.3 Results of the discrimination analysis and classification results within the forests

Within the forests the discrimination of the optical data between the aforementioned stand type classes was better than for the radar data. Especially the high variance within one stand type, the edge effects and the low discrimination power between pure old conifers and mixed stands added up to a less successful classification result for radar data compared to optical data. Even so the Jeffries Matusita Distance points towards compareable classification results, this was only true for the middle part of the image, where most of the training plots were located (Tab. 2).

According to Jeffries Matusita distance radar data showed especially low discrimination between old spruce stands and mixed stands, Landsat data low separability between old spruce stands and mixed stands as well as mixed stands and pine pole stands and Spot-XS data low discrimination power between old spruce stands and young spruce stands as well as spruce pole stands and deciduous stands.

The classification results proved that there are distinct differences between all 3 sensor types (Tab 3). If Landsat-TM are compared with Spot-XS the pole spruce stands and deciduous stands seem to be overclassified with Spot data while mature spruce stands seem to be underclassified. This points towards the fact that with Spot different intensities and distributions for the same objects are measured compared to Landsat-TM. With SPOT data more often young spruce timber were classified as spruce pole wood while with Landsat-data these stands mostly were classified as spruce timber. For Landsat-TM data a slight overclassification of spruce timber on expenses of mixed stands was observed. The high possibility of misclassification

Tab. 2: Best separability using Jeffries Matusita Distance* for the training areas of forest stand type classes based on optical and radar image data

sensor type	Radar	Landsat - TM	Spot - XS
bands	P-band chi 0/0 psi 0/0 L-band chi 0/0 psi 0/0	band 4 band 7	band 2 band 3
forest classes			
spruce timber/pole spruce	1174	1174	1072
spruce timber/ deciduous	1134	1410	1414
spruce timber/ mixed stands	846	868	1123
spruce timber/pine thicket	-	1249	1413
pole spruce/ deciduous	1408	1413	1411
pole spruce/ mixed stands	1265	1172	676
pole spruce/ pine thicket	-	1071	1345
deciduous/ mixed stands	1049	1175	1284
deciduous/ pine thicket	-	1395	1412
mixed stands/pine thicket	1395	921	1094
Ave	1154	1102	1110
Min	846	8680	676

*0-1414

between pole spruce stands and spruce timber with Spot data and the high possibility of misclassifications between spruce timber and mixed stands for Landsat-TM data on the other hand seem to be the reason for the classification differences.

The radar classification shows a strong tendency to overclassify mixed stands and deciduous stands on expenses of young spruce stands and old spruce stands. The overclassification is correlated with the slant range position. With very small incidence angle objects with lower signal response ,according to the signature analysis, get overclassified what is even more distinct with wider incidence angle.That means despite of the slant range grey value corrections applied, there is still a drop in grey values towards the edges, especially in the far range.

The four step classification within the forest proved for radar data better results than the one level classification.within the forest. In the one level classification within the forest the biggest class mixed stands became totally dominating. In the middle parts of the radar image the classification results were compareable to the Landsat respectively Spot classification (Tab.3).

Tab. 3: results of the stand type classification

sensor types	radar	Landsat-TM	SPOT-XS
spruce timber	6,9% 13.93%*	34.93%	18.40%
pole spruce	8.11% 6.46%*	15.00%	25.41%
deciduous	16.57% 16.57%*	6.65%	12.09%
mixed stands	69.02% 60.01%*	42.60%	40.92%
pine thicket	0.1% 0.68%*	0.83%	3.18%

*results from the four level classification

Nevertheless the structures of the features in classified radar image are not enhanced the same way as in the optical data, because structures perpendicular to slant range are enhanced while structures in slant range are depressed.

4 CONCLUSIONS

1. A strong influence with changing range position was observed on the grey values
2. A high standard deviation of the radar data demands a filtering of the data a prior to classification. The Lee sigma[2] filter proved to be appropriate for speckle filtering
3. Roughness seems to be the decisive parameter for radar backscatter, therefore stand types were separated if they differed in their stand structure
4. In this specific test area the multipolarimetric and multifrequent radar image data proved better discrimination power for the defined landuse classes and nearly compareable results for the separability of the defined forest classes than the optical data sets.
5. The most successful radar band proved to be P-band and L-band for stand type classification. For the P-band and L-band the HV followed by the HH polarisation was found to be most succesful in stand type identification.
6. Between P- and L-band high correlation was observed, while optical data are not or negatively correlated with radar data.

5 REFERENCES

Keil, M.. & Raupenstrauch, J. & Tares, T. & Winter, R.1993. Investigations of Forest Information Content of Polarimetric AIRSAR Data in the Harz Mountains.In:*Proceedings of the 25th International Symposium 'Remote Sensing and Global Environment Change'. I-339 ff. 1993* Graz.

Lavalle, C., and Sieber, A.1993. Forest Backscattering Observations by Using Particular Polarimetric States. In: *Proceedings of the 25th Symposium, 'Remote Sensing and Global Environmental Change. I-310 1993* Graz.

Rauste, Y. 1993. Multitemporal Analysis of Forest Biomass Using AIRSAR Data. In: *Proceedings of the 25th Symposium, 'Remote Sensing and Global Environmental Change. I-328 1993* Graz.

Sensors and Environmental Applications of Remote Sensing, Askne (ed.) © 1995 Balkema, Rotterdam, ISBN 90 5410 524 0

Retrieval of forest biomass using SAR

H. Israelsson & J. Askne
Chalmers University of Technology, Göteborg, Sweden

ABSTRACT: The potential of using the radar backscattering coefficient to retrieve information concerning the forest biomass is analysed. The radar backscattering dependence on the bole volumes in a poplar forest is studied in particular. The PLC-bands were examined. Bole volume estimation at C-band was found unrealisable. At L-band very young forest stands could be distinguished from mature poplar stands. At P-band biomass estimation was feasible up to 120 m^3/ha at cross polarisation, which however still is a too small number to be useful in an operational algorithm. The frequency dependence was theoretically validated using MIMICS.

1 INTRODUCTION

The growing number of spaceborne imaging SAR sensors has increased the interest in using the SAR technique for forest mapping. One main objective of the research going on is to invert the radar backscattering coefficient in order to retrieve useful information concerning biophysical parameters. A potential application is estimation of forest biomass, that is important on a global scale as a crucial input parameter to global carbon models. Biomass estimation is of major importance also at a regional level in assessment of forest properties. The objective of this paper is to analyse whether the radar backscattering coefficient contains relevant information about the forest biomass. By examining the SAR data together with relevant ground truth from a poplar forest in the Netherlands, the ability to estimate biomass is determined empirically. The multi frequency polarimetric JPL/AIRSAR is used together with data from the spaceborne SAR systems ERS-1 and JERS-1. The results found from the empirical study is evaluated theoretically using the MIMICS forest canopy backscattering model (Ulaby *et al.* 1990).

2 TEST AREA

The test area that was used was chosen since it was a test site during the MAESTRO-1 campaign in 1989 and the MAC'91 campaign in 1991, when the JPL/AIRSAR was flown over some test sites in western Europe. Hence a large amount of SAR data were available from the area. The forested area that was chosen is a poplar forest in the Flevoland area in the Netherlands, that has been planted in regular row patterns. The stand sizes vary between 5 and 10 hectares. The area is almost completely flat, hence no effect of topography is expected.

2.1 Ground Truth

To study the relation between radar backscattering, *i.e.* the SAR image, and forest biomass, precise biomass values of a number of forest stands are needed. The above ground biomass is however hard to measure and has to be found from empirical models that are limited to certain conditions. In order to use reliable numbers, we consequently have chosen to use bole volumes to characterise the biomass. The bole volume represents the volume of the trunks without branches and is defined,

$$V_b = \frac{fd^2\pi hN}{4} \qquad \left[\frac{\text{m}^3}{\text{ha}}\right] \tag{1}$$

where d is the diameter at breast height, h is the tree height, N is the number of trees per hectare and f is a formfactor taking into account that trunks are not perfect cylinders. The formfactor was in this case put to 0.45. Forty stands in the poplar forest were chosen to be examined. In each stand, the bole volume was estimated at a number of points using a relascope and height measure instruments. The range of bole volumes that was found span from 10 to 315 m^3/ha. The relative accuracy of the measured values was estimated to be around 10 %.

2.2 SAR data

In the study, data from three different SAR systems were used: the airborne JPL/AIRSAR, the ERS-1 satellite and the JERS-1 satellite. Thus a frequency range from 440 MHz to 5.3 GHz was considered.

Table 1. SAR systems that were used in the study.

system	freq.	pol.	inc.ang. (at the test area)
JPL/AIRSAR	440 MHz 1.25 GHz 5.3 GHz	fully polarimetric	30°-55°
ERS-1	5.3 GHz	VV	23°
JERS-1	1.25 GHz	HH	40°

Table 2. JPL/AIRSAR data sets that were used in the study.

date	inc.ang. (at the test area)	leaf grav. moist. [%]	trunk grav. moist. [%]	soil vol. moisture [%]
89-08-16	40-50	65	46	23
91-06-15	42-52	76	59	31
91-07-03	45-54	73	58	37
91-07-12	32-45	no data	no data	28
91-07-28	29-43	68	53	23

The JPL/AIRSAR data were acquired during the MAESTRO-1, in August 1989, and MAC'91, four dates during the summer of 1991, campaigns. They have been polarimetrically and radiometrically calibrated using the responses from point-targets giving the radiometric accuracy of ±1dB. The relatively wide swath of incidence angles shown in Table 1 corresponds to all flights during the two campaigns. Within a single scene, the forest stands are located within a range of around 10°. In Table 2 the data sets are summarised together with some in situ measurements (Droesen 1989), (* 1991).

The ERS-1 scene that was used is a FD product acquired 91-10-25, which is radiometrically calibrated with the accuracy of ±0.3dB (Ulander 1992). The JERS-1 scene that is used is an ESA test image acquired 93-03-28. The accuracy and precision of that calibration factor is not known. Thus only relative analyses of the JERS-1 data were performed.

3. EMPIRICAL RESULTS

The relationship between the radar backscattering and different biomass attributes such as mean stand height, basal area and bole volume look much similar. For all these cases there is a sensitivity to a certain value on the biomass characteristic, where the radar backscattering becomes saturated at a constant level of σ°.

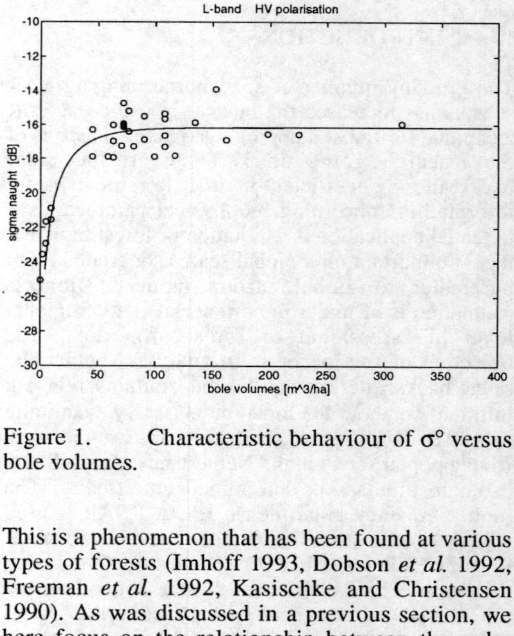

Figure 1. Characteristic behaviour of σ° versus bole volumes.

This is a phenomenon that has been found at various types of forests (Imhoff 1993, Dobson *et al.* 1992, Freeman *et al.* 1992, Kasischke and Christensen 1990). As was discussed in a previous section, we here focus on the relationship between the radar backscattering coefficient and the bole volume. In order to evaluate the sensitivity, the measured radar backscattering coefficients have been matched in the least square sense to an empirical expression,

$$\sigma^{\circ} = \sigma^{\circ}_{sat.}\,(1 - \exp(-\alpha B)) \qquad (2)$$

where B denotes different bole volumes. For all the cases studied, the rms deviations of the empirical fits compared to the measured numbers are less than 1 dB. The equation above is similar to the water cloud model (Attema and Ulaby 1978), where the radar backscattering coefficient was modelled as a function of the depth of the vegetation layer. The parameters of equation (2) differ among various polarisations and radar frequencies. The variations are examined in the following sections.

3.1 Sensitivity saturation

The ability to estimate bole volumes from the radar backscattering coefficient using the empirical relation in equation (2) depends on the magnitude of the attenuation coefficient α. For small α the backscattering coefficient exhibits a nearly linear dependency on the bole volumes, while large α make the dependence saturate at very low values on the bole volumes. We have defined the bole volume sensitivity saturation level B_s as the bole volume level where the sensitivity is 0.1 [dB]/ 100 [m^3/ha].

$$B_s = \frac{1}{\alpha}\ln\left(\frac{10\alpha}{\ln(10^{\frac{0.1}{100}})}+1\right) \qquad (3)$$

The bole volume saturation levels at C-band, Figure 2, are well below 30 m^3/ha at all polarisations and registrations. Thirty m^3/ha corresponds to a very young or intensely thinned forest. Hence C- band cannot be used to estimate forest biomass. It is in many cases even hard to distinguish a mature forest stand from an agricultural field. No significant variation among different polarisations and incidence angles was found.

At L-band, Figure 3 the bole volume saturation levels are higher than at C-band. Cross polarisation exhibits the highest levels between 70-85 m^3/ha, while co polarisation shows lower saturation levels to some extent. The saturation levels at VV are generally somewhat higher than the ones at HH. The results indicate that it is possible at L-band to distinguish very young or thinned forest stands from mature stands, particularly at cross polarisation. The saturation levels are however too low to permit operational biomass assessment.

The P-band bole volume saturation levels are higher than the L-band ones at cross and HH polarisation, Figure 4. At VV polarisation, there is however no apparent difference from L-band. The values at HH polarisation show the largest variation among the different registrations. The results show that P-band cross polarisation is the best configuration to estimate bole volumes of the cases studied. The saturation levels around 120 m^3/ha indicate that a few bole volume classes can be determined. The saturation levels are however still too low to enable bole volume estimations of mature forests.

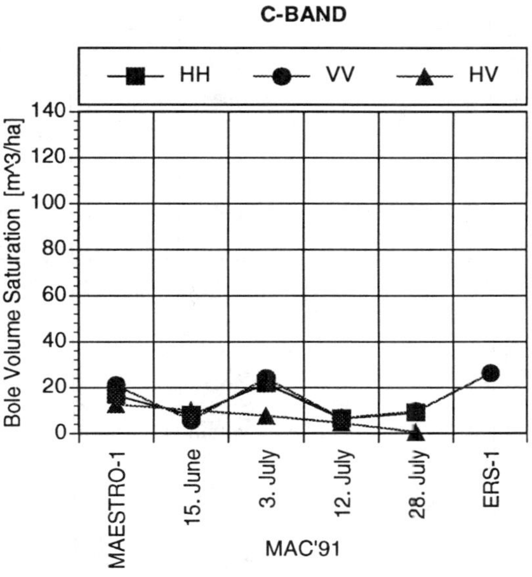

Figure 2. C-band bole volume saturation levels

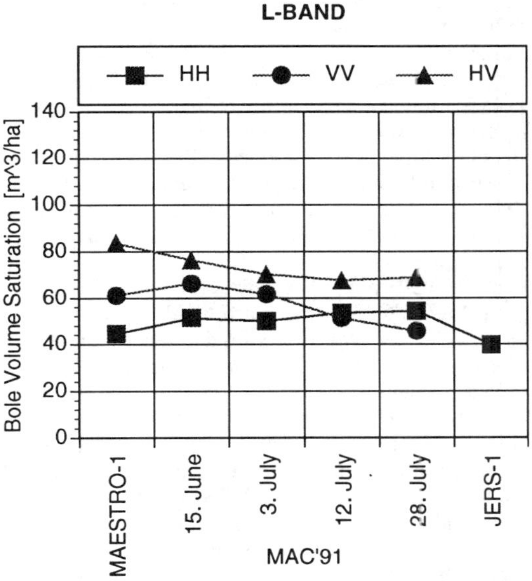

Figure 3. L-band bole volume saturation levels

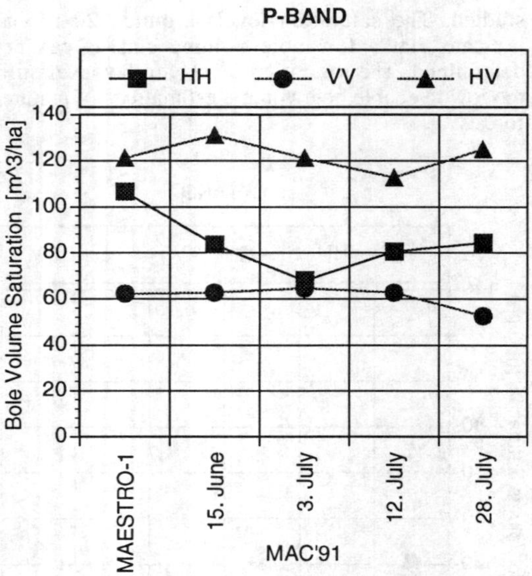

Figure 4. P-band bole volume saturation levels

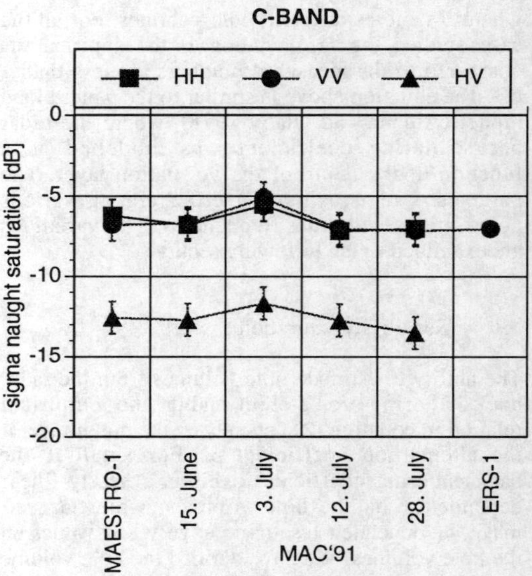

Figure 5. Saturation levels of the backscattering coefficient at C-band.

3.2 σ° saturation

In equation (2), $\sigma^{\circ}_{sat.}$ represents the level that the backscattering coefficient reaches at forest stands with bole volumes above a certain level. In Figure 5-7 those levels are shown for the different data registrations. The error bars denote the radiometric calibration accuracy, that varies among the SAR systems.

At C-band the levels of HH and VV polarisation coincide in all registrations, Figure 5. The backscattering coefficient is quite stable at -7 dB including the ERS-1 image. The registration 91-07-03 exhibits a slightly higher level, that might be due to calibration errors. Cross polarisation is around 7 dB below co polarisation at all registrations.

The HH and VV returns coincide at L-band as well, Figure 6. The levels are fixed between -8 and -9 dB. The HV returns are 5-6 dB lower indicating a slightly higher portion cross polarised backscatter than at C-band.

At P-band the HH return is more 2 dB higher that the VV return, Figure 7. This can be explained by an enhanced ground-trunk backscattering at HH (Israelsson and Askne 1993). The HV backscattering is around 8 dB lower than the HH backscattering.

The variations of the σ° saturation levels among the different registrations indicate that a bole volume estimation is restricted to a certain condition, for which the parameters of equation (2) are well known.

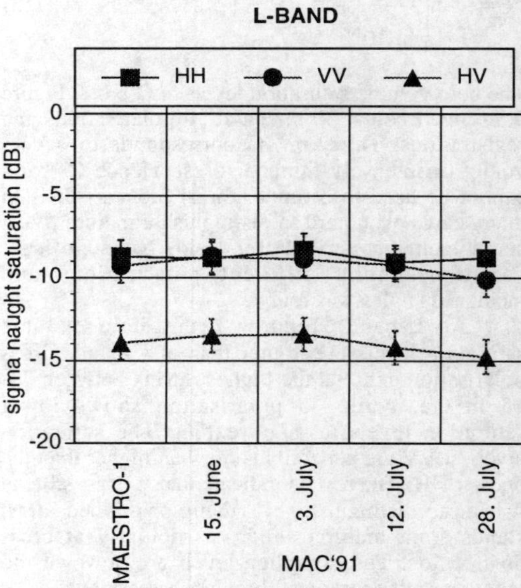

Figure 6. Saturation levels of the backscattering coefficient at L-band.

3.3 HHVV correlation

Using fully polarimetric data the HHVV correlation coefficients may add information to the

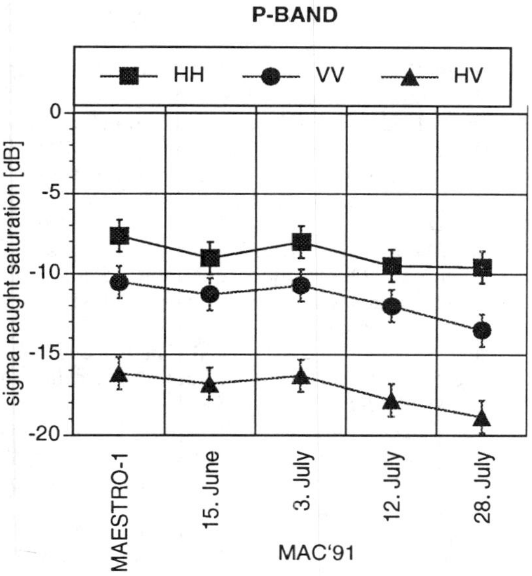

Figure 7. Saturation levels of the backscattering coefficient at P-band.

conventional backscattering coefficients at fix polarisations.

$$\rho_{HHVV} = \frac{\langle S_{hh} S_{vv}^{*} \rangle}{\sqrt{\langle |S_{hh}|^2 \rangle \langle |S_{vv}|^2 \rangle}} \tag{4}$$

The S-parameters in the equation above denote the coefficients of the polarimetric scattering matrix and the brackets denote ensemble averaging.

At C-band the magnitude of the correlation coefficients increases with increasing bole volumes. The spread is however too large to be of use in a bole volume estimation algorithm. For mature forest stands the magnitudes saturate around 0.5. The phases of the coefficient are centred around 0 for all bole volumes at C-band.

At L and P-band the magnitudes decrease slightly with increasing bole volumes and saturate at 0.25 for mature stands. The phases are centred around -10° at L-band and are varying between 0° and -180° at P-band. The ground-trunk backscattering shifts the phase 180°, which explains the large phase variation at P-band.

4 THEORETICAL RESULTS

To understand the results in the previous section, a theoretical analysis of the radar backscattering was performed using the Michigan Microwave Canopy Scattering Model, MIMICS, (Ulaby *et al.* 1990). MIMICS is based on the first order radiative transfer solution, with the forest canopy modelled as two homogenous layers. The different terms of the solution correspond to different scattering paths.

The backscattering coefficient was simulated as function of increasing bole volumes. The geometrical model input parameters were given numbers, that were established from the ground data collection. The parameters were related to varying tree heights. The form of the relation (constant, linear, quadratic or exponential) was chosen for each parameter separately. The moisture parameters were assumed independent on bole volumes and were given characteristic numbers.

Figure 8 shows the bole volume saturation levels according to MIMICS. The saturation levels are determined in the same way as in the empirical study. The lack of bole volume sensitivity at C-band is well predicted. According to MIMICS the C-band backscattering is totally predominated by backscattering from the tree crown, even at low bole volumes. This effect saturates the backscattered signal. At L and P-band the bole volume saturation levels are overestimated at co polarisation, while they are fairly well predicted at cross polarisation. According to the model output, the tree crown dominates the cross polarised backscattering at all the three modelled frequencies. The higher bole volume sensitivity at the longer wavelengths is due

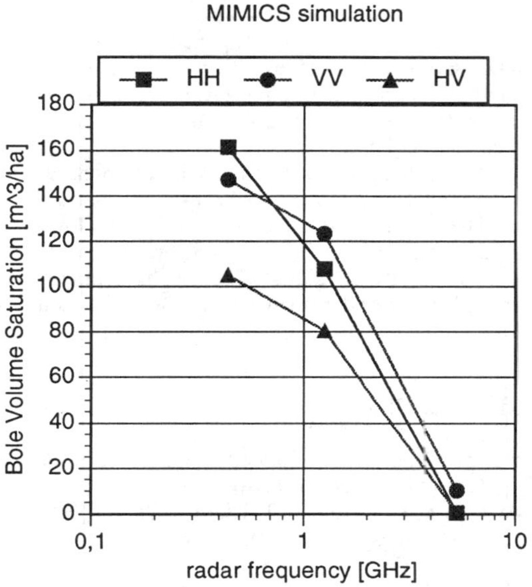

Figure 8. MIMICS simulation of the bole volume saturation levels at PLC-band.

to lower attenuation. At L-band and even more at P-band, in HH polarisation particularly, the trunk-ground backscattering also becomes important. This gives a theoretical sensitivity level that is higher than at cross polarisation. The overestimation compared to the measured values can be explained by the fact that the real forest is not as homogenous as the simulated one.

The theoretical saturation levels of the backscattering coefficient, $\sigma°$, are to some extent dependent on the moisture parameters chosen. At C-band the co polarised backscattering is well predicted, while the cross polarised backscattering is underestimated by more than 10 dB. This is probably an effect of the first-order approximation, which neglects second order effects within a scattering layer. At L and P-band the backscattering coefficients are well predicted at cross polarisation, while the co polarised backscattering coefficients are overestimated by a few dB. This can be partly explained by uncertainties in the input parameters and the approximations being made in the simulations.

5 CONCLUSIONS

The capability to assess bole volumes of a forest using the radar backscattering coefficient has been examined using SAR data from a poplar forest at PLC-band. Retrieval of bole volumes was found impossible at C-band, which was theoretically validated by MIMICS. At L-band bole volume assessment was found feasible up to 70-80 m^3/ha at cross polarisation, which also was theoretically verified. At P-band the empirical levels at cross polarisation increased to 120 m^3/ha. At L and P-band co polarised saturation levels were empirically lower than at cross polarisation, while MIMICS predicted higher levels.

A conclusion of the study is that none of the examined frequencies can be used to assess bole volumes in an operational way. At P-band low bole volume classes can be assessed, but mature forest stands saturate the signal. The results of this paper imply that a radar using still lower frequencies should do a better job.

REFERENCES

* 1991. Ground Data Collection 1991 JPL-SAR Campaign. Heidemij Adviesbureau.

Attema, E. P. W. and F. T. Ulaby 1978. Vegetation modeled as a water cloud. Radio Science. Vol. 13: 357-364

Dobson, M. C., F. T. Ulaby, T. LeToan, A. Beaudoin, E. S. Kaischke and N. Christensen 1992. Dependence of radar backscatter on coniferous forest biomass. IEEE Transactions on Geoscience and Remote Sensing. Vol. 30: 412-415

Droesen, W. J. 1989. MAESTRO 89 Ground Data Collection Horsterwold/ Speulderbos/ Flevoland. Wageningen Agricultural University.

Freeman, A., S. Durden and R. Zimmerman 1992. Mapping sub-tropical vegetation using multi-frequency, multi-polarization SAR data. IGARSS'92 in Houston, Texas, p. 1686-1689.

Imhoff, M. L. 1993. Radar backscatter/biomass saturation: Observations and implications for global biomass assessment. IGARSS'93 in Tokyo, Japan, p. 43-45.

Israelsson, H. and J. Askne 1993. The Effect of Leaning Trunks on Forest Radar Backscattering. IGARSS'93 in Tokyo, p. 57-59.

Kasischke, E. S. and N. L. Christensen 1990. Connecting forest ecosystem and microwave backscatter models. International Journal of Remote Sensing. Vol. 11: 1277-1298

Ulaby, F. T., K. Sarabandi, K. McDonald, M. Whitt and M. C. Dobson 1990. Michigan microwave canopy scattering model. International Journal of Remote Sensing. Vol. 11: 1223-1253

Ulander, L. M. H. 1992. Calibration of the ERS-1 Fast-Delivery Images. First ERS-1 Workshop in Cannes,

Sensors and Environmental Applications of Remote Sensing, Askne (ed.) © 1995 Balkema, Rotterdam, ISBN 90 5410 524 0

Estimation of soil moisture from the MAESTRO-1 SAR data of Flevoland

Jiankang Ji, Henning Skriver & Preben Gudmandsen
Electromagnetics Institute, Technical University of Denmark, Lyngby, Denmark

ABSTRACT: The preliminary results of estimation of soil moisture by applying the L- and C-band SAR data acquired during the MAESTRO-1 Campaign over the Flevoland agricultural areas in The Netherlands are presented. For bare soil case, the calibrated SAR data is examined with classical models. Then the empirical models for co- and cross-polarized ratio developed by Yisok Oh *et al.* [1] are employed. A good linear correspondence between backscattering coefficient σ° and surface rms height σ for all polarizations is observed. Soil moisture content m_v is estimated from L-band data by applying the cross-polarized ratio model and the empirical model of dielectric constant. The estimation accuracy is about 10%. For crop fields, the "Water Cloud" model [2] is applied, where backscatter contribution of vegetation is neglected and only vegetation attenuation is taken into account. Assuming the roughness model for bare soil could be applied, a linear model between $\sigma^{\circ}(dB)$, σ, m_v and LAI(Leaf Area Index) is developed. The model coefficients are regressed with the data. It is applicable for estimation of soil moisture for sparse crop fields and further validation of the model with more experimental data is necessary.

1 Introduction

The MAESTRO-1 experiment undertaken in 1989 has provided an important opportunity to evaluate the potential of multi-frequency polarimetric imaging SAR in agricultural and forestry applications. The NASA/JPL P-, L- and C-band polarimetric airborne SAR mounted on a DC-8 aircraft acquired radar data for four test sites covering essentially typical European forests and agricultural areas scattered in four European countries. During the flight campaign, an extensive ground data collection was accomplished. The significant achievements obtained by the analysis of the SAR and ground data were summarized in the proceedings of the Final Workshop of the Campaign [3].

The data we deal with in this paper were acquired at the Flevoland Test Site in the Netherlands. Within the Flevoland Test Site there were three test areas. One of them was the Flevopolder test site which consisted of a large agricultural area of flat topography and homogeneous soils. Typical crops for this area were potatoes, sugar beet, wheat and corn. Totally 18 fields including 6 bare fields were selected for intensive ground data collection. The SAR data acquisition over the Flevopolder area was performed on August 16, 1989. At the same time, soil moisture, surface roughness and plant physical parameters were collected over the selected fields except for two corn fields.

The estimation of soil moisture for the bare and crop fields from the Flevopolder SAR data is investigated in the paper. Until now, there has been some papers concerning crop discrimination and a paper which mention the possibility of retrieving soil moisture by using the SAR data [4], but apparently there is no publication concerning the soil moisture estimation with this data set. Since P-band SAR data are only weakly sensitive to soil moisture variations [5, 6], only L- and C-band data are applied in the analysis. Before the analysis the data are calibrated with trihedral corner reflectors by use of the integral method [7].

2 Bare Soil Scattering Models

2.1 Classical Models

The Small Perturbation Model (SPM), the Physical Optics (PO) model and the Geometric Optics (GO) model are usually considered as the classical scattering models which are widely used in connection with surface scattering problems [1]. Each of the three models is valid in a certain range which is related to the surface rms height and surface correlation function. To evaluate the applicability of the classical models to the Flevoland SAR data, we plot the validity regions of these models and the measured soil roughness conditions in the $k\sigma - kl$ space, which is shown in Figure 1. In the calculation, we suppose the random surface could be modeled by a Gaussian correlation function. Figure 1 shows that a part of surface roughness conditions is satisfied with SPM at L-band and two soil conditions are valid for PO model at C-band, all the others fall outside the validity regions of all the three models. The figure illustrates that some of the soil surfaces do not satisfy any of the three models, and the classical models can depict surface scattering for only part of agricultural fields, which is consistent with the conclusions drawn in [1]. Therefore the classical models are not adequate for describing the surface conditions of the Flevopolder test site during the Maestro-1 Campaign.

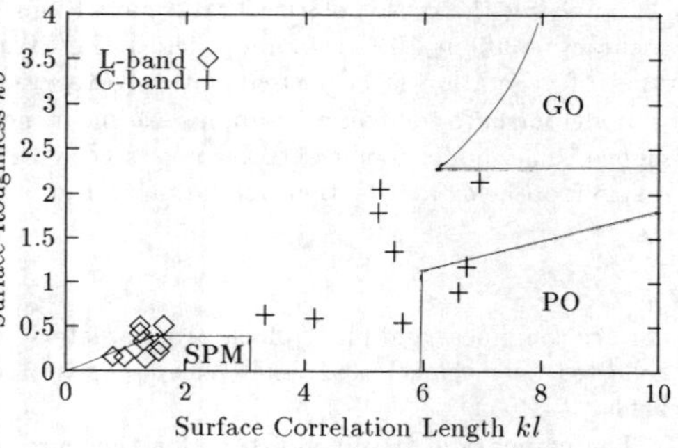

Figure 1: Validity regions for SPM, PO and GO models and the field roughness conditions of the Flevopolder Agricultural Test Site. The incidence angle θ is taken as 45° which is the mean value over all the test fields. The soil surface is characterized by a Gaussian correlation function.

2.2 Integral Equation Model

With the integral equations under Kirchhoff approximation, Fung *et al.* developed a theoretical model to calculate scattering from rough surfaces [8]. The model takes into account multiple scattering and can therefore be applied with cross-polarized scattering. Compared with the classical scattering models, the model is valid in the large regions of surface roughness $k\sigma$ and correlation length kl including all validity regions of SPM, PO and GO models. Under the assumption of $k\sigma << 1$ which is valid for the single scattering approximation, it reduces to SPM. The Flevopolder soil roughness data falls into the validity region of the model so this model is applied to simulate the backscattering coefficients for bare fields. Unfortunately, the model results were not in good agreement with the measurements.

2.3 The Empirical Model for Polarized-ratios p, q

Based on the backscattering data measured at L, C, and X-band, an empirical model is developed by Yisok Oh *et al.* [1] for cross-polarized ratio q and co-polarized ratio p:

$$q = \frac{\sigma^\circ_{hv}}{\sigma^\circ_{vv}} = 0.23\sqrt{\Gamma_0}[1 - exp(-k\sigma)] \quad (1)$$

$$\sqrt{p} = \sqrt{\frac{\sigma^\circ_{hh}}{\sigma^\circ_{vv}}} = 1 - \left(\frac{2\theta}{\pi}\right)^{\frac{1}{3\Gamma_0}} \cdot exp(-k\sigma) \quad (2)$$

where $k = 2\pi/\lambda$ is the wavenumber and λ is the wavelength, σ is the rms surface height, θ is the incidence angle in radians and Γ_0 is the Fresnel reflectivity of the surface at nadir and given by:

$$\Gamma_0 = \left|\frac{1 - \sqrt{\epsilon_r}}{1 + \sqrt{\epsilon_r}}\right|^2 \quad (3)$$

The model is valid in a large range of $k\sigma$, kl, m_v and θ [1]. We checked the measured data of Flevopolder and found that $k\sigma$ and θ are within the validity range, m_v is a little large and kl at L-band is small [7]. But the model is still applicable according to the authors [1].

Equation (3) tells us that Γ_0 is only a function of soil dielectric constant ϵ_r. As ϵ_r is unknown, we do not know the value of Γ_0 for each field. But we know that ϵ_r is basically determined by soil moisture content. From the soil moisture data, it is learned that the variations of soil moisture between fields are small. The maximum difference is about 8% [7]. It may therefore be expected that the variation of Γ_0 between the fields is small. Thus, we assume that Γ_0 is constant as an approximation when we check the model with the roughness data $k\sigma$. With this assumption, the measured and computed p and q (eqs. (1) and (2)) are compared. The results illustrate that the L-band cross-polarized ratio q agrees well with the model, which denotes that q strongly depends upon soil roughness and soil moisture, and therefore can be applied to estimate $k\sigma$ and m_v. The L-band co-polarized ratio p is consistent with the model to some extent but does not match well, C-band measured p is always larger than 0 dB, which is contrary to the model. Because of the poor quality of the C-band hv-polarized image [7], C-band q data are not used.

3 The Inversion of Soil Moisture for Bare Soil

3.1 The Estimation of Soil Textural Components

As mentioned in previous sections, the backscattering coefficient is influenced by soil moisture m_v through the dielectric constant ϵ_r. Therefore, it is necessary to know the relations between ϵ_r and m_v for estimation of soil moisture from backscattering coefficient. It is known that soil moisture and soil texture are two major factors governing the dielectric constant of soil. Dobson *et al.* developed a model to calculate dielectric constant from soil moisture and the soil textural components [10, 11], which is

$$\begin{aligned}\epsilon_r = a_0 &+ (a_1 + b_1 S + c_1 C)m_v \\ &+ (a_2 + b_2 S + c_2 C)m_v^2 \\ &+ (a_3 + b_3 S + c_3 C)m_v^3 \end{aligned} \tag{4}$$

where S and C are, respectively, the weight percent of sand and clay component of a soil, volumetric soil moisture m_v is in $cm^3 \cdot cm^{-3}$, and the coefficients a_i, b_i and c_i $(i = 0, 1, 2, 3)$ are all constants at a given radar frequency. This model is applicable for both the real part ϵ_r' and the imaginary part ϵ_r'' of dielectric constant ϵ_r $(\epsilon_r = \epsilon_r' - j\epsilon_r'')$.

Unfortunately, we have not access to detailed data about the soil textural composition of the Flevopolder Agricultural Test Site. The only information available is that the soil is classified as a fine textured Calcaric Fluvisol and the soil type is Marine Clay, and that it is very uniform with the clay content varying between 30% and 35%. For estimation of soil moisture from the dielectric constant with backscattering models, it is necessary to know soil sand content but no sand content data are reported [9]. Fortunately, we have the clay content data. With this data, we may retrieve the sand content from equations (1), (3) and (4) by use of the cross-polarized ratio and volumetric soil moisture content. The idea is that we use the cross-polarized ratio model to inverse the Fresnel reflection Γ_0 thus the dielectric constant is obtained, then substituting the dielectric constant, volumetric soil moisture and clay content into ϵ_r model (4), with the sand content being the only unknown parameter in the model and therefore retrievable.

As the empirical model q gives a considerable agreement with L-band data, it is applied to estimate the Fresnel Reflectivity Γ_0 and then the complex dielectric constant ϵ_r. Since the imaginary part is much smaller than the real part, we may as an approximation assume the imaginary part of the dielectric constant to be negligible, which is a reasonable approximation for soil material [8, 1, 12]. Thus, only the real part is accounted for in the inversion of dielectric constant. In which case the empirical model for ϵ_r' at L-band is simplified as [10, 11]

$$\begin{aligned}\epsilon_r' = 2.37 &+ (-5.24 + 0.55S + 0.15C)m_v \\ &+ (146.04 - 0.74S - 0.85C)m_v^2 \end{aligned} \tag{5}$$

In this way, soil sand content is retrieved from equation (5). The values of S varies within 12.2% - 24.2% with a mean value of 16.6%. Apparently, large variation of S is obtained. By changing the value of S between 12.2% and 24.2% we find that 10% variation of S yields about 0.9% variation of m_v. Thus, it is acceptable to take the mean value of S as the estimated sand content for all the bare fields with no significant influence on the estimation of m_v with the ϵ_r' model. In the calculation, the clay content of soil is taken as $C = 32.5\%$ which is the mean value of the data. Because the measured q for Field 114 does not fit equation (1) well, the data from this field are not included.

3.2 The Backscattering Model for Soil Roughness

From the measured soil moisture data it is noted that m_v for all the bare fields is greater than 30%, which indicates that the soils are very moist. The variation of m_v between the bare fields is small. As there is no relevant dry fields to compare with it is difficult to examine the validity of inversion model. On the other hand, we may ignore the variation of soil moisture between the fields to examine the variation of backscattering coefficient σ° with soil surface roughness.

The model developed by Ulaby *et al.* is employed to describe bare soil backscattering [13]. It is usually expressed as a linear function between the backscattering coefficient $\sigma^\circ_{soil}(dB)$ and soil moisture m_v:

$$\sigma^\circ_{soil}(dB) = A + Bm_v \tag{6}$$

where B is the moisture sensitivity and A is the dry soil contribution. They are all constants for a given radar configuration and given surface roughness.

As moisture sensitivity B is weakly dependent upon surface roughness, it could be assumed as a first-order approximation to be independent of surface roughness [13, 14]. Thus the roughness influences only parameter A in (6). For evaluating the roughness effect of the simple model, m_v could be approximately taken as a constant for all the bare fields in our case, therefore we could further assume that Bm_v in (6) is a constant for a given radar configuration.

Based on the approximations above we can evaluate the variation of A with surface rms height σ in a certain radar configuration. The variations of backscattering coefficient $\sigma^\circ_{soil}(dB)$ versus surface rms height σ

at L- and C-band with hh, vv and hv polarizations are regressed except for C-band σ°_{hv} because of the poor data quality. The results clearly indicate that there is a good linear relationship between σ° and σ. With the increase of surface rms height, backscattering coefficient of soil increases linearly. Thus, a simple model to connect backscattering coefficient of bare soil with surface rms height can be found from (6)

$$\sigma^\circ_{soil}(dB) = A'\sigma + B' \tag{7}$$

where $B' = Bm_v$, A' is a constant only related to the radar configuration.

The parameters A' and B' may be easily computed from the measured σ°_{soil} and σ. Table 1 gives the values of A' and B' regressed with the measured data for bare fields. High correlation coefficients are obtained for all the available data.

Table 1: The parameters A' and B' in the model $\sigma^\circ_{soil}(dB) = A'\sigma + B'$. B' and A' are the intercept and the slope of the regression equations respectively. r is the correlation coefficient.

	$\sigma^\circ_{soil}(dB) = A'\sigma + B'$					
	L-band			C-band		
	B'	A'	r	B'	A'	r
hh	-21.46	0.528	0.88	-13.89	0.384	0.78
vv	-16.67	0.346	0.73	-13.99	0.368	0.75
hv	-33.22	0.656	0.94	-	-	-

3.3 Retrieval of rms Height from σ°_{soil}

The good linear relations between surface rms height and backscattering coefficient reveals that the rms height of bare soil could be approximately retrieved from backscattering coefficient by ignoring the variations of soil moisture between fields. It is available at least when the soils are very wet and the moisture differences are insignificant. Using the expression (7) and the regressed parameters A' and B' in Table 1, retrieval of surface rms height σ_{est} from the measured σ°_{soil} at all available radar configuration combinations has been made. The results of the estimated surface rms height σ_{est} compared with the measured surface rms height σ_{mea} are illustrated in Figure 2. σ_{mea} is taken as the averages of the measured surface rms heights at two perpendicular directions. Each σ_{mea} corresponds to five σ_{est} incorporated with five estimations at L- and C-band with hh, vv and hv polarizations. We take the mean value $\overline{\sigma}_{est}$ which is the average of the five σ_{est} as the estimated mean rms height. The standard deviation for $\overline{\sigma}_{est}$ is $S_\sigma = 3.0\ mm$. This is the uncertainty in the estimations of surface rms height from the measured backscattering coefficients. The uncertainty comes from the data with large rms height and might be generated from the averages of rms heights measured at two perpendicular directions. Rough fields have significantly different roughness at the two perpendicular directions and a simple average might not be able to represent the surface roughness appropriately.

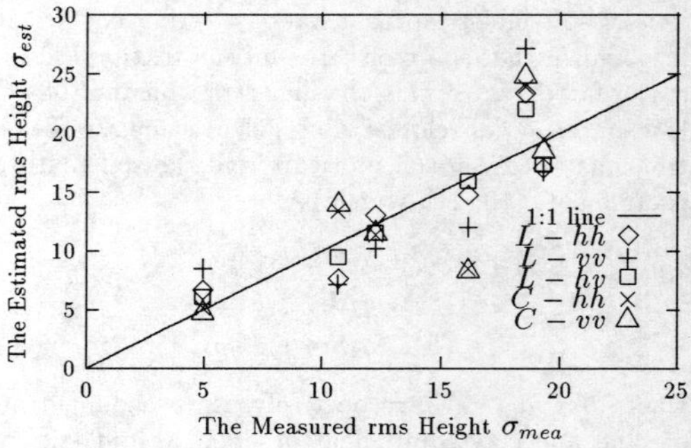

Figure 2: Comparison between the estimated rms height σ_{est} and the measured one σ_{mea}. Each σ_{mea} corresponds to five σ_{est} retrieved from L- and C-band multi-polarized backscattering coefficients σ°.

3.4 Retrieval of Soil Moisture from Cross-polarized Ratio

With the surface roughness being estimated, it is possible to inverse soil moisture content from the cross-polarized model and the dielectric constant model given by equation (1), (3) and (5).

Firstly, Fresnel reflectivity Γ_0 is retrieved by substituting the estimated rms height $\overline{\sigma}_{est}$ and cross-polarized ratio q of L-band into (1). Secondly it is assumed that ϵ_r could be approximated only by its real part ϵ'_r with the imaginary part, ϵ''_r, been small compared with ϵ'_r and therefore omitted. Then ϵ'_r is computed from (3). Finally soil moisture content is retrieved from (5) by substituting the estimated soil textural contents and ϵ'_r into the formula. Figure 3 presents the estimated soil moisture content versus the measured one. The rms uncertainty of the estimated m_v is about 10% for the bare fields.

The estimation uncertainty comes from a few sources. The main uncertainty stems from the estimated rms height $\overline{\sigma}_{est}$. The approximations applied to the linear equation (7) might introduce an uncertainty when σ is estimated from σ°_{soil}. If the soil moisture m_v could be taken into account in the determination of the constant coefficient $B' = Bm_v$, it might reduce the estimation uncertainty of the rms height.

Another source is the elimination of the imaginary part of the dielectric constant ϵ_r. The imaginary part do contribute to ϵ_r even though the contribution is small compared with that of the real part. Actually, $|\epsilon_r| = \sqrt{|\epsilon_r'|^2 + |\epsilon_r''|^2}$, the elimination of ϵ_r'' makes the estimated ϵ_r' larger than it should be, which results in a overestimation of soil moisture. The calculation from model (4) illustrated that $|\epsilon_r''/\epsilon_r'| \approx 19\%$. The third factor influencing the estimation accuracy is the sand content of soil. The estimated value is the average of five bare fields, which may differ from the real one.

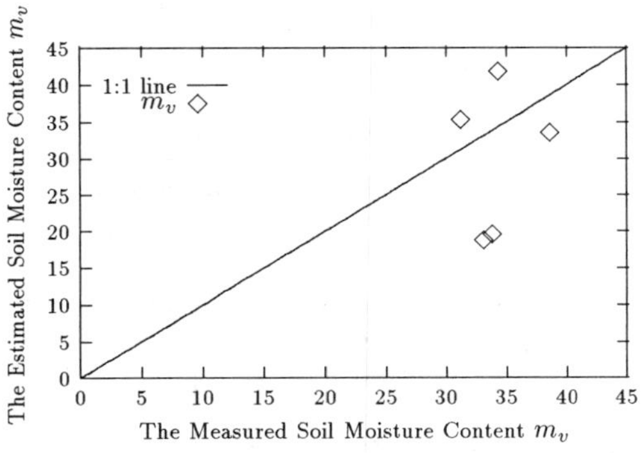

Figure 3: Comparison between the estimated soil moisture and the measured one. No data can be retrieved for Bare Field 114.

4 Backscattering for Crop Fields

Within the 18 test fields at Flevopolder, a total of 12 are crop-covered fields. Four with wheat, three with sugar beets, three with potatoes and two with corn. Unfortunately, only a wheat field, a sugar beet field and two potato fields have the necessary radar and ground data for soil moisture analysis [7].

Since the crops can be treated as a vegetation canopy covering a soil surface, the scattering and attenuation of the crops should be taken into account in the soil scattering study. The basis for radar sensing of soil moisture in this case is that the backscattering signal could carry the information of soil scattering. In other words, the radar wave should be able to penetrate the crop layer and to give an adequate backscatter from the soil surface. In the following analysis, we always assume these conditions are satisfied for all crops at L- and C-band.

4.1 Vegetation Scattering Models

For estimation of soil moisture, we apply the "Water Cloud Model" developed by Attema *et al.* [2] initially and modified by others [15]. It is a semi-empirical model and appears to be suitable for inversion purposes.

The water cloud model represents the power backscattered by the whole canopy σ° as the incoherent sum of the contribution of the vegetation σ°_{veg} and of the underlying soil σ°_{soil}. The latter is attenuated by the vegetation layer.

$$\sigma^\circ_{can} = \sigma^\circ_{veg} + t^2\sigma^\circ_{soil} \tag{8}$$

$$\sigma^\circ_{veg} = E\cos\theta(1 - t^2) \tag{9}$$

$$t^2 = exp(-2DM_v/\cos\theta) \tag{10}$$

or

$$t^2 = exp(-2DL/\cos\theta) \tag{11}$$

where θ is the incidence angle, t^2 is the two-way attenuation through the canopy layer, M_v and L are areal canopy water content (kg/m^2) and leaf area index LAI (m^2/m^2) respectively, D and E are parameters depending on the canopy type. This model corresponds to the first-order solution of the radiative transfer equation through a weak medium, where multiple scattering effects may be neglected.

4.2 The Bare Soil Scattering Model

In Section 3.2, the bare soil scattering model developed by Ulaby *et al.* [13] is modified and a linear correspondence between backscattering coefficient $\sigma^\circ_{soil}(dB)$ and rms height σ is assumed. Accounting for both soil moisture m_v and surface rms height σ, (6) and (7) could be combined and a two-variable linear model for bare soil scattering is derived as:

$$\sigma^\circ_{soil}(dB) = C + A'\sigma + Bm_v \tag{12}$$

where A', B and C are constant coefficients for a given radar configuration.

4.3 The Approximations

The parameters of water cloud model have been fitted to experimental observations by many authors [2, 13, 15, 16]. Based on the results of Prevot *et al.* [15], the fitted parameter E in (9) is zero at C-band, which illustrates that the effect of crops is mainly a two-way attenuation of the backscattering from the underlying soil. Therefore at L- and C-band it is an acceptable approximation to neglect the scattering contribution of the vegetation and only include the attenuation. The scattering model for a vegetation canopy is simplified as follows:

$$\sigma^\circ_{can} = t^2 \sigma^\circ_{soil} \tag{13}$$

Substituting equations (10), (11) and (12) into (13) and expressing σ°_{can} in dB, we get a linear function between σ°_{can} and the "target parameters" $M_v(L)$, σ and m_v:

$$\sigma^\circ_{can}(dB) = C + A'\sigma + Bm_v + D'M_v/\cos\theta \tag{14}$$

or

$$\sigma^\circ_{can}(dB) = C + A'\sigma + Bm_v + D'L/\cos\theta \tag{15}$$

where $D' = -20D/ln10$, and the constants A', B, C and D' depend on crop type and radar configurations and should be experimentally determined.

4.4 The Determination of the Coefficients and Model Results

Models (14) and (15) are applied to the Maestro-1 SAR data. The constant coefficients A', B and C with respect to σ°_{soil} are determined using the bare field data. D' is computed using the crop fields and is fitted with both canopy descriptors L and M_v. Multi-variable least-square linear regression is applied for determination of the parameters. The values of the coefficients are listed in Table 2.

Table 2: The values of the constant coefficients A', B, C and D' in the canopy scattering model, $\sigma^\circ_{can}(dB) = C + A'\sigma + Bm_v + D'M_v(L)/\cos\theta$, where $m_v(\%)$ is the volumetric soil moisture content in the top $0 - 5cm$ layer, σ is soil surface rms height in mm, $M_v(\%)$ is the areal canopy water content in kg/m^2, and L is the Leaf Area Index in m^2/m^2, and r is the correlation coefficient.

Band-Polar.	C	A'	B	r
L-HH	-11.29	0.576	-0.310	0.907
L-VV	-12.51	0.366	-0.126	0.742
C-HH	-17.76	0.364	0.125	0.786
C-VV	-15.43	0.361	0.049	0.756

$D'(L)/D'(M_v)$	Wheat	Sugar Beets	Potatoes
L-HH	-/28.55	0.97/4.56	1.58/6.09
L-VV	-/21.73	0.43/2.03	0.99/3.74
C-HH	-/71.50	0.98/4.61	0.83/3.17
C-VV	-/43.25	0.95/4.46	0.86/3.25

After the coefficients are determined, the model is applied to estimate the backscattering coefficient σ°_{can}. The estimated σ°_{can} is compared with the measured one. Figure 4 illustrates the results for the two potato fields. It is shown that both LAI model and M_v model give good estimations of backscattering coefficients. The rms error of the estimation for the LAI model is $1.49dB$, and for the M_v model is $0.50dB$ respectively. If the rms error of estimation could be taken as the criteria for judging the fitness of the models, M_v model is better than LAI model as may be seen from the figure.

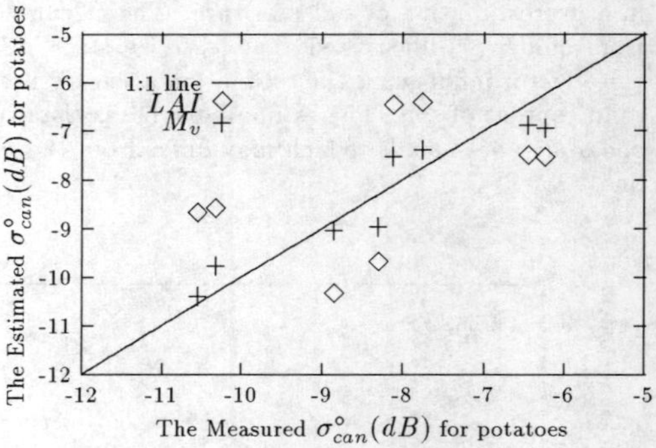

Figure 4: Comparison of estimated $\sigma^\circ_{can}(dB)$ with the measured ones for the two potato fields. L- and C-band, HH and VV polarized σ°_{can} are all combined.

4.5 Some Problems of σ°_{can}

Under certain approximations to the water cloud model, a linear relation is found between $\sigma^\circ_{can}(dB)$ and "target" parameters. The coefficients in the model are determined with Maestro-1 experimental data. In principle, the above approach could be applied to retrieve soil moisture, surface roughness and leaf area index from L- and C-band dual-polarized backscattering coefficients for sparse crops where scattering contribution of vegetation could be neglected. But in fact, there still exist some problems.

In the regression of the model coefficients, only 6 bare fields and 4 vegetated fields (for LAI, only 3 vegetated fields) are used. The reliability of the parameter values of the model is questionable. Further validation of the approach with more experimental data is needed.

It is noted that the soil moisture sensitivity B in Table 2 is negative at L-band, which indicates that the increase of moisture content yields a decrease of backscattering coefficient. This is difficult to explain theoretically, although the consistency with the measured data. An explanation may be that the data set used to estimate B is too small. Totally there are 6 bare fields involved in the calculation and the soil moisture content distributed in a narrow range of $31.2\% - 38.6\%$. Furthermore, only field No.117 has $m_v = 38.6\%$ and all the others are within $31.2\% - 34.3\%$. It is obvious that the soil moisture variation is too small to make the regression a valid one resulting in a large uncertainty in the determination of B.

Table 2 also shows the values of D' that describes the attenuation of vegetation. Large variations of D' are yielded for different types of crops. This result is observed by various authors [15]. The canopy structure has a significant influence on the parameters in the water cloud models. The values of D' for wheat, sugar beets and potatoes are given separately in the table. It is impossible to define a universal algorithm for all types of crops and at all growth stages, but one can expect to define one for each class of vegetation, defined by their geometrical structure [15].

5 Conclusion and Discussion

The paper presents the preliminary results on the estimation of soil moisture of bare fields from Maestro-1 SAR data of the Flevoland Agricultural Test Site. By ignoring soil moisture variations, a simple parameterization of surface rms height is first introduced to extract surface roughness from backscattering coefficient. Applying the empirical cross-polarized ratio model and the dielectric constant model, we present an algorithm to estimate soil moisture from polarimetric SAR data. An accuracy of 0.1 cm^3cm^{-3} is achieved. For crop fields, the "water cloud model" is applied and a linear model between m_v, σ, M_v and $\sigma^o_{can}(dB)$ is proposed.

Parameterization of surface rms height and an algorithm for soil moisture estimation are developed using the limited data available. To improve the algorithm, emphasis should be placed on the following points:

1.) Take into account soil moisture and incidence angle effects in the parameterization of surface roughness.

2.) Take into account the differences in surface rms heights between parallel and transverse directions and thereby the observation geometry relative to the surface structure.

3.) The Maestro-1 data available for soil moisture estimation are limited. It is expected the approaches applied in the paper could be validated with other experimental data.

Experiments carried out by the Danish Centre for Remote Sensing at Electromagnetics Institute using the EMISAR will provide an excellent opportunity to continue the work.

6 Acknowledgment

The work was funded by the Danish Space Board. The authors wish to express their thanks to colleagues at EMI for all the helps at image processing, data calibration and calculation.

References

[1] Yisok Oh, et al., 1992. An Empirical Model and An Inversion Technique for Radar Scattering from Bare Soil Surfaces, *IEEE Trans. Geoscience and Remote Sensing,* Vol. GE-30, No.2, pp.370-381.

[2] Attema E.P.W., et al., 1978. Vegetation Modeled as A Water Cloud, *Radio Science*, Vol.13, No.2, pp. 357-364.

[3] Churchill P.N., et al., 1992. The *MAESTRO*-1 European Airborne Polarimetric Synthetic Aperture Radar Campaign, *The proceedings of the final workshop on MAESTRO-1/AGRISCATT*, pp. 1-11.

[4] Van Leeuwen H.J.C., 1992. Model-based Derivation of Vegetation And Soil Parameters in Multi-frequency Radar Data Sets, *The Proceedings of IGARSS'92,* Vol.1, pp. 54-56.

[5] Jiankang Ji, H. Skriver & P. Gudmandsen, 1994. Estimation of Bare Soil Moisture from MAESTRO-1 SAR Data of Flevoland, *Internal Report, Electromagnetics Institute, Technical University of Denmark.*

[6] Kerekes J.P., 1992. Measurement of Soil Moisture with Synthetic Aperture Radar, *The Proceedings of IGARSS'92,* Vol. , pp. 1751-1753.

[7] Jiankang Ji, H. Skriver & P. Gudmandsen, 1993. SAR Data and Ground Data at Flevoland Agricultural Test Site for the MAESTRO-1 Campaign, *Internal Report, Electromagnetics Institute, Technical University of Denmark.*

[8] Fung A.K. et al., 1992. Backscattering from a Rough Dielectric Surface *IEEE Trans. Geoscience and Remote Sensing,* Vol. GE-30, No.2, pp.356-369.

[9] Churchill P.N. 1990. DATA REPORT of Maestro-1, *Report Number IRSA/MWT/1.90 Version 2.0.* CEC/JRC and ESA/ESTEC.

[10] Dobson M.C., et al., 1985. Microwave Dielectric Behavior of Wet Soil. Part II: Dielectric Mixing Models, *IEEE Trans. Geoscience and Remote Sensing,* Vol. GE-23, No.1, pp. 35-46.

[11] Dobson M.C., et al., 1984. A Reexamination of Soil Textural Effects on Microwave Emission and Backscattering, *IEEE Trans. Geoscience and Re-*

mote Sensing, Vol. GE-22, No.6, pp. 530-536.

[12] Hallikainen M.T., et al., 1985. Microwave Dielectric Behavior of Wet Soil. Part I: Empirical Models and Experimental Observations, *IEEE Trans. Geoscience and Remote Sensing,* Vol. GE-23, No.1, pp. 25-34.

[13] Ulaby F. T., et al., 1982. Effects of Vegetation Cover on The Radar Sensitivity to Soil Moisture, *IEEE Trans. (GE),* Vol. GE-20, No. 4, pp.476-481.

[14] Ulaby F.T., et al., 1978. Microwave Backscatter Dependence on Surface Roughness, Soil Moisture and Soil Texture: Part 1. Bare Soil, *IEEE Trans. (GE),* Vol. GE-16, No.4, pp.286-295.

[15] Prevot L., et al., 1993. Estimating Surface Soil Moisture and Leaf Area Index of a Wheat Canopy Using a Dual-Frequency (C and X Bands) Scatterometer, *Remote Sensing of Environment* Vol.46, pp.331-339.

[16] Prevot L., et al., 1993. Estimating the Characteristics of Vegetation Canopies with Airborne Radar Measurements, *International Journal of Remote Sensing,* Vol.14, No.15, pp.2803-2818.

Sensors and Environmental Applications of Remote Sensing, Askne (ed.) © 1995 Balkema, Rotterdam, ISBN 90 5410 524 0

Use of ERS-1 SAR data for soil moisture assessment

K. Dabrowska-Zielinska, M. Gruszczynska, M. Janowska & K. Stankiewicz
Remote Sensing and Spatial Information Centre, Warsaw, Poland

ABSTRACT: This study has been carried out for grassland and surrounding agriculture in the western part of Poland. Ers-1 SAR data obtained from ESA for Pilot Project PL-4 for the 1992 and 1993 growing seasons have been analyzed and compared to ground measurements collected over the test site. The impact of various ground truth variables on backscatter signatures was examined to develop the method of soil moisture assessment over agricultural crops and grassland. The effectiveness of deriving a backscattering coefficient averaged over different sized blocks of ERS-1 SAR image pixels and subsequent comparison to measured soil moisture was examined. Two-way character of backscatter-soil moisture relationship has been noticed due to succeeding development stages of crops. Good negative correlation between backscatter and soil moisture occurred for stubble and for short grass (LAI from 0.5 up to 1.5) and for early and final stages of root crops and maize. Good positive correlation between backscatter and soil moisture occurred for different crops as: tall grass (LAI above 3) spring wheat, rape and root crops in their full green mass. It was considered that the relationship between backscattering coefficient and soil moisture is strongly dependent on surface roughness which was characterized by leaf area index (LAI) or height of the vegetation values.

1. INTRODUCTION

The investigation of the application of satellite remote sensing data to assess and monitor soil and vegetation variables over large agricultural areas have been undertaken by the Remote Sensing and Spatial Information Centre (OPOLiS) of the Institute of Geodesy and Cartography (IGiK) since 1987. The work has resulted in the establishment of several relationships between information derived from satellite images and the state of vegetation development. In order to accurately derive and predict information about the yield of grassland or other crops information about soil moisture is essential. The method developed by OPOLiS is based on the application of NOAA/AVHRR data. However, the operational use of this method is hampered by the presence of frequent cloud cover which makes it difficult to acquire routine AVHRR observations. The instruments on board the first European Remote Sensing satellite ERS-1 give the opportunity to make operational observations of the agricultural surfaces independently of weather conditions. Of particular interest to the required investigations of soil moisture is the AMI synthetic aperture radar (SAR) instrument.

In order to estimate the utility of the ERS-1 SAR data for soil moisture assessment the analysis of the crop and grass backscatter profiles simultaneously with the profiles of various ground truth variables throughout a growing season was examined to determine their impact on backscatter signatures. The backscatter - soil moisture relationships have been undertaken according to various vegetation surface roughness conditions connected closely to the vegetation development stage. The results and discussion are presented in this paper.

2. TEST SITE DESCRIPTION

The selected test site is located in Western Poland, in the Obra Valley approximately 80 km south-west of Poznan. The area is covered by large areas of grassland which are flooded during the early part of the growing season and by agricultural fields with the following dominant crops: wheat, rye, barley, oat, rape, maize, sugar beet and potatoes. The crop types varied from 1992 to 1993 with the exception of two sites where winter rye and spring wheat were planted in both years. The first ERS-1 SAR images for each year were acquired (towards the end of May) when winter crops were in their heading or flowering stage, spring crops were in their tillering or jointing stage, root crops and maize were in their emergence or germination stage and grass in the flowering stage before harvest. The predominate soil types for crops are sandy and sandy loam and for grassland peat and muck. The growing season of 1992 was affected by drought. The area of the test site is flat with small parts of the upland areas covered by coniferous or mixed forest.

3. GROUND TRUTH DATA

Throughout the growing seasons of 1992 and 1993 and simultaneously to ERS-1 descending overpasses the measurements of soil and vegetation parameters were carried out at 30 points, 15 grassland and 15 crop covered points. The parameters measured were:

- volumetric soil moisture (0-10,10-20.20-30) [%]
- soil water potential (pF) (0-10,10-20,20-30 cm)
- wet and dry biomass [g/m²]
- gravimetric vegetation moisture [%]
- leaf area index (LAI)
- height of the vegetation [m].

These measurements were based on samples taken at the sites which represent the whole field. Also, surveys were undertaken to record crop type with their actual development stage and grass growing condition.

4. ERS-1 SAR DATA

ERS-1 SAR.PRI data acquired during the 35-day repeat orbit and obtained from ESA for Pilot Project PL-4 were available for the following dates with Julian days in brackets:

23.05.1992 (144)
11.06.1992 (163)
16.07.1992 (198)
01.08.1992 (214)
05.09.1992 (249)
27.05.1993 (147)
12.06.1993 (163)
01.07.1993 (182)
17.07.1993 (198)
05.08.1993 (217)
21.08.1993 (233)
09.09.1993 (252)
25.09.1993 (268)

These were 16 bit 3 look images with a pixel size of 12.5 x 12.5 m covering areas of approximately 100 x 100 km (8000 pixels x 8000 lines). From each of the ERS-1 SAR images a sub-scenes corresponding to the area of the test site were geometrically transformed to the topographic map in the scale of 1:50 000. Ground measurement points were located on multitemporal composites using an ERDAS image processing system. Backscatter coefficients for each ground truth points were calculated and averaged for blocks of pixels ranging from 3 x 3 up to 11 x 11 pixels. The ground sample point was placed in the middle of each block. Extracted data were used to investigate the effect of pixel sample size on measurements accuracy of backscatter signature over ground measurement points. This preliminary work indicated that for the crop covered points the block of 6 x 6 pixels and for the grassland the block of 8 x 8 pixels are required to give the minimum variability. Such backscatter signature values have been included in the subsequent data analysis.

5. RESULTS AND DISCUSSION

With data from two consecutive years 1992 and 1993 it was possible to examine the backscatter differencies for the same sites and crop types due to soil moisture differences. For the growing season of 1992 which was affected by drought the backscatter values for crops and grassland were significantly lower than in 1993. For the grassland the backscatter differences varied from 1 to 5 dB between these two years are presented on Figure 1.

Figures 2,3,4 and 5 show the temporal profiles of backscatter and volumetric soil moisture for the both growing seasons for winter rye and spring wheat. Because the crop type, soil type and site is the same, the lower values of backscatter signatures in 1992 were caused by lower values of soil moisture. Hot dry weather during the growing season of 1992 speeded the crop development stages and maturity of about two weeks which was confirmed by the displacements of the temporal profiles of backscatter for both years.(Fig.2 and Fig.4).

During early stage of growing season (up to 163 Julian Day) root crops and maize showed a

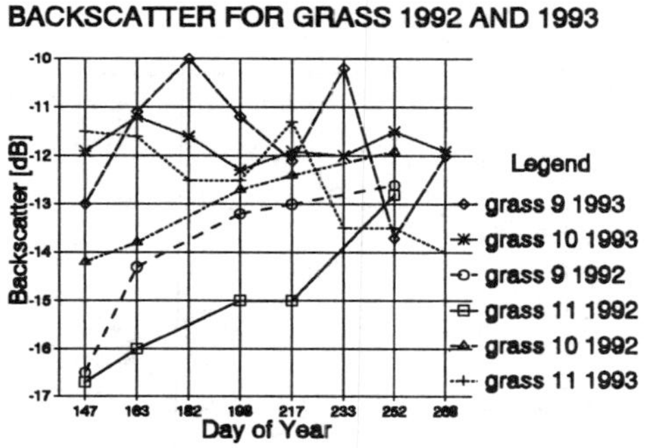

Fig.1. Differences in backscatter for grass in the growing seasons of 1992 and 1993.

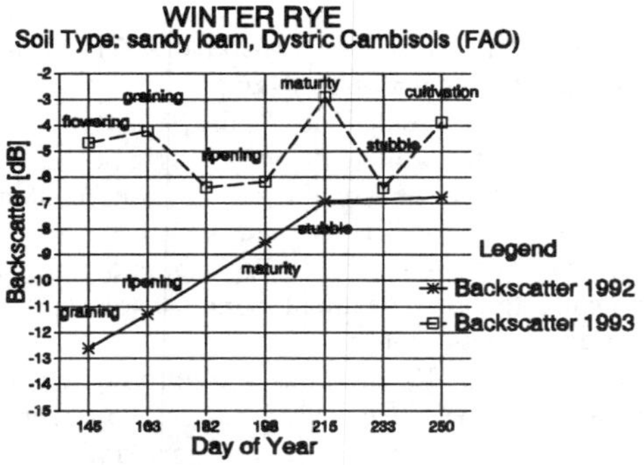

Fig.2. Backscatter profiles for winter rye for 1992 and 1993.

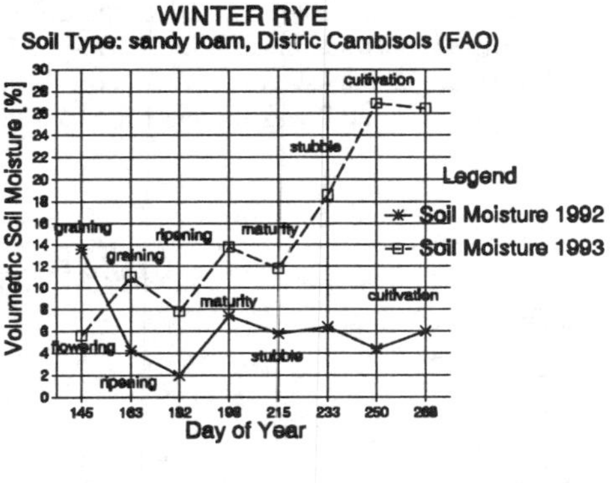

Fig.3. Soil moisture profiles for winter rye for 1992 and 1993

similar profiles of backscatter signatures mainly due to the influence of partially covered soil surface. During these early development stages backscatter

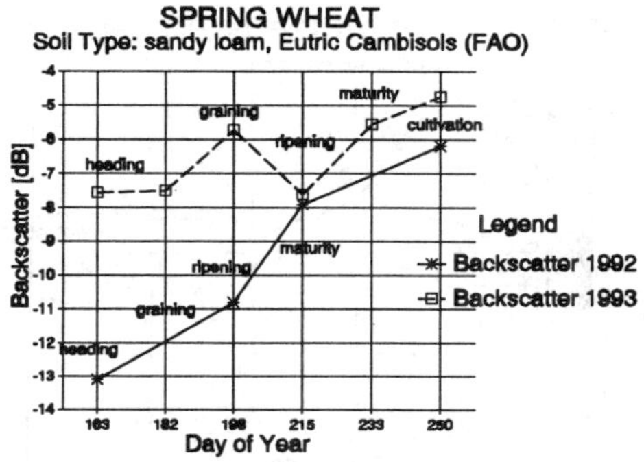

Fig.4. Backscatter profiles for spring wheat for 1992 and 1993.

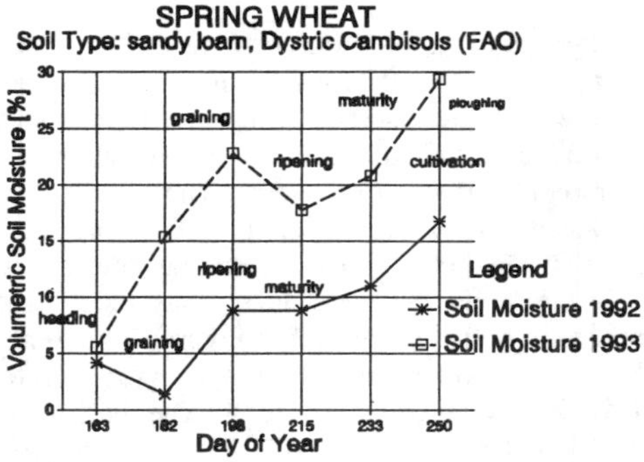

Fig.5. Soil moisture profiles for spring wheat for 1992 and 1993.

decreased when soil moisture increased . For the winter crops and grassland the backscatter increased due to the increase of soil moisture (Fig.6). Two-way character of linear model of backscatter-soil moisture relationship was related to the differences in surface roughness.

The grassland area was well distinquished on ERS-1 SAR image. The smooth surface of grass gave nearly the specular reflection and appeared in dark grey tone. The averaged backscattering coefficients values were from -11 up to -13 dB and differed from the other crops by 5-8 dB. The smallest differences in backscattering coefficient values appeared with spring wheat when the differences were c.a. 2-4 dB.

It was noticed that the relationship between backscattering coefficient and soil moisture is strongly dependent on surface roughness. Even for the same type of vegetation the surface roughness was different. For the considered grassland the height

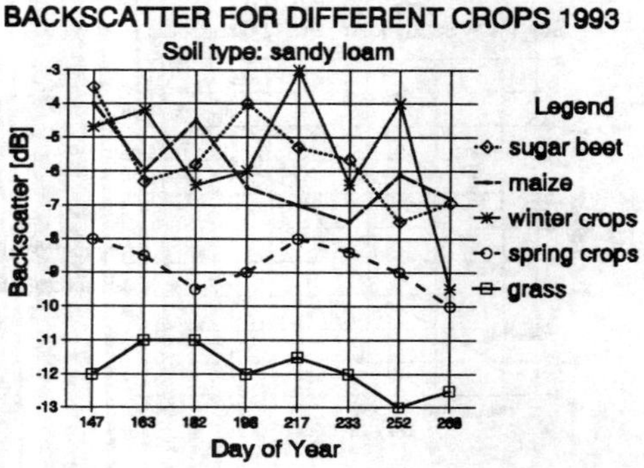

Fig.6. Temporal backscatter profiles for different crops in 1993.

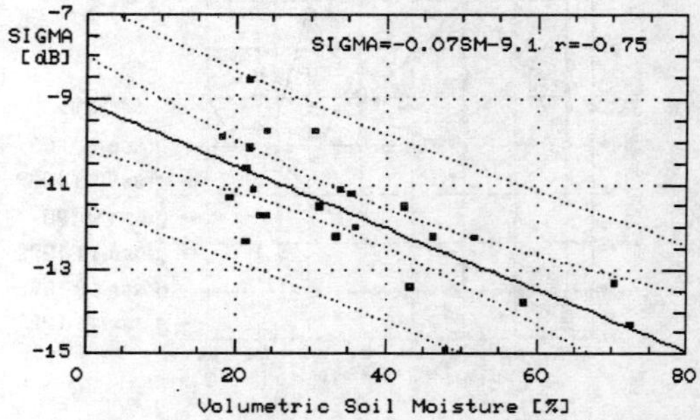

Fig.8. Backscatter-soil moisture (0-10 cm) relationship for grass (LAI 0.5-1.5).

of the grass varied from 0.10 to 0.80 m and LAI from 0.5 to 6.0. Also for the other type of vegetation like wheat the backscattering coefficient strongly depended on the crop growth stage which influenced different surface roughness. The similar results obtained Le Toan et al (1993). Therefore the relationship between backscattering coefficient and soil moisture (0-10 cm level) has been establish within the values of LAI. Figure 7 presents the backscatter-soil moisture relationship for grass for the LAI values between 3 and 4. The backscattering coefficient ranged from -13 dB to -9.4 dB for soil moisture between 30 and 70%. For the increasing LAI values from 4.5 up to 6 higher values of soil moisture dominated. We found that the relationship for grassland was stronger for higher LAI values which corresponded to smoother surface. For the small values of LAI between 0.5 and 1.5 higher values of backscatter corresponded to smaller values of soil moisture (Fig.8). In this case soil moisture was very small for this grassland area. The dominant role played vegetation moisture.

It is presented on Figure 9 that the spring wheat with the same LAI values as grass ie. 2.3 to 3.5 could also be considered for the relationship between the backscatter and soil water potential together with the grass. We applied soil water potential to eliminate the influence of soil type which varied for both plots.

Figure 10 presents the information about backscatter, height of the grass, soil moisture and LAI for various points at the grassland. As it is shown the relationship between backscatter and soil moisture can be considered for the same roughness condition what is represented by different LAI and height values. For the point 5 and 6 when the height of the grass was the same i.e. 0.20 m the difference in soil moisture between these points corresponded to the difference in backscatter. While soil moisture values in points 2 and 7 was the same the backscatter

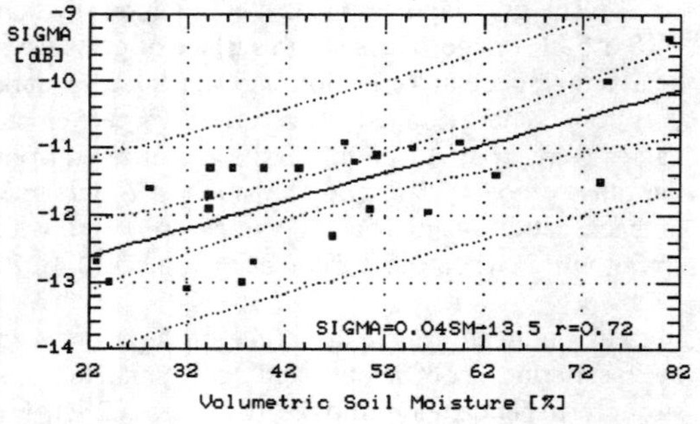

Fig.7. Backscatter-soil moisture (0-10 cm) relationship for grass (LAI 3-4).

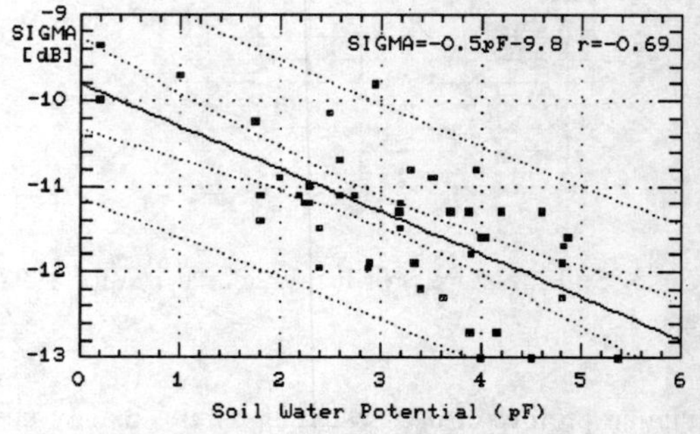

Fig.9. Backscatter-soil moisture (0-10 cm) relationship for grass and spring wheat (LAI 2.3-3.5)

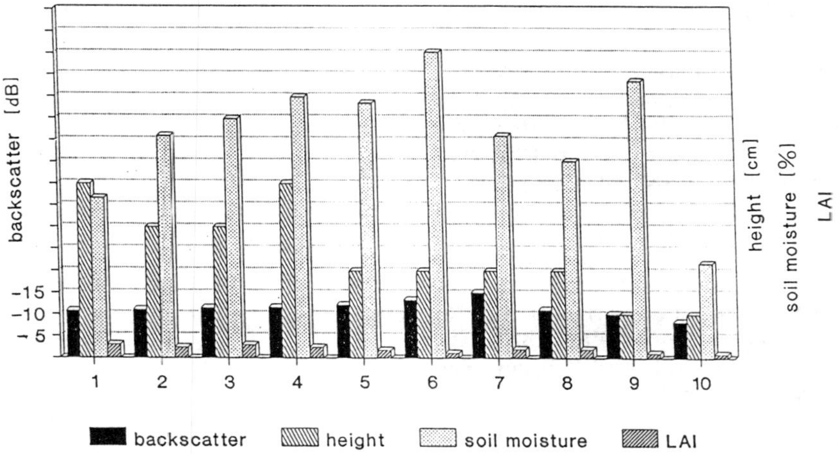

Fig. 10. Different grass conditions.

differed due to the difference in LAI and height of the grass.

Figure 11 represents the relationship between backscatter and soil moisture (0-10 cm) for spring wheat. During the time of ground measurements LAI values varied from 2.5 up to 3.5. With the increasing soil moisture values, the backscatter values increased which gave the good positive correlation.

Figure 12 presents the relationship between backscatter and soil moisture (0-10 cm) for the sugar beet. Backscatter values between -4 and -7 dB represented low values of soil moisture i.e. 12% while soil moisture between 18 and 23% has been represented by backscatter values from -14 to -16 dB. The measurements concerned the first and the last stages of sugar beet growth when the vegetation was sparse and the surface was rough and dry. Such condition gave good negative correlation.

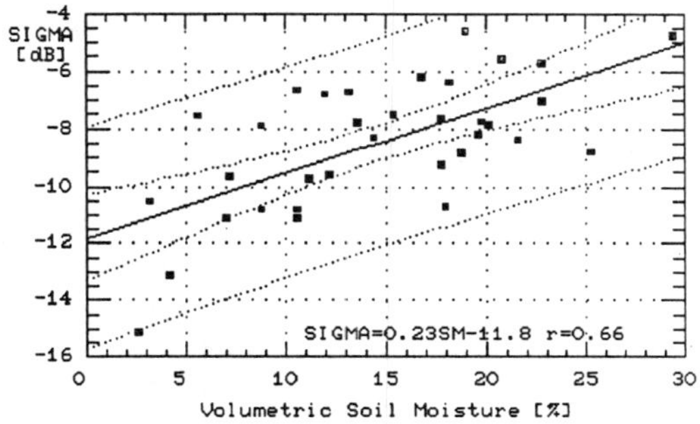

Fig.11. Backscatter-soil moisture relationship for spring wheat.

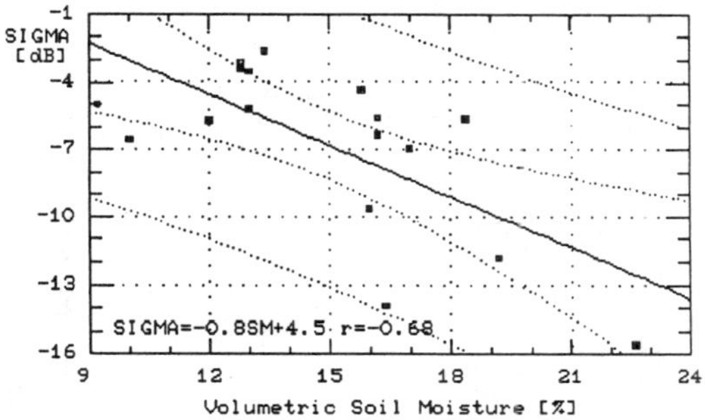

Fig.12. Backscatter-soil moisture relationship for sugar beet.

6.CONCLUSIONS

The surface roughness and soil moisture influence the ERS-1 SAR backscatter signatures. When the soil was covered by vegetation leaf area index (LAI) could represent the surface roughness. For different vegetation roughness during crop developing stages there was not straight relationship between backscatter and soil moisture. The assessment of soil moisture in the depth of 0-10 cm can be considered from backscattering coefficients derived from SAR images for the distinguished vegetation roughness classes.

ACKNOWLEDGEMENTS

The authors want to acknowledge prof. A. Ciolkosz for the scientific discussion, our collegues from the OPOLiS: St. Lewinski, Zb. Bochenek, I.Malek. Zb. Goliaszewski for the assistance and A. Kijowski who provided us with the soil parameters data.

REFERENCES

Le Toan. T., P.Smacchia, J.C. Souyris et al 1993. On the retrieval of soil moisture from ERS-1 SAR data, Proceedings of the Second ERS-1 Symposium, Hamburg, Germany:883-889.

Sensors and Environmental Applications of Remote Sensing, Askne (ed.) © 1995 Balkema, Rotterdam, ISBN 90 5410 524 0

The use of ERS-1 SAR imagery for groundwater assessment

D.Wladis
Geology Department, Chalmers University of Technology, Göteborg, Sweden

M.S.Rosenbaum
Geology Department, Imperial College of Science, Technology & Medicine, London, UK

ABSTRACT: Lineament mapping using remotely sensed data improves the prediction of bedrock groundwater occurrence. However, problems arise from the variable quality of the imagery and the diverse geological and hydrogeological situations affecting the area. Within a Pilot Project for the European Space Agency (ESA), the use of ERS-1 SAR PRI images for lineament interpretation has been evaluated. SAR imagery is particularly useful because of its sensitivity to surface roughness and the ability to obtain multi-temporal scenes of high quality. For the study area, which is situated in south-western Sweden, a comprehensive database has been compiled, enabling both the geology and the hydrogeology to be integrated with the imagery. This allows a more detailed assessment than could have done hitherto. The length of lineament per unit area and proximity analyses have been used as a comparative measure for the success of each remote sensing system.

1 INTRODUCTION

Lineament mapping using remotely sensed data has been used to improve the prediction of bedrock groundwater occurrence for some years (e.g. Greenbaum 1992). However, problems arise due to the variable quality of the imagery and the diverse geological and hydrogeological situations which prevail in the area.

Within a Pilot Project for the European Space Agency (ESA), the use of ERS-1 SAR PRI images for lineament interpretation has been evaluated in the Department of Geology at Chalmers University of Technology in Sweden. SAR imagery is particularly useful because of its sensitivity to surface roughness of the terrain and the ability to obtain multi-temporal scenes of high quality.

For the study area, which is situated in south-western Sweden (Figure 1), a comprehensive database has been compiled. This database enables both the geology and the hydrogeology to be integrated with the imagery, so allowing assessment in more detail than was possible hitherto.

2 LINEAMENT DEVELOPMENT

The crustal rocks of south-western Sweden consist predominantly of Precambrian high-grade metamorphic gneisses. These types of rock are very heterogeneous with respect to hydrogeological parameters. A parameter such as the hydraulic conductivity normally varies by several orders of magnitude within short distances (Gustafson & Krasny 1993). This abrupt variability can be explained by the fact that these rocks have a very small primary porosity and permeability and so groundwater flow mainly takes place along fractures and fissures rather than through the solid rock material. Such cracks often appear on remotely sensed imagery as lineaments, i.e. linear features which represent the intersection of the steeply dipping fracture where it intersects the (almost flat) ground surface. A lineament represents a zone of weakness characterised by extensively fractured rock (Clark 1985). This develops a secondary porosity and permeability, thus enhancing the potential groundwater yield of the aquifer.

Features yielding a tonal change on an image include elongated bogs, streams and geological dykes. In practice, these are often surface representations of an underlying lineament. The topography therefore also generally reflects the location of tectonic lineaments; this tends to be expressed as narrow depressions, from a few metres to a few tens of metres wide.

In humid maritime regions, such as Sweden, water supply has not until recently been regarded as being a potential hydrogeological problem. However, groundwater is now becoming an increasingly valuable resource as the population increases alongside the environmental constraints in an attemt to prevent contamination. Bedrock hydrogeological studies are also of significance for geotechnical and engineering tasks due largely to the control on pore water pressure (and thus effective stress) and on volumes of water flow.

3 LINEAMENT INTERPRETATION USING ERS-1 SAR PRI IMAGES

Lineament interpretation has been undertaken using three different ERS-1 SAR images (acquired 29 Nov 92, 7 Feb 93 and 14 Mar 93). The images were con-

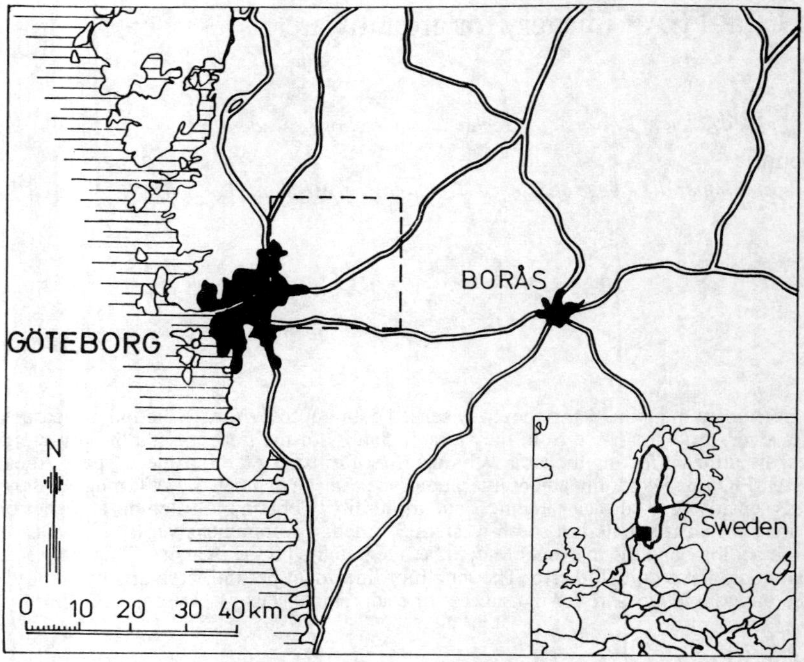

Figure 1. Location map for the study area, east of Göteborg.

trast streched and combined to produce a False Colour Composite (FCC) using the RGB colour space. The FCC supressed any SAR-specific noise (speckle) and also enhanced time-constant features such as the lineament and tectonic structure. Different adaptive filters were tested but did not further assist the structural interpretation. The FCC was considerably better at detecting lineament than use of just a single image.

The lineaments detected using the SAR imagery (Figure 2) have been compared with the published interpretation undertaken by the Swedish Geological Survey (SGU) based on an analysis of aerial photographs (SGU 1978). In order to analyse the spatial distribution of the mapped traces, rose diagrams were plotted at 5° intervals, both with respect to lineament length (Figure 3) and to lineament frequency, which gave a similar pattern.

The interpretation of lineament trace length using ERS-1 SAR shows a directional behaviour which is in very close agreement with that obtained from the aerial photographs, but the total length of lineaments traceable from SAR is approximately one third of that from the aerial photographs. This is probably due to the smaller resolution of the SAR, but structures lying parallel to the SAR illumination angle are also believed to have been supressed. There is also a tendency for lineaments interpreted from aerial photographs to have been subjectively interpolated, at least across lakes and urbanised areas, so yielding a longer total length than if interpreting the remotely sensed imagery.

4 COMPARISON WITH IMAGES PRODUCED BY OTHER SENSORS

Interpretations using images produced by other sensors have also been undertaken for the study area, including SPOT and a Digital Elevation Model (DEM). This comparative study between different sensors is reported in more detail in Wladis & Rosenbaum (1994). In general, it has been found that the interpretation obtained using the large-scale aerial photographs offers the highest level of lineament detection. However, the procedure is time-consuming and aerial photographs cannot provide a regional overview.

Optical sensors such as SPOT require that the sun elevation and sun azimuth be considered since these significantly influence the ability to detect lineaments, as do clouds and atmospheric water vapour. For the SPOT imagery used in this study, a large number of lineaments appear to have been obscured by such moderate atmospheric influences. The sun angle of 170° has tended to enhance those structures which trend east-west at the expense of those which trend north-south.

Lineament interpretation has also been attempted using a DEM. It is known that a surface expression often corresponds to the outcrop of a tectonic lineament, and so it is argued that there should also be some evidence for lineaments in a DEM provided that there is a sufficiently fine resolution. However, the interpolation routines normally used for the creation of a DEM tend to have a smoothing effect which has been found to obscure such tectonic structures.

Figure 2. Map of lineaments from the study area east of Göteborg detected from ERS-1 SAR.

5 CONSIDERING THE INFLUENCE OF LINEAMENTS ON HYDROGEOLOGICAL CHARACTERISTICS

Within the study area there is a large number of well records, primarily obtained from the SGU well archive (Wallroth & Rosenbaum 1994). These records include data such as borehole length, soil thickness, well yield and casing. Since transmissivity data is not recorded as such for most of these wells, the best way of characterising the well capacity is to compute the ratio Q/d, where Q is the yield of the well and d is its length. Q/d gives an conservative estimate for the specific capacity, proportional to the transmissivity. Such computed well capacity values have been used in proximity analyses reported elsewhere (Rosenbaum & Wladis 1994).

Lineaments interpreted using the aerial photographs when considered against well capacity show a positive correlation of spatial position with Q/d, as shown in Figure 4a. However, lineaments derived from the DEM when considered against well capacity show a marked drop in magnitude at short distances (Figure 4b). This is probably due to errors of precision inherent in locating the wells together with errors introduced by the digitising process. A further source of error could be the rather coarse resolution used by the DEM (50 m compared with between 5 and 30 m for the other imagery discussed here). Correlation of well capacity with lineaments analysed using the

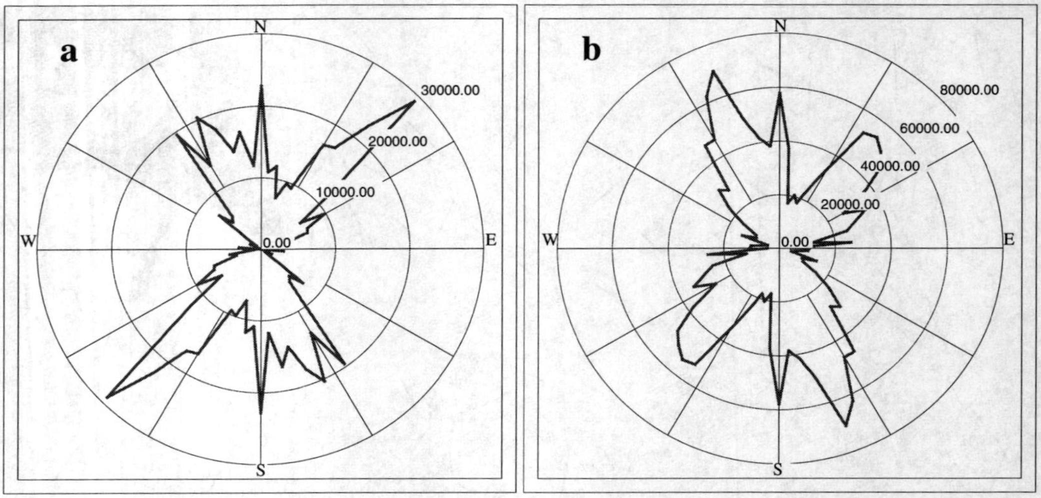

Figure 3. Rose diagrams showing orientations based on lineament lengths detected using remotely sensed imagery (Fig. 3a shows lineaments detected from ERS-1 SAR , Fig. 3b shows lineaments detected from aerial photographs).

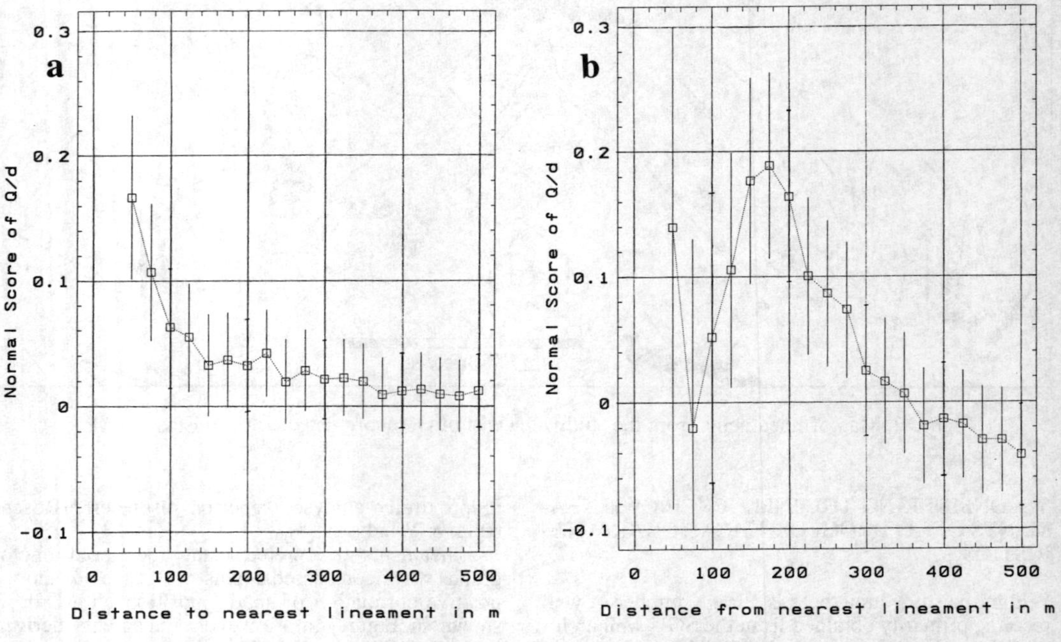

Figure 4. Proximity analysis of the well capacity (Q/d) and its associated standard error at the 95 % level of confidence (determined as the normal score transformation with an overall mean of zero) plotted against the distance to the nearest lineament detected (Fig. 4a shows lineament detected from aerial photograps, Fig. 4b shows lineaments detected from the NLS DEM).

SAR imagery showed a rather similar behaviour to that using the DEM, again because of the relatively low positional accuracy of the SAR PRI images.

6 CONCLUSIONS

Fractures are of considerable importance for locating groundwater in crystalline bedrock and can be used to detect locations likely to yield reasonable supplies of water.

Remotely sensed imagery assists the task of identifying fractures when these are expressed as lineaments, but problems remain regarding the unambigious interpretation of linear structures and locating positions on the ground with sufficient accuracy.

The length of lineament detected per unit area has been used as a comparative measure for the success of each remote sensing system. ERS-1 SAR is currently able to yield better results than can SPOT and DEM. Accuracy enhancement methods such as geocoding and GPS will soon improve the reliability of location. In the future, the best results will probably be achieved by using a combination of different types of sensors, so employing a synergistic approach.

Development of new imaging sensors, combined with a growing availability of digital data, is encouraging research aimed at evaluating how these new sensors and data can be used for hydrogeological modelling. The feasibility of using ERS-1 SAR images for lineament interpretation has therefore been evaluated. Integration of such imagery with spatial data such as well capacity, magnetic field and lithology is necessary to achieve a groundwater model suitable for groundwater production and protection in practice.

REFERENCES

Clark, L. 1985. Groundwater abstraction from basement complex areas of Africa. *Quarterly Journal of Engineering Geology*, 24, 437-451.

Greenbaum, D. 1992. Structural influences on the occurrance of groundwater in SE Zimbabwe. *In*: Wright, E.P. and Burgess, W.G. (Eds.) *Hydrogeology of crystalline basement aquifers in Africa.* Geological Society Special Publication No. 66, 77-85.

Gustafson G. & Krásny J. 1993. Crystalline rock aquifers: their occurrence, use and importance: Memoires of the 24th Congress of the International Association of Hydrogeologists, Ås, Oslo, pp.3-20.

Rosenbaum, M.S. & Wladis, D. 1994. Influence of lineament proximity to groundwater yield in Swedish high grade metamorphic rock aquifers. *Ground Water [In review].*

Wladis D. & Rosenbaum M.S. 1994. Remotely sensed data allowing a multi-indicator approach to groundwater assessment. Proceedings of the 18th Nordisk hydrologisk konference. Thórshavn, Färöarna, 2-4 Aug, 10 pp. [In press].

SGU. 1978. *Tektoniska kartan Göteborg SO*, SGU ser. Af No. 117. Sveriges Geologiska Undersökning, Stockholm.

Wallroth, T. & Rosenbaum, M.S. 1994. Estimating the spatial variability of specific capacity from a Swedish regional data base. *Proceedings of the First Nordic Symposium on Petrophysics in Computer Modelling of Resevoir and Basin Processes*, Gothenburg, Sweden. 1 pp.

Sensors and Environmental Applications of Remote Sensing, Askne (ed.) © 1995 Balkema, Rotterdam, ISBN 90 5410 524 0

Change detection in Swedish land cover from a year of ERS-1 SAR data

D.B.Michelson & J.W.Seaquist
Remote Sensing and GIS Laboratory, Department of Physical Geography, University of Lund, Sweden

ABSTRACT: The information for ten agricultural and seven forest land cover classes are analysed using eight ERS-1 SAR images, covering a full vegetation cycle in southern Sweden. Backscattering coefficients ($\sigma°$) were derived for these classes and compared both visually and statistically. Jeffries-Matusita distances were calculated to determine the statistical separabilities between classes and to determine optimal image combinations for given classification applications. It is shown that ERS-1 data can be used to follow changes in agricultural land cover but not forest land cover, aside from possibly distinguishing fresh clearings over time. Data from the ERS-1 SAR may show potential in effectively classifying agricultural land cover, if applied together with optical data. The data appears to be of limited practical benefit if applied alone. It is not suitable in distinguishing forest land cover classes in Swedish environments.

1 INTRODUCTION

The simple fact that Sweden is located in a region with frequent cloud cover and with very low seasonal natural lighting conditions is reason enough to motivate investigating the potential use of SAR data in remote sensing of terrestrial environments. It is desirable to evaluate the usefullness of SAR data alone and in combination with optical data (Brisco *et al.* 1989, Sandholt *et al.* 1994).

In classifying land cover using SAR data, knowledge of the temporal variation in radar backscatter from defined classes throughout the vegetation cycle is required in order to be able to select appropriate SAR acquisitions. This is made necessary due to the difficulty in separating classes in a single channel of SAR data; the strategy of multitemporal analysis is thus seen as essential.

The objectives of this study are:

1. to characterise SAR backscatter from ten agricultural and seven forest landscape classes over a full vegetation cycle in representative Swedish environments using ERS-1 SAR Precision Image (C-band, VV-polarization) data.

2. to determine which of these classes can be distinguished statistically prior to classification.

3. to determine optimal combinations of acquisitions for a number of given classification applications.

The direction of this study is hence more towards pre-classification activities than the title suggests.

1.1 *Study area and data set*

The study area is the river Genevadsån's drainage basin (250 km^2), located in Halland on the Swedish west coast (figure 1). Both agricultural and forest landscapes are located in the area. Both sandy and clayey soils exist. The area's topography is characterized by flat to undulating terrain with elevations rising to just under 200 m.a.s.l. about 15 km inland.

The ERS-1 data set is presented in table 1. Due to the study area being located in the southern part of the middle of the image swath, the entire area can be imaged only every full repeat orbit cycle, in this case 35 days.

Table 1. ERS-1 SAR.PRI data for the study period.

Acquisition date	Orbit	Frame	PAF
1992-10-22 (Oct92)	6637	2457	UK
1992-11-26 (Nov)	7138	2457	*
1992-12-31 (Dec)	7639	2457	☠
1993-02-04 (Feb)	8140	2457	UK
1993-03-11 (Mar)	8641	2457	UK
1993-04-15 (Apr)	9142	2457	☠
1993-05-20 (May)	9643	2457	UK
1993-06-24 (June)	10144	2457	D
1993-07-29 (July)	10645	2457	UK
1993-09-02 (Sep)	11146	2457	I
1993-10-07 (Oct93)	11647	2457	I

* = not ordered, ☠ = satellite failure

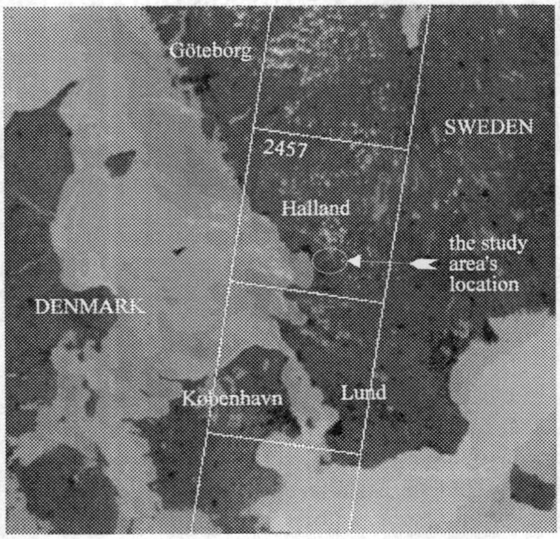

Figure 1. The study area's location and the placement of ERS-1 SAR frame 2457.

2 METHODOLOGY

Each ERS-1 SAR.PRI image was radiometrically calibrated using the method presented by Laur (1992), so as to enable the derivation of σ°. The images, retained in 16-bit resolution, were then geometrically registered to the Swedish national coordinate system using a nearest neighbour algorithm and second order polynomial transformations; this was done using software created by PCI Inc. (1993) and an image-to-image approach where the master image was a corrected Landsat TM acquisition from 1993-04-11. The root mean square error of the control points averaged <50 m for all images.

For each agricultural and forest class, representative fields were selected from field data collected in conjunction with each image acquisition. The sampling scheme involved collecting data from all fields/compartments (which could be located in the imagery) located alongside given stretches of road. This tended to favour the collection of larger volumes of data in the agricultural landscape and smaller data volumes in the forest landscape. Table 2 presents those classes used in the study and the number of polygons used for each class. It should be noted that these fields/compartments were selected to reflect the natural variation in each class and not to be "optimal" in a facilitative sense, i.e. large, flat and homogenous.

All polygons were digitized using 1:20 000 scale orthophotographic maps. These polygons were then interactively edited using the SAR images sequentially as backdrops; this was also done using PCI software. A σ° was derived for each polygon using Laur's (1992) method and in-house software. Descriptive statistics were generated for each class and acquisition.

In order to determine the statistical separabilities among selected classes for a given combination of acquisitions, Jeffries-Matusita distances (Niblack 1985) were calculated using software created by the CHIPS Development Team (1992). This measure of separability produces an index ranging from 0 (no separability whatsoever) to 1.41 (total separability) for given pairs of classes and is thus an indirect indicator of the probability of successful classification of these classes. This routine was applied using a large number of combinations of images for both agricultural and forest landscape classes.

Table 2. Defined classes and the number of polygons used for each class.

Class	No. of polygons
Agricultural landscape	
Unvegetated soil	12
Sugar beet	5
Potato	14
Rapeseed	14
Fallow	6
Wheat	14
Rye	9
Barley	28
Oats	12
Ley pasture	42
Forest landscape	
Deciduous	8
Mixed coniferous/deciduous	8
Mixed coniferous	3
Pine	6
Spruce	12
Clearing	5
Peat bog	7

2.1 *Geocoding*

The data set was also geocoded, using PCI software, in an attempt to correct the geometrical distortions in the SAR imagery created due to topographical surface relief. However the routine is structured in such a way that a simulated SAR image, created using orbital and sensor parameters and a digital elevation model, is used as the master image in an image-to-image correction. The difficulty in identifying common control points between two images, where the master image contains no information on surface features other than pertaining to relief and where the slave image is a SAR image full of speckle, implies that the study area's relief must be more montaine in order for the geocoding routine to be more effective than conventional geometrical registration. The errors obtained when using the routine in Halland were considerably larger than with conventional routines. The results of the latter (see above) were thus used in the analysis.

3 RESULTS

By averaging the individual polygon σ° values for each class and acquisition and then plotting these mean σ° values (figures 2 and 3), the temporal variation in radar backscatter for each class can be illustrated. Figure 4 summarizes class σ° dynamics for the whole vegetation cycle, a descriptive technique previously employed by Cihlar *et al.* (1992). The agricultural classes appear relatively well separated from each other when visualized. The forest classes, however, appear very similar in all respects, with the only exception being the difference between clearings and everything else.

In calculating Jeffries-Matusita distances using all images and all classes for both landscape types, the maximum number of totally separable pairs of classes is revealed along with an overall indicator of how separable all classes are. Table 3 summarizes this and the statistical separabilities of classes for given applications.

3.1 *Temporal variations in σ°*

Agricultural landscape signatures generated during autumn and winter were generally higher than those for spring and summer. Unvegetated and spring-sown fields show relatively high σ°. Ley pasture and fallow fields display low σ° throughout the year.

Towards the end of the vegetation cycle (Sep, Oct-93), signatures are more grouped together. Autumn-sown cereals display the largest annual σ° ranges, followed by rapeseed, spring-sown cereals, ley pasture, sugar beets, fallow land, and potatoes.

Forest landscape signatures show surprisingly little variation throughout the cycle, with the exception of those for clearings. The latter are usually well distinguishable whereas the other classes are almost always clustered together. It is interesting to note, however, that the σ° for mixed coniferous forest is often located between those of pine and spruce monocultures. Deciduous forests have margially higher σ° during the winter. All forest classes together with peat bogs display very low annual σ° ranges, whereas those for clearings are somewhat higher.

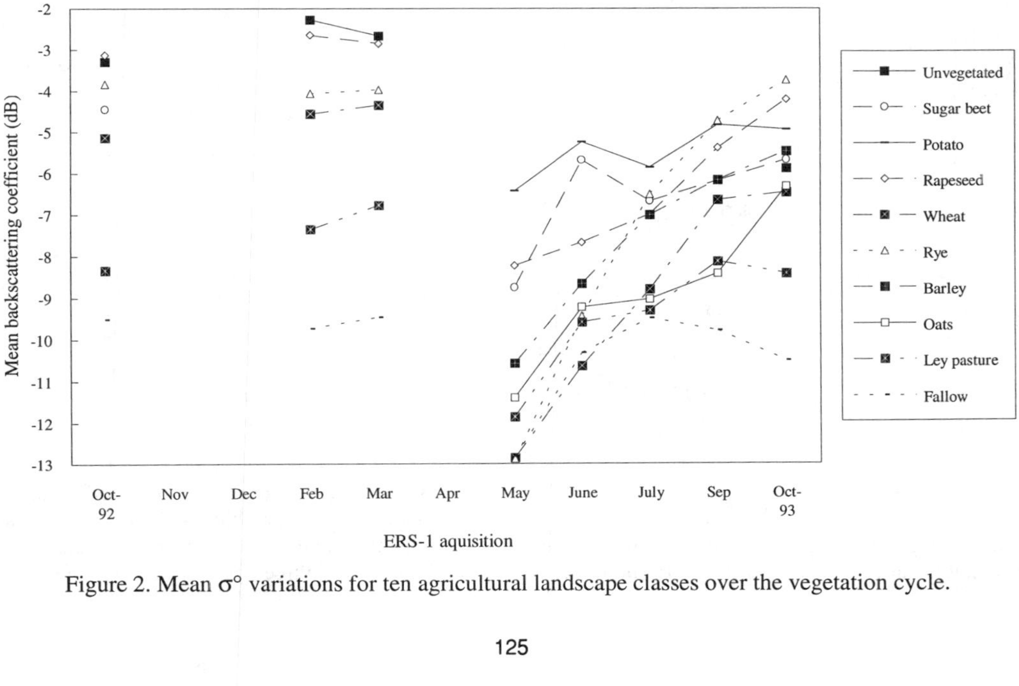

Figure 2. Mean σ° variations for ten agricultural landscape classes over the vegetation cycle.

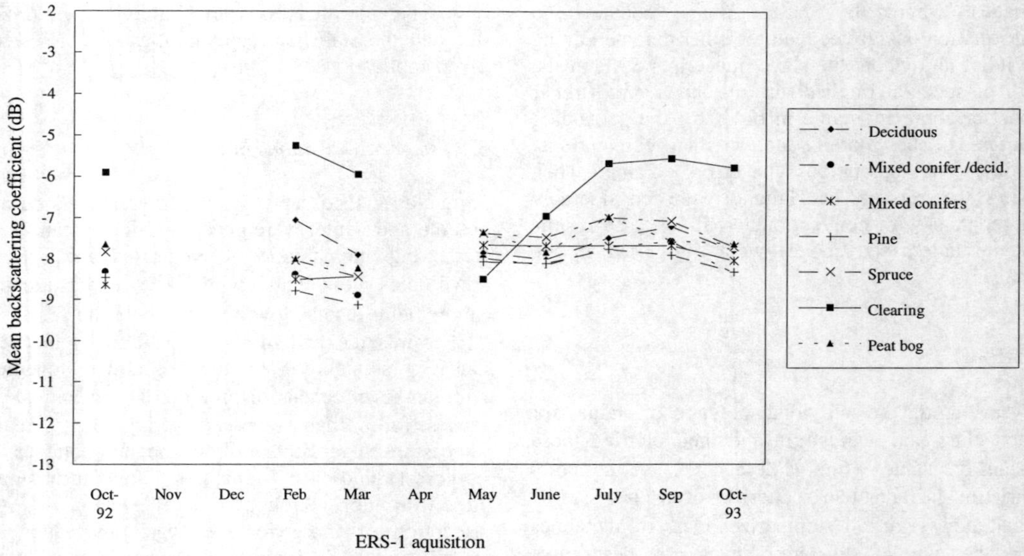

Figure 3. Mean σ° variations for seven forest landscape classes over the vegetation cycle.

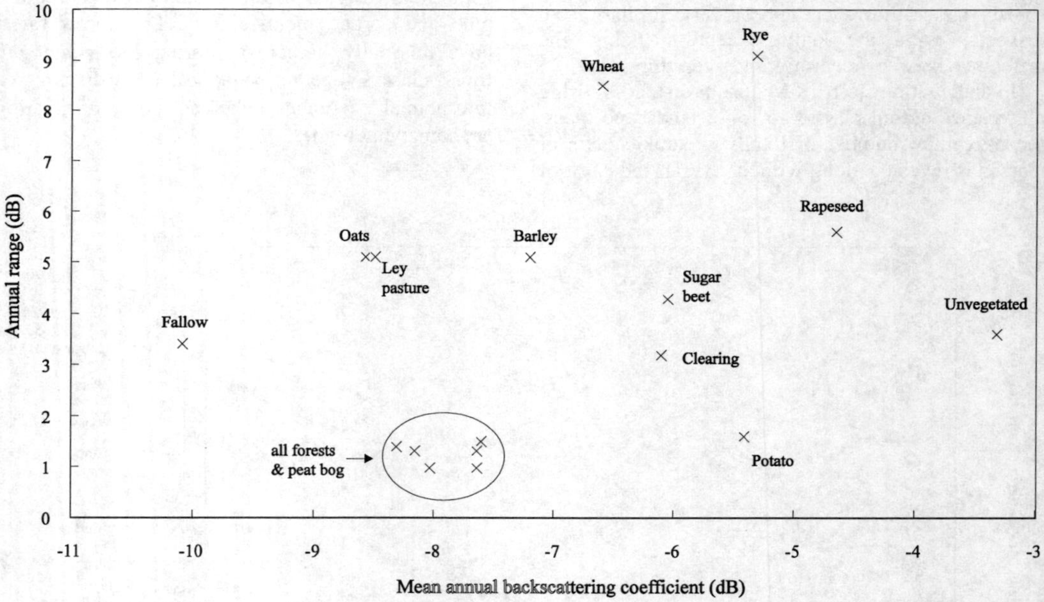

Figure 4. Class σ° dynamics on an annual basis.

3.2 *Signature separability*

When using all images to distinguish between agricultural landscape classes, a very high mean Jeffries-Matusita distance is obtained, but the number of totally separable pairs of classes is less than half the total number of class pairs. When combining the three best images, the mean

Table 3. Statistical separabilities among classes for given classification applications and image combinations, based on calculated Jeffries-Matusita distances.

Application	Optimal acquisitions	Mean J-M dist.	% totally separable pairs of classes
Agriculture	all	1.37	40
Agriculture (3 best acquisitions)	feb, may, june	1.19	2
Agriculture (4 best acquisitions)	feb, may, june, july	1.27	7
Autumn- vs. spring-sown crops	oct92, may, june	1.21	0
Unvegetated winter fields	feb, may, june	1.21	20*
Forests	all	1.31	14
Forests (3 best acquisitions)	oct92, mar, may	1.10	0
Clearing detection	mar, sep, oct93	1.32	0
Single best acquisition	mar	0.93	11

* = does not include spring-sown crops

separability drops considerably and virtually no pairs of classes are totally separable. By adding a fourth image, the mean separability again rises but very few class pairs are totally separable. Using the optimal three image combination, autumn-sown crops are somewhat separated from spring-sown crops. In distinguishing between unvegetated fields and autumn-sown crops and ley pasture/fallow land, the mean separability is reasonably high with 1/5 of the class pairs being totally separable. All optimal image combinations involve the use of an autumn or winter acquisition, a spring acquisition and a summer acquisition.

When using all images to distinguish between forest landscape classes, a high mean Jeffries-Matusita distance is achieved but a very low proportion of class pairs are totally separable. When combining the three best acquisitions, the mean separability drops greatly and no class pairs are totally separable. In attempting to distinguish clearings, the optimal combination of three images yields a high mean separability but no pairs of classes are totally separable. Optimal image combinations were acquisitions from autumn and winter.

The single best acquisition, that from March, gave a low mean separability and few totally separable class pairs.

4 DISCUSSION

The ability to statistically separate agricultural classes was limited. Those classes that exhibited the largest separabilities were unvegetated soil and fallow fields/ley pasture. Differences between other classes were more ambiguous. This could mean that the combination of ERS-1 SAR data and optical data may prove more effective in locating areas with bare fields during the winter and in locating areas with autumn-sown and spring-sown crops. This is a task of great interest in environmental monitoring applications and in the enforcement of land use management legislation. It is also an area where optical remote sensing alone is not entirely suited, in Swedish environments.

The generally higher signatures in the agricultural landscape during autumn and winter may be due to generally wetter conditions. The differences between unvegetated and spring-sown crops and fallow fields/ley pasture may be due to the greater smoothness of the latter's canopies. The grouping of signatures at the end of the vegetation cycle is probably caused by the cereal/rapeseed fields containing stubble or even newly ploughed soil. That the annual σ° ranges of autumn-sown crops are the highest is not surprising since these are very poorly developed during the winter and highly mature by summer; they are thus very similar to unvegetated fields during the winter (very high return) and to spring-sown cereals during summer (low return). That sugar beets and potatoes display low σ° ranges may be due to their canopies developing quickly and scattering strongly.

The optimal image combination of winter-spring-summer in statistically separating agricultural classes is not surprising since fields tend to become more similar when their crops mature (and are harvested), towards autumn.

The difficulty of separating forest landscape classes in this study reveals the unsuitability of applying the ERS-1 SAR to this particular task. These findings perhaps confirm those of Israelsson & Askne (1994) who report on the σ° becoming saturated at relatively low forest biomasses. The higher return from deciduous forest during the winter may reflect its state of defoliation, even though these differences being significant are difficult to confirm through the statistical separability tests. That the annual forest $\sigma°$ ranges are low reflects the relatively slow rate of change in this landscape. The higher return from clearings must be due to both their generally wetter condition and their content of stumps, stones and boulders which can act as effective scatterers. This effect could perhaps be used in locating fresh clearings over time.

The results when using the single best image (March) confirm the necessity of a multitemporal approach when analysing land cover with ERS-1 SAR data.

In deriving $\sigma°$ for each distributed target, Laur (1992) recommends using a minimum of 500 pixels. For ERS-1 SAR.PRI data, this means that fields should be at least 7.8 ha in area. In this study, $\sigma°$ have been derived for some fields that are smaller than Laur's recommended size for the simple reason that these are fully representative southern Swedish fields which cannot be neglected due to their size.

Multitemporal comparison of $\sigma°$ values for individual classes, based solely on Laur's (1992) method, have been shown to be subject to additional sources of error which are correctible but which can lead to variations of up to 2 dB for UK-PAF products (Meadows 1994). The temporal dynamics presented in this study may thus be masked by these errors. The differences between classes in single images remain, however, unaffected.

Quantitative hydrometorological information, that can be used to explain the differences in the radar return from land cover types, were not incoroprated in this study. The analysis of those factors that can be used to explain the dynamics of backscattering from the landscape remains to be undertaken.

5 CONCLUSIONS

The following conclusions can be drawn from the results of this study:

1. The changes in Swedish agricultural land cover, over a full vegetation cycle, can be followed using ERS-1 SAR data, even at a 35-day repeat orbit cycle.

2. ERS-1 SAR data shows potential in classifying Swedish agricultural land cover. It may be beneficial in classifying conventionally "difficult" classes, such as fallow land and autumn-sown crops, if applied together with optical data. It is of limited benefit if applied alone. The optimal combination of images for distinguishing agricultural land cover involves the use of acquisitions from winter, spring and summer.

3. The changes in Swedish forest types, over a full vegetation cycle, cannot be followed confidently using ERS-1 SAR data, probably due to the $\sigma°$ easily becoming saturated.

4. ERS-1 SAR data is unsuitable for forest land cover classification, other than distinguishing between clearings and forest, in Swedish environments. The optimal combination of three images for distinguishing clearings involves the use of acquisitions from autumn and winter.

Aknowledgments: We would like to express our sincere grattitude to: Bjarne Fog at the Institute of Geography, University of Copenhagen, for helping us out with the CHIPS software; Stefan Pinzke in Lund for on the spot software artistry; Henning Skriver at the Electromagnetics Institute, Technical University of Denmark, and Peter Ulriksen, Department of Engineering Geology, University of Lund, for guidance in times of doubt and hesitation. The Swedish National Space Board is gratefully aknowledged for financing this study.

REFERENCES

Brisco, B., Brown, R.J. & Manore, M.J. 1989. Early season crop discrimination with combined SAR and TM data. *Canadian Journal of Remote Sensing*. 15:1, 44-54.

CHIPS Development Team 1992. CHIPS satellite image processing system v. 3.0 User's guide, Routine reference & Technical reference. Institute of Geography, University of Copenhagen, Denmark.

Cihlar, J., Pultz, T.J. & Gray, A.L. 1992. Change detection with synthetic aperture radar. *Int. J. Remote Sensing*, 13:3, 401-414.

Israelsson, H. & Askne, J. 1994. Retrieval of forest biomass using SAR. *Proc.* 14th EARSeL Symposium on Sensors and Environmental Applications, Göteborg, Sweden, 6-8 June. (abstracts volume)

Laur, H. 1992. Derivation of Backscattering Coefficient $\sigma°$ in ERS-1.SAR.PRI Products. Issue 1, Rev. 0, ESA ESRIN.

Meadows, P.J. 1994. The Calibration of the ERS-1 Synthetic Aperture Radar using UK-PAF Imagery. *Proc.* 14th EARSeL Symposium on Sensors and Environmental Applications, Göteborg, Sweden, 6-8 June. (abstracts volume)

Niblack, W. 1985. *An Introduction to Digital Image*

Processing. Birkerød: Strandberg.
PCI Inc. 1993. EASI/PACE 5.3, ImageWorks & GCPWorks reference manuals. Richmond Hill, Canada.
Sandholt, I., Fog, B., Stjernholm, M., Poulsen, J.N. & Skriver, H. 1994. Classification of agricultural crops in Denmark using ERS-1 SAR and SPOT imagery. *Proc.* 14th EARSeL Symposium on Sensors and Environmental Applications, Göteborg, Sweden, 6-8 June. (abstracts volume)

Sensors and Environmental Applications of Remote Sensing, Askne (ed.) © 1995 Balkema, Rotterdam, ISBN 90 5410 524 0

The Eurasian large area ecosystem mapping experiment

S. Folving & J. Mégier
Institute for Remote Sensing Applications, Environmental Mapping and Modelling Group, CEC Joint Research Centre, Ispra (Va), Italy

ABSTRACT: This paper describes a networked project that is being established in order to map and monitor the major ecosystems in the European-Asian Continent. The experiment will especially concentrate on a few longitudinal transections that will be used for validation of AVHRR based results from the whole area. The experiment is presently (May -94) being defined and is expected to be launched by the end of this year.

INTRODUCTION.

During the last few years the large natural and semi-natural ecosystems of the Northern Hemisphere, especially the boreal forests, have become a central theme for many studies. This is due to the fact that it has been generally realized that many environmental problems are of global character, air pollution is transboundary - to mention a single phenomenon. But, it is also due to the fact that we now, thanks to the supply of satellite borne remotely sensed (RS) data, are able to cover large regions in our studies.

The growing awareness of the need to preserve as much genetic variation as possible, biodiversity, and the growing political willingness to maintain a sustainable development has likewise served to turn the scientific community towards studies on global ecosystems. Bio-geography and geo-biosphere studies are once more becoming central. These studies now also benefit from the development of geographical information systems (GIS).

It is of course not just the traditional vegetation geography that is being repeated. Topics like assessment of carbon dioxide emissions, carbon sinks and climatic change are indeed becoming key issues.

Perhaps it should not be excluded that the changed political situation after the "fall of the Wall" has helped to bring the large northern Eurasian regions into focus.

BACKGROUND AND OBJECTIVES.

Studies to provide Eurasian large area ecosystem maps first of all require useful and well defined data. The Institute For Remote Sensing Applications (IRSA) should have both the experience to handle large amounts of data and the experience in pre-processing and formatting data from satellite.

The experiences from the programme on Monitoring of Tropical Vegetation has provided experience in handling large amounts of low resolution optical data. Also the Marine programmes have provided this experience. The MARS (Monitoring Agriculture with Remote Sensing) project has provided the necessary background for handling large area data sets in a GIS format. The studies on atmospheric and topographical correction and satellite data integration in the

Environmental mapping and Modelling Group have given the knowledge for pre-processing large data sets.

The recent development in application of Neural Networks has proven potential for large scale processing - and can even be used for multi-data processing, which will be required, especially for "calibration" purposes, in this mapping experiment. i.e. in the comparison of AVHRR data with high resolution satellite data used in the principal transections.

This experiment is thus very appropriate for IRSA because it allows to use, in an integrated way, the knowledge and experience gained from a broad variety of projects, primarily from the MTV and EMAP units.

The Eurasian large area ecosystem mapping experiment will map the present extent of the main ecosystems and study the development by comparison with historical data sets, and will provide methods for future monitoring of the ecosystems.

Special emphasis will be given to the boundaries between the various ecosystems in order to make it possible to map and monitor changes both in the content *and* in the borders between them. It is evident that any important climatic change should influence the geographical distribution of the sub-ecosystems along the boundaries between the major systems, therefore it is important to concentrate on these boundaries.

The main objectives of the experiment can be summarized as follows, however, it has to be stressed that they are of a provisional character as the very final programme definition is not yet available:

* To develop a large area monitoring system capable of rapidly documenting ecosystem change useful in land management, regional policy making, and in global change and climate studies.

* To promote international cooperation and experience in practical application of Earth resources observation satellite data for regional development and environmental protection.

* To foster exchange of knowledge and experience in the application of remotely sensed data for regional and local land/water management, especially in the less favoured areas having high sensitivity to human impact.

* To identify needed science and application research to effectively fulfill this type of monitoring mission. And to secure that the programme, taken as a whole, should have strong science, operational and social objectives.

<u>AREA AND THEMES.</u>

The area to be covered is situated within approximately 37 to 75 degrees North latitude stretching from the Atlantic coast of Europe eastward to the East Chinese Sea. This area will be covered using NOAA AVHRR data.

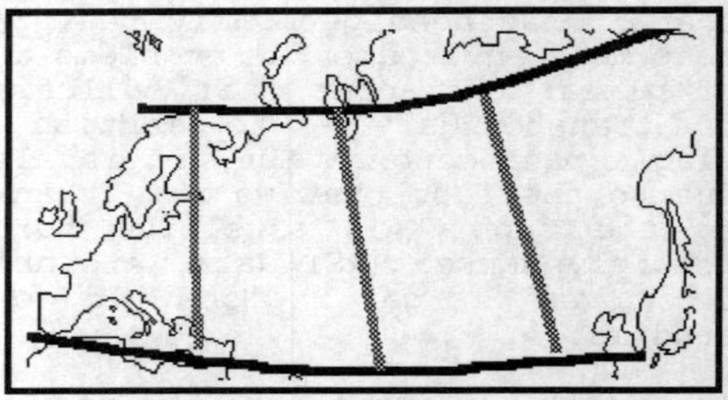

Figure 1. The main study area and the approximatel location of the transects.

Three transections, approximately along the eastern longitudes 30, 70 and 110 degrees, figur 1, have been selected for more intensive studies and for validation of the results obtained with the AVHRR data. High resolution satellite data, primarily from Landsat TM and SPOT, but also ERS-1 and -2 (SAR data) will be used.

This experimental programme is

aimed, as mentioned previously, at providing the areal extension of the main ecosystems, such as the arctic steppes, the Taiga of boreal forest, the mixed and temperate forests and the various types of temperate steppes, figure 2. Mapping of the extension of permafrozen - and partly permafrozen - ground is an other important task that is included in the experiment. Permafrost is believed to be an important environmental indicator for studying climatic change.

Land degradation and landscape dynamics are also considered key topic areas for mapping and for method development. These topics will especially concern the southern steppe regions where for example over-grazing has lead to severe erosion, occasionally in such a degree that it can be considered desertification. Thus a global dimension will be linked to the present EMAP study on Mediterranean land degradation and soil erosion.

CONCLUDING REMARKS.

Maps of ecosystems and ecosystem changes are of course interesting in themselves, but, they will not become useful until they are linked with specific purposes. Therefore the Eurasian Large Area Ecosystem Mapping Experiment from the very beginning has been meant as a network with participation of expertgroups/institutes with a big variety of backgrounds. Presently this network is under definition and agreement, it is therefore somewhat premature to present the details, but, it should be mentioned that socio-economic aspects are having a big importance.

The interest in mapping and assessing biodiversity will not be justified until the results can be included in regional policy making and regional land management. Therefore the experiment is linked to institutes and research groups engaged in studying the development and present stage of the use of the land, e.g. timber production, husbandry and grazing, cropping in the steppe regions etc..

Several West and East European, Russian and Chinese institutes are participating in the network and

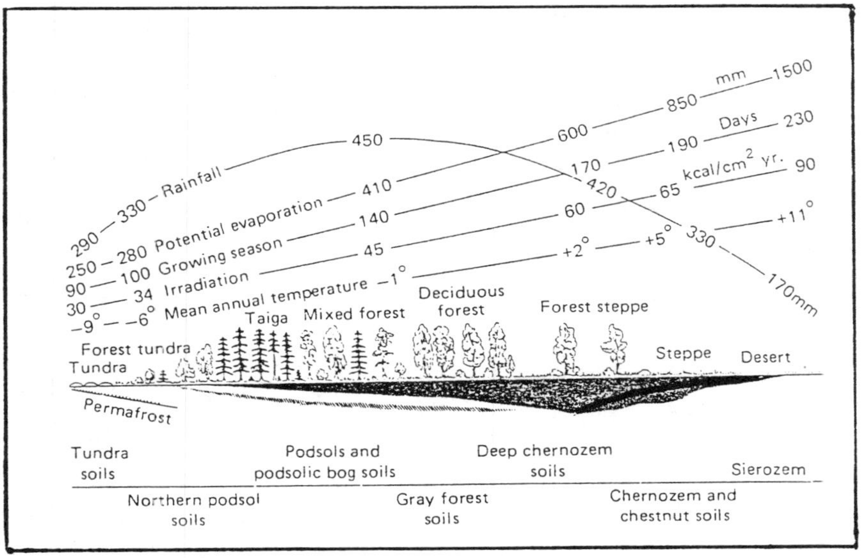

Figure 2. Schematic profile approximately along the 30 degree East transection showing the variation of some geo-biological indicators. (Modified from Walter, 1979).

experiment, especially in the transection areas mentioned above.

REFERENCES.

Walter, H. 1979. Vegetation of the Earth - and Ecological Systems of the Geo-biosphere. Second Edition. Heidelberg Science Library, Springer-Verlag, Heidelberg.

Annual Reports: 1991 (EUR 14833 EN), 1992 (EUR 15407 EN) and 1993 (in print). Institute for Remote Sensing Applications, Joint Research Centre, Commission of the European Communities. ECSC-EEC-EAEC, Bruxelles and Luxemburg.

Sensors and Environmental Applications of Remote Sensing, Askne (ed.) © 1995 Balkema, Rotterdam, ISBN 90 5410 524 0

EC initiatives on European forest ecosystems mapping and forest statistics

S. Folving, P. Kennedy & N. McCormick
Institute for Remote Sensing Applications, Environmental Mapping and Monitoring Group, Joint Research Centre, Ispra (Va), Italy

ABSTRACT: Forestry is becoming an important issue in European environmental and agricultural policy. The need to assess, map and monitor forested areas constitutes a challenge for development of applicable methods based on remote sensing. The paper describes the background for this new development and some of the formal requirements to the information needed. The FIRS (Forest Information from Remote Sensing) Project is described; it is a programme that aims at developing the necessary tools and methods for providing needed information on forest ecosystems. The foundation actions provides the regional units and the nomenclature that is used for the statistical assessment and mapping themes which consists of several applications modules.

1. INTRODUCTION

Until very recently European forestry interests have been over-shadowed by agricultural politics. However, with the increasing recognition that large areas of European forests have suffered from air pollution, fire, disease and storms, the general awareness and concern for forests within the scientific, industrial and public sectors of the European society has grown. Forests and forestry have been areas of conflict between different ministries. The production aspects have naturally been part of the responsibility of the agricultural ministries; the environmental aspects have been in the interests and responsibilities of the environment ministries. A steady growing interest and concern for the environment in Europe has gradually changed the traditional acceptance that agriculture and environment can be considered two different items. Forested areas thus become even more central in the public debate and it seems as if the forestry sector is rapidly becoming a focus for coordinated policy and planning in Europe.

The present trend to concentrate increasingly industrialized agricultural production on an increasingly smaller area leaves large areas for other purposes. As a result of the Council Regulation 2078/92 (CEC, 1992) on agricultural production methods compatible with the requirements of protection of the environment and the maintenance of the country side, and Regulation 2080/92 (CEC, 1992) on instituting a Community aid scheme for forestry measures in agriculture, new challenges for the development and application of remote sensing methods have emerged. The main interest concerns not so much existing forests, but more new forest land and areas set aside for extensive grassland.

As rather large amounts of funds are involved in the set-aside land schemes, it is evident that new methods of control have to be developed and high-resolution remotely sensed data, especially air photos, could become an important tool in the control procedures. For many reasons it is interesting to note that the augmentation of set-aside land, and therefore the afforestation and increasing of grassland area should preferably take place as part of a general landscape planning. For example, as measurements against erosion, protection of watersheds and water courses, for creating landscape diversity etc. Alternative production of energy crops such as fast-growing tree species is also foreseen. This might be considered a challenge for new initiatives in research and development (R/D) in regional and especially local landscape planning.

The agreement on scientific cooperation between the EU and the EFTA countries, together with the possibility of the EFTA countries joining the EU, have brought forestry in general into focus. This, of course, is due to the fact that the EFTA countries are important producers of forestry products.

The general acceptance in Europe of sustainable forest management, conservation of the biodiversity of the European forests, and strategies for long term adaptation of forests in Europe to climatic change, as for example stated in the Helsinki Conference (Min of Agrc., 1993), also constitutes a challenge for development of new methods for monitoring by remote sensing. It is therefore evident that the remote sensing (RS) community should move from pure research and experimental applications towards more direct applications. There will still be a big need for R/D projects. However, some of this R/D should be put within a frame of large scale applications.

The Institute for Remote Sensing Applications has initialized such a large scale application programme, FIRS (Forest Information From Remote Sensing). The FIRS Programme, in fulfilment of EU needs, will assist in defining the required R/D needed for providing means for mapping and monitoring the forest environment and providing the statistical information on production and decline.

2. BACKGROUND FOR THE FIRS PROGRAMME

During the period 1985 to 1992 the Institute for Remote Sensing Applications managed the European Collaborative Programme (ECP) on Application of RS in Local Management (Folving, 1992). The ECP provided through the Theme on Forestry Applications, valuable experience and knowledge on the potential of using high resolution satellite data and airborne RS data for mapping, planning and management of forest and natural vegetation on a local level throughout Europe. A few national programmes were also run under the ECP during the period - namely in Denmark, Ireland, Italy, Germany and Greece.

On a very general level it was found that the EU consisted of three main forest management-types regions (Folving, 1992 and McCormick and Folving, 1992). On the local level, where most of the studies and most of the method development were carried out, the potential of applying RS data was evident. The main drawbacks were economy: both software and RS data were too expensive for the local forest manager. Therefore a rather powerful image processing and GIS software package, PC_IMEGA and IM-X, was developed (Folving et al.,1992 and Hoey et al.,1991) and freely distributed. Note that the availability of data on a local level is still a major problem. The ECP was stopped in 1992; it had fulfilled its aims and the accepted principle on subsidiarity within the EU made it easy to decline the many requests to continue parts of the Programme.

In 1985-86 probably one of the most important projects on application of RS, the CORINE Land Cover Project (CORINE, 1993) was started. The aim was, and is, to map Europe at a scale of 1:100.000 in 44 land use classes.

The CORINE Land Cover map contains three forest classes, coniferous, broadleaved and mixed forest. The methodology, which has been specifically developed for the CORINE Project, requires that the land use classification and generalization is done at the same time. Thus the delineation of forest areas will not always be the same as provided by a pure forest map. Forests smaller that 25 ha are not included as this is the threshold size for the land use units.

It is evident that the use of RS data as in the CORINE Project does not quite harmonize with the possibilities of automatic processing of RS data. Nevertheless, the CORINE Land Cover Project very clearly shows that RS data are useful for general mapping, provided that the methodology is adapted to the aim, as it was in this case.

The European Forest Information and Communication System, EFICS, was established and defined by a Council Regulation in 1989 (EEC, 1989). It was extended by a Regulation in 1994 (EEC, 1994). The objective of EFICS is to collect, coordinate, standardize and process data concerning the forestry sector and its development. EFICS shall take account of existing data, and make use of information available in the Member States of the EU. The background for this decision is that the forestry data available at Community - EU - level is incomplete and only covers part of the information necessary for taking consistent forestry measures. This has been described in the Communication from the Commission Document "Community Strategy and Action Programme for the Forestry Sector" (COM 1988). A draft of the structure of EFICS is shown in Figure 1.

3. FIRS - EUROPEAN FOREST INFORMATION FROM REMOTE SENSING

Building upon the experiences described earlier, as well as upon a study on the needs for forest information in various organizations and administrative entities on regional, national and European level carried out by DG XII for the Institute for Remote Sensing Applications, the FIRS Programme has been defined to develop methods for providing statistical and mapped information on the European forest area (Kennedy et al., 1994). The main aim of FIRS is thus to assist in developing and implementing

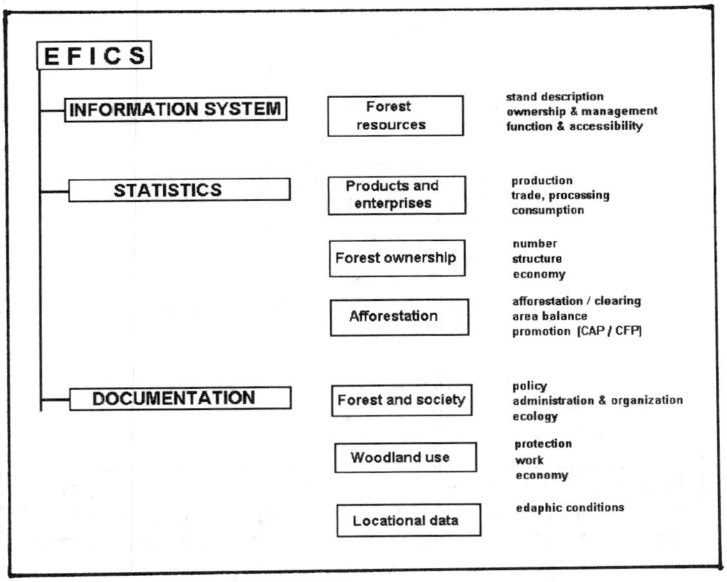

Figure 1. Simplified overview of the EFICS indicators.

methodologies for using remotely sensed data for practical forest and forest environment information assessment on a European and regional scale.

Parts of FIRS are directly aimed at supporting EFICS and other Community needs. Small scale mapping and monitoring of European forest areas is also included and constitutes a part of the Eurasian large area ecosystem monitoring that is aimed at providing a basis for climatic modelling.

It is evident that, in order to include all of Europe, a harmonized terminology and methodology will have to be developed. Therefore the first tasks of FIRS were to create a basis for a coherent Pan-European nomenclature. The first step (i.e. Foundation Action) has been to divide Europe into main, homogeneous forest ecosystem regions. A call for tender was issued in late 1993 for this task, a consortium lead by Scot Conseil, GAF mbH, and Swedish Space Corporation won this tender.

The regions are primarily defined using geophysical and biological variables. Anthropological variables, such as management, will be used for stratifying the primary regions. Within each region representative test areas will be selected. These will be used for the development of mapping methods. An area frame sampling will be performed for each region that will provide the ground units for statistical sampling. The procedure is shown in Figure 2.

A workshop and expert meetings will be used for initializing the development of a common nomenclature on forest mapping and statistics (the second Foundation Action). The nomenclature will be linked closely to the regionalization. It is the aim to provide a general nomenclature that can serve a wide spectrum of purposes, not solely RS based mapping. This action will also define the major variables (e.g. vegetation associations) to be considered in the compilation of the forest information system, and

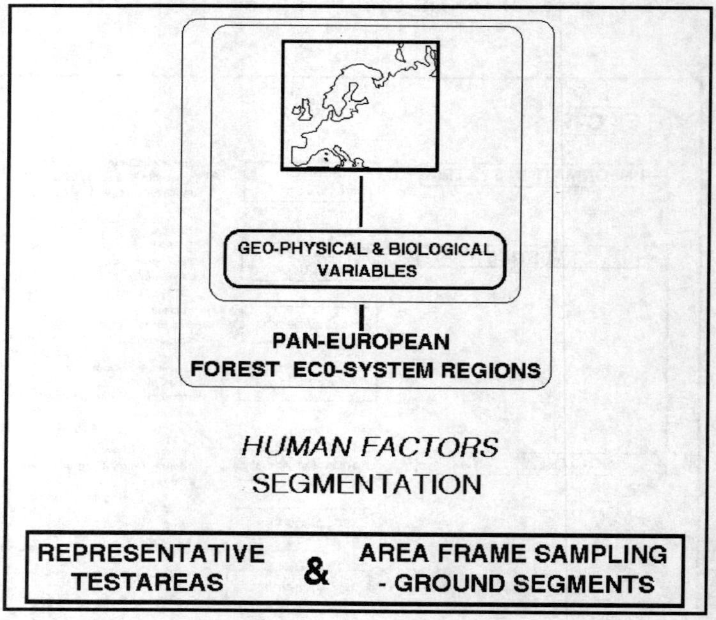

Figure 2. The procedure for defining representative test areas and performing the area frame sampling.

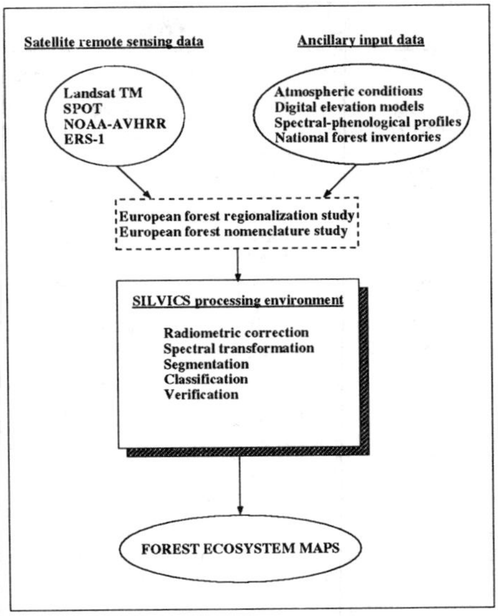

Figure 3. Schematic overview of SILVICS.

selection of the attributes (e.g. species age, hight etc.) required to describe each of these variables.

A third Foundation Action is being undertaken in cooperation with UNEP/GRID in Geneva: Provision of a European Geo-referenced Forest Information Directory. The objective is to investigate and systematically document the availability and quality of forest and forest-related geo-referenced data in Europe.

A software package, SILVICS (Satellite Image Land Vegetation Integrated Clasification System) is being developed (McCormick and Folving, 1992) for mapping the forest ecosystems. SILVICS includes a standardized methodology for processing satellite data, primarily Landsat TM data. The imagery, which may be multitemporal, is corrected for effects of varying sun position, topographic and atmospheric conditions. After this preprocessing the data can be classified, (using either standard classification procedures or neural networks,) based on phenological or seasonal variation in each forest type's spectral characteristics(Figure 3).

The FIRS Programme is planned to have a four year development and evaluation phase and a two to three year implementation and transfer phase. Some of the application modules are, however, foreseen to be applicable already within the first phase. Figure 4 gives an overview of the major themes and application modules.

As it is envisaged that remote sensing data should be used whenever possible, a strong emphasis has been placed on the development of appropriate methodologies and assessment of their capabilities for making FIRS operational.

The main objectives of the major themes of FIRS are as follows:

European Forest Statistics.
To develop methods to obtain statistics on sectorial forest information, e.g. stand composition (species, age, yield, lass, etc.), standing crop volume. Forest ownership and management practices will also be considered.

European Afforestation Monitoring.
To describe the regional variation and to monitor the set-aside

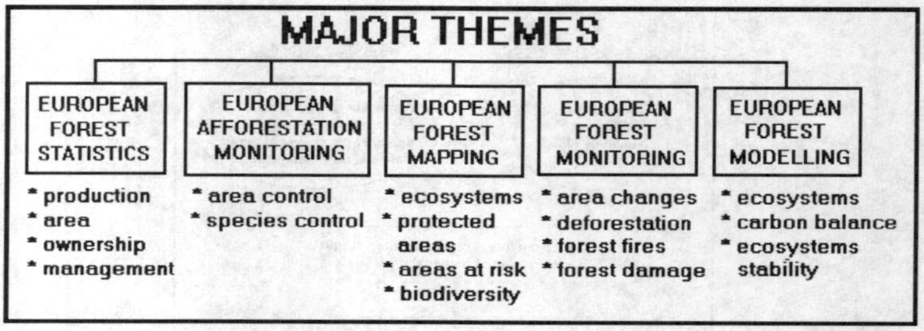

Figure 4. The major Themes and most important Application Modules of FIRS.

afforested land of Europe. This will also include areas situated outside intensive agricultural land. The areal extent, tree species and changes to extensive grasslands will be monitored. The possibility of establishing some kind of control system will also be investigated.

European Forest Mapping.
To develop standardized and operational methods for mapping and monitoring the forests of Europe at a scale of 1:100,000. By adopting an ecosystem approach, emphasis will be placed on ecological forest information. This information will also include geo-referenced data on the geo-physical and bio-physical parameters which determine or influence forest characteristics and growth patterns.

European Forest Monitoring.
To develop methods and models for "fast" monitoring of forest dynamics e.g. area changes, forest damage, forest fire.
European Forest Modelling.
To develop models to monitor changes in ecosystem stability, to assess the fragility of forest systems and to identify those under stress or threat.

4. CONCLUDING REMARKS

It is evident that a bigger public support to the forest sector will lead to a bigger need for information on the forest production and the use of the forested land (e.g. public access, maintenance of areas with broadleaved species). Especially the large support to farmers for afforestation of set-aside land leads to a political requirement for control. The fact that the afforestation is expected to take place as parts of larger landscape plans requires the development of new tools for control and monitoring.

Presently a common forest policy does not exist. However it is evident that this situation is slowly changing so that a common community forest policy can be foreseen soon. RS has been proven to be useful in assessing the EU agricultural production. There is every reason to believe and expect RS to be also one of the main tools, if not the main tool in supplying information needed for a forest policy management.

5. REFERENCES

COM (88) 255 Final. Community strategy and action Programme for the forestry sector. Communication from the Commission CB-CO-88-443-EN. Luxembourg, 11.11.1988.

CORINE. 1993. CORINE Land Cover - Guide Technique. EUR 12585 FR. CECA-CEE-CEEA, Bruxelles & Luxemburg.

EEC Council Regulation 1615/89. Official Journal of the European Communities 15.6.89. No. L. 165/12-13

EEC Council Regulation 2078/92. Official Journal of the European Communities 30.7.92. No. L 215/85-90.

EEC Council Regulation 2080/92. Official Journal of the European Communities 30.7.92. No. L 215/96-99.

EEC Council Regulation 400/94. Official Journal of the European Communities 25.2.94. No. L 54/5.

Folving, S., and J. Mégier. 1991. European Collaboration on Development of new Environmental Thematic Map-types from Remotely Sensed Data. Symp. Int. de Cartographie Thematique derivee des Images Satellitaires. Bulletin du Comite Francais de Cartographie. N 127-128, (pp41-46), Mars-Juin 1991.

Folving, S. 1992. Application of Remote Sensing to forestry mapping - experiences from the European Collaborative Programme for the Less Favoured Areas. Proceeding of the Central Symposium of the "International Space Year" Conference, 30 March - 4 April 1992, Munich. (ESA SB-341, July 1992). (pp 857-861).

Folving, S., M. Hoey and D. Whelan. 1992. Development of a low- cost PC-based image-processing and mapping GIS within the frames of the EEC Joint Research Centre's Collaborative Programme. Comput., Environ. and Urban Systems, Vol 16, (pp 195-201).

Hoey, M., D. Whelan and S. Folving. 1991. A PC based Geographical Information System that integrates image data and attribute data. Second European Conference on GIS, Bruxelles, April 2-5. Vol 1, (pp 596-476).

Kennedy, P., N. McCormick and S. Folving. 1994. Introduction to the FIRS Programme. Proceedings from International Workshop on Satellite technology and geographical information systems for Mediterranean forest mapping and fire management. Thessaloniki 4-6 Nov. 1993.

McCormick, N. L. M. and S. Folving. 1992. Application of remotely sensed data in forestry, grassland and natural vegetation management in European less favoured areas: from case studies to ecosystem mapping. International Symposium "Operationalization of Remote Sensing", 19-23 April 1992, ITC, Enschede, The Netherlands. Vol.8. (pp 1-10).

McCormick, N. L. M., P. Kennedy and S. Folving. 1994. An integrated methodology for mapping European forest ecosystems using satellite remote sensing. Proceedings from International Workshop on Satellite technology and geographical information systems for Mediterranean forest mapping and fire management. Thessaloniki 4-6 Nov. 1993.

Ministry of Agriculture and Forestry. 1993. The Resolutions and the General Declarations of the Second Ministerial Conference on the Protection of Forests in Europe. Ministerial Conference on the Protection of Forest in Europe, 16-17 June 1993, Helsinki.

Sensors and Environmental Applications of Remote Sensing, Askne (ed.) © 1995 Balkema, Rotterdam, ISBN 90 5410 524 0

Improved forest mapping by combining corrections of atmospheric and topographic effects in Landsat TM imagery

Joachim Hill & Wolfgang Mehl
Institute for Remote Sensing Applications, Environmental Mapping and Modelling Unit, Commission of the European Communities, Joint Research Centre, Ispra (Va), Italy

Volker Radeloff
Remote Sensing Department, Faculty of Geography/Geosciences, University of Trier, Germany

ABSTRACT. The spectral reflectance of ground objects in mountainous areas is largely contaminated by second order effects which are due to topographic slope and aspect. Such topographic effects present severe problem for the consistent analysis of optical remote sensing images, in particular for satellite-based forest cover mapping. We have integrated a topographic correction module into a modified 5S atmospheric correction model, where targets are assumed to have lambertian reflectance characteristics. The method was successfully applied to four Landsat Thematic Mapper images with large seasonal differences in solar elevation. Classification methods of increasing complexity (euclidian minimum distance, maximum likelihood, and a backpropagation neural network) have then been used to produce forest stand maps from images which were either corrected for atmospheric effects only or for radiometric distortions due to both atmosphere and topography. It is demonstrated that the topographic corrections provide important improvements when direct and diffuse radiation components are properly accounted for. Differences between the actual bidirectional reflectance properties of forest stands and their approximation by Lambertian reflectance characteristics seem to be less important.

1 INTRODUCTION

It is commonly known that the spectral reflectance of ground objects in mountainous areas is largely contaminated by second order effects which are due to topographic slope and aspect. Particularly in remote sensing images acquired over homogeneous ground cover, reflectance variations due to the topographic effect are perceived as a visual impression of relief. Slopes facing away from the sun appear relatively dark, while the sun-facing slopes appear brighter than horizontally oriented surfaces with the same cover type. It is evident that this effect varies throughout the year as a function of sun elevation and azimuth, and the steepness and orientation of slopes. In extreme cases, effects of self shadowing and shadows cast on opposite slopes may occur frequently, causing further disturbances in the radiometric consistency of satellite images.

The topographic effect presents a severe problem for any consistent analysis of optical remote sensing images, in particular for automated land cover classification (Hill, 1993). This affects especially the production of precise satellite maps over forested areas which tend to be dominant in mountainous areas, being less suitable for other land uses.

The principles of terrain illumination have been described in detail, and various approaches for topographic correction or normalisation were already presented (Justice et al., 1981; Dozier and Frew, 1981; Woodham and Gray, 1987; Leprieur et al., 1988; Kawata et al., 1988; Civco, 1989; Proy et al., 1989, Colby, 1991; Itten et al., 1992; Conese et al., 1993). The most critical issue in correcting the topographic effect seems related to selecting a suitable bidirectional reflectance distribution function (BRDF) for determining how a surface will appear under any condition of viewing. Using the BRDF of a Lambertian reflector and a class of reflectors first identified by Minnaert (1941) have provided partially contradictory results, such that it has recently been recommended to identify empirical correction factors by a regression analysis of the relationship between terrain slope/aspect and corresponding pixel intensities from forested areas (Itten et al., 1992).

We have attempted to integrate the correction of the topographic effect into our implementation of the 5S atmospheric correction model of Tanré et al. (1990). In order to study whether frequently reported shortcomings of topographic corrections (usually attributed to deficient approximations of the surface BRDF) might also be due to improperly balanced direct and diffuse solar radiation terms, our topographic correction was initially implemented under the assumption of Lambertian reflectance characteristics.

Figure 1. Digital elevation model of the Heiderscheid study area, illuminated according to the conditions during the Landsat satellite overpass from 21 September 1989 (top of page is north). Colour plate see page 492

2 STUDY SITE

This study is embedded in a more extensive experiment on automatic updating of the CORINE Land Cover data base on the Grand-Duchy of Luxembourg where we have used multi-temporal Landsat Thematic Mapper data from 1989. Ground reference mapping was obtained within 105 segments (50 ha) during summer 1991 in cooperation with the "Direction des Eaux et Forêts", the "Institut Supérieur de Technologie" (Luxembourg) and the Remote Sensing Department of the University of Trier (Germany). Due to the fact that the reference data were collected 2 years later than the satellite imagery, only the forest-related ground data could be used to directly assess the accuracy of the mapping results. Although an overall precision of 78 % was achieved for the satellite-based mapping of deciduous and coniferous forest stands, this accuracy may locally drop to lower values, particularly in areas of rugged terrain with strong illumination differences.

A smaller study site was selected for analysing possibilities of improving forest mapping accuracy through a correction of the topographic effect in the satellite images. This site covers an area of 52.9 km^2 and is located in the Oesling region of Luxembourg (Ardennes foothills), north-west of the towns Diekirch and Ettelbruck. Although the maximum altitude is only 512 m above sea level, the Heiderscheid test area is characterised by deeply incised, meandering valleys with pronounced illumination differences (figure 1).

In particular the steep slopes of the V-shaped valleys are mostly covered with forests, where coppice-like *Niederwaelder* of oaks are dominant. Climax vegetation would be dominated by mixed beech (fagus silvatica) and oak (quercus robur) forests. Since the early 18th century, fir (abies alba) was introduced, and afforestations with various coniferous species have become popular ever since.

Table 1. Cloud-free Landsat-5 Thematic Mapper images in 1989 (Path 197, Row 25)

Date	Acquisition time (GMT)	Scene centre latitude	Scene centre longitude	Sun elevation	Sun azimuth
16 May 1989	9:56.33	50.28° N	5.93° E	53.47°	139.76°
17 June 1989	9:55.54	50.28° N	5.94° E	56.37°	134.45°
20 Aug 1989	9:54.39	50.28° N	5.92° E	46.44°	141.16°
21 Sep 1989	9:53.47	50.28° N	5.94° E	36.35°	149.82°
10 Dec 1989	9:52.42	50.28° N	5.91° E	13.67°	156.97°

3 DATA SETS

Detailed forestry information for the Heiderscheid study site was obtained from the Luxembourg "Direction des Eaux et Forêts", and has been introduced into a georeferenced data base under ARC-Info and ORACLE. A detailed digital elevation model (elevation, slope, aspect) with a spatial resolution of 10x10 meters was generated from digitised contour lines of the 1:20,000 scale topographic maps, using the ARC-Info TIN module (figure 1).

In 1989 the for central European conditions relatively rare situation occurred that five cloud-free Thematic Mapper scenes were acquired by the Landsat-5 satellite (table 1). All scenes have been system-processed by SSC Satellitbild (Sweden), and were delivered as geocoded products in UTM-projection, being resampled with the cubic convolution scheme. The December image was not used in this study, because the signal dynamics is by far too low.

4 CORRECTING THE TOPOGRAPHIC EFFECT

In case of an inclined surface, the intercepted radiance due to direct solar illumination is proportional to the cosine of the angle γ between local surface normal and the direction of the incoming sunlight

$$\cos(\gamma) = \cos(\theta_0)\cos(\theta_n) + \sin(\theta_0)\sin(\theta_n)\cos(\phi_n - \phi_0), \quad (1)$$

where θ_0 is the solar zenith angle, θ_n the zenith angle of the normal to the surface, ϕ_0 the solar azimuth angle, and ϕ_n the topographic aspect angle. Assuming that the target behaves as a lambertian reflector, also the reflected radiance is proportional to cos (γ) which can be derived for each pixel from registered digital elevation data. Various correction approaches have been proposed most of which attempt to correct the topographic effect through this simple cosine correction, i.e. separately from a correction of atmospheric effects (i.e., Itten et al. 1992).

4.1 *Simple cosine correction*

Assuming lambertian reflectance characteristics, the corrected radiance (which would be observed in case of horizontal orientation) for a pixel on sloping ground is calculated with

$$L_h = L_s \cdot \cos(\theta_0) / \cos(\gamma) \quad (2)$$

where L_h denotes the corrected radiance corresponding to a horizontal surface, and L_s the actually observed radiance over sloping ground; θ_0 is the sun zenith angle (90° - sun elevation angle), and γ refers to the angle between the incident solar radiation and the surface normal on a tilted plane (pixel) which is derived from digital elevation models.

Almost as often as this simple method (cosine correction) has been used, it has also been reported that the approach tends to produce overcorrected images, in particular when the angle between the incoming irradiance and the pixel's surface normal is larger than 55°, implying values for cos (γ) < 0.57 (Civco, 1989; Colby, 1991; Itten et al., 1992). This is usually attributed to the fact that the BRDF of most surface types is not lambertian, and several modifications have been proposed in order to better account for the directional reflectance characteristics of important surface cover types (e.g., Minnaert coefficients, empirically derived correction factors).

However, it must be emphasised that this simple cosine correction applies only to the directly transmitted part of the solar irradiance. Diffuse irradiance terms which, depending on the turbidity of the atmosphere, may be of substantial magnitude are not accounted for correctly. In other words, applying the simple cosine approach (conceptually limited to direct irradiance terms) for correcting an effect which results from the combined direct and diffuse illumination, *must produce* overcorrection phenomena, notwithstanding any assumptions about targets behaving like lambertian reflectors. Corrections of the topographic effect should be directly integrated into atmospheric correction models, because these can provide realistic estimates of direct and diffuse irradiance terms.

4.2 *Topographic corrections as an integral part of atmospheric modelling*

We have therefore tried to incorporate a topographic correction module into a radiative transfer code. The atmospheric correction method is originally based on the formulation of radiative transfer as developed by Tanré et al. (1990), and it was modified to account for atmospheric extinction processes as a function of sensor and terrain altitude (Hill and Sturm, 1991; Hill, 1991). The modified '5S' code makes extensive use of analytical expressions and preselected atmospheric models, resulting in a rather short execution time. It provides corrections for atmospheric absorption, scattering and pixel adjacency effects, where diffusion and absorption processes are assumed to be independent. Upward and downward transmission coefficients are derived by introducing the auxiliary quantity of optical thickness τ which measures the total extinction of a light beam due to molecular and aerosol scattering when passing through an airmass. Multiple scattering is accounted for according to Sobolev's approximate solution (Sobolev, 1963), and absorbing atmospheric gases are assumed to condense at the top of the atmosphere and at the top of the layer between the earth surface and the sensor altitude.
An important element of the code is its ability to estimate atmospheric key parameters, such as the aerosol optical depth, directly from dark objects (e.g., water) in the satellite image (Hill, 1993).

Once the aerosol optical depth τ_a is known* , the downward direct (beam) and diffuse transmittance terms ($t_d\downarrow$, $t_s\downarrow$) for a given wavelength λ are calculated with

$$t_d \downarrow = \exp\left(-(\tau_r + \tau_a)/\mu_0\right) \qquad (3)$$

and

$$t_s \downarrow = \exp\left(-(0.52 \cdot \tau_r + 1/6 \cdot \tau_a)/\mu_0\right) - t_d \downarrow , \qquad (4)$$

where μ_0 denotes the cosine of the solar zenith angle θ_0 (Tanré et al., 1979). The attenuated global irradiance at ground (E_g) in this spectral bandpass is then obtained as the sum of direct and diffuse irradiance terms

$$E_g = E_0' \cdot t_d \downarrow + E_0' \cdot t_s \downarrow . \qquad (5)$$

E_0' denotes the exoatmospherical solar irradiance E_0, being already corrected for ozone absorption.

Assuming lambertian reflectance characteristics, the satellite-measured radiance L_t for a horizontal surface relates to the modelled target reflectance ρ_t according to

$$L_t = E_g \cdot \rho_t \cdot T\uparrow / \pi . \qquad (6)$$

$T\uparrow$ denotes the total upward transmittance in the spectral band. With all other variables but E_g unchanged, the modelled reflectance for an inclined surface is obtained from

$$\rho_t^* = \frac{L_t \cdot \pi}{E_g^* \cdot T\uparrow} , \qquad (7)$$

where E_g^* denotes the slope-aspect-corrected global irradiance. Substituting L_t in equ. (7) by the right-hand term of equ. (6), we obtain

$$\rho_t^* = E_g \cdot \rho_t / E_g^* , \qquad (8)$$

yielding the illumination-corrected target reflectance ρ_t*. The objective is therefore to find a good approximation for the direct and diffuse components of E_g*. In order to do so, we have to examine with greater detail the illumination of mountainous areas where the topography modifies the irradiance in different ways (Proy et al., 1989):

(a) the direct irradiance and also parts of the diffuse radiation are modulated according to the pixel orientation
(b) the topography may also cast shadows on a pixel so that it receives no direct radiation
(c) the topography may reduce the scattered (diffuse) sky radiation by hiding a part of the sky hemisphere
(d) the topography may reflect radiation towards the pixel, thereby providing an additional irradiance source

Neglecting the effect noted under (d), we can state that a pixel receives in

case 1: direct and diffuse radiation when cos (γ) > 0, and no shadow is cast on the pixel by other topographic features
case 2: only diffuse radiation when cos (γ) > 0, but shadow is cast on the pixel by other topographic features
case 3: only diffuse radiation when cos (γ) < 0 , i.e. self shadow.

While in previous work diffuse solar radiation was mostly treated as being isotropic, it has been demonstrated by various authors (e.g., Proy et al., 1989) that it is important to consider its anisotropic distribution. Similar to Conese et al. (1993) we have also used Hay's approximate solution (Hay, 1979; Hay and McKay, 1985; Hay et al., 1986). According to this approach, the diffuse downwelling radiance increases in the circumsolar zone, while it is isotropically distributed for the rest of the sky dome. In Hay's approximation the relative weights of anisotropic and isotropic parts of the diffuse irradiance depend on a wavelength-dependant anisotropy index

$$k = \left(E_0' \cdot t_d \downarrow \right) / E_0 , \qquad (9)$$

* the molecular optical depth τ_r is derived from the classical relation $\tau_r = 0.0088\ \lambda^{(-4.15 + 0.2\,\lambda)}$

which is based on the assumption that "the isotropic and circumsolar components are a linear combination based on the transmissivity of the direct radiation." (Hay et al., 1985). The secondary maximum of diffuse radiation near the horizon which is due to the longer path length (Temps & Coulson, 1977) is neglected in this approximation. It follows from equ. (9) that k also varies with the atmospheric conditions.

Accordingly, the value of E_g* to be used in equ. (8) is calculated either with

$$\begin{aligned} E_g * = \; & (E_0' \cdot t_d \downarrow) \cdot \cos(\gamma) / \cos(\theta_0) \\ & + (E_0' \cdot t_s \downarrow) \cdot k \cdot \cos(\gamma) / \cos(\theta_0) \\ & + (E_0' \cdot t_s \downarrow) \cdot (1-k) \cdot Fr_{sky} \end{aligned} \qquad (10)$$

(i.e., case 1: the sum of direct, anisotropic diffuse (circumsolar), and isotropic diffuse irradiance), or with

$$E_g * = (E_0' \cdot t_s \downarrow) \cdot (1-k) \cdot Fr_{sky} \qquad (11)$$

(i.e., case 2/3: only isotropic diffuse irradiance). For each pixel, Fr_{sky} quantifies the proportion of the hemispherical sky which is not hidden by topographic features in its surroundings. It is determined through an analysis of the digital elevation model, where the local horizon is calculated for 16 directional segments, starting from north (= 0°) in increments of 22.5°. The addition of these visible sky segments provides a good approximation for the hemispherical sky which contributes to the isotropic diffuse illumination of a pixel.

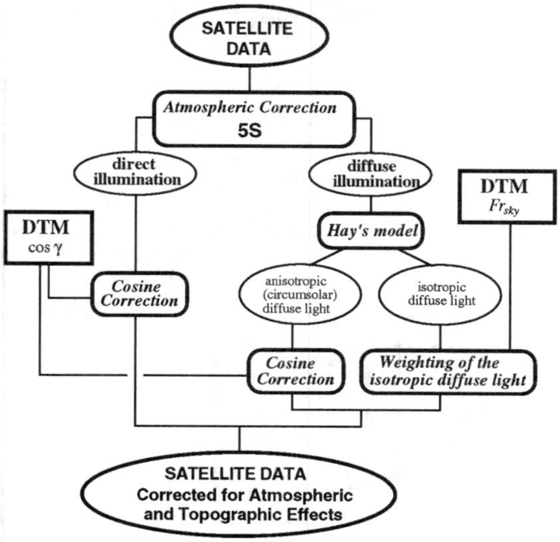

Figure 2. Simplified flow-chart of the processing scheme for combining atmospheric and topographic corrections of satellite imagery.

The flow-chart of Figure 2 summarises this processing scheme for obtaining atmospherically and topographically corrected satellite images, which then will be subject to further thematic analysis.

4.3 *Validation of the topographic corrections*

All Landsat images were processed according to the scheme which was described in the previous section. Atmospheric conditions for each scene were estimated from the satellite-measured radiance over two clear lakes in the vicinity of the Heiderscheid study site (Lac de Haute Sûre, Reservoir de Vianden). The comparison of reference signatures from a horizontal surface with uniform reflectance characteristics (Luxembourg Airport) confirms that these corrections have efficiently reduced the atmosphere-dependant radiometric differences between the images (figure 3).

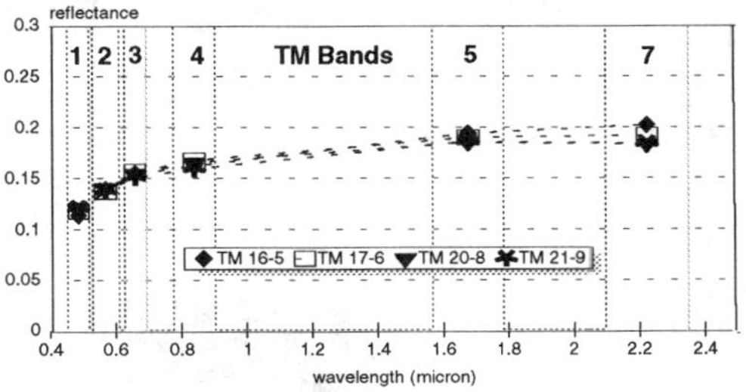

Figure 3. Spectral signatures from a horizontal surface with uniform reflectance characteristics (asphalt, Luxembourg Airport), after atmospheric correction.

Although some noise has been introduced by high frequency terrain features which were present in the digital elevation model, visual examination of the corrected images indicates obvious improvements. Compared to atmospherically corrected images, the additional correction of the topographic effect has substantially reduced the relief-induced illumination differences in all images (figure 4).

A more quantitative validation of the correction method is obtained through a regression analysis of the relationship between cos γ and the modelled scene reflectances for all forest pixels. As anticipated, significant linear relationships are obtained prior to correcting the topographic effect (figure 5). These relationships, the correlation coefficient of which increases with lower sun elevation, collapse after the slope-aspect correction. For all images, the slope of the regression line as well as the correlation between the two variables is then fluctuating around zero, thereby indicating that illumination-dependant variations of pixel intensities have been successfully removed (figures 5, 6).

Figure 4. Colour composite of the Landsat image from 20 August 1989 (band combination is 4-5-3 in red-green-blue): atmospheric correction only (left) and atmospheric/topographic correction (right). Colour plate see page 493

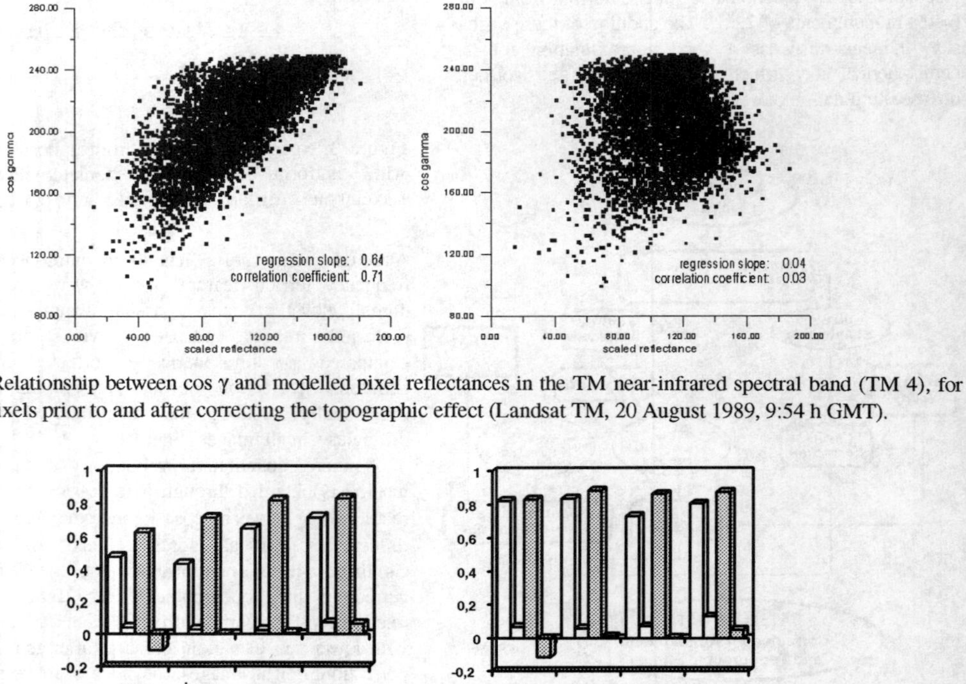

Figure 5. Relationship between cos γ and modelled pixel reflectances in the TM near-infrared spectral band (TM 4), for all forest pixels prior to and after correcting the topographic effect (Landsat TM, 20 August 1989, 9:54 h GMT).

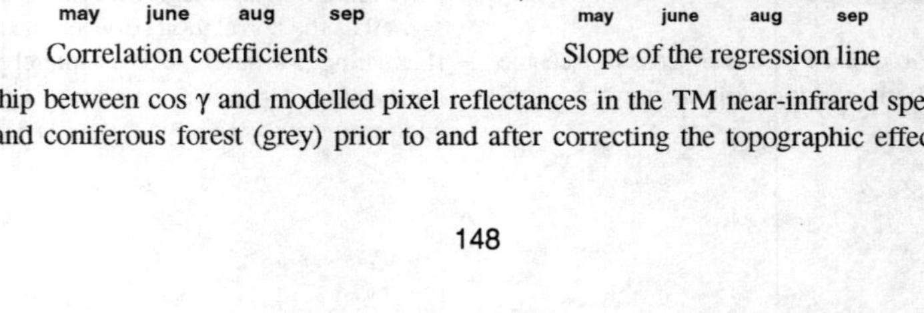

Correlation coefficients

Slope of the regression line

Figure 6. Relationship between cos γ and modelled pixel reflectances in the TM near-infrared spectral band (TM 4) for deciduous (white) and coniferous forest (grey) prior to and after correcting the topographic effect for each acquisition date

Table 2. Percentage of error in mapping deciduous and coniferous forest stands and the relative improvement through the topographic corrections

corrections	May		June		August		September	
	only atmos.	atmos. + topo.	only atmos.	atmos. + topo.	only atmos.	atmos. + topo.	only atmos.	atmos. + topo.
Euclidean Distance	22.4	21.1 (+1.3)	16.2	13.8 (+2.4)	25.9	14.3 (+11.6)	26.3	19.0 (+7.3)
Maximum Likelihood	17.6	11.8 (+5.8)	13.3	14.0 (-0.7)	25.8	14.8 (+11.0)	27.8	19.9 (+7.9)
Neural Network	15.4	11.5 (+3.9)	12.9	12.6 (+0.3)	24.6	14.5 (+10.1)	26.2	19.4 (+6.8)

An examination of the regression slopes is particularly important because these should become negative in case of overcorrecting the topographic effect. This effect, however, occurs only in the May image, which might be related to the high transmissivity of oak leaves during the sprouting period in April/May, and as such correspond to substantial deviations from lambertian reflectance characteristics. For all other dates, no indications of overcorrections are evident which may seriously question the use of lambertian reflectance behaviour as a pragmatic solution for processing optical satellite data over forested areas.

However, severe overcorrection of the topographic effect occurs when the diffuse component of the atmospheric transmittance is not correctly accounted for. This can be demonstrated by comparing a correction with scene-estimated atmospheric parameters and a simulation where the atmosphere is assumed to be unrealistically clear (only molecular, no aerosol scattering). In the latter case, the diffuse radiation component is underestimated by the model such that a too high proportion of the global irradiance is corrected by the cosine correction. This affects in particular the TM bands in the visible part of the spectrum.

It is therefore concluded that an incorrect separation into direct and diffuse radiation components rather than the bidirectional properties of the forest canopy might be the main reason for missing improvements in thematic mapping applications which have been conducted after empirical topographic corrections (i.e., cosine correction).

5 IMPROVEMENTS OF FOREST STAND MAPPING AFTER TOPOGRAPHIC CORRECTIONS

Three classification methods of increasing complexity, such as the classical euclidian minimum distance and maximum likelihood classifiers and a backpropagation neural network (Kanellopoulos et al., 1992), have been used to compare the mapping results obtained from Thematic Mapper images which were either corrected for atmospheric effects only or for both atmospheric and topographic effects. The classification experiment concentrates only on the forested parts of the Heiderscheid study site. All other land cover types have been excluded by applying a corresponding mask which was obtained from the digitised forest stand information of the "Direction des Eaux et Forêts" (see section 3). It is important to mention that an equivalent separation of forested and non forested areas can also be obtained by applying unsupervised classification techniques to the corrected images, thus reducing substantially the omission errors in the uncorrected images (i.e., forested areas on illuminated slopes not mapped as forest but grassland). This already provides an improvement of about 30 % in comparison to the uncorrected images.

Spectral signatures for several forest sub-classes were obtained for each corrected satellite image through unsupervised cluster analysis within the forested areas. The different classifiers have been trained with identical pixel samples, and the mapped sub-classes were then regrouped into deciduous and coniferous stands. About 70 % of the available ground data* have been used to assess the accuracy of the mapping results (table 2).

Our statistical evaluation of the mapping accuracy for the four dates confirms the visual impression of the produced stand maps (figure 7). In addition to the above mentioned reduction of illumination-dependant omission errors, the topographic corrections also provide substantial improvements in the separation between deciduous and coniferous stands, except for the scene from mid June. This was expected because the topographic effect is minimum at this time of the year.

Differences between the various classification methods are small except for the May image. The best results with mapping accuracy close to 90 % are achieved for the images from May (optimum separation between deciduous and coniferous stands during the sprouting period of the deciduous trees), but the results from the June (86 %) and August scenes (85 %) are also acceptable, taking into account the difficult topography and the high fragmentation of the forests. Most improvements are

* forest stands smaller than 3 pixels were discarded.

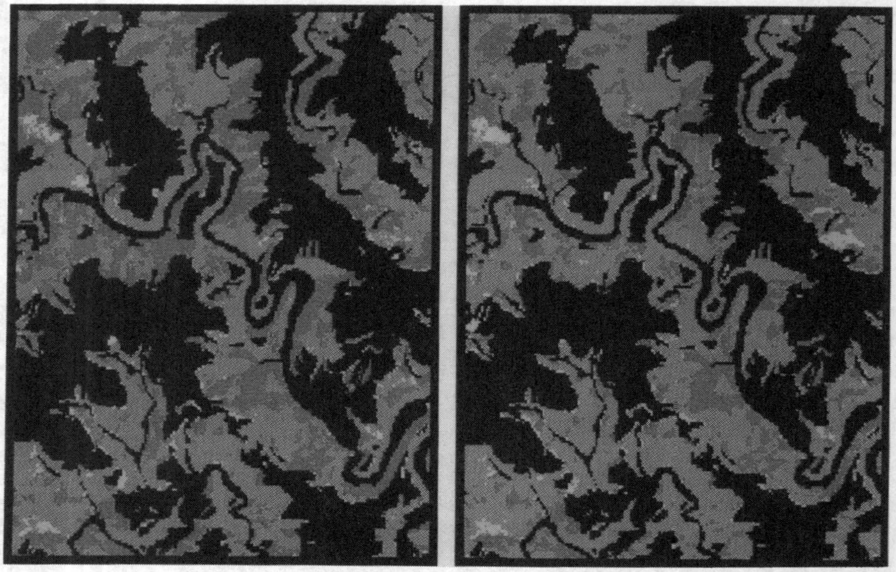

Figure 7. Comparison of forest mapping results for the Landsat TM scene from 20 August 1989, obtained with the neural network classifier before and after topographic corrections (blue - conifers, green - deciduous, yellow - storm damage). Colour plate see page 493

achieved in August and September when the sun elevation is low and leafs are matured.

The remaining errors mainly result from misclassified plantations of young conifers (spectrally not separable from deciduous trees) and from poorly illuminated and shadowed areas which, similar to the results of Kawata et al. (1988), are not fully recovered with our correction method. This explains in particular the reduced mapping accuracy for the September image. It seems that further improvements can only be achieved when the near-horizontal increase of the diffuse irradiance is also accounted for.

These results show that the correction of terrain illumination improves forest mapping also in areas of rugged topography, and it is demonstrated that robust mapping can be achieved when satellite data are pre-processed in an optimum way. None of the classifiers exhibits a particular advantage, except that, for May and June, the maximum likelihood and neural network classifiers perform better than the minimum distance approach. The reasons for this difference are not yet fully understood.

6 CONCLUSIONS

The combination of high quality terrain models, atmospheric and topographic correction modules has shown its suitability to efficiently minimise the effects of illumination differences which are present in optical satellite images of mountainous terrain. Visual examination of the corrected scenes as well as the statistical analysis of mapping results which were obtained with various classification algorithms confirm that reliable separation of deciduous and coniferous stands (accuracy of 85 to 90 %) can be achieved even in highly fragmented forest regions of complex topography.

Separation and correct estimation of direct and diffuse solar radiation components forms the basis for successful corrections of the topographic effect. Differences between the actual bidirectional reflectance properties of forest stands and their approximation by Lambertian reflectance characteristics appeared less important, at least for the Landsat TM scenes which were analysed in this study.

It is concluded that the combination of digital satellite images, topographic data and existing forest information in a GIS-environment opens promising perspectives for operational mapping applications. Further research should be done in other test areas, maybe in alpine regions, in order to understand whether the quality of commercially available DTMs (30 to 50 m resolution) is sufficient for improving forest mapping, so that the method could be applied on larger regions. Whether or not more detailed forest mapping (information on age, density or other ecological parameters) will be possible with such corrections can not be said so far. Finally, the introduction of improved models of the bidirectional behaviour of forest stands could lead to additional improvements.

It is further concluded that optimised radiometric pre-processing of optical satellite imagery is more important for obtaining robust results in forest stand mapping than the use of particular classification approaches.

7 ACKNOWLEDGEMENTS

The authors wish to thank I. Kanellopoulos (JRC) for providing the neural network classifier and his advise, and K. Fullerton (JRC) for her support in using ARC-Info and Oracle. Thanks are also due to the "Direction des Eaux et Forêts" of the Grand-Duchy Luxembourg for providing the forestry information, and to K. Itten and S. Sandmeier (Remote Sensing Laboratories, University of Zurich, Switzerland) for discussions about the topographic corrections.

8 REFERENCES

Civco, D.L., 1989, Topographic normalization of Landsat Thematic Mapper digital imagery, *Photogrammetric Engineering and Remote Sensing*, vol. 55, no. 9, 1303-1309.

Colby, J.D., 1991, Topographic normalization in rugged terrain, *Photogrammetric Engineering and Remote Sensing*, vol. 57, no. 5, 531-537.

Conese, C., M.A. Gilabert, F. Maselli and L Bottai, 1993, Topographic normalization of TM scenes through the use of an atmospheric correction method and digital terrain models, *Photogrammetric Engineering & Remote Sensing*, vol. 59, no. 12, 1745-1753.

Dozier, J. and J. Frew, 1991, Atmospheric corrections to satellite radiometric data over rugged terrain, *Remote Sensing of Environment*, 11, 191-205.

Hay, J.E., 1979, Calculation of monthly mean solar radiation for horizontal and inclined surfaces, *Solar Energy*, vol. 23, 301-307.

Hay, J.E. and D. McKay, 1985, Estimating solar irradiance on inclined surfaces: a review and assessment of methodologies, *Int. J. Solar Energy*, vol. 3, 203-240.

Hay, J.E., R. Perez and D. McKay, 1986, Technical note: Addendum and errata to the paper "Estimating solar irradiance on inclined surfaces: a review and assessment of methodologies", *Int. J. Solar Energy*, vol. 4, 321-324.

Hill, J., and D. Aifadopoulou, 1990, Radiometric comparison of Landsat-5 TM and SPOT HRV-1 sensors for the use in multiple sensor approaches, *Remote Sensing of Environment*, vol. 34, 55-70.

Hill, J. and B. Sturm, 1991, Radiometric correction of multi-temporal Thematic Mapper data for use in agricultural land-cover classification and vegetation monitoring, *Int. J. Remote Sensing*, vol. 12, no. 7, 1471-1491

Hill, J., 1991, Analysis of GER Imaging Spectrometer data acquired during the European Imaging Spectrometry Aircraft Campaign (EISAC '89), *EARSeL Advances in Remote Sensing*, vol. 1, no.1, 64-77.

Hill, J., 1993, *High precision land cover mapping and inventory. The Ardèche Experiment*, EUR 15271, (Office for Official Publications of the European Communities: Luxembourg).

Itten, K.I., P. Meyer, T. Kellenberger, R. Leu, S. Sandmeier, P. Bitter and K. Seidel, 1992, *Correction of the impact of topography and atmosphere on Landsat-TM forest mapping of alpine regions*, Remote Sensing Series, vol. 18, (University of Zurich Zurich, Switzerland).

Justice, C.O., S.W. Wharton and B.N. Holben, 1981, Application of digital terrain data to quantify and reduce the topographic effect on Landsat data, Int. J. Remote Sensing, 2, 213.

Kanellopoulos, I., A. Varfis, G.G. Wilkinson and J. Mégier, 1992, Land-cover discrimination in SPOT-HRV imagery using an artificial neural network - a 20-class experiment, *Int. J. Remote Sensing*, vol. 13 no. 5, 917-924.

Kawata, Y., S. Ueno and T. Kusaka, 1988, Radiometric correction for atmospheric and topographic effects on Landsat MSS images, *Int. J. Remote Sensing*, vol. 9, no. 4, 729-748.

Leprieur, C., J.M. Durand and J.L. Peyron, 1988, Influence of topography on forest reflectance using Landsat Thematic Mapper and digital terrain data, *Photogrammetric Engineering and Remote Sensing*, vol. 54, no. 4, 491-496.

Minnaert, M., 1941, The reciprocity principle in lunar photometry, *Astrophys. J.*, vol. 93, 403-411.

Proy, C., D. Tanré and P.Y. Deschamps, 1989, Evaluation of topographic effects in remotely sensed data, *Remote Sensing of Environment*, 30, 21-32.

Sobolev, V.V., 1963, *A treatise on radiative transfer*, (Van Nostrand: Princeton).

Tanré, D., M. Herman, P.Y. Deschamps and A. de Leffe, 1979, Atmospheric modelling of the background contribution upon space measurements of ground reflectance, including bi-directional properties, *Applied Optics*, 18, 3587-3594.

Tanré, D., C. Deroo, P. Duhaut, M. Herman and J.J. Morcrette, 1990, Description of a computer code to simulate the satellite signal in the solar spectrum: the 5S code, *Int. J. Remote Sensing*, vol. 11, no. 4, 659-668.

Temps, R.C. and K.L. Coulson, 1977, Solar radiation incident upon slopes of different orientation, *Solar Energy*, 19, 179-216.

Woodham, R.J. and M.H. Gray, 1987, An analytic method for radiometric correction of satellite multispectral scanner data, *IEEE Trans. on Geoscience and Remote Sensing*, vol. GE-22, no. 3, 258-271.

Cartographic aspects

Sensors and Environmental Applications of Remote Sensing, Askne (ed.) © 1995 Balkema, Rotterdam, ISBN 90 5410 524 0

Maps and remote sensing in mountainous areas of Third World countries: Example of the upper Mustang district in Nepal

R. Kostka & W. Krämer
Graz University of Technology, Austria

ABSTRACT: In Third World countries there is a big gap between scientific investigations on remote sensing sensors, their development, methodological approaches, data processing and price situation on the one hand and the application to well-defined topics of environmental interest on the other.
Especially in remote mountain areas many difficulties arise when preparing maps on the basis of remote sensing data. Despite these problems image-based maps have many advantages over pure line maps. The Mustang district in northern Nepal will be considered for illustration. Formerly a restricted area, it has lately been confronted with development aid and tourism, and for the past few years NGOs and PPs have also taken interest in this region. Thus, economic development and environmental protection were pushed to the fore. The importance of base maps and thematic information for these purposes has led to the preparation of a whole range of map products for this area. Two special map sheets - SPOT image and a LFC image-based map - are described in more detail and are compared with existing line maps.
Apart from other advantages it can be shown that the confidence in image-based map products is higher than in line maps prepared on the basis of a "quick-and dirty" approach.

1 REMOTE MOUNTAIN AREAS AND TOPOGRAPHIC/THEMATIC MAPPING

There is hardly a region on our planet that has escaped man's attempts at exploration. As a result of his activities there are no white spots (unexplored or unknown areas) to be found on the world map, and basic information is available even on remote and hardly accessible areas of the Earth.

Cartographers, however, are still confronted with problems which mainly concern hardly accessible mountainous border regions in Third World countries. The most frequent shortcomings are the following:

- information is not accurate enough (not only concerning geometry) or simply is not correct,
- information (esp. on developing and hazard regions) is not up-to-date,
- existing information is not available in the form of maps or map derivates.

Therefore, there is a continuing demand for the mapping of topographic/thematic information in order to support human activities in an environmental context. This particularly applies to special areas of land use, for which a wide range of remote sensing products and methods is available on the market.

It has to be noted, however, that in the poorer Third World countries technical difficulties are accompanied by organizational and financial problems. This is often due to the way remote sensing cartography, a useful tool for solving these problems, is conducted on an international level.

Very often the focus is on building up a reputation in the scientific community and on carrying out projects in the framework of development aid programmes whose usefulness can be questioned in many cases. The production of cartographic products which meet the regional demands, however, often seems to be a matter of minor importance.

The existing maps of the Mustang district in northern Nepal will be considered for illustration.

2 MUSTANG DISTRICT AND RELEVANT INTERESTS IN LANDUSE

Until recently, the Mustang district in the north of Nepal was a restricted area. This was due to the political/military situation of this Himalayan region, which extends well into Tibet/People's

Republic of China and provides easy passage to this neighboring area.

Despite this situation, maps were produced in the framework of development projects, which also covered the Mustang district.

In 1985 the Land Utilization Map was published at the original scale of 1:50.000. The index of the map sheets of the Mustang district can be seen in Fig. 1. Apart from topographical data this map also contains significant thematic information, which is divided into a land use and a forestry legend. A reduction of the map section showing the irrigated

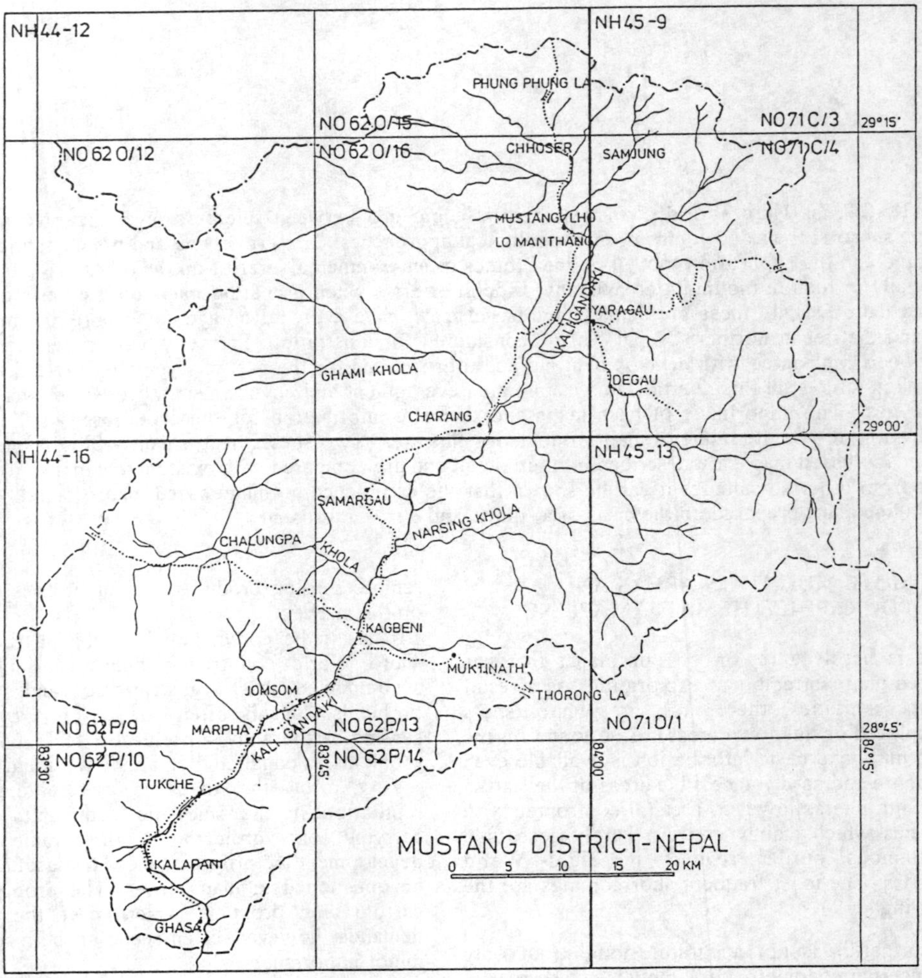

Fig. 1: Mustang District - Index of Map Sheets

1. Land Utilization Map 1:50,000, Mustang District, Nepal. Western Development Region, sheet No. 620/12.......No. 71D/1
Prepared under the Canadian Assistance Program to Nepal by the Topographical Survey Branch, Kathmandu and Kenting Earth Sciences Ltd., Ottawa; printed and published by HMG Survey Department, Topographical Survey Branch, Kathmandu, Nepal 1985

2. Planimetric Map of Satellite Images 1:250.000 Nepal, sheets NH 44-12, NH 45-9, NH 44-16
Compilation, design and supervision: The World Bank, Washington D.C., USA; digital image processing, cartography and printing: IFG - Institute for Applied Geosciences, Offenbach, Germany; published by National Remote Sensing Center, Kathmandu, Nepal, 1986

Fig. 2: Land Utilization Map of the surroundings of Lo Manthang in northern Mustang district: Reduction of the original black-and-white map from scale 1:50,000 to 1:125,000. Topographic information consists of details such as rivers, streams, villages and tracks, spot heights and boundaries. Thematic information is divided into a land use and a forestry part. The most important features are: valley cultivation (F), grazing land (G), rock (R) and sand/gravel/boulders (B).

fields is given in Fig.2.

In 1986 a 1:250,000-scale satellite image map of Nepal was published, which also included the Mustang district. The information content consists of NASA-LANDSAT MSS images with collateral information on boundaries, villages and trails. The index of these map sheets covering the Mustang district is also given in Fig.1.

When the Nepalese government opened Mustang in 1992, the region experienced profound changes, which were marked by outside influences, such as tourism, research and development. Many activities, however, were also aimed at improving the living conditions of the native population. These activities were mainly focussed on the promotion of tourism, the utilization of resources and the search for environmentally sound sources of income, as was demonstrated by the non-governmental organisation (NGO) of ACAP (Annapurna Conservation Area Project).

A region with so few published research results necessarily attracts a number of scientists, who carry out relevant studies in their function as project people (PPs). At the beginning of 1992, for instance, the Nepalese Department of Archeology and the German Research Council initiated a research programme on settlement processes and the formation of states in the High Himalaya. Finally, organized and controlled tourism is to serve as a stable and long-term source of income.

It has to be noted that cartographic information is essential for finding answers to these questions of environmental interest and that neither the line nor the image maps mentioned above can provide the required basic information.

3 RECENT MAP EXAMPLES (1993) OF THE MUSTANG DISTRICT

This demand has led and is still leading to independent mapping projects. In 1993 two map sheets of the Mustang district were published for touristic purposes:

- Trekking Map Mustang, the Salt Route 1:70,000, published by Travel/Trekking Int. Pte. Ltd. Kathmandu, Nepal,
- Mustang - Route from Jomosom to Lo Manthang 1:125,000, published by NEPA Publication, Kathmandu, Nepal.

Fig. 3 shows a black-and-white section of the latter map sheet, which was originally published in a four-colour version. It is not adequate for orientation purposes, as neither the water bodies nor the trails are represented satisfactorily. Apart

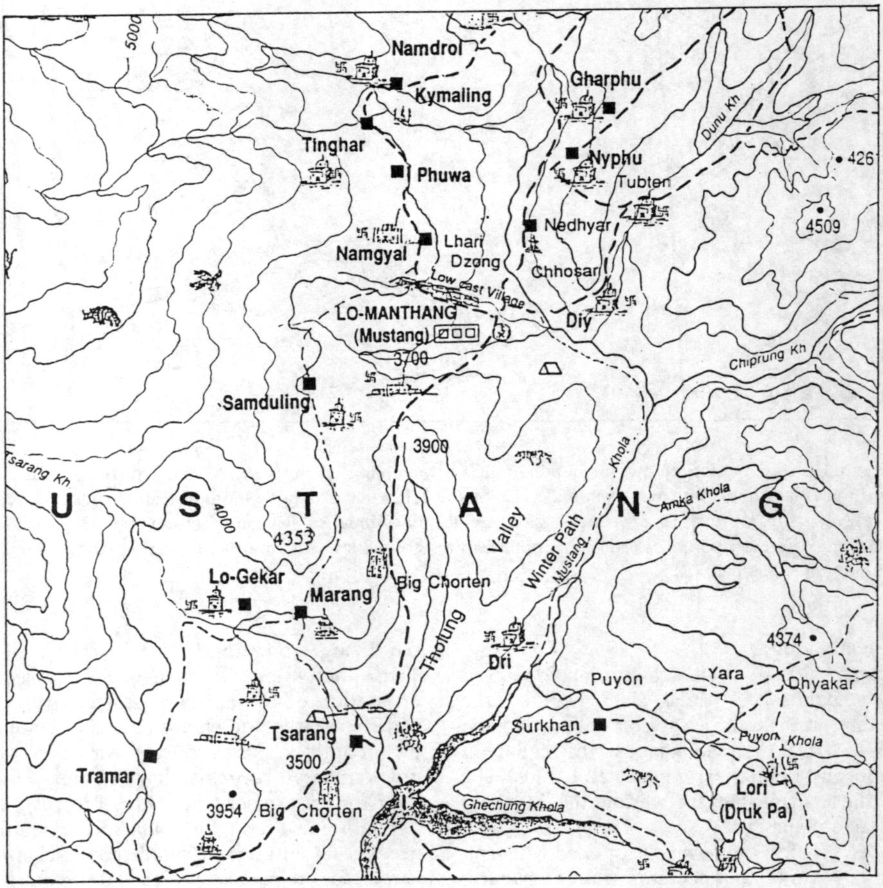

Fig. 3: Black-and-white section of the Trekking Map Mustang - Route from Jomosom to Lo Manthang 1:125,000; Paolo Gondoni, 1993 published by NEPA Publication Kathmandu, Nepal

from that, the villages cannot be located precisely, and the contour intervals of 250 meters are not very informative, either.

Another recent sketch map, which also dates from 1993, may be considered for illustration: It is a small general map designed for the representation of activities carried out in the framework of the Upper Mustang Conservation and Development Project between 1992 and 1993 by ACAP (Fig. 4). Although the aim was not to produce a map for general orientation purposes, important information, such as accessibility of or distances between individual locations, can hardly be derived from this map product.

Investigations of environmental change (e.g. vegetation) and comparative studies (e.g. North - South Mustang) will be based on the Land Utilization Map.

Comparisons show (Fig. 5) that the existing maps do not render satisfactory results. Villages and trails can only partly be located, and the vegetation areas cannot be studied efficiently.
Thus, these map sheets cannot be considered a reliable basis for serious studies and must therefore be classified as "quick-and-dirty" products.

4 SPACEBORNE IMAGES AS BASIC INFORMATION FOR CARTOGRAPHY

The above-mentioned research project on settlement structures and the formation of states in

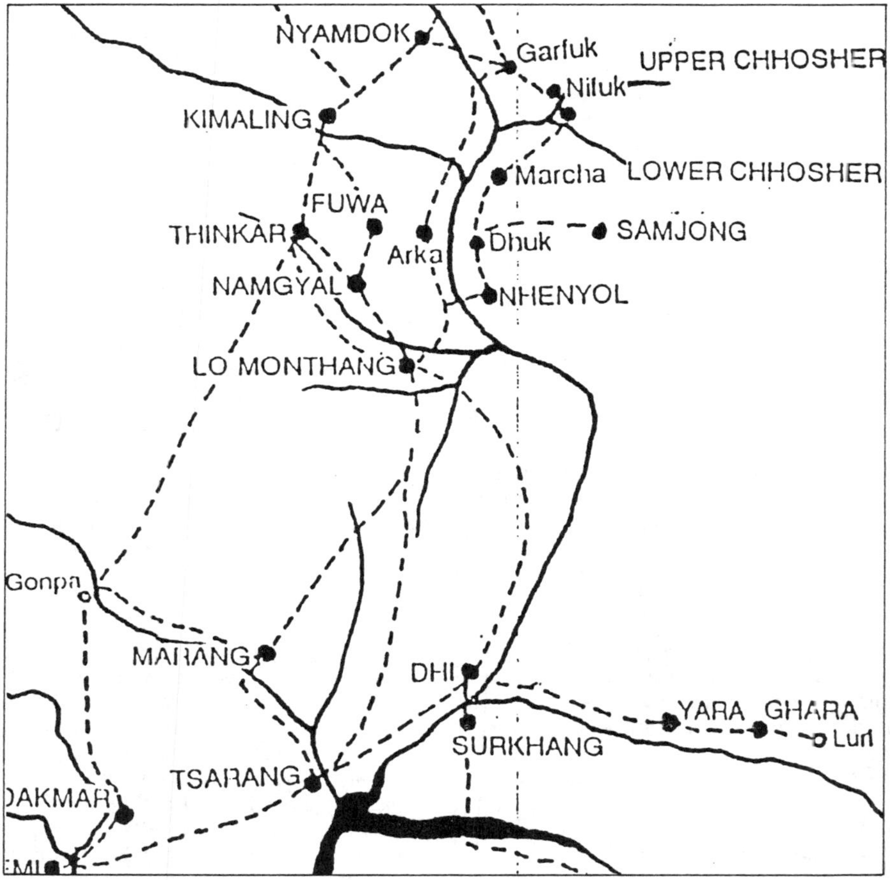

Fig. 4: Sketch map of the Upper Mustang Conservation and Development Project: Enlargement of the section around Lo Manthang. Localisation of activities in the fiscal year 1992/93 through ACAP (Annapurna Conservation Area Project). The scale was changed to 1:125.000 in order to facilitate comparison studies. Drainage patterns, trails and the location of villages should correspond with the map shown in Fig. 3.

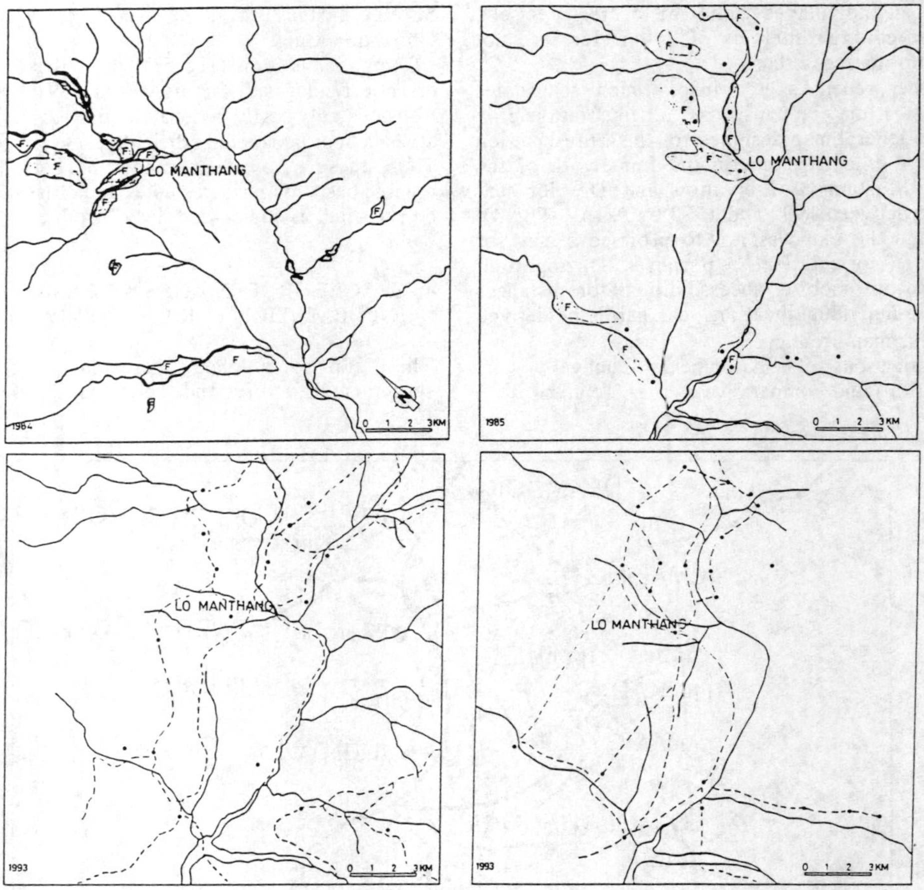

Fig. 5: Map sections of the area around Lo Manthang for comparative studies:
The irrigated fields of the Land Utilization Map (F) from 1985 can be compared with interpretation results of vegetation in the LFC-image taken only one year before. The trails in the surroundings of Lo Manthang are given as broken lines in the map published in 1993. The results of comparison are insatisfactory in both cases, it is not possible to draw reliable conclusions from comparisons.

the Tibetan Himalayas requires independent cartographic studies. As these studies cannot be considered in detail here, two special satellite image products representative of the research work will be described.

The cartographic representation of the Kali Gandaki valley at a scale of 1:50,000/1:125,000 is based on a SPOT satellite image, using a slide of the SPOT-XS image from May 1990 (Fig.6).

Image information for the region around the administrative centre of Lo Manthang was taken from an LFC image from October 1984, which provided data for a cartographic representation at a scale of 1:50,000/1:125,000.

The images were digitized by means of a Vexcel Scanner VX3000 with a resolution of 100 nm, i.e. 2540 dpi. Image processing took place on a Macintosh II CX, Motorola 68030 Processor using the Adobe Photoshop D1-2.5 image processing software. Various filters were used to improve image quality. The output device for the black-and-white image was a HP4 laser printer with a maximum resolution of 600 dpi and a raster frequency of 106 lin/inch.

The work has not been completed yet, the image products available at this stage, however, can already be used as a basis for conclusions and comparative studies.

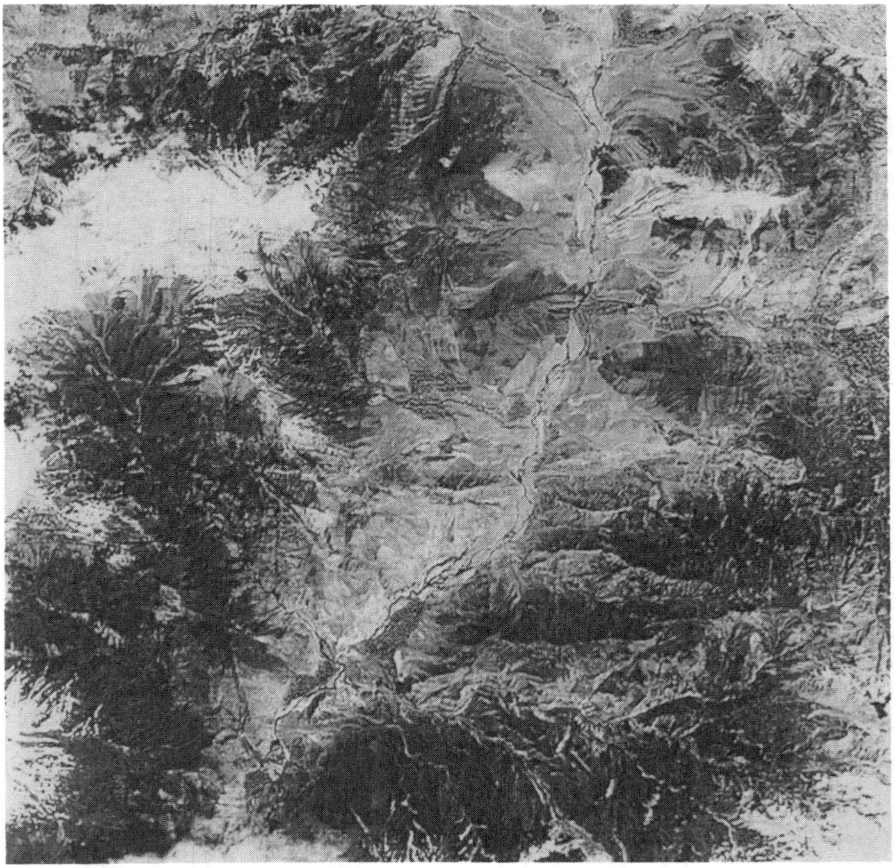

Fig. 6: Black-and-white section of an enlargement of a SPOT2-XS image taken in May 1990; It shows the Kali Gandaki valley in the southern Mustang district. The groundplan is represented in much more detail than in the existing map sheets. Visual interpretation of the relief, however, is not or only partly possible.

The SPOT image (Fig. 6) shows the groundplan, such as river courses or irrigated fields, in much more detail than the above-mentioned line maps. However, it also contains a wealth of additional thematic information which is not essential for the investigation, e.g. vegetation, geology or morphology. This information covers the relief to such an extent that the terrain relief can hardly be distinguished any more: a well-known effect when using remote sensing images in mountain cartography.

The LFC image (Fig. 7) shows fewer details and emphasizes terrain relief. Therefore it is more suitable as a basic map for incorporating thematic information, such as settlements, buildings and trails, from external data sources. In order to keep the orientation of the satellite image north, the image information is represented as as negative, thus rendering the relief more distinct.

When comparing the LFC image with the map examples mentioned earlier on, it becomes obvious that the trails in these maps were drawn without any basic topographic information and that the location of the villages does not correspond with reality.

5 CONCLUDING REMARKS

Pure image maps based on space imagery cannot provide the information content required by NGOs and PPs for their investigations. External data sets are necessary, and cartographic plotting can be a valuable tool for visualizing image information.

Fig. 7: Black-and-white section of an enlargement of the panchromatic LFC-image taken in October 1984; It shows the area round Lo Manthang in the northern Mustang District. Only few details in the groundplan can be interpreted, therefore the relief is emphasized. It is represented as a negative, so as to give the visually correct impression when orienting the image north.

The examples shall demonstrate that image-based maps are nonetheless more reliable than line maps prepared on the basis of a "quick-and-dirty" approach. By systemizing data collection and image processing methods, the potential of remote sensing can already be studied in test areas where excellent reference data are available, and can subsequently be utilized for the mapping of remote mountain regions.

Satellite image products would be of much more use to Third World countries, if the image material, the method of processing and the collection of additional information were coordinated more efficiently. Furthermore, these parameters should be better adapted to the respective task, and more attention should be paid to the non-technical parameters characteristic of the respective countries.

Sensors and Environmental Applications of Remote Sensing, Askne (ed.) © 1995 Balkema, Rotterdam, ISBN 90 5410 524 0

Multisource data integration with an integrated and unified geometric modelling

Thierry Toutin
Canada Centre for Remote Sensing, Ottawa, Ont., Canada

ABSTRACT: This article presents a method to generate ortho-images with Digital Elevation Model (DEM) and a few Ground Control Points (4-8). The method has been integrated to take into account the geometric distortions of the full geometry of viewing (sensor - platform - Earth), and unified to process images from different sensors (VIR and SAR) on various platforms (airborne and spaceborne). Results from eight types of images show an absolute accuracy of 1/3 of pixel for VIR satellite images and one to two pixels for the other images (SAR and airborne), and a relative accuracy of one pixel between the image. Integration (mosaics and colour composites) of these ortho-images with the road network overlaid confirms the relative and absolute accuracies.

1 INTRODUCTION

Cartographic and thematic applications in remote sensing use and process multi-source data (multi-date images from different sensors, topographic and thematic maps, raster and vector data) to take advantage of their complementarity (visible vs radar; airborne vs spaceborne; raster vs vector). The superposition of remote sensing data and their integration with vector data bases require precise and robust geometric processing, as well as user friendly systems to meet these needs and standards. To obtain pixel accuracy or better, the polynomial, spline or kriging functions generally used, cannot meet such results; mainly with images which have large distortions due to off-nadir viewing, to fine resolution, to airborne motion, etc. Furthermore, the methods do not take into account terrain elevation, which is a major factor in these distortions, and only correct locally at the ground control points (GCPs).

Considerable international research has been carried out to develop robust and rigorous mathematical models describing the acquisition geometry related to different types of images (visible and SAR) and of platforms (spaceborne and airborne): for visible satellite images (Masson d'Autume, 1980; Sawada et al., 1981; Friedmann et al., 1983; Guichard, 1983; Toutin, 1983; Salamonowicz, 1986; Gugan, 1987; Konecny et al. 1987; Kratky, 1987; Paderes et al., 1989; Westin, 1990); for SAR satellite images (Rosenfield, 1968; Gracie et al., 1970; Leberl, 1978, Wong et al., 1981; Curlander, 1982; Naraghi et al., 1983; Guindon and Adair, 1992; Toutin and Carbonneau, 1992a); for visible airborne images (Gibson 1984; Ebner and Muller, 1986; Hoffmann, and Muller, 1988); and for airborne SAR images (LaPrade, 1963, Rosenfield, 1968; Gracie et al., 1970; Derenyi, 1970; Leberl, 1972; Konecny and Schurr, 1984; Toutin et al., 1992) and many others.

Some of this research reported promising results for one specific image on one specific platform. To date, no method and system can geometrically correct precisely different images (visible and radar) on different platforms (spaceborne and airborne).

This paper will present a more general approach using an integrated and unified method and a user-friendly system to geometrically correct, raw spaceborne and airborne, visible and SAR images with DEM. The mathematical model is derived from the collinearity equations with photogrammetric method (Toutin, 1983). It integrates the complete viewing geometry with the position and the velocity of the platform, with the parameters of the sensor and with the characteristics of the Earth and of the cartographic projection.

Quantitative and qualitative results will be presented with eight different types of images: MOS-MESSR, Landsat-TM, SPOT-MLA/PLA, ERS-SAR, SEASAT-SAR, airborne SAR and airborne CCD sensor MEIS). The 1:50 000 digital topographic map is used to verify the accuracy.

2 INTEGRATED AND UNIFIED GEOMETRIC MODELLING

This geometric correction modelling was originally developed for SPOT-HRV (Guichard, 1983; Toutin, 1983) and tested on Landsat-TM (O'Brien and Handy, 1991), on MOS-MESSR and SEASAT-SAR (Toutin and Carbonneau, 1992a) and on airborne-SAR (Toutin et al., 1992). This modelling has been now adapted in an operational environment at the Canada Centre for Remote Sensing (CCRS) as an integrated and unified geometric modelling, to geometrically process most of the commonly used remote sensing images.

The geometric modelling is integrated because in the final equations which represents the collinearity condition, it takes into account, the different distortions relatives to the global geometry of viewing i.e.:

- the distortions relative to the platform (position, velocity, orientation);
- the distortions relative to the sensor (orientation angles, instantaneous field of view, line integration time);
- the distortions relative to the Earth (geoid-ellipsoid including elevation); and
- the distortions relative to the cartographic projection (ellipsoid - cartographic plane).

The integration of the different distortions and the derivation of the equations are out of the scope of this paper, and described earlier for the SPOT case by Toutin (1983), and for the SAR case by Toutin et al. (1992).

The geometric modelling is unified because it has been adapted to process indiscriminately spaceborne or airborne data, and visible infrared (VIR) and radar data.

3 PROCESSING STEPS

The key processing steps of the geometric correction are:

1. acquisition and pre-processing of the remote sensing data (images and if they exist, ephemerid and attitude);
2. acquisition and management of ground coordinates (cartographic and image);
3. computation of the geometric correction model(s) using few GCPs (4-8); and
4. ortho-image generation and mosaicking with a DEM.

In relation to these four steps, and the geometric modelling, the software SRIT <<Système de Rectification des Images de Télédétection>>, has been developed. This was done originally for satellite VIR data (Toutin et Carbonneau, 1992b), and later adapted to process airborne and satellite SAR data.

Different modules have been developed for each step of the processing, previously mentioned:

1. to read the images, and (if they exist) the ephemeris and the attitude. As the data organization on the tape is different for each sensor and platform, each one needs a specific module. In addition to the image files, there is a parameter file which describes the geometry of viewing and an attitude file (if the attitude exist: e.g. for SPOT, airborne MEIS). If a new sensor and platform is added to the method, a new module has to be developed only in this step;
2. to interactively visualize, edit and manage the reference points (control, planimetric, altimetric, homologous, check). This module enables the user to find the best set of data points and the best configuration of images (if more than one image is being processed), minimizing the residual values on the

ground control points and/or the errors on check points (if there are any);

3. to compute the geometric correction models. An interactive least square adjustment process is used to solve the equations of observations. As more than one image can be solved at a time, each group of equations is weighted as a function of the accuracy of the image data and the cartographic data.

Some parameters which are specific to the airborne images or to the SAR images are not taken into account for the other images. The SRIT system decides for each image how many and which parameters have to be resolved in this step 3, according to the image acquired in step 1; and

4. to generate ortho-images and mosaics; the map projection system can be any conformal projection (UTM, MTM, Lambert, etc.). The user can use a DEM or a mean reference altitude, depending on the variation in relief and the viewing angles at which the image was acquired.

The system is user friendly because:

- it is the same to operate for any image;
- throughout the processing steps, the user does not have to worry about the type of images. The relevant information on the images(s) are taken directly from the tape; and they are transferred to the other steps;
- the user can process and tie one, two or more images together in steps 2 and 3, giving a better relative accuracy between the different ortho-images.
- other satellite and airborne data can be added easily in the future; and
- the user can control and select the data and the method (GCP and DEM accuracies, image viewing angle, ...) as a function of the accuracy required for his application.

4 DATA SET

The topographic data is the Sherbrooke Data Set for the topographic applications of remote sensing (Lassere et Lemieux, 1990). The study area is made up of two half of map sheets of the province of Quebec (Canada), thus representing a land coverage of approximately 26 x 40 km, with an altitude variation of about 350 metres. The data used in this experiment are:

- ground control points which have been obtained using a STK-1 stereocomparator. The root mean square error of the cartographic coordinates are better than three metres. These points are mainly the intersections of expressways, highways, roads, streets or railroads.
- a DEM with a 50-m grid, generated from the 1:50 000 contour lines (10-m intervals). The accuracy is in the order of five metres
- the vector data of the digital topographic map. All the elements are positional data, as observed on the surface of the Earth in X, Y and Z coordinates and without movement of the element due to cartographic generalization. The accuracy is in the order of five meters.

The remote sensing data are summarized in Table 1. They consist of eight images:

- six spaceborne (MOS, Landsat, SPOT, SEASAT, ERS) and two airborne (Falcon, Convair) ; or
- five VIR (MESSR, TM, MLA, PLA, MEIS) and three SAR.

5 RESULTS

5.1 Quantitative results

The number of points acquired on each image is different because of the difference in target shape and appearance due to spectral response. The image coordinate accuracies are 1/3 to 1/2 of pixel for VIR data and one to two pixels for SAR data.

Table 2 gives the full results of two modelling tests with the root mean square (RMS) and maximum residuals on the GCPs, and the RMS, maximum and bias errors on the check points (CP). The residuals reflect the modelling accuracy, and the errors the restitution accuracy. As the plotting error on the CP is included, the ortho-image accuracy is better than the restitution accuracy.

In the reverse to polynomial methods, which are sensitive to the number and spatial distribution of the GCPs and their elevation, this model is not affected by these factors because it respects the

Table 1. Characteristic of the image data set.

Platform	Sensor	Level of Processing	Acquisition Date	Supplemental Data	Ground Coverage (km)	Pixel Size (m)	Viewing Angles (deg)
Convair	SAR-C	slant range	90/09/11	yes	16 x 40	4 x 4.31	45 / 76
Falcon	CCD-MEIS	raw	90/06/11	yes	6 x 40	5 x 4	-20 / +20
SPOT	HRV-PLA	raw	87/06/20	yes	70 x 60	12 x 10	27.2/31.3
SEASAT	SAR-L	ground range	78/09/17	no	100 x 100	12.5 x 12.5	19 / 25
ERS-1	SAR-C	ground range	92/06/16	no	100 x 100	12.5 x 12.5	21 / 25
SPOT	HRV-MLA	raw	87/08/02	yes	60 x 60	22 x 20	-10.7 / -6.6
LANDSAT	TM	bulk	89/08/27	yes	88 x 94	30 x 30	0 / 7
MOS	MESSR	raw	89/06/22	no	100 x 100	50 x 50	-3 / +3

Table 2. Results of the geometric correction.

Test	Image		MESSR		TM		MLA		SEASAT-SAR		ERS1-SAR		PLA		MEIS		SAR	
1	GCP / CP		89	0	95	0	154	0	24	0	25	0	139	0	69	0	44	0
	Res.		X	Y	X	Y	X	Y	X	Y	X	Y	X	Y	X	Y	X	Y
	GCP	RMS	18.3	17.2	13.1	9.5	7.1	7.0	13.7	11.7	22.3	21.1	3.4	3.1	6.1	5.3	6.6.	5.6
		MAX	-45.1	37.3	-32.8	26.7	-20.9	-18.8	-30.8	-22.4	-42.5	-50.8	-8.7	-9.0	15.4	-12.4	13.1	-16.9
2	GCP / CP		12	77	17	72	19	135	10	14	10	15	19	120	12	57	12	32
	Res / Error		X	Y	X	Y	X	Y	X	Y	X	Y	X	Y	X	Y	X	Y
	GCP	RMS	16.2	13.2	10.7	6.6	6.3	7.0	15.1	7.8	12.2	25.0	2.7	3.1	4.7	9.0	6.3	4.9
		MAX	-38.4	25.2	-17.6	-16.0	12.1	16.3	-30.8	-14.2	23.1	32.3	-5.0	7.6	-8.3	-17.5	-10.2	-9.9
	CP	**RMS**	**23.4**	**20.5**	**12.5**	**10.1**	**8.0**	**8.2**	**18.8**	**14.7**	**39.9**	**20.1**	**3.8**	**3.4**	**7.3**	**7.5**	**10.9**	**6.0**
		MAX	-55.8	45.9	23.8	22.9	23.7	-23.6	35.5	-29.3	-75.8	32.3	-8.7	-9.9	-15.0	-18.1	-24.4	-20.2
		BIAIS	-9.2	-3.9	-1.9	4.6	3.5	-4.5	7.5	-1.4	-1.7	5.3	1.4	-0.1	0.4	0.0	0.3	-0.3

complete geometry of viewing, including the terrain given that there is no extrapolation in planimetry and also in altimetry (Toutin and Carbonneau, 1989). The same research has shown that eight GCPs is a good compromise to get the same accuracy, as shown in Table 2, because the geometric modelling corrects globally the image, when the polynomial modelling corrects locally.

In the first test, as the RMS residuals are on the same order of magnitude than the plotting error (1/3 - 1/2 pixel for satellite VIR images and 1 - 2 pixels for the others) and as the maximum errors is less than three times the RMS residuals (this remark is also valid for the second test on the GCPs and CPs), the mathematical model describes properly the viewing geometry of images acquisition, but it is also stable and robust for the full image without generating local errors. Furthermore, as the bias errors on the CP for the second tests are small, the mathematical model does not introduce trends or systematic errors. These results give a good level of confidence for the geometric correction and rectification of all these types of images with different terrain and geometric configurations.

In summary, the general results are:

- one third of pixel accuracy for the satellite VIR data; and
- one to two pixel accuracy for the other data (SAR and airborne)

For the MOS-MESSR image, the results are half pixel because no ephemeris and attitude were available; and in optical system, angles are one of the important keys. With attitude, the same one-third of pixel accuracy could be achieved.

For the SAR satellite images, 12.5 metres is the pixel spacing, but the pixel resolution is 25-30 metres. This results in one-half to one pixel resolution accuracy to keep consistency when compared with VIR images accuracy. Furthermore, no ephemeris data was available for the SAR images.

For the airborne images, the difference in results may be accounted for the airborne motion which is less stable than the satellite. High frequency distortion cannot be corrected without precise attitude and position data.

5.2 Qualitative results

Finally, with the last SRIT module, ortho-images have been generated for the different images, using a DEM. That enables more complex products as mosaics or colour composite images to be created.

Figure 1 is a mosaic of the eight ortho-images previously described. The image is 39 x 29 km large with a common 10-m pixel size. From west to east, or north to south, there are the airborne SAR, airborne CCD-MEIS sensor, SPOT-PLA, ERS-SAR, SEASAT-SAR, SPOT-MLA, Landsat-TM and MOS-MESSR. The mosaic becomes fuzzy as moving from a 4-m (airborne data) to a 50-m (MOS-MESSR) pixel size, resampled at 10 metres.

Figure 2 is an enlargement (17 x 15 km) in the North-West part of the Figure 1 with the road network overlaid.

Figure 3 is also an enlargement (10 x 10 km) in the North-West of a colour composite ortho-image of SPOT-PLA and two airborne SAR from opposite-side flight lines using the intensity-hue-saturation coding. The roads network has been overlaid. The pixel size of this image is five metres.

The qualitative viewing of these ortho-mosaics and colour composite images confirms the quantitative results of 1/3 to 1/2 pixel accuracy for VIR satellite images and one to two pixel accuracy for the other images. In some areas, maximum errors appear (20 m for airborne CCD sensor and SAR; 50-60 m for spaceborne SAR) in accordance with the statistics of Table 2. There is no strong local distortion, greater than the tolerance. That is one of the qualities of this geometric modelling which corrects the images globally. The topographic file has an accuracy of five meters, which is in the same order of magnitude than errors on some images (SPOT and airborne). Therefore, the topographic file cannot be considered as an absolute reference, when compared to SPOT data or airborne data.

Since the radiometric quality of the ERS-SAR was poor, the integration

Figure 1. Ortho-mosaic of eight spaceborne and airborne images (39 x 29 km; 10 m pixel size)

Figure 2. Enlargement of the ortho-mosaic (17 x 15 km; 10 m pixel size) Colour plate see page 494

Figure 3. Enlargement (10 x 10 km; 5 m pixel size) of an IHS colour composite SPOT-P and airborne SAR Colour plate see page 495

of ERS and Landsat data in the Charlevoix area (Quebec, Canada) along the St. Lawrence River was done. This integration was carried out with the Université du Québec à Montréal for structural geomorphological application (DeSève et al., 1994). The altitude variation is in the order of 1,000 metres. The same geometric processing and SRIT system with DEM was used for the two images. Figure 4 (67 x 52 km; 30-m pixel size) is the result of a colour composite of TM7-TM4-SAR with RGB coding, respectively. The relative and absolute accuracies have been checked to be 30 metres. Visual evaluation on some cartographic features (power line clearcuts, lakes, etc.) confirms the relative accuracy of the integration.

6 CONCLUSIONS

This paper has shown a method and a system to geometrically correct with DEM raw images. The mathematical model is based on the collinearity

Figure 4. Integration of a RGB colour composite Landsat-TM (7,4) and ERS-1-SAR (67 x 52 km; 30 m pixel size) Colour plate see page 496

conditions and represents the physical realities of the full viewing geometry, platform- sensor-Earth.

The method was already tested at CCRS on different images (SPOT-HVR, Landsat-TM, MOS-MESSR, SEASAT-SAR, airborne VIR and SAR) in the past years. The method and the mathematical model are now integrated and unified to process images from different sensors (VIR and SAR) on different platforms (spaceborne and airborne).

The operational SRIT software, based on this method, enables the generation with DEM of ortho-images and mosaics of the previously mentioned images. The user does not have to consider the type of images, except for the GCP acquisition step, because the information relative to the platform and the sensor are directly extracted from the tape and transferred to the other processing steps. One of the major advantages of the method is that the users need to operate only one system.

Processing different images MOS-MESSR, Landsat-TM, SPOT-PLA/MLA, SEASAT-SAR, ERS-SAR, airborne SAR and CCD sensors, this method has achieved accuracies of:

- 1/3 of pixel for VIR satellite images; and
- 1-2 pixels for the other images.

The difference in the result can be accounted for:

- a better definition and quality of the GCPs on VIR images; and
- a better knowledge of the spaceborne path than the airborne path.

Using a DEM, different ortho-images and their integration (mosaics, colour composites) have been generated to qualitatively check the relative accuracy between the different ortho-images. The road network has been overlaid to check the absolute accuracy. It confirms the quantitative results on check points, but for some data (SPOT and airborne), the road network cannot be used as an absolute reference, because it has an accuracy of five metres, in the same order of magnitude the SPOT ortho-image accuracies (3 m for PLA; 7 m for MLA) or to the airborne ortho-image accuracies (7.5 m for CCD-MEIS; 6 - 10 m for SAR).

7 ACKNOWLEDGEMENTS

The author would like to thank the Canada Centre for Geomatics for providing the digital topographic data and for their participation in the SRIT development. He also thanks Ms. Danielle DeSève and Dr. Robert Desjardins of the <<Departement de géographie de l'Université du Québec à Montréal>> for providing the ERS-SAR and Landsat-TM of Charlevoix (Canada), and for their participation in the data processing.

8 REFERENCES

Curlander, J.C., 1982, Location of Spaceborne SAR Imagery, IEEE Transaction of Geoscience and Remote Sensing, GE-22 (2), pp: 106-112.

Derenyi, E.E., 1970, An Exploratory Investigation into the Relative Orientation of Continuous Strip Imagery, Research Report No. 8, University of New Brunswick, Canada.

DeSève, D., Desjardins R. et Toutin, Th., 1994, Contribution des données radar d'ERS-1 dans l'appréhension de la géomorphologie structurale de l'Astroblème de Charlevoix, submitted to the Canadian Journal for Remote Sensing.

Ebner H., and Muller F., 1986, Processing of Digital Three Line Imagery Using a Generalized Model for Combined Point Determination, International Archives of Photogrammetry and Remote Sensing, 26(311), pp. 212-222.

Friedmann, D.E., Friedel, J.P., Magnussen, K.L., Kwok, R and Richardson, 1983, Multiple Scene Precision Rectification of Space-borne Imagery with very few Control Points, Photogrammetric Engineering and Remote Sensing, 49(12), pp 1657-1667.

Gibson, J., 1984, Processing Stereo Imagery from Line Imagers, 9th Canadian Symposium on Remote Sensing, (Canada) pp. 471-488.

Gracie, G., Bricker, J.W.,

Brewer, R.K., and Johnson, R.A.., Stereo Radar Analysis, U.S. Engineer Topographic Laboratory, Report No. FTR-1339-1, Ft. Belvoir, VA, U.S.A.

Gugan, D.J., 1987, Practical Aspects of Topographic Mapping from SPOT Imagery, Photogrammetric Record, 12 (69), pp 349-355.

Guichard, H., 1983, Etude théorique de la précision dans l'exploitation cartographique d'un satellite à défilement : application à SPOT, Société Française de Photogrammétrie et de Télédétection, 90(2), pp. 15-26.

Guindon B., and Adair, M., 1992, Analytic Formulation of Spaceborne SAR Image Geocoding and Value-Added Products Generation Procedures Using Digital Elevation Data, Canadian Journal of Remote Sensing, 18 (1), pp. 2-12.

Hoffmann, O., and Muller, F., 1988, Combined Point Determination Using Digital Data of three line Scanner Systems. International Archives of Photogrammetry and Remote Sensing, 27 (B11) pp. III / 567 - III / 577.

Konecny, G., and Schuhr, W., 1984, Practical Results of Geometric SAR-580, Image Evaluation, International Archives for Photogrammetry and Remote Sensing, 25 (A3).

Konecny, G., Lohamann, P., Engel, H., and Kruck, E., 1987, Evaluation of SPOT Imagery on Analytical Instruments, Photogrammetric Engineering and Remote Sensing, 53(a), p.p. 1223-1230.

Kratky, V., 1987, Rigorous Stereophotogrammetric Treatment of SPOT Images. Compte rendus du Colloque International sur SPOT-1 : utilisation des images, bilans, résultats, Paris, France, pp. 1281-1288.

La Prade, G.L., 1963, An Analytical and Experimental Study of Stereo for Radar, Photogrammetric Engineering, 29(2), pp. 294-300.

Lassere, M., and Lemieux, J.P., 1990, Sherbrooke Data Set for Topographic Applications of Remote Sensing. Final Report, EMR Canada, 40 pages.

Leberl, F., 1972, On Model Formation with Remote Sensing Imager, Oesterr, Zeitschrift fur Vermessurgswesen und Photogrammetric, 60(2) pp. 43-61.

Leberl, F., 1978, Satellite radargrammetric, Seutsche Geodactische Kommission, Munich, BDR, Serie C, 239, 156 pages.

Masson d'Autume, M. G., 1980, Le traitement géométrique des images de télédétection, Annales des Mines, Paris, France, 2, pp. 53-62.

Naraghi, M., Stromberg, W., and Daily, M., 1983, Geometric Rectification of Radar Imagery using Digital Elevation Models, Photogrammetric Engineering and Remote Sensing, 49(2), pp. 195-159.

O'Brien, D., and Handy, J., 1991, Recent Image Mapping Experiments in the Surveys, Mapping and Remote Sensing Sector, National Conference on GIS, Ottawa, Ontario, Canada, pp. 458-470.

Paderes, F.C.., Mikhail, E.M., Fagerman, J.A., 1989, Batch and On-line Evaluation of Stereo SPOT Imagery. Proceedings of the ASPRS - ACSM Convention, Baltimore, MD, U.S.A., Vol. 3, pp. 31-40.

Rosenfield, G.H., 1968, Stereo Radar Techniques, Photogrammetric Engineering, 34(6), pp. 586-594.

Salamonowicz, P.H., 1986, Satellite Orientation and Position for Geometric Correction of Scanner Imagery. Photogrammetric Engineering and Remote Sensing, 52(4), pp. 491-499.

Sawada, N., Kikode, M. Shinoda, H, Asada, H., Iwanaga, M., Watanabe, S., and Mori, K., 1981, An Analytic Correction Method for Satellite MSS Geometric Distortion, Photogrammetric Engineering and Remote Sensing, 47 (8), pp. 1195-1203.

Toutin, Th., 1983, Analyse mathématique des possibilités cartographiques du satellite SPOT,

Mémoire de DEA, Ecole Nationale des Sciences Géographiques, France, 74 pages.

Toutin Th., et Carbonneau, Y., 1989, La multi stéréoscopie pour les corrections d'images SPOT-HRV, Journal canadien de télédétection, 15(2), pp.. 110-119.

Toutin, Th., and Carbonneau, Y., 1992 (a), MOS and SEASAT Image Geometric Corrections, IEEE Transactions on Geoscience and Remote Sensing, 30(3), pp. 603-609.

Toutin, Th., and Carbonneau, Y., 1992b, La création d'ortho-images : description d'un nouveau système, Journal canadien de télédétection, 18(3), pp. 136-141.

Toutin, Th., Carbonneau, Y., and St. Laurent L., 1992, An Integrated Method to Rectify Airborne Radar Imagery using DEM, Photogrammetic Engineering and Remote Sensing (58(4), pp. 417-422.

Westin, T., 1990, Precision Rectification of SPOT Imagery, Photogrammetic Engineering and Remote Sensing, 56(2), pp. 247-253.

Wong, F., Orth, R., and Friedmann, D., 1981, The Use of Digital Terrain Model in the Rectification of Satellite-Borne Imagery, 15th International Symposium on Remote Sensing of Environment, Ann Arbor, Michigan, U.S.A., pp. 653-662.

Oceans

Sensors and Environmental Applications of Remote Sensing, Askne (ed.) © 1995 Balkema, Rotterdam, ISBN 90 5410 524 0

Mesoscale and submesoscale atmospheric phenomena studied by the synthetic aperture radar aboard the ERS-1 satellite

Werner Alpers
Institute of Oceanography, University of Hamburg, Germany

ABSTRACT: Radar images acquired over the ocean by the synthetic aperture radar (SAR) aboard the First European Remote Sensing Satellite ERS-1 delineate not only features in the ocean, but also phenomena in the marine boundary layer. These phenomena include katabatic wind fields, convective cells, atmospheric boundary layer rolls, internal gravity waves in the atmosphere, and atmospheric undular bores. Examples of ERS-1 SAR images showing sea surface manifestations of these atmospheric phenomena are presented.

1 INTRODUCTION

An imaging radar is capable of delineating mesoscale and submesoscale atmospheric phenomena because they are associated with variations of the wind stress at the sea surface. The wind stress depends on the wind speed at the sea surface and on the stability of the air-sea interface which is a function of the temperature difference between the water and the air (Keller et al. 1989). Changes in the wind stress at the sea surface disturb the small-scale sea surface roughness and thus give an "imprint" on the sea surface which is visible on radar images. The synthetic aperture radar (SAR) aboard the First European Remote Sensing Satellite ERS-1 is much more sensitive to changes in the wind stress than the SAR aboard the Seasat satellite because the ERS-1 SAR operates at C band (wavelength: 5.7 cm) and not at L band (wavelength: 24 cm) like the Seasat SAR. For ERS-1 SAR (incidence angle: 23 degrees; VV polarization), the slope of the wind scatterometer model function is 1.1 while for Seasat SAR (incidence angle: 23 degrees; HH polarization), it is only 0.5 (Alpers and Brümmer 1994). Thus the ERS-1 SAR should be more suitable for studying atmospheric phenomena than the Seasat SAR. This can be seen by comparing ERS-1 and Seasat SAR images acquired over the ocean.

2 KATABATIC WIND FIELDS

Katabatic winds are cold winds blowing in the evening and night down a sloping terrain ("gravity flow") and, at a coast, over the adjacent sea surface. These winds are generated because in the evening and night the air near the surface cools off faster over the land than over the sea. Over mountain slopes a horizontal density difference develops between the cooled air at the slope surface and the free air in the same altitude over the lower ground or the sea. This results in a "down-hill" flow of cold air. Sea surface manifestations of katabatic wind fields have often been identified in the summer on ERS-1 SAR images of coastal regions which are adjacent to mountainous areas. Fig. 1 shows an example of a katabatic wind field which extends from the Calabrian coast near the Italian town of Gioia into the Tyrrhenian Sea. Also patterns arising from katabatic winds which are channeled through valleys in northern Sicily can be delineated north of the Sicilian coast.

It is also possible to extract the distribution of sea surface wind velocity in the katabatic wind tongues from ERS-1 SAR images by using an ERS-1 wind scatterometer model function (The wind direction has to be inferred from the orientation of the valley and from the form of the katabatic tongue).

3 CONVECTIVE CELLS

Convective atmospheric cells are generated over the sea when the water temperature is higher than the air temperature (unstable marine boundary layer) and when the wind speed is low. The cellular features visible on Fig. 1 between the island of Stromboli and the Sicilian coast are very likely sea surface manifestations of such convective atmospheric cells. Such features can often be delineated on ERS-1 SAR images taken over the Mediterranean Sea on calm days during summer. These quasi-regular features are absent in a coastal sector adjacent to the north-

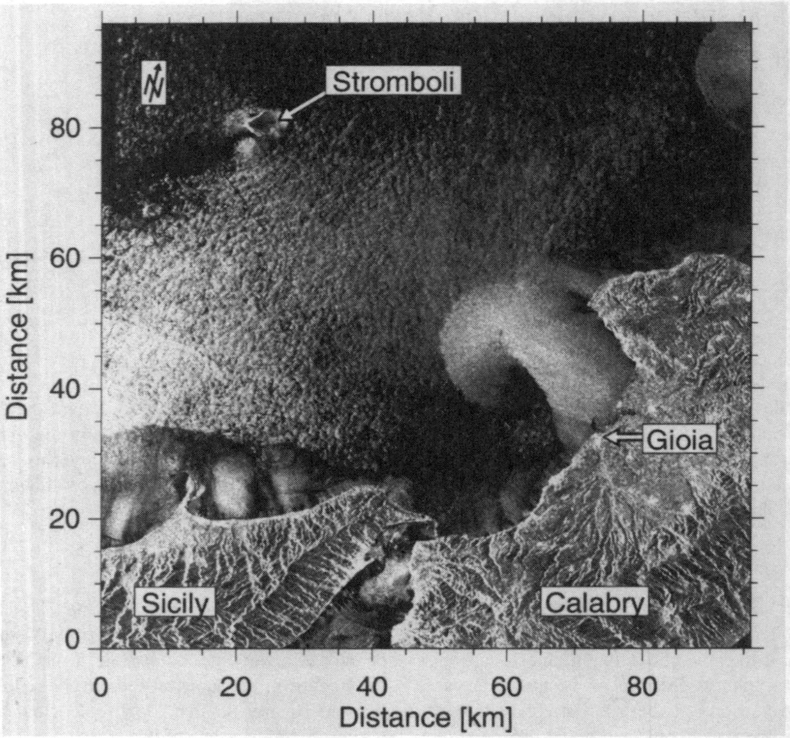

Figure 1. ERS-1 SAR image (100 km x 100 km) of the Mediterranean Sea (Tyrrhenian Sea) north of the Strait of Messina (Italy) acquired on Sept. 8, 1992, at 21:13 UT (orbit 6014, frame 765). It shows north-west of Gioia sea surface manifestations of a katabatic wind tongue (bright blob in the image). Furthermore, between the island of Stromboli and the Sicilian coast, a granular pattern can be delineated, which we believe are sea surface "imprints" of atmospheric convective cells. This cellular structure is destroyed in the vicinity of the Sicilian coast by the katabatic wind blowing from the mountains onto the sea. In the lower section of the image an oceanic internal wave train propagating southwards in the Strait of Messina can be delineated.

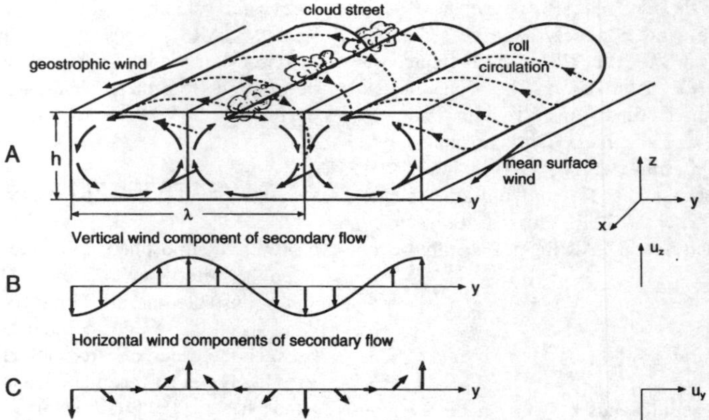

Figure 2. Schematic plot of secondary flow pattern associated with atmospheric boundary layer rolls. Panel A: Perspective view of the three-dimensional flow; Panel B: Variation of the vertical component u_z of the wind velocity along the y direction; Panel C: Variation of the horizontal components u_x and u_y (in the x-y plane).

ern Sicilian coast. We suspect that there the organized cellular circulation pattern in the unstable marine boundary layer is destroyed by coastal processes associated with the katabatic wind field blowing from the northern Sicilian mountains through the valleys.

4 ATMOSPHERIC BOUNDARY LAYER ROLLS

Atmospheric boundary layer rolls are helical circulation patterns in the atmospheric boundary layer which are superimposed on the mean wind field, i.e., the primary flow (see Fig. 2). They can be generated either by thermal instability (Rayleigh-Bénard instability) when the layer is heated from below or cooled from above, or by dynamic instabilty (inflection point instability) when the wind velocity changes with height in such a way that an inflection point occurs in the wind component normal to the roll axis. The axes of the boundary layer rolls are oriented between the directions of the mean surface wind and the geostrophic wind above the boundary layer. Usually the boundary layer is capped by an inversion so that the depth of the boundary layer and the roll layer coincide. In the case of a thermal instability, the aspect ratio, i.e., the horizontal wavelength of the roll pattern, λ, divided by the roll height, h, is 2.8 according to the linear Rayleigh-Bénard convection. However, the most frequently observed values range between 2 and 4.

If the moisture conditions are favorable, cloud streets may be formed in the upward rising branches of the roll circulation. Atmospheric boundary layer rolls contribute significantly to the vertical exchange of momentum, heat, and moisture in the atmosphere. At higher altitudes their relative contribution to the total vertical fluxes in the atmosphere can even be larger than the turbulent fluxes.

Sea surface manifestations of atmospheric boundary layer rolls have first been noted on L band SAR images which were acquired over the Atlantic ocean off the coast of Florida during the Marineland Experiment in 1975 from a NASA airplane (Thompson et al. 1983). Later such features have also been observed in L band SAR images from the Seasat satellite (Fu and Holt 1982), and from Space Shuttle Columbia during the Shuttle Imaging Radar-A (SIR-A) experiment (Ford et al. 1983), and in X band real aperture radar (RAR) images from various Russian satellites (Mitnik 1992). However, all these radars were not well calibrated, and therefore it was not possibe to extract quantitative information on sea surface wind variations from the radar images. This is, however, possible with the ERS-1 SAR since it is a calibrated instrument.

Fig. 3 shows sea surface manifestations of atmospheric boundary layer rolls acquired over the Greenland Sea when a cold wind was blowing off the ice over the warmer sea surface. The streaks are aligned approximately in wind direction. The horizontal separation of the streaks increases with increasing distance from the ice edge. This increase is linked to an increase of the height of the atmospheric boundary layer with distance from the ice edge.

Other examples of sea surface manifestations of atmospheric boundary layer rolls visible on ERS-1 SAR images can be found in the paper by Alpers and Brümmer 1994.

5 INTERNAL WAVES IN THE ATMOSPHERE

Atmospheric internal gravity waves exist in layered atmospheres. They either occur as quasi-periodic waves or as solitons. They are often generated behind mountain ranges in which case they are called lee waves. In the steady state lee waves are stationary with respect to the terrain feature, but they are propagating relative to the mean air flow above the earth surface. Lee waves are very common in visible remote sensing imagery where they manifest themselves as wave-like cloud patterns. However, they also can manifest themselves on the sea

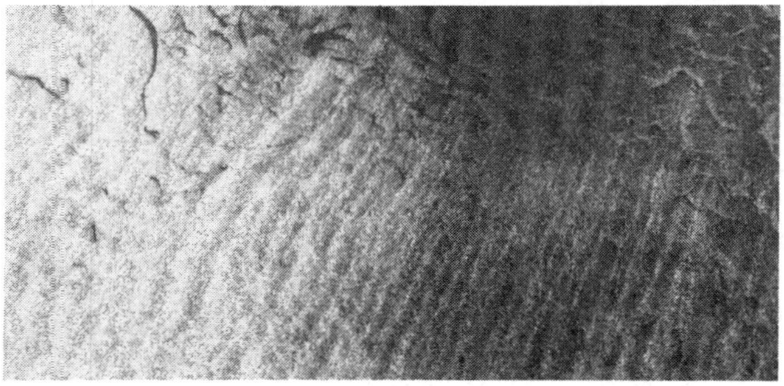

Figure 3. ERS-1 SAR image (100 km x 50 km) of the marginal ice zone west of Spitsbergen in the Greenland Sea acquired on March 5, 1993, at 19:41 UT (orbit 8561, frame 1629). It shows sea surface manifestations of atmospheric boundary layer rolls generated by a strong cold wind blowing off the ice over the warm sea (cold air outbreak). The ice edge is located in the upper part of the image.

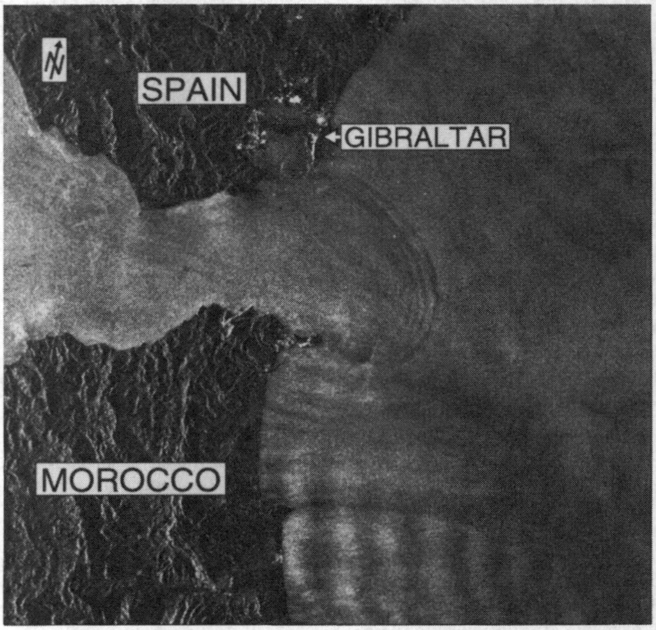

Figure 4. ERS-1 SAR image (100 km x 100 km) of the Strait of Gibraltar acquired on Sept. 3, 1993, at 22:39 UT (orbit 11168, frame 711). It shows in the center sea surface manifestations of an oceanic internal wave packet generated in the Strait of Gibraltar and in the lower right hand section sea surface manifestations of atmospheric internal waves (lee waves) generated by an eastward blowing wind over the 600 m high mountain range Sierra del Hauz in Morocco.

surface since they are associated with a varying surface stress which modulates the small-scale sea surface roughness. An example of an atmospheric lee wave visible on an ERS-1 SAR image in the Mediterranean Sea is shown in Fig. 4. The atmospheric lee wave is generated by the interaction of an eastwards blowing wind (approx. 7 m/s) with Sierra del Hauz mountain range in Morocco.

Fig. 5 shows an image intensity (grey level) scan normal to the eastern Morrocan coast line through the lee wave pattern. The image intensity has been converted into normalized radar cross section (NRCS) by using the conversion (calibration) factor provided by ESA. Furthermore, the normalized radar cross section can be related to the wind speed by using a C band wind scatterometer

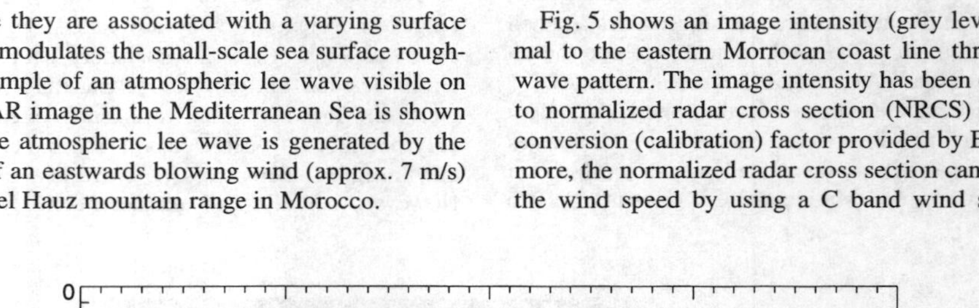

Figure 5. Image intensity scan from west to east through the internal lee wave pattern east of the Moroccan coast shown in Fig. 4. On the left-hand vertical coordinate axis the normalized radar cross section (NRCS) is plotted and on the right-hand vertical coordinate axis the wind speed at a height of 10 m above sea level (under neutral wind conditions).

model. The wind speed scale calculated from the CMOD 4 model (Stoffelen et al. 1992) is plotted on the right-hand vertical coordinate axis.

6 ATMOSPHERIC UNDULAR BORES

Fig. 6 shows an ERS-1 SAR image of the German Bight of the North Sea which was acquired on March 18, 1992, at 10:25 UT. The wave-like pattern visible east of the island of Heligoland is not a sea surface manifestation of an oceanic internal wave, but of an atmospheric undular bore propagating eastwards. The identification of this feature as an atmospheric undular bore is possible because simultaneous in-situ meteorological measurements from the island of Heligoland exist. At Heligoland the German Weather Service had erected a 80 m high meteorological mast at which wind and temperature measurements were carried out every minute at height levels of 10 m, 30 m, 50 m, and 80 m. A wind with a mean wind speed of 4.0 m/s was blowing at the time of this ERS-1 overflight from 140° N. Twenty-seven minutes before the ERS-1 SAR image was taken, a pronounced periodic wind fluctuation in east-west direction was observed at Heligoland. The period of the fluctuations was 5.6 minutes, and the wind speed fluctuations at 10 m height ranged from -1.2 ms^{-1} to 1.7 ms^{-1} (Alpers and Stilke 1994).

Atmospheric undular bores are solitary wave disturbances propagating on a stably-stratified layer in the lower atmosphere. The most spectacular examples of atmospheric undular bores are the "Morning Glories" which often occur near the southern coast of the Gulf of Carpentaria in Northern Australia (Smith 1988). The name "Morning Glory" derives from an impressive roll cloud, or series of roll clouds, which occurs early in the morning, most frequently at the end of the dry season. They are associated with surface wind gusts and with pressure jumps on the order of few hectopascals.

However, atmospheric undular bores have also been observed in other parts of the world (Doviak and Ge 1984). One example are the "Berlin Fog Waves of October 11, 1969", reported by Scherhag (1969) and Egger (1985). There exist several reports by pilots of small aircraft and crew members of small sailing boats who encountered a suddenly occurring succession of strong wind squalls or wind gusts of varying direction, the period of which is typically 5-10 minutes. These unexpectedly occurring strong wind squalls associated with atmospheric undular bores can become dangerous to small landing aircraft (Christie und Muirhead 1983).

The generation of atmospheric undular bores is usually

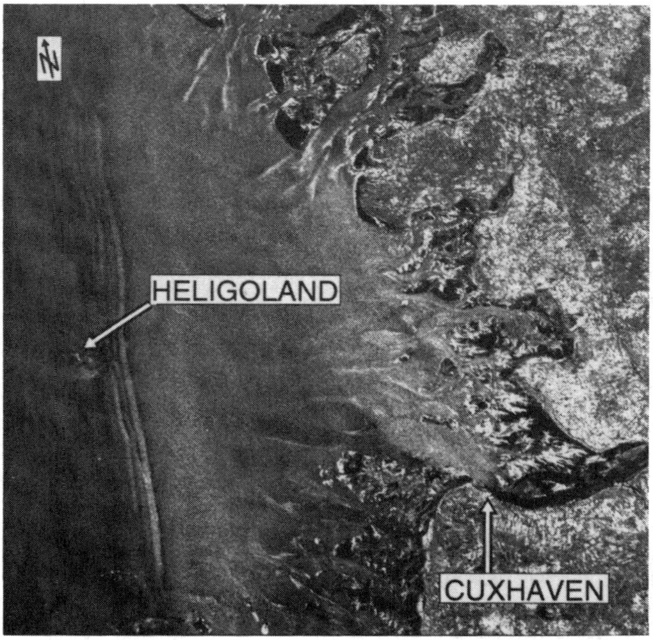

Figure 6. ERS-1 SAR image (100 km x 100 km) of the German Bight of the North Sea acquired on March 8, 1992, at 10:25 UT (orbit 3373, frame 2511). It shows an elongated wave-like pattern east of the island of Heligoland which is an imprint on the sea surface of an atmospheric undular bore generated by an atmospheric front. The bright streaks near the coast are sea surface manifestations of underwater bottom topographic features.

linked with the intrusion of colder, denser air into a stably layered atmosphere. On March 8, 1992, a cold front was moving from west to east over the German Bight. Like atmospheric lee waves, atmospheric undular bores are associated with a varying wind stress at the sea surface and thus with a varying small-scale sea surface roughness which is detectable by radar. By applying a wind scatterometer model, variations in image intensity can be converted into variations of wind speed.

ACKNOWLEDGMENTS

The author thanks ESA for providing the ERS-1 SAR images and D. Gamarnik and J. Heybrock for preparing the figures. This work has been supported by the Bundesministerium für Forschung und Technologie, Bonn, under contract numbers 01 QS 9014 and 50 EE 9413.

REFERENCES

Alpers, W. and B. Brümmer (1994). Atmospheric boundary layer rolls observed by the synthetic aperture radar aboard the ERS-1 satellite. *J. Geophys. Res.* **99**: 12,613-12,621.

Alpers, W. and G. Stilke (1994). Observation of internal solitary waves in the atmosphere by the synthetic aperture radar aboard the ERS-1 satellite and by in-situ measurements. Manuscript, University of Hamburg.

Christie, D.R. and K.J. Muirhead (1983). Solitary waves: A hazard to aircraft operating at low altitudes. *Aust. Meteorol. Mag.* **31**: 97-109.

Doviak, R.J. and R.S. Ge (1984). An atmospheric solitary gust observed with a Doppler radar, a tall tower, and a surface network. *J. Atmos. Sci.* **41**: 2,559-2,573.

Egger, J. (1985). Die Berliner Nebelwellen. *Meteorol. Rdsch.* **38**: 103-107.

Ford, J.P., J.B. Cimino, C. Elachi (1983). Space shuttle Columbia views the world with imaging radar: The SIR-A experiment. Jet Propul. Lab., Calif. Inst. of Technol., Pasadena, Publ. 82-95.

Fu, L.L. and B. Holt (1982). SEASAT views oceans and sea ice with synthetic aperture radar. Jet Prop. Lab., Calif. Int. of Technol., Pasadena, Publ. 81-120.

Keller, W.C., V. Wismann, W. Alpers (1989). Tower-based measurements of the ocean C-band radar backscattering cross section. *J. Geophys. Res.* **94**: 924-930.

Mitnik, L.M. (1992). Mesoscale coherent structures in the surface wind field during cold air outbreaks over the Far Eastern seas from the satellite side looking radar. *Extrait de La mer (Bulletin de la Société franco-japonaise d'océanographie)* Tome **30** (3): 287-296.

Scherhag, R. (1969). The Berlin fog waves of Oct. 1969. Supplement to the Berlin Weather Map.

Smith, R.K. (1988). Travelling waves and bores in the lower atmosphere: The "Morning Glory" and related phenomena. *Earth-Science Reviews* **25**: 267-290.

Stoffelen, A., D.L.T. Anderson, P.M. Woiceshyn (1992). ERS-1 scatterometer calibration and validation activities at ECMWF: B. From radar backscatter characteristics to wind vector solutions. Proceedings of the workshop *"ERS-1 Geophysical Validation"*, held at Penhors, Betagne, France, April 27-30, 1992, published by the European Space Agency, Paris, France, ESA publication WPP-36 (Aug.92): 89-93.

Thompson, T.W., W.T. Liu, D.E. Weissman (1983). Synthetic aperture radar observations of ocean roughness from rolls in an unstable marine boundary layer. *Geophys. Res., Lett.* **10**: 1,172-1,175.

Sensors and Environmental Applications of Remote Sensing, Askne (ed.) © 1995 Balkema, Rotterdam, ISBN 90 5410 524 0

An operational, satellite-based, European oil slicks monitoring system

D.R. Sloggett & I.S. Jory
Earth Observation Sciences Ltd, Farnham, Surrey, UK

ABSTRACT: Recent large scale oil slick spills and an increasing public awareness of the environmental damage that they can cause have heightened political interest in potential solutions based upon satellite based observations, particularly in cases of deliberate spilling. This paper describes a proposed operational infrastructure that would be based upon three ground stations in Europe. The system envisaged provides a service to a number of end users organisations, such as the coastguards and local coastal defence authorities, in Europe in countries such as Holland, Norway, Germany, Turkey, the UK and to the European Environmental Agency (EEA).

The oil slick detection infrastructure is based upon an Oil Slicks Detection Workstation (OSDWS) developed by Earth Observation Sciences (EOS) for the Defence Research Agency (DRA) at Farnborough and the British National Space Centre (BNSC). This system is able to ingest ERS-1/2 imagery and information derived from meteorological sources.

1. INTRODUCTION

Increasing levels of oil pollution in European marine basins is a source of concern for environmental groups and government agencies working at the national, regional and local levels. Recent accidents, such as the now-famous Braer grounding in the Shetlands, have raised public concern at the potential environmental impact such accidents can cause. The Exxon Valdiz accident in Alaska and several other incidents have all filled the media with horrifying pictures of the consequences of such a spill. The biggest incident recorded so far is the Ixtoc blow-out in 1979 off the coast of Mexico - where over 400,000 tonnes of oil leaked out over a 10 month period. Even small scale spills can have a dramatic effect upon local ecosystems. An incident in the Wadden Sea in 1969, where only 200 metric tons of oil was spilled, killed 40,000 seabirds. Every year around 4 million metric tons of oil contaminate the sea (Lean and Hinrichsen, 1992). Half of this total originates from ships, the rest comes from land based pollution sources. More than 1.1 million metric tons of oil is deliberately discharged every year from tankers pumping out their tanks before taking on new cargoes.

Currently, within Europe, many countries operate aircraft, helicopters and ships to monitor oil slicks that are notified to authorities, such as the United Kingdom Marine Pollution Control Unit (MPCU) in Southampton. In Germany, Italy, Norway and the Netherlands similar capabilities exist to detect oil slicks.

Oil slicks in the sea arise from many sources. Lean and Hinrichsen (1992) estimate these sources as follows:

- operational discharges from tankers (22%)
- municipal waste (22%)
- tanker accidents (12.5%)
- atmospheric sources – (industry) (9.5%)
- bilge and fuel oils (9.0%)
- natural seeps (7.5%)
- non-refining industrial wastes (6.0%)
- urban run-off (3.5%)
- coastal refineries (3.0%)
- offshore production (1.5%)
- river run-off (1.0%)
- other sources (2.5%).

Evidence suggests that the cost of cleaning up the results of oil spills outweighs the cost of prevention. But a key question is how best to ensure that the above causes can be reduced to a minimum. In the case of monitoring deliberate discharges, then the axiom that the polluter must pay would appear to apply. The key issue in the short term is how to catch the polluter. In the medium term this emphasis on prosecution shifts to a mode of deterrence – where ship's masters no longer take the risk as the cost of being caught is too high.

A key question is, how to monitor or police the vessels that might be polluting our seas? Aircraft have been the traditional solution – but they lack range and search area. Helicopters are equally disadvantaged. Ships are only capable of rapid response to a specific locality once information on the location of a potential slick has been gleaned from other sources.

Satellites offer one potential solution to this problem of providing regional coverage of seas such as the Baltic, the Mediterranean, the English Channel and the North Sea. Visual sensor systems are limited in their ability to detect oil slicks on the surface of the water and have to contend with cloud cover. In contrast active Synthetic Aperture Radar (SAR) systems have an all weather capability which enables them, within a limited swath region, to monitor the state of the oceans in detail.

Research, reported by Wahl et al. (1992), has demonstrated the potential detection of slicks using ERS-1 SAR data and the difficulties associated with minimising false alarms. These are due to natural surface phenomena which exhibit reduced levels of radar backscatter similar to those of an oil slick. Wahl et al. (1993) concluded that more sophisticated algorithms would be required to ensure that false alarms, arising for example from oceanographic features in the water, could be eliminated from the list of potential slicks. Research undertaken at Earth Observation Sciences (EOS) devised and validated automated techniques for oil slick detection based upon the analysis of edges of segmented regions within ERS-1 imagery. This work has been continued and an approach based upon evidential reasoning proposed. This attempts to take into account additional *contextual* information within an overall reasoning approach.

This paper summarises the development of an algorithm that automatically detects oil slicks in ERS-1 SAR data. Section 2 of the paper describes the target phenomena created by oil slicks on the water - in terms of the reduced radar cross section. Section 3 outlines the algorithm that is used to detect the slicks and Section 4 defines the evidential reasoning processes at the core of the processor. The algorithm runs entirely automatically using features of the slicks in the water and local contextual information to discriminate between real oil slicks and many different potential false alarms. Section 5 considers an operational framework in which the techniques developed within this research could be used and Section 6 has some conclusions. Section 7 acknowledges the participants in the programme and Section 8 provides a series of references.

2. THE CHARACTERISTICS OF OIL SLICKS IN ERS-1 IMAGERY

In the course of this research programme EOS has collected a number of scenes that contain known slicks. We have also been allowed to view the extensive library of potential slicks held by Professor Werner Alpers and his team in Hamburg. We have noted several underlying features of slicks that might be exploited as discriminants in an automated processing system. The first of these, the reduction in the radar backscatter, was noted by Wahl (1993) and in several papers by Alpers and his team. The reduction in backscatter levels arises due to the damping effect that the oil has on the oceans surface. However this effect is known to have limitations when the wind speed and sea surface state rise above certain levels when the reduction in the radar backscatter is lost in the complex radar returns from the ocean.

A second feature exhibited by the oil slicks considered within this research was the relatively smooth edges associated with slicks. After segmentation potential slicks were identified using a number of criteria. The most important of these was an assessment of the fractal nature of the edges of the slicks.

A third feature of the slicks that has been noted is linked to the dispersal of the oil from the initial point at which it was released and the way in which the release occurs. In those cases where a large amount of oil is released over a short period of time the thickness of the oil tends to vary from the initial point. The dispersal pattern also, in very low sea-states, tends to follow prevailing tide, current and wind conditions. It is therefore possible to predict, given knowledge of the above parameters, the likely dispersal pattern for a specific type of oil on the surface of the sea at a given sea-state.

These potential observables have been noted in a number of images collected by EOS and its research partners. What has also been noted is a number of features in the water that exhibit similar signatures to those observed in real oil slicks. These areas in the sea with low levels of radar backscatter are often caused by a number of natural phenomena, such as sea which is sheltered from the wind by a local part of the coastline, oceanographic features, features arising due to shallows, and many other potential effects. Natural slicks, such as plankton blooms and surface films, can also reduce the radar backscatter but these are not oil slicks. These natural phenomena often result in large numbers of potential slicks being identified in segmented ERS-1 imagery. On some occasions several hundred potential slicks are identified in each ERS-1 scene. What is required are techniques that complement the direct attributes of features in the water, such as the radar cross section and an assessment of the fractal nature of the edges. This research is using contextual information to enable the possibility of a feature in the water being a slick or some other from of natural phenomena.

3. AN AUTOMATED DETECTION ALGORITHM

The algorithm arising from this research can be viewed as three logically separate, but linked, processing chains that use an ERS-1 image as their input. The structure of the processing chain envisaged is illustrated in Figure 1.

The ERS-1 image is initially analysed by the *direct* processing chain. This chain, which includes the segmentation algorithms, aims to detect features in the water that might be a slick and then derive attributes of those features that can be used within an overall framework based upon evidential reasoning. The *direct* attributes of a feature can include: average radar cross section, standard deviation, higher statistical measures, an assessment of the fractal nature of the edges of the feature, texture measures and topological measures that reflect the shape and compactness of the feature.

The second processing chain is the *indirect* or inferred processing chain. This is the element of the processor that links real world observations with models to infer other attributes about a feature. In the case of the oil slick the *indirect* processing chain has to look at information on meteorological conditions, tide and current data and predict a likely dispersal pattern for various forms of oils. This processing chain is essentially providing models to enable the overall framework to reason about parameters that it cannot directly observe.

The third element of the processing chain is the *contextual* processing chain. This seeks to provide additional sources of evidence that might be relevant to the overall reasoning process. Of particular interest in this case, to assist in the elimination of potential false alarms, is *contextual* information on the location of sewage outfalls shallows in the water the coastline meteorological parameters and tide and current data. This information, stored in a Geographic Information System (GIS) -like facility, can be used within the reasoning process to decide which features in the water are most likely to be a slick.

The fourth element of the framework is the *evidential* reasoning chain. This has to read all of the attributes derived from the direct, indirect and contextual processing chains and use them in a process of balanced reasoning about each feature in the water. The approach undertaken in this research is to propose one of several possible solutions for each feature in the water. These hypotheses provide a view as to what each feature in the water might be. The evidential reasoning process seeks to find, through its deductions, evidence that supports or conflicts with each hypothesis.

In the course of reasoning about each feature the process will weight the sources of evidence. In particular viewing conditions the significance of a specific attribute might need to be reduced as it does not convey the same value of information as it would in other conditions, such as when the wind speed exceeds some value.

The evidential reasoning system proposed has already been discussed in detail at a workshop held in April 1994 at the JRC at Ispra. At this workshop many examples of real oil slicks in ERS-1 imagery and oceanographic features that exhibited features similar to oil slicks where debated at length by a team of visiting scientists. The framework envisaged by EOS was debated and the outline of the reasoning process discussed in detail. Prototypes of the evidential reasoning system, using fuzzy logic techniques, have been developed by TNO in the Netherlands in collaboration with EOS. These are undergoing detailed validation over the next few months in collaboration with UK pollution monitoring agencies and the DRA ground station at West Freugh.

4. A POTENTIAL OPERATIONAL FRAMEWORK FOR OIL SLICKS MONITORING

The development of a reliable automated means of detecting oil slicks in ERS-1 SAR imagery provides a basis for considering the implementation of an operational service providing information, in the short term, on the location and size of oil slicks. In the medium term with access to Radarsat and Envisat this service might be extended, providing estimates of the type of oil and the volume in the slick.

Within Europe several ground stations, such as those operated by DRA at West Freugh in Scotland, the Tromso station in Norway and the ESA station at Fucino in Italy could be used as reception sites for ERS-1 data covering European Maritime basins. It is envisaged that the raw SAR data would be processed into an image at the ground station. The image would then be geocoded prior to being passed into the oil slick detection algorithm where it would be processed to detect any slicks within the scene. The resulting list of slicks, estimates of their size and location would then be stored within the information service. It is envisaged that the service would provide several levels of information to potential end users. At one level this would analyse the locations of the slicks and notify the agencies in those countries whose coastal waters were affected. At the highest level the service would record statistics on

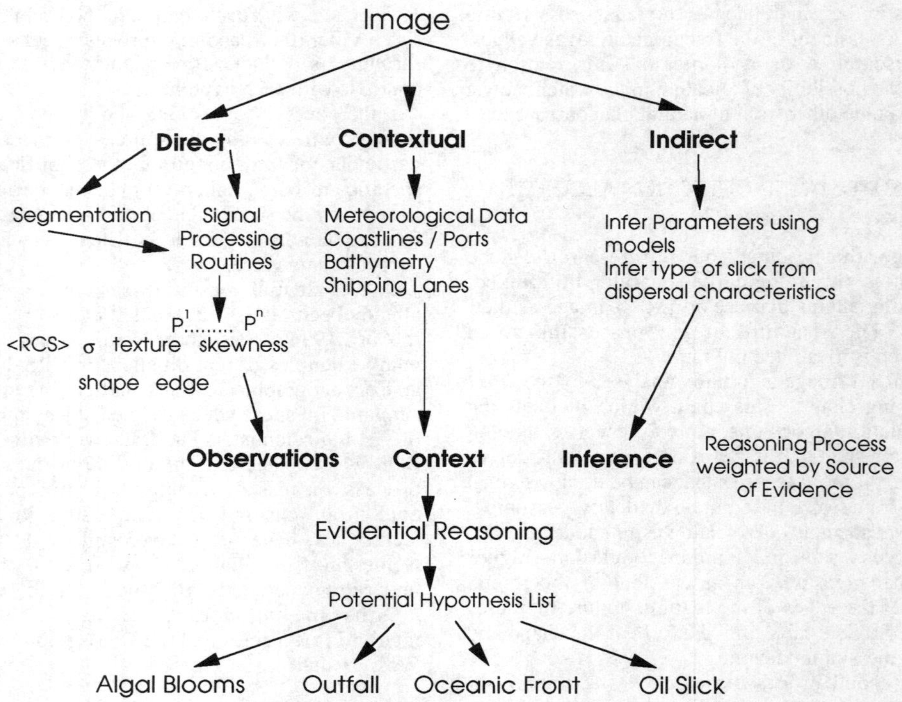

the distribution of slicks in order to provide statistical summaries to regional agencies, such as the European Environmental Agency.

In order to manage local incidents it is envisaged that pollution monitoring agencies will integrate the information on the location and size of slicks with local tide, current and meteorological data to provide a nowcasting service indicating its likely dispersal and the area affected by the slick. This type of service would be of great benefit to organisations such as insurance agencies who might face claims for compensation from a number of potential sources as a direct result of an incident.

5. CONCLUSIONS

The research reported within this paper has demonstrated that it is feasible to define an algorithm that automatically detects oil slicks within ERS-1 SAR imagery. The algorithm uses a combination of direct attributes derived from the image after segmentation and contextual information to rank order features in the water thought most likely to be oil slicks.

The key development outlined is this paper is the automation of the process. With reliable automatic detection it is feasible to envisage the implementation of an oil slicks monitoring service being established in Europe for all European maritime basins. This would provide information on the location and size of slicks detected by ERS-1. It is envisaged that this data would be used by pollution control agencies working at a regional, national and local scale within Europe. At the regional scale the information would be used to generate assessments of the scale of the pollution problem and to assist in policy formulation to combat the problem. At the national and local scale the information would also be used to assist in the management of incidents that have occurred, such as the grounding or sinking of a supertanker. In this case the information on the location and size of the oil slick would be combined with local tide, current and meteorological information to provide a short term forecast of the dispersal of the slick. This would be used by pollution control agencies to decide upon areas that were under threat from the slick, such as marine aquaculture or tourist beaches, and select the most correct course of action to minimise the effects of the slick.

In the future, with the advent of more advanced SAR instruments on Radarsat and Envisat, it may also be possible to provide some form of qualitative assessment of the type of oil detected, such as heavy crude, light oil etc.. This may allow an additional item of information to be added to the service envisaged which is an estimate of the total volume of oil on the surface.

6. ACKNOWLEDGEMENTS

The authors wish to acknowledge the work of the staff of the Defence Research Agency (DRA) at Farnborough for their support and advice in the course of developing the oil slicks workstation. Funding from the British National Space Centre (BNSC) is also gratefully acknowledged. The authors also wish to acknowledge the work of the Oil Slicks team at EOS, notably the efforts of Simon Tackley and Dr Ken Sullivan. The research has continued due to funding provided by the Advanced Techniques Group of the Institute for Remote Sensing (IRSA) at the Joint Research Centre at Ispra led by Dr Alois Sieber. Dr Sloggett also wishes to express his thanks to Professor Werner Alpers, Dr Volkmar Vissman and Dr Erik Halsema for many helpful discussions in the course of the recent research activities sponsored by JRC.

7. REFERENCES

Lean, G., Hinrichsen, D. (1992) WWF Atlas of the Environment, Helicon Publishing Ltd, Oxford, England.

Sensors and Environmental Applications of Remote Sensing, Askne (ed.) © 1995 Balkema, Rotterdam, ISBN 90 5410 524 0

Gravity profiles from satellite altimetry compared with ice gravity profiles in the Gulf of Bothnia

J. Noréus
Department of Radio and Space Science, Chalmers University of Technology, Göteborg, Sweden

K. Hayling & M. Nyborg
Department of Geology, Chalmers University of Technology, University of Göteborg, Sweden

ABSTRACT: This paper compares gravity profiles from space altimetry with *in situ* gravity field measurements provided by the Swedish Geological Survey. The processing of repeated altimeter data is described. Averaged altimetric height profiles are converted to gravity profiles, which provides accurate determinations of the short wavelength part of the gravity field spectrum. We present results with the accompanying *in situ* measurements interpolated into profiles along orbit segments from the Geosat, ERS-1 and Topex/Poseidon satellites. The altimetric gravity shows excellent agreement with the finer details (≥8 km) size in the short wavelength part of the spectrum. However the fit is not as good for gravity anomalies with an extension between 30 km and 100 km. Discrepancies near the coast lines also have to be considered.

1 INTRODUCTION

The altimeter transmits a short electromagnetic pulse in the microwave range, receives the reflected echo and measures the time of two-way propagation. The distance from the Earth's sea surface to the satellite is then calculated, and by independent knowledge of the satellite orbit, the height of the sea surface above a reference surface can be determined. As a reference surface is a standardised ellipsoid usually used. The altimetric measurements have to be corrected for propagational and systematic effects using models or measurements. Further details about space altimetry is found in *e.g.* Seeber (1993).

Space altimetry has since 1975 continuously increased its importance as a tool for determining the geoid and gravity field of the Earth (*e.g.* McGoogan 1975, Stanley 1979, Sandwell & McAdoo 1988). Focus have primarily been made on global geoid models or gravity field (Rapp & Pavlis 1990). Recent interest in the details of the gravity field in local areas have emphasised on resolution determination and resolution improvement of altimetry.

The evaluation of altimetry can be done by either indirect methods, using *e.g.* power spectra and coherence analysis (Marks & Sailor 1986, Noréus 1993), or direct methods of comparison using ground truth of the gravity field (Small & Sandwell 1992). The most frequent data used as such a ground truth are ship gravity measurements, which however often suffer from accuracy limitations of the vessel's measurement conditions. This paper presents a comparison of satellite altimetry with a gravity data set of extraordinary quality. This data set comprises measurements conducted on the sea ice in the Gulf of Bothnia.

2 THE DATA SETS

2.1 *General description*

The *in situ* gravity measurements (from now and on called ice gravity data) in the Gulf of Bothnia which are used in this paper were conducted by Swedish Geological Survey on the sea ice during a number of cold winters since the 70's. Location of the data is shown in Fig. 1. The sampling distance is approximately 5 km.

The ice gravity measurements represent a very accurate and consistent data set of the gravity field in the Gulf of Bothnia, with a measurement accuracy of better than 0.1 mgal (1 mgal = 10^{-5} ms^{-2}). This should be compared with the accuracy of ship measurements, which yield 0.5 mgal accuracy at ideal conditions, but generally are accurate within 2-5 mgal.

The combined altimeter data set of the Geosat, ERS-1 and Topex/Poseidon altimeters in the Gulf of Bothnia and the Northern Baltic Sea is shown in Fig. 2. ERS-1 orbit segments are identified as those with the highest inclination, and Topex/Poseidon gives orbit segments with the lowest inclination. The mixing of different satellite data give rise to irregularly shaped data gaps. The sizes of those gaps vary between 5-40 km (across).

All three satellites were designed with exactly repeated orbits (+/– 1 km), yielding several measurements at each orbit segment location. Topex/Poseidon is characterised with a retrograde orbit, while the orbits of ERS-1 and Geosat both are sun synchronous.

2.2 *Data processing*

Three sub-areas (one in the Bothnian Sea and two in the Bothnian Bay) were defined using the characteristics of the ice gravity data set. Using applied geostatistics (kriging) the ice gravity data set was characterised with three different variogram models, describing the spatial coherence of the sub-sets of data. The uncertainty of estimation varies between 0.5 mgal and 2.8 mgal.

The satellite altimetry data were edited and thoroughly inspected for outliers and bad data. The number of passes in a typical pre-processed orbit segment was 35 (Geosat ERM), 8 (ERS-1 35-day), 23 (ERS-1 3-day) or 28 (Topex/Poseidon). The individual height profiles diverge usually within some metres, because of uncertainties in the orbit. The geographically correlated orbit error is also included which can not be traced by inspecting cross-overs or average biases. However, since the absolute level of the height above the reference ellipsoid is out of interest we do not have to be concerned about this. What do concern us, however, are the dissimilarities between subsequent altimeter measurements, and possibilities to reduce them.

The height profiles were averaged using an iterative polynomial adjustment method described in detail in Noréus (1993). First, linear overlapping were made of all repeats with a length of 75% or more of the longest profile belonging to a particular orbit segment, then an average height profile was determined for that orbit segment. The individual residuals to this average profile were then modelled with polynomials up to degree three. The models were then subtracted from the original data. This procedure reduced medium and long wavelengths discrepancies, which include orbit errors, satellite perturba-

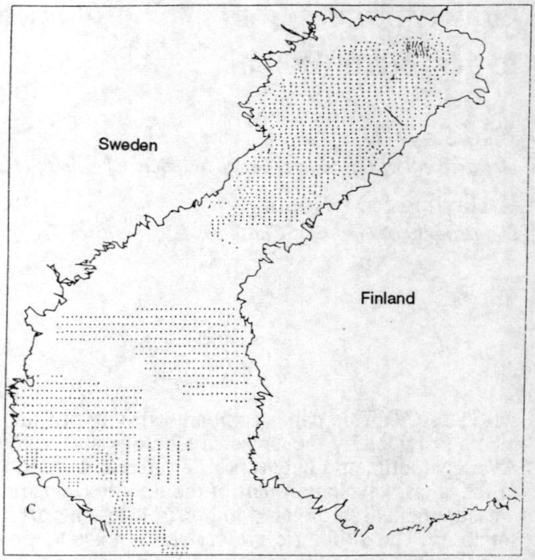

Fig. 1. Location of the ice gravimetric measurements in the Gulf of Bothnia. Sampling density is approximately 5 km.

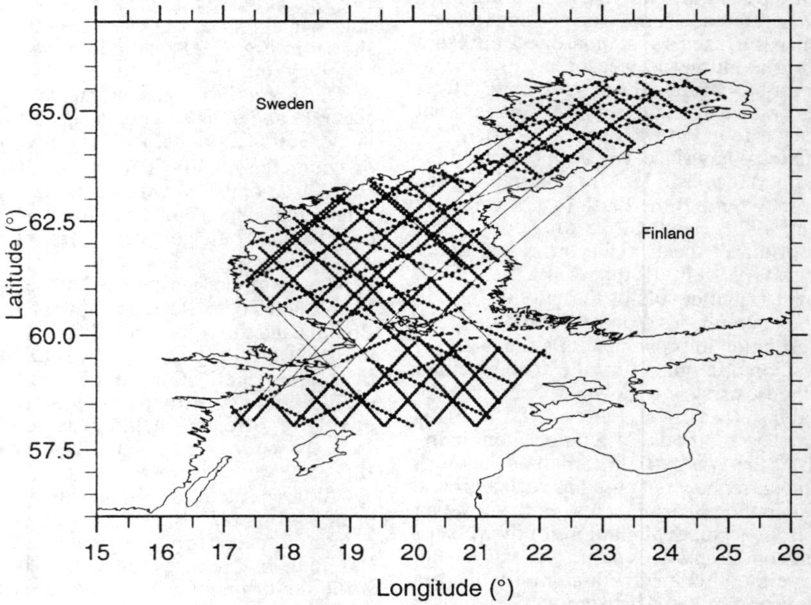

Fig. 2. Location of the altimetry data from the three altimeters Geosat, ERS-1 and Topex/Poseidon in the Gulf of Bothnia and Northern Baltic Sea.

tions, or propagation and dynamic effects that were not removed by correction terms.

One-dimensional gravity profiles were then computed from the averaged altimetric height profiles, using the relation between the potential field and the gravity field (Eq. 2.1)

$$\Delta g = \frac{2\pi g_0}{\lambda} \Delta H \qquad (2.1)$$

(Heiskanen & Moritz, 1967), where g_0 is the normal acceleration due to gravity (~9.80 ms^{-2}) and λ is the wavelength of the geoid undulation ΔH. Special at-

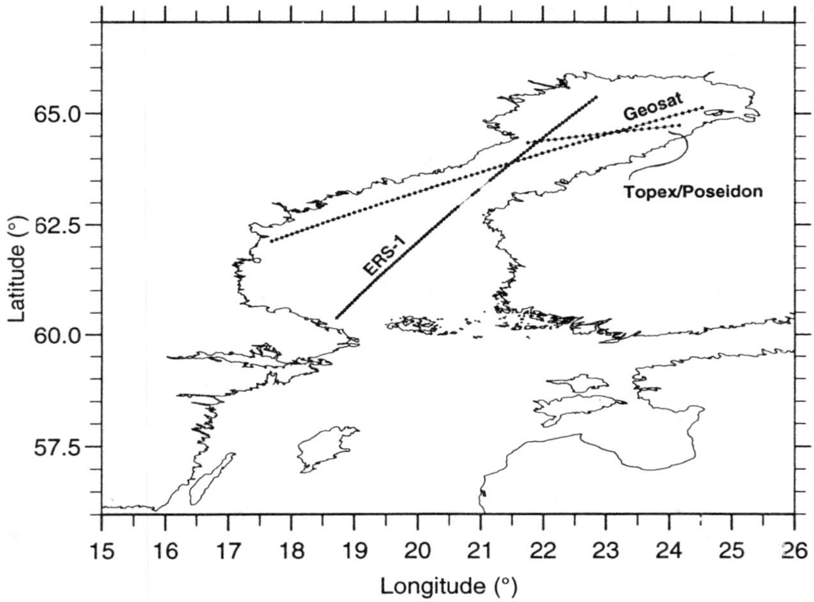

Fig. 3. Location of the three selected altimeter tracks presented as examples in this paper.

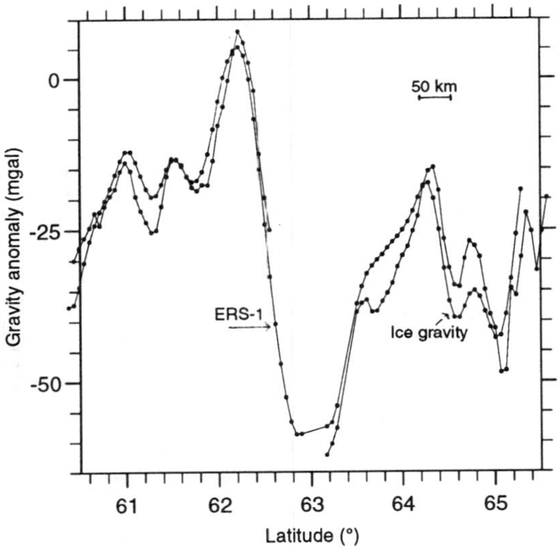

Fig. 4a. Gravity profile from an ERS-1 3-day track and the corresponding ice gravity profile.

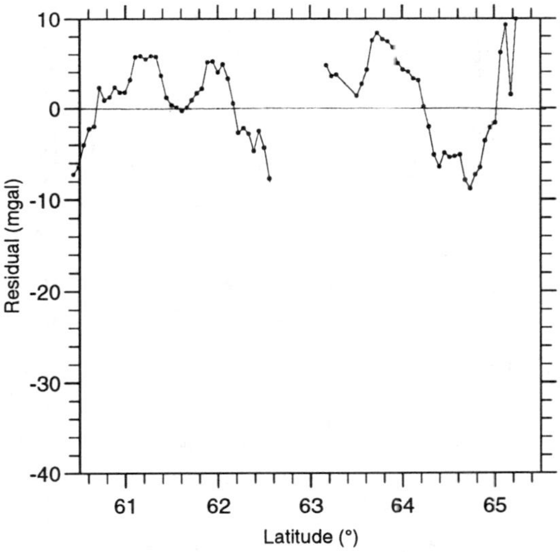

Fig. 4b. Residual between the ERS-1 altimetric gravity profile and the ice gravity profile.

tention was given to height profiles with data gaps; a spline interpolator simulated missing points, in order to obtain complete data series with equidistant points. The interpolated values were used in the transformation but not written in the resulting gravity profile. Thus, data gaps remained in the resulting gravity profile, also yielding a smooth approximation of the gravity field near the data gaps.

We also compared altimetric gravity profiles calculated from the standard 1 Hz-average data format with 10 Hz-average data. The 10 Hz-average data set was obtained implicitly from the geophysical data records. The "unpacking" procedure was successful for Geosat and Topex/Poseidon, but not fruitful for ERS-1. This is probably due to the re-tracking of the ERS-1 altimeter waveforms that is made in a prior ground processing step at the French-PAF facility (J. Lillibridge, NOAA, personal communication).

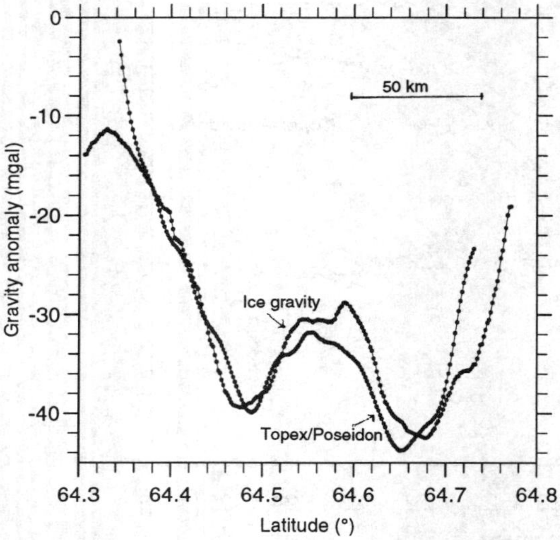

Fig. 5a. Gravity profile from a Topex/Poseidon track and the corresponding ice gravity profile.

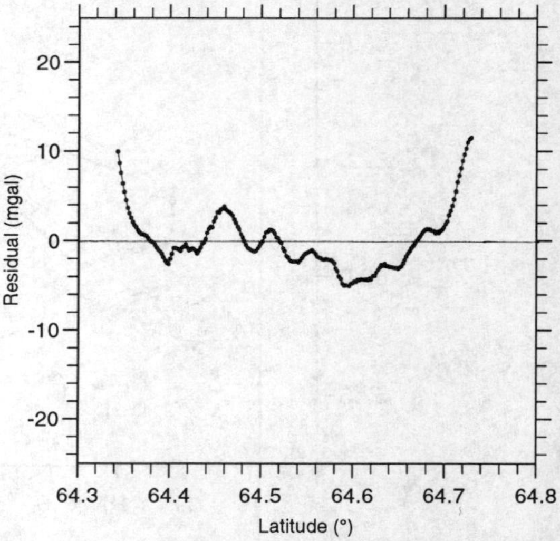

Fig. 5b. Residual between the Topex/Poseidon altimetric gravity profile and the ice gravity profile.

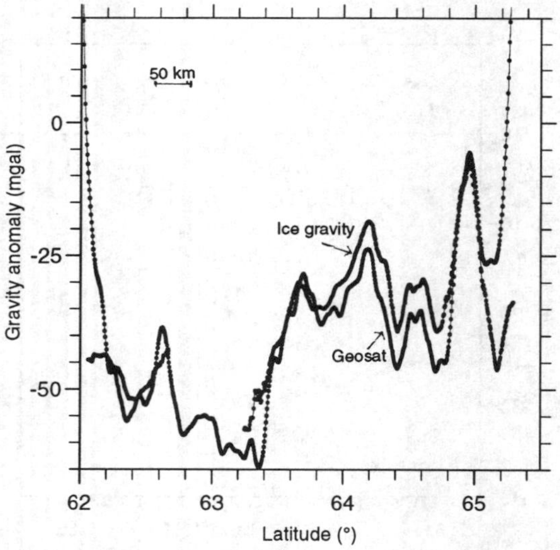

Fig. 6a. Gravity profile from a Geosat 10 Hz-sampled track and the corresponding ice gravity profile.

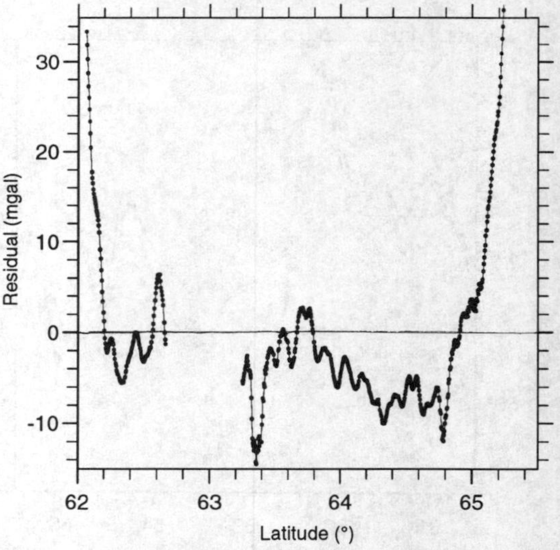

Fig. 6b. Residual between the Geosat 10 Hz-sampled altimetric gravity profile and the ice gravity profile.

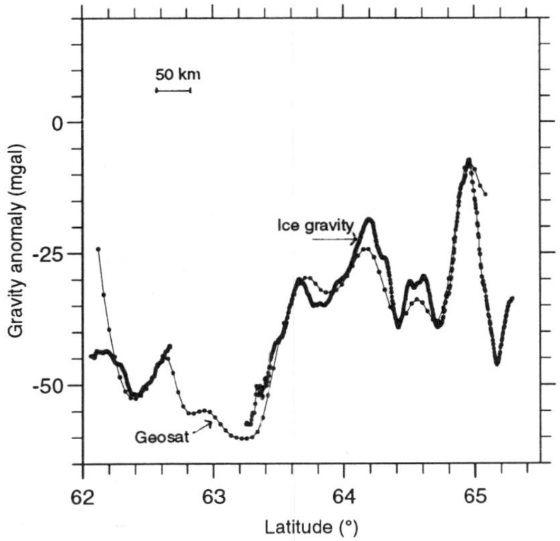

Fig. 7a. Gravity profile from a Geosat 1 Hz-sampled track and the corresponding ice gravity profile.

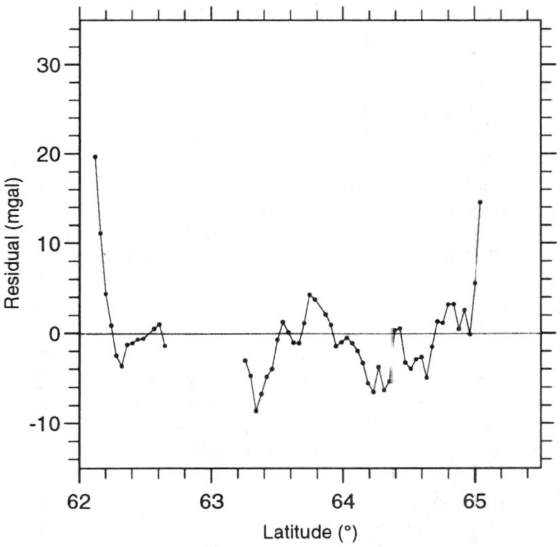

Fig. 7b. Residual between the Geosat 1 Hz-sampled altimetric gravity profile and the ice gravity profile.

3. RESULTS

The results of the gravity profile comparison is presented here in the form of three examples; one from each of the ERS-1, Geosat and Topex/Poseidon satellites. A map showing the location of the selected tracks is given in Fig. 3. Figs. 4-7 present the gravity profiles from altimetry together with the ice gravity profiles, the residuals between the two, and comparative figures of the Geosat profile for the 1 Hz-average density and 10 Hz-average density formats.

The profile from the denser 10 Hz-average format shows more details of the gravimetric field. This is especially noticed in the region at 64.5° latitude in Fig. 6a.

4. CONCLUSIONS AND DISCUSSION

The "unpacking" of the standard 1 Hz-average data format into a 10 Hz-average data set yields a recovered gravity field with more short wavelength anomalies than do 1 Hz data only. Details of the gravimetric field is accurately recovered by the altimeter in the wavelength region 8-30 km, showing a high degree of correlation with the ice gravity. In the wavelength region 50-100 km the correlation is not as good. Discrepancies of wavelengths from 300 km and longer were removed with a least squares bias and tilt adjustment of the altimetric gravity profile to the ice gravity profile.

Part of the discrepancies that can be observed in the residuals between the altimetric and ice gravity profiles can be caused by positioning discrepancies between the ice gravity and altimetric data sets. One example is given for the Topex/Poseidon profile in Fig. 5a. This has to be further examined in more detail.

Two-dimensional gravity maps from altimetry can be constructed in the sub-areas after calibrating the altimetric gravity profiles relative to the ice gravimetry, and interpolating onto a regular grid. Compared with the ice gravimetry, the altimetric gravity data set will be under-sampled in the cross track direction. From this point of view, a denser track spacing is desired, for example that of Geosat GM phase (classified as secret above 40°S latitude) or the current ERS-1 168-day cycle. However, an important aspect of the design of altimeter missions, is to what extent a better gravity field anomaly determination is obtained with a denser track spacing at the cost of fewer repeats.

The possibility of combining satellite gravimetry with other sources of measurements, implies a very high potential of using altimetry in accurate determinations of the gravity field.

ACKNOWLEDGMENTS

Swedish Geological Survey provided the ice gravity measurements. Swedish National Space Board provided financial support.

REFERENCES

Heiskanen, W.A., & H. Moritz 1967. *Physical Geodesy*, San Francisco-London: W.H. Freeman & Co.

Marks, K.M. & R.V. Sailor 1986. Comparison of

Geos-3 and Seasat Altimeter Resolution Capabilities, *Geophysical Research Letters* 13, No. 7: 697-700.
McGoogan, J.T. 1975. Satellite Altimetry Applications, *IEEE Trans. on Microwave Theory and Techniques* MTT-23, No. 12: 970-978.
Noréus, J.P. 1993. Resolution of Geosat Altimeter Data using Averaging and Short Spacing, submitted to *International Journal of Remote Sensing*.
Rapp, R.H. & N.K. Pavlis 1990. The Development and Analysis of Geopotential Coefficient Models to Spherical Harmonic Degree 360, *Journal of Geophysical Research* 95, No. B 13: 21885-21911.
Sandwell, D.T. & D.C. McAdoo 1988. Marine Gravity of the Southern Ocean and Antarctic Margin from Geosat, *Journal of Geophysical Research* 93, No. B9: 10389-10396.
Seeber, G 1993. *Satellite Geodesy*, Berlin-New York: Walter de Gruyter.
Small, C. & D.T. Sandwell 1992. A Comparison of Satellite and Shipboard Gravity Measurements in the Gulf of Mexico, *Geophysics* 57, No. 7: 885-893.
Stanley, H.R. 1979. The GEOS-3 Project, *Journal of Geophysical Research* 84, No. B8: 3779-3783.

Sea ice

Sensors and Environmental Applications of Remote Sensing, Askne (ed.) © 1995 Balkema, Rotterdam, ISBN 90 5410 524 0

Sea ice – A remote sensing challenge

Preben Gudmandsen
Technical University of Denmark, Lyngby, Denmark

ABSTRACT: This largely qualitative exposition is describing problems related to observation of sea ice by means of satellite remote sensing. It addresses the operational application of data carried out routinely to assist shipping in ice-frequented water concluding that the most pressing problems to be solved is the frequency of observation of a given area and the timely dissemination of data to the "end user" on board the ship. Furthermore, it reviews present state of ice dynamics modelling and discuss the role of remote sensing in this context. Also, it reports on an investigation of the model sensitivity described recently in the literature which points at the most important parameters of the model including ice concentration (minimum lead fraction), sensible and latent heat exchange coefficients and the atmospheric and oceanic drag coefficients. Improvements in retrieval accuracy of these parameters are briefly discussed. The importance of ensuring proper calibration of sensors in a long-term monitoring involving successive missions by overlap in time is stressed.

1 INTRODUCTION

Sea ice covers large parts of the oceans of the Earth. About 13% and 10% of the oceans in the northern and the southern hemispheres, respectively, are covered by ice every winter with large variations from year to year and during the winter. In the Arctic Ocean a large part of the ice survives the summer melt, thereby forming second and multiyear ice which may have thicknesses up to six meters. Since sea ice effectively insulates the relative warm ocean from the atmosphere it has an impact on the regional weather and the Earth's climate by reducing the heat flux.from the warmer ocean to the atmosphere. In climate change research large-scale observation of sea ice is therefore of great importance. Sea ice is a very dynamic medium with large variations within few hours under influence of ocean current, wind and waves, broadly described as Air-Sea-Ice interaction processes and great efforts have therefore been invested in the development of methods for deriving parameters such as extent of sea ice canopy, ice concentration, ice classification and ice motion.

In economical important zones such as the Gulf of Bothnia between Finland and Sweden, sea ice is often an hindrance to shipping and fisheries. This is also the case on the east coast of Canada, in the Bering Sea, the Northeast Passage along the coast of Siberia, and at the south and west coast of Greenland. For operational and for safety reasons ice reconnaissance by aircraft and satellite is essential in support of icebreakers and ships. Also, in recent years scientific field campaigns in the Arctic and the Antarctic sea ice regions have been supported by satellite observations by transmission of up to date ice charts or by own receiving capability such as an AVHRR receiving station installed on board to give data in real time.

Thus, sea ice presents a challenge to the remote sensing techniques ranging from observations and derivation of various important parameters to the timely dissemination of data in the operational case. For climate research the accuracy of the parameter retrieval presents another challenge to the scientists. It is the purpose of this note to review briefly these challenges and to indicate ways of approaching them wherever possible.

2 REMOTE SENSING TECHNIQUES

Literature shows that all remote sensing techniques are exploited in observation of sea ice - passive and active techniques - covering the entire useful electromagnetic spectrum. For large-scale observations in connection with climate research and numerical modelling passive microwave radiometer data are being used with advantage to scales of tenths of kilometres. For detailed studies, fine-scale active microwave data including synthetic aperture radar (SAR) and altimeter data are important. Due to the frequent cloud cover in polar and sub-polar regions microwave techniques has and will play a major role, but experience has shown that also visual/infrared techniques are useful. Thus, NOAA AVHRR data are regarded by some users the backbone of sea ice observation due to the frequent coverage of large areas and due to the the fact that clouds often changes in time and space. The AVHRR data at one-kilometer spatial resolution may be used for retrieval of a number of the parameters mentioned in the introduction although it has been found from comparison with floating buoy and radar data that in a number of cases important information is lost in connection with special weather conditions (with cloud cover) and that radar data often complement the visual/infrared data. Thus, sea ice observation has become a multi-sensor and thereby a multi-mission undertaking and this will certainly also be the case in the future.

3 OPERATIONAL ASPECTS

Sea ice observation is used operationally or semi-operationally to support shipping in ice-frequented regions as mentioned in the introduction. Of special interest are parameters such as ice extent, ice classification and detection of leads and ridges. In order to be useful data shall be available at relative short time delay on the bridge of a ship preferably in the format of interpreted data and with recommendations as to routing based on a prediction process. It is a very demanding task which presently cannot be performed fully in spite of the fact that studies have been carried out in many regions and during many years. One reason is that fine-resolution, all-weather data like SAR data have not been available regularly until the launch of ERS-1. Also. it is only recently that numerical modelling has reach a state that it may be used routinely to assist prediction of sea ice conditions - and still it is a technique that requires improvements.

The requirements to an observation system for sea ice are summarized in Table 1 (Ramseier & Strübing, 1994), listed according to different time scales of operation and planning of ship routes i.e. Close Tactical Planning to take place within less than three hours, Tactical Planning within 24 hours etc. The Medium and Long Range Planning approaches the tasks of climate research which is often handled on a monthly basis. The table also lists the spatial scale of the observation area associated with the planning where Close Tactical Planning needs observation within a distance of 30 km and with a resolution of less than 100 m, for instance. Finally, the resolution in time is given which means that data, i.e. interpreted data, should not be more than 3 hours old to satisfy Tactical Planning, for instance.

Table 1 Requirements to Ice Information Support Systems

It is clear that for Close Tactical, Tactical and Strategic Planning rather stringent requirements are placed on the observation system(s). With the present techniques it may be concluded that Close Tactical Planning requirement may only be fulfilled by aircraft and helicopter observations, but also that Tactical Planning may not be satisfied with any of the present satellite systems except with the NOAA AVHRR system with its 1-km resolution and very wide swath of about 1 300 km. However, cloud cover often limits the utility of these data as already mentioned. Experience has shown that ERS SAR data may ease the situation since they are rather frequent at high latitudes which compensates for the narrow swath of the system (100 km) and the orbit configuration (35-day repeat). However, the future ASAR on Envisat with its wide swath of 400 km may come close to the requirements although the time delay of three hours for data processing and dissemination planned for presently, still is too long considering the subsequent process of interpretation.

Strategic Planning may be fully supported by data from passive microwave radiometer systems such as the current SSM/I with a spatial resolution of retrieved parameters of 25 km and a swath of about 800 km. The Multichannel Imaging Microwave Radiometer (MIMR) to be carried by Metop 1 and 2 and the NASA Polar Platform EOS PM may find applications in Tactical Planning in spite of the fact that the spatial resolution exceeds the 1-km requirement. With the finest resolution of about 5 km obtained at 88.5 GHz detailed studies of the ice canopy may be carried out including determination of large-scale ice drift patterns. The data might replace SAR data complementing AVHRR data during cloudy conditions.

Recently it has been shown that ERS scatterometer

TABLE 1 Summary of Scales and Resolutions for Ice Information Support

	Close Tactical	Tactical	Strategic	Medium Range	Long Range
TIME SCALE	< 6 hrs	< 24 hrs	< 7 days	< 1 month	> 1 month
Space Scale	< 30 km	< 100 km	< 1000 km	> 1000 km	> 1000 km
RESOL. SPACE	< 100 m	< 1 km	< 30 km	< 30 km	< 30 km
RESOL. TIME	< 1 hrs	< 3 hrs	< 12 hrs	< 3 days	< 1 week

data at 5.3 GHz and a spatial resolution of 25 km are useful for determining large-scale ice drifts in the Arctic Ocean. This type af data is complementary to microwave radiometer data and may therefore be exploited in the context of Strategic Planning, but further studies are needed to fully ascertain the potential of this proposition.

In addition to the functional characteristics also the parameter retrieval should be considered. Present satellite techniques determine the ice extent i.e. the boundary between the ocean and the ice canopy at sufficient accuracy in all planning cases except for Close Tactical Planning. Also, ice concentration is determined with good accuracy while ice classification seems to be less useful but fortunately of less importance. In Close Tactical Planning information about leads, size and direction, and of ridging are important parameters which may only be obtained from SAR data at finest resolution and still, detection and determination of ridges may be considered a development area.

4 CLIMATE RESEARCH AND MODELLING

Climate research and climate change research have a component of sea ice mainly due to the influence of sea ice on the heat flux from the ocean to the atmosphere. Considering that during winter there may be a temperature difference between the ocean and the atmosphere of more than 20 to 30 degrees and that sea ice reduces the heat flux very drastically it will be understood that the formation of sea ice in an area will have a major impact on the atmosphere. Often it is observed that during ice formation large areas, measured in thousands of square kilometres, may be covered by sea ice within few days due to a positive feedback mechanism with cooling of the air over the ice and consequently a cooling of the adjacent open water.

Table 1 may also serve the purpose of discussing the requirements to remote sensing in the context of climate research. To some extent the columns of Strategic, Medium and Long Range Planning are describing the situation related to modelling at different scales. Thus, Strategic Planning describes the requirements related to so-called Local Area Models or High-Resolution Local Area Models (HIRLAM), the latter with a grid size of 15-20 km. Presently, these models may be served by AVHRR and SAR data. In the future, data from the passive MIMR with a spatial resolution of 5 km may be very useful as a supply of the important parameters of ice extent and concentration. Although ice is very dynamic the resolution time of less than 12 hours may be relaxed to 24 or 72 days in the case of modelling. Medium Range and Long Range Planning relate to global models which will be useful in global climate research although there seems to be a trend of reducing the grid size of these models with the advent of more and more advanced and faster computers.

Long-term observations from satellite is dependent upon the life time of the sensors and the satellite and since there is a tendency that follow-on mission takes advantage of the latest technical development to improve accuracy or life time there is a problem of continuity - well-known in meteorology for example. Overlap in time between different missions is therefore a must and suitable means should be applied to "calibrate" one sensor by its predecessor. An example of this is the two instruments SMMR and SSM/I (Scanning Multichannel Microwave Radiometer on Nimbus 7 and Special Sensor Microwave Imager on DMSP satellites) launched in sequence whereby a data set of more than 15 years is now available. Using a two-month overlap in 1987 data were compared over a stable region like the Antarctic Ice Cap. A difference in brightness temperature of as much as 14 K was observed and corrected for to establish continuity (Zabel & Jezek 1994) - of importance also for sea ice observations.

4.1 Ice dynamics modelling

In addition to long-term monitoring of geophysical parameters climate research has a strong component of numerical modelling partly to understand the observed variations and to be able to predict the climatic trend and partly to study geophysical processes that are part of the climate system. In the polar regions modelling is generally made to describe the air-sea-ice (A-S-I) interaction which may be described qualitatively by the now classical Figure 1 showing a number of exchange processes occurring during both winter and summer periods.

A full model description of an ice covered region includes by tradition three models: an ice dynamics model, an ice-ocean model and an ice-atmosphere model. These models are coupled to describe the total

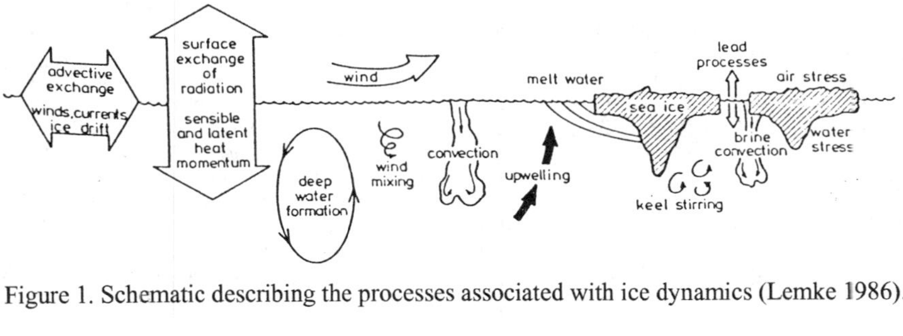

Figure 1. Schematic describing the processes associated with ice dynamics (Lemke 1986).

physical situation. A thorough review of these models is given by Preller et al. (1992) including an extensive list of references. Here, we shall only give a brief account of relevance to remote sensing.

The dynamic formulation describes the momentum balance for ice mass, i.e. the total mass of ice in a resolution cell of the model, influenced by ice motion, ice growth and decay. It includes parameters such as ice mass and velocity, the stress of wind and ocean on the upper and lower ice surfaces, the influence of the surface tilt of the ocean and a term, internal ice stress, taking deformation rate and ice strength into account. The ice strength which is related to the movement of the ice is a function of the ice thickness and (exponentially) of the ice concentration or more precisely by the (complementary) minimum lead fraction i.e. the amount of open water and thin ice (in the resolution cell) which is continually created by deformation and freezing.

The formulation includes a thermodynamic component which determine the rate of ice growth and melt. It comprise parameters such as surface albedo (for snow and bare ice), solar shortwave radiation describing the flux of energy into the ice surface, longwave radiation from clouds and outgoing from the surfaces, atmospheric and ocean heat fluxes including the temperature of the mixed ocean layer and ice surfaces, and a number of physical constants of the media involved, ocean, ice and atmosphere.

4.2 Ocean-ice and Atmosphere-ice models

Ice-ocean models have been developed in which the ice dynamics model is coupled to a multi-level ocean model which includes temperature and salinity profiles and in one case also the forcing of the ocean obtained from a global atmospheric model.

Thus, there is a link between the ocean model and the atmospheric model which give the geostrophic wind retrieved from global or regional atmospheric models developed to the grid size (resolution cell) of the model in question. For studies of local and regional cases Local Area Models or High-Resolution Local Area Models (HIRLAM) are applied.

The three coupled models which are developed initially to study specific processes in polar regions may be used with some success for short-term predictions of the ice conditions. However, they need a number of parameters for initialization and validation as will be understood from the above descriptions. Presently, satellite remote sensing may supply some of them, i.e. the ice extent, the ice concentration or lead fraction, coarse ice classification into first-year and multiyear ice and recently also thin ice, and ice velocity vector fields. In fact, since in situ measurements are not generally available from these regions, satellite data become crucial for running the models.

4.3 Model sensitivities

The success of operating a model is - of course - dependent inter alia on the accuracy of the parameters used for initialization of the model. With present remote sensing techniques and parameter retrieval methods the accuracy of the above-mentioned parameters seems largely sufficient for large-scale modelling whereas improvements are necessary when dealing with meso-scale modelling. However, sensitivity studies of the models have been carried out in order to ascertain which parameters of the many are the most critical with respect to accuracy or quality. Whereas previous studies were made as single-parameter sensitivity investigations, a recent study (Chapman et al. 1994) has dealt with the model sensitivity by varying a number of parameters simultaneously using an advanced statistical procedure. It investigates the ice dynamics model as described above where 13 parameters are varied simultaneously to study their effect on issues such as

Ice mass
Ice area
Mean drift speed
Ice range of area

the latter describing the difference between the ice area in winter and summer. The study addresses the entire Arctic region at a coarse scale of 110 km and so the results describe the total ice mass in this region, for instance.

The parameters which are varied are

Atmospheric drag coefficient
Oceanic drag coefficient
Ice strength parameter
Minimum lead fraction
Snow albedo
Ice albedo
Open-water albedo
Exchange coefficient, surface sensible heat
Exchange coefficient, surface latent heat
Snowfall rate
Cloud reduction of shortwave solar flux
Cloud enhancement of longwave flux
Ocean heat flux

Each of these parameters are varied randomly in prescribed ranges in 157 statistical experiments including cross-validation tests. From these experiments the following observations were made:

- The ice mass depends strongly on the two parameters, the minimum lead fraction and the exchange coefficient for surface sensible heat whereas the other parameters play a minor role with a small influence of clouds on the longwave radiation as an exception. It is shown that the two variables modify the surface heat balance in the way that the effect of an increase of the minimum lead fraction is greater if the heat transfer coefficient is large and visa versa. A large fraction of leads increase the heat flow from the ocean to the atmosphere and so does a larger heat transfer coefficient, and thereby larger ice masses are formed during winter due to increased cooling of the water surface. It is worth mentioning that the ice mass is almost linearly dependent upon the two parameters in the ranges of 0 - 6% and 0 - 6, respectively, in an almost one-to-one relationship. Thus, the ice mass increases from 20 to 60 10^{15} kg with a change in the lead fraction changing from 1 - 5%, for instance.

- The ice area is for the greater part dependent upon the sensible heat exchange coefficient and to a smaller degree upon cloud cover,.but also the latent heat exchange coefficient has some influence. A large exchange coefficient for sensible heat will increase the ice formation and the ice thickness so that the ice area becomes larger with time. Similarly, the evaporation "controlled" by the latent heat exchange coefficient will cool the ocean surface and thus increase the ice formation. The snow albedo plays a minorrole with the other parameters including the lead fraction having little or no influence.
- The mean ice drift speed is controlled almost entirely by the two drag coefficients but also the lead fraction and the ice strength play roles with the sensible heat transfer coefficient having a small influence. It is interesting to note that the surface drags on the floes in general are counteracting. The ice is largely wind driven with the ocean drag retarding the floe movement. However, this might not apply in some areas like the Greenland Sea where the drift is largely determined by the south-going East Greenland Current with the wind having an advancing or a retarding effect with dominant northern and southern winds, respectively. The study shows that the dependence of the atmospheric drag is weaker when the oceanic drag coefficient is large.
- Ice range of area or the difference between winter and summer sea ice extent is dependent on the two drag coefficients at almost the same rate as well as the ice strength and the minimum lead fraction with a smaller influence by the snow albedo and snow fall rate. The study shows that there is a strong interdependence among the drag coefficients but also between the atmospheric drag coefficient and the ice strength.

The conclusions we may draw from this study - and other one or two-dimensional parameter studies vis à vis remote sensing is that the minimum lead fraction or the ice concentration is the most important factor which therefore has to be determined with great accuracy. The numerical example given above points to the fact that major improvements of the present retrieval accuracy are needed. It is suggested that a goal of 2% should be aimed at. The literature refers to accuracies that range between 5 and 15% for global retrieval of the ice concentration by means of passive microwave radiometer data whereas 3% has been quoted when algorithms are tuned to regional conditions. A combination of passive and active data may lead to a solution. In regional studies SAR data may be the way forward indicated by recent studies using ERS-1 SAR data.

Presently, the atmospheric drag coefficient may only be determined by field measurements but it is suggested that better accuracy of actual mean coefficients may be obtained when such measurements are related to simultaneous observations by fine-resolution SAR, but the accuracy is debatable. In this context, the distinction between first-year and multiyear ice is of interest as an indicator of ice thickness and surface smoothness. But so far the multiyear ice fraction, for instance, is not retrieved by means of passive microwave data to better accuracy than 10 to 15% . During summer melt conditions it is not possible to distinguish between the two ice classes. Therefore, the observation of the on-set of melt and freezing is important. SAR data are likely to give a better retrieval, and accuracies of 5 to 10% have been quoted and from pattern.recognition it may be possible to distinguish between the classes even during summer conditions. The oceanic drag coefficient may only be determined by field measurements but the ice classification may also be essential in selecting proper (regional) numbers for the coefficients.

Heat fluxes cannot be measured by satellite except for large-scale observations of the outgoing longwave radiation fluxes. However, the fluxes are modified by clouds so that a reliable cloud cover observation may improve the situation considerably compared to that where only sporadic field measurements are available. The sensitivity study showed that cloud cover had a small influence on the determination of ice mass and ice area. This is clearly a research area. In the same context, the temperature of open water and ice surfaces are of importance. Presently, they may be determined from satellite at a spatial resolution of about 1 km and is thus useful for measurements over large floes and large open-water bodies whereas the interesting situation with smaller leads during winter conditions cannot be approached.

The sensitivity study is a large-scale study and it is not clear to what extent it is representative for regional modelling at smaller scales. The Greenland Sea case referred to above points to some discrepancies. However, the general conclusions are probably of significance in this context. This should be studied further.

5 CONCLUSIONS

Ice observation is of great importance for shipping in ice-frequented waters and may serve the purpose of guiding ships and icebreakers through the ice canopy if the data are received on board at a short delay. The requirements are different for different ranges of planning ship routes, distinguishing between Close tactical, Tactical or Strategic planning with 6 and 24 hours and 7 days terms. The requirements from the first-mentioned range may only be fulfilled by helicopter observations whereas the other may be satisfied by remote sensing assuming that the data are available to the user within three hours, analyzed and interpreted. Tactical planning require frequent observation i.e. every 24 hour at a minimum.

Remote sensing may assist modellers in contributing important parameters such as ice extent, ice concentration, ice formation and melt, ice motion, wind (in larger open-water areas only), ice surface temperature and cloud cover fraction, some of them used as input data to initialize models and others as control data. A number of other parameters may not be retrieved by remote sensing or only at large scales and often they are obtained from field measurements carried out in various regions and at different times of the year, etc.

Present A-S-I models serve the purpose of studying the interaction processes with advanced versions coupled to three-dimensional meteorological and oceanographical models. These coupled models may serve as prediction models and coupled to models for lower latitudes as a means of studying global climate change.

A multi-dimensional sensitivity study of the ice dynamics model has demonstrated that the so-called minimum lead fraction is the most important parameter of the 13 parameters studied. It is concluded that this parameter, or the ice concentration should be determined at an accuracy of at least 2% and that therefore additional research is needed first of all on a regional scale. Other important parameters are the sensible heat exchange coefficient and the atmospheric and oceanic drag coefficients. Still other parameters are the ice strength, the snow albedo and the snow fall rate. These parameters may only be retrieved from satellite data with great uncertainty but coupled with surface measurements carried out under field campaigns general or average numbers may be arrived at,.that may improve the model performance.

The result of any model describing geophysical phenomena or regional or global ice conditions depend upon the accuracy of the parameters used to initialize the models. The challenge to remote sensing is therefore to improve the retrieval techniques possibly by combining data of different origin into a multimission, multi-data and multitemporal system. In operational services the challenge lies more on the operational aspects where the frequency of observation and the dissemination delay times are the crucial problems.

REFERENCES

Chapman, W.L.,W.J. Welch, K.P. Bowman, J. Sacks, J.E.Walsh 1994. Arctic sea ice variability: Model sensitivities and multidecadal simulation. J. Geophys. Res.,99, C1: 919-935.

Lemke, P. 1986. Stochastic description of the atmosphere-sea ice-ocean interaction, Chapter 13 in N. Untersteiner (ed.) The geophysics of sea ice, NATO ASI Series B: Physics, Plenum Press, Vol 146, 785-823.

Preller, R.H., J.E. Walsh, J.A. Maslanik, 1992. The use of satellite observations in ice cover simulations, Chapter 22 in F.D. Carsey (ed.), Microwave Remote Sensing of Sea Ice, Geophysical Monograph 68, 385-404.

Ramseier, R. & K. Strübing, 1994. Personal communication.

Zabel, I.H.H. & K.C. Jezek 1994. Consistency in long-term observations of oceans and ice from space. J. Geophys. Res. 99. C5: 10,109-10,120.

Sensors and Environmental Applications of Remote Sensing, Askne (ed.) © 1995 Balkema, Rotterdam, ISBN 90 5410 524 0

Arctic sea ice climate system parameters as observed by synthetic aperture radar

J. Askne
Department of Radio and Space Science, Chalmers University of Technology, Göteborg, Sweden

ABSTRACT: As part of the ARCTIC-91 expedition August - October 1991 the first field expedition to the high Arctic took place with simultaneous coverage by satellite SAR. The aim was to study ice properties of importance for climate aspects. During the advanced melt period the ice concentration and melt pond concentration are important quantities and during the early freeze up the variability of the sea ice signatures and the new ice formation are of high interest. This paper will give the general background and results based on ARCTIC-91. In particular the possibilities to use SAR to study effects due to melt ponding will be considered.

1 INTRODUCTION

Models of the earth as a system indicate an increase in global temperature due to the green house effect. The effects are taking place over time scales of decades and early warning signs are important to identify. The models predict that effects due to a global temperature rise will be amplified in the Arctic (Stouffer et al. 1989) due to the various feedback mechanisms associated with a strong coupling between the Arctic sea ice cover, the atmosphere and the ocean.

Modelling is a very complex process and each one of the different models for the atmosphere, sea ice, and ocean has to be further developed. Recently a model of the Arctic sea ice was analysed (Chapman et al. 1994). Different parameters of the ice model were varied. Ice mass, ice area, mean ice drift, and winter-summer range of area were used as response parameters of the model. The strongest sensitivities arise from the minimum lead fraction, the sensible heat exchange coefficient and the atmospheric and oceanic drag coefficients.

Consequently there is a need to observe the various parameters of interest as indicators of a global temperature change and several remote sensing sensors are used to follow the ice properties and the cloud situation, (Gloersen and Campbell 1991), (Barry et al. 1993). However, we can not use a single remote sensing sensor to study the important properties of the Arctic ice cover but have to combine observations from various sensors and develop new techniques to observe some of the properties.

2 SEA ICE MODEL

We will follow (Chapman et al. 1994), who used a two-level dynamic formulation going back to (Hibler 1979) and a thermodynamic formulation from (Parkinson and Washington 1979) with various extensions. The dynamic formulation is based on momentum balance for a mass of ice which is affected by the air and water stresses prescribed by drag coefficients, Coriolis force, acceleration due to sea surface tilt and internal stress related to ice deformation and ice strength. The thermodynamic model is based on the surface energy budget, where the flux of the solar radiation, the downcoming flux of longwave radiation from the atmosphere, the transfer of sensible and latent heat between the ice and the atmosphere, the surface emission of Planck radiation, and the thermal conductivity to the ocean are balanced.

The temporal variability is based on a prescribed cloud fraction varying between winter and summer, monthly snowfall rates and daily grids of surface winds and air temperature from 1979 onward. The sensitivity test of effects of variations of the 13 different parameters require simulations of about five years in order to determine the equilibrium response, which limits the evaluation of the interdependencies of the parameters.

Ice mass, ice area, mean drift speed, and ice range (winter to summer) of area are determined as function of the various parameters. Results for mass area and ice area are concluded to be more robust than the others.

The ice mass is most sensitive to the minimum lead fraction, while the ice area is most sensitive to the turbulent exchange of sensible heat, cloud depletion of longwave flux, the turbulent exchange of latent heat, and the snow albedo. The model demonstrates a negative trend in the wintertime maxima of ice covered area, which decrease by approximately 5% from the mid-1970s to the mid-1980s, consistent with observations by (Gloersen and Campbell 1991). However it is warned that the model can easily be adjusted to produce trends of either sign in the simulated ice mass or ice area and the model was indeed tuned to give a temporally stable time series by varying the cloud depletion of solar flux. It is also stressed that the correlation between simulated and observed variations of ice coverage in various parts of the Arctic is such that much of the observed variance is unexplained by the model.

Although models of the kind described above are hampered by the complexity of the phenomena, the limitations of computer simulations and the uncertainties of parameter variations, the models can give information on important parameters to monitor, the need for accuracy, the need for sampling frequency etc. E.g. the total ice mass varies by nearly a factor of 2 as the minimum lead fraction varies from 1 to 4%.

The advance of the melt process during the summer should be important for global climate models. The ice floe melts from the surface, sides, and bottom (Steele 1992). The lateral melt is significant only for small floes while the surface and bottom melt dominates are nearly equal for large floes. One aspect of the melt process is the melt ponding. The variability of the melt pond area should be of interest to determine in order to characterise early stages in the degree of melting. The snow melt is typical for the Arctic (in contrast to the Antarctic) and the Arctic old ice may be covered in the summer by 10-60% of melt ponds (Gow and Tucker 1990). Increasing melt pond coverage serves to accelerate melt rates as more heat is absorbed from incoming short-wave radiation.

3 REMOTE SENSING OF THE ARCTIC

The Arctic has been covered regularly independent of clouds and weather conditions by passive multichannel microwave radiometry since 1978, (Gloersen et al. 1992). Radar altimeters have also been used since a long time and has been used for sea ice studies, (Laxon 1990). Synthetic aperture radars onboard Seasat 1978 for three months and ERS-1 since 1991 and JERS-1 since 1992 provide high resolution information of Arctic ice. With the swath limited to 100 km the sensor can be used to analyse test areas of the Arctic, while sensors like SSM/I are more appropriate for large coverage. RADARSAT and ASAR on Envisat will provide information over swaths of the order of 400/500 km and the SAR will be more useful for coverage of large areas of the Arctic.

The technological developments are putting constraints on what we can achieve by remote sensing and we are forced to develop the information obtained from available and planned sensors while we at the same time develop sensors which can measure the wanted parameters with sufficient accuracy. The sensitivity analysis described above (Chapman et al. 1994) illustrates the importance of a large number of parameters, only some of them may be possible to obtain by means of present day remote sensing techniques. Such properties include the ice area, and drift speed and it should be possible to relate these properties to e.g. the minimum lead fraction. However the modelling does not include a very detailed analysis of all possible phenomena and their variation by season, and there may be other properties as well of interest to determine by means of remote sensing.

Passive microwave sensors have been the standard method for long time observations of Arctic and Antarctic sea ice, see (Gloersen et al. 1992). The Antarctic sea ice extent mapped by radar altimeter, (Laxon 1990), reveals excellent overall agreement but show differences in the ice extent mapped near the end of the melting period. This was also shown to be the case in a study of Arctic sea ice areas, (Laxon and Askne 1991). The differences may be attributed to effects of e.g. melt ponds or ice floes in combination with relatively large open water areas. High resolution sensors are necessary to study the discrepancies.

With the high resolution of the SAR we can observe other parameters than with low resolution sensors such as SSM/I, and of the above mentioned properties we may observe the minimum lead fraction and the ice concentration with high accuracy, we can determine ice drift, we can observe properties of the wind and may relate these to the atmospheric drag coefficient of the sea ice. From high resolution SAR images we may also identify different ice classes and then indirectly various types of ice properties. By combination with information from the observed ice motion and possibly also with thermodynamic modelling further constraints can be obtained in order to identify ice thickness, and ice mass. However this is a complex procedure.

In this paper we are studying in particular the possibilities of the SAR sensor. One unique property is the possibility to determine ice motion in spite of the cloud cover. The high resolution of the SAR is also important to determine melt pond concentration, minimum lead concentration etc. The lead size is important for the melting process, and the reduced ice strengths during melting coupled to the ice motion accelerates the melt process by producing smaller ice floes, more sensitive to lateral melt. The SAR is mainly used for limited areas of the Arctic but by statistical sampling of properties over different parts of the Arctic the entire Arctic may be characterised. The SAR can also be used to derive local tie point values for the passive microwave sensors. The combined use of data from different sensors is necessary to overcome the limitations of each one with the goal to provide data for the Arctic sea ice.

4 RESULTS FROM ARCTIC-91

Only field expeditions can provide the information to validate the capacity of a sensor, and ARCTIC-91 provided for the very first time during the life time of ERS-1 information on sea ice properties in the high Arctic together with simultaneous ERS-1 coverage. The ARCTIC-91 expedition took place August - October 1991 during the advanced melt period and the early freeze-up period. The evaluation program has coupled satellite data with *in-situ* data and electromagnetic modelling (Carlström and Ulander 1993) (Ulander et al. 1993) In order to combine different sources of information comparison of SAR and SSM/I data has taken place (Pettersson et al. 1994). Ice motion has illustrated phenomena on different time scales from 100 minutes and up (Sun 1993) (Sun and Askne 1994).

Ships observations (Ulander et al. 1992) provided hourly information such as ice concentration, melt pond concentration, snow fall, wind etc. The remote sensing and sea ice program provided information on properties of importance for remote sensing of sea ice such as properties of the ice surface, temperature and salinity of ice cores, properties of snow layer, etc.

5 OPEN WATER SIGNATURES

The basis for ice concentration analysis is the contrast in the SAR image between the ice and the open leads. The open water leads are wind roughened and can show quite varying radar backscatter. The open water signature is not established completely for a SAR, but the signature for the ERS-1 scatterometer is illustrated by Fig. 1.

The scatterometer signatures derived for areas of 50 $\times 50$ km^2 have been verified within the estimated accuracy of the scatterometer of ± 2 m/s, but for the SAR with its higher resolution more local conditions may be studied.

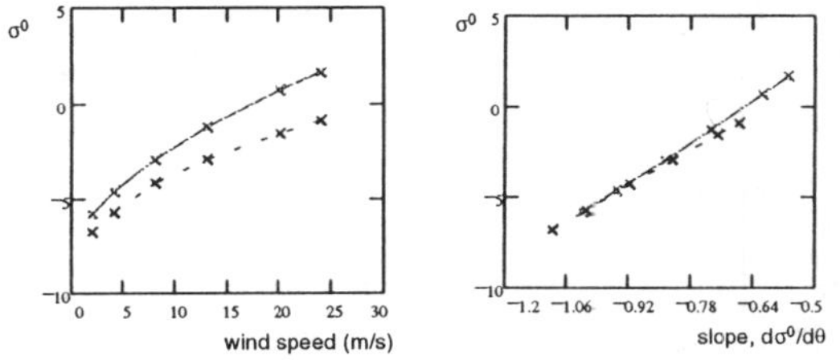

Fig. 1 Radar backscattering coefficient versus wind velocity and slope. The marked points correspond to 2, 4, 8, 13, 20, and 24 m/s wind speed. Solid line marks upwind and dotted line cross wind conditions. (From (Long 1992) (Askne 1993).)

Comparison of the local wind information from Oden with the backscatter averaged over a large area we have obtained reasonable agreement with the scatterometer expressions taking into account their uncertainties.

6 ADVANCED MELT PERIOD

During this period we have the minimum ice extent and various properties contribute to this minimum value, including the maximum melt pond coverage

6. 1 SAR signatures during the melt period

During the advanced melt period we only have old ice covered by melt ponds and open water. The old ice has a very low backscatter and with a wind speed of only a few meters wind per second a contrast relative the open water. Ice floes can be identified by their form and typical backscatter values for old ice and open water can be identified over the image. An illustration is given in Fig. 2 as function of incidence angle. If we have no systematic changes of the backscatter conditions along track we may fit a line to the identified values and these lines then represent the best estimate of the ice and open water signatures which may be compared to the along track mean backscatter value to determine the ice concentration, see (Askne et al. 1994). The variability of the old ice illustrated in Fig. 2 is caused by variability in melt pond properties and in the bare ice conditions, while the variability in the open water signature is mainly caused by wind variability but also by e.g. melt water runoff etc.

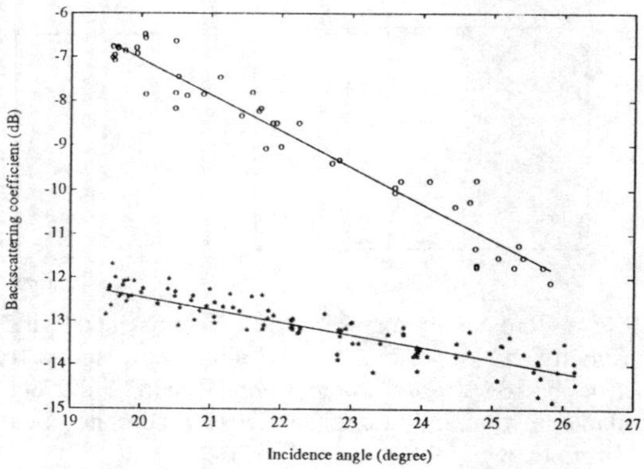

Fig. 2 Illustration of backscattering coefficient values obtained over the swath of ERS-1 SAR on 18 August 1991. Backscatter values over areas typical for open water or sea ice have been determined and the regression line is an estimate of the mean properties.

6. 2 The SAR melt pond signature

Under some conditions we may determine the melt pond coverage by means of the SAR. Let us make the following assumptions:

- We assume we have melting conditions (determined from meteorological charts)
- We assume that the melting ice consists only of melt ponds and old ice in between
- We assume that the melt ponds have the same backscattering signature as open water.

The backscattering coefficient for the old ice, σ_{ice}, would then be determined by an area average

$$10^{\frac{\sigma_{ice}}{10}} = \mu 10^{\frac{\sigma_{ow}}{10}} + (1-\mu)10^{\frac{\sigma_{bi}}{10}} \qquad (6.1)$$

where σ_{ice} is the backscatter value for ice including melt ponds, σ_{ow} is the value for open water and σ_{bi} is the value for bare old ice, all values expressed in dB. We then obtain for the melt pond concentration μ

$$\mu = \frac{10^{\frac{\sigma_{ice}}{10}} - 10^{\frac{\sigma_{bi}}{10}}}{10^{\frac{\sigma_{ow}}{10}} - 10^{\frac{\sigma_{bi}}{10}}} \qquad (6.2)$$

σ_{ice} and σ_{ow} are determined from the image and if we can identify values on σ_{bi} either from the SAR image or by modelling we can determine the melt pond coverage μ. Fig. 2 illustrated values for σ_{ice} and σ_{ow} derived by averaging along track the SAR image, resulting in mean values for different incidence angles. In (Askne et al. 1993) values for σ_{bi} were identified using video images of SAR covered ice areas to -15.1 dB for ERS-1 SAR from 20 August. The question is how stable this value is or if σ_{bi} can be determined by other methods. This may be the case if we have overlapping scenes from consecutive orbits. We may then expect with the above assumptions that the only difference in backscatter from the old ice is due to the difference caused by the wind effect on the melt ponds, assumed to be the same as that on the open water signature. In that way the bare ice value σ_{bi} can be determined from the image pairs. This has been done i 3 cases, and resulted in values consistent with the value above.

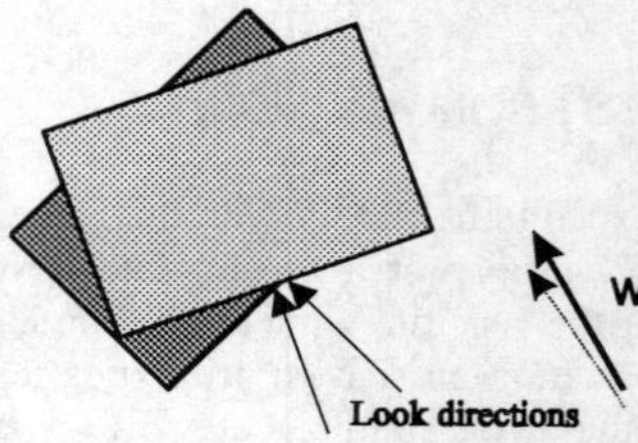

Fig. 3 Illustration of SAR imaging from two consecutive orbits with some possible change of wind conditions.

- Assume now we have the same melt pond coverage over the entire scene. If so we expect the same melt pond coverage, μ, from both the two images. (In principle we should identify the same area in the two images which are not co-located.)

In Fig. 3 we have illustrated the relation between σ_{ice} and σ_{ow} for different values of μ (7, 13, and 20%) assuming σ_{bi} = -15.1 dB. The curves converge to the point $\sigma_{ice} = \sigma_{bi} = \sigma_{ow}$ and the sensitivity for uncertainties in σ_{bi} is depending on σ_{ow} as indicated by the figure. Experimental values have been included from three pairs of the same area acquired 17, 18, and 19 August 1991. The pairs indicate a melt pond coverage of 13, 20, and 20%, indicated be "diamond"-signs in Fig. 3. The pairs fall on one and the same melt pond concentration curve, in agreement with the assumed σ_{bi} value. From the icebreaker Oden we have observations from the area corresponding to the SAR scenes from 18 and 19 August. Oden reports on 20 August a mean of 20% varying from 10 to 40% and with an outlier of 60%. Although the results are very limited in number the results are consistent.

A basic assumption in the analysis is that the melt ponds yield the same backscatter as open water. It is expected that the water surface of the melt ponds

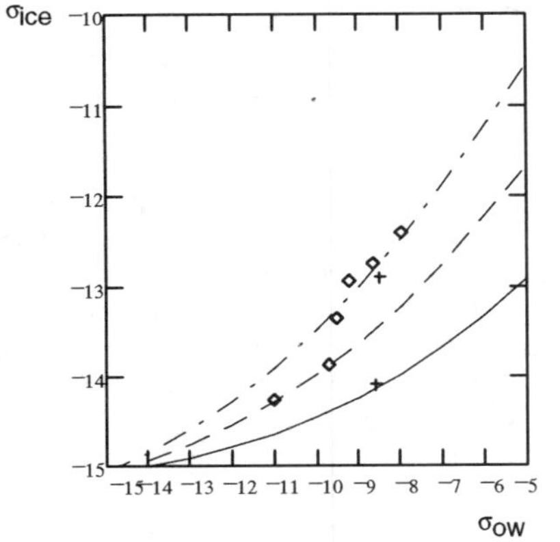

Fig. 4 Illustration of backscattering values 17 August - 20 August 1991 and curves corresponding to 7, 13, and 20% melt pond coverage. For explanation see text.

would give a lower value on σ due to wind shielding and/or fetch effects as well as depth effects. The SAR derived figure would then rather be a lower limit. The wind over ice may also differ somewhat from the wind over open water in particular when the ice concentration is low. The results from the image pairs on 18 and 19 August and the corresponding results from Oden indicate that the difference may be small under the conditions on these days.

In Fig. 3 the upper plus sign in Fig. 3 is derived from a single SAR image on 20 August. The analysis was based on the assumption of non-frozen melt pond surfaces. On 20 August we had a temperature transition region affecting the radar backscatter according to Fig. 4, cf. (Askne et al. 1993). The backscatter values in Fig. 4 is illustrated together with the modelled value from a surface with a linear temperature decrease with increasing distance assuming the σ_{bi} value changing from - 14.7 to -10.0 dB

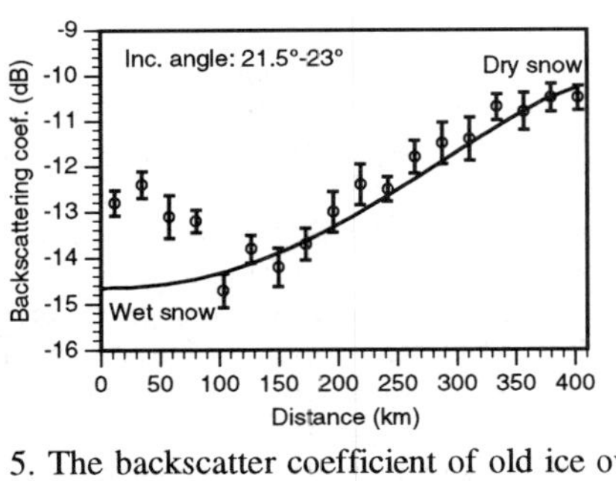

Fig. 5. The backscatter coefficient of old ice over an area where temperatures change from above to below 0 °C, from (Askne et al. 1994).

mainly due to volume scattering and decreased surface attenuation. The 0°C transition zone is expected somewhere at 150 km. The value of 20% melt pond coverage was derived over the 0 - 100 km area. The SAR scene covering the 100 - 200 km range yields a result representing the lower "plus" in Fig. 3 at 7%. From single scenes over the area 300 - 400 km we would have derived a completely wrong value on melt pond coverage due to the freezing of the melt pond surface. It is therefore important either to have image pairs over the same area or temperature charts ensuring that we have melting conditions.

6.3 The SSM/I melt pond signature

The SSM/I ice concentration value for the position of Oden (courtesy Don Cavalieri) is illustrated in Fig. 6 together with daily mean values of ice concentration and melt pond concentration along the route of Oden. (Values for day 232 missing as Oden was not moving this day.)

We see a difference between field observations and SSM/I observations of ice concentration which may be attributed to the melt pond coverage of the old ice, cf. (Comiso and Kwok 1992). This difference is decreasing when the temperature decreases possibly due to freezing of the melt ponds. Such freezing

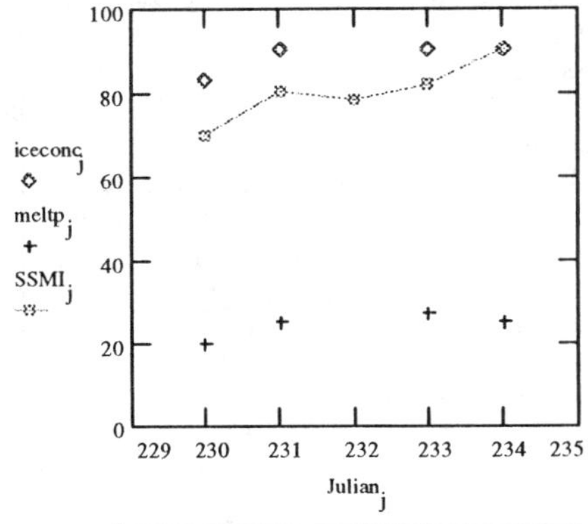

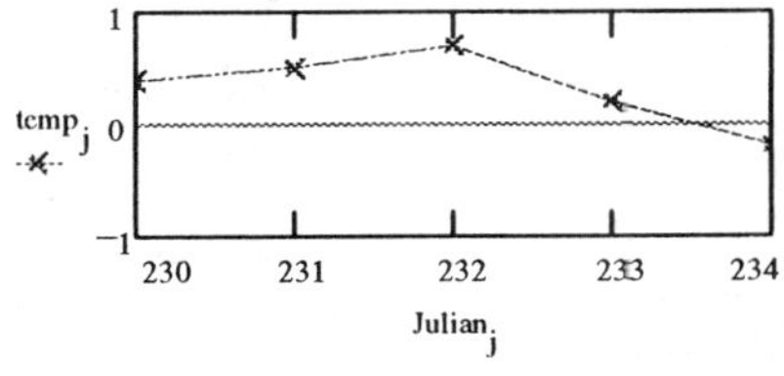

Fig. 6. NASA team SSM/I ice concentration (courtesy Cavalieri) compared with daily averages of ice and melt pond concentration along the Oden route. Daily mean temperatures illustrated below.

has been observed in the field (courtesy L. Ulander). When the ice is covered by 20% of melt ponds we expect the ice concentration measured by SSM/I to decrease by 20% of the true ice concentration, in line with Fig. 6. It should be noted that the exact co-location of satellite and field information is important but less so for SSM/I with a 50 km resolution than for the SAR. For day 230 the ice concentration obtained from a SAR image is 61% while the Oden report from the corresponding area approximately corrected for ice drift is 60% (Askne and Ulander 1992).

7 EARLY FREEZE-UP PERIOD

The early freeze-up period studied during ARCTIC-91 is from 21 September - 4 October when Oden was on its southbound route and covered by ERS-1 south of latitude 84.6°. For details we refer to (Pettersson et al. 1994) in this volume. During this period we have a decrease in temperature from zero to -15 with some variability and the precipitation is in the form of snowfall. New ice is formed, sometimes combined with frost flower formation on the new ice surface, (Ulander et al. 1993). The backscatter of the multi year ice stabilises at a level of -5 to -8 dB while the new ice formation yields signatures at a lower level as long as the transient frost flower phenomenon does not take place. Areas covered by frost flower and saline wet snow may as well as the open water areas yield backscattering coefficients of the same order as multi year ice. Thin ice is important for the salt production and for the thermal exchange between the atmosphere and ocean and can be recognised by its low signature. However the ice concentration, which is very high, can only be determined with high accuracy if the wind is sufficiently high, of the order of 5 m/s and higher. The variability of the signatures causes problems for automatic routines for ice class segmentation. However the problems are not sufficiently serious to be a great problem for ice motion recognition by identifying major similarities between the images, (Sun 1993). During this period the ice concentration is very low and we are approaching the period with the minimum lead extent. It is important to combine the information from the high resolution SAR sensor with the passive microwave sensors with high radiometric resolution in order to identify features of importance to characterise parameters of importance for the climate system description.

8 FUTURE NEEDS

The resolution of the SAR allows us to study details like ice ridges, leads and features associated with new ice formation. The SAR provides sufficient resolution in order to determine ice drift and can then provide information with details hereto impossible using buoys. However the single channel information from the SAR limits the classification properties. We encounter different conditions and fast changing conditions between the seasons and even within a specific season. For that purpose further field experiments are very important in combination with simultaneous satellite coverage. In order to provide support from scattering models it is important to study the ice with multi-frequency, multi-polarisation scatterometer where the different channels provide constraints on the interpretation and determines the sensitivity to various ice parameters. Such information is crucial for the interpretation of the single channel information from ERS-1 and is of importance for future satellite sensors providing more information.

ERS-2 (to be launched 1995) will hopefully extend the ERS-1 dates with high radiometric and geometric resolution data over a 100 km swath, while RADARSAT (to be launched 1995) will provide wide swath information of lower radiometric accuracy but expected to be of high importance for understanding spatial variations. The combined use of the two sensors would be very useful together with the field experience. SSM/I would similarly provide information of very high value.

For the field experiment it is important to stress some investigations relative those carried out in ARCTIC-91, dielectric properties of the snow and ice surface, variability of properties, air bubble inclusions in the upper layer of the sea ice to mention some.

9 CONCLUSION

The sensitivity of the Arctic cover to a global temperature increase means that it is important to develop the information extraction from the available satellite sensors. The addition of the SAR has given new possibilities to study details of relevance in them selves as well as providing information for large scale sensors. Today we have good possibilities to study ice motion and related effects on ice formation and deformation. We have increased possibilities to study parameters like melt ponds, ice concentration, leads and wind in leads, ice types. This information from the SAR will clearly be useful for estimation of geophysical parameters of importance for studying the

Arctic sea ice climate system. Properties based on the geometric resolution such as ice motion can be derived from the SAR operationally while properties based on the radiometric resolution such as ice classification is more of a problem. With the help of the geometric resolution, properties such as leads, ice floes and even melt ponds can be recognised during the melt period and ice concentration and melt pond concentration determined with high accuracy. For the freeze-up period the spatial resolution yields possibilities to follow the change of backscatter but the single channel information limits the interpretation.

The analysis in this paper has been based on experience from field observations in conjunction with ARCTIC-91. Field experiments yield the possibility to derive the relation between field conditions and the observations by remote sensing to some detail but only for the cases studied. By necessity these observations are limited and should be generalised by studies of great amounts of satellite data under relatively constant field conditions. The observations during ARCTIC-91 took place mainly in the high Arctic, far from the ice boundaries with more dynamic conditions. The melt pond concentration was studied during the advanced melt period and in areas with relatively small variations of the back scatter relative the incidence angle dependence, cf. Fig. 2, ensuring relatively stable conditions. The results were also shown to be very sensitive to the temperature. The melt pond surface must not be partly frozen or covered by snow if the radar signature should resemble that of open water. Therefore the results can only be generalised to other conditions with great care and should also be further tested by SAR data from the same period during other years.

ACKNOWLEDGEMENT

SSM/I data obtained from Don Cavalieri, NASA is gratefully acknowledged. ERS-1 SAR data was provided by ESA within the PIPOR AO-projects and financial support for the evaluation was obtained from the Swedish National Space Board. Discussions within the Remote Sensing Group at Chalmers University of Technology are gratefully acknowledged.

REFERENCES

Askne, J. (1993). Wind derived from leads in ice covered areas as seen in SAR images. Abstract volume, Third Scandinavian SAR Symposium, Göteborg.

Askne, J., A. Carlström, et al. (1994) ERS-1 SAR backscatter modeling and interpretation of sea ice signatures. to be presented at IGARSS-94,L os Angeles,

Askne, J. and L. M. H. Ulander (1992). "Remote sensing of Arctic sea ice using the ERS-1 SAR." *Central Symposium of the International Space Year* ESA **SP-341**: 129-133.

Askne, J., L. M. H. Ulander, et al. (1993). Accuracy of ice concentration derived from ERS-1 SAR images during the late melt period in the Arctic. Microwave Remote Sensing of Ice, Lyngby, to be published in EARSeL Advances in Remote Sensing.

Askne, J., L. M. H. Ulander, et al. (1994). Validation of ERS-1 SAR for Arctic ice properties from ARCTIC-91. Second ERS-1 Symposium, Hamburg, ESA **SP-361**, 257-262.

Barry, R. G., M. C. Serreze, et al. (1993). "The Arctic Sea Ice - Climate System: Observations and Modeling." *Reviews of Geophysics* **31**: 397-422.

Carlström, A. and L. M. H. Ulander (1993). "C-band backscatter signatures of old sea ice in the central Arctic during freeze-up." *IEEE Trans. Geoscience and Remote Sensing* **31**: 819-829.

Chapman, W. L., W. J. Welch, et al. (1994). "Arctic sea ice variability: Model sensitivities and a multidecadal simulation." *J. Geophysical Research* **99**: 919-935.

Comiso, J. C. and R. Kwok (1992). "Summer Arctic ice concentrations and characteristics from SAR and SSM/I data." *Proc. first ERS-1 Symposium* : ESA **SP-358**

Gloersen, P. and W. J. Campbell (1991). "Recent Variations in Arctic and Antarctic Sea-Ice Cover." *Nature* **352**: 33-36.

Gloersen, P., W. J. Campbell, et al. (1992). *Arctic and Antarctic sea ice, 1978-1987*. Washington, DC, NASA SP-511.

Gow, A. J. and W. B. Tucker (1990). Sea-ice in the polar regions. *Polar Oceanography*. San Diego, Academic Press. 47-122.

Hibler, W. D. (1979). "A dynamic thermodynamic sea ice model." *J. Phys. Oceanogr.* **9**: 815-846.

Laxon, S. (1990). "Seasonal and inter-annual variations in Antarctic sea ice extent as mapped by radar altimetry." *Geophysical Research Letters* **17**: 1553-1556.

Laxon, S. and J. Askne (1991). Comparison of Geosat and SSM/I mapping of sea ice in the Arctic. IGARSS'91, Helsinki, ESA.

Long, A. E. (1992). C-band V-Polarized Radar Sea-Echo Model from ERS-1 Haltenbanken Campaign. ERS-1 Geophysical Validation, Penhors, ESA **WPP-36**: 113-118

Parkinson, C. L. and W. M. Washington (1979). "A

large scale numerical model of sea ice." *J. Geophys. Res.* **84**: 311-337.
Pettersson, M., J. Askne, et al. (1994). Classification of ice properties in the Arctic during freeze-up using ERS-1 SAR and SSM/I. *Proc. EARSeL 14th Symposium*. Balkema (to be published).
Steele, M. (1992). "Sea ice melting and floe geometry in a simple ice-ocean model." *J. Geophys. Res.* **97**: 17729-17738.
Stouffer, R. J., S. Manabe, et al. (1989). "Inter hemispheric Asymmetry in Climate Response to a Gradual Increase of Atmospheric CO2." *Nature* **342**: 660-662.
Sun, Y. (1993). A new correlation technique for the ice motion analysis. EARSeL workshop on sea ice, Lyngby, to be published in EARSeL Advances in Remote Sensing.
Sun, Y. and J. Askne (1994). Ice motion in the high Arctic on different time periods by a fast algorithm of motion estimation. EARSeL'94, Göteborg, Balkema (to be published).
Ulander, L., A. Carlström, et al. (1992), ARCTIC-91: Remote Sensing and Sea Ice Data Report, Dep. of Radio and Space Science, RSG 1993:1.
Ulander, L. M. H., A. Carlström, et al. (1993). ERS-1 backscatter from Nilas and Young Ice during Freeze-up. Microwave Remote Sensing of Ice, Lyngby, to be published in EARSeL Advances in Remote Sensing.

Sensors and Environmental Applications of Remote Sensing, Askne (ed.) © *1995 Balkema, Rotterdam, ISBN 90 5410 524 0*

Classification of ice properties in the Arctic during freeze-up using ERS-1 SAR and SSM/I

M.I.Pettersson & J.Askne
Department of Radio and Space Science, Chalmers University of Technology, Göteborg, Sweden

D.J.Cavalieri
Laboratory for Hydrospheric Processes, NASA Goddard Space Flight Center, Greenbelt, Md., USA

ABSTRACT: We have studied the ice concentration derived from ERS-1 SAR and SSM/I during the freeze-up period of the Arctic-91 expedition. During this complex period we have examples of melting conditions and new ice formation with frost flowers. The ice concentration from the SAR was derived by an averaging method and a segmentation method, and was compared with the SSM/I ice concentration (NASA team algorithm) and ship observations. It was found that the SSM/I values may underestimate the ice concentration by up to 25% compared with the (local) ship observations. ERS-1 SAR images may overestimate the ice concentration. The SAR classification results were also compared with helicopter video observations and it was found that the segmentation method differs up to 10% in the classification results after a period of frost flower formation. Reasons for these discrepancies are discussed. It is concluded that it is important to compare results from different sensors and different methods in combination with temporal sequences and meteorological information and to use only the consistent results.

1 INTRODUCTION

The Arctic area is an important region in the Earth's climate system, especially as an indicator of global warming (Stouffer et. al. 1989). Remote sensing techniques are an efficient tool in monitoring the Arctic. Sea ice concentration is one of the parameters of interest. SSM/I is today the only sensor which gives the ice concentration over of the whole Arctic, but future SAR sensors will cover the major part of the Arctic. Several data processing techniques are used to obtain concentration estimates from SSM/I and SAR data. In this work we compare results of different methods applied to SAR and SSM/I data with observations from the Arctic'91 expedition during the freeze-up period.

2 WEATHER AND BACKSCATTER CONDITIONS

The analyses are made using satellite data during a period from Sept. 21 to Oct. 5 1991. Under this period the temperatures as observed during the Arctic'91 expedition dropped from 0° C to -16°C (see Fig. 1), and the ERS-1 SAR backscatter signature changed rapidly (Ulander et. al. 1993). Especially between Sept. 26 to Sept. 29 there was a dramatic change in the radar backscatter from nilas and young ice areas. Backscatter from multi year ice was stable under the studied period.

As can be seen in Fig. 1, the temperature only increased on Sept. 27. The increase in temperature and the snowfall this day may have caused the large change in the thin ice backscatter signatures between Sept. 26 and Sept. 29 (Ulander et. al. 1994) .

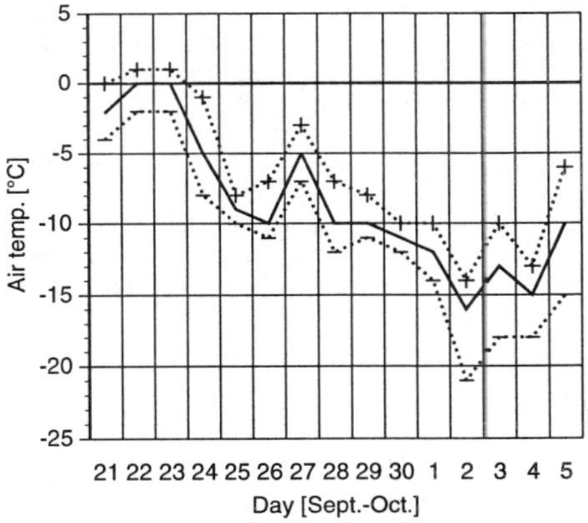

Figure 1. The maximum (+) and minimum (-) temperatures observed during on the Arctic'91 expedition during the freeze-up period. Daily averaged temperatures are shown as a continuous line.

Normally the radar backscatter changes with incidence angle (ground range), and the angle dependency is different depending on ice types and weather conditions. Therefore we have analysed the backscatter as a function of ground range using all SAR images available over the areas where Oden was located. In Figure 2 radar backscatter

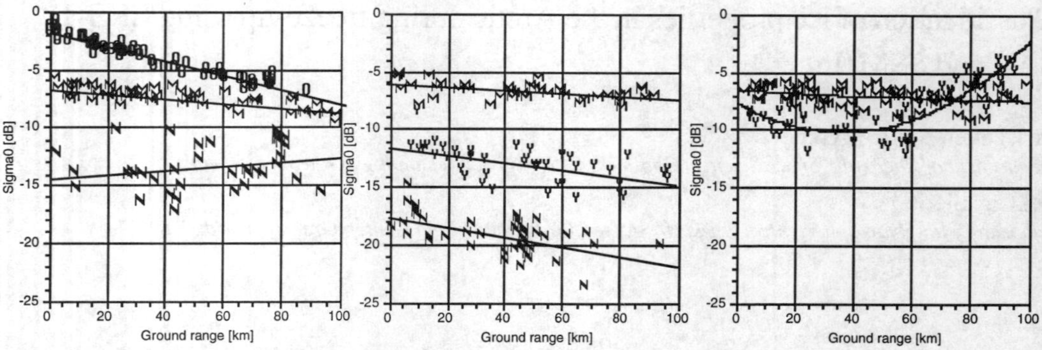

Figure 2 Backscatter as a function of ground range for (M) multi-year, (Y) FY2 and (N) FY1 for Sept. 22, Sept. 26 and Sept. 29.

coefficients are plotted for Sept. 22, Sept. 26 and Sept. 29, as a function of ground range. In the plots we have defined three different classes "First year 1" (FY1) which is probably nilas, "First year 2" (FY2) probably young ice, and multi year ice (MY). The backscatter from MY is stable over the entire period, only the standard deviation is smaller on Sept. 22. On the other hand the backscatter coefficients from FY2 and FY1 change dramatically between Sept. 26 and Sept. 29.

According to Figure 2 the ice classes can be separated by the backscatter coefficient on Sept. 22 and Sept. 26. Only in the far range the open water areas are difficult to distinguish from MY on Sept. 22. However, the situation has changed completely on Sept. 29 and it is almost impossible to do any classification in the far range between FY2 and MY. We believe the change in the backscatter for FY2 is caused by frost flowers. The range dependence of FY2 areas on Sept. 29 is probable due to the weather conditions.

3 METHODS FOR ICE CLASSIFICATION

We have used two different SAR classification methods a local average method (LAM) (Pettersson 1994) and a segmentation method (SEG) (Sun et. al. 1992). The LAM takes the average of N pixels and the classification is done by thresholds between the different classes. The SEG first detects borders between different areas. These borders then form segments and the classification is done by setting thresholds in the segmented image.

3.1 Local Average Method (LAM)

The LAM needs two parameters: the number of pixels N used in local averaging and the thresholds between backscattering coefficient of the different ice classes. To select these parameters we used the pixel intensity distributions of the different ice classes.

The statistics for the different ice classes were determined by adding together histograms from a large number of identified areas belonging to the same ice type.

Normally there is an overlap between the distributions of the different ice classes. To eliminate this overlap pixel averaging was performed over N pixels. To determine N we computed the average distribution (Kendell et. al. 1958) :

$$f_{\Sigma}(x) = \frac{\int_{0}^{\infty} e^{-iwx} \phi^{N}(w) dw}{N} \tag{3.1}$$

where Φ is the characteristic function of the distribution. We used this distribution to compute N in such a way that the overlap between the two most separated classes will be small. We have also used the distributions to compute the thresholds. Details are given in (Pettersson 1994).

We have also tested thresholds as given by (Rignot et. al. 1993) and we found just a small difference between the threshold levels.

3.2 Segmentation method (SEG)

The segmentation method used is described in (Sun et. al. 1992). It requires the selection of the edge detector window size, the threshold value for edge detection and the thresholds for ice classification in the segmented image. The edge detector window size depends on the ice floe size distribution and the separability of the different ice classes. Empirically a 7x7 pixel window was found to be a good edge detector window.

To determine a threshold value for the edge detected image the pixel distribution for the edge detector image for the different classes was computed. The edge detector image distribution is given by the ratio between two stochastic variables and according to (Papoulis 1991) this distribution is given by:

$$f_z(z) = \int_0^{\infty} w f_x(wz) f_y(w) dw \qquad (3.2)$$

where f_x and f_y are given by the distribution of the averaged pixel distribution in Eq. (3.1). We used f_z to compute the threshold for the edge detected image in the maximum likelihood sense (Pettersson 1994) . The segmentation method resulted in a segmented image, and for classification thresholds were computed using the distributions in Eq. (3.1).

3.3 Problems with LAM and SEG

Classification with these two methods cause problems which are unique for each method.

In the SEG method it is very difficult to get an error estimation because of borders not being detected. This is in particular a problem when backscattering coefficient changes gradually between different ice areas. Because of the lack of border detection the two areas will be classified as belonging to the same class.

The LAM will not result in large homogeneous classified areas because we will get error pixels within a homogenous area. However, error pixels will not cause any error in the classification due to the maximum a posteriori estimation of the thresholds. This means that the error pixels of one class in another class are as many as the other way around and they should cancel. One problem with the LAM method is to estimate the error caused by border pixels between different areas.

4 HELICOPTER VIDEO - SAR IMAGE COMPARISONS

To estimate the accuracy of the SAR image classification we have used video recordings made during Arctic'91 to compare with SAR images. In Figure 3d a helicopter video strip from Sept. 28 is shown. Figures 3c and 3e are SAR images from Sept. 26 and 29. Figures 3a-b, 3f-g are classified SAR images using LAM and SEG. In the SAR images there is some ice drift between the ice floes in between Sept. 26 and 29 and we have a change in σ^0 for FY1 and FY2 between the two days. The difference between FY1 and FY2 can not be detected in the video image and therefore these two classes are combined as class TI (Thin ice).

Looking into details of the video strip, we recognise the white areas of the multi-year ice floes. However, there are also some small almost white areas of TI (Thin ice) due to a thin snow layer. All ice floes marked MY only consists of MY, the other white areas are both MY and TI. The grey and black areas are TI and there is no open water on the video. (The difference between the MY and the snow covered TI does not show up fully in the print.)

Segmentation of the video strip was done by using a threshold in the image and obtained 19% of thin ice and 81% of multi-year ice. A problem in the video classification is that the camera's exposure changes between different photos, which causes problem to differentiate between the white TI and the MY. However, we believe that this error is small and that the classification accuracy is about one per cent.

Figures 3c and 3d are the SAR images from Sept. 29 and Sept. 26 over the same area as the helicopter strip. For comparison with the video images we have to keep in mind that there are some internal motions during the periods between the SAR images and the video image. The figures illustrate brightness changes between the two SAR images. The area of MY is stable but the areas of TI change between the two days. We believe this is due to the snowfall on Sept. 27 affecting in particular TI which had frost flowers on top.

The two SAR images were classified by the LAM method. Because the strip is so narrow a big learning area was chosen outside the image in order to derive the window size N and the threshold. The LAM estimate of MY concentration over the studied strip is 79% and 84% for Sept. 26 and Sept. 29 respectively compared to 81% according to the video.

The two SAR images were also classified by the SEG method and the result is illustrated in Fig. 3a and 3g. The segmentation seems to work well on Sept. 26 but not as well on Sept. 29. The MY concentrations is 73% and 71%, respectively. If we compare with the helicopter video this method underestimate the MY ice concentration but the error is smaller than 10%. This error seems to be high but we have to keep in mind that the segmentation area is small, it consists of small ice floes and there is some drift between Sept. 26 and Sept. 29.

5 SSM/I-SAR CLASSIFICATION COMPARISON

The most important sensor for ice classification today in the Arctic is SSM/I. We have made a comparison of the ice concentration derived from the SAR and SSM/I data along the route of Oden during the Arctic-91 expedition. We also compared the classification results with the ship observations.

In the SAR case we have made classification over areas of 25x25 km. We have used four different classes: open water (O), multi-year ice (MY), first year 2 (FY2), and first year 1 (FY1).

In the SSM/I case we have used the NASA team algorithm and we have in this case two different classes: ice and open water.

5.1 SAR classification

As mention before we used two different methods for ice classification, LAM and SEG. For these methods we have to choose pixel size and threshold limits. In Figure 4 are shown the LAM threshold for each classification day. The area under, between and over the thresholds define the different classes

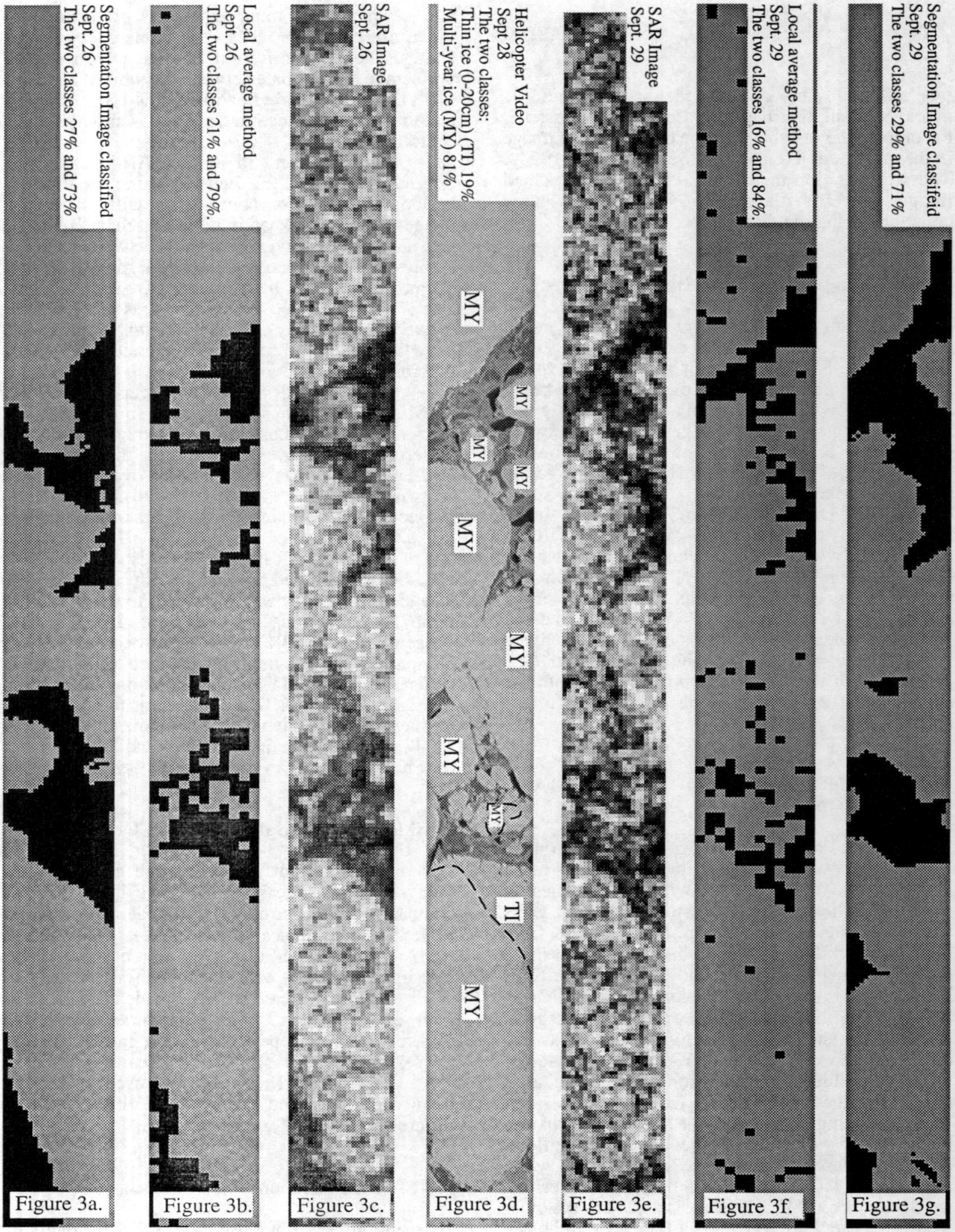

Figure 3. Comparison helicopter video (middle), SAR images and classified images from Sept. 26 and Sept. 29.

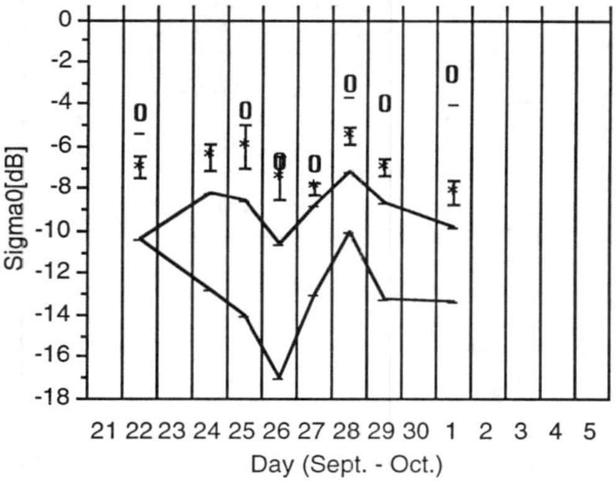

Figure 4. Threshold levels during the freeze up period. The thresholds are shown as continuous lines. The figure also shows the observed multi year average radar backscattering (*) and the backscatter from open water (O) calculated by an empirical model

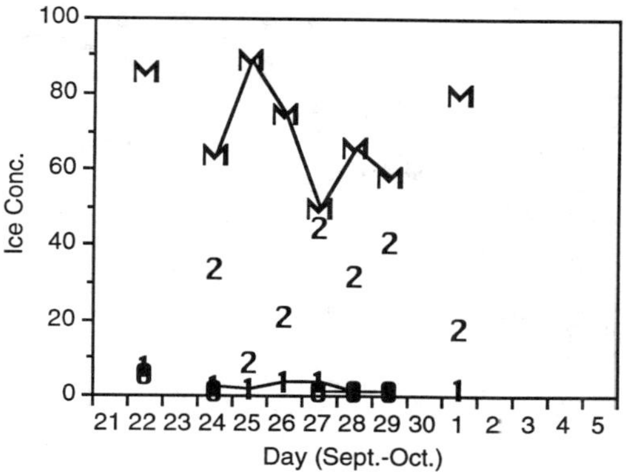

Figure 5. The concentrations derived by LAM.

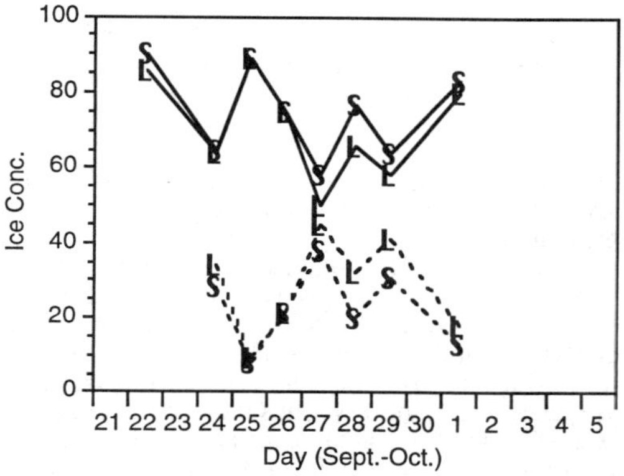

Figure 6. Changes of MY and FY2 ice concentrations during the freeze-up period. SEG is marked (S) and LAM is marked (L).

and they are marked with FY1, FY2, MY and O. From manual inspection of the SAR images open water areas have been identified but not for all days. This may be due to similar backscatter from open water and ice. For that reason the estimated backscatter from open water has been calculated using the empirical model (Long 1992) derived for the ERS-1 scatterometer. Wind speed and wind direction from the ship observations were used and the model gave a backscatter signal which was not more than 1 dB from the measured in the SAR images in the cases when open water areas were identified. According to Fig. 4, the expected open water backscatter will be higher than the MY backscatter except for Sept. 24 when the wind speed was under 1m/s.

Under the period the backscatter from MY was stable, but the backscatter from the first year ice was unstable. The ground range variation is the reason why the MY mean value is fluctuating between -6 and -8 dB in Figure 4. The change in FY ice backscatter affects the thresholds between the different classes for different days. The thresholds are also influenced by the variation of the backscatter with ground range.

In Figure 4 error bars on the MY mean value are included indicating the standard deviation of the MY ice backscatter coefficient. The standard deviation is different for different days because the window size N depends on the classification conditions.

In Figure 5 the result of the classification using LAM is shown. The MY ice fraction dominates followed by FY2. The concentration of FY1 and open water is low during the whole period.

5.2 Comparison between LAM and SEG

In Figure 6 the fractions of MY ice and open water derived by LAM (L) and SEG (S) are compared. Large differences between both methods are caused by problems in the ice classification as discussed in section 3.3. Figure 6 shows that the SEG yields a higher concentration of MY ice than the LAM. If we can not identify a border we may connect a small young ice area with a larger MY ice area and this area will be classified as MY.

5.3 Results from the comparison between SAR and SSM/I

In Figure 7 we see the ice concentration observed by SSM/I, SAR and ship observations. The ship observations where made every hour from Oden (Liljeström 1992) by visual observations over a

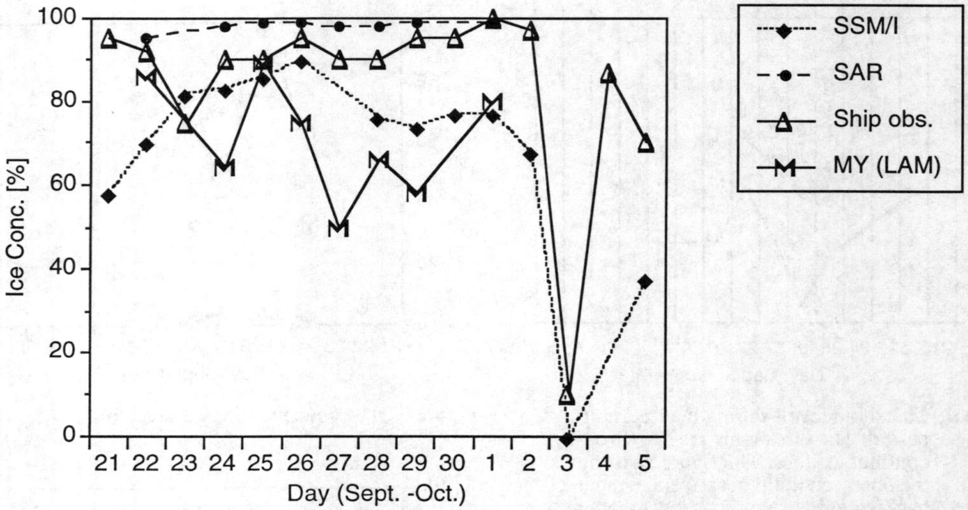

Figure 7. Total ice concentration derived from SSM/I and SAR as compared to the ship observations. The concentration of multi year ice as derived from SAR is also marked.

roughly 6 nautical mile wide area along the track of Oden. Because the ship takes the easiest track through the ice the ship observation may be an underestimation of the total ice extent. The accuracy of the visual observations is not better than ±10%.

The SAR ice concentration was derived by the LAM, and the result indicates a possible overestimation of the ice concentration. The difference to ship observations may be partly due to the SAR resolution, small leads are not detected. In case of low wind speeds it is difficult to distinguish between open water and ice (see Figure 4). Particularly if the leads are small the texture difference between open water and ice can not be detected. However, the difference between the SAR and the ship observations are within the accuracy of the ship observations (note there are no SAR observations from Sept. 23).

The SSM/I seems to underestimate the ice concentration during the entire period. During the first days the SSM/I yields a very low value. This maybe due to the fact that the surface temperature of the ice is close to melting(see Figure 1) and the snow layer is partly melting. After Sept. 27 SSM/I seems to underestimate the ice concentration by 20%. If we compare the SSM/I ice concentration with the MY concentration we see that the SSM/I gives an estimation of the ice concentration which is normally in between the MY fraction and the ship observations or the SAR observations of the ice concentration. This is expected according to (Cavalieri 1994) as the new ice will be between MY and open water in GR PR space. Ambiguities resulting from mixtures of open water, thin ice and MY ice are unresolved in the NASA Team algorithm.

6 CONCLUSIONS

It was shown that the multi-year ice areas are stable in radar backscatter but areas of nilas and young ice areas may change rapidly by as much as 13 dB. It was also shown that the backscatter from the young ice may change 7 dB over the ERS-1 ground range due to varying weather condition. These effects make the SAR classification more difficult .

There are today many different methods for ice type classification using SAR images. It is therefore important to verify the differences and the accuracies of these methods. We have used two different methods LAM and SEG. It was found that when we have contrast in backscatter between different classes we obtain the same results with the two methods, but when we have smaller contrast the methods yield different results. When we compared the two methods with a classified helicopter video image we found that the LAM gave results for the MY concentration within 3%. The SEG method underestimated the MY fraction by 10%.

Ice concentration in the Arctic is today derived mainly by SSM/I. We have compared the ice concentration derived from the SSM/I and from ERS-1 SAR. We found that during the freeze-up period it seems like the SSM/I underestimate the ice concentration. The error in ice concentration may be as large as 25%. The error is probably caused by the presence of thin ice when the temperature is low, or by a melting snow layer when the temperature is close to 0° C. The SAR, on the other hand, possibly over estimates the ice concentration, as open water may be mistaken for ice.

We have noticed problems in the determination of sea ice concentration with both SSM/I and SAR during the complex freeze up period. For that purpose the combined use of the sensors is important. The sensors may be improved in the future, e.g. we will obtain more information if we use multifrequency polarimetric SAR observations (Drinkwater et. al. 1991). Until then it is important to determine under what conditions there are risks for misinterpretation.

7 REFERENCES

Cavalieri, D.J. 1994. A microwave technique for mapping thin sea ice. *Journal of Geophysical Research* . In press

Drinkwater, M.R., *et al.* 1991. Multifrequency Polarimetric Synthetic Aperture Radar Observations of Sea Ice. *J. Geophys. Res.* 96(11): 20,679-20,698.

Kendell, M.G. and A. Stuart, *The advanced theory of statistics,* . 1958, Charles Griffin & Company: London. p. 188-189.

Liljeström, G. 1992. *Icebreaker Oden trafficability data*, The Swedish Polar Research Secretariat.

Long, A.E. 1992. C-band V-Polarized Radar Sea-Echo Model from ERS-1 Haltenbanken Campaign. in *URSI, Microwave Signature -92 Conference "Terrestrial Remote Sensing with Microwaves: Signatures, Techniques and Systems"*. 113-118. IGLS-Innsbruck, Austria:

Papoulis, A. 1991. *Probability, Random Variables, and Stochastic Processes*. Third ed. 1991, McGraw-Hill International Editions.

Pettersson, M. 1994. Internal report, Department of Radio and Space Science, to be published.

Rignot, E.J.M. and J.J.v. Zyl 1993. Change Detection Techniques for ERS-1 SAR Data. *IEEE Transactions on Geoscience and Remote Sensing* 4: 896-906.

Stouffer, R.J. and S. Manabe 1989. Inter hemispheric Asymmetri in Climate Responce to a Gradual Increase of Atmosheric CO2. *Nature* 342: 660-662.

Sun, Y., A. Carlström, and J. Askne 1992. SAR image classification of ice in the Gulf of Bothnia. *Int. J. Remote Sensing* 13(13): 2489-2514.

Ulander, L.M.H., A. Carlström, and J. Askne 1993. ERS-1 SAR Backscatter from Nilas and Young Ice during Freeze-up. in *EARSeL'93. Microwave Remote Sensing of Ice*. Lyngby:

Ulander, L.M.H., A. Carlström, and J. Askne 1994. Effect of frost flowers, rough saline snow and slush on the ERS-1 SAR backscatter of Arctic thin ice. *Int. J. Remote Sensing* . Submitted for publication

Sensors and Environmental Applications of Remote Sensing, Askne (ed.) © 1995 Balkema, Rotterdam, ISBN 90 5410 524 0

On the accuracy of estimation of ice type concentration from ERS-1 SAR images

Henning Skriver
Electromagnetics Institute, Technical University of Denmark, Lyngby, Denmark

ABSTRACT: A segment-based parameter retrieval algorithm for estimating the ice type concentration from ERS-1 SAR data is presented. The algorithm is based on an edge detection, region growing approach together with segment characterization by backscatter coefficient and texture parameters. A refined method for edge thinning is presented. The algorithm runs fully automatic and it performs well in pack ice images, whereas it needs some tuning for ice edge images. The ice concentration may be obtained by an accuracy of at least 10 %. Comparison with SSM/I data shows fairly good agreement.

1 INTRODUCTION

There is no doubt that microwave remote sensing is an essential tool for studying large-, medium-, and small-scale sea ice phenomena in polar regions (e.g. Carsey, 1989). This is due to the fact that microwave remote sensing is largely independent of light and weather conditions. The synthetic aperture radar (SAR) is essential due to its fine spatial resolution that enables detailed imaging of the scene of observation, and data from ERS-1 have already shown great potential for sea ice mapping. A number of parameters, which is of great importance for research and application purposes may be derived from the radar data. This applies to ice concentration, ice motion, lead structure, floe size distribution, etc. (e.g. Holmes et al., 1984; Lyden et al., 1984; Burns et al., 1987, Fily and Rothrock, 1987; Carsey, 1989; Skriver, 1991; and Nystuen and Garcia, 1992).

The basis of parameter retrieval of any kind is the ability to discriminate between the targets in question. In SAR images the basis of discrimination is the properties of the backscatter coefficient, i.e. the mean backscatter coefficient and the spatial variations in the backscatter coefficient - also called texture. Discrimination based on these properties is disturbed by the speckle noise inherent in SAR images. To reduce the influence of speckle on the discrimination it is necessary to use as large areas in the image as possible to estimate the mean backscatter coefficient and the texture parameters, and thereby reducing the variance of the estimates of the parameters. Two approaches can be used to perform this task, i.e. window-based methods or segment-based methods. Both approaches may be preceded by or combined with a filtering process to reduce the speckle. The disadvantage of the filtering is that fine details in the texture that may be important for the discrimination are lost.

In a window-based approach a moving-window technique is used, and for each position of the window the mean backscatter coefficient and texture parameters are computed. These parameters can then be used for the discrimination. In a segment-based approach the SAR image is segmented prior to the discrimination, which is then performed using the mean backscatter coefficient and texture parameters of the segments. The advantage of the window-based method is that it is simple and thereby fast. However, problems at borders between different target areas reduce the window sizes that can be effectively used, and thereby reduce the number of pixels that can be used for the estimation. The segment-based methods are, on the other hand, often complex and relatively slow. However, they maximize the number of pixels used for the estimation using only the pixels covering a specific target area and in most cases without severe border problems.

In this paper, results of an evaluation of the accuracy of using a segmentation approach are presented. The segmentation algorithm is based on an edge detection and a region growing technique. This algorithm is able to cope with low contrast edges and missing edge points due to speckle and/or very low contrast. The segmentation algorithm is well suited for SAR images of sea ice, but it is also applicable to other applications of SAR data. The segments are then discriminated using the mean backscatter coefficient and texture parameters. Finally, the classified segments can be used to compute the ice type concentration, ice edge, lead structure, etc. In this paper only the ice type concentration will be studied.

In the following, the parameter retrieval algorithm

is outlined in Section 2, and the ERS-1 SAR data used in the paper are presented in Section 3. In Section 4 the results of the segmentation algorithm and the analysis of the discrimination parameters are given, together with the results of estimation of the ice type concentration, including a comparison with SSM/I-data. Finally, in Section 5 the conclusions are given.

2 PARAMETER RETRIEVAL

The segmentation algorithm and parameter retrieval approach were originally proposed by Skriver (1989), but since then it has been further developed and adapted for ERS-1 SAR data. The purpose of the segmentation is to perform a spatial discrimination of objects from different and identical target classes (such as different ice types and water). By means of segmentation an image is subdivided into segments, where each segment represents a spatially limited area of pixels from the same target type. The segments can be identified and accessed one by one at later steps of processing.

The segmentation algorithm is a region growing algorithm and it runs in a number of steps:

- Edge detection
  - Edge enhancement
  - Edge thinning and detection
- Region growing
  - Distance transform
  - Segment kernels
  - Linking of pixels
- Segment merging
- Segment interpretation
  - Segment parameters
  - Segment classification
  - Parameter estimation

The individual steps are explained in details below. For the individual steps it is outlined how they are made more or less automatic.

2.1 Edge detection

The purpose of the edge detection is to delineate the segments, and due to the multiplicative nature of the speckle noise a requirement of the edge detector is that its false alarm rate is independent of the average intensity of the bordering areas. This is fulfilled by the gamma ratio edge detector (Madsen, 1986, and Touzi et al., 1988). Edge detection usually runs in two steps: edge enhancement and thresholding. For the gamma ratio edge detector the edge enhancement is performed by testing in appropriately positioned subwindows and thereby an edge image and a gradient direction image are formed. The edge image is the test statistic for the maximum likelihood test of equal means of independent samples from two gamma pdf's.

Due to the speckle noise the window size has to be large to obtain estimates of the test value with sufficiently low variance. In the present case, the window size observed to give sufficient estimates is 21 x 21 pixels. Of course, this large window size results in very broad edges, which are not very well suited for the segmentation because of the poorly defined edges.

Therefore, a new combined detection and thinning algorithm that yields a detected edge width of approximately 1 pixel has been developed. The algorithm runs as follows. For each pixel in the image the pixel value of the edge enhanced image is compared with the two neighbours in the edge gradient direction. If both of the neighbours are smaller than the pixel in question, the pixel is a potential edge pixel. If the pixel value of this potential edge pixel exceeds a specific threshold it is labelled as an edge-detected pixel. The result is that only the pixels in the middle of the broad edge from the edge enhancement are labelled as edge-detected pixels, and very fine, thin edges are produced. An appropriate value of the threshold is by manual inspection found to be 2.2. The algorithm is outlined in Fig. 1.

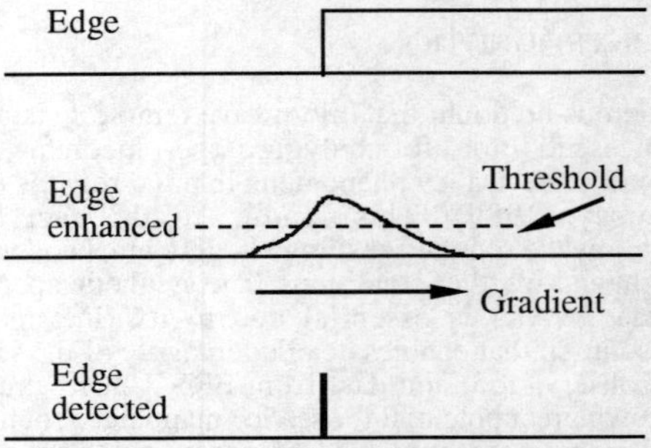

Fig. 1 The edge detection and thinning algorithm.

2.2 Region growing

The segmentation approach used in the present work belongs to a large group of segmentation algorithms called region growing. A region growing algorithm runs basically as follows: A number of seed points is spread in some way over the image. From all these seed points regions are grown "simultaneously". The growing for the individual regions is stopped when the border of the area is reached, where the border is defined in some appropriate way. The most difficult parts of the algorithm are to determine the seed points and the border conditions.

The segments are "delineated" with the edge points from the edge detection as sparsely determined border points. Based on these edge points both the seed points and the border conditions are determined. The centres of these segments are found by assigning for each non-edge pixel a value equal to the distance from the nearest edge pixel. For a closed contour the distance transform will have one or more maxima in the middle of the contour. These maximas make up the segment kernels. To avoid

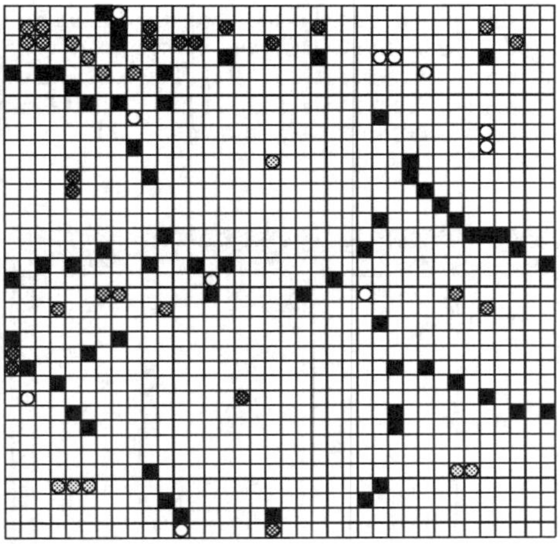

Fig. 2 The local maxima for a set of synthetic edge points. Neighbouring local maxima with the same shade of grey form a segment kernel.

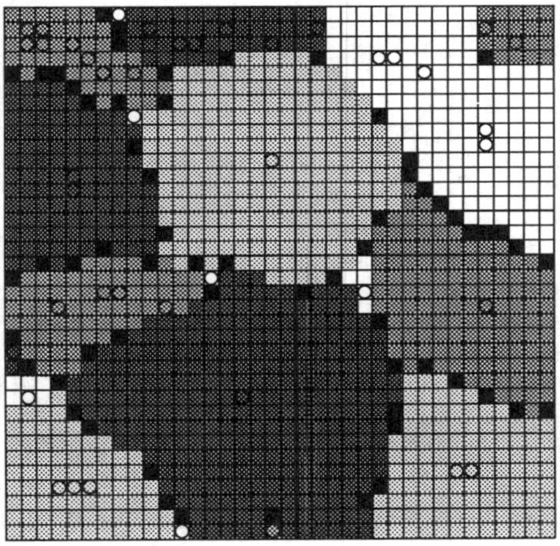

Fig. 3 The result of the segmentation of the set of synthetic edges in Fig. 2.

multiple segment kernels for the same segment, neighbouring maximas are merged to form a single segment kernel. The distance transform is generated by an approximate method using window operations. It approximates the Euclidean distance with a maximum error of 3% (Skriver, 1989). An example of artificial edge points together with the maximas is shown in Fig. 2.

Regions are then grown from the seed points as explained above, until the edge points are reached. The non-edge and non-maxima pixels are linked to the segment kernels by using the gradient direction of the distance transform. The gradient points in the direction of the appropriate segment kernel. Hereafter, the edge pixels are linked to the segment kernels using the gradient directions from the edge detector. (Skriver, 1989). The segments corresponding to Fig. 2 are shown in Fig. 3.

The region growing part of the retrieval algorithm runs fully automatically. One of the advantages of this algorithm is that it produces segments irrespectively of the actual intensity level, i.e. it produces segments of bright areas as well as black areas. Furthermore, it can cope with missing edge pixels.

2.3 Segment merging

The segmentation algorithm generates a large number of small segments. To reduce the computational load of any postprocessing step and to be sure that the segment size is sufficiently large for estimation of parameters, small segments have been merged to neighbouring larger segments. For small segments only the mean intensity can be used to discriminate between these areas. Therefore, the test of whether to merge a small segment with a larger one or not is if the ratio of the mean intensities of the two segments is lower than a threshold. The ratio is used because it is the test for equal means in two gamma pdf's.

The segment merging runs in three steps. First, all small segments (i.e. size < 50 pixels) are merged with larger ones. Hereafter, the somewhat larger segments (i.e. size < 500 pixels) are merged but only if the ratio is smaller than 2.5 dB. Finally, all segments are merged if the ratio is smaller than 1 dB. The number of segments after these steps is shown in Table 1 for two examples, in the pack ice with many and smaller segments and at the ice edge with few and larger segments.

Table 1. Number of segments after the steps in the segment merging

	Pack ice	Ice edge
After segmentation	230.000	40.000
Merging, size < 50	77.000	15.000
Merging, size < 50, ratio < 2.5 dB	28.000	7.500
Merging, ratio < 1 dB	12.000	3.500

The above-mentioned stepwise approach is found to give the best results for both merging all small segments and merging very large segments if they are very close in the mean intensity. The segmentation and merging algorithm produces only very few mixed segments, at the same time as the number of segments after the merging is manageable.

2.4 Segment interpretation

As mentioned earlier, the basis for the parameter retrieval is classified segments. The segments are classified using parameters determined from the pixel values of the pixels in the segments. The most important parameter is the mean backscatter coefficient, which is determined by averaging the intensity values in the segment.

In some cases ambiguities exist between the backscatter coefficient for different ice/water classes. Calm open water may have the same backscatter coefficient as smooth new ice, wind-roughened water may be confused with rough new ice or multiyear ice, rough new ice and multiyear ice may have the same backscatter coefficient, etc. In order to resolve these ambiguities parameters describing the spatial variation of the backscatter coefficient (i.e. texture) may be included.

In the present work, three texture parameters have been tested with the ERS-1 data. The power-to-mean ratio (PMR) is a well-known first-order texture parameter (1), that takes the speckle noise into account (Skriver, 1989), giving comparable results in high and low intensity areas.

$$\mathrm{PMR} = \frac{E\{x^2\}}{E\{x\}^2} \tag{1}$$

where x is the intensity pixel value.

The other texture parameters are second-order parameters of the co-occurrence matrix type (Haralick et al., 1973). The co-occurrence matrix is an estimate of the second-order pdf in the stationary case. These parameters have been used previously with fairly good results (Skriver, 1989 and 1991, and Nystuen and Garcia, 1992). In the present work two parameters are used, a contrast measure (CON), and an entropy measure (SENT, Sum Entropy):

$$\mathrm{CON}(\mathbf{r}) = \iint (x - y)^2 \, p(x, y; \mathbf{r}) \, dx dy \tag{2}$$

$$\mathrm{SENT}(\mathbf{r}) = \int P_{x+y}(z; \mathbf{r}) \log(P_{x+y}(z; \mathbf{r})) \, dz \tag{3}$$

where x and y are intensity pixel values with relative positions given by **r**, and $P_{x+y}(z;\mathbf{r})$ is the sum matrix averaged over all z = x+y. In the present work the length of **r** is fixed to 3 pixels.

Based on one or more of the above-mentioned parameters the segments may now be classified as one of a number of classes. The ice type concentration, for instance, may then be computed for the entire image or for subsets of the image by determining the number of pixels (and thereby the area) of each ice class within the image or image subset.

3 ERS-1 DATA

Six ERS-1 images from the Greenland Sea from the winter 1993 have been used in the investigations. The data product used is the Precision Image product and the backscatter coefficient is computed using the approach given in Laur (1992). The location of the images is shown in the map in Fig. 4, and the time and centre coordinates are shown in Table 2.

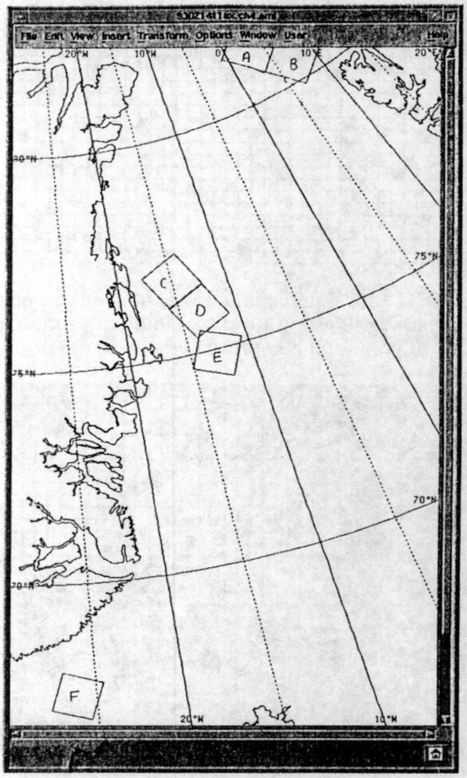

Fig. 4 Map showing the location of the ERS-1 images.

Table 2 Time and location of the ERS-1 data

	Time	Coordinates	Location
A	93-03-12	82.2N 3,5E	Fram Strait
B	93-03-12	80.5N 7.2E	Fram Strait
C	93-02-07	76.8N 14.9W	Greenland Sea
D	93-02-07	76.0N 13.3W	Greenland Sea
E	93-02-02	74.9N 12.8W	Greenland Sea
F	93-02-11	67.3N 26.1W	Denmark Strait

The images A and B from the Fram Strait contain mostly multiyear ice, but also a number of different new ice classes and wind-roughened water are present. The images C and D from an ascending pass from the Greenland Sea contain multiyear ice and various new ice types. Both the E image from the Greenland Sea and the F image from the Denmark Sea contain approximately 50% new ice and 50 % wind-roughened water.

The ERS-1 data have been segmented by the algorithm described in Section 2. A large number of segments in the images has been classified manually for reference. The classification has been by visual inspection and is based on the experience of the author and general knowledge about the weather conditions and the potential ice types.

4 RESULTS

Four types of results are presented in this section, the evaluation of the segmentation algorithm, the results of the analysis of the parameters for the various classes, the classification results and the results of the comparison with the SSM/I data.

4.1 Segmentation results

The segmentation algorithm runs with fixed set up, and thereby without any manual interference. The algorithm works very well in the pack ice images (A-D). Relatively few segments are produced, and they delineate very accurately patches of different classes. Only very few of the segments contains mixed classes. For the ice edge images (E - F) the performance is somewhat worse. A relatively large number of segments contains mixed classes. In this case the standard set up which is determined from pack ice images does not work very well. The reason is probably that the edges between the classes are poorly defined, so adjusted parameters for the edge detection part of the algorithm may improve the results.

4.2 Parameter results

In Figs. 5 and 6 are shown the backscatter coefficient for the manually classified segments in the B image from the Fram Strait and the F image from the Denmarks Strait, respectively. Also, in Table 3 is shown the standard deviation for segment sizes between 4500 and 5500 and for segment sizes larger than 25000. The ice type codes are MY for multiyear ice, NI 1 for rough new ice, NI 2 for medium rough new ice, NI 3 for relatively smooth new ice, and WA for open water. ICE is probably new ice.

We observe that the variance of the estimates of the backscatter coefficient is somewhat larger for small segments than for larger segments. However, even for very large segments we see some variation

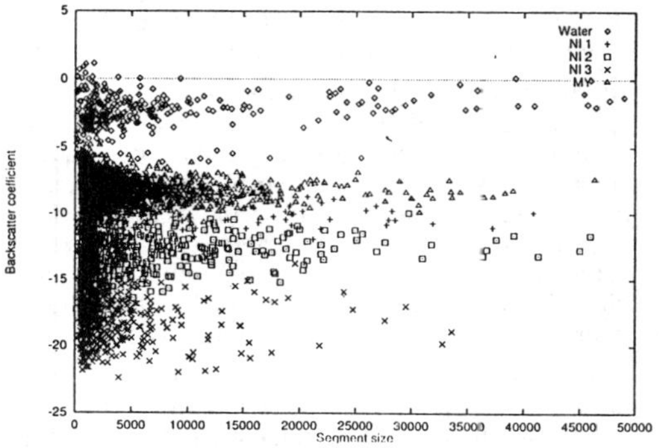

Fig. 5 Backscatter coefficient vs. segment size for the B image from the Fram Strait.

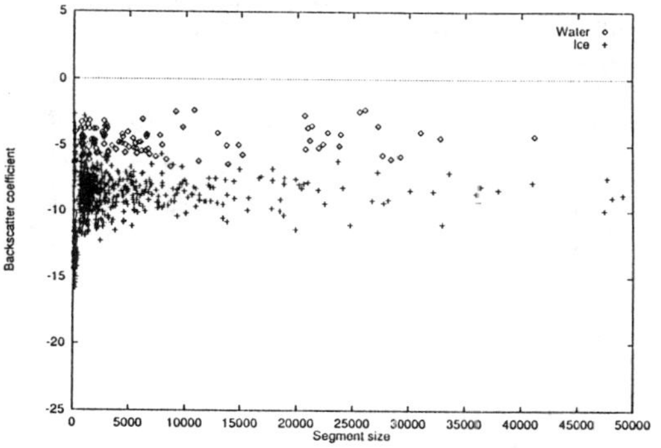

Fig. 6 Backscatter coefficient vs. segment size for the F image from the Denmarks Strait.

Table 3 Mean backscatter coefficient and standard deviation for segment sizes between 4500 and 5500 pixels (STD 1) as well as for segment sizes larger than 25000 pixels (STD 2) for the classes present in the B and F images.

Im.	Ice type	$<\sigma^0>$ [dB]	STD 1 [dB]	STD 2 [dB]
B	MY	-8.2	0.9	0.6
B	NI1	-9.9	1.1	0.5
B	NI2	-13.2	1.3	0.9
B	NI3	-17.7	1.5	0.9
B	WA	-2.5	1.2	1.1
F	WA	-4.5	0.5	1.1
F	ICE	-9.2	1.0	1.0

of the backscatter coefficient. This may to some extent be explained by the missing correction for the variation of σ^0 with incidence angle.

The average of the backscatter coefficient is an unbiased estimator, therefore we can use all segments to estimate the overall mean backscatter coefficient. In Table 3 is shown the mean backscatter coefficient for the classes present in the images estimated on the basis of a large number of manually classified segments. The strong backscatter from the open water corresponds to a relatively strong wind of more than 10 m/s. The σ^0 found for multiyear ice corresponds well with other results for winter conditions (e.g. Onstott and Shuchman, 1992, and Stern et al., 1993) The new ice signatures, on the other hand, have high variability and the interpretation is somewhat difficult. The manual discrimination between the new ice types is largely based on the intensity differences corresponding for the saline new ice types to surface roughness differences.

In Figs. 7 and 8 are shown the PMR for the B image

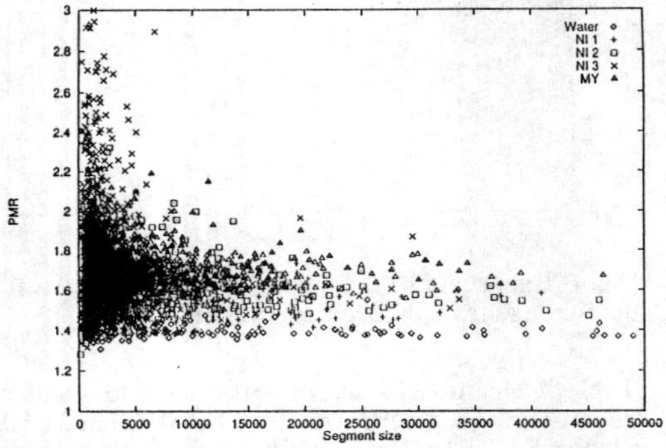

Fig. 7 PMR vs. segment size for the B image from the Fram Strait.

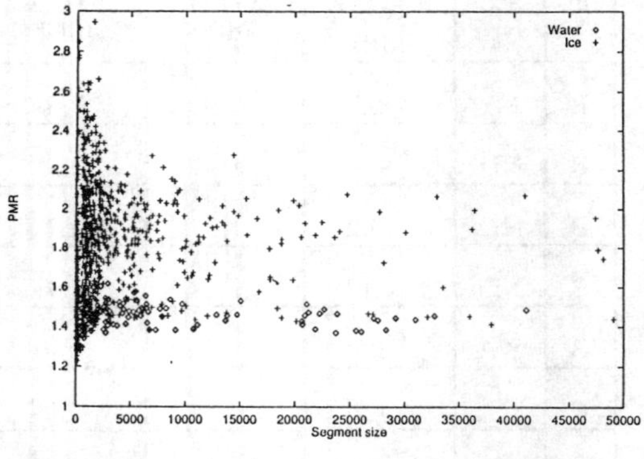

Fig. 8 PMR vs. segment size for the F image from the Denmark Strait.

and the F image, respectively. Also, in Table 4 is shown the mean and standard deviation of PMR. The estimator of the PMR is asymptotically unbiased (Skriver, 1989), therefore the mean PMR is computed on the basis of large segments only. We observe in this case a large decrease in the variance of the estimate of PMR for large segments. Furthermore, we see a large overlap between the classes, except for water, which is rather homogeneous, i.e. low PMR.

From Table 4 we see that the water areas have significantly lower PMR compared to the other classes. This is important, because the PMR may be used to delineate the water areas independently of the backscatter and thereby the wind speed. This requires further studies under different conditions.

Table 4 PMR for segment sizes larger than 25000 pixels and standard deviation of PMR for segment sizes between 4500 and 5500 pixels (STD 1) as well as for segment sizes larger than 25000 pixels (STD 2) for the classes present in the B and F images.

Im.	Ice type	PMR	STD 1	STD 2
B	MY	1.67	0.1	0.06
B	NI1	1.51	0.08	0.05
B	NI2	1.54	0.11	0.04
B	NI3	1.61	0.27	0.14
B	WA	1.39	0.03	0.03
F	WA	1.41	0.06	0.03
F	ICE	1.65	0.18	0.22

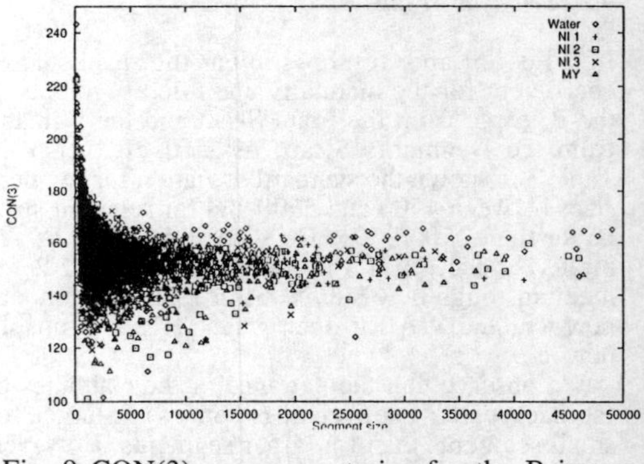

Fig. 9 CON(3) vs. segment size for the B image from the Fram Strait.

Table 5 CON(3) for segment sizes larger than 25000 pixels and standard deviation of CON(3) for segment sizes as in Table 4. b

Im.	Ice type	CON	STD 1	STD 2
B	MY	151	7.3	3.9
B	NI1	152	7.5	4.6
B	NI2	153	9.3	3.9
B	NI3	154	6.8	1.4
B	WA	159	6.9	7.1
F	WA	162	9.1	4.3
F	ICE	158	7.7	6.4

The CON(3) parameter is shown in Fig. 9 for the B image and the mean and standard deviation is given in Table 5. The discrimination potential of the CON parameter is seen to be rather poor. This is also the case for the SENT parameter.

4.3 Classification results

The segments in the B and the F images are used for a classification study. Between 50 and 80 % of the segments (measured by areal coverage) have been used in the study. A small number of segments from each class is picked for the training set. The statistics are computed and the rest of the segments is classified. The hit rate (measured in areal coverage) is then determined.

Table 6 Classification accuracy (measured in areal coverage) for the individual parameters.

Im.	Ice type	σ^o	PMR	CON	SENT
B	MY	88.7	76.7	33.6	37.9
B	NI1	77.0	44.1	28.4	20.5
B	NI2	8.9	27.4	7.9	5.0
B	NI3	97.3	21.1	12.6	26.9
B	WA	99.4	88.6	68.8	48.9
F	WA	99.3	98.7	63.6	64.6
F	ICE	99.6	86.0	82.4	82.8

Table 7 Classification accuracy (measured in areal coverage) for the backscatter coefficient combined with the other parameters. The classification accuracy for σ^o alone is also shown.

Im.	Ice type	σ^o	σ^o + PMR	σ^o + CON	σ^o + SENT
B	MY	88.7	89.7	71.3	71.3
B	NI1	77.0	94.2	79.9	78.6
B	NI2	8.9	5.0	8.7	8.5
B	NI3	97.3	74.8	78.6	67.3
B	WA	99.4	96.6	99.5	99.4
F	WA	99.3	93.9	98.3	98.5
F	ICE	99.6	99.9	99.9	99.5

The classification has been performed for the parameters outlined in Section 2.4 and the backscatter coefficient combined with the other parameters. From the analysis in Section 4.2 it is clear that the backscatter coefficient has a large discrimination potential. Therefore, it must always be included, whereas the other parameters may be added to resolve the ambiguities that exist for the backscatter coefficient as mentioned in Section 2.4.

The classification accuracy for the individual parameters is shown in Table 6 and for the backscatter combined with the other parameters in Table 7. As expected from the analysis in Section 4.2 the hit rate is high for the backscatter coefficient and low for the co-occurrence matrix parameters CON and SENT. The PMR is somewhat worse than σ^o. It is observed that the best performance for the PMR is for the water classes. In general, the hit rate for the open water is rather high, indicating that the accuracy of a total ice concentration retrieval will be high. The new ice types are somewhat confused, the manual classification is in this case however relatively uncertain.

The classification using the backscatter coefficient combined with the texture parameters shows that only the addition of PMR results in a slight increase in the hit rate for some of the classes.

4.4 Comparison with SSM/I

Preliminary results from a comparison between estimation of ice concentration from SSM/I data and ERS-1 SAR data have been obtained. SSM/I data closest possible in time have been used for the comparison, and the time difference between SSM/I and SAR data acquisitions is between 40 minutes and 5 hours 45 minutes. The total ice concentration is estimated from the area covered by the SAR image in the SSM/I data, and the total ice concentration is estimated in the SAR image using the segment method and the backscatter coefficient as discriminating parameter. The results are shown in Table 8.

Table 8 The ice concentration (in %) estimated from SSM/I data and from ERS-1 SAR data.

Im.	SSM/I	SAR
A	74	87
B	92	95
E	98	100
H	43	37
G	60	64

We see that the ice concentration estimated from the two data sets agrees fairly well with the largest deviation being for the A image from the Fram Strait where the difference is almost 20 %.

5 CONCLUSIONS

A segment-based approach for parameter retrieval applied to ERS-1 SAR-data has been outlined. It is an automated algorithm, and it shows fine results for pack ice images, whereas it needs tuning for ice edge images. The parameters for the algorithm may have to be adapted for the two different cases. The backscatter coefficient and texture parameters (PMR, CON, and SENT) have been studied for ice type discrimination. The backscatter coefficient shows very good results, and is the most important parameter for discriminating between ice types (and water). PMR shows also good performance for discriminating water from ice, and it may help when ambiguities exist for the backscatter coefficient. The co-occurrence matrix parameters (CON and SENT) show poor performance. It is expected that the ice concentration may be determined with an accuracy of better than 10 %. A comparison between SSM/I data and ERS-1 SAR data showed fairly good agreement for the total ice concentration.

ACKNOWLEDGEMENTS

The work is funded by the Danish Space Board. The author would like to thank Leif Toudal Pedersen, Electromagnetics Institute, for processing of the SSM/I data.

REFERENCES

Burns, B. A., D. J. Cavalieri, M. R. Keller, W.J. Campbell, T. C. Grenfell , G. A. Maykut, and P. Gloersen, 1987, "Multisensor Comparison of Ice Concentration Estimates in the Marginal Ice Zone", J. Geophys. Res., vol. 92, pp. 6843-6856.

Carsey, F., 1989, "Review and Status of Remote Sensing of Sea Ice", IEEE J. Oceanic Eng., vol. 14, pp. 127-138.

Fily, M., and D. A. Rothrock, 1987, "Sea Ice Tracking by Nested Correlations", IEEE Trans. Geosc. Rem. Sens., vol. GE-25, pp. 570-580.

Haralick, R. M., K. S. Shanmugan, and I. Dinstein, 1973, "Textural Features for Image Classification", IEEE Trans. Syst., Man, Cybern., vol. SMC-3, pp. 610-621.

Holmes, Q. A., D. R. Nuesch, and R. A Shuchman, 1984, “Textural Analysis and Real-Time Classification of Sea-Ice Types Using Digital SAR Data", IEEE Trans. Geosc. Rem. Sens., vol. GE-22, pp. 113-120.

Laur, H., 1992, "Derivation of Backscatter Coefficient in ERS-1.SAR.PRI Products", ESA Document.

Lyden, J. D., B. A. Burns, and A. L. Maffett, 1984, "Characterization of Sea Ice Types Using Synthetic Aperture Radar", IEEE Trans. Geosc. Rem. Sens., vol. GE-22, pp. 431-438.

Onstott, R. G., and R. A. Shuchman, 1992, "Arctic Sea Ice Microwave Signature and Geophysical Processes Study", Proc. First ERS-1 Symposium, ESA SP-359, pp. 329-332.

Madsen, S. N., 1986, "Speckle Theory, Modelling, Analysis and Applications Related to Synthetic Aperture Radar Data", Electromagnetics Institute, Ph. D. thesis, LD 62.

Nystuen, J. A. and F. W. Garcia, 1992, "Sea Ice Classification Using SAR Backscatter Statistics", IEEE Trans. Geosc. Rem. Sens., vol. 30, pp. 502-509.

Skriver, H., 1989, “Extraction of Sea Ice Parameters from Synthetic Aperture Radar Images", Electromagnetics Institute, Ph. D. thesis, LD 74.

Skriver, H., 1991, “Sea ice type concentration from MIZEX-87 SAR data", International Geoscience and Remote Sensing Symposium, pp. 415-418.

Stern, H. L., D.A. Rothrock, R. Kwok, and B. Holt, 1993, "The Geophysical Processor System: Automated Analysis of ERS-1 SAR Imagery", Proc. Second ERS-1 Symposium, ESA SP-361, pp. 281-286.

Touzi, R., A. Lopes, and P. Bousquet, 1988, "A Statistical and Geometrical Edge Detector for SAR Images", IEEE Trans. Geosc. Rem. Sens., vol. GE-24, pp. 235-245.

Sensors and Environmental Applications of Remote Sensing, Askne (ed.) © 1995 Balkema, Rotterdam, ISBN 90 5410 524 0

Active microwave signatures of sea ice in the Greenland Sea during summer 1993

Bjørn Bavnehøj Thomsen, Leif Toudal Pedersen & Henning Skriver
Remote Sensing Unit, Electromagnetics Institute, Technical University of Denmark, Lyngby, Denmark

ABSTRACT: During a 3-month period in the summer 1993 a great number of ERS-1 SAR scenes were acquired in connection with a field campaign in the Northeast Water polynya at the Greenland coast. Most data were received as low-resolution fast-delivery data (UILR) but also some precision data (PRI) were acquired for comparison. The data were used for identifying and tracking floes; in total 80 scenes were analyzed with 149 individual observations of 41 different floes. This data set has subsequently been used for studying the microwave signatures of the floes as a function of time. The data have been referred to measurements of temperature and wind acquired on a ship and an automatic weather station in the area. The calibration processes of the two types of SAR data have been examined.

1 INTRODUCTION

In connection with a field campaign in the Northeast Water Polynya at the northeast corner of Greenland satellite data were collected and analyzed in support of other activities by ships in the area. The interest was concentrated on the period July-August 1993 but data were also acquired from the preceding period although at less intensity. An annotated catalogue of the data including AVHRR and ERS-1 SAR data has been worked out (Gudmandsen and Thomsen, 1994).

One of the analyses was concerned with derivation of the motion of the floes in the area based on ERS-1 SAR data which were acquired rather frequently in the period. In spite of the relatively narrow swath of the SAR and the change in the backscatter coefficient through the period it was found possible in several cases to track floes through long periods using an interactive method and thereby obtain a good impression of the flow pattern. The result of this work is shown in Figure 1 as an illustration, where floe tracks are drawn as straight lines between positions at which individual floes have been identified.

On the basis of this work an analysis of the microwave signatures of the floes has been carried out by determining the backscatter coefficient at each observation, i.e. as a function of time and position. This is the subject of the paper. In order to obtain good accuracy the work also included a study of the calibration of the data used, including low and high-resolution SAR data.

Figure 1 may also serve to briefly describe the area of research, since the tracks have been drawn on an AVHRR scene for a particular day, 10 August 1993. It is seen that the polynya - a stretch of open water or area of low ice concentration in an ice canopy - has developed along the coast between a protrusion of shore-fast ice extending far into the sea at about 79.5°N and another protrusion of shore-fast ice in the form of a peninsula extending from the northeast corner (Nordøstrundingen) of Greenland to about 10°W. It is seen that the polynya at this late summer time extends far towards the east where it meets the pack ice drifting southwards from the Arctic Ocean carried by the East Greenland Current.

2 DESCRIPTION OF DATA

ERS-1 scenes were received as UILR-images (Low Resolution Fast Delivery Products) via the ESA Broadband Data Dissemination Network (BDDN) at the Danish Meteorological Institute (DMI) in Copenhagen, and transferred to the institute (EMI) at the Technical University of Denmark (DTU).

The UILR-images are low-resolution SAR images with a resolution of about 110 meters, produced at the ESA station at Salmijarvi (Kiruna) by averaging and rescaling the Fast Delivery Images (UI-16). The averaging process reduces the image from 5000 x 6300 pixels to 1000 x 1000 pixels, and the rescaling process reduces the radiometric resolution from 16 bit to 8 bit using a histogram equalization.

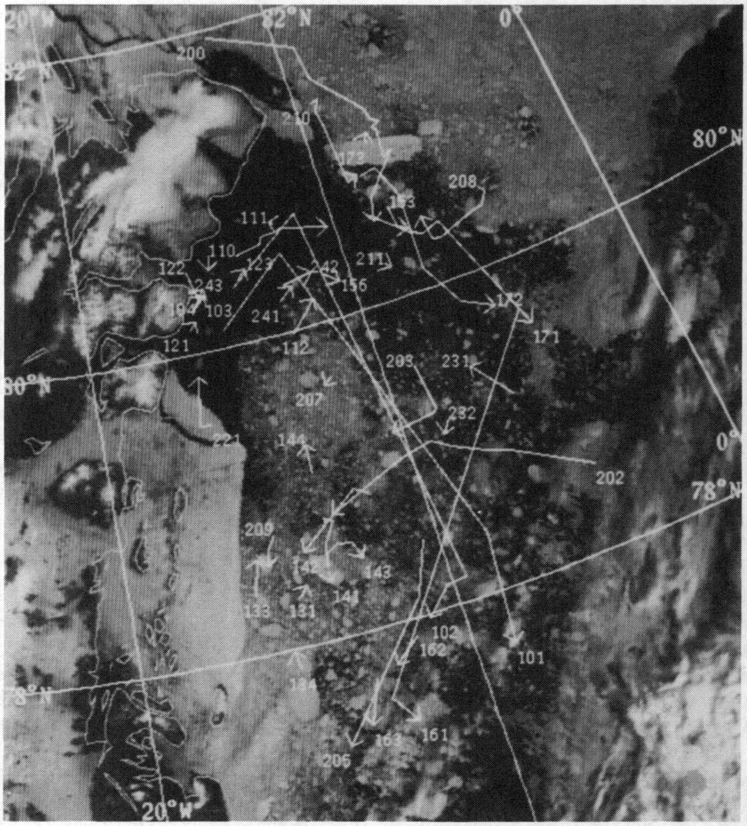

Figure 1: *Ice floe tracks in the North-East Water area.*

A calibration was performed on all UILR images received to invert the histogram equalization and correct for range spreading loss, antenna gain pattern and illumination.

The UILR images were then compared with Precision Images (PRI) in order to evaluate the calibration. The PRI images were mainly acquired from the UK PAF (Processing and Archiving Facility).

Floes are recorded by their center position. Around this position (coordinate) an octagon is defined, constituting a mask for derivation of statistical properties. The size of the octagons is 296-335 pixels, representing an area of about 3 km^2. Due to the averaging process, no significant correlation between neighbouring pixels is expected.

In addition to the floes, also test areas of open water were defined in order to evaluate the calibration process.

3 CALIBRATION METHOD

The PRI and UILR SAR images should be corrected for range spreading loss, antenna pattern and illumination. In UILR images none of these corrections have been performed, while PRI images acquired from the UK-PAF and processed after 1 September 1992, have been corrected except for the illumination. Corrections not performed in the received images, have been performed at EMI before converting them to an internal format.

The scenes were all internally stored as 16-bit images using a gain value equal to the post-1.9.92 UK-PAF processor. Due to the difference in gain between processors, scenes in general need also to be corrected for the processor gain.

The correction of PRI images is straight forward as described by Laur (1992), and the total correction in the amplitude image is

$$\frac{\sqrt{\sigma^o}}{\sqrt{\sigma^o_{uncor}}} = \frac{g_{proc}}{g_{ref}} \frac{g(20.35^o)}{g(\xi)} \sqrt{\frac{sin\theta}{sin23^o}} \sqrt{\left(\frac{R}{847km}\right)^3} \quad (1)$$

where g_{proc} is the processor gain; g_{ref} is the reference processor gain; ξ is the look-angle; 20.35^o is the reference look-angle; $g(\xi)$ is the antenna gain; θ is the incidence-angle; 23° is the reference incidence-angle; R is the range; and 847 km is the reference range. Equation (1) denotes correction for processor gain, antenna pattern, illumination, and range spreading loss.

The information in the UILR images is found to be insufficient to make a similar calibration of the image [Ulander, 1993], due to the lack of incidence angle information. Investigations of 30 PRI images (Orbit 8375-15101, Frame 1449-2475) support former observations [Ulander, 1993] of the center incidence angle being in the range 22.9^o - 23.1^o. Hence, the center incidence angle is approximated to 23.0^o, and the start incidence angle θ_0 can be calculated as

$$\begin{aligned} \theta_0 &= \theta_c - \Delta\psi - \\ & sin^{-1}\left(\frac{R}{r_0}(sin\theta_c - sin(\theta_c - \Delta\psi))\right) \quad (2) \\ \Delta\psi &= \frac{l}{2R} \end{aligned}$$

where θ_c denotes the center incidence angle (assumed to be 23.0^o); R is the radius of earth; r_0 is the slant range distance to the first pixel; and l is the image swath (100 km).

From this point the calibration of UILR images is similar to the calibration of PRI images.

The minimum and maximum incidence angles are stored in the internal image format at EMI. Hence, the incidence angle for an area in the image is determined by

$$tg\theta = \frac{n - n_0}{n_1 - n_0}(tg\theta_{max} - tg\theta_{min}) + tg\theta_{min} \quad (3)$$

where n is the center point sample number in the image line; n_0 is the sample number at near-range; n_1 is the sample number at far-range; θ_{min} is the incidence angle at near-range; θ_{max} is the incidence angle at far-range. Using (3) to determine θ results in a maximum error of 0.02 degrees for the ERS-1 SAR images.

4 COMPARING UILR AND PRI IMAGES

Using the UILR images instead of PRI calls for an investigation on the deviation introduced in backscatter, stemming from the averaging and rescaling processes.

To increase the number of on-line available PRI images, they were all scaled by a factor of 8 in both directions, resulting in the same pixel spacing as in UILR images, 100 m. The averaging process performed in intensity, affected the measured values of backscatter by less than 0.01 dB.

In a selection of 7 scenes, a total of 21 homogeneous test areas of open water have been selected, and compared in UILR and PRI images with respect to σ^o. The test areas contains open water sized between 13 and 53 km^2.

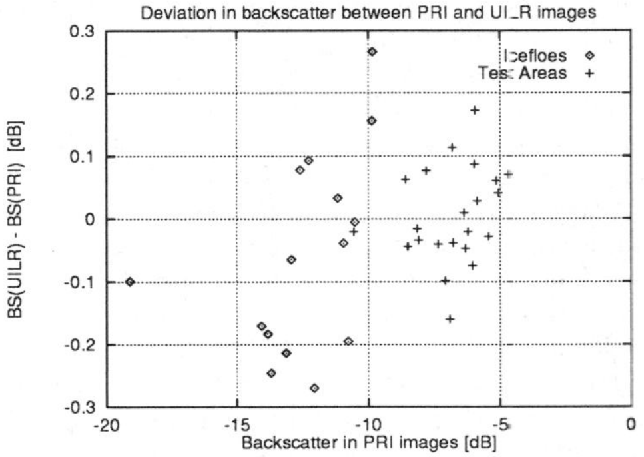

Figure 2: *Deviations in backscatter values larger than -20 dB, as derived from UILR-images and PRI-images.*

In addition to this, all tracked ice floes appearing in a PRI image received at EMI, have also been investigated.

Figure 2 shows the difference in measured backscatter values derived from the UILR-images and PRI-images. It is seen that the spread of points is smaller for the test areas (open water) than for the ice floes describing a higher degree of homogeneity of the former.

The difference is less than 0.2 dB for the test areas, but slightly higher for the ice floes, where the difference is 0.3 dB. Defining the standard deviation (STD) as

$$STD = \sqrt{\frac{1}{n}(\sigma^o_{\mathrm{UILR}} - \sigma^o_{\mathrm{PRI}})^2} \quad (4)$$

results in a STD of 0.074 dB for the test areas, and 0.164 dB for the ice-floe data.

It has been found that large deviations can occur in areas with σ^o below -20 dB, probably due to the histogram equalization method applied.

The variance-to-mean-squared-ratio (VMR) of the calibration areas is in the range of 0.03 to 0.05. Assuming uncorrelated pixels and a VMR of 0.05, the relative standard deviation on the estimated σ^o for a homogeneous area is

$$\frac{\mu}{\sigma} = \frac{1}{\sqrt{N}}\sqrt{VMR} = \sqrt{\frac{0.05}{300}} = 0.013 \quad (5)$$

which corresponds to a speckle noise of 0.06 dB.

By changing the size of the test areas to the same size as the ice-floe octagons, STD increases from 0.074 dB to 0.097 dB. Adding STD from speckle (0.060 dB) with STD from the large test areas (0.074 dB) and assuming they are independent, results in a total STD of 0.094 dB, which is closed to the speckle noise previously calculated.

Investigations have shown that there is a difference in pixel positions in PRI and UILR images. The problem is that different orbit data are used so that different areas are processed. Since the ice floe tracking is based on pixel position, floe tracking shall be performed seperarately in the UILR and PRI images. Unfortunately, the ice floe tracking is very sensitive to the positioning of the octagon mask, so this also results in a large deviation (0.3 dB) mainly due to problems of hitting the same location in the PRI and UILR images.

Due to their homogeneity the test areas are less sensitive to positioning deviations, compared with the floes, which may contain ridges, melt ponds etc. on the edge of the polygon mask.

5 WEATHER CONDITIONS

In the period from 23 July 1993 to 15 August 1993, the weather in the NEW area was observed and reported hourly at the U.S. Coast Guard ship, USCG Polar Sea. Temperatures are found in the range of -2°C to +10°C, and are plotted in Figure 3, together with wind speeds and directions.

The wind speeds show large fluctuations, in the range 0 m/s to 19 m/s (0-36 knots). In the major part of the period, the wind direction is south, but also winds from north are common. The winds from north are usually combined with high wind speeds as was the case around 13 August. Note that wind directions are defined as zero when no wind occurs. Figure 4 shows a stick diagram of the wind. As already shown in Figure 3, the wind is mainly coming from south. However the northern winds are often stronger.

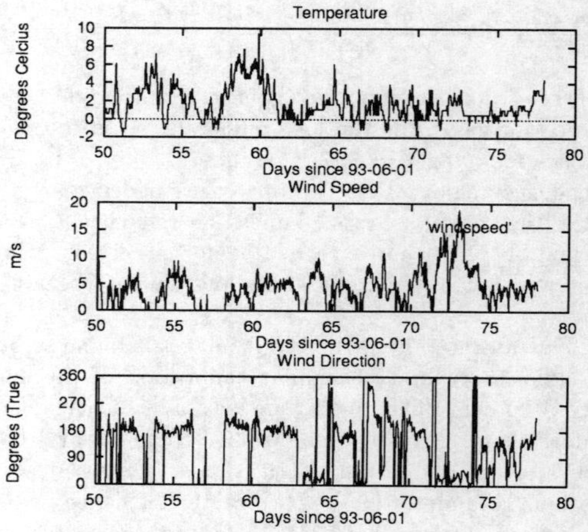

Figure 3: *Temperature, wind speed and wind direction measured on board the USCG Polar Sea. Note that the wind direction is defined as zero when the wind speed is zero.*

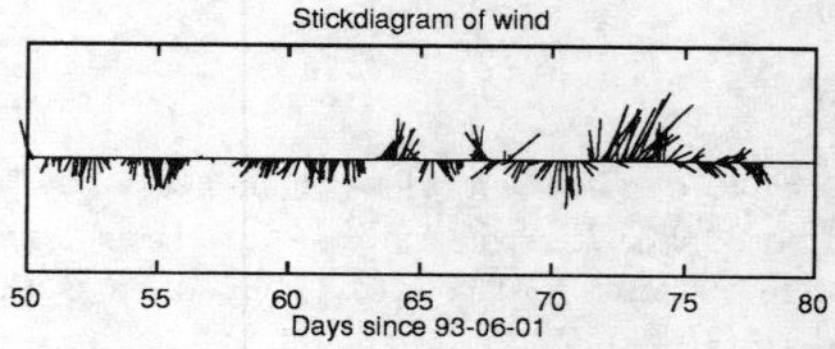

Figure 4: *Stick diagram derived from the wind speed and direction measured on USCG Polar Sea. Sticks pointing down represents winds coming from south.*

6 SHORE-FAST ICE

Two areas of shore fast ice were investigated to derive information on different ice types. The first one selected is connecting Krøyers Holme (three small islands about 16 km east of the Greenland coast at 80.651°N, 13.722°W) with the shore. This ice breaks usually off every year and constitutes thus first-year ice. However, it cannot be stated with confidence that it did brake off in its entirety during the summer 1992, wherefore some of the area may constitute second-year ice. (The ice disappeared in late July 1993).

The second area of shore fast ice selected is located east of Nordøstrundingen (81.34°N, 11.40°W). It constitutes the peninsula extending over shallow water about 40 km to the east, and has been observed as a permanent feature during more than 10 years. The main part is sustained by small stranded floebergs, while the outer regions break off during the summer both on the northern and the southern sides.

Hence the first area will represent first-year/second-year ice and the latter area multiyear ice.

Observations in August by Mortensen (1993) report large areas of melt water ponds on the ice at Nordøstrundingen.

6.1 Analysis method

Assuming that the backscatter value in dB for a specific area or floe has a linear relationship with a number of factors and that σ^o has been observed under different conditions at different moments in time, σ^o

can be described as

$$\sigma^o(i,j) = \sigma_i^o + \beta_{X_1,i} X_1(i,j) + \beta_{X_2,i} X_2(i,j) \qquad (6)$$

where i is an index for the area/floe; j is an index for the observation time; X_1 and X_2 are the dependence parameters; β_{X_1} and β_{X_2} are the dependence of σ^o to the dependence parameters; σ_i^o is the backscatter value when all dependence factors are zero.

The β-parameters in (6) are estimated by linear regression using the statistical analysis system SAS (1985). In general the values of the dependence parameters (X_i) will depend of both area/floe and time of observation. However, weather parameters are assumed to be only dependent on the observation time.

6.2 Krøyers Holme

In 7 different ERS-1 scenes, and at 6 different locations of the shore-fast ice at Krøyers Holme, investigations were made of the backscatter signal.

Unfortunately the weather data from USCG Polar Sea do not cover this early period, but temperature data from a weather station located on Krøyers Holme have been acquired for the entire 3-month period.

These data show that the air temperature for the first time goes above 0°C on 930616 (Day 15) and that the temperature is constantly above 0°C in the period 930703 (Day 32) to 930727 (Day 56), except for two observations on 930714 (Day 43).

In order to derive the variation of backscatter with time, the following model was applied to the data

$$\sigma^o(i,j) = \sigma^o + \Delta\sigma^o_{time,j} + \Delta\sigma^o_{target,i} + \beta_\theta \theta(i,j) \qquad (7)$$

where j is indexing the observation time and i the target area; θ is the incidence angle; β_θ is the dependence of backscatter signal to incidence angle.

Using *SAS* to solve (7), results in a regression coefficient $R^2 = 0.79$, and β_θ = -0.80±0.33 dB/°. The standard deviation on the residuals are 0.7 dB in terms of backscattering coefficient.

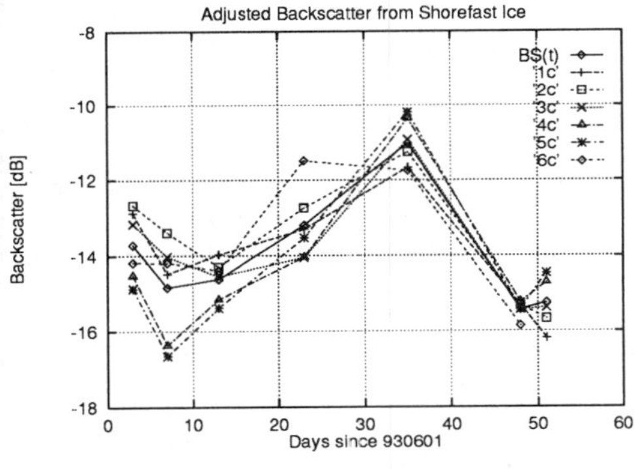

Figure 5: *Average backscatter $\sigma^o + \Delta\sigma^o_t ime$ of shore-fast ice at Krøyers Holme with the backscatter from target areas adjusted to the incidence angle of 23.0° by +0.8 dB/degree.*

Figure 5 shows $\sigma^o + \Delta\sigma_j^o$ and the backscatter values of the 6 different areas as a function of time. The backscatter values are adjusted to 23° by +0.8 dB/degree, before plotting. The figure shows the effects of first-year ice during warm-up as reported by Gogeneni (1984). The snow layer on top of the ice begins to melt as the air temperature reaches 0°C, but when the melt water drains through the snow layer and meets the cold ice at the ice boundary, it freezes again, resulting in a layer of superimposed ice which is very rough. This is only a temporary effect, and as the temperature rises above 0°C and the snow layer melts away, the superimposed layer also begins to melt. The effect of the superimposed rough layer is clearly seen in the figure as increased backscatter values from -15 dB to -11 dB during the first period, followed by a drop in backscatter values during the latter.

6.3 Nordøstrundingen

At the ice protrusion located east of Nordøstrundingen three different areas were selected and found in 14 different SAR scenes from 930607 to 930816. The 10 images in the period 930722-930813, where weather data from USCG Polar Sea were available, were used to determine the influence of the incidence angle, wind speed and temperature on σ^o. The areas are locates as A: 81.34°N 11.05°W, B: 81.27°N 10.62°W and C: 81.20°N 10.19°W.

Incidence angle, wind speed and temperature dependence are modeled by

$$\sigma_j^o = \sigma^o + \beta_\theta \theta_j + \beta_{WS} WS_j + \beta_T T_j \qquad (8)$$

where j is indexing the observation time; β_θ is the incidence angle dependence; β_{WS} is the wind speed dependence; β_T is the temperature dependence; θ is the incidence angle; WS_j is the wind speed at the observation time j; T_j is the temperature at the observation time j; and σ^o is the backscatter when incidence angle, wind speed and temperature are zero.

The results of applying (8) to the three selected areas are shown in Table 1.

There was found no significant dependence of inci-

Table 1: *Regression coefficients and dependency estimates for 3 areas at the ice protrusion east of Nordøstrundingen.*

Area	R^2	$\beta_\theta \left[\frac{dB}{degree}\right]$	$\beta_{WS} \left[\frac{dB}{ms^{-1}}\right]$	$\beta_T \left[\frac{dB}{^oC}\right]$
A	0.05	-0.005±0.248	0.058±0.099	0.050±0.279
B	0.63	-0.027±0.250	0.259±0.099	-0.212±0.280
C	0.78	0.213±0.185	0.295±0.086	-0.345±0.243

dence angle in any area, and area A showed no significant dependence of any of the parameters. Wind speed dependence was very significant with values of $\beta_{WS} = 0.25 - 0.30 \frac{dB}{m/s} (\pm\ 0.10\ \frac{dB}{m/s})$ for areas B and C. Area C showed also a less significant dependence of temperature with $\beta_T = -0.35 \pm 0.25 \frac{dB}{^oC}$.

A similar model like (8) without the terms of incidence angle and temperature dependences showed regression coefficients of $R_B^2 = 0.59$ and $R_C^2 = 0.66$. This shows that the backscatter from the area (B and C) were mostly dependent on wind speed.

In model (8) area A only had $R_A^2 = 0.05$, and hence the variations in σ^o could not be explained by incidence angle, wind speed or temperature. However, the standard deviation of σ^o was also much less for area A.

STD of the residuals, which represents the variation not explained by the model, were approximately equal to 0.7 dB for all three areas, which also was the results from Krøyers Holme in Section 6.2.

7 ICE FLOES

The ice floes observed and tracked in June are mainly multiyear ice, and they are easy to identify due to the large difference in backscatter compared with the surrounding first-year ice. However, in July and August ice classification is hampered by the fact that the melting surfaces produce equal backscatter from all types of floes.

7.1 Incidence angle

Like the shore fast ice, also the floe backscatter signal shows a dependence of incidence angle.

This is seen in Figure 6, where backscatter is plotted against incidence angle. There is a trend of decreasing backscatter for increasing incidence angle.

7.2 Temporal variations

The temporal variations as showed in Figure 7 are very large during the period. In the warm-up period during June, large drops in the backscatter from multiyear ice is seen, exceeding 12 dB in the course of 10 days. This is the effect of melting, cutting off the volume scattering from multiyear ice. A similar effect is reported by both Holt et al. (1993) and Gogeneni (1984).

In July the backscatter level rises again, but more slowly than the previous drop. Large variations still occur in the backscatter signal; both short term variations of the same floe and variations between different floes observed at the same moment in time. These variations can be up to 5 dB.

A similar trend of slowly increasing backscatter is

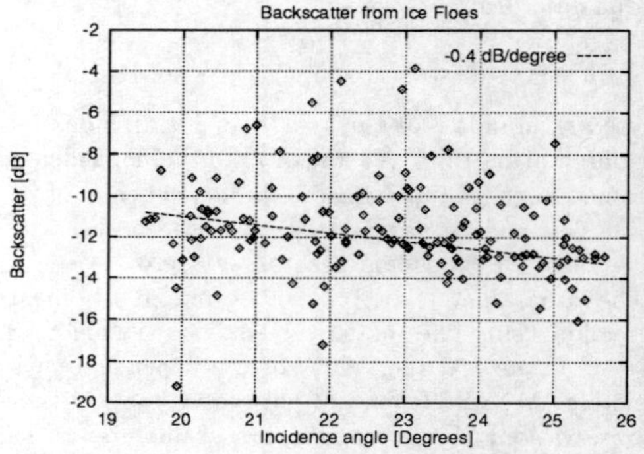

Figure 6: *Backscatter of ice floe as function of incidence angle.*

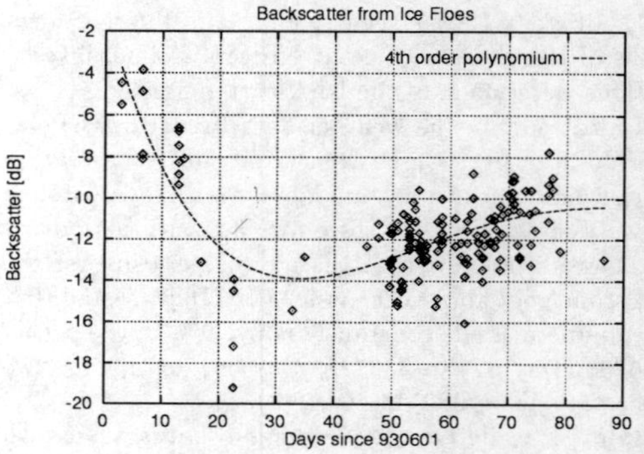

Figure 7: *Backscatter of ice floes as function of time.*

also reported by Gogeneni et al. (1992) and explained by cooling, drying and drainage of the ice surface. The increase that is reported by Gogeneni is 10 dB in a 2-month period at an incidence angle of 40°, where as an increase of 3-4 dB during a one-month period was found in the present work it, but with an angle-of-incidence in the range 19°–26°.

7.3 Dependence factors

Assuming a linear relationship between backscatter and a series of different factors, like incidence angle, wind-speed, temperature, time and squared time, a

Table 2: *Dependence estimates for ice floes observed between day 52 and day 75.*

β_θ	-0.379±0.082	dB/degree
β_{WS}	0.098±0.048	dB/ms^{-1}
β_T	-0.192±0.083	dB/°C
β_t	-0.211±0.603	dB/day
β_{t^2}	0.002±0.005	dB/day^2

model can be established as :

$$\sigma^o_{dB} = \sigma^o_i + \beta_\theta\theta + \beta_{WS}WS + \beta_T T + \beta_t t + \beta_{t^2} t^2 \quad (9)$$

where σ^o_{dB} is backscatter in dB; i is an ice floe index; θ is incidence angle; WS is the wind speed; T is the temperature; t is the moment in time; and β_X are the dependencies to the parameters.

The model results in a regression coefficient R^2 of 0.53 and the estimates of the β-parameters are listed in Table 2.

It is noted from Table 2 that the dependence of incidence angle is very significant. The value of -0.4 dB/degree is, however, only half as large as that observed from the shore-fast ice at Krøyers Holme.

The dependence of windspeed and temperature are both significant. The former is positive, but only half as large as the one observed from the shore-fast ice at Nordøstrundingen, which may stem from differences in the total area of melt-water ponds. The latter is also smaller than the observation from the shore-fast ice at Nordøstrundingen.

An explanation for the lower dependences inthe case of floes, may very well be the mixture of first-year (FY) and multiyear (MY) floes having different characteristics. This may also explain the low value of R^2.

Time dependence is modelled as a second-order effect, and reaches 0.11 dB/day at the end of the period, which corresponds to an increase of 3 dB during a one-month period.

8 DISCUSSION

The investigations showed that the calibration of UILR images made a usable alternative for PRI images in the study of backscatter from ice floes. The difference between PRI and UILR images was less than 0.3 dB for values in the range -15 dB to -5dB, but may be shown to be even smaller, i.e. less than 0.2 dB.

After preliminary studies of shore-fast ice, evident temporal variations were observed for both FY and MY ice. Especially in the melting period large variations occurred for the same ice type at the same time as large differences between the different ice types. After the melting period the difference in σ^o for FY and MY ice types was very small, due to surface wetness.

A temporal trend in σ^o is observed as an increase of 3 dB during one month, which is explained by cooling, drying and drainage of the ice-floes as also reported by Gogeneni (1984).

Large deviations in the backscatter of more than 3 dB from the same floes were observed even though the duration between the observations was only a few hours. These deviations can, however, be explained by changes of the conditions, i.e. incidence angle, wind speed or temperature.

The summer period provides totally different conditions for the ice floes compared to the winter period. However, an extension of the investigated period to September, may be useful for showing how and at what level the backscatter stabilizes, increasing our knowledge of the backscatter characteristics of sea ice.

9 ACKNOWLEDGMENT

This analysis could not have been made without the good understanding and assistance of the ERS-1 Order Desk planning the extensive data acquisition over the area in question. Also, the Danish Meteorological Institute is acknowledged for operation of the ERS-1 BDDN station on behalf of Danish users and for the AVHRR data acquired at the receiving station of the Institute. Dr. Peter Minnett of the Brookhaven National Laboratory, Long Island, provided the meteorological data recorded on board the American icebreaker USCG Polar Sea. Thanks are extended to Preben Gudmandsen for support and valuable discussion in the course of the work.

10 REFERENCES

Gogeneni, S.P. (1984),
Radar Backscatter from Summer and Ridged Ice, and the Design of Short-Range Radars, Ph.D. Dissertation, The University of Kansas, Lawrence, KS.

Gogeneni, S.P., R.K.Moore, T.C.Grenfell, D.G.Barber, S.Digby, and M.Drinkwater (1992),
The Effects of Freeze-Up and Melt Processes on Microwave Signatures, Microwave Remote Sensing of Sea Ice, Geophysical Monograph 68, American Geophysical Union, pp.329-340.

Holt, B., G.Cunningham, and R.Kwok (1993),
Sea Ice Radar Signatures from ERS-1 SAR During Late Summer and Fall in the Beaufort and Chukchi Seas, Proceedings First ERS-1 Symposium, ESA SP-359, pp.339-342.

Laur, H. (1992),
Derivation of Backscatter Coefficient σ^o in ERS-1.SAR.PRI Products, ESA, 16 pp.

Mortensen, H.B. (1993),
Private communication.

SAS Institute Inc. (1985),
SAS User's Guide: Statistics, Version 5 Edition, Cary NC, pp.1-13.

Ulander, L.M.H. (1993),
Calibration of the ERS-1 SAR Fast-Delivery Images, Proceedings First ERS-1 Symposium, ESA SP-359, pp.173-178.

Sensors and Environmental Applications of Remote Sensing, Askne (ed.) © 1995 Balkema, Rotterdam, ISBN 90 5410 524 0

Influence of ice and snow properties on ice ridge mapping by SAR

A. Carlström, L. M. H. Ulander & W. Dierking
Chalmers University of Technology, Göteborg, Sweden

ABSTRACT: Effects of different ice and snow properties on the signature of ice ridges in ERS-1 SAR images are investigated. Scattering models based on the physical optics approximation are applied to surface backscatter from ice ridges and frozen brash ice. *In situ* ice properties are used as input and the results are compared with SAR image signatures and radar scatterometer measurements. The results indicate that C-band backscattering from ice ridges composed of blocks is mainly dependent on the slope distribution of the ice blocks. The agreement between model and measurements is good for cases with slopes less than 0.4. The observed backscattering coefficients of frozen brash ice are almost as high as for ridged ice. The scattering model for brash ice is sensitive to the shape of the surface height correlation function and to the dielectric properties of the snow layer.

1 INTRODUCTION

Remote sensing of ice ridges is of interest both for ship navigation and for climatological studies of sea ice processes. The 5.3 GHz (C-band) synthetic aperture radar (SAR) onboard the European satellite ERS-1 is a promising sensor for mapping deformed ice surface areas due to its sensitivity to the roughness of the surface and its ability to work during all weather conditions. Even single ice ridges may be observed in the imagery during certain conditions, due to the high spatial resolution (30 x 30 m) of the SAR.

A major problem with interpretation of ice ridge parameters is that the signature of ice ridges may be confused with returns from other ice types with rough surfaces. A typical example is refrozen brash ice, which consists of fragments of ice that are frozen on to each other and thus form a rough surface with a backscatter of the same order as of an ice ridge in a level ice background. The widths of ridges are usually smaller than the resolution of the SAR. In those cases the image contrast to the surrounding ice is reduced and the probability for misinterpretations is increased. The snow layer may also affect the SAR signatures so that the detectability of ridges is reduced.

A first-year ice ridge is composed of blocks of broken ice that are forced to pile up by pressure or shear. The size and tilt angle distribution of the blocks determine the main characteristics of the surface roughness spectrum. The first attempt to model the backscattering from ice ridges was made by Johansson and Askne (1987). They proposed a model with contributions from specular reflections and scattering from the rough surfaces of ice blocks randomly tilted in the plane of incidence. This model was later extended (Manninen 1992) to allow for a second degree of freedom of the block tilt and a size distribution of the blocks. Manninen used actual measured distribution functions and found the rough surface scattering from ice block facets to dominate the scattering, while specular reflections were concluded to be a second order effect. Scattering from deformed ice has also been modelled using the conventional rough surface scattering formulation of physical optics (Drinkwater 1989) which showed a good fit to SAR measurements of deformed ice in the marginal ice zone (Livingstone and Drinkwater 1991).

Based on results from field experiments in the Baltic sea we investigate the major physical parameters determining the backscatter from ice ridges and brash ice. First, the different measurement techniques are described. Then, we compare *in situ* measurements with radar data and establish an empirical relationship. In order to determine the dominating scattering mechanisms, we compare theoretical scattering model results with SAR and scatterometer data. Finally, we discuss implications of the results.

2 MEASUREMENT TECHNIQUES

2.1 *Ice and snow property measurements*

As part of the Baltic Experiment for ERS-1 (BEERS), field campaigns have been carried out during the winters of 1992-94 (Thompson et al. 1992; Thompson et al. 1993). Extensive field measurements of surface roughness and other relevant snow and ice parameters were conducted simultaneous with ERS-1 SAR overflights. The measurement sites were mostly chosen so as to have similar properties over an area of at least 10 x 10 resolution cells in the SAR image. The reason for this was to avoid the problem of having a mixture of different ice types in the same resolution cell and to reduce the effect of speckle. Hence, the ice types we investigate here do not include narrow ice ridges, but rather distributed rubble fields or jammed brash barriers. The 1993 and 1994 experiments also included narrow ridges, and results will be presented elsewhere.

During BEERS-92, a manual profiling technique was used for measuring ten metre long height profiles of deformed ice, such as ice ridges, rubble fields, and jammed brash barriers. A 10-m string was used as a reference when sampling the surface height regularly every 5 or 10 cm. The sampling distance was chosen so that the smallest ice blocks were sampled with at least two points. The deflection of the string was accounted for by subtracting a least-squares fit parabola from the obtained height profile. The precision of the height measurements was estimated to 3 mm (RMS) by measuring a flat surface.

For BEERS-93-94, a more accurate laser-based profiling instrument was developed. Cross sectional as well as longitudinal 10-metre profiles of ice ridge sails were measured with horizontal and vertical resolutions of 20 mm and 3 mm, respectively.

The ice density, temperature, and salinity were measured in the surface layer and size distributions of air bubbles were measured for selected samples. Snow density and wetness were measured by a dielectric sensor.

2.2 *Scatterometer measurements*

The Chalmers Shipscat as used during BEERS-94 consisted of an HP8720B vector network analyser, a dipole-fed 0.6 m parabolic dish antenna and a control computer. The network analyser was used as a transceiver transmitting continuous waves in vertical (V) or horizontal (H) polarization and coherently receiving amplitude and phase of the backscattered signal at V and H polarisation. The centre frequency of the radar was 5.4 GHz and the range resolution 0.3 m.

The antenna was mounted on the rail of the icebreaker *Oden* at an elevation of 23 m above sea level. The backscattering coefficient was measured for different incidence angles ranging from 20° to 65° in steps of 5°. For each incidence angle, data for 14 different azimuth angles were obtained by rotating the antenna around the vertical axis in steps of 5°. External calibration of the Shipscat was performed by recording the radar response of a 47 cm trihedral.

2.3 *Calibration of SAR images*

For each measurement area, 50-500 pixels were extracted from ERS-1 SAR fast delivery (FD) images processed by ESA at the Kiruna receiving station. The backscattering coefficient σ^0 can be determined according to (Ulander 1992)

$$\langle \sigma^0 \rangle = \frac{f}{K}\left(\langle DN^2 \rangle - \langle DN_n^2 \rangle\right) \tag{1}$$

where < > denotes spatial averaging, K is the calibration constant, f is a range-dependent correction, and DN and DN_n are the pixel values in the area of interest and noise, respectively. The definition of the range correction f for all FD images is according to

$$f = \left(\frac{R}{R_{ref}}\right)^3 \frac{\sin\theta}{\sin\theta_{ref}} \frac{G^2(\vartheta_{ref})}{G^2(\vartheta)} \tag{2}$$

where R is the slant range ($R_{ref} = 847$ km), θ is the local incidence angle ($\theta_{ref} = 23°$) and G is the antenna gain pattern as a function of elevation angle from nadir ϑ ($\vartheta_{ref} = 20.35°$ is the antenna boresight angle). The FD images are not provided with sufficient information to calculate θ and ϑ. Instead, they are estimated assuming their nominal values, i.e. $\theta = 23°$ and $\vartheta = 20.35°$ at mid-swath. The elevation gain pattern of the ERS-1 SAR antenna has been measured by ESA in flight and the calibration constant K has been determined to 72.6 dB from analysis of the BEERS-92 calibration experiment (Ulander 1992). The noise pixel value was determined to RMS $DN_n \approx 226$ based on a noise-equivalent backscattering coefficient of -25.5 dB at mid-swath. The calibration accuracy using this procedure is estimated to better than ±1.0 dB in the mid-80 km swath.

3 RESULTS

3.1 *Scattering as a function of RMS height*

Comparisons of backscatter models with SAR measurements of Baltic sea ice at C-band (Ulander et al. 1992) showed that scattering from the ice surface is the principal backscattering mechanism for temperatures below 0°C despite a low salinity. Surface scattering is dependent on the RMS height and the height correlation function of the surface. The characteristics of the acquired roughness profiles were investigated by deriving the power spectrum of each profile. The spectral density was mostly decaying as an inverse power-law of spatial frequency with exponents varying between 1.3 and 2.2. This implies that the surface has fractal properties and that the measured RMS heights and correlation lengths are dependent of the profile length. It also means that the statistical fluctuations of the roughness parameters are not decreasing with increasing profile length (Church 1988). Hence, we high-pass filter the profiles before computing the roughness. This corresponds to averaging measurements from shorter sub-profiles with lengths corresponding to the cut-off wavelength and in this way we get a larger number of independent samples. However, the cut-off wavelength must be chosen so that the radar scattering components of the surface are preserved. It was found that it was possible to high-pass filter with a cut-off at 1 m without changing the slope distribution significantly. This agrees with observed block dimensions mainly being less than one metre.

The measured backscattering coefficients of deformed ice are shown in Figure 1. Based on these data we propose an empirical relation between roughness and backscatter of deformed ice according to

$$\sigma^0 = a\left(s/s_0\right)^b \cos^2(\theta)/\cos^2(\theta_0) \tag{3}$$

where s is the RMS height of surface Fourier components with wavelengths shorter than one metre, $s_0 = 0.040$ m, and $\theta_0 = 21.6°$. A least-squares fit in log-log space gives $a = -6.5 \pm 0.4$ dB and $b = 0.44 \pm 0.16$ with a correlation coefficient of 0.77, see Figure 1. The incidence angle dependence assumed in (3) is according to a Lambertian surface. The relatively good fit of the empirical relation indicates that it may be used in an inversion procedure to estimate roughness from SAR images (Carlström and Ulander 1994; Carlström et al. 1994).

3.2 *Scattering model for ridged ice*

Scattering models for ice ridges which average the scattering from randomly oriented ice blocks were suggested in (Johansson and Askne 1987; Manninen 1992). We will here compare the results of a similar but simplified block scattering model with the measured backscattering coefficients of the deformed ice.

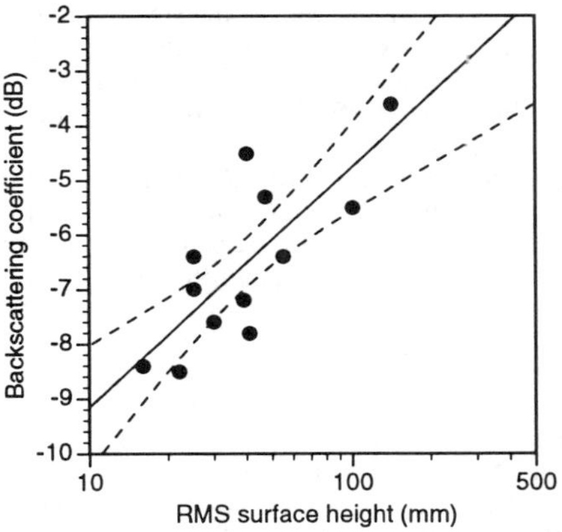

Figure 1. Fit of empirical backscatter model to measurements of deformed ice, the dashed lines indicate the ±1 standard deviation of the model fit.

If we consider an ensemble of randomly oriented and equally sized facets with the average ground projected area A, the mean backscattering coefficient is

$$\sigma^0(\theta) = \frac{1}{A}\iint p\left(z'_x, z'_y\right)\sigma(\theta_i)\,dz'_x dz'_y \tag{4}$$

where σ is the radar cross section of the facet at the local incidence angle θ_i, z'_x and z'_y are the slopes of the facets along the x and y directions, respectively, and $p(z'_x,z'_y)$ is the joint probability density of the facet slopes. We may evaluate (4) approximately for flat facets using the fact that the probability density function is slowly varying compared to the scattering pattern of the facet, which is evaluated under the physical-optics approximation. The validity conditions are that $r \gg \lambda$ and $m \gg \lambda/4r$, where m is the RMS slope, λ is the radar wavelength, and r is the equivalent circle radius of the block facet. The result is (Carlström and Ulander 1994)

$$\sigma^0(\theta) = |R(0)|^2 \frac{\pi}{\cos^3\theta} p(\tan\theta, 0) e^{-4k^2s^2} + \iint p\left(z'_x, z'_y\right)\sigma_s^0(\theta_i)\,dz'_x dz'_y \tag{5}$$

where R is the Fresnel reflection coefficient and $k = 2\pi/\lambda$ is the radar wave number. The first term represents specular (or coherent) reflections from the block facets. A small-scale roughness on the facet is accounted for by a mean rough surface backscattering coefficient σ^0_s, and the exponential factor which reduces the magnitude of specular reflections. For simplicity we may neglect the polarisation dependence since we are only interested in like-polarised scattering. We see that the backscatter is independent of the size of the facets as long as $r >> \lambda$, and is mainly determined by the facet slope distribution. Hence, (5) is a good approximation also for an ensemble of blocks with different sizes.

The surface slope distributions of the measured ice surfaces were estimated by differentiating the unfiltered profiles. As a model function, we choose the Cauchy probability density function,

$$p(z') = \frac{m}{\pi} \frac{1}{m^2 + z'^2} \tag{6}$$

where z' is the surface height derivative. This model agrees better with the data than a Gaussian distribution (see Figure 2). An extreme case of the slope distribution is $m = 1$, which is equivalent to a uniform angular distribution of the surface normal. The RMS slope is not defined for this distribution and the slope parameter m is instead a measure of the half-value width of the distribution. A least-squares fit of the Cauchy function to the distribution of height increments was carried out for each profile in order to estimate the slope parameter m.

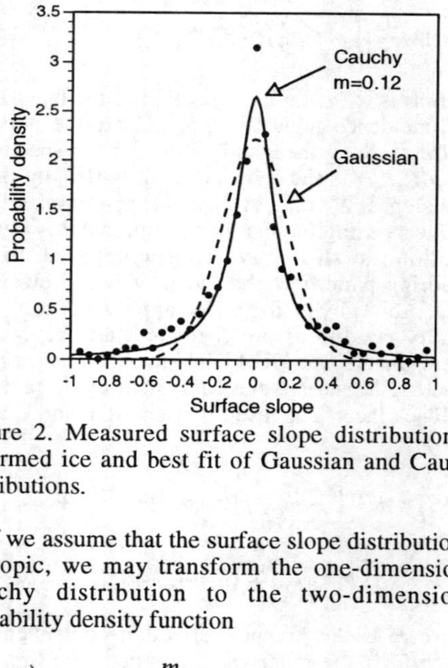

Figure 2. Measured surface slope distribution of deformed ice and best fit of Gaussian and Cauchy distributions.

If we assume that the surface slope distribution is isotropic, we may transform the one-dimensional Cauchy distribution to the two-dimensional probability density function

$$p(z'_x, z'_y) = \frac{m}{2\pi\left(m^2 + z_x'^2 + z_y'^2\right)^{3/2}} \tag{7}$$

in which $m = 1$ corresponds to a uniform distribution of the facet normal solid angle. Figure 3 shows numerical results of (5) for both a Cauchy-type and a Gaussian (with RMS slope m.) slope distribution. The measured average incidence angle of 21.6° was used. Surface scattering from the small-scale roughness was computed using the integral equation model (Fung et al. 1992). The results show that the difference is small between flat facets and facets with a small-scale roughness of $s = 3$ mm and $l = 70$ mm. These values represents relatively rough blocks, since laser profiler measurements of ice blocks in one of the ridges gave $s = 2.0 \pm 0.5$ mm and $l = 69 \pm 17$ mm. The reason for the small difference is that an increase in the rough surface scattering is compensated by a decrease in the specular component. We conclude that the specular component dominates and that the slope distribution of the blocks is the most important parameter.

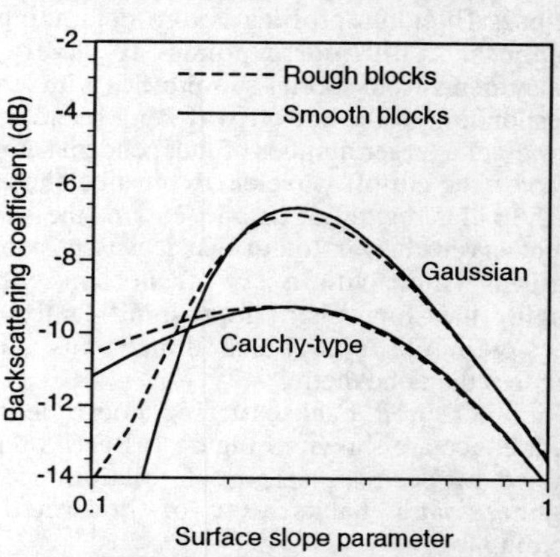

Figure 3. Comparison of model predictions for ice blocks with different slope distributions and surface roughness.

So far we have neglected the part of the wave that is transmitted into the ice block and reflected against the far side of the block. The effect of finite thickness may be incorporated by replacing the reflection coefficient with an effective reflection coefficient for a plane-parallel layer. In this case, a realistic assumption is that the ice blocks are loss-less (low salinity) and that the shadow side of the block is towards an air filled space. We also neglect volume scattering within the ice. If the blocks are of slightly varying thickness, phase interactions will average out and the mean effective reflectivity is given by the incoherent model for a dielectric slab (Ulaby et al. 1986)

$$|R|^2 = \frac{2|R_1|^2}{1 + |R_1|^2} \tag{8}$$

where R_1 is the Fresnel reflection coefficient of the air-ice interface. This corresponds to an increase in the backscattering coefficient of 2.7 dB, compared to the case with infinite block thickness. For rough blocks, the reflection and transmission coefficients are modified by the small-scale roughness and the specular term is reduced. On the other hand, the diffuse term is increasing due to rough surface scattering from the far side of the block and the effect of small-scale roughness is small also in this case.

If we use the effective reflectivity together with the model (5) and the Cauchy-type distribution we get the result shown in Figure 4, which are compared with the SAR data for the deformed ice with large block sizes. The agreement between the model and SAR measurement is good for slope parameters less than 0.4. For larger slope parameters, the measurements show an increasing trend, while the model is decreasing with increasing slope. Possible explanations for this are effects like multiple scattering and edge diffraction, which become important for large slopes. Diffraction effects are especially important for small block sizes where physical-optics theory is not valid. These scattering effects should be considered in further developments of scattering models for deformed sea ice surfaces.

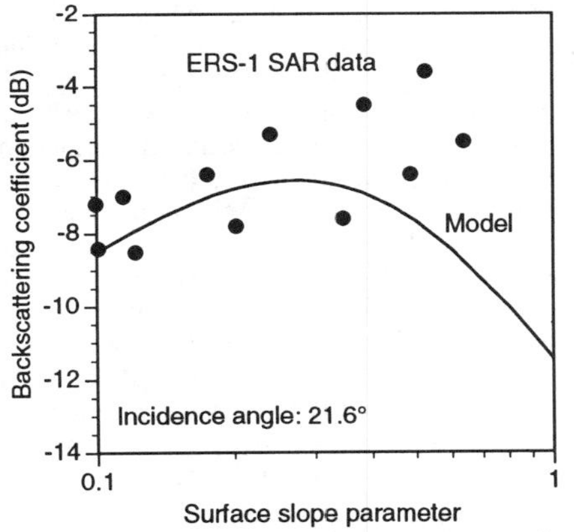

Figure 4. Comparison between model prediction, including effect of finite block thickness, and SAR data for deformed ice with large block sizes.

3.3 *Scattering model for brash ice*

During BEERS-94, backscatter measurements were carried out on an area of frozen brash using the C-band scatterometer which was operated on the icebreaker *Oden*. In addition, surface roughness data were acquired. The ice surface was rough and covered by a 2-5 cm layer of dry snow.

The backscattering coefficients σ^o of frozen brash as observed by the scatterometer are presented for different incidence angles and different polarisations in Fig. 5. Since the differences between the VV-HH and VH-HV-polarised backscatter are very small, only VV and HV are shown. The error bars are a function of the number of independent samples and of the signal-to-noise ratio. The strong decrease of σ^o with increasing incidence angles indicates the dominance of surface scattering contributions. The cross-polarised return can probably be explained by multiple scattering processes either on the ice surface or within the ice volume, or both (Fung and Eom 1981a). Since the ice cover consisted of frozen brash, the presence of volume inhomogeneities is likely. Thus a volume scattering contribution can not be totally excluded. The sudden drop of σ^o at incidence angles around 35-40 degrees may indicate a change in the surface and/or volume characteristic.

The average correlation function of surface heights as measured by the 10-m laser profiler is shown in Figure 6 together with 95% confidence limits. The RMS height is 24 mm and the 1/e correlation length is 0.34 m. The result was obtained by averaging the correlation functions of eight profiles and the uncertainty is caused by the statistical fluctuations that arise from the finite profile length. The autocorrelation resembles an exponential function for displacements lower than 0.6 m. Close to the origin, however, it approaches a Gaussian with a correlation length of about 0.1 m. This implies that the average radius of curvature of the surface is about 0.2 m. The reason for the Gaussian shape is partly due to that the profile is smoothed by the 20 mm footprint of the laser but the larger correlation length of the Gaussian implies that the actual surface profile is smooth on the surface-projected radar wavelength scale.

For the given roughness paramters, the physical-optics solution (i.e. tangent plane approximation) as described in (Fung and Eom 1981b) can be applied in order to calculate the surface scattering contribution. The backscattering coefficient is

$$\sigma^0(\theta) = \frac{2k^2|R(0)|^2}{\cos^2\theta}\exp\left(-4k^2s^2\right)\times$$
$$\times\int_0^\infty J_0(2kr\sin\theta)\left[\exp\left(4k^2\cos^2\theta\, s^2\rho(r)\right)-1\right]rdr \quad (9)$$

where ρ is the measured autocorrelation function and J_0 is the zeroth order Bessel function. Our model consists of a layer of dry snow above a rough ice surface. The effect of the dry snow cover is to reduce the dielectric contrast, the wavelength, and the incidence angle on the ice surface. Results of model computations are shown in Figure 5 for snow-covered ice, and for comparison, for bare ice. The main difference is that the snow-cover reduces the backscattering at the low incidence angle range. For RMS heights of the order of half a wavelength, the model is very sensitive to the shape of the correlation function close to the origin, which is related to the RMS slope and average radius of curvature of the surface. Simulations show that the model computation at larger incidence angles is sensitive to the effect of surface smoothing by the laser profiler. A reason for this is that the laser footprint filters out undulations of the surface that radar waves with incidence angles above 45° interacts with, i.e. Fourier components with wavelengths larger than the Bragg resonant wave. This may also explain the difference between model and measurements at larger incidence angles.

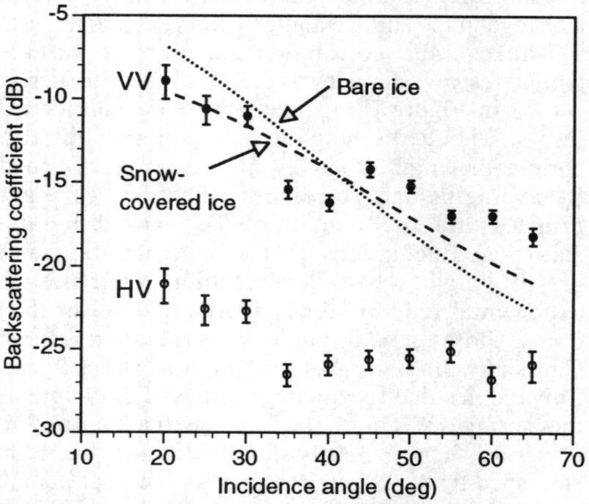

Figure 5. Ship-scatterometer measurements and model result. The result for snow-covered ice was obtained by assuming a dielectric constant of 1.5.

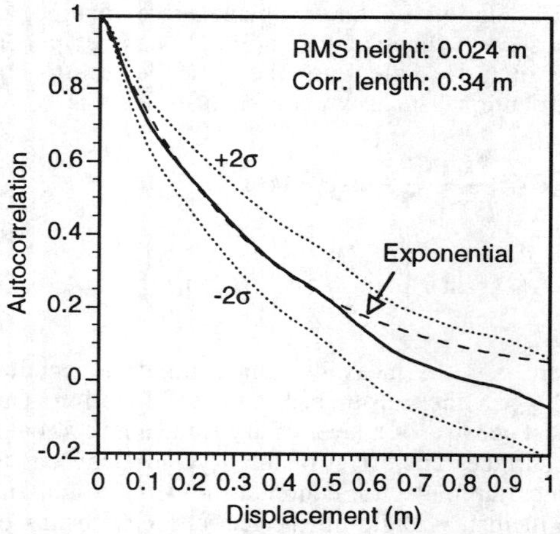

Figure 6. Average surface height autocorrelation function of frozen brash.

4 CONCLUSIONS

Radar signatures of ridged and rough ice have been investigated during several field campaigns in the Baltic Sea. The main contribution to C-band radar scattering at lower incidence angles from deformed sea ice is due to the surface structure on the cm-dm scale. When the snow is dry, the ice surface is dominating the scattering, whereas if the snow is wet, scattering from the snow surface is also important.

The scattering model for tilted ice blocks indicates that scattering from ice ridges is dominated by specular point scattering and thus depends mainly on the slope distribution of the ice blocks, while the small-scale roughness of the blocks is less important for the incidence angle range of the ERS-1 SAR. The single scattering contribution is independent of the block size as long as the blocks are considerably larger than the wavelength. The agreement between model and measurements is good except for cases with large slopes, where it is suspected that multiple-scattering and edge diffraction effects are important.

The frozen brash surface consists of smaller fragments of ice and may not be described by a block model. Instead, we used the RMS height and the surface height autocorrelation function as input to the physical-optics model. For RMS heights of the order of half a wavelength, the model is mainly sensitive to the shape of the correlation function close to the origin, which is related to the RMS slope and average radius of curvature of the surface.

Dry snow affects the backscattering signature of rough ice, since it changes the dielectric contrast, wavelength, and incidence angle on the ice surface. Modelling results indicate that a dry snow-cover lowers the scattering at lower incidence angles. If the penetration depth of the radar wave is substantially reduced due to snow wetness, the dominant scattering source is the snow surface. Since the snow surface roughness is not necessarily correlated to the roughness of the ice, rough ice areas may be masked and detection becomes difficult.

5 REFERENCES

Carlström, A. & L.M.H. Ulander 1994. Validation of backscatter models for level and deformed sea ice in ERS-1 SAR images. Submitted to *Int. J. Remote Sensing.*

Carlström, A., L.M.H. Ulander & B. Håkansson 1994. Model for estimating surface roughness of level and ridged sea ice using ERS-1 SAR. *Proc. of Int. Geoscience and Remote Sensing Symp., IGARSS '94, Pasadena, California.*

Church, E.L. 1988. Fractal surface finish. *Applied Optics.* 27: 1518-1526.

Drinkwater, M.R. 1989. LIMEX'87 ice surface characteristics: implications for C-band SAR backscatter signatures. *IEEE Trans. Geosci. and Rem. Sens.* 27: 501-513.

Fung, A.K. & H.J. Eom 1981a. Multiple scattering and depolarization by a randomly rough Kirchhoff surface. *IEEE Trans. Antennas Propag.* AP-29: 463-71.

Fung, A.K. & H.J. Eom 1981b. Note on the Kirchhoff rough surface solution in backscattering. *Radio Science*. 16: 299-302.

Fung, A.K., Z. Li & K.S. Chen 1992. Backscattering from a randomly rough dielectric surface. *IEEE Trans. Geosci. Remote Sens.* 30: 356-69.

Johansson, R. & J. Askne 1987. Modelling of radar backscattering from low-salinity ice with ice ridges. *Int. J. Remote Sensing*. 8: 1667-1677.

Livingstone, C.E. & M.R. Drinkwater 1991. Springtime C-band SAR backscatter signatures of labrador sea marginal ice: measurements versus modeling predictions. *IEEE Trans. Geosci. and Rem. Sens.* 29: 29-41.

Manninen, A.T. 1992. Effects of ice ridge properties on calculated surface backscattering in BEPERS-88. *Int. J. Remote Sensing*. 13: 2469-2487.

Thompson, T., B. Håkansson, L. Ulander & A. Carlström 1992. Experiences from the Swedish sea ice programme during BEERS-92. *Proc. of First ERS-1 Symposium, Cannes, France, 4-6 Nov.*: 313-318. ESA SP-359.

Thompson, T., B. Håkansson, L. Ulander, A. Carlström, A. Gustavsson & B. Larsson 1993. The use of ERS-1 SAR data for sea ice mapping and icebreaker operations in the Gulf of Bothnia. *Proc. of Second ERS-1 Symposium, Hamburg, 11-14 Oct.*: 271-276. ESA SP-361.

Ulaby, F.T., R.K. Moore & A.K. Fung 1986. *Microwave remote sensing: Active and passive, 3 vols.* Dedham, MA: Artech House.

Ulander, L.M.H. 1992. Calibration of the ERS-1 SAR fast-delivery images. *Proc. of First ERS-1 Symposium, Cannes, France, 4-6 Nov.*: 173-178. ESA SP-359.

Ulander, L.M.H., R. Johansson & J. Askne 1992. C-band radar backscatter of Baltic sea ice: theoretical predictions compared with calibrated SAR measurements. *Int. J. Remote Sensing*. 13: 2447-2468.

Sensors and Environmental Applications of Remote Sensing, Askne (ed.) © 1995 Balkema, Rotterdam, ISBN 90 5410 524 0

An analysis of daily ice motion over the Yermak Plateau

Y. Sun & J. Askne
Department of Radio and Space Science, Chalmers University of Technology, Göteborg, Sweden

ABSTRACT: The ERS-1 SAR images from cross orbit or neighbouring orbit provide an image sequence with short time interval (e.g. one day). In this paper, such an image sequence has shown great potential for studying the ice motion with daily variations due to wind, currents and tides. Two algorithms, the fast correlation and the optical flow algorithms, are applied to an image sequence over the north part of the Yermak Plateau which has been recognised as a region having complex ice drifts due to the integrated affects of wind, tides and mixing of the surface polar water and underlying warmer Atlantic water. The early results show that besides the wind forced drift the ice motion has the components with circular property that may be the affect of the tidal ellipse and the ice deformation has some relation with the topography of the plateau.

1 INTRODUCTION

The Fram Strait is the largest opening of the Arctic ocean through which about 20% of total ice area of the Arctic Basin annually exits (Barry et al. 1993), via the Transpolar Drift Stream away from the Siberian coast, across the pole over the Yermak Plateau to the Fram Strait. The ice export through Fram Strait has important consequences for the overall Arctic ice cover and oceanography, which has been recognised for many years and confirmed by numerous observations in this region (Untersteiner 1988). Use of ERS-1 SAR images has great flexibility for spatial and temporal coverage. It is very appreciated for studying the property of the Transpolar Drift Stream (TDS) near the Fram Strait. Kergomard et al. (1993) has reported the study of the TDS using ERS-1 SAR images. Their results showed that the north part of the Yermak Plateau has a rather strong deformation of the ice cover which implies that the ice packs in this area may be derived by a integrated force (e.g. wind, currents and tides). To investigate the effects of different forces, we use in this paper, a relatively long sequence of images (from September 20 to 2 October, 1991) with a short time interval (one day) covering the north part of the Yermak Plateau. Meanwhile the deformation grids atop the plateau are also studied using the optical flow method. Both ice motion measurements show that the ice motion along the north and north-west flanks of the Yermak Plateau is a strong indicator of the variations in wind, diurnal tidal current and the gradient of the plateau topography.

2 A TIME SERIES OF ICE DRIFTS

The ERS-1 SAR images acquired during the 3-day repeat orbit are commonly used for the ice motion study. However the ice drift over the Yermak Plateau may have significant variations during a three day period due to the variations of the ice motion forces. Therefore cross orbital and neighbouring orbital

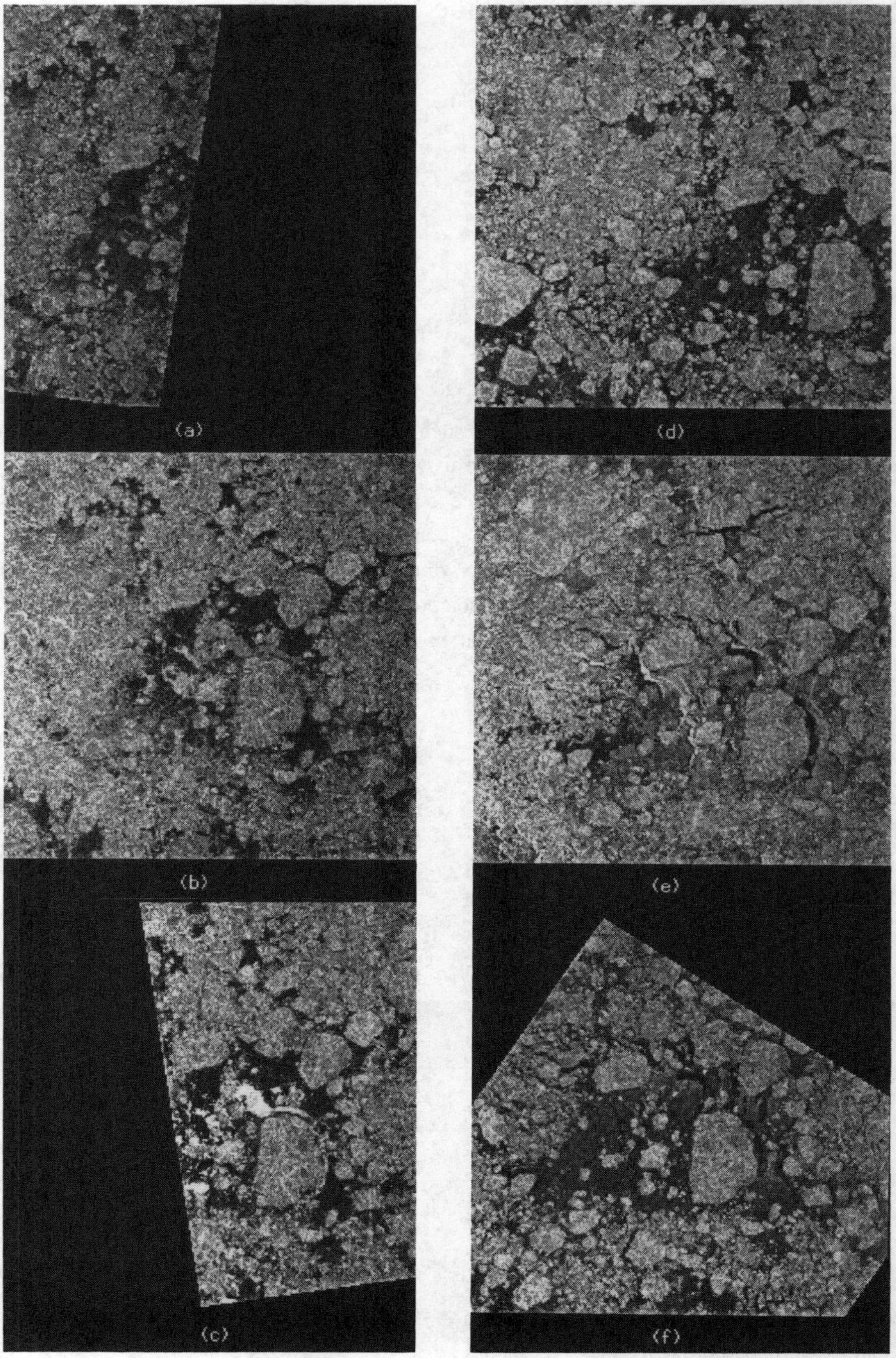

Fig. 1 An example of the images from cross orbits and neighbouring orbits, where the images (a), (b), (c), (d), (e), and (f) are acquired on 20, 21, 22, 24, 27, and 28 Sept. 1991, and the scene centre is 83.46° N, 14.35° E.

images are used to obtain an image sequence which has short time interval (one day) and covers the approximately the same area. The important aspect for generating such an image sequence is to transform the images from individual image co-ordinates into one with geographic co-ordinates. In this work we use bilinear transformation whose parameters are determined from the four corner co-ordinates of an image found in the header of the image. The rms. error introduced by the transformation is about 80 meters for the four corners. Fig. 1 is an example of such an image sequence. To obtain the mean drift a fast cross-correlation algorithm (Sun, 1993a) is applied to the image sequence acquired during a 14-day period (from 20 September to 3 October, 1991). The locations of these images are shown in Fig. 2, where the centre of the area is about 83.0° N and 13.5° E.

With the cross-correlation method, a mean drift of ice packs is obtained by finding the maximum in a cross-correlation function of two images. To deal with uniform scale change and rotation between two images the Fourier analysis is implemented to determine the rates of rotation and scale changes, and finally to compensate them. The basic idea is briefly described following. The Eq. (1) presents the relation of the power spectra of two images which differ by a rotation angle θ_0 and scale factor a.

$$P_d(u,v)=\frac{1}{a^4}P_r\left(\frac{u}{a}\cos\theta_0+\frac{v}{a}\sin\theta_0,\ -\frac{u}{a}\sin\theta_0+\frac{v}{a}\cos\theta_0\right) \quad (1)$$

P_r and P_d are the power spectra of the reference image and deformed image respectively. Applying polar co-ordinates on the Eq. (1) we get

$$P_d(r,\theta) = \frac{1}{a^4}P_r\left(\frac{r}{a}, \theta - \theta_0\right) \quad (2)$$

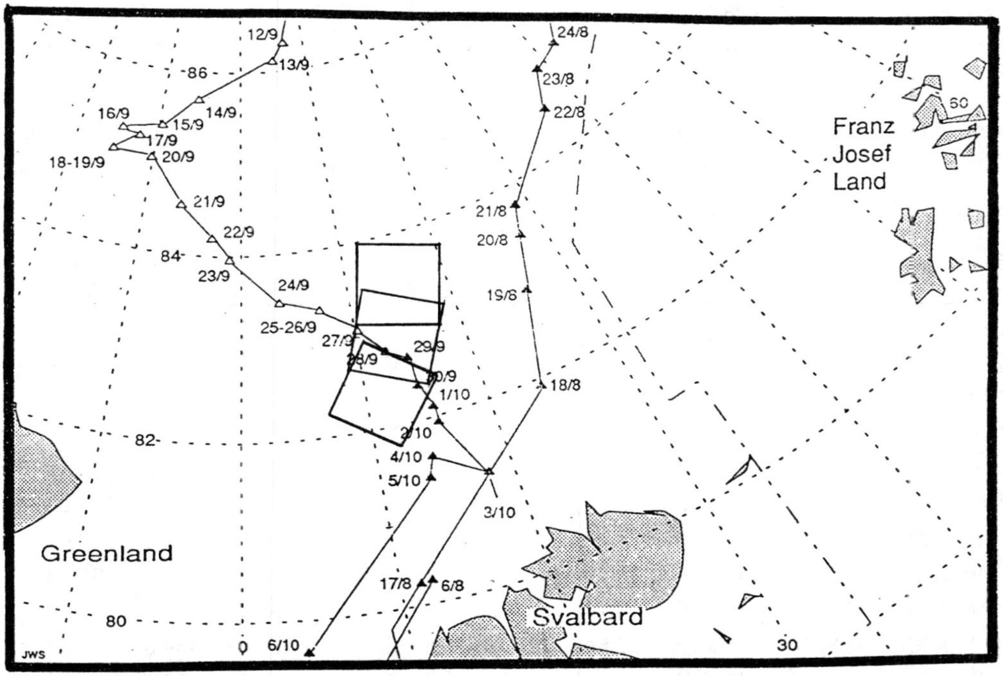

Fig. 2 The location of the study area, as well as the ship track during the ARCTIC-91 expedition. The wind velocities are obtained along the ship track.

Consider a projection which maps (r, θ) plane into (ξ, η) plane, where $\xi = \ln(r)$ and $\eta = \theta$. The power spectra, P_r and P_d, are expressed in (ξ, η) co-ordinates as follows.

$$P_d(\xi, \eta) = \frac{1}{a^4} P_r(\xi - \ln(a), \eta - \theta_0) \quad (3)$$

In Eq. (3) the scale factor a and the rotation angle θ_0 become translation parameters which can be easily determined using the cross-correlation function of P_r and P_d in (ξ, η) co-ordinates. The mean drifts obtained by this algorithm are then, converted to motion velocities and are listed in Table 1.

3 EXAMINE THE ICE MOTION DATA

The ice motion data obtained in the preceding section has been examined by the wind-driven linear model (Thorndike and Colony 1982), we then assume

$$\mathbf{V} = \mathbf{A}\ \mathbf{W} + \mathbf{C} + \mathbf{E} \quad (4)$$

where ice velocity **V**, wind velocity **W**, constant **C**, and a residual **E** are complex numbers. A turning angle θ in the complex multiplier, $\mathbf{A}=|\mathbf{A}|(\cos\theta - j\sin\theta)$, is measured clockwise positive form **W** to **V**.

The wind velocities used for this study are obtained during the ARCTIC-91 expedition which are listed in Table 1. A complex regression is carried out on the two velocities (ice motion and wind) and the results are listed in Table 2.

Table 2. Ice motion Statistics

$\lvert \mathbf{A} \rvert$	θ_a	$\lvert \mathbf{C} \rvert$ cm s^{-1}	θ_c	$\lvert \mathbf{E} \rvert_{rms.}$ cm s^{-1}	$\lvert \mathbf{V} \rvert_{rms.}$ cm s^{-1}	r^2
0.0146	12.7°	2.02	175.6°	5.5	10.4	0.73

From table 2. it is clear that wind is a major force (about 70%) of ice motion and the constant current, **C**, is southward. The correspondence between wind and ice motion velocities is illustrated in Fig. 3, as a plot of two time series vectors. However the residual term **E** is not negligible. In order to verify **E**, we convert all velocity vectors of **E** which is the difference between the estimated ice motion velocity

Table 1. The mean ice motion velocities and the corresponding wind velocities, as well as the locations for both measurements, positive angle of a velocity is defined from north clockwise to the velocity.

Time period	Ice motion velocity: Location Latitude	Location Longitude	Speed cm s^{-1}	Direction* degree	Wind velocity: Location Latitude	Location Longitude	Speed cm s^{-1}	Direction* degree
20/9 - 21/9	83.2° N	13.5° E	13.46	1.4	85.0° N	10.0° W	668.3	-22.2
21/9 - 22/9	83.5° N	14.5° E	7.05	40.9	84.5° N	6.5° W	855.6	14.0
22/9 - 23/9	83.2° N	16.5° E	12.64	48.1	84.2° N	2.5° W	1127.1	12.9
23/9 - 24/9	83.1° N	15.5° E	10.56	40.4	84.0° N	1.0° W	704.0	44.2
24/9 - 25/9	81.9° N	15.0° E	3.11	-80.7	83.6° N	3.5° E	130.8	146.8
25/9 - 26/9	82.8° N	11.5° E	8.64	-165.3	83.5° N	7.5° E	356.0	-172.3
26/9 - 27/9	83.2° N	13.5° E	9.44	-140.8	83.5° N	7.5° E	394.6	-158.9
27/9 - 28/9	83.4° N	15.5° E	20.68	-129.4	83.1° N	10.0° E	604.1	-137.2
28/9 - 29/9	83.0° N	11.5° E	9.74	-129.3	82.9° N	12.5° E	598.5	-134.2
29/9 - 1/10	82.8° N	11.5° E	2.64	-167.4	82.8° N	14.5° E	339.6	-91.7
1/10 - 2/10	82.8° N	11.5° E	7.18	-166.9	82.1° N	15.5° E	575.6	-151.5
2/10 - 3/10	83.2° N	13.5° E	10.24	136.2	81.9° N	15.8° E	470.1	-187.3

* The reference north is taken from the point (82.9° N, 7.1° E) instead of the local north for the individul measurements.

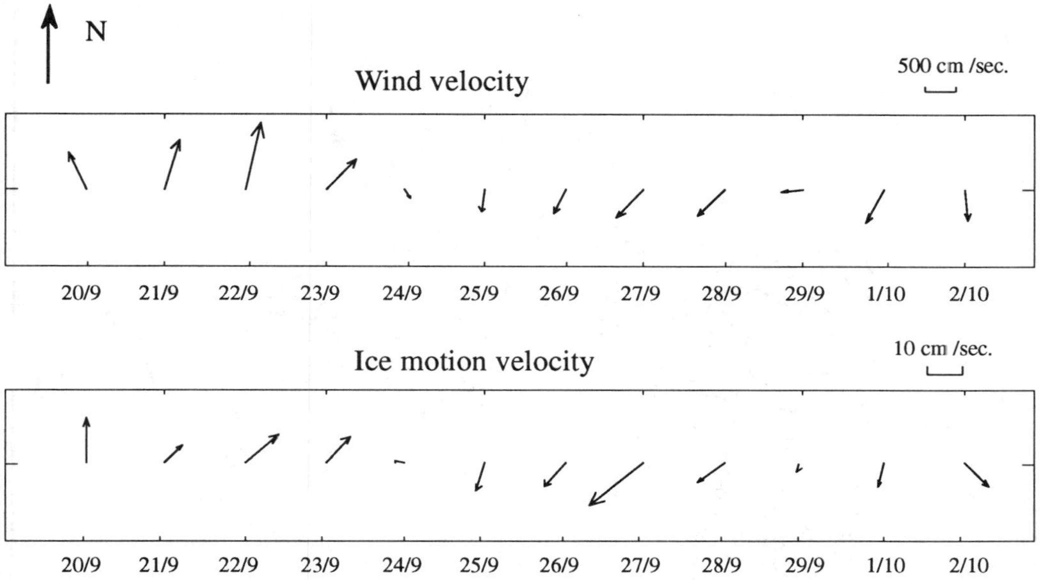

Fig. 3 The plot of two time series vectors, wind velocities and ice motion velocities.

and the true velocity, into distance vectors and plot them in Fig. 4 as a motion track. As we see the motion track forms a circle during a 14-day period. This could be the effect of a tide cycle, (see comment by Kergomard et. al. 1993).

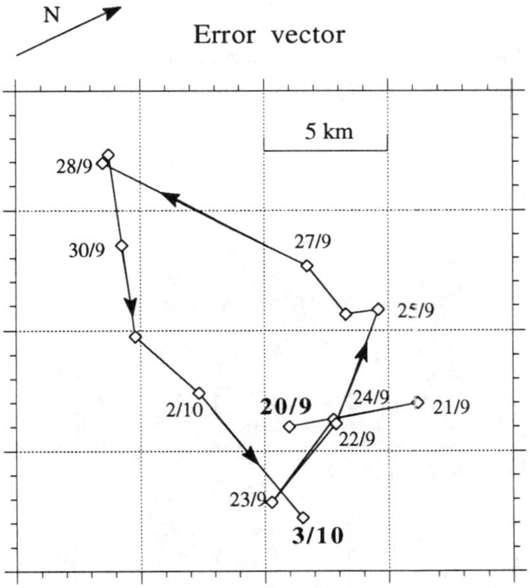

Fig. 4 The plot of the **E** vectors.

4 DEFORMATION GRIDS

Besides the mean ice drifts, the details of the ice motion in the spatial domain are derived by the optical flow algorithm (Verri et. al., 1990). This algorithm calculates the ice motion velocities at each pixel by solving following equation:

$$\frac{\partial I(x,y,t)}{\partial x \partial x}u_1 + \frac{\partial I(x,y,t)}{\partial x \partial y}u_2 + \frac{\partial I(x,y,t)}{\partial x \partial t} = 0$$
$$\frac{\partial I(x,y,t)}{\partial y \partial x}u_1 + \frac{\partial I(x,y,t)}{\partial y \partial y}u_2 + \frac{\partial I(x,y,t)}{\partial y \partial t} = 0 \qquad (5)$$

where u_1 and u_2 are two components of the ice motion velocity, I(x,y,t) is the image intensity, the partial derivatives of the intensity are calculated in a moving window. A pyramid structure and refinement steps are also required by the algorithm. For further details see (Sun 1993). This algorithm is applied on two image pairs, one acquired from orbits 1056 and 1099 (frame 1701) of 28 September and 1 October, the other one acquired from orbits 1070 and 1113

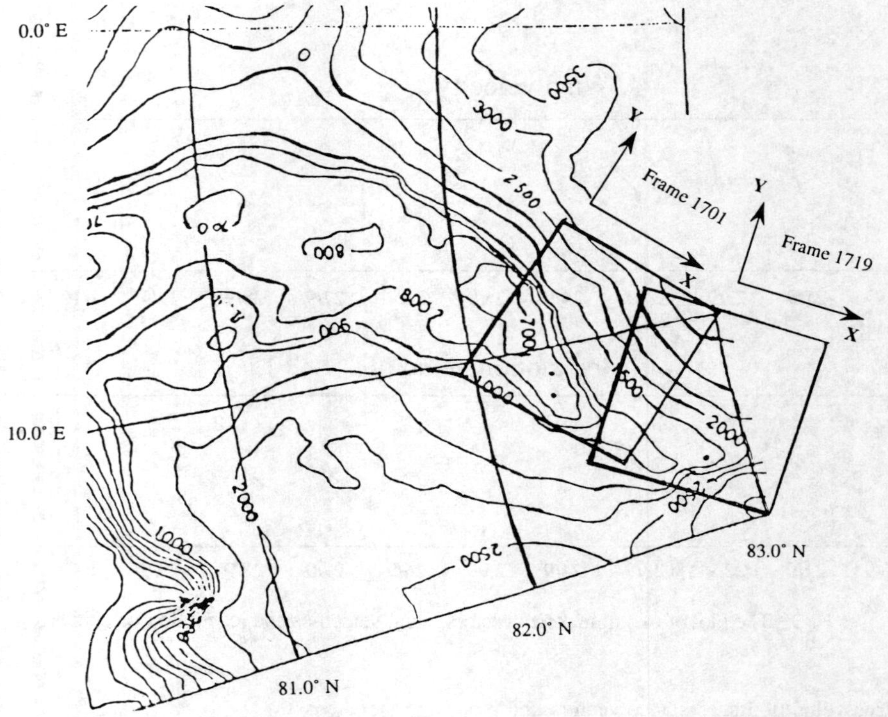

Fig. 5 Bathymetry of the north part of the Yermak Plateau and the area covered by the deformation grids, where the X-axis is in range direction and Y-axis in azimuth direction.

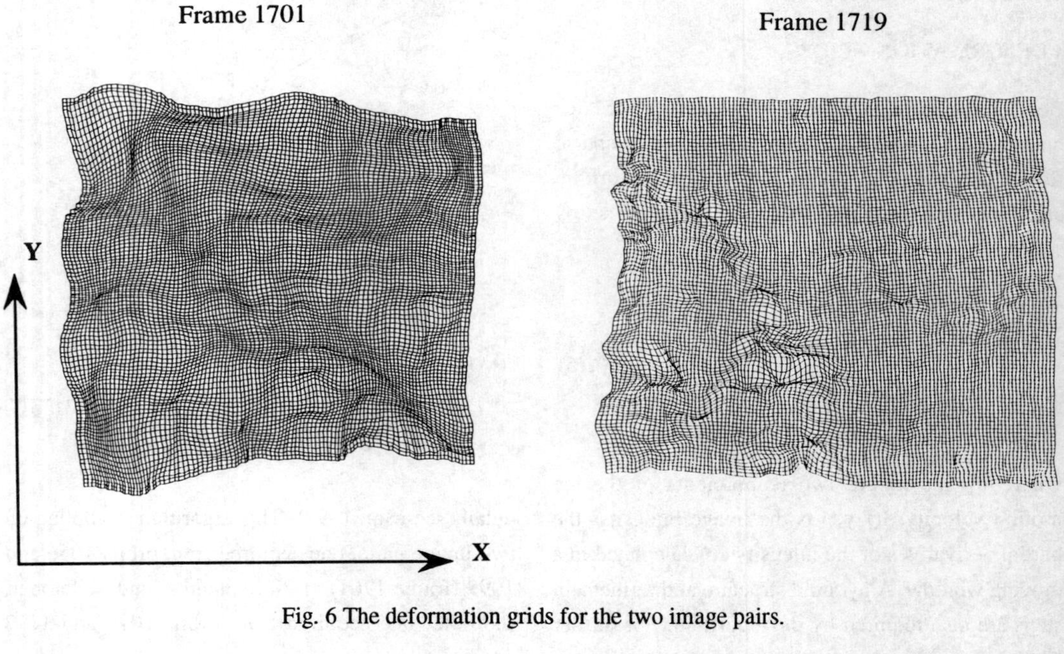

Fig. 6 The deformation grids for the two image pairs.

(frame 1719) of 29 September and 2 October 1991. The coverage of the two image pairs is just atop the Yermak Plateau see Fig. 5. The results are displayed as deformation grids (see Fig. 6). Since within the image scenes some areas do not have correspondence between two images. The deformation grids are only plotted on an area with approximately 80 x 80 km^2, (as marked in Fig. 5) instead of entire coverage of an ERS-1 image scene (100 x100 km^2). The basic idea of the deformation grids is that at the beginning of the motion the ice cover has an uniform grids. After a period of motion the grids will be changed, a non-rigid motion velocity field will cause the grids to be irregular or deformed. The deformation grids illustrates the property of a velocity field and is not affected by the mean velocity.

In Fig. 6 the deformation grids for both cases contain a shear zone which corresponds to the gradient of topography of the plateau. This phenomena manifests that the topography enhanced current plays major role in the deformation of ice cover since for both cases the wind is not strong about 400 cm s^{-1}.

5 DISCUSSION AND CONCLUSION

In this paper we present the results of daily ice motion over the Yermak Plateau by using the ERS-1 SAR images acquired from cross orbits and neighbouring orbits. The question naturally arises whether the accuracy is acceptable. The accuracy of a motion drift derived by our algorithm is mainly depends on three aspects, the error in corner co-ordinates, the error introduced by the linear transformation, and the pixel spacing. The former two errors are within 100 meters respectively for the PRI (precision image) image and the pixel spacing used in our algorithm is 200 meters. The total error could be limited in 0.5 km. In general, the daily drift of ice packs is about 8 km. In this study the maximum and minimum values are 18 and 3 km respectively. Therefore accuracy of 0.5 km is acceptable for the study of daily ice motion.

In section 3, we examine the daily motion data and find that besides the wind forced ice drifts the residual term **E** shows circular property (Fig. 4). This phenomena might be related to the tidal ellipse which is a tidal *seiche,* or standing wave oscillation in channels, embayments and boxes (Apel 1987). As we see in Fig. 4 the cycle of the ellipse is approximately 14 days (i.e. the neap-spring tidal cycle). Its major axis is about 72° from north, which is along the north-west flank of the Yermak Plateau. Therefore the **E** term has stronger components along the flank of the plateau than that cross the flank. From Fig. 4 we also found that the first four vectors of **E** (i.e. from 20/9 to 23/9) are quite random. As described in section 3 the **E** term is obtained by subtracting the wind forced ice drift (including the constant term) from the true ice drift. Hence the errors in wind velocities will affect the **E** term. From Fig. 2 and Table 1, it can be seen that the locations of wind measurements for 20, 21, 22, and 23, September, are quite far away (over 100 km) from the area for the ice motion study. The wind velocities for these four days are biased. This could be the reason why the **E** vectors are random on those four days. For the future work the accuracy wind information is desired.

Based on the analysis in this paper, we conclude that

- The ice motion over the Yermak Plateau is a strong indicator of the variations in wind, diurnal tidal current and the gradient of the plateau topography.
- The ERS-1 SAR image sequence with a short time interval has great potential for studying the ice motion with daily variations due to wind, currents and tides
- The fast correlation algorithm is suited for analysing daily ice motion.

ACKNOWLEDGEMENT

This work is supported by Swedish National Space Board.

REFERENCE

Apel J.R. 1987, Principles of Ocean Physics, (London, San Diego, New York, Boston, Sydney, Tokyo: Academic Press).

Barry, R.G., Serene, M. C., and Maslanik, J. A., 1993, "The ARCTIC sae ice - climate system: Observations and modelling," *Reviews of Geophysics*, **31, 4,** pp 397-422.

Kergomard C., Gascard J.C., Fily M., Delacourt Ch., Lefort Ch., and Darr D., 1993, The Transpolar ice drift north of Fram Strait studied from ERS-1 SAR imagery, Proceedings Second ERS-1 Symposium - Space at the Service of our Environment, Hamburg, Germany, ESA SP-361.

Sun Y., 1993a, Anew correlation technique for ice motion analysis, *Proceedings of EARSeL workshop on ice*, Lyngby, Denmark, to be published.

Sun Y., 1993b, Using ERS-1 SAR images for ice motion estimation, *Research Report No. 164, Depart. of Radio & Space Science, Chalmers University of Technology, Sweden.*

Thorndike, A. S., and Colony, R., 1982, "Sea ice motion in response to geostrophic winds, *J. Geophys. Res.*, **87**, pp. 5845-5852.

Untersteiner N. 1988, On the mass and heat balance of Fram Strait, J. Geophys. Res., 93, pp. 527-531.

Verri A., Girosi F., and Torre V., 1990, Differential techniques for optical flow, *J. Opt. Soc. Am.* , A/Vol. **7**, No. **5**, pp. 912-922.

Cryosphere

Sensors and Environmental Applications of Remote Sensing, Askne (ed.) © 1995 Balkema, Rotterdam, ISBN 90 5410 524 0

Real time runoff forecasts for two hydroelectric stations based on satellite snow cover monitoring

Klaus Seidel, Walter Brüsch, Charlotte Steinmeier & Jaroslav Martinec
Communications Technology Laboratory, Swiss Federal Institute of Technology, ETHZ, Zürich, Switzerland

Jürg Wiedemeier
Northeastern Power Company of Switzerland (NOK), Baden, Switzerland

ABSTRACT: Seasonal and short-term runoff forecasts for two hydroelectric stations in the upper Rhine basin are carried out in real time based on snow cover monitoring by Landsat and SPOT satellites. Evaluation of snow reserves on 1 April 1993 from satellite data reveals uncertainties in estimates using point measurements on the ground as index. Runoff is computed by the SRM model with snow covered areas as well as temperature and precipitation forecasts as input variables. A SRM menu system has been installed for operational data acquisition and management. The runoff forecasts can be exploited, among other purposes, for optimizing the hydropower production and for timely decisions on the electricity market.

1 INTRODUCTION

Many hydroelectric stations in Switzerland are situated in mountainous regions where snow is a major runoff factor. The satellite remote sensing of the snow cover therefore emerged as a useful tool in runoff forecasts which are needed to optimize the electricity production. This paper deals with seasonal and short-term forecasts for two hydroelectric stations of the Northeastern Power Company of Switzerland in the upper Rhine basin. Runoff is computed by the Snowmelt Runoff Model (SRM) which uses the snow covered area, temperature and precipitation as input variables. Periodical snow cover mapping is based on Landsat and SPOT data. Quantitative precipitation forecasts and temperature forecasts are taken into account in real time.

2 HYDROELECTRIC STATIONS AND THEIR CATCHMENT AREAS

The Rhine basin at Felsberg (3250 km^2, 560-3614 m a.s.l.) was divided into 9 partial areas as shown in Figure 1 in order to evaluate the regional distribution of snow (Martinec et al., 1991). The catchment areas of the powerplants *Sedrun* and *Tavanasa* are situated in the western part (area no. 1), where the snow accumulation is normally higher than in the eastern parts. As required by the Snowmelt Runoff Model (SRM), the catchment areas of Sedrun and Tavanasa are divided into elevation zones which are listed in Table 1.

Sedrun is an accumulation power plant with three artificial reservoirs totalizing 152·10^6 m^3 of storage capacity and an installed output of 150 MW.

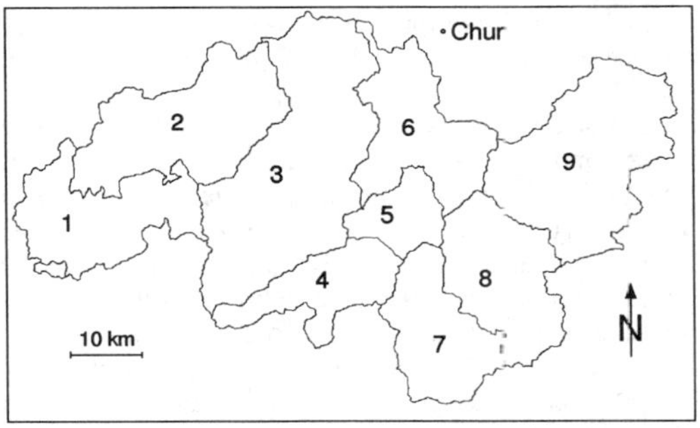

Figure 1: Division of Rhine-Felsberg basin (3250 km^2, 560-3614 m a.s.l.) into 9 regions

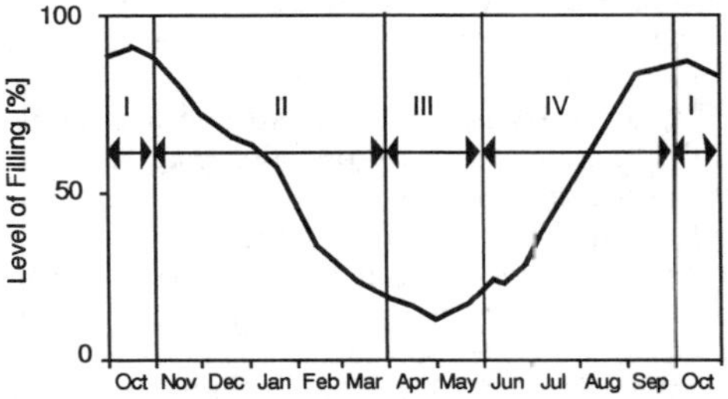

Figure 2: Operation phases of a reservoir

Table 1: Elevation zones of the catchment areas of the stations Sedrun and Tavanasa

	Sedrun			Tavanasa		
zone	area [km^2]	elevation range [m a.s.l.]	mean hypsometric elevation	area [km^2]	elevation range [m a.s.l.]	mean hypsometric elevation
B				19.8	1277-1600	1492
C	14.0	1840-2100	2000	77.9	1600-2100	1877
D	63.2	2100-2600	2380	85.9	2100-2600	2323
E	30.3	2600-3210	2726	31.8	2600-3210	2750
total	107.5	1840-3210		215.4	1277-3210	

Tavanasa is a run-of-the river plant situated down stream with a compensating reservoir of $0.4 \cdot 10^6$ m^3 volume. It uses the outflow from the Sedrun station and an additional runoff from its catchment area. The installed capacity is 180 MW.

3 MERITS OF SEASONAL AND SHORT-TERM RUNOFF FORECASTS

In Switzerland, the electricity consumption is higher in winter than in summer, reversely to the runoff regime which is influenced by snowmelt. In order to reduce this discrepancy, the hydroelectric station Sedrun accumulates water in the summer and improves runoff in the winter by emptying the reservoirs. This mode of operation is illustrated by Figure 2 (Seidel et al., 1990).

In phase 1, the reservoir operation aims at filling the reservoir as late as possible and simultaneously avoiding an overflow. In phase 3, the lowest water level in the reservoir is attained but a necessary reserve must be maintained until the snowmelt season sets in. In these months, short term forecasts may help to achieve these goals. Seasonal runoff forecasts are important for accumulation hydroelectric stations mainly in the phase 3 and 4. They also help to plan the maintenance work on turbines to avoid overflow.

The run-of-the river station Tavanasa has only a very small storage reservoir to avoid situations when the inflow exceeds the capacity of turbines so that water overflows the weir unused. Quick decisions must be made to reduce such losses and therefore short-term runoff forecasts are needed.

Thanks to runoff forecasts, advanced information on the future electricity production may be obtained. As a result, demands on base and peak electricity supply in the national grid can be satisfied while achieving favourable prices on the electricity market. A numerical example has been published earlier (Seidel et al., 1990): Stations Sedrun and Tavanasa produce in summer about $300 \cdot 10^6$ kWh. If $50 \cdot 10^6$ kWh of excessive summer energy can be sold contract based for a price higher just by 0.005 sFr/kWh, a gain of 250'000 sFr results for the cost/benifit analysis of forecasts, considering just these two stations.

4 REAL TIME SEASONAL FORECASTS IN 1993

The target of forecasts are natural inflows into reservoirs of the station Sedrun and the natural runoff from the catchment area of the station Tavanasa in the months April through September.

The seasonal runoff originates from two components: Snow reserves on 1 April and precipitation in the subsequent months. The first component is more reliable because snow is actually accumulated in the basin at the time of forecast. Precipitation can only be estimated from historical records and subsequently updated if actual amounts differ from the estimate.

However, to evaluate the water equivalent of the snow cover in altitudes indicated in Table 1 is no easy task either. The only snow gauging station in the region, Sedrun (1420 m a.s.l.), measured only snow depths: 45 cm on 16 March and 20 cm on 1 April. As the only station in an adequate altitude, Weissfluhjoch (2540 m a.s.l.) measured a water equivalent of 684 mm on 16 March and 725 mm on 1 April. Normally, the region no. 1 in Figure 1 has more snow than the Weissfluhjoch area no. 9 (Martinec et al., 1991). The data of 1993 thus appeared to indicate snow conditions similar as in 1985 but with a higher altitude gradient. In 1985, areal average water equivalents for all elevation zones listed in Table 1 were evaluated from periodical snow cover mapping by Landsat-TM and SPOT-XS recordings with the use of the so-called modified depletion curves (MDC) of the snow coverage as described previously (Martinec et al., 1991).

Based on this evaluation and snow measurements of 16 March, a provisional forecast was issued on 1 April 1993 and also reported at the ITC Symposium in Enschede 1993 (Seidel et al., 1993). Later, the results have been presented in a much more informative and detailed manner as shown in Figure 3. The first row in the figure refers to the initial forecast values derived for TAVANASA and SEDRUN. The numerical values computed by SRM are summarized in Table 2. Monthly temperature and precipitation values have been taken from historical long-term mean measurements. In addition, the actually measured runoff is shown for comparison.

5 UPDATING OF RUNOFF FORECASTS WITH ACTUAL SATELLITE SNOW COVER MAPS

Whenever the snow coverages of successfully evaluated satellite recordings became available an update of the respective MDC's was possible.

In Table 3 a list of satellite recordings is given which were acquired during the runoff season 1993.

Based on the data of 21-APR-93 and 16-JUN-93 a revision of the MDC's for the respective elevations was possible and the initial monthly distribution curves have been updated for 25 April 1993 and 20 June 1993 as shown in Figure 3. The numerical values are also given in Table 2. Evidently, the snow cover in the Sedrun-Tavanasa region was in 1993 not deeper than in the Weissfluhjoch area as would correspond to the long term average trend. This experience casts doubts on using the station Weissfluhjoch as index of snow reserves for Tavanasa and Sedrun before a proper evaluation of satellite data. This example shows that satellite snow cover mapping is more reliable than point measurements in a normal snow gauging network.

In spite of the overestimation of snow reserves, the initially given forecast was accurate but this was due to a massive precipitation excess in September.

6 REAL TIME SHORT-TERM FORECASTS IN 1993

Similar as in a previous study of simulated short-term forecasts with historical data (Seidel et al., 1990), the SRM was used for the computation of daily flows. In the morning of the forecast day n, the following data were available:

Yesterday's daily runoff	Q_{n-1}
Forecasted temperatures	$T_{max,n}$ ... $T_{max,n+3}$
	$T_{min,n}$ ... $T_{min,n+3}$
Forecasted precipitation	P_n ... P_{n+3}

Snow covered areas S_n ... S_{n+3} for the respective elevation zones, derived from the modified depletion curves of the snow coverage with the use of temperature forecasts as explained elsewhere (Martinec, 1985).

Considering one elevation zone, the runoff Q_n is computed as follows:

$$Q_n = (c_{sn} a_n T_n S_n + c_{rn} P_n) A \frac{10000}{86400} (1 - k_n) + Q_{n-1} k_n \quad (1)$$

where

c_{sn}	runoff coefficient for snow
c_{rn}	runoff coefficient for for rain
a_n	degree-day factor [cm °C^{-1} d^{-1}]
T_n	$= \frac{T_{max,n}+T_{min,n}}{2}$ forecasted average temperature at the mean hypsometric elevation of the zone [°C d]
S_n	ratio of snow covered area to total area
P_n	forecasted precipitation [cm]
A	area of zone [km^2]
$\frac{10000}{86400}$	conversion from [cm km^2 d^{-1}] to [m^3 s^{-1}]
k_n	$= xQ_{n-1}^{-y}$ recession coefficient with parameters x, y to be determined for the given basin

Recalling Table 1, the model input must be computed separately for the respective elevation zones.

7 ASSESSMENT OF RESULTS

The accuracy of short-term runoff forecasts for Sedrun and Tavanasa can be assessed by taking reference to the results of a recent WMO Project (WMO, 1992) entitled *Simulated Real-Time Intercomparison of Hydrological Models.*

The accuracy of the forecasts for lead times from 1 to 4 days can be evaluated by the root mean square error:

$$RMSE_n = \sqrt{\frac{1}{N} \sum_{i=1}^{N} (Q_{f,n,i} - Q_{n,i})^2} \quad (2)$$

where

N	total number of forecasted periods
$Q_{f,n,i}$	forecasted flow for n-th day ahead for i-th forecast period
$Q_{n,i}$	measured flow for n-th day ahead for i-th forecast period

Table 2: Runoff forecasts 1 April - 30 September 1993 for Sedrun and Tavanasa (in 10^6 m^3)

	Tavanasa			Sedrun		
Runoff from:	1-APR-93	25-APR-93	20-JUN-93	1-APR-93	25-APR-93	20-JUN-93
forecasted snowmelt	129.1	88.6	82.8	97.0	63.8	62.8
forecasted precipitation	123.7	134.5	145.1	64.7	73.3	84.1
total runoff	252.8	223.1	227.9	161.7	137.1	146.9
measured runoff	246.3			166.9		

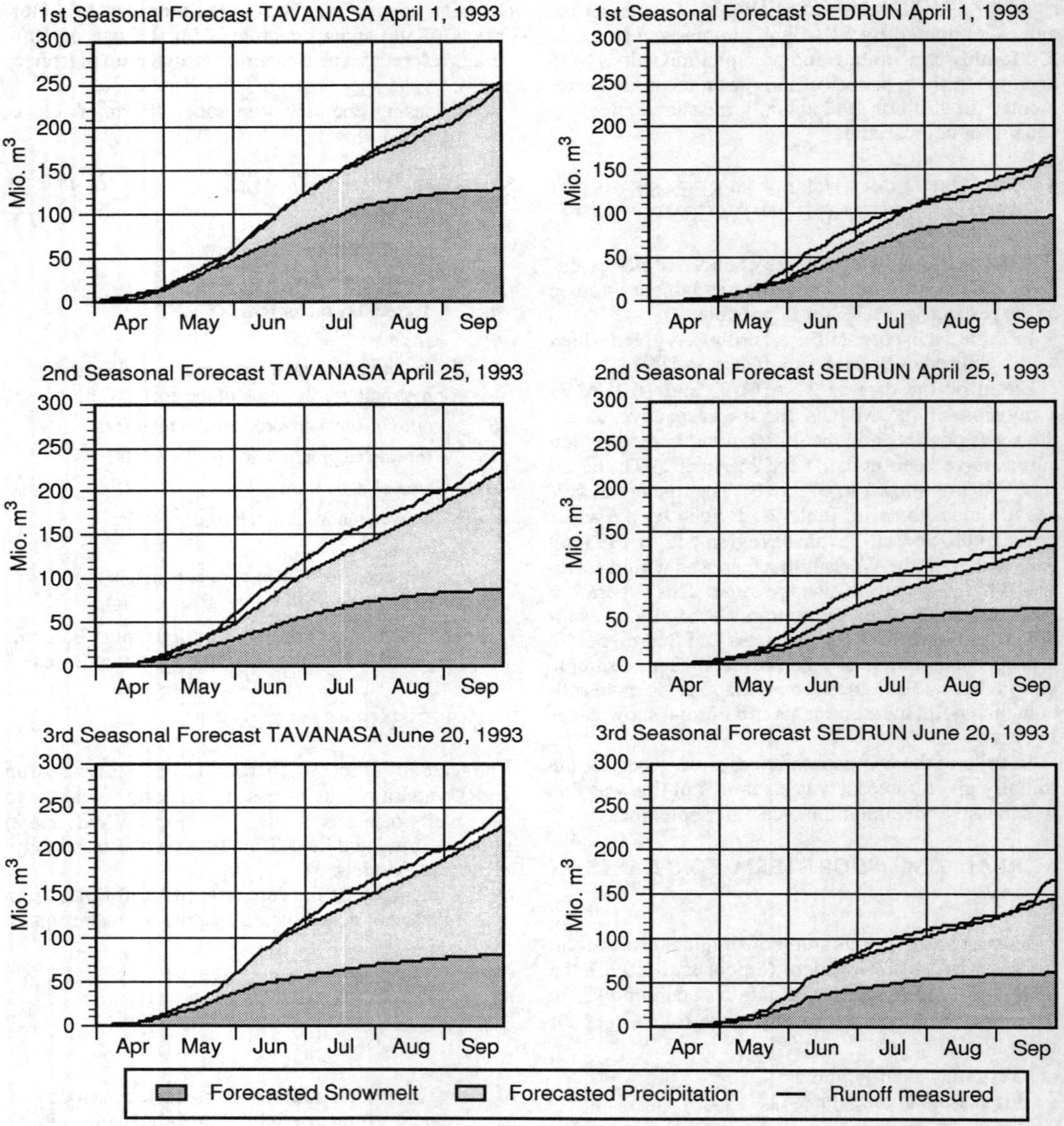

Figure 3: Runoff forecast 1993 for SEDRUN and TAVANASA separately for snow and precipitation components

In Table 4 the achieved accuracies for Tavanasa are summarized. The percentage is with 35% rather high. One part of the deviations is certainly caused by the errors in the temperature and precipitation forecasts.

8 SRM Data Acquisition and Management System

The previously described approach for runoff simulations and forecasts has been installed as the *SRM Data Acquisition and Management System.* An overview of the system is outlined in Figure 4.

The aim is to support runoff simulations and forecasts in an operational manner. The kernel of the menu system is a Relational Data Base Management System (RDBMS) which integrates the results of the snow cover mappings and the online collection of meteorological data such as temperature and precipitation inputs which are supplied as measured and forecasted

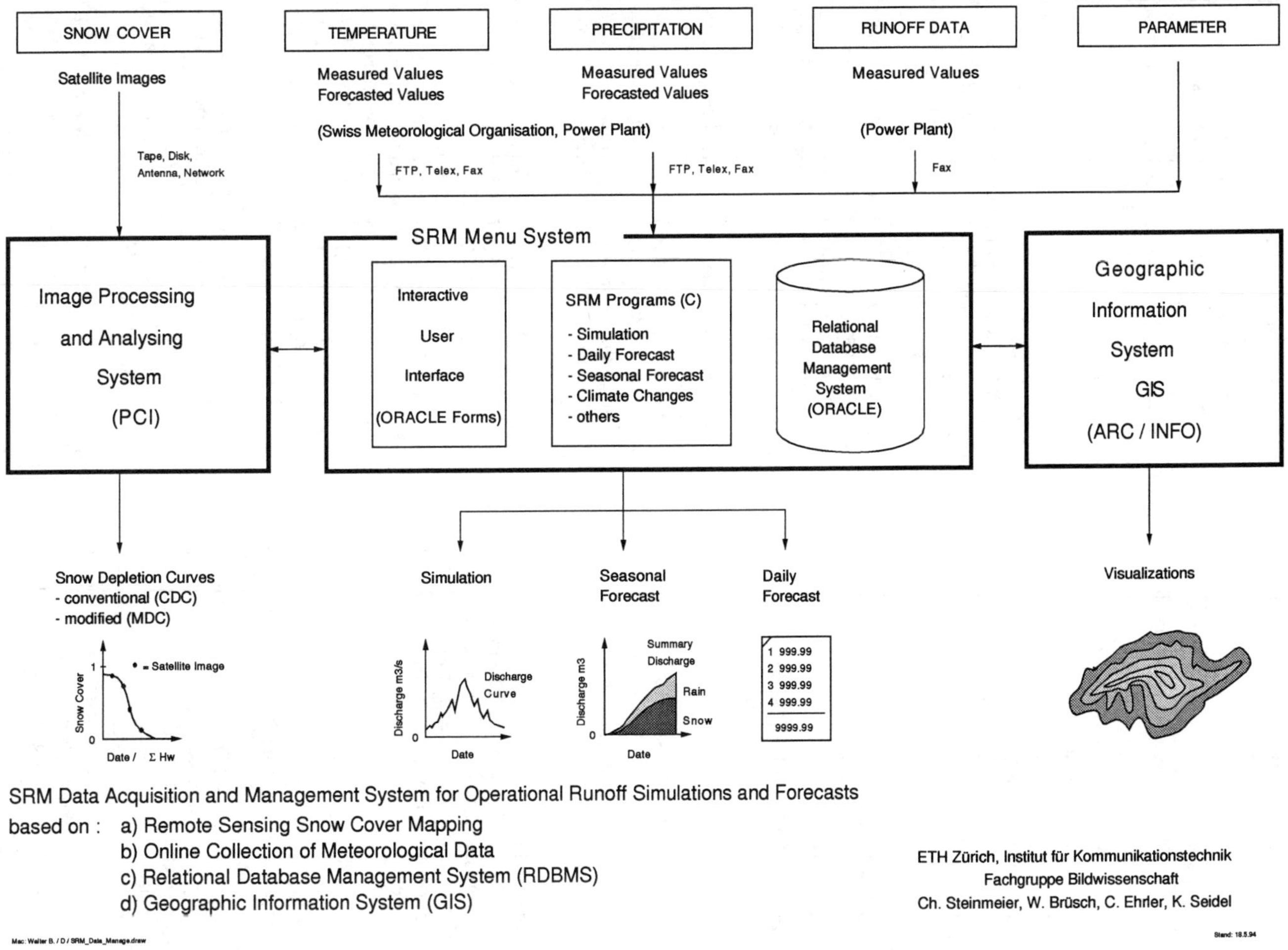

Figure 4: SRM Data Acquisition and Management System

Table 3: List of satellite recordings for the Sedrun and Tavanasa area in 1993

Date	Sensor	Quality - Remark
30-MAR-93	XS	excellent, evaluated
20-APR-93	MSS	excellent
21-APR-93	XS	good, evaluated
22-MAY-93	MSS	cloud coverage $>30\%$
1-JUN-93	XS	excellent, evaluated
7-JUN-93	MSS	cloud coverage $>30\%$
16-JUN-93	XS	excellent, evaluated
12-AUG-93	XS	good

Table 4: Accuracy for daily runoff forecasts for Tavanasa: Root mean square errors for step ahead forecasts from 1 to 4 days

Tavanasa	RMSE $[m^3s^{-1}]$	RMSE [%]	Discharge $[m^3s^{-1}]$
1	4.75	26.0	18.17
2	6.61	36.2	18.21
3	6.82	37.4	18.26
4	7.37	40.4	18.38
1-4	6.39	35.0	18.26

values by the Swiss Meteorological Services.

An additional link to a Geographic Information System (GIS) has been installed to allow easy data access related to different regions for comparative studies. The GIS incorporates a powerful visualization tool for input data and results.

9 CONCLUSIONS

Periodical satellite mapping of the seasonal snow cover together with appropriate hydrological models can improve the reliability of short-term as well as seasonal runoff forecasts. These forecasts can be exploited, among other purposes, for optimizing the hydropower production and for timely decisions on the electricity market.

A data acquisition and management system supports operational utilization and advanced data handling for scenarios on regional basis.

As has been recognized in other countries, for example in the USA (Castruccio et al., 1980), in India (Kumar et al., 1991) and in Japan (Kawata and Kusaka, 1988), this application of remote sensing is a promising example of a favourable cost/benifit relation.

10 ACKNOWLEDGEMENTS

This paper presents results from an ongoing research project which is supported by the *Commission for the Promotion of Scientific Research (KWF)* and the *Swiss Association of Producers and Distributors of Electricity (VSE)* under contract no. 2331.1.

References

Castruccio, P., Loats, H., Lloyd, D., and Newman, A. (1980). Cost/benefit analysis. in operational applications of satellite snowcover observations. In *Workshop, Sparks, Nevada, 1979*, pages 239–254. NASA Conference Publication 2116.

Kawata, Y. and Kusaka, T. (1988). Snowmelt runoff estimation using snow cover extent data and its application to optimum control of dam water level. In *IGARSS '88 Symposium in Edinburgh, Scotland*, pages 439–440, Paris. ESA SP-284. Vol. 1.

Kumar, V., Haefner, H., and Seidel, K. (1991). Satellite snow cover mapping and snowmelt runoff modelling in Beas basin. In *XX. General Assembly IUGG in Vienna 1991, IAHS-IUFRO Symposium Snow, Hydrology and Forests in High Alpine Areas*, pages 101–109. IAHS Publication No. 205.

Martinec, J. (1985). Snowmelt runoff models for operational forecasts. *Nordic Hydrology, Munksgaard, Copenhagen*, 16:129–136.

Martinec, J., Seidel, K., Burkart, U., and Baumann, R. (1991). Areal modelling of snow water equivalent based on remote sensing techniques. In *XX. General Assembly IUGG in Vienna 1991, IAHS-IUFRO Symposium Snow, Hydrology and Forests in High Alpine Areas*, pages 121–129. IAHS Publication No. 205.

Seidel, K., Martinec, J., Steinmeier, C., and Bruesch, W. (1993). Remote sensing of snow cover for operational forecasts. In *Intern. Symposium "Operationalization in Remote Sensing", vol. 4*, pages 71–79. ITC Enschede NL.

Seidel, K., Wiedemeier, J., and Martinec, J. (1990). Operationelle Überwachung von Schneeschmelzvorgängen: Möglichkeiten zur Prognose von Wasserabflussmengen. *Bulletin SEV/VSE*, 81(10):9–17.

WMO (1992). Simulated real-time intercomparison of hydrological models. Operational Hydrology Report 38, World Meteorological Organization, Geneva.

Sensors and Environmental Applications of Remote Sensing, Askne (ed.) © 1995 Balkema, Rotterdam, ISBN 90 5410 524 0

Snow cover monitoring in a high altitude mountain using SPOT and ERS data

F.Cauneau, L.Castagnas & M.Albuisson
Centre d'Energétique, Groupe Télédétection & Modélisation, Ecole des Mines de Paris, France

I.Pons
Société Geosys, Ramonville, France

ABSTRACT: ERS-1 SAR data and SPOT-HRV data are combined into a Geographical Information System in order to evaluate the snow cover in a high altitude mountain, near the valley of the Durance, in the French Alps. The basic difficulty is to achieve the geocoding of the SAR image, as the geometrical effects due to the ranging on sharp ridges degrade the lisibility of the image and the modelisation of the acquisition. The results presented here show that the lisibility of the SAR image is enhanced through the geocodinc, and specific radiometry can be observed in the areas where ice and snow are expected to be present..

1 INTRODUCTION

Monitoring snow cover during the summer period is of a key interest for the management of hydroelectric plants. Classical methods make use of *in situ* measurements, such as snow gauges which present a strong limitation mainly due the lack of their spatial covering.

Space borne sensors offer the possibility to detect the snow cover in both the optical and microwave domain. Significant results have been obtained by application of optical data in this domain, leading to the possibility of a mapping of snow cover. Similar applications of microwave data are basically limited by specific problems of both radiometric and geometrical effects. Thus, to get the full benefit of the possibility offered by the microwave sensors of an all-weather access, there is a need to study the synergy between both optical and microwave sensors.

2 PRESENTATION OF THE METHOD

The aim of the present work is to combine the informations coming from the French sensor SPOT 2 - HRV and the European ERS-1 Active Microwave Imager, to achieve a proper evaluation of the snow cover in the high valley of the Durance, in the French Alps.
The specific radiometric signatures of the ice and snow, in flat land cases, have been investigated by previous authors (see, *e.g.* Bredow *et* Cogineni, 1990; Karam *et al.* 1992; Murthy *et* Haykin, 1987). The basic problem is that such models may not be applied to the mountain case, the radiometric corrections to be applied need informations on both the geometry of the signal/target interaction, and the nature of the ground cover. Our combination method is based on a classification of the SAR data, achieved in a domain defined by the backscatter coefficient, and the local true incidence angle. Part of the SPOT data is jointly used with *in situ* data for the training phases of the classification. The other parts of the data are to be used to evaluate the accuracy of the method.

Starting with the estimation offered by a Geographical Information System (GIS) using both *in situ* and SPOT data (latest acquiered on 18 June 1993), two couples of ERS-1 scenes have been acquiered (11 June, and 18 July 1993). Each couple consists in a descending and an ascending pass, to improve the spatial covering.

The ERS-1 scenes, which are Single Look Complex (SLC) products are first reduced from the speckle effects : each image is analyzed and decomposed through a wavelet analysis, and several classical filters (Lee filter, Wiener filter ...) are applied in the first wavelet coefficient planes (Cauneau, Ranchin 1993). The filtered wavelet transform is used to reconstruct the filtered image. When using the wavelet transform, the efficiency of a given filter is generally improved. In this case study, the application of a filter at the first level of processing leads to a crucial choice : most of the image contrast is due the combined effects of the incidence angle and the foreshortening effects. As the speckle reduction may be not applied after the step of the geometrical corrections, the filter has to be chosen in such a way that the variations of the radiometries on each side of the mountains have to be conserved. A Wiener filter has lead to better reults than a Lee filter in the present case. After this step, icy areas become visible on the images.
The correction of the terrain effects on SAR images

Figure 1. Representation in grey levels of the raw SAR image of the basin of the Durance.

in this alpine zone leads to specific problems that are not to be encounterd in more moderate reliefs. Especially some disconnected parts of the terrain may be acquiered at similar ranges, resulting in a blurring efect on the image. Using a Digital Elevation model (DEM), the geometrical effects of the ranging are computed and corrected. The SAR image is thus geocoded in the frame of the GIS. At this step, the local incidence angle is also computed, and integrated in the GIS as a new parameter. The radiometric corrections on the image can also be achieved using a modified version of the algorithm defined for the flat land case (Laur, 1992).

3 RESULTS AND DISCUSSION

The firsts results are shown on the case of the scene of the 30 June 1993 (descending path). On figure 1, the raw image is shown, using a negative look-up table and a grey levels representation. One can notice that most of the information at this step is mainly the shading effect of the SAR (see, *e.g.* Guindon, 1990). The DEM is shown on figure 2, the horizontal resolution is 30 m. It consists in a frame of 1950 lines and 1150 columns of pixels, in a Lambert 3 geographical frame. The valley of the Durance can be recognized, from the top to the bottom of the image. The *Barrière des Ecrins*, reaching more than 3000m of altitude is at the left of the image.

The figure 3 shows the image of the simulation of the SAR data acquisition over the terrain, as modeled in the DEM. The inverse operation is used to build the geocoding frame, which is applied on the SAR image, and used to compute the local incidence angles.

The figure 4 shows the result of the geocoding. The lisibility of the image is drastically improved. The shape of the landscape appears as clearly as in optical images, and one can notice some features over the *Barrière des Ecrins* that can be related with the presence of ice, or snow. The sides of the mountains that are situated in the foreshortening zone of the acquisition cannot be properly retrieved. This problem can be solved using the corresponding pass in the cross direction.

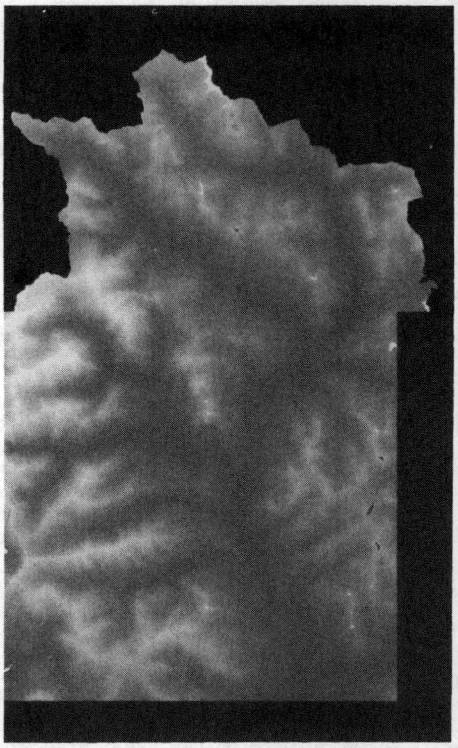

Figure 2 . Representation in grey levels of the DEM. The low altitude zones (valleys) are encoded in black; The high altitude zones are encoded in white.

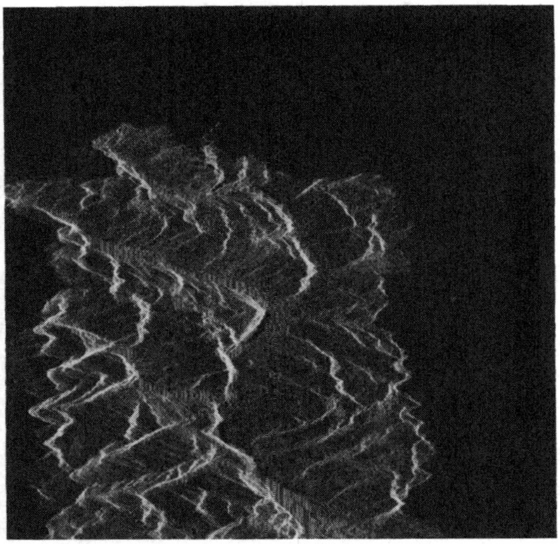

Figure 3. Representation in grey levels of the synthetic SAR image, computed with the help of the DEM. See the text for details.

Figure 4. Representation in grey levels of the geocoded SAR image presented in figure 1. See the text for the details.

This study was achieved with the grant of the *Centre National d'Etudes Spatiales, Electricité de France*. The ERS-1 data was provided by the European Space Agency, as a Pilot Project.

REFERENCES

Bredow, J.W., S. Gogineni, 1990 : Comparison of measurement and theory for backscatter from bare and snow-covered saline ice. *IEEE trans. on Geosci. and Remote Sensing*. Vol. 28, n° 4, p. 456-463.

Cauneau , F., T. Ranchin, 1993 : Speckle removal in SAR images using the wavelet transform. In : *Proc. of the 12th EARSeL symposium : Remote sensing for monitoring the changing environment of Europe*. Winkler Ed. Rotterdam, 1993, p. 97-104.

Guindon, B., 1990 : Development of a shape from shading technique for the extraction of topographic models from individual spaceborne SAR images. *IEEE trans. on Geosci. and Remote Sensing*. Vol. 28, n° 4, p. 654-661.

Karam, M.A., A.K. Fung, R.H. Lang, N.S. Chauhan, 1992 : A microwave scattering model for layered vegetation. *IEEE trans. on Geosci. and Remote Sensing*. Vol. 30, n° 4, p. 767-784.

Laur, H., 1992 : Derivation of backscattering coefficient σ_0 in ERS-1 SAR PRI products. ESA /ESRIN report, Issue 1, Rev. 0, 17th october 1992. 16 p..

Murthy, H.A., S. Haykin, 1987 : Bayesian classification of surface-based ice radar images. *IEEE journal of oceanic engineering*. Vol. OE12, n°3, p. 493-502.

Sensors and Environmental Applications of Remote Sensing, Askne (ed.) © 1995 Balkema, Rotterdam, ISBN 90 5410 524 0

The discrimination of glacier facies using multitemporal ERS-1 SAR data

G.J. Marshall & W.G. Rees
Scott Polar Research Institute, University of Cambridge, UK

J.A. Dowdeswell
Centre for Glaciology, Institute of Earth Sciences, University of Wales, Aberystwyth, UK

ABSTRACT: Remote-sensing methods are able to provide regional glacier mass balance information; Equilibrium Line Altitude (ELA) can be estimated from the location of particular glacier facies margins. Sensors using visible/near-infrared wavelengths may be constrained by cloud cover and certain surface conditions, whereas SAR has the potential to provide guaranteed annual mass balance information. This study examines the ability of ERS-1 SAR to discriminate the different glacier facies on a sub-polar mountain glacier using a multitemporal dataset and *in situ* measurements from two field campaigns. Results, interpreted through simple backscatter models, indicate that two SAR images are needed to estimate the ELA for sub-polar glaciers; one acquired at the end of the ablation season and the other at any time outside this period.

1 INTRODUCTION

The increasing concentration of atmospheric CO_2 and other greenhouse gases has led to predictions of a warmer climate and concern about the consequent melting of terrestrial ice sheets and glaciers. These ice masses respond to changes in local climate by altering their total mass, which initiates a modification in flow regime and eventually causes a relocation of their margins. Mountain glaciers, as distinct from ice sheets, respond fastest to climatic fluctuations because of their rapid movement and high rates of accumulation and ablation. The change in mass for a given year is called the net mass balance and may be correlated with various climatic parameters (e.g. Lefauconnier & Hagen 1990). Thus, suitably interpreted, such glaciers are 'sensitive indicators of climatic change' (Untersteiner 1984).

Satellite remote sensing offers a valuable monitoring technique for mountain glaciers because the scale and remoteness of glaciated regions often restricts the widespread undertaking of *in situ* measurements. For example, field mass balance measurements are only routinely acquired for a very small percentage of the Earth's glaciers (Meier 1984). A survey of regional mass balance changes using reconnaissance methods, such as remote sensing, may often be of more value than detailed measurements on one or two glaciers (Paterson 1981).

The ability of visible and near-infrared (VIS/NIR) sensors to provide data for mass balance studies can be severely restricted by cloud cover. In north-west Spitsbergen cloud cover is greatest at the end of the ablation season, the period when imagery for mass balance estimation is needed (Marshall et al. 1993), whereas the magnitude of cloudiness for parts of the Russian High Arctic is always such that very few useful VIS/NIR images of these areas have been acquired in total (Marshall et al. in press). For such regions ERS-1 Synthetic Aperture Radar (SAR), which can penetrate through cloud, is a sensor with a sufficiently high spatial resolution (nominally 12.5 m for ERS-1) to be used as a possible alternative source of regional mass balance information.

The altitude of certain glacier-facies margins at the end of the ablation season may be directly related to net mass balance; this paper describes the ability of SAR imagery to discriminate these glacier facies and hence provide the required mass balance data. Four SAR scenes of Ayerbreen, a small mountain glacier in Svalbard, are analysed, two of which were acquired during validating field campaigns. A preliminary interpretation of the results is provided by simple modelling of the interactions between microwave radiation and the glacier. SAR imagery acquired at the end of the ablation season is compared with corresponding Landsat Thematic Mapper (TM) imagery to determine their relative merits for facies discrimination.

2 THE GLACIER-FACIES CONCEPT AND MASS BALANCE ESTIMATION

The glacier-facies concept has developed from pioneering field work in Greenland by Benson (e.g. Benson 1959), and has since evolved into the terminology used by Williams et al. (1991), which is employed in this paper, with one further revision. Although they may occur on smaller ice masses the percolation and dry-snow facies are generally

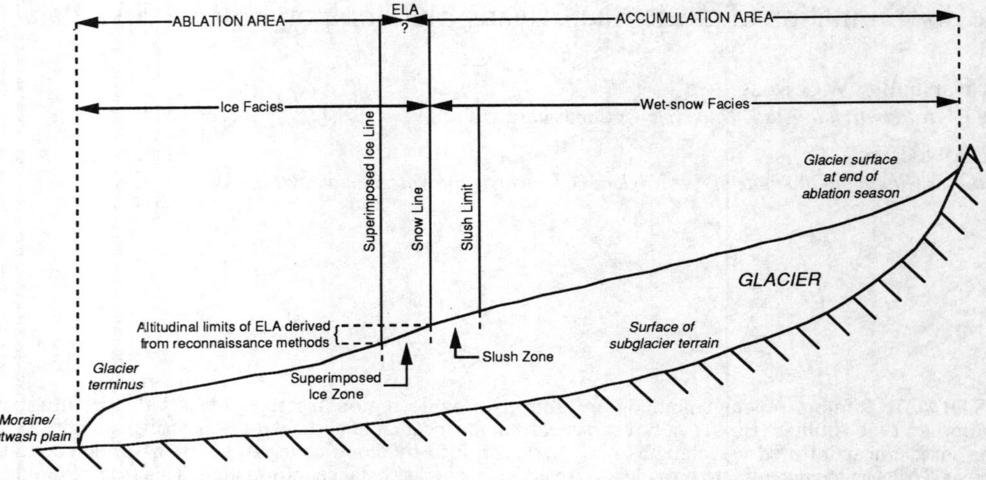

Figure 1. Cross-section of a sub-polar glacier showing the location of the different glacier facies and margins discussed, and their use in estimating the ELA position using remote-sensing techniques (modified from Williams et al. 1991).

restricted to ice sheets and are not discussed here. Mountain glacier facies are the result of differential diagenesis within the overlying snowpack caused by altitudinal variations in the magnitude of ablation and accumulation processes. This dependence on local meteorological conditions means that the spatial distribution of glacier facies is temporally variable, and they retreat up-glacier as the ablation season progresses. Unless observed at the end of the ablation season the facies margins are given the qualifying adjective 'transient'. A cross-section of a glacier illustrating the different glacier facies and margins described in this paper is shown in Figure 1.

For mass balance studies using remote sensing methods the ideal parameter to locate is the Equilibrium Line Altitude (ELA), where the mass balance of a glacier is zero. Its height at the end of the ablation season can be directly related to the net mass balance of the glacier (Østrem 1975). Many Arctic glaciers, including Ayerbreen, can be defined as sub-polar (Ahlmann 1935) and may have a superimposed ice zone above the glacier ice. Such glaciers undergo surface melting accompanied by the formation of fluid water; superimposed ice forms by the refreezing of this meltwater from the snowpack onto the glacier surface at its base (Koerner 1970), and so constitutes part of the accumulation zone. Although smooth when first exposed this ice eventually assumes a characteristic white broken crystalline ablation surface. The spatial distribution of internal snow, firn and ice temperatures means that the ELA lies somewhere within the superimposed ice zone (Østrem & Brugman 1991). Thus, the superimposed ice line, herein defined as the lowest elevation of superficial superimposed ice, and the snow line, the highest elevation to which the snow cover has retreated, are the lower and upper limits of the ELA, respectively. The height difference between these two margins determines the accuracy of an ELA estimated using remotely sensed data. In temperate glaciers, without a superimposed ice zone, the ELA may be approximated by the snow line.

Above the snow line are the wet-snow facies, where all the snow deposited since the end of the previous summer has been raised to 0°C (Paterson 1981). At the lower margin of the wet-snow facies there may be a slush zone which is completely saturated with water; its upper margin is the slush limit; the highest point at which any material is lost by runoff (Paterson 1981). Within the snowpack horizontal layers of ice, known as ice lenses, are formed as a consequence of summer melting events when free water percolates into the snowpack until encountering a relatively impermeable layer where it spreads out laterally and refreezes.

3 MICROWAVE BACKSCATTER FROM SNOW AND ICE

The various glacier facies and zones have different physical properties, because of their differing diagenesis, that may allow them to be discriminated using SAR imagery. The nature of backscatter from the different glacier facies and the theoretical backscatter models, with which the SAR data are compared, are now briefly discussed. The interaction of microwaves with snow is given a comprehensive review by Mätzler (1987).

Microwave backscatter from a semi-transparent medium like a glacier snowpack consists of three contributions from a) the snow-air interface, b) the snow volume, and c) the underlying ice surface. Following (Stiles & Ulaby 1980) the backscatter coefficient per unit surface area σ^0 may be written as

$$\sigma^0(\theta) = \sigma^0_{sa}(\theta) + \sigma^0_s(\theta') + \frac{T^2_{sa}(\theta)}{L^2(\theta')}\sigma^0_{gs}(\theta') \qquad (1)$$

where θ is the incidence angle, θ' is the transmission angle, $\sigma^0_{sa}(\theta)$ is the backscatter at the snow-air interface, $\sigma^0_s(\theta')$ is due to volume scatter from the snowpack, $T_{sa}(\theta)$ is the power transmission coefficient across the snow-air boundary, $L(\theta')$ is the one-way loss factor of the snowpack, and $\sigma^0_{gs}(\theta')$ is due to backscatter coefficient from the underlying glacier surface. The power reflection coefficient of dry snow is very small and the surface scattering contribution may be neglected excepting specular reflection.

Volume scattering takes place from snow grains and water inclusions; if these scatterers are small relative to the wavelength such that

$$k_0 r < 1 \qquad (2)$$

where the wavenumber $k_0 = 2\pi/\lambda$ and r is the scatterer radius, then the Rayleigh approximation may be used to model the volume scattering; e.g. Rott et al. (1985) and Drinkwater (1989).

For a given incident wavelength the penetration depth of a snowpack is controlled by scattering and absorption losses dependent on the physical structure and dielectric properties of the medium. The presence of liquid water alters the latter; it increases the dielectric constant and particularly the dielectric loss factor causing a significant reduction in the penetration depth. For dry snow, the snowpack is almost transparent to radar, and the penetration depth is of the order of tens of metres, but with 5% volume liquid water this is reduced to approximately one wavelength (Ulaby et al. 1982) and surface scattering dominates.

Surface backscatter may be modelled using the Kirchhoff formulation; it is applicable to surfaces with undulations larger than the incident wavelength, and assumes plane-boundary reflection at the surface. If a surface is characterised by its rms height variation (σ), rms slope (s) and correlation length (l), usually derived from surface profiles with sampling interval $\lambda/10$ (Ulaby et al. 1982), the criteria for the use of the Kirchhoff formulation are that

$$k_0 l > 6 \quad \text{and} \quad l^2 > 2.76\sigma\lambda \qquad (3)$$

Depending on the magnitude of s the scalar approximation for smooth-surface backscatter ($s < 0.25$ rad) or the stationary-phase approximation for rough-surface backscatter ($s > 0.25$ rad) is used. The former is often applicable to wet-snow surfaces while the latter, suitable for glacier ice, requires a further assumption (Rott 1984) that

$$(2k_0\sigma\cos\theta)^2 > 10 \qquad (4)$$

so that for typical values of σ for glacier ice the limit of validity occurs at $\theta = 35°$ for C-band ($\lambda = 57$ mm). For a more detailed treatment of the Kirchhoff formulation see e.g. Ulaby et al. (1982) and Rott (1984).

4 FIELD MEASUREMENTS

Two seasons of validating fieldwork were undertaken on Ayerbreen, a small mountain glacier located in the Nordenskiöld Land region of Spitsbergen, the largest island in the Svalbard archipelago. Ayerbreen descends from approximately 950 m to 380 m a.s.l. along its 6.5 km length and covers an area of 4.5 km^2. The fieldwork was carried out during July-August 1992 and April-May 1993 and a coincident SAR image was acquired during each season. Ayerbreen was chosen because it has a relatively constant and shallow gradient; thus, an initial analysis could be undertaken without considering topographic effects. In addition, data loss from terrain distortions such as layover and shadowing, typical for SAR imagery in mountainous regions (e.g. Rott et al. 1985), is not significant because the lower two-thirds of the glacier are relatively open towards the south-east, the SAR look direction.

Measurements of a wide variety of surface and subsurface parameters known to influence microwave interaction with snow and ice were obtained. Such data include micro (mm) and macro (cm) surface roughness profiles, snow wetness, density, grain size, temperature and stratigraphy profiles, meteorological data and regular surveys of the facies margins during the ablation season. This information was used to aid interpretation of the SAR imagery and as input data for the backscatter models.

5 INTERPRETATION OF MULTITEMPORAL DATA

Multitemporal variations of σ^0 along Ayerbreen long-profile (6.5 km) are illustrated in Figure 2. Pixel values from the SAR PRI product may be converted to dB values using the following equation (Laur 1992)

$$\sigma^0_{(dB)} = 20 \cdot \log_{10} DN - K_{(dB)} + \beta_{(dB)} \qquad (5)$$

where DN is the PRI digital number, K is the calibration constant (59.49 dB for the UK-PAF) and β is the angular correction factor (taken as 0 dB because Ayerbreen lies in the swath centre of the images acquired). The data have been median filtered to reduce the effect of image speckle. *In situ* measurements are available for 1st August 1992 (Figure 2a) and 8th May 1993 (Figure 2c).

In the 1st August image (Figure 3a) there is a clear boundary between a generally homogenous zone of high backscatter ($\overline{\sigma}^0$ = -7 dB) and a zone of low backscatter ($\overline{\sigma}^0$ = -15 dB). Ground survey of the

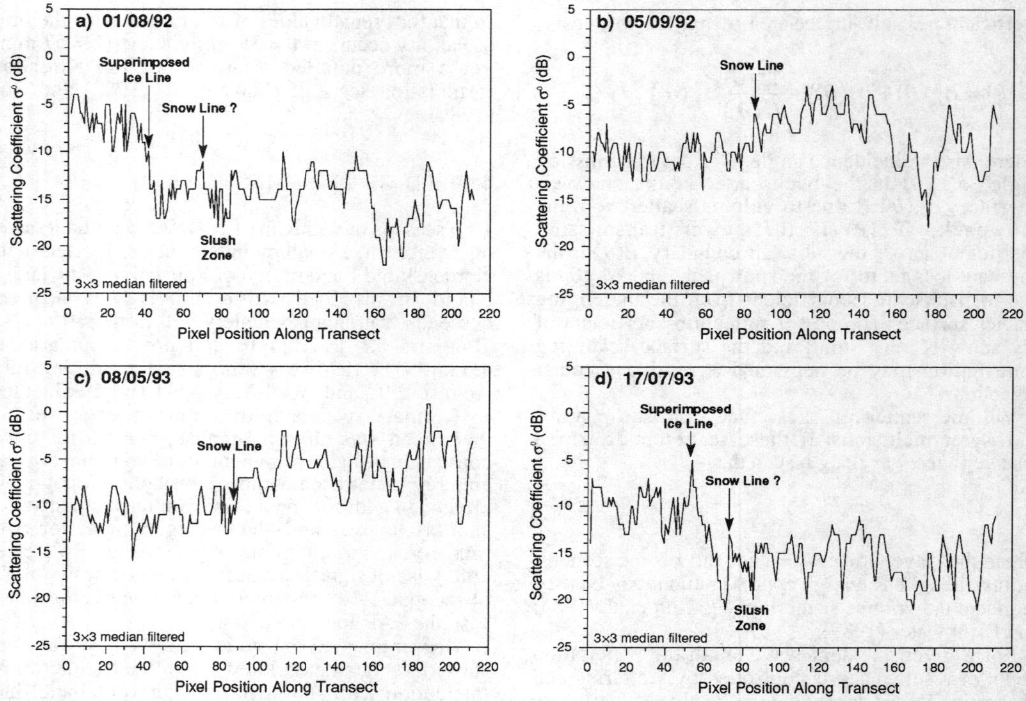

Figure 2. Multitemporal variation of the backscatter coefficient along the 6.5 km Ayerbreen profile showing the location of the glacier-facies margins.

facies margins two days later (Ayerbreen was shrouded in fog on 1st August) indicates that this is the superimposed ice line; glacier ice (high backscatter) is discriminated from superimposed ice (low backscatter). Modelled results for glacier ice using the stationary-phase approximation give the expected backscatter as -8 dB. When $s \rightarrow 0.25$ rad, as with the glacier ice surface on 1st August, the effect of incidence angle on model backscatter becomes significant; the southern part of the glacier terminus which faces the SAR at a relatively steep angle has a backscatter coefficient as high as 0 dB.

Evidence of the slush zone is seen on the profile (Figure 2a); it corresponds to the more up-glacier of the two zones where σ^0 drops to -17 dB, but is difficult to distinguish in the image (Figure 3a). The slush zone may be expected to have very low values of σ^0 because the high water content causes significant dielectric losses. The 15-20 m width of the slush zone on Ayerbreen is probably too narrow to enable it to be easily identified in the presence of speckle. However, the slush zone is recognisable in Seasat L-band imagery of Iceland ice caps (Rott & Mätzler 1987). As the Ayerbreen slush zone is not discernible, the similar backscatter from superimposed ice and wet snow means that the transient snow line could not be located on the 1st August image. In regions where the superimposed ice zone is not significant the snow line is easily identified using C-band because of the marked differences in surface roughness between the ice and wet-snow facies (e.g. Dowdeswell et al. 1994), although for the longer wavelength of L-band (λ = 235 mm) the two surfaces may not be sufficiently dissimilar (Hall & Ormsby 1983). The first zone of low backscatter (-17 dB) adjacent to the glacier ice is possibly the very smooth and wet surface of superimposed ice that has almost melted away.

At C-band for $\theta \leq 30°$ surface scattering is the dominant scattering mechanism for wet snow (Rott 1984). The modelled results for wet snow using the scalar approximation compare favourably with the SAR data, giving an expected surface scattering coefficient of -15 dB for wet snow with the average roughness and dielectric properties measured in the field. Backscatter variation within the wet-snow facies may result from changes in surface roughness and liquid water content because of differential exposure to ablation processes.

The 5th September 1992 image (Figure 2b) was acquired at the end of the ablation season; the air temperature was below 0°C and there had been a snow fall which had settled and frozen in the topographic lows of the glacier ice reducing its surface roughness. This can be seen in Figure 2b; σ^0

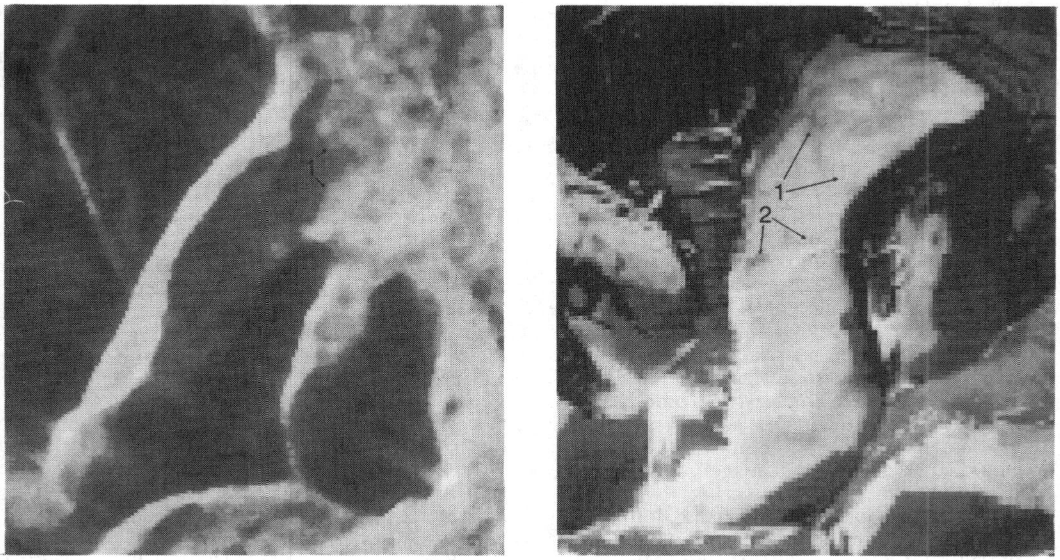

Figure 3. Images of Ayerbreen acquired during the 1992 ablation season from a) ERS-1 SAR (01/08/92) and b) Landsat TM (28/08/92). The superimposed ice line is indicated by 1 and the snow line by 2.

from the glacier ice surface has dropped to -10 dB and there is no indication of a superimposed ice line. To be certain that a SAR locates this margin at the end of the ablation season the satellite revisit interval needs to be about five days. However, the increasing overlap between adjacent suborbital tracks towards the poles means that this is possible at high latitudes. Marshall et al. (in press) showed that ERS-1 provided 31% daily coverage for a point close to Ayerbreen during its 35-day orbit.

The profile displays a distinct rise in backscatter up-glacier, assumed to be the location of the snow line. Above it the wet-snow facies has a high σ^0 of up to -4 dB. A Rayleigh scattering model from snow grains would severely underestimate the scattering strength; fieldwork in Greenland (Jezek 1993) has shown that such values of σ^0 are a result of enhanced scatter from the rough tops of ice lenses. The lower half of the wet-snow facies has higher values of σ^0 because it has experienced more melting events during the ablation season and hence the series of ice lenses has developed further. More than ten ice lenses at different depths were observed in shallow snow pits dug during the summer field season.

Figure 2c, the 8th May 1993 profile, indicates the seasonal reversal of the relative magnitudes of backscatter from the glacier ice and wet-snow facies, which had commenced in the 5th September image, and has also been reported from Greenland (Fahnestock et al. 1993). The only significant change between the two images is the marked rise in σ^0 from the upper wet-snow facies so that there exists an up-glacier increase in backscatter within this facies. Field measurements indicated only minimal surface melting at the acquisition time so this change must be caused by an internal structural alteration.

On 8th May the snowpack above Ayerbreen was approximately 2 m in depth with the lower 300 mm consisting of depth hoar. The latter is formed by sublimation (Paterson 1981) and had grain sizes of 0.9 mm radius as compared to the overlying snow with grains of mean radius 0.1 mm. The radius of the depth hoar is still sufficiently small for the Rayleigh approximation to be valid, and the snowpack was modelled as two distinct layers. Results indicate that the layer of depth hoar is responsible for almost the entire volume scattering from the snowpack. However, model calculations also show that the snowpack is virtually transparent to the radar with most of the backscatter coming from the underlying glacier ice surface.

The snow line can still be located in the 8th May 1993 image at approximately the same pixel position as in the 5th September 1992. Other images, not described here, indicate that this feature may be discriminated at any time of year outside the ablation season; i.e. when there is no melting within the snowpack. Similar findings have been made for Black Rapids Glacier in Alaska (Ahlnäs et al. 1992).

The 17th July 1993 image (Figure 2d) displays similar features to the 1st August image from the previous year. The superimposed ice line is, however, already at a higher altitude than on 1st August the previous year and the glacier ice shows greater σ^0 variation. If the second zone of low backscatter is assumed to be the slush zone it has a similar location to 1992 so the superimposed ice zone is reduced in extent. Further imagery acquired during the 1993 ablation season indicate the ELA to be higher than in 1992.

6 ANALYSIS OF LANDSAT TM IMAGE

Sensors operating at VIS/NIR wavelengths are often able to locate the snow line because the spectral reflectance, or albedo, of snow and ice are markedly different at such wavelengths; for example, 0.85 and 0.32, respectively, at 800 nm (Zeng at al., 1984). Actual values for albedo may be derived from Landsat DN values and modelled atmospheric corrections (e.g. Winther 1993). Difficulties may be caused by snow contamination, which reduces the snow reflectance, or a fresh snow fall, which gives both the ice and wet-snow facies similarly high reflectances. The superimposed ice zone has been discriminated on another Svalbard glacier, Austre Brøggerbreen, using both Landsat TM imagery (Winther 1993) and SPOT imagery (Parrot et al. 1993).

A histogram-equalised TM band 4 image of Ayerbreen, acquired on 28th August 1992, is illustrated in Figure 3b. This was the only Landsat image of Ayerbreen obtained during the 1992 ablation season in which the glacier was not obscured by cloud cover.

The snow line on Ayerbreen can be identified clearly in TM bands 1-4 because of the presence of the slush zone. This has a very low albedo, due to water saturation, making it distinct from the surrounding snow and ice surfaces. With the aid of terrestrial photography, taken a few days prior to image acquisition, the superimposed ice line may also be located. Superimposed ice appears more homogeneous and has a higher albedo than glacier ice, which is partly debris-covered although still distinct from the terminal moraine. Analysis of training areas indicates that the two ice types are statistically separable at a 95% confidence interval for TM bands 1-4. The relatively low sun elevation of 23° enhances surface morphology but considerably widens the range of DN values associated with a particular facies. Therefore, as with SAR data, slope corrections would improve facies discrimination.

7 ESTIMATING THE 1992 ELA OF AYERBREEN

The 1st August is too early in the ablation season to obtain useful mass balance information, and unfortunately the superimposed ice line, clear in this earlier image, was not apparent in the 5th September image. Consequently, the elevation of this facies margin was derived from the TM image and located at 520 m. The snow line altitude was found to be at 560 m in both SAR and TM imagery. Therefore, the 1992 ELA of Ayerbreen was 540±20 m. Assuming an ELA of 540 m, the accumulation area ratio (AAR), the accumulation area at the end of the ablation season divided by the total glacier area, for Ayerbreen is 0.58. For other glaciers a value of 0.7 has been found to correspond to a net balance of zero (Glen 1963) so it is likely that Ayerbreen underwent a slight decrease in mass during 1992. The 40 m altitudinal range of the superimposed ice zone is the same as that reported by Parrot et al. (1993) for Austre Brøggerbreen.

8 CONCLUSIONS

The principal conclusions from this study are that:

- For a sub-polar glacier with a superimposed ice zone the accuracy of a mass balance estimate derived from remote sensing methods is determined by the ability to discriminate two facies margins; these are the superimposed ice line and snow line, the lower and upper limits of the ELA, respectively.

- ERS-1 C-band SAR has been shown to be able to discriminate the superimposed ice line. Marked differences in surface roughness between glacier ice (rough) and superimposed ice (smooth) cause σ^0 to be approximately 8 dB higher for glacier ice.

- The slush zone may be too narrow to be identified in the presence of image speckle. Consequently the similar σ^0 from superimposed ice and wet snow means that the transient snow line cannot always be identified by SAR during the ablation season. However, at any other time enhanced backscatter from the rough surfaces of ice lenses, formed in the wet-snow facies, causes a reversal in the magnitude of σ^0 between glacier ice and wet snow and provides the means for locating the snow line.

- Landsat TM data could easily detect the snow line on Ayerbreen because of the very low albedo of the slush zone. Using validating terrestrial photography, the superimposed ice line could also be discriminated.

- SAR has the potential to provide guaranteed annual estimates of glacier mass balance. For sub-polar glaciers two SAR images are needed; one acquired at the end of the ablation season and a second acquired at any time prior to the onset of melting the following year. Only the latter is required for temperate glaciers.

ACKNOWLEDGEMENTS

We thank ESA for providing the ERS-1 SAR data for project A0 Int 8-9. G.J. Marshall acknowledges the support of a research studentship from the U.K. Natural Environment Research Council who also kindly supplied the Landsat image.

REFERENCES

Ahlmann, H.W. 1935. Contribution to the physics of glaciers. *Geogr. J.* 86(2):97-113.

Ahlnäs, T.K., C.S. Lingle, W.D. Harrison, T.A. Heinrichs & K.A. Echelmeyer 1992. Identification of late-summer snow lines on glaciers in Alaska

and the Yukon Territory with ERS-1 SAR imagery (abstract). *EOS* 73(43):204.
Benson, C.S. 1959. Physical investigations on the snow and firn of north-west Greenland: 1952, 1953, and 1954. *SIPRE Res. Rep.* 26.
Dowdeswell, J.A., W.G. Rees & A.D. Diament 1994. ERS-1 SAR investigations of snow and ice facies on ice caps in the European High Arctic. *Proceedings of the 2nd ERS-1 Symposium: Space at the Service of our Environment, held at Hamburg, Germany, 11-14 October, 1993.* ESA SP-361(2):1171-1176.
Drinkwater, M.R. 1989. LIMEX '87 ice surface characteristics: implications for C-band SAR backscatter signatures. *IEEE Trans. Geosci. Rem. Sens.* 27(5):501-513.
Fahnestock, M., R. Bindschadler, R. Kwok & K. Jezek 1993. Greenland Ice Sheet surface properties and ice dynamics from ERS-1 SAR imagery. *Science* 262:1530-1534.
Glen, J.W. 1963. *Int. Assoc. Sci. Hydrol. Bull.* 8(2):68 (contribution to discussion).
Hall, D.K. & J.P. Ormsby 1983. Use of Seasat synthetic aperture radar and Landsat multispectral scanner subsystem data for Alaskan glaciology studies. *J. Geophys. Res.* 88(C3):1597-1607.
Jezek, K.C. 1993. Spatial patterns in backscatter strength across the Greenland Ice Sheet. *Proceedings of the 1st ERS-1 Symposium: Space at the Service of our Environment, held at Cannes, France, 4-6 November, 1992.* ESA SP-359(1):269-272.
Koerner, R.M. 1970. Some observations on superimposition of ice on the Devon Island ice cap, N.W.T., Canada. *Geogr. Ann.* 52A(1):57-67.
Laur, H. 1992. Derivation of backscattering coefficient σ^0 in ERS-1.SAR.PRI products. *Issue* 1, *Rev.* 0. ESA-ESRIN.
Lefauconnier, B. & J.O. Hagen 1990. Glaciers and climate in Svalbard: statistical analysis and reconstruction of the Brøggerbreen mass balance for the last 77 years. *Ann. Glaciol.* 14:148-152.
Marshall, G.J., W.G. Rees & J.A. Dowdeswell 1993. Limitations imposed by cloud cover on multitemporal visible band data sets from polar regions. *Ann. Glaciol.* 17:113-120.
Marshall, G.J. J.A. Dowdeswell & W.G. Rees in press. The spatial and temporal effect of cloud cover on the acquisition of high quality Landsat imagery in the European Arctic sector. *Rem. Sens. Environ.*
Mätzler, C. 1987. Applications of the interaction of microwaves with the natural snow cover. *Rem. Sens. Rev.* 2:259-387.
Meier, M.F. 1984. Contribution of small glaciers to global sea level. *Science* 226(4681):1418-1421.
Østrem, G. 1975. ERTS data in glaciology - an effort to monitor glacier mass balance from satellite imagery. *J. Glaciol.* 15(73):403-414.
Østrem, G & M. Brugman 1991. Glacier mass-balance measurements. A manual for field and office work. *NHRI Sci. Rep.* 4. National Hydrological Research Institute, Saskatoon, Saskatchewan and Norwegian Water Resources and Energy Administration, Oslo.
Parrot, J.F., N. Lyberis, B. Lefauconnier & G. Manby 1993. SPOT multispectral data and digital terrain model for the analysis of ice-snow fields on Arctic glaciers. *Int. J. Remote Sensing* 14(3):425-440.
Paterson, W.S.B. 1981. *Physics of glaciers. 2nd edn.* Oxford: Pergamon Press.
Rott, H. 1984. The analysis of backscattering properties from SAR data of mountain regions. *IEEE J. Oceanic Eng.* 9(5):347-355.
Rott, H., G. Domik, C. Mätzler, H. Miller & K.G. Lenhart 1985. Study on use and characteristics of SAR for land snow and ice applications. *ESA Contract Report* 5441/83/D/IM(SC). Paris: European Space Agency.
Rott, H. & C. Mätzler 1987. Possibilities and limits of synthetic aperture radar for snow and glacier surveying. *Ann. Glaciol.* 9:195-199.
Stiles, W.H. & F.T. Ulaby 1980. Radar observations of snowpacks. *Proceedings of the Microwave Remote Sensing of Snowpack Properties Workshop held at Fort Collins, CO, U.S.A.* NASA C.P. 2153).
Ulaby, F.T., R.K. Moore & A.K. Fung 1982. *Microwave Remote Sensing: Active and Passive. vol. II.* Reading, MA: Addison-Wesley.
Untersteiner, N. 1984. The cryosphere. In J.T. Houghton (ed), *The global climate:* 121-140. Cambridge: Cambridge University Press.
Williams, R.S.Jr., D.K. Hall & C.S. Benson 1991. Analysis of glacier facies using satellite techniques. *J. Glaciol.* 37(125):120-128.
Winther, J-G. 1993. Landsat TM derived and in situ summer reflectance of glaciers in Svalbard. *Polar Res.* 12(1):37-55.
Zeng, Q., M. Cao, X. Feng, F. Liang, X. Chen & W. Sheng 1984. A study of spectral reflection characterisics for snow, ice and water in the north of China. *Proceedings of the Symposium on Hydrological Applications of Remote Sensing and Remote Data Transmission, held at Hamburg, Germany, August 1983.* IAHS Publ. 145:451-462.

Marine and inland waters

Sensors and Environmental Applications of Remote Sensing, Askne (ed.) © 1995 Balkema, Rotterdam, ISBN 90 5410 524 0

Remote sensing and water quality in the Netherlands

R. Allewijn & J. M. M. Kokke
Survey Department of Rijkswaterstraat, Delft, Netherlands

A. G. Dekker
Institute for Environmental Studies, Amsterdam, Netherlands

R. J. Vos
Delft Hydraulics, Netherlands

ABSTRACT: A Dutch framework programme on remote sensing and water quality was initiated, named "REWANET". The aim of this framework programme is to create operational applications of remote sensing for water quality management and policy decision making in the Netherlands. REWANET should lead to (1) the definition and implementation of a structured approach for research and operationalization of remote sensing techniques, in conjunction with in situ monitoring networks, numerical models and (geographical) information systems for water quality management and decision making; and (2) the establishment of a permanent national platform for coordination of remote sensing research and applications related to water quality, placed within an international context. The general outline of the REWANET framework is described in this paper. A number of aspects concerning remote sensing of water quality will be discussed in more detail: (1) inventory of general user-requirements, (2) application-oriented research on atmospheric correction procedures, (3) standardized methods for the calculation of water quality parameters, and (4) integration of remote sensing with numerical water quality models.

1 INTRODUCTION

Remote sensing offers possibilities to derive synoptical spectral and spatial information which can be valuable in adding to water quality information from point observations and numerical water quality models. Present developments in remote sensing and related technology have their implications on development strategy of methods and algorithms. A number of both airborne and spaceborne systems for remote sensing of water quality are available now, others will become available in near future. The development of imaging spectroscopy systems, offering the possibility of acquiring full spectral information as opposed to (integrated) band information with traditional systems, and the definite trend towards airborne remote sensing require flexibility and completeness of data processing algorithms.

Since the start of the Dutch National Remote Sensing Programme in 1986 a number of projects on remote sensing and water quality has been performed in the Netherlands. Part of these projects concentrated on fundamental research, concerning:

- Atmospheric correction procedures (e.g. De Haan et al., 1991; Kokke & Bakker, 1993).
- High resolution optical remote sensing techniques with the emphasis on laboratory, field and airborne experiments (e.g. Dekker et al., 1992; Dekker, 1993; Krijgsman, 1994).

In these projects software has been developed for atmospheric correction and sensitivity analysis. A number of sophisticated water quality algorithms has been defined. However, most software is not implemented in an operational computer environment.

Other water quality projects in the Netherlands concentrated on applicability and operationalization, such as:

- Landsat TM imagery for water quality concentration mapping (Buiteveld et al., 1993).
- Operationalization of NOAA imagery for water quality determination (Harkink, 1991).
- Thermal airborne techniques for enforcement of environmental legislation (Roeters et al., 1993).

Two main conclusions have been drawn from these studies:

- As Landsat and NOAA have limited possibilities for water quality applications, future research and operationalization should be focused on imaging spectroscopy airborne and spaceborne systems (e.g. Van Stokkom et al., 1993).

- For most water quality applications of remote sensing, integration of remote sensing data with point observation networks, numerical models and geographical information systems is needed (Allewijn, et al., 1993; Bijlsma, et al., 1991; Van Tol, et al., 1992).

Aspects concerning data-acquisition, data-processing and data-distribution are generally treated in individual projects. Such a project is the current project on ERS-1 oil spill detection (Pellemans, 1993). With respect to NOAA imagery the Netherlands Royal Meteorological Institute has implemented an operational procedure for data-processing and -distribution (Roozekrans & Prangsma, 1992). However, as part of the total national infrastructure for remote sensing data, an overall national data-management policy for remote sensing water quality information is lacking.

A number of Dutch institutes are involved in European remote sensing projects. However, as cooperation and coordination of activities on the European level is a complex matter, most national projects are performed in a more or less isolated manner, which may lead to duplication of work.

Based on this historical background three main problems can be identified:

1. At this moment a clear *long-term overall strategy* for remote sensing water quality research and applications is missing in the Netherlands. The path to the final goal of operationalization and implementation is poorly defined. Continuity from fundamental research to application-oriented research and operationalization should be the aim. A structured approach for implementation of (advanced) remote sensing techniques, in conjunction with in situ monitoring networks, numerical models and (geographical) information systems for water quality management and decision making is required.
2. An overall national *data-management policy* for remote sensing derived water quality information is lacking, as part of the total national infrastructure for remote sensing data. For operational use a well-defined procedure for receiving, (pre-)processing, analyzing and distribution of remote sensing data to the end-user has to be implemented.
3. Due to the lack of a structured national approach the *international position* of Dutch remote sensing activities on water quality applications is less strong than desired, for instance in relation to projects and programmes of the European Community (EC) and the European Space Agency (ESA).

In order to solve these three problems, a national framework for water quality research and application has been defined (REWANET), which serves as an umbrella for a number of projects. The general outline of this framework programme is presented in figure 1. The main objectives of this framework are:

1. Stimulation of involvement of the end-user by user consultancy meetings and pilot projects.
2. User-guided research on (advanced) remote sensing techniques, in preparation of future (satellite) systems (e.g. ENVISAT/MERIS).
3. Definition and implementation of procedures for integration of remote sensing data with in-situ monitoring networks, numerical models and (geographical) information systems; embedding remote sensing data in existing and new models and information systems; definition and implementation of a national infrastructure for data-acquisition, data-processing and data-distribution of remote sensing water quality information.
4. Increasing the involvement of the Netherlands in international/European remote sensing projects and programmes (EC and ESA).

In this paper four projects will be discussed which are performed within the framework of REWANET:

- Identification of user-requirements (section 2).
- Operationalization of atmospheric correction procedures (section 3).
- Standardized methods for the calculation of water quality parameters from atmospherically corrected remote sensing data (section 4).
- Remote sensing information for improved calibration of water quality models (section 5).

2 USER-REQUIREMENTS

The starting point of the activities of REWANET is the identification of the general needs of water quality managers and policy decision makers. These end-users have been consulted in workshops, by inquiries and by personal interviews. The main application fields and parameters needed were identified, as well as the information needs in terms of availability, accuracy and format of remote sensing data.

For the marine waters (North Sea and the Dutch estuaries) the following applications and parame-

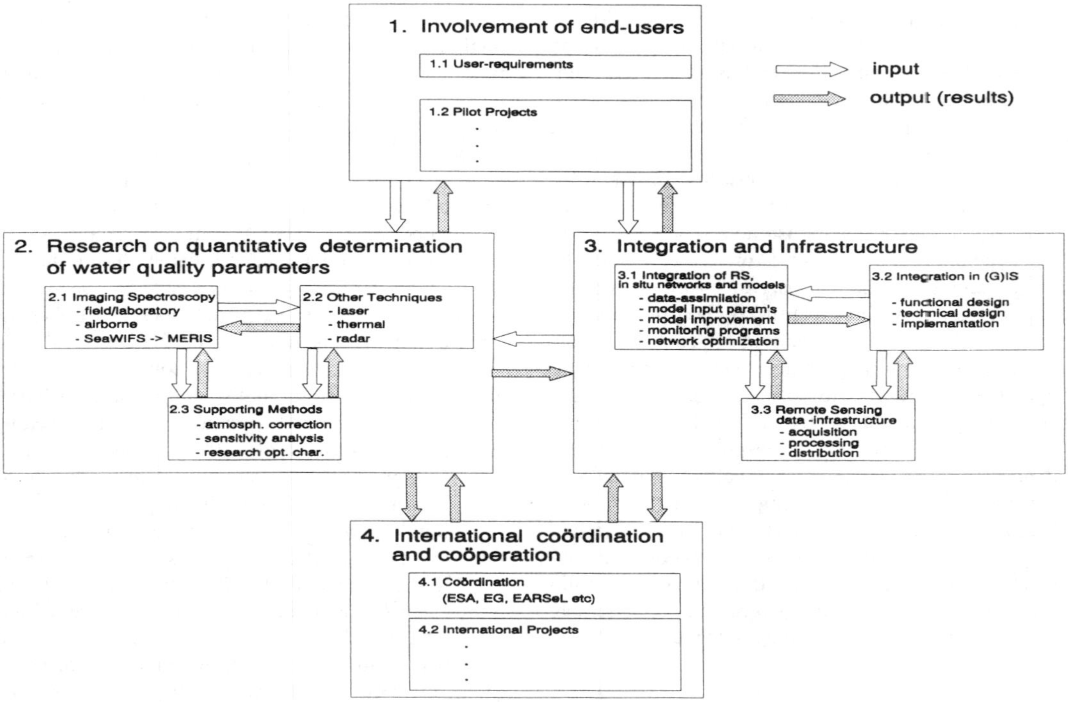

Figure 1. A framework for Remote Sensing and Water Quality in the Netherlands (REWANET)

ters were identified by the end-users:

- Eutrophication (primary production, chlorophyll-a, total biomass, algae-species, temperature).
- Disturbance of habitats, caused by dredging, fishery, etc. (sediment concentration and spatial and temporal distribution of suspended sediment).
- Pollution (oil spills, sea surface temperature and currents, spatial and temporal distribution of suspended sediment, micro-pollutants attached to the fine fraction).

For the fresh waters (inland lakes, rivers and harbours) the following applications and parameters were identified by the water quality managers:

- Eutrophication and biological monitoring (chlorophyll-a, transparency, suspended sediment, algae-species, floating layers of blue-green algae, duckweed in ditches, coverage of water plants).
- Pollution (sediment in rivers during high flows, micro-pollutants attached to the fine fraction).

For marine and fresh waters an increasing effort is put into the acquisition of water quality information. The end-users stress the importance of spatial information obtained by remote sensing. However, for operational use in monitoring programmes attention should be given to the following aspects:

- *Availability of data*: For operational water management remote sensing products have to be available on a regular basis (daily to monthly basis, depending on the application). Especially for smaller water systems the acquisition of airborne remote sensing data has to be more flexible. Application of remote sensing for smaller water systems is hampered by the lack of operational airborne systems which can be used on a routine-basis at relatively low costs.
- *Accuracy of data*: For remote sensing to be used in routine water quality monitoring programmes a well-defined accuracy of data has to be guaranteed. Quantitative remote sensing, providing information of a standard quality, is a prerequisite for remote sensing to be used in operational monitoring programmes.
- *Processing and archiving of data*: Standardized methods for processing of remote sensing and

good archiving facilities are needed. Processing and archiving of data have to be done in cooperation with the end-user and in conjunction with conventional measuring techniques and existing data-bases. By bringing remote sensing closer to the end-user and by integrating it with in situ observation networks, the added value of remotely sensed information can be demonstrated more easily. In general remote sensing products should be distributed to the end-user in a format which is compatible with standardized GIS software.

Main obstacles for quantitative use of remote sensing in monitoring programmes are the lack of operational atmospheric correction procedures and standardized methods for calculation of water quality parameters from atmospherically corrected data. Determination of water quality parameters is generally based on (semi-)empirical methods. Analytical procedures and methods for integrating remote sensing with point measurements and numerical models have to be developed. The main steps in quantitative and integrated use of remote sensing are described in the next sections.

3 ATMOSPHERIC CORRECTION

The main objective of remote sensing of marine and fresh waters is to provide spatial information of water quality and quantity. This requires operational remote sensing algorithms for water quality parameter estimation. An operational algorithm requires two main components:

1. A removal of atmospheric influence from the remotely sensed signal (section 3).
2. The estimation of water quality parameters from the -atmospherically corrected- remotely sensed signal (section 4).

The reflectance derived from the remotely sensed radiance may be increased or decreased as compared to the reflectance at the earth's surface. This is due to variable effects like viewing angle, spectral band characteristics, target reflectance, reflectance of the environment and atmospheric characteristics. Correction of the disturbing effect of atmospheric scattering and absorption on satellite and airborne remote sensing data can hardly be circumvented when the purpose is either quantitative analysis or monitoring of certain phenomena at the earth's surface. Especially in remote sensing of water surfaces the atmosphere can influence the radiance as measured by the remote sensor quite dramatically. The useful signal (i.e. the light emerging from the observed part of the water surface) can easily be less than 20% of the total signal. Of course the atmospheric effect is time dependent, but it also varies in space, which becomes particularly important when satellite imagery is being used.

There has been put an enormous effort into the development of atmospheric correction procedures. The radiative transfer in clear atmosphere is well understood and accurate models have been developed. Well known and widely used models are Lowtran 6 (Kneizys et al., 1988) and 5S (Tanré et al., 1990). In recent years also a number of remote sensing institutes in the Netherlands have addressed the problem of correction/removal of the atmospheric effect on both satellite and airborne remote sensing data. Fundamental research activities (e.g. de Haan et al., 1991) as well as more practical approaches (Kokke & Bakker, 1993) have been chosen to tackle the problem.

Although some of the work was reported in international journals and symposia (e.g. Kokke et al., 1989; de Haan et al., 1991), there is no overall impression of the status of research and the availability of methods. In fact there is a quite common tendency of atmospheric correction work not becoming fully operational and widely available for the remote sensing community.

In order to operationalize existing knowledge in the field of atmospheric correction and also to enhance availability of existing techniques to those who are interested, the following phased approach has been chosen. First an inventory will be performed of available national and international operational methods for atmospheric correction of remote sensing data. Secondly, after inclusion of user requirements, a small number of methods will be selected for further evaluation with respect to user-friendliness, circumstances under which to be used (possibilities/limitations), usefulness for application on satellite or airborne data, possibilities with respect to application to data of both existing and forthcoming systems (e.g. SeaWiFS), flexibility in usage with respect to position of spectral bands (e.g. airborne imaging spectrometry systems such as CASI and ROSIS), accuracy and required additional input. Finally, on the basis of the evaluation results, tools will be developed for atmospheric correction of both airborne and satellite data. These tools will be implemented in an operational computer environment.

4 WATER QUALITY ALGORITHMS

Remote sensing for water management purposes

will greatly increase its potential when it is no longer required to carry out intensive simultaneous in situ measurement campaigns for proper analysis. Besides atmospheric correction, the current state-of-the-art still requires in situ measurements for sensor calibration and correction of air-water surface and interface effects.

Multitemporally valid algorithms for the estimation of water quality parameters from remote sensing data are scarce. Dekker (1993) developed such algorithms for chlorophyll-a and cyanophycocyanin for Dutch inland waters. These algorithms are valid for high concentration ranges (e.g. from 0 - 200 μg l^{-1} pigment), but are less suited for low concentrations (between 0 and 20 μg l^{-1}). For the estimation of seston dry weight (suspended matter), Secchi depth transparency and vertical attenuation coefficients essential information on the inherent optical properties of particulate matter is lacking (Dekker, 1993). Therefore only semi-empirical algorithms are available, requiring in situ measurements each time a remote sensing image is acquired. The estimation of aquatic humus (dissolved organic matter) is currently not possible for inland waters. Only through adequate knowledge of the inherent optical properties of all the other substances present in the water analytical algorithms for aquatic humus determination may become available.

Going from inland waters to marine waters several changes (mainly a decrease in concentrations) occur in the presence and abundance of the various components within the water. When (semi-) empirical methods are used in coastal zones, errors occur in the estimation of water quality parameters, amongst others due to changes in the water surface state.

Methods for derivation of water quality parameters from remote sensing data should not only be multitemporally valid, but also be applicable to different sensors. In view of the present developments towards airborne and in the near future even spaceborne imaging spectrometry systems, data processing techniques must be spectrally flexible and cover the whole spectral range and not only a limited number of bands. Applicability of algorithms based on empirical and semi-empirical methods is generally very limited: i.e. the algorithm is only applicable to the data of a specific experiment and is only valid for the particular sensor used in that experiment. An analytical approach with respect to the retrieval of water quality information from remote sensing data offers the perspective of having the disposal of a 'universal' method, which in the ideal case is applicable to data of any remote sensor at any time in any place. Therefore within REWANET the analytical approach to retrieve water quality parameters from remotely sensed data has been chosen.

This analytical method should meet the following requirements:

- sensitive enough to estimate low concentrations;
- robust enough to estimate a large range of concentrations;
- sufficiently analytical to enable sensitivity analysis;
- as independent as possible of extraneous factors such as water surface state and solar angle;
- intrinsically suited to correct and assess influence of extraneous factors;
- straightforward to implement in a digital image processing system; and
- applicable to or easily adaptable to remote sensing instruments with differing specifications.

If these criteria are combined with the identified user-requirements (see section 2) of availability, accuracy, processing and archiving of data, the rationale for standardizing the methods for retrieval of water quality parameters from remote sensing data (often from different instruments) is obvious. Therefore, within the REWANET programme a toolkit of methods and algorithms will be developed. In first instance the algorithms will be based on (simple) radiative transfer theory using the inherent optical properties of the optical components of the waters as the baseline information, conforming to recent scientific research results (Dekker & Donze, 1994). This toolkit should eventually contain all tools necessary to arrive at any water quality parameter, that can be derived from remotely sensed radiance. In our case the toolkit has to be dedicated to the retrieval of water quality parameters and should contain tools that are necessary to achieve this goal. Once having the disposal of a toolbox, it should be easy to identify which tools are missing or have to be renewed or updated. A user can pick the best suited tools for a specific aim and build his own modular method.

In conclusions, the following guidelines for future research and application of remote sensing algorithms for determination of water quality parameters can be defined:

- Only if no alternative exists, empirical (statistical) models should be used.
- Use semi-empirical (spectral knowledge-based) models as long as essential knowledge or

information for applying an analytical model are absent.

- Concentrate on research that will contribute to the further development of analytical (radiative transfer based) models.

5 WATER QUALITY MODELS

Nowadays the use of water quality and ecological models plays an important role in water management of various water systems. In addition the use of modelling techniques may contribute to a better understanding of the influence of human activities (e.g. waste water discharges) on the environmental state of the water system and the causes of ecological degradation. If well developed and calibrated models allow the simulation of future situations under different waste load scenarios and conditions. In this way the environmental benefits of potential managerial measures to improve and safeguard the ecological values of a water system can be quantified and used in the environmental planning process.

The reliability and accuracy of the modelling results highly depend on the accuracy of input data (such as pollution inputs) and on the knowledge of various processes. During the development and calibration of the model, the comparison with field data is crucial. Often, and especially when dealing with large water bodies, the availability of field data is restricted by the number of monitoring stations and the sampling frequency applied.

Therefore, the use of remote sensing data can give a significant contribution in the comparison between model results and field observations. So far, only scarce examples exist where remote sensing data have been used to improve the accuracy of mathematical models for water quality. As part of the REWANET programme a pilot project is started to investigate the potential use of remote sensing data in the development, the calibration and validation of water quality and ecological models for the North Sea. In this project the main focus will be given to an improved simulation of suspended solids including inorganic and organic components. For the organic components also the simulation of phytoplankton, in relation to nutrients, will be considered by the model. The project, called RESTWAQ (Remote Sensing as a Tool for improved knowledge on Water Quality and ecology in large water bodies) will be carried out in 2 phases. Phase 1 includes the combined use of remote sensing data and field data to improve the calibration of water quality models, while phase 2 intents to use the results from modelling, remote sensing and field monitoring in an integrated way, to optimize the knowledge on water quality and ecological constituents in the water system.

The stepwise approach for phase 1 includes the following activities:

1. Set up and first calibration of a new generation of water quality and ecological models for the North Sea. This work concerns the development a (high resolution grid) water quality model for the coastal waters of the Netherlands and the southern part of the North Sea. Substances to be modelled include suspended sediment, nutrients, phytoplankton and toxic substances as heavy metal and organic micro-pollutants.
2. Acquisition and preprocessing of NOAA-AVHRR (and possibly SeaWiFS) satellite data. Available images from NOAA-AVHRR for the southern North Sea will be used and processed into weekly, monthly and yearly composites to enable the comparison of modelling results and field data for constituents as suspended solids and chlorophyll.
3. Preliminary use of remote sensing imagery to test the water quality models for the North Sea and identification of the needs for improvement. This step will focus on the identification of gaps in data and processes that might occur from spatial and temporal comparison of model results with remote sensing images.
4. Development of a generic method for comparison of model results with remote sensing data and field data, using existing tools for data assimilation and making optimum use of a GIS environment.
5. Application of the generic method to improve the calibration of the water quality models and evaluation of the contribution of remote sensing data for calibration and improvement of the performance of water quality models.

As the project aims the development of generic tools and methods, their benefits will not be restricted to the applications for the North Sea. Potential regions for similar application in this respect include the Yangtze Estuary and Hangzhou Bay (China), the Gulf of Finland and Neva Bay (St Petersburg, Russia), the Bay of Gdansk (Poland), the Adriatic Sea (Italy) and the Gulf of Azov and Black Sea (Ukrainian).

6 CONCLUDING REMARKS

For implementation of remote sensing in water quality management it is necessary to identify the problems and information needs of water quality managers and decision makers. Derived from the information requirements quantitative and integrated procedures for data-acquisition and data-analysis should be developed. By processing remote sensing imagery into basic digital information as input for a GIS and by defining rule-based procedures for integration of remote sensing with other spatial information in a GIS environment, remote sensing can be a useful technique for water quality management.

Remote sensing experts should cooperate, amongst others, with ecologists and model builders to combine remote sensing techniques with conventionally applied observation techniques and numerical models. Furthermore the involvement of the end-user in the application of remote sensing have to be stimulated. Finally, coordination of activities at an European level is needed. Network structures (e.g. CEO) should be the main driving force for standardization of remote sensing and GIS tools for water quality management in Europe.

In order to reach these objectives a Dutch framework programme on remote sensing and water quality has been defined (REWANET). This REWANET framework should lead to:

- The definition and implementation of a structured approach for research and operationalization of remote sensing techniques, in conjunction with in situ monitoring networks, numerical models and (geographical) information systems for water quality management and decision making.
- The establishment of a permanent national platform for coordination of remote sensing research and applications related to water quality, placed within an international context.

REFERENCES

Allewijn, R., H.P.J. Mulder, H.H.J. van Dijk & A.J. de Leeuw, 1993. Integration of remote sensing, GIS and numerical models for management of Dutch tidal waters. *Proc. of EGIS'93 Symposium, Genoa, Italy, March 29 - April 1, 1993:* 932-941.

Bijlsma, A.C., H.F.P. van den Boogaard & A.C. de Smet, 1991. *The assimilation of satellite and in situ data in a temperature model of the North Sea.* BCRS report 91-24, Delft, The Netherlands (in Dutch).

Buiteveld, H., P.B. Roeters, T.H.L. Claassen & K. Appelman, 1993. Water quality monitoring and management with remote sensing, from regional scale to national scale. *Proceedings of International Symposium Operationalization of Remote Sensing, Enschede, The Netherlands, 19-23 April 1993,* Vol 3: 99-108.

Dekker, A.G., T.J. Maltus, M.M. Wijnen & E. Seyhan,1992. The effect of spectral bandwidth and positioning on the spectral signature analysis of inland waters. *Remote Sensing of Environment,* Vol. 41: 211-225

Dekker, A.G., 1993. *Detection of optical water quality parameters by high resolution remote sensing.* Thesis, Free University, Amsterdam.

Dekker, A.G. & M. Donze, 1994. Imaging spectro-metry as a research tool for inland water resources analysis. In J. Hill (ed.) *Eurocourses Remote Sensing: Imaging Spectrometry as a Tool for Environmental Observations,* Kluwer AP, Dordrecht, The Netherlands (in press).

Haan, J.F. de, J.W. Hovenier, J.M.M. Kokke & H.T.C. van Stokkom, 1991. Removal of atmospheric influences on satellite-borne imagery; A radiative transfer approach. *Remote Sensing of Environment,* Vol. 37: 1-21.

Harkink, L., 1991. *Towards the operationalization of NOAA/AVHRR products for the purposes of marine ecology research.* BCRS report 91-09, Delft, The Netherlands.

Kneizys, F.X., E.P. Shettle, L.W. Abreau, J.H. Chetwynd, G.P. Anderson, W.O. Gallery, J.E.A. Selby & S.A. Clough S.A., 1988. *User's Guide to LOWTRAN-7.* Report AFGL-TR-88-0177, E.R.P. 1010.

Kokke J.M.M., H.T.C. van Stokkom, J.F. de Haan& J.W. Hovenier, 1989. Validation of an atmospheric correction method for satellite borne imagery. *Proceedings of the 4th International Colloquium on Spectral Signatures of Objects in Remote Sensing, Aussois, France, ESA-SP-287: 99-104.*

Kokke, J.M.M. & H.C. Bakker, 1993. Operational test of an atmospheric correction method for sequential quantitative water quality assessment with Landsat TM. *Proceedings of International Symposium Operationalization of Remote Sensing, Enschede, The Netherlands, 19-23 April 1993,* Vol. 3: 203-214.

Krijgsman, J., 1994. *Optical remote sensing of water quality parameters.* Thesis Delft University.

Pellemans, A.H.J.M., 1993. Operational application of 'real time' ERS-1 SAR data for oil spill detection. *Proceedings of International Symposium Operationalization of Remote Sensing, Enschede, The Netherlands, 19-23 April 1993*, Vol. 7: 173-180.

Roeters, P.B., H. Buiteveld, K. Appelman & H.C.Bakker, 1993. Enforcement of environmental legislation using thermal infrared remote sensing, from research towards implementation. *Proceedings of International Symposium Operationalization of Remote Sensing, Enschede, The Netherlands, 19-23 April 1993,* Vol. 3: 109-120.

Roozekrans, J.N. & G.J. Prangsma, 1992. *Observation of the earth-atmosphere system with the NOAA satellite*. BCRS report 92-02, Delft, the Netherlands (in Dutch).

Stokkom, H.T.C. van, G.N.M. Stokman & J.W. Hovenier, 1993. Quantitative use of passive optical remote sensing over coastal and inland water bodies. *Int. J. Remote Sensing*, Vol. 14, no. 3: 541-563.

Tanré, D., C. Deroo, P. Duhaut, M. Herman, J.J. Morcrette, J. Perbos, & P.Y. Deschamps, 1990. Description of a computer code to simulate the satellite signal in the solar spectrum: the 5S code. *International Journal of Remote Sensing*, Vol. 11: 659-668.

Tol, A.C. van, A.J. van der Meer, S.B. Hoek & A.J. de Leeuw, 1992. *Integration of remote sensing with the water quality monitoring network of Lake IJssel and Lake Marken for analysis and optimization of the network.* BCRS report 92-21, Delft, The Netherlands (in Dutch).

Sensors and Environmental Applications of Remote Sensing, Askne (ed.) © 1995 Balkema, Rotterdam, ISBN 90 5410 524 0

Estimation of water quality by passive optical remote measurements

Tiit Kutser, Helgi Arst & Sirje Mäekivi
Estonian Marine Institute, Tallinn, Estonia

ABSTRACT: Problems faced in interpreting data of passive optical remote sensing obtained by telespectrometric measurements on board a research vessel (or helicopter) are discussed. Results of the estimation of correlative relationships between remote sensing data and other water characteristics (Secchi disk transparency, chlorophyll, suspended matter and dissolved organic carbon concentrations and values of the beam attenuation coefficient) of the waters of the Baltic Sea and Estonian lakes are presented.

Introduction

Passive optical remote sensing from satellites has become one of the main methods of ocean research. It is obvious that for global-scale investigations satellite measurements are inevitable. However, the interpretation of satellite data is often difficult, not only because of the disturbing influence of the atmosphere and clouds, but also because of some specific features of the objects under investigation. This is particularly noticeable in the case of marginal seas and inland waters, where rather drastic spatial and temporal changes take place in the type and concentration of the ingredients in the water. Consequently, remote measurements from aircraft, helicopters and ships are also of value and deserve further development. Note that these measurements are also of great importance in the calibration of satellite methods.

Equipment and measurements

The telespectrometric measurements were carried out with a telespectrometer Pegasus. It is a scanning device which uses the same photomultiplier for recording sea and sky radiance. The technical characteristics of Pegasus are the following:

Number of measuring channels	3
Spectral interval	400-700 nm
Spectral resolution	6 nm
Measuring speed	4 spectra/sec
Field of view to nadir	± 4 degrees

The telespectrometer Pegasus enables to measure 1) the upwelling radiance from the direction of nadir, L_u; 2) the downwelling radiance from the zenith point of the sky, L_z; 3) the downwelling irradiance, E_d. Then the remote sensing reflectance (or radiance factor) $r(\lambda)$ at the wavelength of light λ can be calculated by the following formula: $r(\lambda)=L_u(\lambda)\pi/E_d(\lambda)$. In case of telespectrometrical measurements directed to the nadir point on board a ship or aircraft the influence of the air layer between the water surface and the receiver on the measured signal is small and the upwelling radiance $L_u(\lambda)$ consists of two main components: 1) solar radiance reflected from the wavy water surface $L_R(\lambda)$; 2) solar radiance backscattered from the depths of the water mass $L_D(\lambda)$. We can estimate the component $L_R(\lambda)$ by an approximate formula $L_R(\lambda)\approx 0.02L_z(\lambda)$ and determine separately two components of the radiance factor:

$$r(\lambda)=\frac{\pi L_u(\lambda)}{E_d(\lambda)}=\frac{\pi L_R(\lambda)}{E_d(\lambda)}+\frac{\pi L_D(\lambda)}{E_d(\lambda)}= \quad (1)$$

$$=r_R(\lambda)+r_D(\lambda).$$

In the present paper some results of telespectrometric and contact measurements carried out in some small Estonian lakes, Lake Peipsi (in Estonia) and the Baltic Sea are considered (see Table 1).

In addition to these results we have also data on total phosphorus concentrations in Lake Peipsi obtained during four expeditions.

Table 1. Regions and dates of the measurements of the concentrations of chlorophyll (CHL), Secchi disk depth (SD), dissolved organic carbon (DOC), suspended matter (SM), mean beam attenuation coefficient ($\bar{c}$) and beam attenuation coefficients (c(λ)). Remote sensing measurements were carried out during all these expeditions.

Region	Date	CHL	SD	DOC	SM	$\bar{c}$	c(450)	c(550)
Lake Peipsi	05.90	*	*					
Estonian lakes	08.90	*	*					
Lake Peipsi	08.90	*	*					
Estonian lakes	05.93	*	*	*	*	*	*	*
Estonian lakes	06.93	*	*	*	*	*	*	*
Estonian lakes	08.90	*	*	*	*	*	*	*
Estonian lakes	10.90	*	*	*	*	*	*	*
Pärnu Bay	06.92	*			*			
Pärnu Bay	04.93	*	*	*	*	*	*	*
Pärnu Bay	06.93	*	*	*	*	*	*	*
Pärnu Bay	07.93	*	*	*	*	*	*	*
Baltic Sea	06.89	*	*					
Baltic Sea	06.90	*	*					

Results and discussion

1. Correlation method

Estimation of the concentrations of optically active substances in water by optical remote sensing data represents the following inverse problem: above the water mass, spectra of the solar radiation reflected from the water surface and backscattered from the water mass are measured by a telespectrometer; on the basis of the numerical values of the results obtained the characteristics describing the location of the extreme values and the shapes of the spectra are calculated. Series of telespectrometrical measurements and simultaneous contact measurements must be made in order to find algorithms connecting the remote sensing and characteristics of the water. These algorithms are determined by investigating correlative relationships between the different characteristics.

Using this method a number of chlorophyll-retrieval algorithms have been constructed by different authors. The task is relatively easy if there is only one substance in the water (e.g. phytoplankton) as in the Case I water (by Morel water types). But if there are several optically active substances in the water, the problem may be complicated. Absorptive and scattering properties of optically active substances cause changes in reflectance spectra and the influence of one substance may overshadow the influence of other substances. Consequently, in case of turbid, multicomponental waters, the problem of the interpretation of optical remote sensing data needs special attention.

We have considered about 20 characteristics (from simple color indices to more complicated integral characteristics) with the purpose of finding out the most suitable characteristics for building the remote sensing algorithms (Miller et al. 1988; Kutser et al. 1994). We used these characteristics in regression analyses to estimate the properties of the water mass (chlorophyll concentration, beam attenuation coefficient of the water etc.).

It must be noted that our data set is far from homogeneous: different weather (and illumination) conditions, different regions (from the open Baltic to small lakes), different distances from the water surface (from boat to helicopter) are involved. For this reason it is practical to consider separately the results obtained on the basis of a single expedition and a whole data set. Some examples of the results of the estimation of the water characteristics from remote sensing data are presented below.

1.1 Secchi disk depth

Rather a great number (about ten) of the characteristics of remote sensing spectra give the correlation coefficients $R>0.7$ for the Secchi disk transparency (SD). A comparison of measured and calculated (using the remote sensing characteristic SF) Secchi disk depths is shown in Fig. 1. Here $SF=r(440)/r(480)+r(480)/r(600)+[r(480)/r_{max}]^A$, were $A=1$ if r_{max} is in spectral region $\lambda>480$ nm

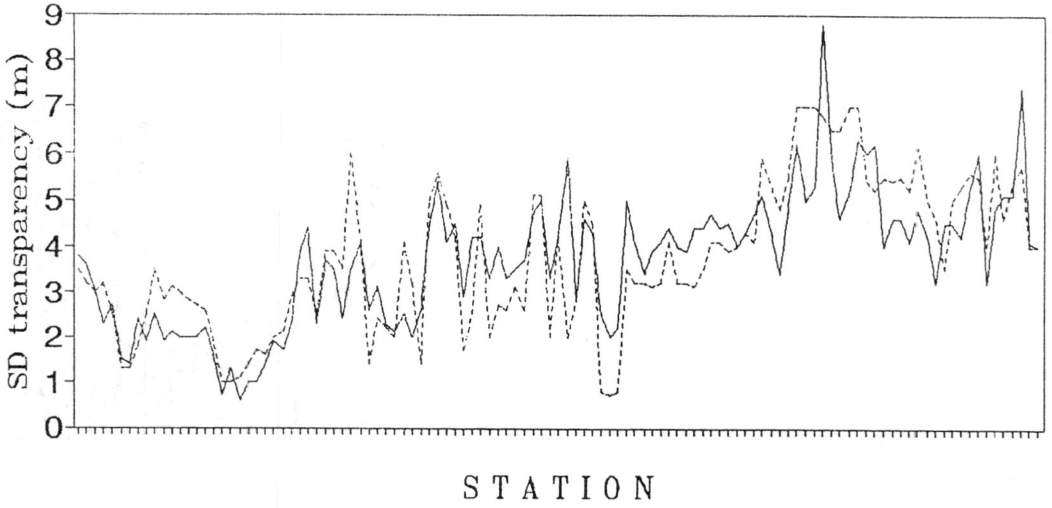

Fig.1 Secchi disk transparencies measured in situ (solid line) and calculated using algorithm SD=2.86SF-3.12 (dotted line).

and A=-1 if r_{max} in spectral region $\lambda < 480$ nm. Correlation coefficient R=0.85. It must be noted that each station in Fig.1 (and the following figures) corresponds to one small lake or one measuring station in the Baltic Sea or Lake Peipsi.

The correlation coefficient is usually higher if we consider the data of some group of measurements or use the data of only a single expedition. For instance, we got R=0.97 for the data in Pärnu Bay in summer 1993.

1.2 Chlorophyll *a* concentration

For chlorophyll only a small number of remote sensing characteristics gave $R > 0.7$ in case we used our whole data set. The values of the correlation coefficient are often higher for single expeditions. For example, Fig.2 presents results obtained during the expedition at Lake Peipsi in May 1990 (R=0.86). We got the highest correlation coefficient R=0.94 using the data of the expedition at 11 small Estonian lakes in May 1990.

1.3 Dissolved organic carbon

We started the measurements of the dissolved organic carbon (DOC) suspended matter (SM) and the beam attenuation coefficient only in 1993. Therefore, the corresponding data set is smaller than for SD and chlorophyll. Up to now we have not found satisfactory correlations between DOC and remote sensing characteristics using the whole data set. One possible reason here is the low sensibility of our telespectrometer for the wavelengths $\lambda < 420$ nm (as known, the dissolved organic carbon influences the light absorption mainly in the region of wavelength $\lambda \leq 450$ nm). Nevertheless, we have found rather good correlation for some separate expeditions (for example, for the June 1993 expedition to 11 small Estonian lakes R=0.92). The comparison of some results of measured and calculated DOC values is shown in Fig.3.

1.4 Suspended matter

Some remote sensing characteristics are applicable for the estimation of the concentration of suspended matter in turbid waterbodies. An example of measured and estimated concentrations of suspended matter in Estonian lakes and coastal regions of the Baltic Sea in 1993 is given in Fig.4. Better correlations occurred when we separated lake and marine measurements. The correlation coefficient was 0.90 for the coastal waters.

1.5 Beam attenuation coefficient

The spectra of the beam attenuation coefficient were measured by means of a spectrophotometer Hitachi in wavelengths range 350-700 nm using

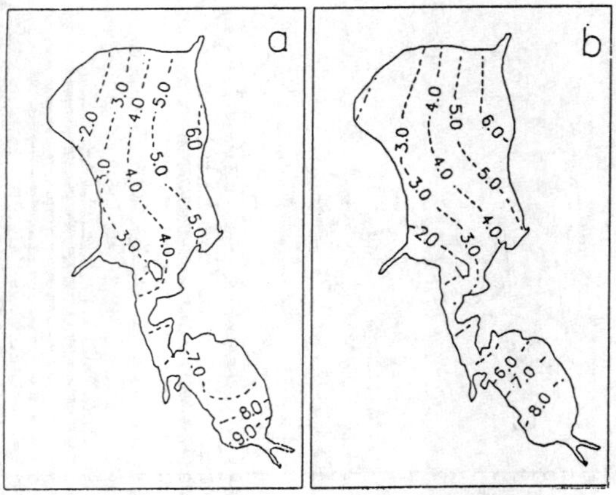

Fig.2 Isolines of chlorophyll *a* concentrations, measured in situ (a) and calculated from remote sensing data (b) in Lake Peipsi in May 1990. The corresponding algorithm is $CHL=1439T_1-3$, where $T_1=[L_u(460)-L_u(410)]/2-L_u(435)$.

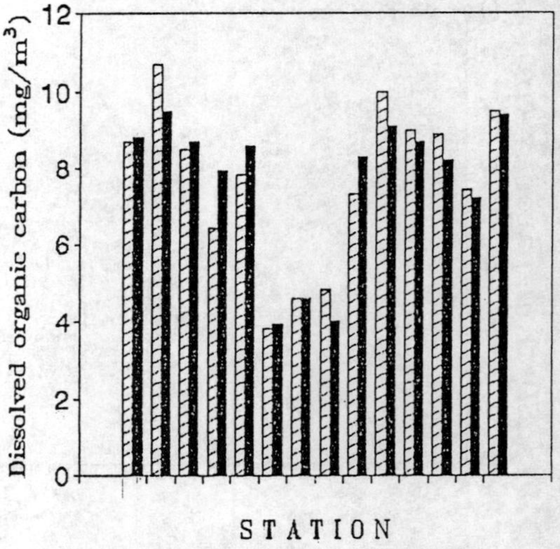

Fig.3 Concentration of the dissolved organic carbon measured in Pärnu Bay and in the Gulf of Riga in summer 1993 (filled bars) and calculated by remote sensing data (hatched bars). The corresponding algorithm is: $DOC=-11.42L_1+15.98$, where $L_1=r(480)/r(620)$, correlation coefficient $R=0.93$.

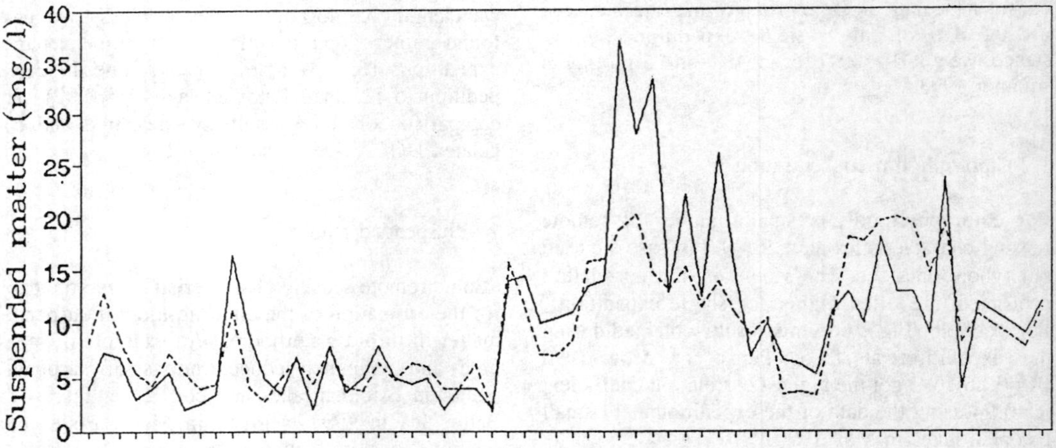

Fig.4 Concentrations of suspended matter measured in situ in some small Estonian lakes and coastal regions of the Baltic Sea (solid line) and calculated using remote measurement data (dotted line). Remote sensing algorithm is $\ln(SM)=-1.54SFM+5.71$, where $SFM=r(430)/r(530)+r(530)/r(620)+[r(520)/r_{max}]^A$. $A=1$ if r_{max} in spectral region $\lambda > 480$ nm, elsewhere $A=-1$; correlation coefficient $R=0.80$.

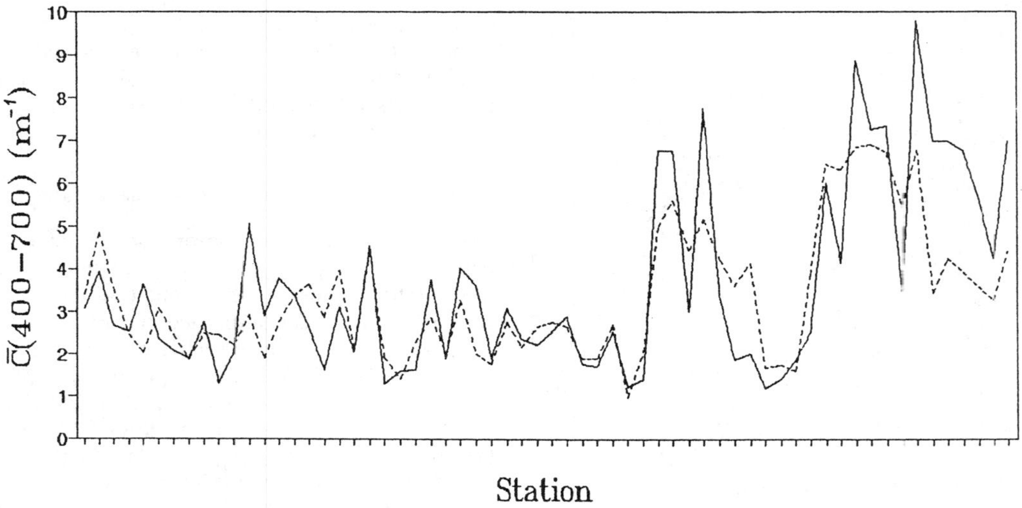

Fig.5 The values of $\bar{c}$(400-700), determined on the basis of water samples taken from Estonian lakes and coastal regions of the Baltic Sea in summer 1993 (solid line) and the corresponding values calculated from remote sensing data (dotted line). Remote sensing algorithm is $\ln(\bar{c})=-1.26SFM+4.16$ (see formula for SFM in Fig.4); the correlation coefficient $R=-0.79$.

water samples taken from the surface layer of the waterbodies. An attempt was made to estimate the averaged over the range 400-700 nm beam attenuation coefficient $\bar{c}$(400-700) and spectral beam attenuation coefficients in two wavelengths (450 and 550 nm) using remote sensing data. Fig.5 presents some results of these estimates.

As earlier, the correlation coefficients for single expeditions (and for lakes and marine measurements separately) are higher than for the whole data set. For c(450) and c(550) the correlation coefficients between measured and calculated values do not differ significantly from $\bar{c}$(400-700) data.

1.6 Total phosphorus concentration

Lindpere et al. (1987) have shown that there is a linear correlation between the relative transparency of the water (measured by Secchi disk) and the total phosphorus concentration in it. As we had some success in estimating the Secchi disk depth using passive remote sensing data then we had the idea to estimate also total phosphorus using remote sensing data. We have total phosphorus data only for Lake Peipsi and here we tried to estimate total phosphorus concentrations from remote sensing data (Kutser et al. 1994). The correlation coefficients obtained for phosphorus are mostly smaller than for Secchi disk depth. The best coincidences between measured and calculated phosphorus was obtained for in August 1990 (Fig.6).

Our results confirm that it is difficult to find an universal algorithm for estimating the properties of different waterbodies using the correlation method. Different characteristics may have good correlation in case of some single expedition, but in most cases the correlation coefficients are remarkably lower if we use the whole data set. However, there exist some cases where the correlation coefficients for a single expedition are lower than for the whole data set. It seems that coefficients in remote sensing algorithms have local and seasonal character, depend on the measurement conditions and also on illumination conditions during the measurements. Our conclusion concerning the seasonal and temporal character of the remote sensing algorithms is supported by the results of some other researchers (Aarup et al. 1990; Bukata et al. 1991; Mitchell & Hansen 1991).

2 Similarity method

The results, presented above, were obtained by means of the classical correlation method. We call the second, alternative method for the interpretation of the remote sensing data "similarity method".

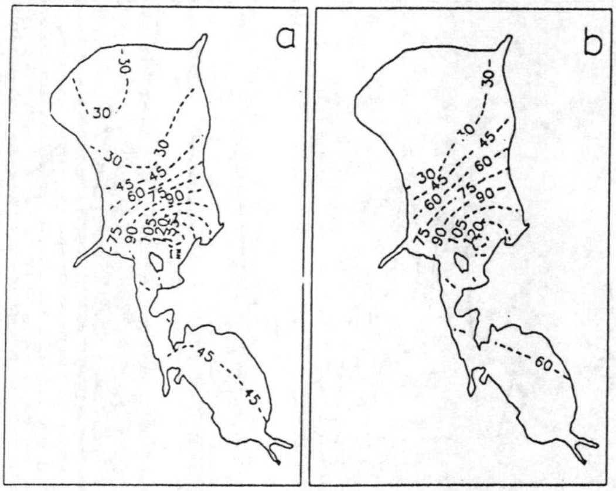

Fig.6 Total phosphorus concentrations (P) measured in situ (a) and calculated from remote sensing data in Lake Peipsi in August 1990 (b). The remote sensing algorithm is $P=279T_4-25.7$, where

$$T_4=\left[\int_{415}^{455} r_d(\lambda)d\lambda * \int_{655}^{685} r_d(\lambda)d\lambda\right] / \int_{405}^{605} r_d(\lambda)d\lambda.$$

The correlation coefficient $R=0.88$.

The basis of this method is the comparison of the measured full spectrum of $r(\lambda)$ (or $r_D(\lambda)$) with the multitude of the corresponding spectra determined by means of model calculations. From this group of spectra we try to identify the one which is the closest to the experimental spectrum. Thereafter we can determine the concentrations of optically active substances in the water for the experimental spectrum under consideration, assuming them to be the same as the corresponding initial data for computing the "most similar" theoretical spectrum. Note that for this kind of comparison normalized spectral curves may also be used. The efficiency of this method depends essentially on the validity of the mathematical model describing the spectra of $r(\lambda)$ or $r_D(\lambda)$. The principal difficulty is with the detailed description of the absorptive and scattering properties of optically active material in the water (e.g. different types of phytoplankton pigments, organic and inorganic substances etc.), especially the determination of the backscattering coefficients. Our estimations show that the scattering (and backscattering) coefficients presented in handbooks and monographs (mainly for the open ocean waters) are unsuitable for turbid coastal and inland waters.

To build up our model we assumed that there are only three optically active substances in the water: chlorophyll a, dissolved organic carbon and suspended matter. Under these conditions the total spectral absorption coefficient of the water, a_λ, is described by the following formula:

$$a_\lambda=a_{w,\lambda}+a_{chl,\lambda}C_{chl}+a_{DOC,\lambda}C_{DOC}+a_{SM,\lambda}C_{SM}, \tag{2}$$

where $a_{w,\lambda}$ is the spectral absorption coefficient of pure water; $a_{Chl,\lambda}$, $a_{DOC,\lambda}$, $a_{SM,\lambda}$ are respectively the spectral specific absorption coefficients of chlorophyll, DOC and suspended matter; and C_{Chl}, C_{DOC} and C_{SM} are their respective concentrations in the water. The values of the specific absorption coefficients of pure water, chlorophyll and hydrosol were obtained from Plass, Humphreys & Kattawar (1978). The values of $a_{DOC,\lambda}$ were calculated by the formula taken from Topliss, Miller & Irvin (1989):

$$a_{DOC,\lambda}=0.565\exp[-\gamma(\lambda-380)], \tag{3}$$

where $\gamma=0.013\pm0.005$, λ is given in nm, and $A_{DOC,\lambda}$ in l/(mg*m). The values of C_{DOC} will be in mg/l, C_{Chl} in mg/m^3 and C_{SM} in mg/l.

The total spectral backscattering coefficient, b^*_λ is given by

$$b^*_\lambda=0.5\,b_{w,\lambda}+\beta\,b_{SM,\lambda}\,C_{SM}, \tag{4}$$

where $b_{w,\lambda}$ is the spectral backscattering coefficient of the pure water and $b_{SM,\lambda}$ is the specific scattering coefficient of suspended matter, β is the ratio of the hydrosol backscattering coefficient to its scattering coefficient. The values of $b_{SM,\lambda}$ and β were estimated from physical models of scattering properties of seawater by Kopelevich (1983).

On the basis of the values of a_λ and b^*_λ we can calculate the diffuse reflectance of the sea just below ($z=-0$) the water surface using the formula by Kirk (1989). Taking into account that according to Gordon et al.(1975) $r_D(z=+0)\approx0.5r_D(z=-0)$ we get:

$$r_{D,\lambda}(z=+0)\approx0.5[(0.975-0.629\mu_0)\frac{b^*_\lambda}{a_\lambda}]. \tag{5}$$

where μ_0 is the cosine of the direct solar beam in the water (after refraction). As different combinations of concentrations of optically active matters may give similar spectra of r_D, we used average results obtained by ten closest spectra.

Until last year we had no experimental data on DOC and SM concentrations and therefore we were able to use this method for estimating only the chlorophyll concentration. The measured chlorophyll *a* concentrations varied between 0.3 and 47 mg/m^3. The correlation coefficient between computed and measured chlorophyll concentrations is 0.82, but for chlorophyll concentration less than 4 mg/m^3 the

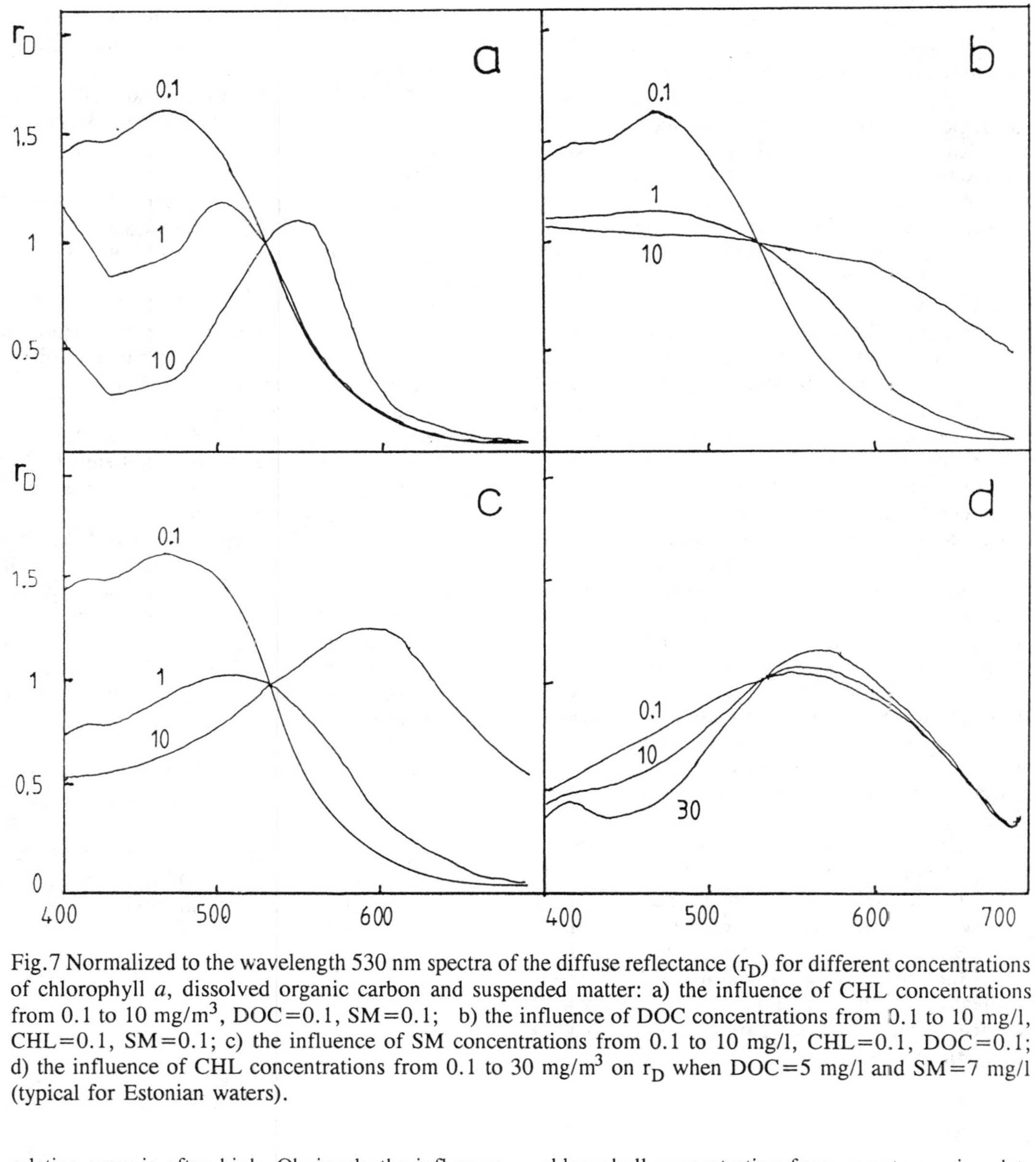

Fig.7 Normalized to the wavelength 530 nm spectra of the diffuse reflectance (r_D) for different concentrations of chlorophyll *a*, dissolved organic carbon and suspended matter: a) the influence of CHL concentrations from 0.1 to 10 mg/m^3, DOC=0.1, SM=0.1; b) the influence of DOC concentrations from 0.1 to 10 mg/l, CHL=0.1, SM=0.1; c) the influence of SM concentrations from 0.1 to 10 mg/l, CHL=0.1, DOC=0.1; d) the influence of CHL concentrations from 0.1 to 30 mg/m^3 on r_D when DOC=5 mg/l and SM=7 mg/l (typical for Estonian waters).

relative error is often high. Obviously the influence of chlorophyll on the remote sensing spectra is here screened by the stronger influence of DOC and SM. As a rule, the relative difference between measured and calculated C_{Chl} decreases with increasing chlorophyll concentration.

Fig. 7 shows the influence of different concentrations of the optically active substances on the spectra of diffusely backscattered radiation above the water surface. How difficult it is to retrieve the chlorophyll concentration from remote sensing data may be seen in Fig.7c.

Conclusions

Preliminary investigations of two methods (correlation and similarity) for the interpretation of passive optical remote sensing data show that both methods are applicable for the estimation of the concentrations of optically active substances in the natural turbid waters. The correlation method (simple regression algorithm) is easier to apply in monitoring of the water state. The drawback side of this method is the dependence of the algorithms on the region, season and weather conditions. We must have different algorithms for different waterbodies or groups of waterbodies and information on the seasonal changes in those algorithms caused by biological activities in the water. A large number of field measurements are needed to elaborate remote sensing algorithms for the waterbodies we are interested in.

If the elaboration of a theoretical model successful, the similarity method will be much more universal than the correlation method. At present there are difficulties in the detailed description (values and spectral distribution) of the absorbing and scattering properties of the optically active substances mainly due to the lack of respective experimental data from different stages of the biological activity in the water. More information is needed especially on the type, size distribution and backscattering properties of living and non-living suspended matter. There are also problems connected with the determination of the chlorophyll specific absorption coefficients: due to the package effect those coefficients may differ more than twice for the same phytoplankton species growning under different illumination conditions and we have to know the typical assemblage of the species of algae (their specific absorption and backscattering coefficients) in the waterbody under investigation.

In spite of the difficulties connected with the interpretation, our results confirm the understanding that the correlation of the optical remote sensing data with water transparency and the concentrations of the optically active substances in the water is not a random phenomenon and that it is possible to get qualitatively authentic results of spatio-temporal changes of several characteristics in the water bodies by means of optical remote methods.

REFERENCES

Aarup, T., S. Groom & P.M. Holligan 1990. The processing and interpretation of North Sea CZCS imagery. *Netherlands Journal of Sea Research* 25:3-9.

Bukata, R.P., J.H. Jerome, K.Y. Kondratyev, D.V. Pozdnyakov 1991. Satellite monitoring of optically-active components of inland waters: an essential input to regional climate change impact studies. *J. Great Lakes Res.* 17:470-478.

Gordon, H.R., O.B. Brown & M.M. Jacobs 1975. Computed relationships between the inherent and apparent optical properties of a flat homogenous ocean. *Appl. Opt.* 14: 417-427.

Kirk, J.T.O. 1989. The assessment and prediction of optical water quality. *Australian Water and Wastewater Association 13th Federal Convention* 504-507.

Kopelevich, O.V. 1983. Few-parameter model of the optical properties of the seawater. In A.S. Monin (ed.) *Optics of the ocean*: 208-235. Moscow: Nauka (in Russian).

Kutser, T., H. Arst, T. Miller, L. Käärmann & A.Milius 1994. Telespectrometrical estimation of water trasparency, chlorophyll a and total phosphorus concentrations on Lake Peipsi. *Int. J. Remote Sensing* (submitted).

Lindpere, A., H. Statast, A. Milius & A. Pihlak 1987. Water characteristics of Lake Peipsi-Pihkva and their correlations with nutrients. *Proc. Est. Acad. Sci., Biology* 36:146-155.

Miller, T., H. Arst, T. Torp, T. Lukk & J. Meitern 1988. Estimation of the transparency of the Baltic Sea waters and Estonian lakes by optical measurements. *Proc. of 16th Conference of the Baltic Oceanographers* 2: 711-716.

Mitchell, B.G. & O.H. Hansen 1991. Bio-optical properties of Antarctic Peninsula waters: differentiation from temperate ocean models. *Deep-Sea Res.* 38:1009-1028.

Plass, G.N., T.J. Humphreys & G.W. Kattawar 1978. Color of the ocean. *Appl. Opt.* 17:1432-1446.

Topliss, G.N., J.R. Miller & B. Irwin 1989. Ocean optical measurements - 1. Statistical analysis of data from the Western North Atlantic. *Contin. Shelf Res.* 9:113-131.

Sensors and Environmental Applications of Remote Sensing, Askne (ed.) © 1995 Balkema, Rotterdam, ISBN 90 5410 524 0

Operational uses of ocean colour data – Perspectives for the OCTOPUS Programme

D.R.Sloggett
Earth Observation Sciences, University College London, UK

M.Srokosz
RSADU of the BNSC

J.Aiken
Plymouth Marine Laboratory, UK

S.Boxall
University of Southampton, UK

ABSTRACT: Research conducted within Europe has highlighted several potential operational uses of satellite-based ocean colour data. This is of particular interest to the European Commission with its mandate to encourage the development of operational applications of satellite based data as a means of ensuring that taxpayers obtain a return on their investment in space. This paper discusses these applications and considers the operational infrastructure required to support the delivery of geophysical information derived from the ocean colour images to user organisations, such as fisheries agencies, global change research groups and agencies responsible for monitoring the coastal zone. The paper also addresses the potential implications of contemporary European data policy perspectives for the development of operational applications. In particular consideration is given to the need to encourage the value added sector to enter the service sector and provide services based upon information derived from satellite based instrumentation and other sources.

1. INTRODUCTION

Several potential applications of ocean colour data has been demonstrated in a number of projects using the Coastal Zone Colour Scanner (CZCS) instrument on the Nimbus satellite; see Barale and Schlittenhardt (1993). Originally scheduled to last for a single year its operation between 1979 and 1986 has given the oceanographic community access to an important source of data. Whilst spectrally and radiometrically limited, the CZCS sensor was able to pioneer many uses of ocean colour data in open oceans and in the important coastal regions of the world. Scientists were also able to conduct detailed research into the algorithms required for accurate atmospheric correction. This is a vital precursor to successful operational applications as the water leaving radiance is only a small percentage of the total radiance levels received at the satellite.

In the next 15 years there are a number of ocean colour based instruments that are scheduled to be launched. These sensor systems represent a second generation of ocean colour sensors with improved spectral and radiometric resolution and, in some cases, improved spatial resolution. This second generation of ocean colour sensors include the SeaWiFS sensor system, the Japanese OCTS, the North American Space Administration (NASA) MODIS and the European Space Agency (ESA) MERIS system. This range of missions and the continuity of access it provides gives ocean scientists and other users of ocean colour data a unique opportunity to investigate operational applications of the data.

Within Europe, the European Commission are keen to establish practical benefits to mankind as a return on the investment in space based instruments. To stimulate investigation of the potential operational uses of ocean colour data the EU have placed a study programme with Earth Observation Sciences Ltd (EOSL), and its science team, to investigate obstacles to applications achieving an operational status. The study programme has investigated existing research and its potential to form the basis of operational uses of the data, the data management infrastructure and ground segment required to support operational applications and associated data policy issues, such as the cost of data, and ownership.

The applications analysis activities have highlighted the potential operational use of ocean colour data in a number of areas. These are discussed in more detail in Section 2. These include: monitoring toxic blooms in European maritime basins; making a contribution to recruitment modelling and managing fisheries, monitoring major riverine discharges, and coastal processes. The study has also addressed the potential use of primary production measurements on a daily, monthly and seasonal basis derived from ocean data in climate change programmes, such as those being undertaken at the Max-Planke Institut and the Hadley Centre in the UK. The study has recommended a number of pilot programmes that should be undertaken to demonstrate the operational use of the data. These projects are designed to enlist the active cooperation of end user communities such as the fisheries and coastal monitoring authorities.

Section 3 of the paper looks at the data management infrastructures required to support the development of operational applications and ensure their effective operation on a day-to-day basis. Section 4 considers the implications of the development of operational uses of the data on contemporary European data policy research. Section 5 draws some conclusions on those applications which are likely to achieve near-term success. It also makes specific recommendations on the data management architectures that should be implemented within the JRC Centre for Earth Observation (CEO) project.

2. CONTEMPORARY OCEAN COLOUR RESEARCH AND POTENTIAL OPERATIONAL APPLICATIONS

Research, see Sloggett et al (1994), has shown that three areas merit serious consideration as potential operational uses of satellite-based ocean colour data. This paper focuses upon two of these in detail. These are applications in: the coastal region, where monitoring pollution and the transportation of sediment is of specific interest; and fisheries, where the monitoring of Harmful Algae Blooms (HABs), the combination with Sea Surface Temperature (SST) information to locate fish schools, and the potential to contribute to the development of policies that will encourage the sustainable management of fisheries have been proposed.

In trying to establish whether or not a specific application is able to mature to a fully operational capability there are a number of issues to be considered. These are:

- what information is required by users?
- how accurate must it be?
- how frequently is it required?
- do proven techniques exist for extracting it?
- how localised are the phenomena being observed?
- are other data sources required to ensure that it can be generated reliably?
- what obstacles exist in establishing an infrastructure to delivering it to the user?

The research also considered obstacles, on an application-by-application basis that might arise to the development of operational uses of ocean colour data, such as the:

- frequency of coverage of an area
- spectral resolution of the instrument on the satellite
- spatial resolution of the instrument
- ability to access the data within the time period required by the application
- need to complement the satellite data with a comprehensive network of in-situ observations
- need for more accurate atmospheric corrections
- stability of the instrument
- complexity of the signatures involved with highly localised phenomena dominating the observations
- need to better understand the phenomenologies involved and their associated geophysical parameters.

2.1 Coastal Applications

The coastal region is an area of specific concern in Europe and the world. Currently over 40% of the world's population live in this region which is equivalent to 5% of the total land surface of the Earth. It is a focus of trade and of the main interfaces between the ocean, land and atmosphere. It is an area which is most heavily influenced by anthroprogenic factors, such as waste disposal and the by-products of agriculture and industry. The coastal region is dynamic with erosion being a particular feature in many European maritime basins. It is a region where the environment needs to be carefully monitored to ensure the development of policies that encourage sustainable development of the region. This is where ocean colour data can play an important role.

The international focus on the coastal region is one recognised within this study due to its increasing importance socially, ecologically and economically. From a social perspective the coastal zone currently supports over 40% of the world's population. This is forecast to increase to over 60% by the early part of the next century increasing the pressures in coastal regions.

From an ecological viewpoint several factors are effecting the coastal zone. Increased levels of pollution are occurring from the discharge of untreated waste into the Mediterranean Sea, in the Baltic pressures are occurring from increased levels of nitrates being discharged from riverine sources and in the North Sea increased levels of metallic waste has been noticed as a result of discharges from industry on rivers such as the Rhine, the Humber and the Elbe. The Black Sea is a specific area of concern ecologically with many of its fisheries in a state of dramatic decline as a result of increased levels of pollution being discharged from, for example, the Danube.

In addition to these pressures the potential consequences on low lying coastal areas of changes in sea level could be catastrophic for many existing communities. In the UK, the National Rivers Authority (NRA) and the local authorities have to work closely together to decide upon major coastal defence programmes. Similar planning is carried out in The Netherlands. The cost of such schemes are often significant. Indeed in recognition of the future costs of maintaining current levels of defences a new idea is being discussed in the UK called 'managed retreat'. This ignores the existing axiom that all land shall be protected. Cost benefit studies are undertaken which consider the costs of coastal defence work against the implications of allowing the sea to invade certain areas of land. Economically the coastal zones is this region through which over 80% of the world's trade passes and is also where over 95% of the world's fish is caught.

Concerns are also being voiced over the potential linkage of increased levels of nutrients and other potentially toxic particulates in the water resulting from discharges from land sources, as a result of agricultural practises, and sewage outfalls. There is a great deal of interest in the physical and chemical means by which local sources of such particulates are dispersed, especially in enclosed basins such as the Black Sea and the Mediterranean Sea, where there is virtually no tidal range. This can result in a long term build-up of concentrations of specific constituents in the water with unknown consequences for local fisheries as pollutants are introduced into the food chain and the zooplankton at the base of the food chain are destroyed.

It is important to know the areas that are affected by such discharges in order to decide upon the likely implications of, for example, fisheries. European statute requires that before the construction of new outfall facilities that Environmental Impact Assessment (EIA) studies are conducted to assess the implications of siting a specific facility in a certain location. EIA studies require observations and models to be used in a combined effort to quantify the implications of a specific siting decision.

Given these concerns it is appropriate to consider the use of ocean colour data in coastal areas. The study has specifically addressed the potential of ocean colour data, both from airborne and satellite sources, to provide information to help in quantifying the nature of the problems and establishing appropriate management responses to the changes that are occurring. One of the key features of the analysis is to develop applications using satellite-based instrumentation that increase the effectiveness of the resources used in the coastal region.

The research has identified a number of problem areas in establishing operational applications of ocean colour data. The requirements for data, for example, are often different and sometimes exclusive from those of other marine areas, particularly when considering time and space scales. The options of ocean colour sensors range from CZCS, providing frequent data but at poor spatial resolution, through the Landsat series, providing good spatial resolution but at poor radiometric and temporal resolution, to aircraft, providing good spatial and temporal resolution but poor global coverage (temporally and spatially).

Where satellite-based observations are spectrally limited, airborne sensor systems, with their high resolution spectral measurement capabilities, can make a significant contribution as part of an overall operational capability. Although satellite based observations may, in the case of AVHRR and SeaWiFS be spatially and spectrally limited, they still have a role to play in identifying phenomena within the water. This is also the case for MERIS were the proposed 300 metre resolution will provide access to improved spatial and spectral information in the coastal regions. An example being the interest in monitoring the transport of trace metals in the channel. This processes is closely linked to the transport of the suspended sediments as many of the pollutants are speciated with clay and other inorganic and organic particles.

The agencies tasked with monitoring coastal processes and devising coastal protection schemes will also be keen to know information on the background levels of suspended sediment before any construction work starts, such as seasonal fluctuations and post storm surge events. This will enable them to predict the consequences of a particular coastal defence scheme. Where large-scale public funded is required to defend a specific part of the coastline satellite-based observations linked with models of local processes could be an important contributor to the decision-making process.

The use of ocean colour data to monitor coastal processes such as sediment transportation and riverine discharges has operational potential for authorities tasked with monitoring coastal erosion and overall water quality within European maritime basins, such as the North Sea, the Baltic and the Mediterranean Sea. This is of particular interest to countries such as The Netherlands, France, Germany, Italy, and the UK. It is of particular importance to enclosed basins, such as the Black Sea where the Danube is a major source of sediment and nutrients. A key factor that must be recognised however is the local nature of such predictions. The sediment transport and deposition processes vary, for example, from estuary to estuary. The tidal range of the estuaries is a key factor in this respect. It is possible to envisage a service, being provided from research stimulated within the OCTOPUS project, that combines satellite-based observations and models of tidal flows and local bathymetry to predict the transportation and the deposition processes. Harbour authorities would be interested in such a service as they need to look at potential problems with silting to focus dredging operations.

It is possible to envisage a distributed monitoring system that would be established within the context of the CEO project, such as a Coastal Information System (CIS). This would use existing centres of excellence in key locations in Europe. Each centre would use its local knowledge of the physical and chemical processes involved to provide a nowcasting service to European and national agencies. With major European riverine discharges occurring from the Po, Elbe, Rhine, Gironde, Humber, Danube and Rhone estimated at 250 tons per km^2, per year a coastal monitoring system linking models and observations from satellites and in-situ sensors could be of great interest to the EEA.

In the course of the study a number of potential applications of ocean colour data have been identified in the coastal region. These include:

- determination of levels of sediment in the water to give a clear indication of rates of coastal erosion which would be of interest to agencies responsible for coastal defence planning
- determination of mechanisms of sediment transport which would be of use to harbour authorities
- determination of the dispersal from major effluent discharge points along the coast to support EIA analysis
- determination of the types of nutrients being discharged from major estuaries which would be of interest to environmental monitoring agencies.

These applications require that ocean colour observations are available on a routine basis, such as daily, over the areas of interest. Ideally they should be able to use spectral analysis to discriminate the types of particles in suspension in the water and quantify the total load of each of these particle types over a grid of around 100 metres.

2.2 Fisheries and Harmful Algae Blooms

The problems affecting the management of fisheries in European are well known. Fishing levels in recent years have, in a number of circumstances, resulted in the almost total exhaustion of species in some areas. In response to this national Governments and the European Commission have had to implement difficult measures in order to try to protect certain species through, for example, the introduction of quotas. These measures have naturally been unpopular with fishermen who traditionally have been able to operate largely free of statutory controls. Policy formulation in this area requires access to the most comprehensive analysis of the environmental factors which might affect and influence recruitment in each fisheries region. This study suggests that ocean colour data could provide important insights into environmental events that can alter the levels of recruitment. This is an area requiring more research. The study also points out the potential uses of ocean colour data in monitoring the occurrence of HAB and notifying the authorities of the location, extent and type of HAB with a view to them establishing an effective response strategy to minimise the risks to public health.

Historically fishing is one of the major sources of employment and protein for the population of Europe. Since the Second World War many fisheries organisations have used airborne and satellite based observations of ocean colour and temperature to locate areas where fish congregate to feed. The operational use of ocean colour data in catching fish is, in some areas of the world, an established use of the data. In Europe however the story has changed. Over the past 20 years the levels of fishing that can be carried out within Europe have become increasingly controlled due to the need to conserve stocks and provide a sustainable source of food. In order to promote conservation fisheries agencies have been forced to consider several ways of limiting the catch. Measures have varied across the European Union from the UK government 'tie-up' policy to attempts at controlling the numbers of fish caught by increasing mesh sizes. This latter approach is often referred to under the title 'technical measures'.

Management of the fish stocks is achieved through establishing firm quotas and monitoring the numbers of fish landed at key fishing ports. Throughout Europe fishermen are restricted in the Total Allowable Catch (TAC) that they can land with consequent effects upon their livelihood and earnings potential. European Commission policy makers, working in concert with International Council for the Exploration of the Sea (ICES), have to decide when it is appropriate for the total landed catch of specific species to be decreased or increased. These decisions are critical to the fishermen and to their support industries and are often the source of controversy. Indeed, in a situation analogous to the set aside subsidies offered to farmers, fishermen are asking for compensation from Governments to help them. This situation in Europe is one of a number of applications were the development of a sustainable approach is required. This is one of the key themes emerging from the European Commission in its space policy; in essence how can space-based observations assist users in the formulation of policies that are based upon the sustainable exploitation of renewable resources, such as fisheries. What is needed is an effective and accurate means of forecasting the recovery of stock levels as the management of the existing reserves takes effect to enable fishermen to return to sea as quickly as possible. Such a forecasting capability is a vital element of an effective fish management strategy.

Shepherd (1993) has reviewed the problems of managing fish stocks and has rejected technical measures, on their own, as a panacea. In his analysis he highlights the problem that technical measures only affect the composition of the catch not the quantity of fish caught. He cites the need to ensure that technical measures should ensure that fish span at least once. For large fish, such as cod, this would require minimum mesh sizes of 200 mm which are far greater than those currently in use. The key issue is to maintain a Spawning Stock Biomass (SSB) of sufficient size to ensure that fishing levels and any other natural events do not increase the mortality levels to the point where a stock collapses. In the North Sea excessive fishing of several species has led, in the last ten years, to their virtual collapse. The important point here is to define an acceptable operating point where sustainable resource management can occur. Shepherd (1993) concludes that realistic technical conservation means, including the closure of some fisheries during spawning, on their own are not enough to ensure effective conservation. They have to be accompanied by direct conservation means such as limits on catches and/or fishing effort.

The Food and Agricultural Organisation (FAO) of the United Nations is another important contributor to the field of fisheries and research into their sustainable exploitation. Garcia (1991) has reviewed the total annual catch in the world and estimated that this could peak at 100 million tons. His report highlights the early recognition of the problems of sustainability; which date back to the identification of problems in the late 1940s. The report describes the so-called four phases of development of contemporary fish exploitation starting with the rapid rise which followed the end of the Second World War. The FAO look at these issues from a worldwide perspective and have to focus upon issues such as: policy; planning; data collection; research; enforcement and regional cooperation. The report highlights that fish and fish-related products contributes 40% of the protein intake to two thirds of the world's population. This reliance upon a single source of protein for such a large percentage of the human population underlines the increasing importance of techniques which emphasize conservation of fish stocks rather than exploitation.

The research has identified several areas in which ocean colour data could make a contribution in the efficient management and exploitation of fisheries. These are:

- commercial fisheries exploitation
- fisheries research and management
- monitoring the environment factors that influence fisheries and aquaculture
- monitoring HAB.

2.2.1 Commercial Fisheries Exploitation

In these times of financial hardship there is a need to catch the allowable levels of fish with the minimum of effort and cost; see Santos (1994). Ocean colour sensors and observations of SST enable fishermen to understand where fish stocks are likely to congregate to feed. If this information can be rapidly disseminated to the fishermen via satellite and shore based communications systems then it is possible to steer fishermen to the likely locations of fish schools. In Portugal such an operational service is already being provided by the SATOCEAN project sponsored by the NATO Science for Stability Programme, see Santos and Fiuza (1992).

The European Commission fisheries report published by the Nansen Centre (Pettersson, 1989) provides an excellent review of the successful use of satellite based observations in fisheries applications. The report summarises the range of satellite-based instruments that can contribute to deriving geophysical parameters of interest to the fisheries management authorities, such as SST, estimates of primary productivity, and meteorological observations, such as wind field measurements. The report describes the marine fishery ecology system, defines their associated life stages, outlines research linking primary production to fish yields and highlights the importance of coastal upwelling zones. The report also addresses the issue of the requirements for access to data to support operational uses of satellite based measurements. Summaries are provided of the use of satellite based data and infrastructures in Japan, the US, Canada, Australia, Brazil, Norway, France and several other countries. The report notes the significant efforts made in Japan to establish a national infrastructure for exploiting measurements from in-situ sources and satellites such as AVHRR.

Fiuza (1990) has discussed the applications of satellite remote sensing to fisheries. He cites the geophysical parameters that must be considered in fisheries research where the local environment plays such an important role. Knowledge of atmospheric parameters, such as incident solar radiation and cloud cover, wind speed and direction, as being important to the levels of primary productivity that will be generated in the upper layers of the ocean. Knowledge of salinity levels, the vertical temperature profile, the colour of the water, and the location of fronts and eddies is also important. The distribution of chemicals such as nitrates and phosphates alongside the acidity levels of the water is also important as fish have certain tolerance levels that effect their survival. Finally knowledge of the predator-prey relationships that exist, levels of pollution and the types and location of plankton also have an effect on the productivity of a fishery. This long list of environmental parameters illustrates the complexity of the relationships that have to be understood in fisheries research. Satellite based remote sensing can measure some of these parameters. It can for example provide insights into the surface temperature of the water, the colour of the water and the location and strengths of currents and eddies.

2.2.2 Fisheries Research and Management

Research has highlighted the potential operational use of ocean colour data in support of fisheries management and exploitation (Santos, 1994 and Sloggett et al 1993) and the potential use of ocean colour data to monitor the evolution and spatial extent of toxic blooms, such as red tides. These events introduce harmful toxins into the food chain and can, in extreme situations, cause severe illness in human beings. Detection of the spatial and temporal extent of such blooms, or the likelihood of their occurrence - given certain constituents being present in the water - would enable fisheries managers the potential to avoid catching fish that may have been in the area where the toxic bloom occurred.

The overall levels of fish stocks in a particular maritime basin can be affected by a number of environmental events. Fish spawn in known areas of the maritime basins. If their larvae are not correctly transported by local currents and tides to regions of the maritime basin where nutrients exist in the water a population can be decimated in a single year. Satellite data can, with the correct level of automatic and manual analysis, provide important complementary information on the likely successful movement of the larvae. Parameters derived from this in conjunction with knowledge of the occurrence of episodic events, would provide important additional information for effective fisheries management.

Shepherd (1993) has considered the need for fisheries to be managed. He argues that management practises are essential to prevent the total collapse of certain stocks due to an over-capacity in the levels of fishing effort undertaken. In his analysis Shepherd considers a relatively simple model of what he refers to as 'bio-economic equilibrium'. Figure 2.1 illustrates this model. Bio-economic equilibrium is achieved where the costs of fishing and the earnings intersect. Shepherd points out that a classic example of a non-regulated bio-economic equilibrium is the near collapse of the stock, illustrated in Figure 2.2. In the case of the North Sea cod the SSB has been reduced to just 1% to 2% of the unfished level. He states that it is little short of a miracle that stocks at these levels are able to maintain any level of reproduction.

A valid question for the analysis of the SSB is how low is too low? This question was posed by Shepherd (1993). He cites two measures which are often used based upon average levels of recruitment. These are:

- twenty percent of the unfished level, widely regarded as a minimum desirable level
- the level corresponding to the biological reference point (F_{med}).

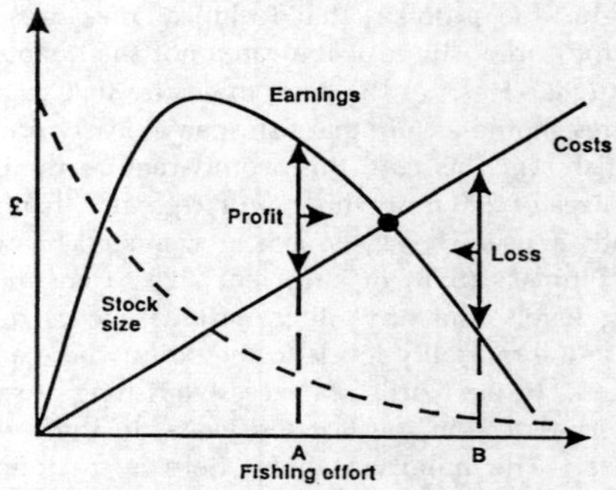

Figure 2.1 The balance of cost and earnings and the 'bio-economic equilibrium' (Courtesey of MAFF)

In his analysis Shepherd uses examples of the experience of the US agencies on the Georges Banks to illustrate how mesh size alone has not protected stock sizes. Figure 2.3 shows that the mesh size would have to rise to around 180 mm. With fishing levels staying around current levels the mesh size of 140 mm would still keep the fish population well below what is regarded as a desirable level. The need to raise mesh sizes to over 180 mm to ensure that fishing effort levels above that in 1990 could be sustained is not practical. The levels of investment required by the fishermen being a key factor in such a consideration.

In his report Shepherd has considered why the bio-economic equilibrium may not be stable. He argues that there is a time lag that operates in the 'system'. This time lag leads to fishing levels initially overshooting the equilibrium point. This results in punitive conservation measures where the fish stock is allowed to recover. Once a recovery is detecting economic pressures build up to release more fishing effort. This leads to a repeat of the cycle. Figure 2.2 illustrates this effect. In addition to this we have to consider the potential effects of environmental factors on the SSB. This can cause major impulses to occur in the cycle. Organisations

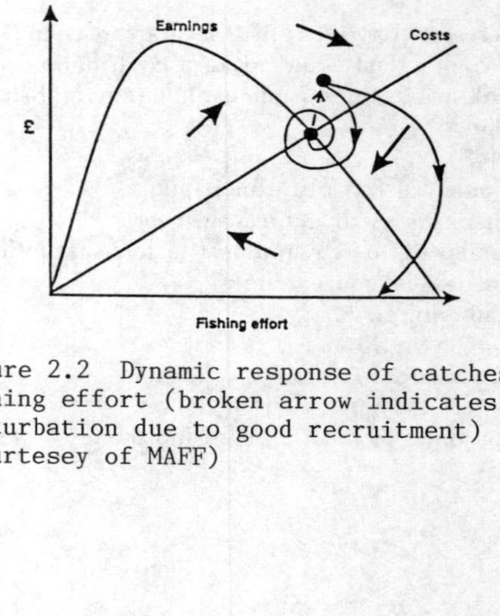

Figure 2.2 Dynamic response of catches and fishing effort (broken arrow indicates a perturbation due to good recruitment) (Courtesey of MAFF)

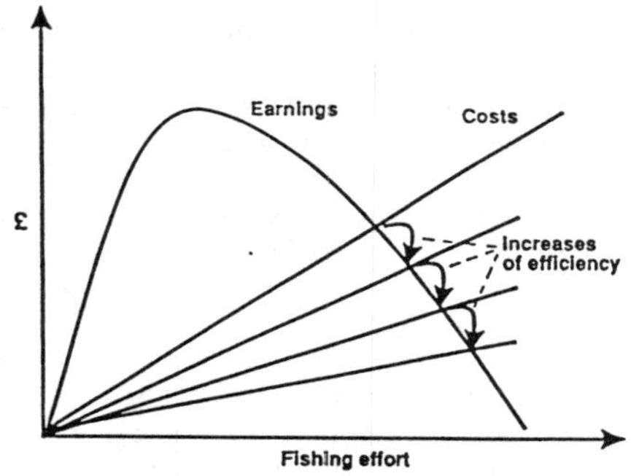

Figure 2.3 Dynamic response to increasing efficiency (Courtesey of MAFF)

such as ICES deal with this by providing themselves with a margin when setting quota levels.

In reports, such as ICES (1992), the organisation recommends the TAC in each of a number of management units in the Baltic. This report uses information on the total landed catches, and acoustic surveys, to decide upon future quotas for the pelagic (Herring and Sprat), demersal (cod) and Salmon fisheries. Of particular note is the decline in the SSB of the Baltic cod from 800,000 tons in 1984 to 200,000 tons in 1990; the lowest value observed in the recorded data from 1970 to 1990. In the same period the catches of cod have declined from 441,000 tons in 1984 to 171,000 tons in 1990. Similar reports are generated by ICES for all of the other fishing areas in Europe.

These reports highlight, inter alia, local environmental factors that are likely to influence recruitment levels. In the Baltic the combination of low salinities and poor oxygen conditions in deep waters are foreseen to limit spawning levels in the areas of the Gdansk Deeps and Gotland. Oxygen deficiencies were also noted in the Bornholm Basin. Interestingly the report links improvements in recruitment to an influx of water from the North Sea into the central Baltic. It suggests that improved conditions for spawning will arise some six months after an influx of North Sea water into the Bornholm Basin. It is worth noting that just such a major influx of North Sea water occurred for the first time in ten years in the early months of 1993.

In recognition of the difficulties currently being experienced in maintaining fishing levels and deciding upon methods of controlling catch levels the study has considered the potential role of ocean colour data in the effective management of fish stocks. A key to this is to know the levels of recruitment to each stock on a year-by-year basis. Sloggett and Aiken (1994) have suggested that it is possible to envisage ways in which parameters derived from ocean colour data might be combined with existing models of fish recruitment to provide more accurate assessments of the SSB.

The issue of recruitment modelling is the subject of on-going research by ICES (1992). The report of the study group meeting on these processes held in Paris in 1990 reviews the problems of modelling recruitment and studies individual approaches to the problems. Their remit was too clarify conceptual ideas and develop existing and new theories. A number of group sessions explored ideas that were subjected to debate within a wider forum.

The report by ICES (1992) highlights recruitment as a dynamic interplay between factors that regulate population, such as predation, and environmentally driven factors were physical and biological processes that determine if, for example, larvae can locate food.

The ICES report considers several key, but interrelated, objectives of recruitment research. These are:

- prediction of annual recruitment
- understanding the factors that stabilize recruitment
- understanding of the factors that cause variability in recruitment
- anticipating long-term changes in recruitment, allied to global change.

The report noted several processes that are of importance when considering modelling recruitment. These included:

- advective processes
- diffusive processes
- processes regulating primary production (upwelling, nutrient supply etc.)
- the dynamics of ocean fronts (surface convergence, tidal-mixing, etc.)
- physiological effects (temperature, salinity, toxins etc.).

It is possible to envisage ocean colour data and information derived from other instruments, such as ATSR, being used to make a direct contribution to many of the processes identified in the ICES report. In an operational context this information would be derived from the satellite-based instruments and disseminated to fisheries agencies for assimilation within their fish recruitment models. In time, as the data collected increased and the overall accuracy of the predictions of SSB improved, the need for fish surveys to be undertaken would be reduced. It is possible to envisage in the medium term satellite data being used with models and ancillary data sources, such as meteorological data, to provide an accurate forecasting service for fish stocks on a European scale.

Remotely sensed observations of ocean colour can be used to monitor the environmental conditions which may affect commercial fisheries, fish farming in sea lochs (salmon fisheries) and other forms of aquaculture (shellfish, kelp etc.). Environmental pollution can arise from a number of sources: direct anthropogenic effects such as oil spills, chemical spills and waste discharges of noxious or toxic substances; natural effects such as harmful algal blooms of dinoflagellates or other toxic algae; eutrophication in coastal waters, enriched by anthropogenic nutrients (nitrates and phosphates from agriculture fertilisers which run-off into rivers, estuaries and coastal waters) which can lead to anoxia (without oxygen), as the plants respire (rot), and use up the oxygen in the water leading to fish mortality.

Santos (1994) outlines the complex physical and biological cycles that affect the recruitment processes. These have been reviewed by a working group established by ICES. The report of the first study group meeting (ICES, 1992) reviewed the problems associated with modelling recruitment, such as the dynamic interplay between population regulatory factors - such as predation - and environmentally driven variability.

This link to environmental variability is of particular importance if the climate is changing as a result of anthropogenic activities. The joint ICES, SCOR, IOC and PICES Global Ocean Ecosystem Dynamics (GLOBEC) project has been investigating the potential effects of climate change on the population of cod (GLOBEC, 1993). The research activities have recognised that several different scales of investigation need to be considered. These range from the effects of small scale turbulence on encounter rates between fish larvae and their prey, to large-scale effects of inter-decadal changes in windfields on circulation and transport of heat and young fish. Despite the complexity of the processes involved, in terms of the range of forcing functions acting upon the population, effects related to climate change have been documented for several stocks. Examples include: periods of low temperature are observed to result in the decline of stocks at the northern boundaries of cod distribution; particular hydrographic and wind conditions that result in unusual transportation of eggs and larvae; or events such as flushing of enclosed basins, such as the Baltic.

2.2.3 Monitoring Harmful Algae Blooms

Another potential application of satellite-based ocean colour observations is in monitoring HAB This topic has been covered in depth in a recent ICES report (1992). The report listed potentially harmful species of phytoplankton, such as Alexandrium, highlighting the emerging evidence of the four most common associated human illnesses. These are Paralytic Shellfish Poisoning (PSP), Diarrheic Shellfish Poisoning (DSP), Neurotoxic Shellfish Poisoning (NSP) and Venerupin Poisoning. Over the past fifteen years there has been a considerable increase in fish and shellfish farming throughout northern Europe. Harmful phytoplankton have caused this developing industry considerable problems. One of the most notable examples being the May 1988 bloom where over 800 tons of rainbow trout, Atlantic salmon and cod were lost along the Swedish and Norwegian coasts. This event was followed by another in the summer of 1988 where a total of 290 tonnes of farmed salmon were killed as a result of phytoplankton blooms in the Faroe Islands. The estimated cost of this event was in the region of 1 million pounds sterling.

There is an extensive literature on HABs which affect mainly shellfish, but can propagate up the food chain causing fish mortality, affecting fisheries, fish farms etc. and even the health of humans who feed on affected animals; fatalities have been known to occur, especially in the sick, old or infirm. HABs are more commonly called 'red tides' or 'brown tides' on account of the red or brown coloration given to the water by the high concentrations of dinoflagellates which characteristically contain the accessory pigments peridinin, and carotenoids (fucoxanthyin and derivatives) all of which absorb at yellow wavelengths. This characteristic colour offers the possibility of detection by airborne or satellite remote sensors for water colour. Although these algae are not unique in their ability to absorb light at green or yellow wavelengths they are the species which most commonly form high concentration phytoplankton blooms. In combination with a Knowledge-Based Expert System or a GIS management tool, populated with historical data base information the risk could be assessed and action initiated. The analysis of SeaWiFS imagery might ask the question: is the bloom absorbing at 510 and/or 555 nm? Knowledge of the existence of a bloom within the locality at the same time of the year and during similar meteorological conditions could well form the basis of a reliable HAB warning service. In 1988 in Norway just such an operational service was established to monitor the rapid progress of the Gyrodinium aureloum bloom which started in the Skagerrak and moved north under the influence of the Norwegian Coastal Current resulting in severe levels of fish mortality in many areas. At the height of the event over 130 fish farms were towed away to saftey within northern fiords. The final extent of the bloom is illustrated in Figure 2.4 where the major blooms in Norwegian waters are shown between 1966 and 1990; quite clearly the event of 1988 has been the largest recorded event in this area to date, with the event of 1981 resulting in the most recorded levels of mortality in fish stocks. A similar bloom event in 1983, whilst being of a lesser extent, occurred in July in the Skagerak. It is illustrated in Figure 2.5. In this image the yellow colouration is used to illustrate medium levels of pigment concentration.

Recent work published by Subramaniam and Carpenter (1994) has reported on the development of a protocol that is capable of detecting the cyanobacterium Trichodesmium. This is a nitrogen-fixing

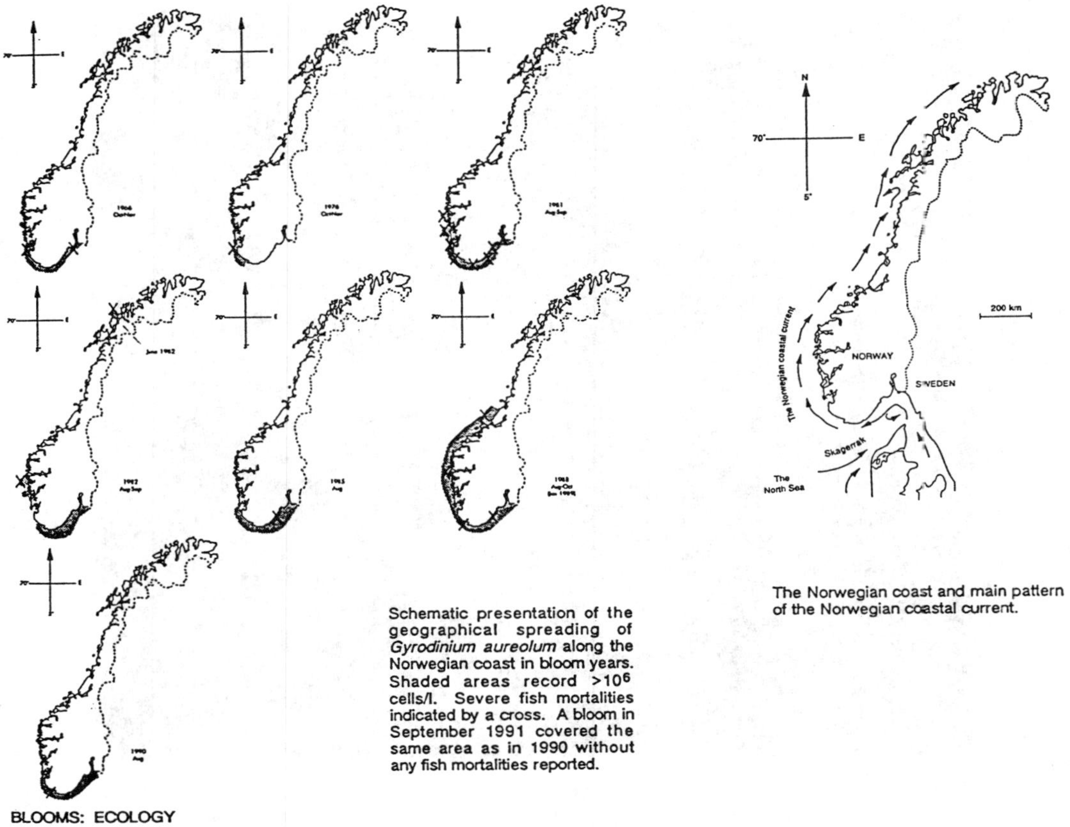

Figure 2.4 Example of Norwegian Algal Blooms

cyanobacterium that is frequently the most abundant phytoplankon element in tropical oligotrophic waters. The protocol takes into account two optical characteristics of Trichodesmium blooms - a high reflectivity due to the presence of gas vacuoles and an absorption feature at 550nm due to the accessory pigment phycoerythrin. The paper suggests that the combination of these two features is unique to Trichodesmium blooms. The research has been validated using CZCS imagery derived from sites in the Pacific, Atlantic and Indian oceans along with sea truth data.

A longer term and more specific objective for satellite imagery would be the detection of precursors to HABs. This objective could only be achieved after a number of years of study and research into the physical, biological and bio-optical precursors to HABs and their bio-optical signature which might be used as an indicator. We can foresee that an analysis/forewarning strategy could be developed using satellite imagery, but not in the short term.

The third environmental pollution topic concerns algal blooms, not necessarily toxic, but having consequences for fish life, through the impact they have on the oxygen concentration of the water. The approach needed has some similarities to the problems associated with HABs discussed above, but the major precursor to the terminal "anoxia" condition, for fisheries is the bloom itself. This should be readily detected by satellite remote sensing methods by virtue of its high biomass, rather than any distinct bio-optical signature arising from its accessory phytoplankton pigments. There may of course be some bio-optical signature which may serve as a precursor to any phytoplankton bloom whether an HAB or otherwise; these may be indicators in the phytoplankton seasonal succession which differentiate the case in question from the normal non-blooming condition, even if the actual mechanism of bloom formation is physical, rather than biological. As mentioned above, such subtleties of the biological succession will only be understood after considerable research, which may combine field and laboratory studies and be linked closely to satellite remote sensing

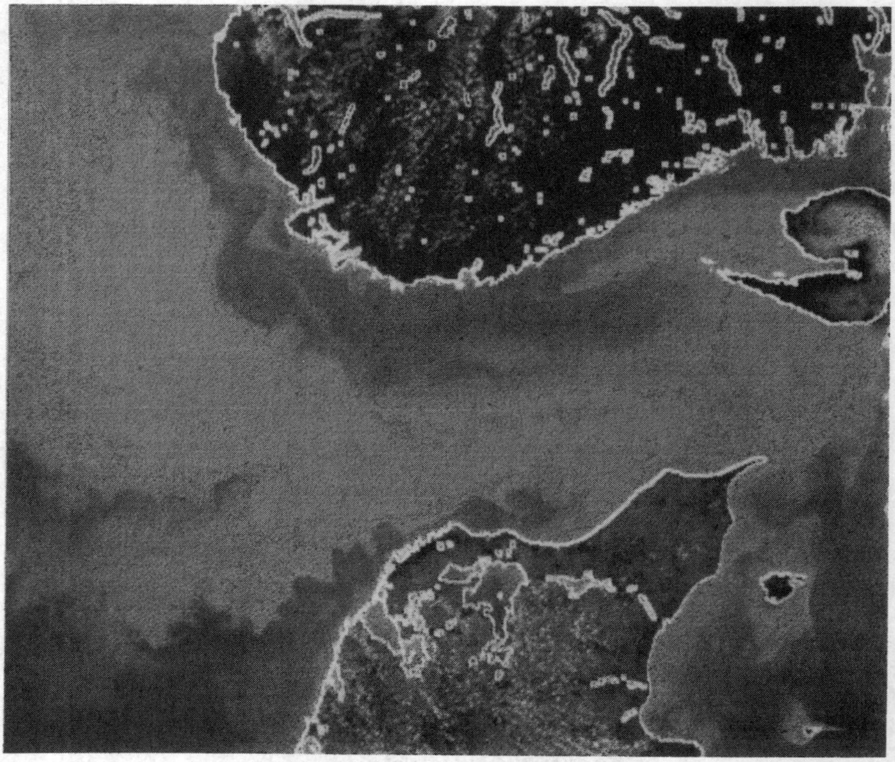

Figure 2.5 CZCS Image of Coccolithophores in the Skagerak on 6 July 1983. Land and cloud boundaries are masked with a white line. In the water green correspondents to a high, yellow to a medium, blue to a low pigment concentration (Courtesey of DLR) Colour plate see page 497

observations. To be useful, the observations must originate from an operational sensor producing data which has been acquired over a number of biological seasons, each of which may have significantly different, distinctive, phenomenological characteristics.

Whilst monitoring a bloom once it has been detected is important it would also be desirable to be in a position to be able to forecast its occurrence from evidence that could be derived from ocean colour and related instrument systems, such as ATSR. If this were feasible a nowcasting service could be established that would provide information on the likely occurrences of blooms for all European maritime basins.

2.2.4 Applications to Commercial Fishery Exploitation

There have been a number of studies, going back many years, reporting the applications of satellite remote sensing to the operational exploitation of commercial pelagic (mid-water) fisheries. The link to fish stocks has been through the surrogates of water colour (phytoplankton biomass, chlorophyll concentration) and SST.

The basic premise is: observations of ocean colour from space provide a measure of the areas of enhanced biological production and plant biomass which can accumulate at shelf-sea fronts, topographic structures (banks and ridges), shelf edge and other upwelling zones, and at physical features (current systems, eddies, spurts, jets, etc.) where zooplankton and fish populations are known to accumulate for feeding, spawning and early life development. Remotely sensed measurements of SST provide an additional measure of these physical structures where biological processes are enhanced, note SST measurements (skin temperature only) need careful interpretation and can be misleading e.g. solar heating of the surface layer can obliterate water structure at fronts or eddies which will reduce the utility of SST imagery. Thus imagery of ocean colour and SST provide both a measure of the primary production and biomass at the base of the food chain and the location of the areas where the higher trophic levels will thrive. Demersal (bottom - living) fish are equally likely to thrive at zones of high pelagic production, since their benthic food resources are directly enhanced by the high primary production in the euphotic zone.

While a new colour sensor is awaited, SST imagery from AVHRR data has proved a useful substitute for commercial fisheries exploitation of the established association of fish schools with thermal fronts. There is no doubt that the synergistic combination of remotely sensed SST and colour imagery would greatly enhance this exploitation; the utility and feasibility of the methods have been demonstrated adequately by the use of SST imagery alone; charts of fronts, likely fish schools etc; can be radio-faxed to ships at sea, adopting the methods which have been used for weather forecast charts for several decades. The current model for radio-faxing SST maps and weather charts seem adequate for colour imagery issued by a central fisheries agency (akin to the meteorological offices uses of weather charts). There would seem to be no need for specialised satellite communications not generally available on fishing vessels. It is ironic to think, that in the current concern for conservation rather than exploitation, satellite imagery might be used to define exclusion zones where commercial fishing is banned during periods of fish spawning and early life history development.

In Portugal the SATOCEAN project has, since 1989, been providing an operational service to the Portuguese tuna and swordfish fishermen, see Santos and Fiuza (1992). The fishermen are provided with charts with the location of thermal fronts and isotherms derived from AVHRR data. Preliminary results from the research activities that are also being carried out indicate that sardines concentrate in moderately cool, relatively old, upwelling water on the inner shelf off western Portugal in temperatures around 14°C. In contrast swordfish and tunas are preferentially found seaward of the semi-permanent thermal front separating coastal waters from the warmer open ocean. The sardine fishery has the largest volume and hence the highest value of the Portuguese fisheries. It is located along the coastal of continental Portugal with a localised concentration around the north-western coast. The sardine schools are found at depths between 20-50 meters, generally near the sea bottom. Santos and Fiuza report that 50% of the schools were found at depths of less than 10 meters from the bottom, and more than 70% in less than 20 metres from the floor of the ocean.

2.2.5 Fisheries Research

There are close links between commercial fishery management and fisheries research through cooperation at national and international level. This is directed towards the estimation of fish recruitment, and stock assessment, for the formulation of policy (government and intergovernmental) on total allowable catch and enactment of legislation for management and conservation of fish stocks.

Historically, fisheries research has been focused on the inter-annual variability of measurements of recruitment and year-class strengths of fish eggs, larvae and juveniles in response to physical and biological forcing. Generally, this approach has been unsuccessful in explaining the observed changes year-to-year with little success towards the development of predictive tools (models). The failures of this approach are largely a result of intra-annual (seasonal) variability in the forcing functions, e.g. timing and interactions. Thus, to understand the fundamental year-to-year variability, it will be necessary to study the seasonal processes, such as the duration and spatial extent of episodic events (upwelling, etc.) which may be vital to the survival of fish eggs and larvae. Ship-board methods cannot adequately measure the extent of variability of physical and biological processes either temporally or spatially and satellite remote sensing of water colour will be vital in this context to fisheries research. Fishing policy for commercial fisheries is determined by management issues which are derived from fisheries research. However, exploitation will be decided by policy rather than any direct benefits which satellite imagery can offer fishermen, i.e. commercial fishing will be controlled by management decisions and efforts to limit over fishing.

The modelling of fish stock levels is described by Doucet and Sloep (1992). In their book they derive mathematical formulae that model populations dynamics. They consider models of the stock levels of single and multiple species and evolve methods for predicting fish stock levels. They focus on the optimisation of harvesting under specific constraints, such as those arising from economic factors. Our aim would be to enhance the performance of these models through the combination of the existing models of the stock levels with geophysical parameters derived from the satellite based observations.

Complementary research has also been reported by Roozekrans (1993). He has shown the successful application of ocean colour and AVHRR data to the detection of Coccolithophore blooms. He has also shown that research is being carried out by The Netherlands Institute for Fisheries Research into correlating sea surface temperature with the locations of spawn of herring in the North Sea. Preliminary results from the project suggest that a potential operational network could be established easily at minimal cost.

3. DATA MANAGEMENT ARCHITECTURES SUPPORTING OPERATIONAL APPLICATIONS

3.1 Historical Perspectives

Historically European ground segment developments have been designed to support a relatively small user community. It is important to reflect upon the nature of that user community and its requirements for access to data. In the early 1980s when the ground segments for the SPOT and ERS-1 systems were being commissioned many of the applications of the data and information derived from their sensor suites were relatively immature. In the case of the ERS-1 mission the priorities were to research the potential of the suite of sensors on board the spacecraft to measure a range of geophysical parameters. Research groups were supported with access to the data to conduct fundamental studies into the physics, chemistry and biology of the processes involved in radar based imaging.

In contrast the development of the SPOT ground segment was driven by commercial priorities. These still required the active involvement of several research groups to examine the applications of the data to studies of inter alia monitoring vegetation, coastal zones and use of the stereo viewing capabilities to support the generation of Digital Elevation Maps (DEMs). The potential SPOT community was primed with simulations of the data from the satellite sensors prior to launch date in order to familiarise themselves with the potential of the data. Gradually operational users of the data and information have emerged from these research activities, such as the Monitoring Agriculture by Remote Sensing (MARS) project, which has so successfully demonstrated the routine use of satellite-based observations. This is seen by many to be a role model for the OCTOPUS programme.

In the course of the ERS-1 mission several operational uses of instruments, such as the wind scatterometer have emerged. Several meteorological centres in Europe have found benefits in using the wind scatterometer data as an input to their forecasting models. To support this ESA have developed and commissioned a fast delivery service for key data streams to ensure they are delivered within three hours of an overpass.

The fast delivery service for ERS-1 is an interesting role model for potential operational uses of ocean colour data where the time taken to deliver the data or information to the end user is of specific importance, such as with monitoring HAB. However other operational users of ocean colour data do not always require data or information within hours of the satellite viewing an area in question, such as those studying climate change, coastal erosion, fisheries and water quality based applications. In these areas users require access to data and information within a month of the overpass in order to integrate the results within other planning and decision making processes.

3.2Forthcoming Ground Segment Programmes

In reflecting upon contemporary ground segment programmes it is appropriate to also consider several initiatives that have been started within ESA, the European Commission and other organisations that have the potential to be linked to a European operational infrastructure arising from the OCTOPUS Project. These projects are called the Centre for Earth Observation (CEO), the Envisat ground segment, the Global Ocean Observing System (GOOS) and the Global Climate Observing System (GCOS). In addition to these projects activities are also being undertaken in Japan, with their Earth Observation Information System (EOIS) activity and in the United States by NASA with their ESDIS Core System (ECS) project.

Each of these developments have their own special requirements. It is important to understand some of these in order to understand the key drivers that might emerge for the architectures proposed. One contrast that is particularly important is the emphasis on research users in the United States on the ECS programme - notably in the climate change field, and the focus in Europe on encouraging operational uses of the data and the impact this has on the way in which the CEO project will be undertaken.

These are important influences on the two programmes. They may well result in somewhat different solutions. In the United States the focus is on data integration between archiving centres. Their aim is to create an interoperable architecture in which research scientists can explore linkages and feedback between various elements of the Earth's ecosystem. In contrast in Europe the short term focus is on demonstrations of real world applications of EO data. Current plans for the CEO project envisage a series of pathfinder projects being undertaken in certain key applications areas. The criteria envisaged for pathfinder projects includes bringing together specific user communities with data providers. A classic role model would include a government agency, a value added concern, a systems development company, a data provider and a number of research groups. In this model the government agency would act as the final user, the value added concern would act as a supplier of the information required by the user, the data provider would ensure the availability of data, the research groups would develop the algorithms and techniques required to extract the required geophysical parameters, and the systems development company who would take the algorithms from the research groups and commission these within an operational infrastructure.

A major difference between the ECS and CEO projects is the emphasis on individual pathfinder projects within CEO. The current CEO programme does not, within the Fourth Framework Programme, attempt to design some all-embracing architecture into which all of the users requirements will fall. In our view this approach is quite correct. The diverse range of users needs prohibits a classical systems engineering approach. What is needed is to explore, through the pathfinder projects, a number of classical solutions to connecting users to data and information. Out of this exercise will come several important insights into the architecture and common system elements that might be required in an overall CEO system in the future. It is into this current thinking of the way in which the CEO project will be carried forward that we have to consider the operational infrastructures that would be needed to support the OCTOPUS project.

4. THE IMPLICATIONS FOR EUROPEAN DATA POLICY

The study also considered the implications of developments in data policy for the operational use of ocean colour data. Its clear conclusion is that provision should be made for the reception and processing of SeaWiFS data in Europe. The use of SeaWiFS, both in its own right to encourage research and the application of ocean colour data, and as a precursor to the ESA MERIS instrument will be invaluable.

The data policy considerations also raised some important points in respect of the commercial operation of the SeaWiFS instrument. In order to encourage the use of the SeaWiFS data NASA and Orbital Sciences Corporation (OSC) have established an arrangement whereby nominated Principal Investigators (PIs) will have, in many cases, direct access to the data from the satellite. Many of the PIs are hoping to undertake research work that is linked to the primary mission of the satellite which is climate change research. It is axiomatic that this in itself, per se, will not lead to the development of many commercial uses of the data. It is important to recognise that in the US the primary mission of the satellite and its PIs is directed at climate change research. In Europe the main rationale for accessing the satellite changes depending upon the interest of the user. Many PIs will be using the data for climate research. Indeed the study has confirmed the interest of the European climate research centres in the United Kingdom and Germany in accessing this data for use within their global change models. By contrast, the European Commission requires access to the data to promote research into applications that might mature when the MERIS instrument becomes available on the Envisat programme. This range of interests requires a data policy that can respond to both research uses of a commercial data source for purely research needs and those uses of the data designed to establish the knowledge upon which operational uses of the data might build.

The study has concluded that research and development organisations that are not currently recognised as PIs on the programme will have to be assisted in the short term to secure access to the data. The findings of the applications element of this study show that more research is needed, especially in the field of data fusion, to develop reliable operational applications of ocean colour data. In order to support the research and to encourage the development of applications access to SeaWiFS data by scientific users must, in the short to medium term, be on the basis of the data being free, or at the most, the cost of reproduction. In this respect ESA have an important role to play during their on-going discussions with the satellite operator both over the initial licence that could be granted for access to data and how that access might develop as operational uses of the data emerge from ongoing research activities.

In the short term the data derived from SeaWiFS could be used to establish some pathfinder projects within the framework of the Fourth Framework Programme of the European Commission. These would demonstrate the feasibility of the applications areas thought likely to mature into operational uses of the data. If these pathfinder projects were successful then ESA would have to address the data policy framework required to support operational use of MERIS data in continuing the services. One key element of these considerations would be the policy to be adopted in charging users for data when the main users would be government agencies or departments working in for example fisheries and coastal defence. Another important element would be the potential restrictions that might be placed upon commercial organisations working alongside research groups to develop the infrastructure to support operational uses of the data. The report postulates that commercial concerns could be given the same access to the data enjoyed by research groups whilst they are developing an application and its associated infrastructure. Once developed, the data would have to be paid for using the normal commercial arrangements with the licence holders. The need to address these issues is well understood by ESA as it features as a key point of their considerations for missions beyond Envisat where operational satellites or 'pathfinders' will be one element of the strategy for Earth observation. Equally the European Commission have, through several of their own initiatives, already started to address this important issue.

5. CONCLUSIONS

This paper has considered ways in which contemporary European research into the extraction of geophysical parameters from ocean colour data might be exploited in operational applications. The study has concluded that three areas, monitoring HAB, the supply of global estimates of phytoplankton activity, and monitoring coastal erosion are the applications most likely to have short term operational potential. The study has also highlighted several applications where, with some further research and development, additional operational uses might be discovered, such as in the management of fisheries.

The study has also made recommendations on the architectural concepts for the CEO programme to support a system that is adaptable, evolvable and maintainable. The system concept is based upon the notion of a federated architecture. This federation of data providers and consumers operates on the basis that anybody can join the system and advertise their datasets for other people to use. This includes ground stations with local archives, large central archives - such as those operated by ESRIN, research centres, specialist data centres, processing and archiving centres and many more.

In the course of the study several key points became clear. The most important of these is that it is unlikely that any operational use of ocean colour data will arise using this source alone. The reliability of using ocean colour as a single data source will not be sufficient for operational users. Real operational uses of satellite-based data occur when data derived from the satellite is combined with data and information drawn from other sources, such as Geographic Information Systems (GIS), meteorological data, aircraft data, in-situ observations and observations from other space-based instruments. Through this process of combination, or fusion, of data and information derived from other sources the accuracy of the final information generated for the user will be higher. A key point in this analysis is the local nature of many phenomena that occur in the water. In order to detect these phenomena reliably operational systems will require access to local meteorological data, such as sea surface temperature, salinity levels, wind direction and speed and atmospheric pressure. These observations will enable an information extraction system to minimise any error that might arise in the atmospheric correction process and hence maximise the signal received from the ocean. For this reason the study team have concluded that any operational infrastructure required to support the supply of information derived from space-based ocean colour instruments should be distributed and not centralised. Given that some operational applications will only be of benefit if they can demonstrate the delivery of information to the user within a set period of time it is also important not to introduce any unnecessary delays in the infrastructure.

The study has also noted several important issues that will contribute new perspectives to the debate on data policy within the European Commission. With its short term emphasis on gaining access to ocean colour data through the commercially operated SeaWiFS satellite the OCTOPUS programme will be breaking some important new ground when it is considering near-real time access to ocean colour data.

6. ACKNOWLEDGEMENTS

The EOSL team would like to acknowledge the important contribution made to our study programme by Dr Vittorio Barale of the Joint Research Centre in Ispra and Dr Alan Cross of DGXII, Space Policy Unit, in Brussels. In the course of the study programme two workshops were held at the JRC. They were both chaired by Dr Roland Doeffer of GKSS in Germany. The EOSL team are most grateful to him for his excellent work in the course of these two workshops in bringing together the views of the study team and the visiting scientists. Dr Sloggett would also like to acknowledge the work of Catherine Stock in organising the second workshop in Ispra.

7. REFERENCES

Barale, V., Schlittenhardt, P. (1993) Ocean Colour: Theory and applications in a decade of CZCS experience, Euro-Courses, Remote Sensing, Volume 3, Kluwer Academic Publishers.

Doucet, P., Sloep, P.B., (1992) Mathematical modelling in the Life Sciences, Ellis Horwood Ltd.

Fiuza, A.F.G., (1990) Applications of satellite remote sensing to fisheries, in Operations research and management in fishing, Kluwer academic publishers.

Garcia, S.M., (1992) Ocean fisheries management: The FAO programme, in Ocean Management in Global Change, Edit Fabbri, Elsevier Applied Science, pp 381-418.

GLOBEC (1993) Cod and climate change, Report No. 4.

ICES (1992) Effects of harmful algae blooms on mariculture and marine fisheries, Report No. 181.

Petterson, L.H., (1992) Applications of remote sensing to fisheries, Final project report to the JRC at Ispra, Nansen Research Centre.

Roozekrans (1993), The operational use of NOAA AVHRR image-products in the marine environment, Proc. Int. Symp. Operationalization of Remote SEnsing, 19th-23rd April.

Santos, A.M.P., Fiuza, A.F.G., (1992) Supporting the Portuguese fisheries with satellites, Symposium of the International Space Year, Munich, Germany.

Santos, A.M.P., (1994) The use of ocean colour data in fisheries, in Global Ecology: Volume 1 - Coastal Zones, Praxis Publishing, In print.

Shepherd, J. (1993) Why Fisheries need to be managed and why technical conservation measures on their own are not enough?, Ministry of Agriculture Fisheries and Food, Directorate of Fisheries Research, Laboratory Leaflet No. 71.

Sloggett, D.R., Aikin, J. (1994) The potential for contemporary ocean colour satellite data to play a role in the effective management of fish stocks, Oceanology International '94, Brighton, UK.

Sloggett, D.R., Aiken, J., Boxall, S., Srokosz, M., (1993) OCTOPUS Study Interim Report.

Sloggett, D.R., Aikin, J., Boxall, S., Srokosz, M., (1994) OCTOPUS Study Final Report.

Subramaniam, A., Carpenter, E.J., (1994) An empirically derived protocol for the detection of blooms of the marine cyanobacterium Trichodesmium using CZCS imagery, IJRS, Vol. 15, No. 8, May, pp. 1559-1570.

Sensors and Environmental Applications of Remote Sensing, Askne (ed.) © 1995 Balkema, Rotterdam, ISBN 90 5410 524 0

Use of fluorescence based estimates of phytoplankton production to validate primary production estimates based on remote sensing techniques

M. Olaizola, D.W. van der Linde & P.M. Schlittenhardt
Joint Research Centre, Institute for Remote Sensing Applications, Ispra (Va), Italy

ABSTRACT: Remote sensing of ocean colour offers the possibility of investigating short-term (daily to seasonal) as well as long term (yearly to decadal) changes in phytoplankton primary production over large geographical areas. However, validation of such estimates is difficult since it is based on their comparison with spotty primary productivity measurements made at sea, based on sample incubations with ^{14}C. Fluorescence-based techniques offer the possibility of extending the temporal and geographical coverage of oceanic primary production studies. These estimates may, thus, be more useful in validation studies of primary production models based on remote sensing techniques. Here, we present preliminary results of investigations carried out to determine whether primary production measurements based on fluorescence techniques can be used to validate primary production estimates based on remote sensing algorithms.

1. INTRODUCTION

The usefulness of remote sensing of ocean colour in global change studies resides in the speed and geographical coverage possible in measuring biogeochemically relevant parameters. Because light absorption by phytoplankton pigments (among other things) modulates the spectral shape of ocean reflectance, algorithms have been developed which utilise this spectral information to derive chlorophyll-like pigment concentration (chl_{sat}) for surface waters of the ocean (Fig. 1). Although satellite-based sensors may provide measurements of biogeochemically relevant pools, such as chl_{sat}, on large geographical scales, the sensors can not provide measurements of the reactivity of the measured chlorophyll pool, i.e. carbon flux or primary production. Therefore, the reactivity of the chl_{sat} pool must be modeled.

Early attempts to model primary production using remotely sensed data pointed out some of the difficulties concerning the use of chl_{sat} estimates without the benefit of localised, *in situ*, ancillary data (Eppley et al. 1985). Platt et al. (1991) have developed the concept of biogeochemical provinces which stipulates that average estimates of photosynthetic parameters can be calculated for specific areas of the ocean based on historical data (productivity measurements and vertical chlorophyll distribution). Thus, a catalogue of phytoplankton physiological characteristics can be produced which lists the specific parameters that can be used as input

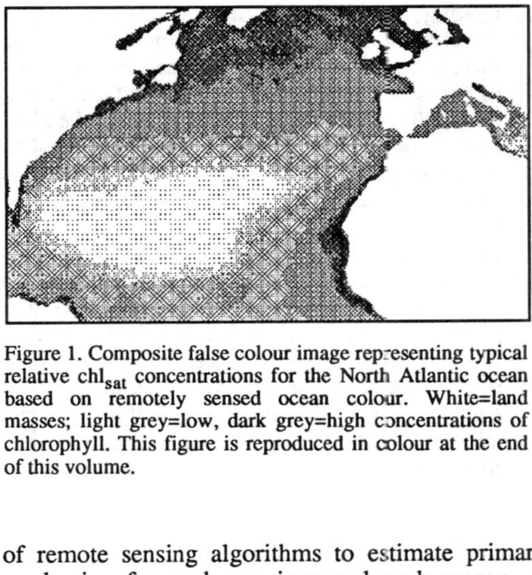

Figure 1. Composite false colour image representing typical relative chl_{sat} concentrations for the North Atlantic ocean based on remotely sensed ocean colour. White=land masses; light grey=low, dark grey=high concentrations of chlorophyll. This figure is reproduced in colour at the end of this volume.

of remote sensing algorithms to estimate primary production for each province and each season. A shortcoming of this approach is that it assumes that surface values of photosynthetic parameters are vertically and horizontally homogeneous for a specific province. However, vertical variations in physiologically relevant parameters have been documented (Kolber et al. 1990, Falkowski et al. 1991, Geider et al. 1993a). Also, the partitioning of the ocean realm into provinces may not be sensitive enough to small scale, transient, changes in oceanographic conditions which may produce changes in phytoplankton physiology and, thus, primary production (Falkowski et al. 1991, Geider et al. 1993a).

Other attempts to estimate primary production based on remote sensing algorithms involve the calculation of new primary production using compound remote sensing techniques (Dugdale et al. 1989, Sathyendranath et al. 1991) which take advantage of the relationship between surface water temperature and nitrate concentration. Nitrate, a necessary phytoplankton nutrient, is generally supplied to the mixed layer from below the thermocline. Thus surface layer temperature works as a tracer of nitrate concentration. A shortcoming of this approach is that it may work only in upwelling areas (Dugdale et al. 1989) or other vertically dynamic areas (Sathyendranath et al. 1991) were a clear temperature signal may be detected and related to influx of deep, nutrient-rich water.

Independently of the model used to estimate primary production from remote sensed data, it must be validated with *in situ* data. However, validation of such estimates is difficult since it is based on their comparison with spotty primary productivity measurements made at sea. Until now, measurements at sea have been obtained by either conducting time consuming incubations of phytoplankton samples with ^{14}C *in situ* (or under simulated *in situ* conditions) or by modeling primary productivity based on photosynthesis *versus* irradiance (P *vs.* I) measurements done in the ship's laboratory, which are faster but not *in situ*. Both of these methods have the added inconvenient of artefacts produced by containment of natural samples.

Here we report results of oceanographic campaigns in which extensive measurements of phytoplankton fluorescence were carried out. Our objective was to determine if estimates of photosynthetic parameters based on measurements of phytoplankton fluorescence are comparable to estimates obtained using the standard ^{14}C incubation technique and whether these estimates can be used in validation efforts of remote sensing algorithms.

2. METHODOLOGY

2.1 *Phytoplankton fluorescence*

The theoretical and practical aspects of the application of fluorescence techniques to measure phytoplankton photosynthesis have been recently discussed by Kiefer and Reynolds (1992) and Kolber and Falkowski (1993). In short, phytoplankton fluorescence represents one of three pathways (fluorescence, thermal de-excitation and photosynthesis) available for energy dissipation in the antenna of photosystem II (Fig. 2). The basic premise is that by measuring changes in fluorescence yield, and with basic knowledge of the thermal loss term (also called non-photochemical quenching), changes in the quantum yield of photochemical energy conversion can be estimated. Specifically, we have measured changes in the so-called F_v/F_m fluorescence ratio, which reflect changes in the quantum yield of photochemistry. These measurements can be done *in situ* very rapidly (within a few seconds) and results obtained in near real time. Because of the lack of a time constrain, as is the case for incubation-based techniques, increased geographical coverage as well as increased sampling frequency are possible.

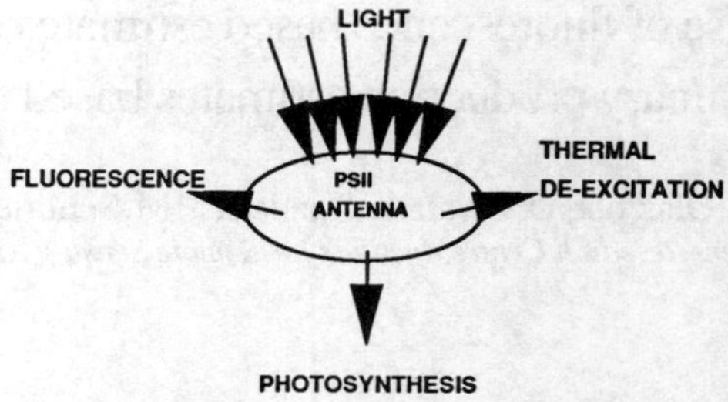

Figure 2. Schematic representation of the three competing paths that lead to dissipation of absorbed excitation energy in photosystem II.

2.2 *Instrumentation*

The results reported here were obtained using two different types of fluorometers: a pump-and-probe (P&P) fluorometer and a pulse amplitude modulated (PAM) fluorometer. Both instruments have been described in the literature (Kolber et al. 1988, Schreiber et al. 1986).

3. RESULTS

During the past year we have conducted several investigations in which phytoplankton primary productivity estimates based on fluorescence techniques have been obtained in parallel with conventional ^{14}C measurements. Here we present preliminary results of two of these campaigns, in the North Atlantic ocean and in the Southern Baltic sea.

3.1 *North Atlantic Fluorescence Experiment*

The North Atlantic Fluorescence Experiment (NAFE) was conducted between 13 May and 9 June,

1993. It was a collaborative project between the Institute for Remote Sensing Applications (Ispra) and the Bedford Institute of Oceanography (BIO, Canada). The cruise consisted of two parts, a detailed study of the upwelling area off NW Africa (to be reported elsewhere) and a transect across the North Atlantic ocean (Fig. 3). The transect lasted 11 days, from 28 May to 6 June 1993. Thus, we have obtained a quasi synoptic map of vertical and cross-ocean variations in the quantum yield of photochemistry.

We measured the F_v/F_m ratio of phytoplankton using a prototype P&P fluorometer. Because this was a benchtop model, our measurements could not be done *in situ* but on samples (dark adapted, >1 hr) collected with a rosette from throughout the euphotic zone. Parallel ^{14}C incubations were carried out to obtain primary production estimates with which to compare the fluorescence data. As of this time, the ^{14}C data has not been worked out so a comparison between the two methods is not available and only the fluorescence data will be presented here.

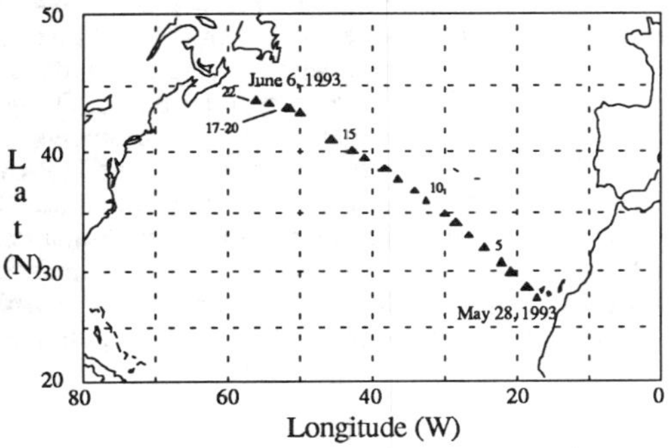

Figure 3. Chart of the North Atlantic ocean showing the location of the stations occupied during the North Atlantic Fluorescence Experiment (black triangles).

The eastern basin of the North Atlantic is dominated by the return circulation of the North Atlantic Gyre while the western basin is dominated by the circulation of the Gulf and Labrador currents. Although we found no evidence of the Gulf current during our campaign we did come across the cold water of the Labrador current (Fig. 4) near the Canadian continental shelf. We found a strong basin-wide longitudinal gradient in nitrate and chlorophyll concentrations (Fig. 4). The eastern side of the basin had lower nitrate and chlorophyll concentrations than the western basin (see also Fig. 1). The depth of the region of maximum chlorophyll concentration closely followed the depth of the nitracline, the depth at which the nitrate

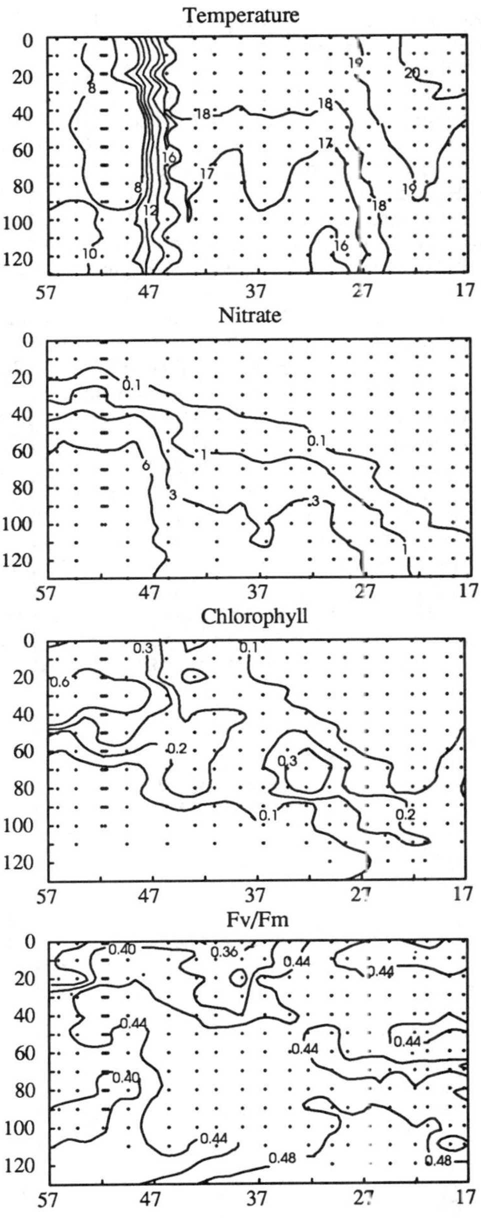

Figure 4. Vertical distribution of water temperature (°C), nitrate concentrations (μM), chlorophyll concentration μg/L), and the quantum yield of photochemistry (as F_v/F_m) measured during the transect across the North Atlantic ocean.

concentration increases rapidly, from about 100 m (east) to 35 m (west).

The range of measured values of F_v/F_m ranged between 0.25 and 0.50. These are typical values as obtained using pump-and-probe fluorometry (Kolber et al. 1990, Falkowski et al. 1992, Geider et al. 1993a). Based on the geographical distribution of F_v/F_m values (Fig. 4) we can divide the North Atlantic into three separate areas. The eastern basin, from 17° to about 32° W, is characterised by a weak vertical gradient in F_v/F_m values. Higher values are encountered at depth, near the nitrate source for the photic zone. This relationship between photosynthetic efficiency and nitrate concentration has been already described for algal cultures (Kolber et al. 1988, Geider et al. 1993b) and natural phytoplankton communities (Kolber et al. 1990, Falkowski et al. 1991, Geider et al. 1993a). The western basin, from 47° to 57° W near the Canadian shelf, can be, itself, divided into a shallow and a deep region. The shallow region, from the surface down to the bottom of the chlorophyll maximum depth (~50 m) is also characterised by a direct relationship between F_v/F_m values and nitrate concentration. Below the chlorophyll maximum depth, however, F_v/F_m values decreased although the nitrate concentration continued to increase. It is expected that phytoplankton cells encountered below the chlorophyll maximum depth are sinking out of the photic zone and probably represent, at least in part, detrital (dead) material producing the decrease in F_v/F_m. Between the two, previously described, basins a third area can be described from 35° to 45° West. A relationship between F_v/F_m and nitrate supply was also apparent in this area. The surface F_v/F_m values measured here were the lowest of the whole North Atlantic although there was not an apparent change in hydrographic conditions (Fig. 4). Phytoplankton light absorption measurements carried out in this area suggested the presence of detrital material (increased absorption between 400 and 450 nm) which may represent a decaying surface bloom (Olaizola et al. 1994). Our results from this campaign support the hypothesis that nutrient supply may control the photosynthetic performance of phytoplankton in the North Atlantic ocean, as previously suggested by Platt et al. (1992).

3.2 *Baltic Sea Fluorescence Experiment*

The Baltic Sea Fluorescence Experiment (BASFE) was conducted between March 1 and 10, 1994. It was a collaborative project between the Institute for Remote Sensing Applications (Ispra) and the Institute for Baltic Sea Research (Germany). During this cruise, we made extensive measurements of phytoplankton fluorescence using the PAM fluorometer. As this is also a benchtop instrument, the measurements were again made on dark-adapted (>1 hr) samples collected with a rosette. Measurements were done along transects as well as at fixed positions over a diel cycle. A limited

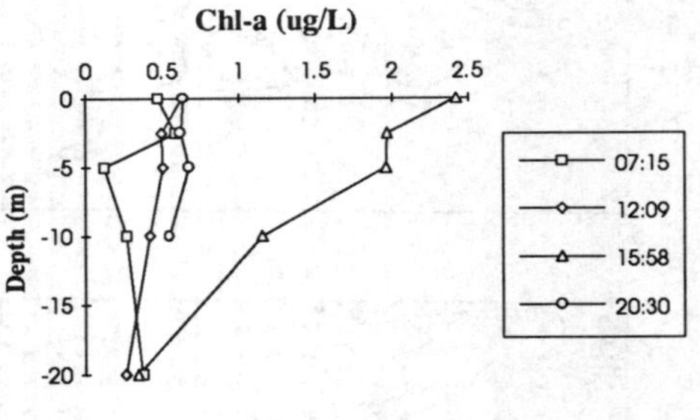

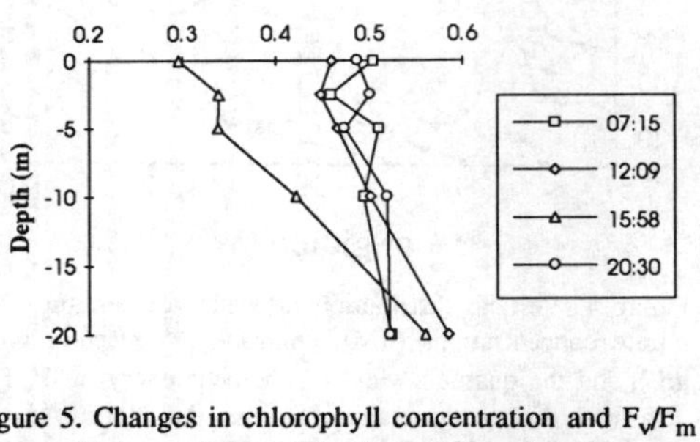

Figure 5. Changes in chlorophyll concentration and F_v/F_m over a 14 hr period on March 4, 1994. Samples were collected along a transect east of Bornholm Island from 55°28.63N 15°28.57E to 54°51.82N 16°18.01E.

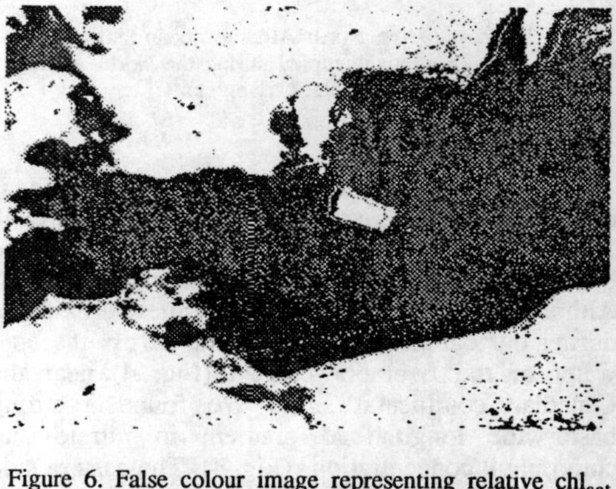

Figure 6. False colour image representing relative chl_{sat} concentrations for the Southern Baltic on March 13, 1983, based on remotely sensed ocean colour. White=land masses, light grey=low concentrations, dark grey=high concentrations of chlorophyll. This figure is reproduced in colour at the end of this volume.

amount of data is presented here to show that the fluorescence measurements provided evidence of rapid changes in the photosynthetic efficiency of phytoplankton (within a few hours).

On March 4 we run a transect east of Bornholm Island from 55°28.63N 15°28.57E to 54°51.82N 16°18.01E. We made fluorescence measurements at four different stations along the transect. We found large changes in both biomass (5 times) as well as F_v/F_m values (Fig. 5). In fact, we found more variability in F_v/F_m within a few kilometres in the Southern Baltic as in the whole North Atlantic (Fig. 4). We have looked for evidence of such small scale variability in biological parameters in the southern Baltic Sea in CZCS images of the Ocean Colour European Archive Network (OCEAN). Because of cloud cover, few good images of this area exist. However, we have found evidence for north-to-south filament-like structures in ocean colour of an appropriate scale to explain the variability that we measured in chlorophyll concentration (Fig. 6). The lowest F_v/F_m value was measured in the surface sample that had the highest chlorophyll concentration. This could be evidence for localised nutrient limitation produced by phytoplankton uptake or light-induced photoinhibition since these

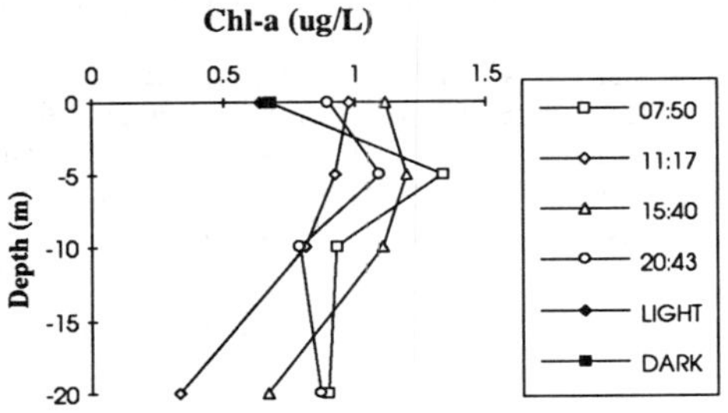

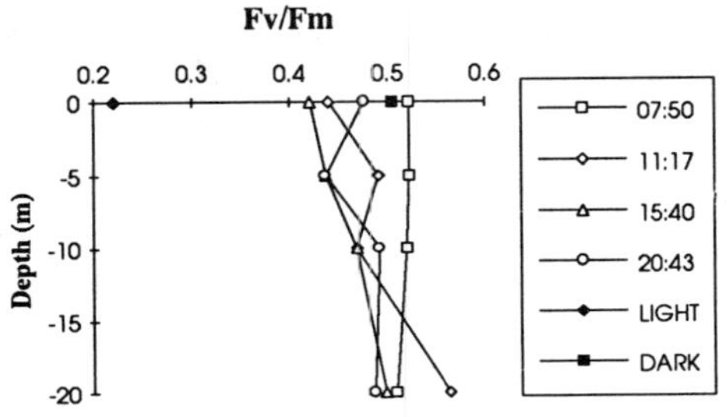

Figure 7. Changes in chlorophyll concentration and F_v/F_m over a 13 hr period on March 9, 1994. Samples were collected while the ship was anchored at 54°55.66N 13°31.15E.

samples were collected in mid afternoon.

On March 9 we occupied a diel station, while the ship was anchored. Samples were collected on 4 different occasions during the day in the early morning, noon, mid-afternoon and evening (Fig. 7). Surface samples collected in the morning were also incubated floating in the surface, either under ambient light (LIGHT) or wrapped in black plastic (DARK) and recovered at the same time as the samples collected in mid-afternoon. Our results indicated changes in chlorophyll concentration of up to 100% during the course of the station, presumably due to horizontal movement of the water mass past the ship. This would indicate that the finger-like structures in surface chlorophyll detected in the CZCS images (Fig. 6) may move horizontally, perhaps due to tidal or wind forcing. The surface F_v/F_m values ranged between 0.42 and 0.52, showing a slight decrease in the mid-afternoon samples. The F_v/F_m measured in the sample incubated under ambient light (0.22, Fig. 7, black diamond) indicated photoinhibition produced by the high illumination. Presumably vertical mixing in the water column did not permit such strong photoinhibition to develop in the absence of containment. Previous investigations have also shown differences in photoadaptation parameters in natural populations of phytoplankton held in bottles versus freely mixing populations (Olaizola et al. 1992). If we consider that bottle incubations for ^{14}C productivity measurements are usually inoculated with samples collected in the morning and placed at fixed depths for several hours, we can deduce that these incubations do not reproduce the true *in situ* conditions of the phytoplankton nor, thus, the *in situ* primary production of the phytoplankton community.

4. CONCLUSIONS

With the advent of commercially available *in situ* fluorometers that are capable of making the necessary measurements, estimates of primary productivity *in situ*, in real time, will soon become routine. We have shown here that fluorescence-based techniques can be used to provide estimate of changes in primary productivity parameters over large geographical scales (NAFE) as well as over short temporal scales (BASFE). Thus, proper validation studies of primary productivity models based on remote sensing techniques will be possible when new platforms become operational (i.e. SeaWIFS, OCTS, MERIS).

5. ACKNOWLEDGEMENTS

This work was supported in part by the Commission of the European Communities (CEC). The CZCS-derived composite image was obtained from NASA, Goddard Space Flight Center (Fig. 1). The CZCS-derived image of the Southern Baltic was provided by the Ocean Colour European Archive Network (OCEAN) Project (Fig. 6). The temperature, nitrate and chlorophyll data for the North Atlantic was provided by the Bedford Institute of Oceanography. Chlorophyll analysis (HPLC) for the Southern Baltic were conducted by M. Anasagasti. The authors thank P.G. Falkowski and Z. Kolber for the loan of the prototype pump-and-probe fluorometer.

Colour plate Figure 1 see page 497
Colour plate Figure 6 see page 498

6. REFERENCES

Dugdale, R.C., A. Morel, A. Bricaud, and F.P. Wilkerson. 1989. Modeling new production in upwelling centers: A case study of modeling new production from remotely sensed temperature and color. *J. Geophys. Res.* 94: 18119-18132.

Eppley, R.W., E. Stewart, M.R. Abbott, and U. Heyman. 1985. Estimating ocean primary production from satellite chlorophyll. Introduction to regional differences and statistics for the Southern California Bight. *J. Plank. Res.* 7: 57-70.

Falkowski, P.G., R.M. Greene, and R.J. Geider. 1992. Physiological limitations on phytoplankton productivity in the ocean. *Oceanography*, 5: 84-91.

Falkowski, P.G., D.A. Ziemann, Z. Kolber, and P.K. Bienfang. 1991. Role of eddy pumping in enhancing primary production in the ocean. *Nature,* 352: 55-58.

Geider, R.J., R.M. Greene, Z. Kolber, H. MacIntyre, and P.G. Falkowski. 1993a. Fluorescence assessment of the maximum quantum efficiency of photosynthesis in the western North Atlantic. *Deep-Sea Res.* 40:

Geider, R.J., J. LaRoche, R.M. Greene, and M. Olaizola. 1993b. Response of the photosynthetic apparatus of *Phaeodactylum tricornutum* (Bacillariophyceae) to nitrate, phosphate, or iron starvation. *J. Phycol.* 29: 755-766.

Kiefer, D.A. and R.A. Reynolds. 1992. Advances in understanding phytoplankton fluorescence. In, P.G. Falkowski & A.D. Woodhead (eds), *Primary Productivity and Biogeochemical Cycles in the Sea*: 155-174. New York: Plenum Press.

Kolber, Z. and P.G. Falkowski. 1993. Use of active fluorescence to estimate phytoplankton photosynthesis in situ. *Limnol. Oceanogr.* 38: 1646-1665.

Kolber, Z., K.D. Wyman, and P.G. Falkowski. 1990. Natural variability in photosynthetic energy conversion efficiency: A field study in the Gulf of Maine. *Limnol. Oceanogr.* 35: 72-79.

Kolber, Z. J. Zehr, and P.G. Falkowski. 1988. Effects of growth irradiance and nitrogen limitation on photosynthetic energy conversion in photosystem II. *Plant Physiol.* 88: 923-929.

Olaizola, M. P.K. Bienfang, and D.A. Ziemann. 1992. Pigment analysis of phytoplankton during a Subarctic spring bloom: xanthophyll cycling. *J. Exp. Mar. Biol. Ecol.* 158: 59-74.

Olaizola, M., G.M. Ferrari, and M. Anasagasti. 1994. Phytoplankton physiological status affects water column light absorption: Applications to remote sensing. *Ann. Geophysic.* 12 (Sp. II): 353.

Platt, T., S. Sathyendranath, O. Ulloa, W.G. Harrison, N. Hoepffner, and J. Goes. 1992. Nutrient control of phytoplankton photosynthesis in the Western North Atlantic. *Nature,* 356: 229-231.

Platt, T., C. Caverhill, and S. Sathyendranath. 1991. Basin-scale estimates of oceanic primary production by remote sensing: The North Atlantic. *J. Geophys. Res.* 96: 15147-15159.

Sathyendranath, S., T. Platt, E.P.W. Horne, W.G. Harrison, O. Ulloa, R. Outerbridge, and N. Hoepffner. Estimation of new production in the ocean by compound remote sensing. *Nature,* 353: 129-133.

Schreiber, U., U. Schliwa, and W. Bilger. 1986. Continuous recording of photochemical and non-photochemical chlorophyll fluorescence quenching with a new type of modulation fluorometer. *Photosynth. Res.* 10: 51-62.

Joint Research Center-Institute for Remote Sensing Applications-Marine Environment Unit

Sensors and Environmental Applications of Remote Sensing, Askne (ed.) © 1995 Balkema, Rotterdam, ISBN 90 5410 524 0

Bathymetry measurement in shallow water areas using airborne spectrometers

J. Bijaoui, S. Jullien & F. Cauneau
Centre d'Énergétique, Groupe Télédétection & Modélisation, École des Mines de Paris, France

ABSTRACT: Multispectral data of shallow water areas are used in this study to obtain bathymetry in moderate turbid waters and with various sea bottom reflectances. A new algorithm make use of *a priori* knowledge of different classes of water contents and bottom types. The retrieved classes appear to be in accordance with *in situ* data.

1 BATHYMETRY FROM SPACE

For seas whose bottom presents a high spatial and temporal variability, it is of a great interest to get at regular intervals accurate measurements of the water depth. It is generally quite difficult and of a high cost to get *in situ* data, and to obtain by this way a high spatial coverage. The basic interests to make use of remote sensing data for such applications are a lower cost and a great spatial covering with a high spatial resolution.

Optical radiometers like SPOT-HRV or Landsat-TM offer good possibilities for monitoring shallow waters. The main limitation is due to both the small number of spectral bands and the limited radiometry accuracy.

Because of their greater number of layers in the optical band, and especially in the blue, spectrometers are more adapted to analyse bathymetry and sea bottom. Actually, some airborne spectrometers like the Geographical Environmental Research (GER) instruments, used in this study, offer the possibility to evaluate the bring of such sensors.

Most of the algorithms developed for the bathymetry retrieval in remote sensing images are limited to the clear water cases. More generally, the basic difficulty for this kind of applications is to detect the presence of turbidity, and, in the case of moderately turbid water, to apply an appropriate correction.

In the present work, we develop an algorithm based on the modelisation of the inverse of the radiative transfert in the water column. Different classes of water contents, bottom reflectances and depths try to obtain a relevant solution. This approach, which is reserved to spectrometers, could offer good synergy with the other models using radiometers.

After a short review of the bathymetry algorithms using space borne radiometers, we present our method tested in the lagoon of Venice.

2 BATHYMETRY FROM SPACE-BORNE RADIOMETERS

Polcyn *et al.* (1980) computes the bathymetry of shallow waters with the help of a spectral band ratio. The layers must be (R_i) chosen between the blue and green wavelength (because of the deepest light penetration in water). Also, this algorithm needs *in situ* measurements of the attenuation coefficients (c_i) and the bottom reflectances (R_{bi}) for each spectral bands. If we suppose these coefficients constant over the study area, for each image pixel the depth is given by:

$$z = \frac{1}{2(c_1 - c_2)} \ln \frac{R_2 R_{b1}}{R_1 R_{b2}}.$$

Over a great area, this algorithm gives poor results because of the high variability of the bottom reflectance and of the attenuation coefficient.

Bierwith *et al.* (1993) developed an algorithm to retrieve the sea bottom reflectance. Let us consider Z as a mean depth computed over N spectral bands:

$$Z = \frac{1}{N} \sum_{i=1}^{N} \frac{\ln R_i}{-2c_i},$$

where R_i are the observations corresponding to an image pixel. If we suppose the attenuation coefficient

c_i for each band constant over the study area, the bottom reflectance R_{bi} for a spectral band i is given for each pixel by:

$$R_{bi} = R_i e^{2c_i Z}.$$

This correction applied to the Polcyn algorithm leads to important improvements of the accuracy. When waters present some turbidity, this method should be used with precautions. If no information is available on the nature and concentration of suspended matters, we advise to evaluate the occurrence of turbidity by the means of a turbidity index (see, *e.g.* Pages *et al.* 1990).

The limitation of such approach is that there is no objective criterion to assess the actual optical depth of the water through radiometric measurements. The application of the previous algorithms may be particularly wrong when the turbidity is highly variable.

3 THE BRING OF SPECTROMETERS

The basic improvements brought by spectrometers reside in both the increasing number of independent spectral measurements for each image pixel and their high radiometric accuracy. For each point of the study area, it is necessary to consider the turbidity vertically constant.

For a given area, the radiative transfert equation is then inversed in the following way. We build a data base containing the attenuation coefficients corresponding to different classes of suspended matters. The spectral reflectances corresponding to different cases of sea bottom are added, as different classes of bathymetry. For each combination of this data base, the algorithm compares a theoretical reflectance (from data base) to an observed one (from image).

The method is tested with the data of the EISAC 1989 campaign, over the lagoon of Venice. The spectrometer used is a GER instrument. The acquisition date is the 25 May 1989. *In situ* data were also provided over six stations. An extensive bathymetry mapping (Alberotanza, 1989) is also taken.

The plate 1 shows the first band of the spectrometric image (wavelength: 477*nm*, spectral band: 12.3*nm*, size: 1024 lines × 416 columns). The pixel size is not exactly defined, because of the wide acquisition geometry. At the top of the image, we can see the coast near Mestre, and part of Venice along the right side. At the bottom, the Lido island is clearly visible.

The approximate water depths in the area are weak (1*m* - 2*m*), except for the navigation channels, where it can reach 10*m* (Plate 2). Despite the fact that the waters of the lagoon are turbid, we can identify some bottom features like the navigation channels at the bottom right of the image.

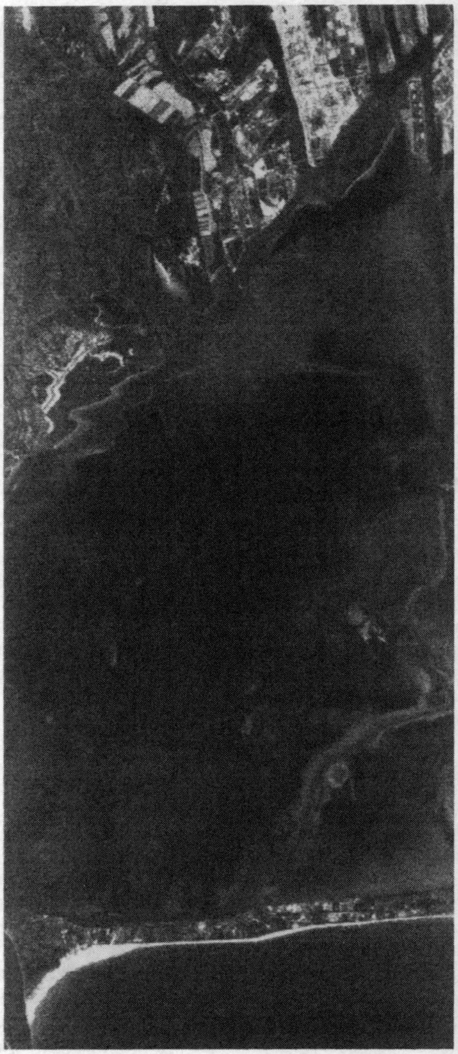

Plate 1. Spectrometric image of the lagoon of Venice.

According to the *in situ* informations, an image classification using the first 31 spectral bands (wavelengths from 477*nm* to 848*nm*) of the GER instrument is achieved. A data base is built, using three classes of sea bottom (sand, mud and algae), three classes of water content (high turbidity, sestons with chlorophyllians, clear water) and five classes of bathymetry ([0.0*m* ; 0.3*m*], [0.3*m* ; 0.6*m*], [0.6*m* ; 1.0*m*], [1.0*m* ; 2.0*m*], [2.0*m* ; 3.0*m*]).

The result of the classification is shown in Plate 3. The algorithm found five combinations (see Table 1 for explanations).

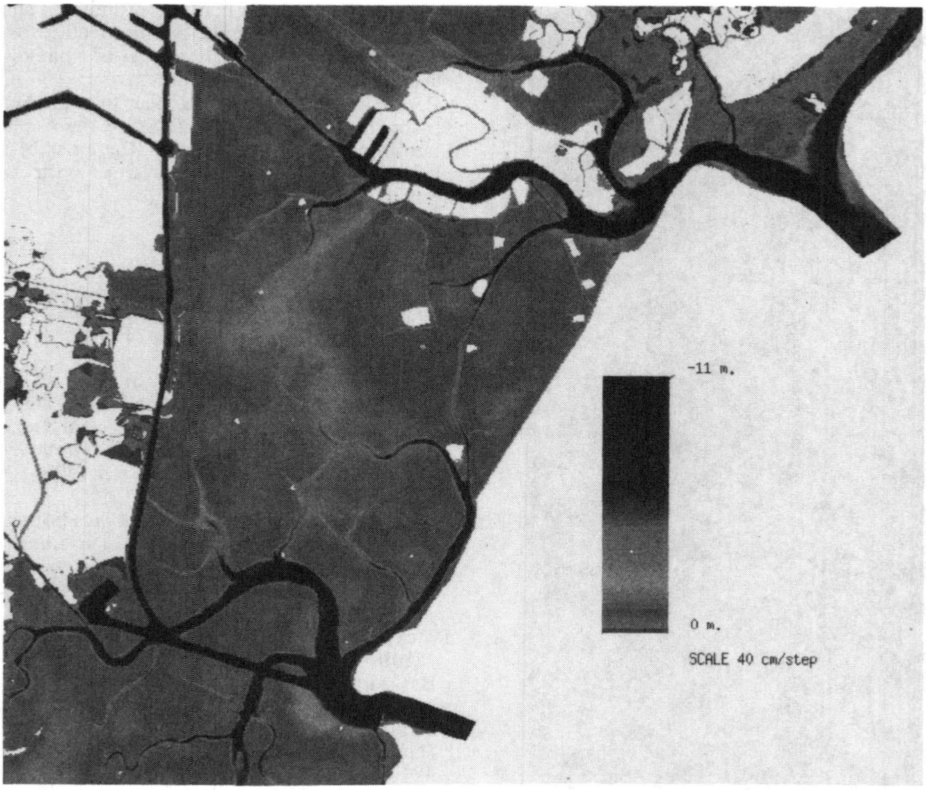

Plate 2. Representation of the bathymetry in the study area.

Table 1 : Description of the retrieved cases. The given codes correspond to those of Plate 3.

Code	Bottom	Water content	Water depth
3	algae	chlorophyllians sestons	[1.0*m* ; 2.0*m*]
1	sand	chlorophyllians sestons	[1.0*m* ; 2.0*m*]
2	mud	highly turbid	[0.3*m* ; 0.6*m*]
4	mud	chlorophyllians sestons	[1.0*m* ; 2.0*m*]
5	sand	highly turbid	[0.3*m* ; 0.6*m*]

Despite the lack of extended *in situ* data, the retrieved water contents are in accordance with the data of the stations, as well as the bottom reflectances. The retrieved bathymetry is roughly in agreement with the reference chart, mainly in the eastern part of the image. In the remaining area, the classification leads to errors (Plate 3, class 4 on the left of the image, in white), mainly due to the fact that water is opaque. This point should be improved by taking in account the reflectance of the water column.

4 CONCLUSION AND PERSPECTIVES

We have presented a method which use the images of an airborne spectrometer tc retrieve the bathymetry in turbid waters. As previous algorithms

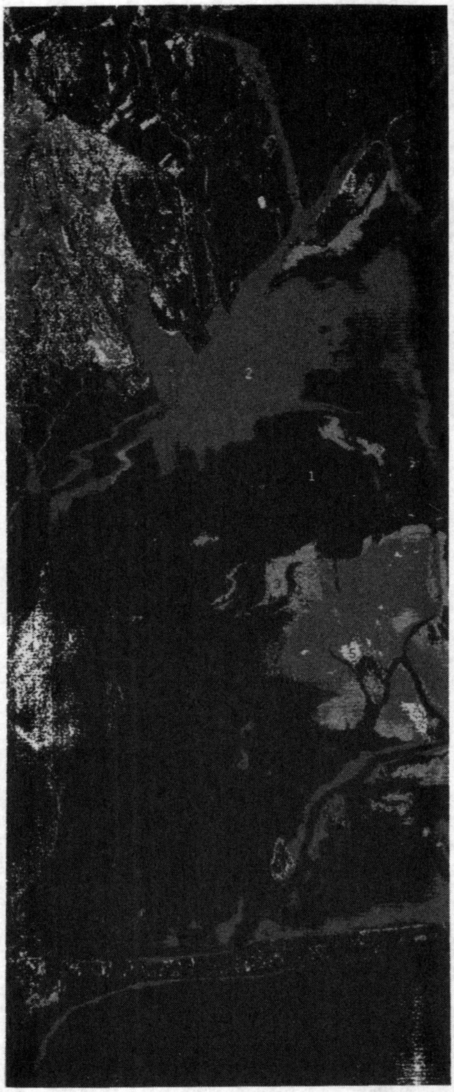

Plate 3. Classification of the spectrometric image. Five classes are retrieved by our algorithm. See Table 1 for the significations of each class.

do not allow to take in account the effects of the high variability of both water turbidity and bottom reflectance with radiometers, we show that great improvements can be expected by the use of spectrometers in such applications. Airborne spectrometers, such as GER instruments, provide interesting results.

Our algorithm is based on the *a priori* knowledge of the spectral characteristics of bottom substrats, water contents and the range of the expected bathymetry.

The results obtained here are promising, showing that this algorithm would be improved, especially with a proper modelisation of backscatter phenomenons in the water column.

Perspectives offered by the near to come SeaWifs sensor, and the future MERIS are then enlarged. Such sensors would provide the possibility to monitor large shallow water areas with a great accuracy.

REFERENCES

Alberotanza L. "Sea-Truth Data Catalogue." *EISAC Campaign, North Adriatic Sea,* 1989.

Bierwirth, N.P., T.J. Lee and R.V. Burne. "Shallow Sea-Floor Reflectance and Water Depth Derived by Unmixing Multispectral Imagery", *Photogrammetric Engineering & Remote Sensing*, vol. 59, no. 3, pp. 331-338, March 1993.

Pages, J., B. Mougenot, J. Citeau. "Radiométrie de terrain pour le satellite SPOT sur un estuaire tropical", *Bulletin de l'Institut Océanographique de Monaco,* special no. 6, pp. 97-122, 1990.

Polcyn, F.C., D.R. Lysenga, R.A. Schuman. "Multisensor Approach to Remote Sensing Bathymetry", *Proceedings of the Fourteenth International Symposium of Remote Sensing of Environment. Environmental Research Institute Publisher, Mi. Po BOX 8618, Ann Arbor Mi.* 48107, vol. 1, p. 525, 1980.

Sensors and Environmental Applications of Remote Sensing, Askne (ed.) © 1995 Balkema, Rotterdam, ISBN 90 5410 524 0

Interannual variations in the extent of surface cyanobacterial accumulations in the Baltic Sea detected by satellites

Ove Rud & Mati Kahru
Department of Physical Geography, Stockholm University, Sweden

ABSTRACT

In the Baltic Sea regularly occuring cyanobacterial accumulations (blue-green algae blooms) pose a major environmental and even health hazard. Satellite data between 1982 and 1993 has been processed and analysed for the occurence of surface cyanobacterial accumulations. During the investigated time period an increase in the area covered by accumulations has been detected, in 1992 reaching over 62 000 km^2. Conspicuous surface accumulations were also present in the early 1980s, coincident with a period of sunny and calm summers. As cyanobacteria play a major role in the Baltic ecosystem, a significant increase in the amount of cyanobacterial accumulations can represent a structural shift in the planctonic communities. For example, increased cyanobacterial blooms in the Baltic mean increased input of atmospheric nitrogen (via N_2 fixation by cyanobacteria) and the consumption of deep water supplied phosphorus, leading to higher levels of eutrophication. The question which environmental factors have caused the increase in the area covered by cyanobacteria remains open. Obvious candidates for the causing factors are increased riverine phosphorous input and/or changes in the phosphorous to nitrogen ratio.

1. INTRODUCTION

Although the first reports of dense cyanobacterial blooms date back to the middle of the 19th century (Wallström 1991), it has been suggested that the extent and intensity of the blooms has increased due to anthropogenic eutrophication (Horstmann 1975). However, due to the inherent spatial and temporal variability, the long-term changes are very difficult to prove using data from conventional shipboard monitoring.

At some stage of the bloom cyanobacterial filaments become positively buoyant and flocculate near the sea surface in a process called inverted sedimentation (Horstmann et al. 1986). At low wind speeds these cyanobacterial agglomerates accumulate at the surface and become clearly visible even on low-sensitive satellite imagery (Öström 1976). Despite the photo-oxidation of chlorophyll at high solar irradiance, the cyanobacterial surface accumulations still indicate increased water column chlorophyll concentration (Kahru et al. 1993). Various satellite imagery of the Baltic Sea (Landsat MSS and TM, CZCS, NOAA/AVHRR) is available since the mid-1970s and could be used for detecting the dynamics of cyanobacterial blooms during the last 15-20 years. Especially useful in that respect is the AVHRR which has wide swath width, frequent coverage, has been flown on the NOAA series of satellites with little modifications since 1979, and is comparatively inexpensive to process - all these characteristics are important in creating a high-volume, long-term data base.

SATELLITE DATA

A total of 135 suffiently cloud free noon passes were digitally processed, and for each year maps of the cumulative area as well as the total area covered by the accumulations were compiled. The main part of the analysis have been based on data from the Advanced Very High Resolution Radiometer (AVHRR) flown on the NOAA series of satellites. Data from the Coastal Zone Colour Scanner (CZCS) and Thematic Mapper (TM) sensors, have also been used to identify periods of accumulations and to study them in more detail. The algorithm for detecting cyanobacterial accumulations relies on their increased reflectance with high spatial structure. A number of procedures using AVHRR channels 1,2,4 and 5 were used to eliminate clouds and other disturbances.

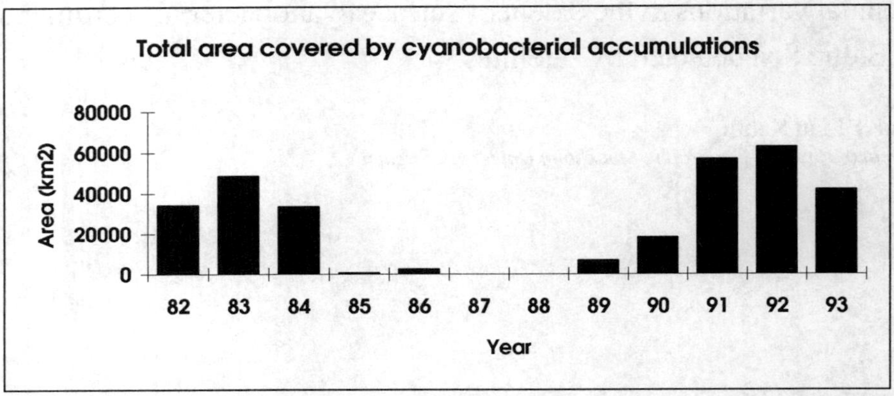

Fig. 1. Interannual dynamics of the total area covered by satellite detected cyanobacterial accumulations in the Baltic Sea.

INTERANNUAL VARIATIONS

The result is based on a twelve year time serie of AVHRR imagery from 1982 to 1993. Analysis of pre- 1982 data and other sensors (CZCS, 1980-1985) is in progress.

As shown in Fig. 1 the extent of cyanobacterial accumulations has varied drastically during the twelve years. Two periods of increased cyanobacterial coverage can be detected: from 1982 to 1984 and from 1990 to 1993. As the weather during July-August 1993 was unfavourable for the accumulations to appear at the surface, the 1993 value can probably be regarded as a fraction of the potential area that would have been affected in the case of extended calm and sunny weather. Although the coverage of years 1982 to 1990 in terms of the number of processed images is not as good as of the last 3 years, and more processed images may add to these yearly total areas, the large variations is evident and may indicate significant shifts in the structure and functioning of the Baltic Sea ecosystem. The question what causes the variations in the area covered by cyanobacterial accumulations remains open.

Another shift that can be noted is the appearance of the accumulations in the Gulf of Riga and in the western Gulf of Finland in 1992 an 1993 (Fig. 2). Whereas the accumulations in the open Baltic Sea are formed by Nodularia spumigena, in the Gulf of Riga they are dominated by Aphanizomenon and Microcystis (A. Andrushaitis, pers. comm.).

In 7 years during our time series the accumulations have been recorded in the Hanö Bight but only twice in the Bay of Gdansk. In controlling distributions like these besides the macronutrients of P and N, the distributions of trace metals leaching from terrestrial areas may be involved.

CONCLUSIONS

Historical NOAA/AVHRR data is a good data source when studying large scale, rapidly changing cyanobacterial blooms in the open sea. Together with CZCS data enough registrations can be compiled to reveal the annual variations of surface cyanobacterial coverage. A full understanding of the mechanism behind the algal bloom variations requires an integrated approach between remote sensing and marine biological and ecological sciences. Satellite based monitoring of the distribution of cyanobacterial accumulations can be a valuable operational tool for environmental monitoring, especially when adequate knowledge about the causing mechanisms becomes available.

During the 1994 summer a test will be conducted where locally received AVHRR images will be processed and interpreted, and then transfered as accumulation maps to different information centres both in Sweden and Finland.

REFERENCES

Horstmann, U. 1975. Eutrophication and mass occurrences of blue-green algae in the Baltic. *Merentutkimuslait. Julk./Havsforskningsinst. Skr.* 239: 83-90.

Horstmann, U., H. van der Piepen & K.W. Barrot 1986. The influence of river water on the southeastern Baltic Sea as observed by Nimbus 7/CZCS imagery. *Ambio* 15: 286-289.

Kahru, M., J.-M. Leppänen,& O. Rud 1993.

Cyanobacterial blooms cause heating of the sea surface. *Marine Ecology Progress Series* 101: 1-7.
Wallström, K. 1991. Ecological studies on nitrogen fixing blue-green algae and on nutrient limitation of phytoplankton in the Baltic Sea. Ph.D. Thesis, Uppsala University, 23pp.
Öström, B. 1976. Fertilization of the Baltic by nitrogen fixation in the blue-green algae Nodularia spumigena. *Remote Sensing of Environment* 4: 305-310.

Sensors and Environmental Applications of Remote Sensing, Askne (ed.) © 1995 Balkema, Rotterdam, ISBN 90 5410 524 0

Thermal plume monitoring from ATM data

R.A.Vaughan, Chan-Chuan Wen & W.Mansell
Department of Applied Physics and Electronic & Mechanical Engineering, University of Dundee, UK

ABSTRACT: Data from the Daedalus AADS-1268 Airborne Thematic Mapper (ATM) was used to study the hydrodynamic characteristics of a thermal plume discharged by the Central Termica Litoral de Almeira power plant located on the southern coast of Spain. Several sets of images taken from different heights were analysed, and two hydraulic models were applied in order to calculate the mixing coefficients of the thermal plume in the transverse direction. The total extent of the affected surface area was also calculated. The hydraulic characteristics of the plume under different sea conditions were studied and the analysis revealed that, in the near field, the mixing process and the dispersion of the plume were quite similar under different sea conditions, whereas in the far field the mixing process in rough sea conditions was several times stronger than in a calm sea state. The plume was dispersed to a large extent, but was trapped around the coast of Carbonaras village in the presence of a south westerly wind.

1. INTRODUCTION

Coastal waters constitute useful and convenient heat sinks for once-through power plant cooling systems. In such systems, cooling water extracted from the water body is passed through the plant condenser system and returned to the receiving water at an elevated temperature. Away from the source region, the temperature excess in the discharged effluent is eventually dissipated through diffusive mixing and far-field dispersion in the receiving waters (Davies and Mofor, 1993), but the detailed nature of the dispersal process is highly site-specific. If one assumes that the generating station has an overall efficiency of 30-40%, then the condenser cooling water carries away nearly twice as much energy as is transmitted along the high voltage lines and, with today's increasing awareness of environmental issues, it is important to try to understand the impact that this thermal pollution may have on the surroundings.

Plume behaviour has been studied conventionally using mathematical models involving parameters such as wind, tide, current and temperature and salinity distributions obtained from *in situ* boat surveys. Spaceborne and airborne remote sensing, however, provides better synoptic, spatial, temporal and spectral data for observing such a highly temporally- and spatially-varying coastal water process and for providing input into hydrodynamic models of the mixing process. Since it is usually the temperature excess of these discharge waters which provides the spectral contrast between the effluent and ambient waters, the thermal waveband is by far the most useful for such studies. However, correlated images in other wavebands will also provide useful information, particularly on sea state conditions.

A previous study of the discharge from this power plant (Wu, 1990, Vaughan and Wu, 1992) was based on data acquired incidentally to another flying campaign and was not optimal for such an investigation. So, in 1992, four sets of imagery and photographs were intentionally taken over the study area. Unfortunately, one set of imagery was unusable. The principle objectives of this present study were to use these images to track the surface motion of the buoyant thermal discharge, to use these observations to estimate a mixing coefficient and to calculate the total areal extent of the affected region.

2. DATA

The Central Termica Litoral de Almeira power plant is situated near the village of Carboneras in south east Spain. It is a coal-burning plant which started commercial operations in 1985. Sea water is pumped from an inlet located to the south of a dyke, in a harbour, and is discharged through an outlet about 100 metres on the other side of the dyke. The inlet and outlet are therefore separated by this wall about 1100 metres in length (see figure 1). The discharge channel is 4.2 x 2 metres in cross section. The outlet is situated at the sea/shore interface, and the jet can therefore be treated as a surface horizontal one. At the time of the overflight, the cooling water was discharged at a rate of about 7000 cubic metres per hour with a temperature of about 30.6^{0}C. The temperature of the ambient water was about 17^{0}C measured from ATM data. There is very little tidal range in the Mediterranean, but there is a weak south westerly current parallel to the coast at this point.

Four flights were made on 6 June 1992, the first three at a height of 1000m and the last at 2000m, but only three of these provided usable data. The analysis presented here is based only on this data which was acquired with S-bend correction applied, and the spatial resolutions were about 2m and 4m respectively. Because the outfall was located near the centre of the scan line, no geometric correction was made. Data was analysed using two image processing packages, PC_IMEGA and R-CHIPS. Other software for retrieving and analysing the data was developed using Quick C. Radiometric correction and was applied using the calibration data provided to estimate brightness temperatures.

The visible band images appeared to be very clear and, since the flying height was so low, no attempt was made to correct this data for the effects of the atmosphere. Lowtran7 was used for correcting the thermal data using reasonable values for concentrations of water vapour and carbon dioxide since model values appeared to give unrealistic results. An almost constant atmospheric correction factor of about 2.6^0 was obtained for temperatures ranging from 16 to 35^{0}C measured from 1000m. However, from the difference between the temperatures measured at 1000 and 2000m, an atmospheric gain of only 0.75^0 was obtained. But the effect of either correction on the calculated mixing coefficients was in any case negligible.

Ambient conditions during the flights were supplied by the Spanish Weather Bureau as follows:

Figure 1. *Channel 1 image of test site from 2000m showing the location of the dyke. This image was from the 1989 dataset (Wu, 1990) in which the plume turns south (to the right) due to the prevailing wind and current, and follows the line of the dyke.*

- Temperature: 17.2^{0}C
- Solar radiation: 203 W/m^2
- Humidity: 71.9%
- Wind speed: 5.3 m/s
- Wind direction: 167^0

These were, however, average values over the whole day, and more precise values were inferred from the images and airphotos taken during the flights. The first flight took place at 0900 hours when the sea was calm and the prevailing current small. By the time of the fourth flight, at 1330 hours, wind speed had increased to about 9 knots which was calculated from the sea state observed on the airphoto. Wind direction could also be inferred from the wavelets present in both airphotos and visible channels of the ATM data.

3. QUANTITATIVE ANALYSIS

Solving practical dilution problems involving jets is generally not a trivial matter requiring knowledge of momentum, buoyancy, density stratification, flow boundaries, loss of buoyancy, cross flows,

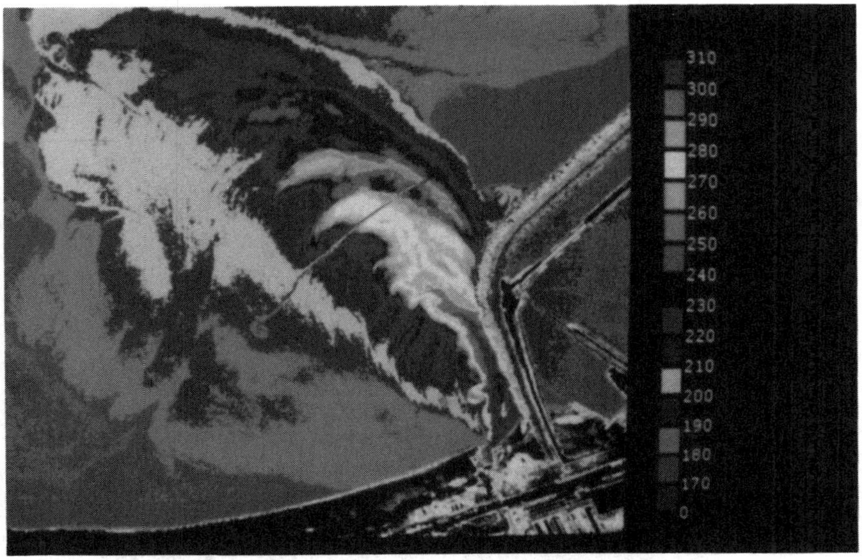

Plate 1. Density sliced image of the thermal plume at 0900h (1^0 per slice). The plume originates at the bottom right and clings to the wall before dispersing off shore. The sea was calm at this time of day and the prevailing alongshore current was small. Note the track of a boat diagonally across the centre of the image demonstrating poor vertical mixing.
Colour plate see page 498

bottom friction, internal friction, internal hydraulic jumps and ambient turbulence. In this case, no ground measurements were made, and also the presence of the dyke generated strong geometric effects. No model adequately describes the present situation adequately, so two were tried, the Gaussian normal spreading model and one that takes account of the boundary effect.

Two-dimensional coordinate systems were set up for both the near-field and far-field parts of the plumes for each set of data. Normally only one coordinate set is used, but in this case the plumes are noticeably curved and no single set adequately fits. In the near-field, the plume is in the shelter of the dyke. In the far-field, the effect of the on-shore breeze and stronger current velocity in the afternoon was to turn the plume along the coast, whereas in the morning the curvature was much less (see Plate 1 and figure 2). It should be noted that in the images obtained in 1990 the plume turned south rather than north and followed the line of the dyke because of the different prevailing conditions (see figure 1).

The theory on which the modelling of outfall plumes is based assumes the vertical mixing takes place very much more quickly than does horizontal mixing. In the absence of *in situ* measurements, it is usual to assume that full vertical mixing is accomplished relatively close to the source and that the average water depth can be used in the calculations. In this case, it was evident that vertical mixing was very weak. A long white line in one of the thermal images, associated with the track of a small vessel observed in the visible images and airphoto, indicated that colder water, about 5^0C cooler than ambient, was being disturbed. Assuming a draft of about 2m for the boat, this suggested that the effluent was not fully mixed even at a distance of 400m away from the source. A value of 2m for average water depth was therefore used in the calculations.

Software was written to calculate the mixing coefficients from the equations developed in Davies and Mofor (1993) for the various conditions encountered. In the near field, the plume remained attached to the dyke, and the model taking account of the boundary effect was used. Even in the far field, the effect of the dyke was still evident from the non-symmetrical temperature profile. In this case, only the data derived from the left hand side of the profile was used.

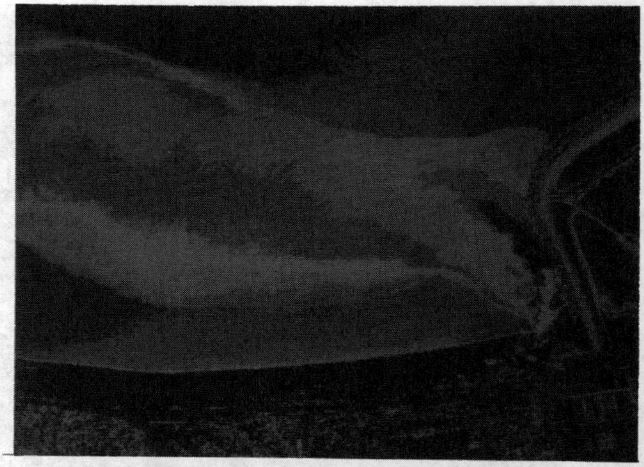

Figure 2. *Plate 1 shows the form of the thermal plume at 0900h when the sea was calm. As the day wore on, an increasing on-shore breeze turned the plume northwards along the coast as seen above (upper image, 1030h, lower image, 1330h).* Colour plate see page 499

Temperature transects were extracted and attempts made to fit curves to these.

Mixing coefficients were then calculated in the near field for each of the transects for two input values of current velocity, namely 0.05 and 0.1 ms^{-1}, values considered realistic in relation to the estimated source flow velocity. Tables of these, together with the graphs of the temperature profiles can be found in Wen (1993).

In the far field, however, values of current velocity were unknown, and so an alternative method was used. From the analysis in Davies and Mofor (1993), the logarithm of the excess temperature profile, $\ln\Delta T(x,y)$, should vary as $-y^2$, where y is the transverse distance from the axis of the plume at a distance x from the source. Graphs of these quantities were plotted, and reasonably linear curves were obtained particularly for the morning flight, allowing values for both the mixing coefficient and the current velocity to be estimated.

4. SUMMARY OF RESULTS

Table 1 summarises the results of the above calculations, showing the average values of mixing coefficients and current values for each transect. In the near field, values of the mixing coefficient decreased slightly

Table 1

Mixing coefficient in the near-field

Distance to source m	Mixing coefficient m^2/s			
	U=0.05 m/s		U=0.1 m/s	
	image 115	image 416	image 115	image 416
25	1.595	0.705	3.160	1.211
50	0.834	0.413	1.653	0.765
75	0.655	0.455	1.297	0.901
100	0.611	0.316	1.219	0.627
125	0.558	0.416	1.121	0.824
150	0.513	0.391	1.031	0.771
175	0.483	0.344	0.961	0.690
200	0.438	0.336	0.871	0.675
225	0.423	0.311	0.849	0.621
250	0.403	0.303	0.802	0.611

Mixing coefficient in the far-field

Distance to source m	Mixing coefficient m^2/s		Current velocity m/s	
	image 115	image 416	image 115	image 416
350		0.267		0.012
400	0.065	0.336	0.007	0.013
450		0.316		0.018
500	0.050	0.573	0.008	0.014
550		0.404		0.018
600	0.073	0.367	0.007	0.019
700	0.073		0.007	
800	0.107		0.006	
900	0.072		0.009	

Table 2. *Results for sample transect*

Transect No. 3
Distance from boundary to transect : 15.00 m
Distance to source : 74.53 m

Point	Temperature °C	Y-dist m	Mix. Coef V.=0.05m/s	Mix. Coef V.=0.1m/s
15	17.4	152.4	1.03000	2.06000
14	17.5	140.7	0.94000	1.87000
13	17.2	131.4	0.66000	1.32000
12	17.0	122.3	0.34000	0.50000
11	16.9	110.5	0.28000	0.50000
10	17.1	101.3	0.35000	0.68000
9	17.7	89.6	0.43000	0.87000
8	18.7	83.7	0.52000	1.04000
7	20.1	72.0	0.52000	1.03000
6	21.9	62.8	0.52000	1.10000
5	23.9	53.6	0.51000	1.01000
4	26.1	41.9	0.42000	0.90000
3	28.2	32.7	0.36000	0.74000
2	30.0	20.9	0.24000	0.48000
1	31.1	11.8	0.14000	0.28000
0	31.4	0.0	0.02000	0.04000
		average =	0.45500	0.90125

with distance from the source after the first 25m and increased with the estimate used for current velocity. Average values of 0.5 m^2s^{-1} (with a cross-flow of 0.1 knot) and 0.9 m^2s^{-1} (with a cross flow of 0.2 knot) were obtained. For the first two or three transects the water depth was shallower than the average value used in the calculation, so results from these were ignored. An inspection of the images showed that eddies were more prominent in the morning image when the state of the sea was calm than they were in the afternoon image. This could account for the different values obtained for the mixing coefficients in these two cases. In theory, also, the coefficient should be constant along each transect, but in this study it was found to increase gradually from the centre of the plume to the edge (see table 2). This was probably caused by taking average values of current velocity rather than actual ones at each point.

In the far field, less variation was found from transect to transect. Average values of 0.07 m^2s^{-1} and 0.35 m^2s^{-1} for the mixing coefficient in the two images were obtained, the first nearly ten times smaller than in the near field and the second about three times smaller. The corresponding value of current velocity were 0.007 and 0.05 ms^{-1} respectively, both surprisingly small.

The greater mixing in the afternoon can certainly be attributed to the prevailing weather conditions, but the absolute values must be treated with some scepticism. It has been noted already that gaussian spreading almost certainly does not occur, and also that vertical mixing is taking place far too slowly for the model to strictly apply. These factors, together with all the errors introduced by not having *in situ* data, make it surprising that a fit as good as this was obtained.

Using the near infrared band, channel 8 (0.91-1.05μm), to mask out land pixels, a count of those water pixels with temperatures exceeding that of the ambient water was made enabling areal extents to be obtained. These values are shown in table 3. It can be seen that the afternoon image has a much greater area corresponding to a small temperature excess than does the morning

Table 3. *Total extent of area affected by effluent*

Temp. °C bright.	Temp. °C excess	Extent-115 m^2	Extent-416 m^2
17.5	0.5	399944	1286807
18.0	1.0	429470	706301
18.5	1.5	123688	342251
19.0	2.0	133040	96742
19.5	2.5	153169	27072
20.0	3.0	72912	21483
20.5	3.5	65986	14088
21.0	4.0	42173	8288
21.5	4.5	37560	5853
22.0	5.0	31125	5572
22.5	5.5	28036	5082
23.0	6.0	23257	3207
23.5	6.5	15310	3207
24.0	7.0	12402	2716
24.5	7.5	8678	1998
25.0	8.0	7250	2138
25.5	8.5	4705	1542
26.0	9.0	4407	1822
26.5	9.5	4981	1419
27.0	10.0	5318	1612
27.5	10.5	5410	1892
28.0	11.0	3925	1980
28.5	11.5	2252	2366
29.0	12.0	1958	1384
29.5	12.5	2147	2015
30.0	13.0	1901	2961
30.5	13.5	2891	3680
Total	=	1623894	2555477

image, reinforcing the above conclusion that the sea state helps disperse the plume.

5. QUALITATIVE ANALYSIS

Even though the results of the quantitative analysis above were rather disappointing, it was found that many of the conclusions drawn above could be inferred from inspecting density sliced thermal images and longitudinal temperature profiles. The effects of prevailing weather conditions, poor vertical mixing and bathymmetry as well as the areal extent of the spreading could all be inferred from these and from the colour airphotos. This is not to say that a quantitative study is not necessary, but it does indicate the dangers of trying to apply a not very sophisticated model to a complex situation. One might almost assume that the present models may almost never apply!

As a general conclusion, however, it would seem that these results demonstrate that the major spreading behaviour is due to ambient weather competing with the small along-shore current flow under the influence of the dike. A better model is required which takes account of these varying conditions together with poor vertical mixing.

6. ACKNOWLEDGEMENTS

This work was undertaken in partial fulfillment of the degree of MSc at the University of Dundee. C-CW would like to thank the Chung Shan Institute of Science and Technology (Taiwan) for supporting his stay in the UK.

REFERENCES

Davies, P A and Mofor, L A. 1993. *Remote sensing observations and analysis of cooling water discharges from a coastal power station.* Int J Remote Sensing, **14** 2 253-273

Vaughan, R A and Wu, Q. 1992. *The study of a coastal plume using airborne thematic mapper data.* In Proceedings of 11th EARSeL Symposium, Graz (Austria) July 1991. (ESA, Paris)

Wen, C-C, 1993. *A study of coastal plume using airborne remotely sensed data.* MSc Thesis, University of Dundee.

Wu, Q, 1990. *A Study of coastal plume using airborne thematic mapper data.* MSc Thesis, University of Dundee.

Atmosphere

Sensors and Environmental Applications of Remote Sensing, Askne (ed.) © 1995 Balkema, Rotterdam, ISBN 90 5410 524 0

Determination of true surface temperature using TOVS and AVHRR data

M.Arbelo, F.Herrera & F.J.Expósito
Communication & Remote Sensing Group, Department of Physics, University of La Laguna, Canary Islands, Spain

V.Caselles
Department of Thermodynamics, University of Valencia, Spain

ABSTRACT: The determination of true surface temperature from satellite is performed by means of multichannel algorithms with channels 4 and 5 of AVHRR-NOAA or using radiative transfer models and radiosounding profiles of temperature and humidity. In this work, an alternative to the current algorithms is established. A new method combining the information supplied by sensors of TOVS and AVHRR systems on board NOAA satellites is proposed.

1 INTRODUCTION

Accurate determination of sea surface temperature is essential for ocean and climate studies. The thermal structure of sea surface plays a critical role in regulating the exchanges between the atmosphere and oceans, and a knowledge of sea surface temperature is important for verifying circulation models. Satellite measurements of the upwelling radiance from the Earth's surface and atmosphere have been carried out in a routine manner for many years (Chaine, 1980). However, the developed algorithms to obtain SST do not consider the environment characteristics of a zone as peculiar as Canary Islands, being necessary a particular study taking into account those conditions.

Present multichannel algorithms use the 4 and 5 channels data of AVHRR. The problem of these procedures is the local character of them coefficients, because they are inadequate and inaccurate for Canary Islands zone. In the same way, the algorithms that use radiative transfer models and radiosondes data, create doubts about the validity of used atmospheric profiles to make corrections to the SST, because they are launched from land surface and it exists, principally over the oceans, a great variability of water vapor that the radiosondes do not measure. There is another problem that we can not obviate if we wish obtaining the true surface temperature, that is, the difference in time between satellite pass and the radiosonde launching.

In this work, an alternative to the algorithms using directly or indirectly traditional radiosoundings is established, and a new method, combining the obtained water vapor data from TOVS with the information of thermal-infrared channels of AVHRR to improve split-window equation, is proposed.

2 METHODOLOGY

2.1 *Split-window equation*

The employed method for the determination of SST is the split-window. The theorical basis of this technique can be found in McMillin and Crosby (1984). This algorithm make use of measured radiances by 4 and 5 channels of AVHRR.

If T_4 and T_5 are the registered brightness temperatures by each channel, the true surface temperature, T, could be calculated by means of the equation:

$$T = T_4 + A(T_4 - T_5) + B \tag{2.1}$$

The A and B coefficients, for sea surface temperature, are given by:

$$A = \frac{1 - \tau_4}{\tau_4 - \tau_5} \tag{2.2}$$

$$B = -\frac{(1 - \tau_4)(1 - \tau_5)}{\tau_4 - \tau_5}(T_{a4} - T_{a5}) \tag{2.3}$$

where τ_4 and τ_5 are the transmitances due to water vapor, and T_{a4}, T_{a5} are the mean radiative temperature of water vapor column in the channels 4 and 5 of AVHRR (Maul, 1983).

2.2 *Calculation of water vapor with TOVS data*

The water vapor, W, is an atmospheric component which is not evenly distributed on the global scale. Humidity and cloud distribution over the oceans are functions of sea surface temperature. Humidity may changes from 0.5 gr/cm² for high latitudes to about 6 gr/cm² for the equatorial regions (Szekielda, 1988). All that makes impossible to know its distribution with only radiosounding data, because they are limited to a fixed zone. Then, it would be interesting, having data over the oceans for characterizing correctly the water vapor.

NOAA satellites, with TOVS system on board, make possible the study of total atmospheric water vapor with an accurate between 5 and 10% on clouds free zone (Young, T.L. 1992). For it, the proposed method by Smith et al. (1985) is employed. This procedure calculates the precipitable water total amount, temperature profiles and surface skin temperature. It consists in a physical solution of radiative transfer equation making use of all channels of HIRS-2 (High Resolution Infrared Radiation Sounder mod. 2) and MSU (Microwave Sounding Unit) of TOVS to solve all parameters simultaneously, thus increasing the accuracy and numerical efficiency of the retrieval process with regards to the iterative methods. Briefly, the method entails:

1.-Integration by parts of radiative transfer equation and it express in perturbation form.

2.-Deviations in surface skin temperature, atmosphere temperature profile and water vapor mixing ratio profile are expanded in terms of the spectral band weighting functions $d\tau/d\ln P$.

3.-A direct matrix solution is accomplished.

3 AVHRR and TOVS combination

The most current algorithms, that combine data of AVHRR and TOVS sensors to obtain the surface temperature, use associations of the apparent temperatures of both systems (Schluessel, 1987), (Ottlé, 1992), or making use of humidity and temperature profiles from TOVS for to correct the atmospheric effect on 4 channel of AVHRR with a radiative transfer model (Reutter, 1991).

Our procedure propose the synergy of the calculated water vapor from TOVS channels radiances with the split-window algorithm of coefficients depending of this atmospheric component.

The coefficients A and B of split-window have been calculated for a big enough set of atmospheres using the radiative transfer model LOWTRAN 7. A lineal correlation is observed between A and water vapor, while B showed a quadratic behaviour with W (Coll et al., 1993c),

$$A = A_0 + A_1 W \tag{3.1}$$

$$B = B_0 + B_1 W + B_2 W^2 \tag{3.2}$$

These relations let us to obtain the temperature knowing the apparent temperatures of 4 and 5 channels of AVHRR and the water vapor (gr/cm²) from TOVS data. The followed procedure can be observed on the next figure:

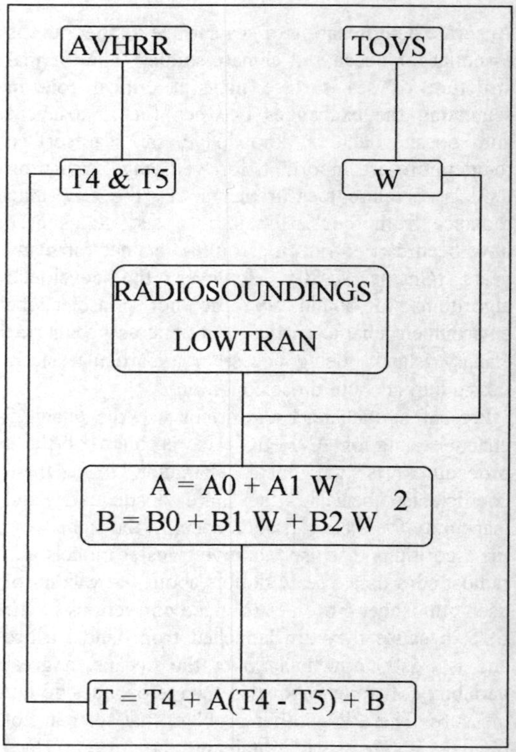

Figure 1. AVHRR and TOVS combination.

4 RESULTS

The equipment of reception, storage, processing and representation of HRPT telemetry data of NOAA satellites is a TeraScan System. This instrumentation allow us to extract information from the next sensors on board NOAA satellites:

AVHRR: Multispectrals images with 1 Km resolution.

TOVS: Vertical profiles of atmospheric humidity-temperature.

ARGOS: Data of small transmitters on board buoy, globes, wild animals ...

Antenna server operates of independent manner and it is linked to a local area network .

The system is operative, receiving the information that the satellites automatically send, during 24 hours.

The used image, for application of proposed algorithm, is a partial scene from NOAA 11 pass at 15:03 GMT, 18 of June of 1991. It is centered to 28.5° North Latitude and 15.6° West Longitude. Radiosounding of 12:00 GMT launched from Santa Cruz de Tenerife by Instituto Nacional de Meteorología is available too.

The obtained results with our algorithm, are compared with data of other authors and the disagreements between both are analyzed. Used mask for eliminating clouds and land zones, has been the same in all cases. So, calculation of SST has been realized only for the "cleared" pixels.

One point situated to 27° 36.30 N and 16° 3.03 W presents, after application of different algorithms, the next results:

1.- (McClain et al., 1985): 288.9 K
2.- (Sobrino et al., 1991): 293.1 K
3.- (Coll et al., 1993b): 293.4 K
4.- (Arbelo et al., 1993): 293.9 K

1.- McClain equation is of global character and it is not extrapolable to zones as particular as Canary Islands. The difference with the rest of algorithms is obvious.

2.- In this case, A coefficient of split-window has been obtained by interpolation of total water vapor data, 1.09 gr/cm², calculated from vertical profile of humidity and temperature of that day, with Sobrino values. The problem of this method is the validity of radiosounding launched over the land to obtain SST, and the difference in time between that and satellite pass. Three hours can be too much time in hot months with high evaporation.

3.- The particularity of Coll method is the quadratic term, but the coefficients have been calculated taking into account water vapor amounts less 2 gr/cm², then, in principle, it is not valid for a subtropical region like Canary Islands.

4.- The proposed method in this work uses the split-window equation with the coefficients as a functions of W like it has been described. The calculation of both coefficients, A(W) y B(W), has been done choosing four free-clouds TOVS soundings in the studied image. All data are situated to the South of Occidental Islands. The atmospheric water vapor values for each point are: 2.14, 2.05, 1.86 y 2.03 gr/cm². The used value into the coefficients expressions to obtain them, is the average of four data.

The proposed local split-window equation is:

$$T = T_4 + 2.61\,(T_4 - T_5) + 0.35 \qquad (4.1)$$

CONCLUSION

The combination of TOVS and AVHRR data must give us the best results of sea surface temperature because it guarantee the simultaneity in time and space of T and W measurements.

The validity of proposed method will be proved with in situ measurements and satellite images during the next months.

REFERENCES

Chaine, M. T. 1980. Infrared remote sensing of sea surface temperature. *Remote Sensing of Atmospheres and Oceans*. Edited by A. Deepack. Academic Press.

McMillin, L. M.. and Crosby, D. S. 1984. Theory and validation of the multiple window sea surface temperature technique. *Journal of Geophysical Research* 89, C3: 3655-3661.

Maul, G. A. 1983. Zenith angle effects in multichannel infrared sea surface sensing. *Remote Sensing of Environment* 13: 439-451.

Szekielda, K.H. 1988. *Satellite Monitoring of the Earth*. Edited by Ji Au Kong. Wiley Interscience.

Young, T. L. 1992. *Terascan Reference Manual*. SeaSpace, San Diego, California, USA.

Smith W. L. et al. 1985. The simultaneous retrieval export package. In *II ITSC*, Igls, Austria.

Schluessel, P. et al. 1987. Comparison of satellite derived sea surface temperatures with in situ skin measurements. *J. Geoph. Res.* 92: 2859-2874.

Ottlé, C. and Vidal-Madjar, D. 1992. Estimation of land surface temperature with NOAA-9 data. *Rem. Sens. Environ.* 40, 27-41.

Reutter, H. and Olesen, Folke-Soren 1991. Atmospheric correction of AVHRR IR-DATA with TOVS sounding. In *5th AVHRR Data User's Meeting*, Tromso.pp 143-148.

Coll, C y Caselles, V. 1993c. Analysis of the atmospheric and emissivity influence on the split-window equation for sea surface temperature. *International Journal of Remote Sensing.* (Forthcoming).

McClain, E. P. et al. 1985. Comparative performance of AVHRR based multichannel sea surface temperatures. *Journal of Geophysical Research* 90, C6:11587-11601.

Sobrino, J. A. et al. 1991. Atmospheric correction for land surface temperature using NOAA-11 AVHRR channels 4 and 5. *Remote Sensing of Environment.* 38: 19-34.

Coll, C. et al. 1993b. On the atmospheric dependence of split-window equation for land surface temperature. *International Journal of Remote Sensing.* (forthcoming).

Arbelo, M. et al 1993. Utilización de los datos TOVS y AVHRR para obtener la temperatura de la superficie terrestre de Canarias. In *XXIV Bienal R.S.E.F.* Jaca. Tomo II, MT-51.

Sensors and Environmental Applications of Remote Sensing, Askne (ed.) © 1995 Balkema, Rotterdam, ISBN 90 5410 524 0

Comparison between the aerosols optical depth data obtained by the NOAA satellites and the Optronic OL752 spectroradiometer

F.J. Expósito, M. Arbelo & F. Herrera
Communication & Remote Sensing Group, Department of Physics, University of La Laguna, Canary Islands, Spain

J.P. Díaz & A. Díaz
Atmospheric Physics & Environment Group, Department of Physics, University of La Laguna, Canary Islands, Spain

ABSTRACT: The aerosol optical depth data obtained by two ways are compared. The first one uses solar radiance measurements, by means of the spectroradiometer Optronic OL752 placed over the ground. The second one takes the solar radiation scattered by the atmosphere using the AVHRR radiometer channels on board NOAA meteorological satellites. A correlation about 85% between both procedures was obtained.

1 INTRODUCTION

The study of atmospheric aerosol has a special interest in a wide set of important topics: radiative mechanics in the Earth-Atmosphere system, urban pollution, correction of satellite images, measurements of atmospheric gases: ozone, NO_X,... (Hansen 1992, Stowe 1992).

In general, it is necessary to know three parameters in order to describe adequately the aerosol extinction in the atmosphere: the complex refraction index of the particles, the size distributions and the particle shapes. A knowledge of the first two quantities is sufficient if it can be assumed that the particles are spherical and the Mie theory is applicable.

We will present the improvements done in an Optronic OL752 spectroradiometer. Initially, this equipment was thought as a global radiation measuring instrument. We developed it adding a collimating tube in order to measure direct solar radiation. This tube was designed following the rules dictated by World Meteorological Organization (WMO 1971). In a second step, and to get a best tracking of the sun, we connected the output of the tube to the input port of the monocromator, via optical fiber bundle.

The TeraScan system is an automatic equipment which can process the NOAA satellites data. With the radiometer on board these ships and several aproximations we can give the aerosol optical depth for a wide zone around the Canary Islands. The result to compare this values versus the OL752 show a correlation about 85%, which permits us to calibrate the satellite data and compute a best results for this area.

2 RETRIEVAL OF AEROSOL OPTICAL DEPTH

2.1 *Optronic algorithm*

The incident solar beam transmitted directly by the Atmosphere can be written as (Lambert-Beer-Bouguer law):

$$L(\lambda) = L_0(\lambda)\exp\left(-\left[\tau_R + \tau_{abs} + \tau_A\right]m\right) \qquad (2.1)$$

where $L_0(\lambda)$ and $L(\lambda)$ are the monochromatic intensities of the incident and transmitted radiation, respectively, τ_R is scattering Rayleigh coefficient, τ_{abs} is the coefficient of absorption, τ_A is the aerosol optical depths and *m* is the air mass.

In order to use the equation (2.1), it is necessary to measure direct solar radiation, so we change the initial design of the Optronic (originally developed to calculate absolute solar global radiation, from 250 to 800 nm). We build a collimating tube following the WMO rules. In fig. (1) we show the diagram of this dispositive, which has the following features: $a=R/r=1$, $b=L/r=64.7$ and $z_0=\tan^{-1}(a/b)=0.88$.

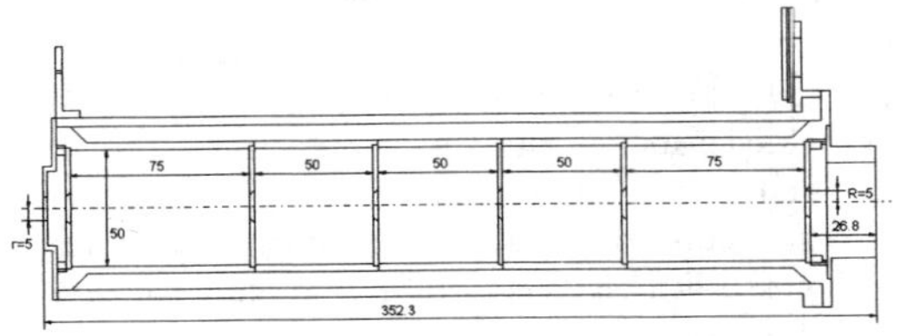

Figure 1: Collimating tube diagram (distances are in mm).

The coefficients τ_R and *m* were calculated using the

next expressions (WMO 1978, Iqbal 1978):

$$m = \left[\cos\theta_0 + 0.15(93.885 - \theta_0)^{-1.263}\right]^{-1} \quad (2.2)$$

and

$$\tau_R = 0.00838\lambda^{-(3.916 + 0.074\lambda + 0.05/\lambda)} \quad (2.3)$$

To correct the measures did at the Izaña GAW station (altitude: 2360 m), we use the factor: P(mb)/1013.25 (Iqbal 1978), and the ozone value for this day (300 DU), which was obtained by the Brewer of this station.

The fig. (2) shows the Langley's calibration to the wavelength of 630.0 nm (Díaz 1992). The fit to these data gave the regression line: y=-14.861-0.308x with a correlation factor of 98%. So, the extraterrestrial constant for the equipment was of 3.515E-7A .

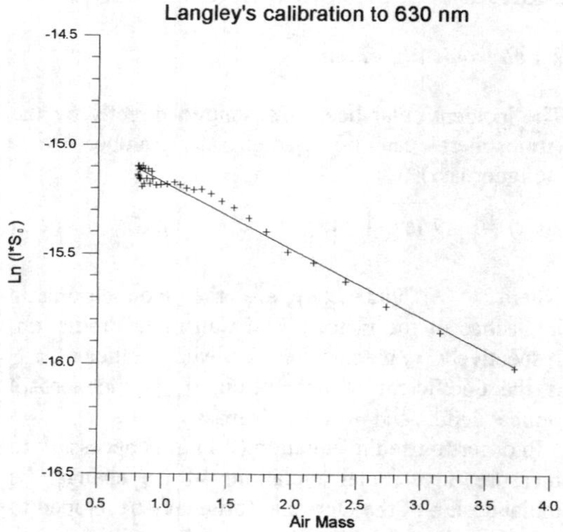

Figure 2. Langley's calibration to 630 nm.

The design of the bundle used to connect the monocromator to the collimating tube was negotiated with Polymicro Technologies, Inc., trying to resolve the handicap of the high attenuation and the low solar radiance in the UV region. The principals characteristics are:

Total length: 2m, diameter: 1cm.
Silica core (for one optical fiber): 200±8μm, doped silica cladding: 220±5μm, buffer coating: 240±5μm.
Numeric aperture:0.22 ±0.02.
Step refraction index.
Temperature range: -50 to 400ºC.

2.2 *TeraScan algorithm*

To obtain the aerosol optical depths by the NOAA satellites, we used a TeraScan system that can capture and process this kind of data. The system block diagram is shown in the fig. (3).

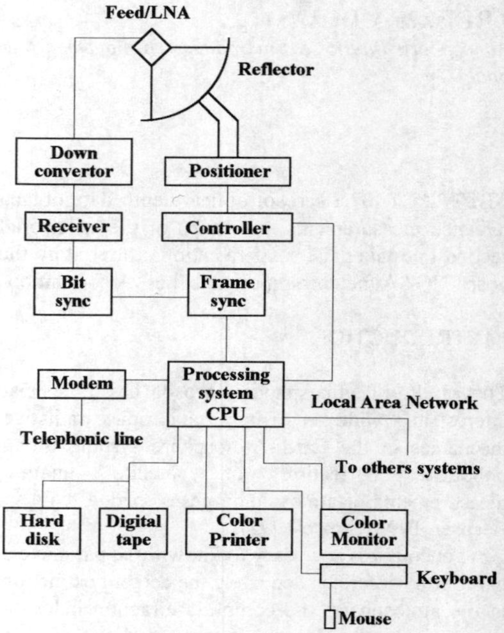

Figure 3. HRPT system block diagram.

The subroutine employed begins with the radiative transfer equation, which will be applied to every clear pixel of the image (Liou 1980, Kaufman 1993):

$$\mu \frac{dL(0;\mu,\phi)}{d\tau} = L(\tau,\Omega) - \frac{\omega_0}{4\pi}\int_{4\pi} L(\tau,\Omega')P(\Omega,\Omega')d\Omega' - \frac{\omega_0}{4\pi}\pi E_0 P(\Omega,-\Omega_0)\exp(-\tau/\mu_0) \quad (2.4)$$

where L is the diffuse intensity, τ the optical depth, the single scattering albedo is ω_0, μ the $\cos\theta$ (θ = satellite zenith angle), μ_0 the $\cos\theta_0$ (θ_0 = sun zenith angle), Ω the solid angle (θ,ϕ) (ϕ = azimuth angle), $P(\Omega,\Omega_0)$ the phase function (maritime, because we have these environmental conditions) and πE_0 the incoming radiative solar flux. A cloudless pixel (or clear pixel) must not have any of the next conditions:

satellite zenith angle > 50°
solar zenith angle > 50°
brightness temperature CH4 < 0°
CH1/CH2 ratio <1.5
CH2 albedo >3
CH2 albedo variation of a pixel with his neighbours > 0.125

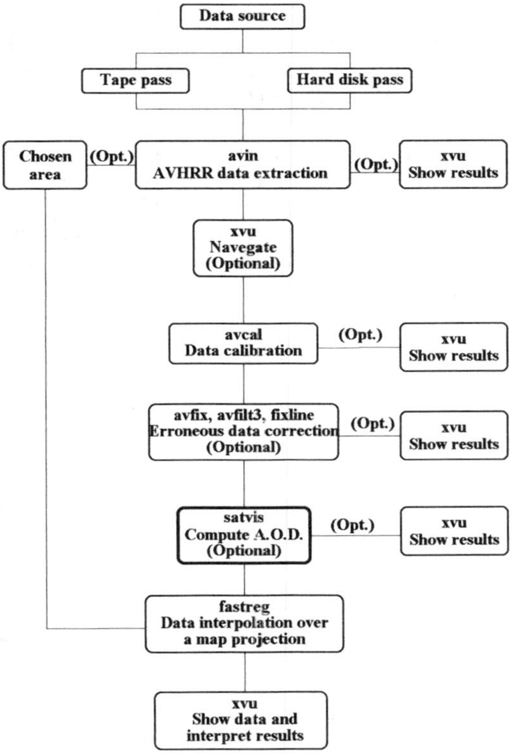

Figure 4. Satellite algorithm to compute the A.O.D.

The maritime air verifies several conditions, and one of the more important is that the single scattering dominates the multiple scattering. So, we can write the solution of the above equation as:

$$L(0;\mu,\phi) = \left[\frac{\omega_0 \mu_0 E_0}{4(\mu + \mu_0)}\right] P(\Omega,\Omega_0) \cdot \left\{1 - \exp\left[-\tau\left(\frac{1}{\mu} + \frac{1}{\mu_0}\right)\right]\right\} \quad (2.5)$$

If we suppose an optically thin atmosphere, i.e., $\tau(1/\mu + 1/\mu_0) << 1$, the equation (2.5) become:

$$L(0;\mu,\phi) \cong (\omega_0 E_0 / 4\mu) P(\Omega,\Omega_0)\tau \quad (2.6)$$

Due to the fact that the sea contribution is negligible for the signal detected by the satellite, to the wavelength of the CH1 and CH2, we obtain that (Durkee 1986, Frost 1988):

$$\begin{aligned} \tau_A &= (4\mu / \omega_0 E_0) P(\Omega,\Omega_0) L_A \\ L_A &= L_* - L_R \end{aligned} \quad (2.7)$$

where L_A is the radiance due to aerosols, L_* the radiance detected by the satellite and L_R the radiance due to scattering Rayleigh.

The satellite algorithm flux diagram to compute the aerosol optical depth is drawn in fig. (4).

2.3 *Comparison between both algorithms*

The instrument placed over ground was installed at sea level at the Nautical School inside the harbor of Santa Cruz. We use a half band width of 2 nm and an ozone value of 300 DU. The NOAA channel 1 is centered on 630 nm and has a wide band width, about 100 nm. The data taken was obtained over the sea, very close to the coast line and in a cloudless area. The results are showed in the fig. (5).

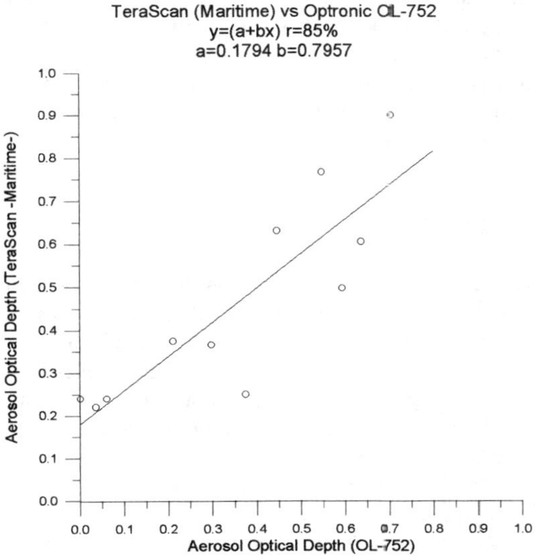

Figure 5. TeraScan data vs. Optronic data.

The fit gave the straight line: y=0.1794+0.7957x with a coefficient of correlation of 85%.

3 CONCLUSION

The good conditions at Izaña GAW station for the days 8th September and 18th November 1993, permitted us to get very high correlation in the Langley's plot calibration (about 98%), such as it's shown in the fig. (2).

We have used a correction to eliminate of the measurements over ground the absorption due to the ozone. Although the recorded aerosol optical data are not so much, and the correlation between them is not high (r=85%), they can give us an idea, in a first approximation, of the behavior of the employed instruments. So, we can see that the TeraScan has a shift about 0.18, which could be explained by different factors, like the wide bandwidth in the AVHRR channels and the several approximations done to resolve the radiative transfer equation (equation (2.4)). On the other hand, the data are included in a wide range of optical depths, which permit us to check the response of the instruments for a complete set of the different conditions.

AKNOWLEDGEMENTS

The authors gratefully thank the Instituto Nacional de Meteorologia, and specially to Emilio Cuevas (Chief of the Izaña Observatory) and all the observers of this station. Many thanks also go to the Centro Superior de Náutica y Estudios del Mar for use of the installation.

REFERENCES

Hansen J., A. Lacis, R. Ruedy & M. Sato. Potential climate impact of Mount Pinatubo eruption. *Geophys. Res. Letters,* 1992, 19 (2), 215-218.

Stowe L.L., R.M. Carey & P.P. Pellegrino. Monitoring the Mt. Pinatubo aerosol layer with NOAA-11 AVHRR data. *Geophys. Res. Letters,* 1992, 19 (2), 159-162.

World Meteorological Organization. *Measurements of radiation and sunshine. Guide of meteorological instrument and observing practice.* 4th edition. WMO No.8, T.P.3. Geneva, Switzerland 1971.

Iqbal M. *An introduction to solar radiation.* Academic Press, Canada 1978.

World Meteorological Organization. *Report of second expert meeting on turbidity measurements.* October 24-27, Boulder, Colorado, U.S.A. 1978.

Díaz J.P., F.J. Expósito and A. Díaz. Espesor óptico de aerosoles a la longitud de onda de 500 nm. desde la estación BAPMoN de Izaña. *Rev. Acad. Canar. Cienc.* 1992. IV (núms. 1 y 2), 159-168.

Liou K.N. *An introduction to atmospheric radiation.* Academic Press, New York, N.Y., U.S.A. 1980.

Kaufman Y.J. Aerosol optical thickness and atmospheric path radiance. *J. of Geophys. Res.* 1993. 98 (D2), 2677-2692.

Durkee. P.A., D.R. Jensen, E.E. Hindman and T.H. Vonder Haar. The relationship between marine aerosol particles and satellite-detected radiance. *J. of Geoghys. Res.* 1986. 91 (D3), 4063-4072.

Frost E.M. *Global scale estimates of aerosol particle characteristic.* Master's Thesis. Naval Postgraduate School. Monterey. California. U.S.A. 1988.

Sensors and Environmental Applications of Remote Sensing, Askne (ed.) © 1995 Balkema, Rotterdam, ISBN 90 5410 524 0

An algorithm for the reconstruction of columnar aerosol size distribution from spaceborne spectrometer data

A. Bartoloni & M. Mochi
TELESPAZIO SpA, Rome, Italy
G. Milillo
ASI, Centro di Geodesia Spaziale, Contrada Terlecchia, Matera, Italy

ABSTRACT: The proposed retrieval algorithm for the reconstruction of columnar aerosol size distribution is based on two steps procedures: aerosol spectral thickness determination from spaceborne spectrometer measured radiance; columnar aerosol size distribution reconstruction from spectral thickness determined in step one. In this paper the second step has been approached using an ad hoc regularization method in which an explicit positivity constraint has been posed on the solution.

1. INTRODUCTION

The GOME instrument that will be on board the ERS-2 satellite has been designed in order to collect radiation over the entire wavelength region from 240 to 790 nm, in which several atmospheric species and also aerosols and clouds can be observed. Atmospheric aerosols are produced both in nature and by man. Naturally occurring materials include volcanic dust, meteoric dust, spores and seeds, soil particles, sea salt, ash from burning vegetation, and sulfates formed from photochemical and catalytic oxidation. Man's contribute materials such as fly ash from smoke stacks, mineral particles from industrial source, soil particles due to the agricultural activity and particles produced photochemically from gaseous pollutants.

The proposed retrieval algorithm is based on two steps procedures: aerosol spectral thickness determination from GOME measured radiance; columnar aerosol size distribution reconstruction from spectral thickness determined in step one. In this paper the second step has been approached using an ad hoc regularization method in which an explicit positivity constraint has been posed on the solution.

2. THE MATHEMATICAL MODEL

The aerosol particles are characterized by shape, size and refractive index (real part and imaginary part, the latter related to the absorption); they are generally polydispersed around a mean size, and various semi-empirical laws have been proposed to express the size distribution. Aerosols interact with optical radiation in a way that can be described using the electromagnetic theory. The solution to Maxwell's equations for electromagnetic radiation interacting with spherical bodies of arbitrary size and composition was first formulated by Mie in 1908 (see [1]).

Mie's theory is used for solving problems of light scattering by aerosols, though it has to be remarked that it applies in the strict sense only to spherical particles, which real aerosols often are not. But provided the aerosols themselves are nearly spherical (not disklike or rodlike, for example) and provided they do not have preferential orientations in space, Mie theory provides an acceptable approximation for the optical scattering and attenuation from a cloud of heterodispersed particles.

In the case where there are many particles, the radiation fields or attenuation coefficients are obtained by summing the contribution from each particle.

Assuming that the atmosphere particles can be modeled by equivalent spheres of known refractive index to a sufficient degree of accuracy, the integral equation which relates optical thickness, designated with $\tau(\lambda)$, to an aerosol size distribution can be written as:

$$\tau(\lambda) = \int_0^\infty \pi r^2 Q_{ext}(r,\lambda,m)n(r)dr \qquad (1)$$

where:

- λ is the wavelength of the incident illumination;
- m is the complex refractive index of aerosol particles;
- $Q_{ext}(r,\lambda,m)$ is the extinction efficiency factor from Mie theory given by

$$Q_{ext}(r,\lambda,m) = \frac{2}{\tilde{x}}\sum_{l=1}^{\infty}(2l+1)Re\left[{}^{e}B_l + {}^{m}B_l\right] \quad (2)$$

- n(r) is the columnar aerosol size distribution, i.e., the number of particles per unit area, per unit radius interval in a vertical column through the atmosphere.

The inference of the columnar size distribution n(r) from the spectral aerosol optical thickness $\tau(\lambda)$ is based on the numerical solution of (1); this is a Fredholm equation of the first kind that can be written in the form:

$$g(\lambda) = \int_0^\infty K(r,\lambda)f(r)dr \qquad (3)$$

with

- $K(r,\lambda) = \pi r^2 Q_{ext}(r,\lambda,m)$;
- $f(r) = n(r)$;
- $g(\lambda) = \tau(\lambda)$.

If we take into account that $g(\lambda)$ is the results of an experimental measurement procedure and if we look for a piecewise-linear solution f(r), we have to consider the following discretization of (3)

$$g_i + \varepsilon_i = \sum_{j=1}^{n} A_{ij} f_j \qquad i = 1,\dots,m \quad (4)$$

where:

- $g_i = g(\lambda_i)$ for $i = 1,\dots,m$;
- $f_j = f(r_j)$ for $j = 1,\dots,n$, are the n values of the particle size spectrum at the different rays r_j;
- ε_i for $i = 1,\dots,m$, is an error term resulting from the experimental uncertainty in the estimation of g_i;

$$A_{ij} = \begin{cases} a_{i1} + a_{i1}\dfrac{r_1}{\Delta r_1} - b_{i1}\dfrac{1}{\Delta r_1} & j = 1 \\ -a_{ij-1}\dfrac{r_{j-1}}{\Delta r_{j-1}} + b_{ij-1}\dfrac{1}{\Delta r_{j-1}} + a_{ij} + a_{ij}\dfrac{r_j}{\Delta r_j} - b_{ij}\dfrac{1}{\Delta r_{j-1}} & j = 2,\dots,n-1 \\ -a_{in-1}\dfrac{r_{n-1}}{\Delta r_{n-1}} + b_{in-1}\dfrac{1}{\Delta r_{n-1}} + a_{in} & j = n \end{cases}$$

- $a_{ij} = \int_{r_j}^{r_{j+1}} K(r,\lambda_i)dr \quad b_{ij} = \int_{r_j}^{r_{j+1}} rK(r,\lambda_i)dr$

  $i = 1,\dots,m \qquad j = 1,\dots,n$

We have modeled the error vector $\boldsymbol{\varepsilon}$ as zero-mean gaussian noise with standard deviation σ.
In practice errors introduced into the equations by approximating the integral by quadrature expansion are usually small in comparison with random errors associated with experimental measurements.
Equations (4) may be expressed in matrix notation:

$$\mathbf{g} + \boldsymbol{\varepsilon} = \mathrm{A}\mathbf{f} \qquad (5)$$

where $\mathbf{g}$ and $\boldsymbol{\varepsilon}$ belong to R^m (m-dimensional Euclidean space), $\mathbf{f}$ belongs to R^n and A belongs to R^{mxn}.
There are some inherent limitations and difficulties when looking for the solution of (5):

- the fundamental equations are unstable; attempts at direct solution will generally give large, oscillating errors;
- any successful method of solution must in some way discriminate against high frequency components and can therefore only provide a filtered (smoothed) version of the true but inaccessible solutions.

In order to solve the linear system (5) we have to use a regularization method that is to incorporate in the reconstruction procedure some a priori knowledge about the size distribution.

This is usually made searching for the smoothest solution compatible with the measured data (see [6]).

Hence we have to minimize the following functional

$$\min_f \sum_i (f_{i-1} - 2f_i + f_{i+1})^2 + \gamma^{-1}\sum_i \varepsilon_i^2 \qquad (6)$$

where the regularization parameter γ is determined by the relation

$$\sum_i \varepsilon_i^2 = E^2 \qquad (7)$$

where E is a constant. After having differentiated with respect to the f_i we have

$$A^T\varepsilon + \gamma H f = 0 \qquad (8)$$

with A^T we indicate the transpose matrix of A, and $H \in R^{n \times n}$ is the following matrix

$$H = \begin{pmatrix} 1 & -2 & 1 & 0 & & & & & & \\ -2 & 5 & -4 & 1 & 0 & & & & & \\ 1 & -4 & 6 & -4 & 1 & 0 & & & & \\ 0 & 1 & -4 & 6 & -4 & 1 & 0 & & & \\ & \cdot & \cdot & \cdot & \cdot & \cdot & \cdot & \cdot & & \\ & & 0 & 1 & -4 & 6 & -4 & 1 & 0 & \\ & & & 0 & 1 & -4 & 6 & -4 & 1 & \\ & & & & 0 & 1 & -4 & 5 & -2 & \\ & & & & & 0 & 1 & -2 & 1 & \end{pmatrix} \qquad (9)$$

hence from (8) and (5) we have the regularized solution:

$$\mathbf{f}_{REG} = (A^TA + \gamma H)^{-1}A^T\mathbf{g} \qquad (10)$$

In the application of this technique to our reconstruction problem we note the presence, in the solution, of a large negative and clearly unphysical parts.

This suggest a modification of the usual technique in order to incorporate also the constraint of positivity on the required solution.
Hence we transform (6) in the following quadratic programming (QP) problem (see [7]):

$$\min_f \frac{1}{2}\alpha(\mathbf{f},H\mathbf{f}) + \frac{1}{2}\beta(A\mathbf{f}-\mathbf{g},A\mathbf{f}-\mathbf{g}) \qquad (11)$$

with

$$\mathbf{f} \geq \mathbf{0} \qquad (12)$$

where $(\cdot,\cdot)$ is the scalar product, α is a regularization parameter that has to be chosen and β is chosen as

$$\beta = \frac{1}{\sigma^2} \qquad (13)$$

The problem (11), (12) can be solved through the Lagrange multipliers method, that is searching for the saddle point of the following lagrangian function:

$$L(\mathbf{f},\mathbf{l}) = \frac{1}{2}(\mathbf{f},K\mathbf{f}) + (\mathbf{c},\mathbf{f}) + (\mathbf{l},-\mathbf{f}) \qquad (14)$$

where $\mathbf{l} \in R^n$ is the Lagrange multipliers vector and

$$K = \alpha H + \beta A^TA \qquad (15)$$

and

$$\mathbf{c} = -\beta A^T\mathbf{g} \qquad (16)$$

We have to determine $\mathbf{f}^*$ and $\mathbf{l}^* \in R^n$ which satisfy the following condition

$$\begin{cases} K\mathbf{f}^* + \mathbf{c} - \mathbf{l}^* = 0 & \\ \mathbf{f}^* \geq 0 & \\ l_j^* f_j^* = 0 & j = 1,\dots,n \\ l_j^* \geq 0 & j = 1,\dots,n \end{cases} \qquad (17)$$

So we write the solution of (11) (12) as follows:

$$\mathbf{f}_{QP} = \beta(\alpha H + \beta A^TA)^{-1}A^T\mathbf{g} + (\alpha H + \beta A^TA)^{-1}\mathbf{l}^* \qquad (18)$$

Now we give a method to choose the regularization parameters α and β of (18). In order to do this we compare the two solutions of (5) that we have found, $\mathbf{f}_{REG}$ and $\mathbf{f}_{QP}$ given by (10) and (18) respectively; if we make the position

$$\gamma = \frac{\alpha}{\beta} \qquad (19)$$

in (18) we have:

$$\mathbf{f}_{QP} = \mathbf{f}_{REG} + (\alpha H + \beta A^TA)^{-1}\mathbf{l}^* \qquad (20)$$

The formula (20) shows that the QP solution is obtained from the usual regularized solution adding a term that cures the unphysical negative part of that solution. This fact suggests a method for the determination of the regularization parameter α, based on the idea of minimizing the L_2 norm of this added term; that is:

$$\min_\alpha \|\mathbf{f}_{REG} - \mathbf{f}_{QP}\| \qquad (21)$$

3. NUMERICAL RESULTS

In order to test the proposed retrieval algorithm we generate a data sets of simulated data starting from equation (1) and choosing the size distribution n(r) as the following modified gamma (see [4])

	a_1	a_2	b_1	b_2
Haze M	5.333×10^4	1	8.9443	1/2
Haze L	4.97×10^6	2	15.1186	1/2
Haze U	100	2	12	1/2

For the computation of the simulated the optical thickness we have calculated the extinction efficiency factor $Q_{ext}(r,\lambda,m)$ using the Bessel functions supplied by NAG library (see [2]).
The integral (1) has been evaluated in a particle ray range from ray_min to ray_max using the routine D01AJF of NAG Library (see [3]) and dividing the range of integration into subintervals equally spaced in logarithmic scale.

To the integral result a normally distributed random numbers (zero mean and fixed standard deviation) has been added in order to simulated the measurements errors.

The following quantity

$$err = \frac{\sum_{j=1}^{n}\left(f_j - f_j^{true}\right)^2 \Delta_j}{\sum_{j=1}^{n}\left(f_j^{true}\right)^2 \Delta_j}$$

has been used to evaluate the obtained reconstructed size distribution, with $\Delta_j = r_{j+1} - r_j$.

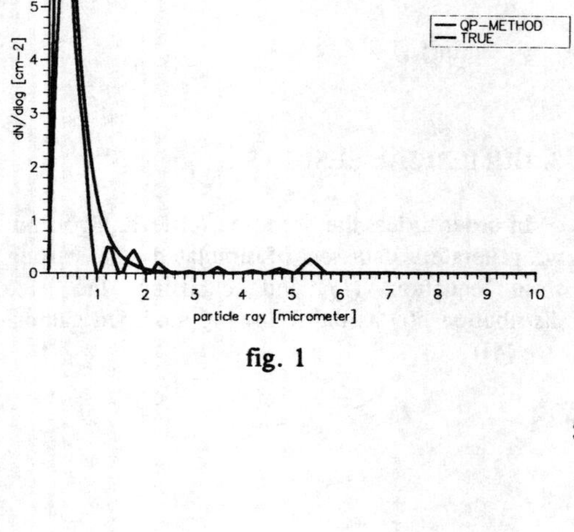

fig. 1

In figs 1, 2, 3 examples of the obtained results are shown; while in figs 4, 5, 6 the shapes of the corresponding expression (21) in the vicinity of the selected of the regularization parameter are represented.

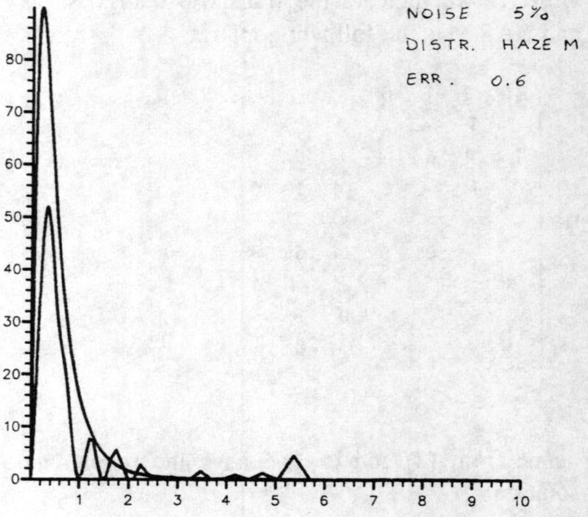

fig. 2

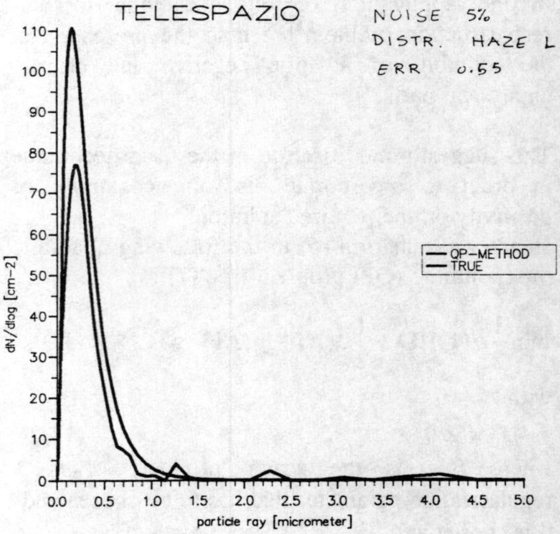

fig. 3

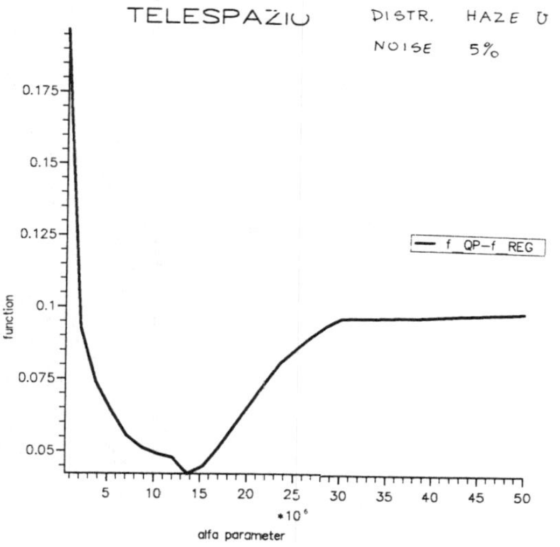

fig. 4

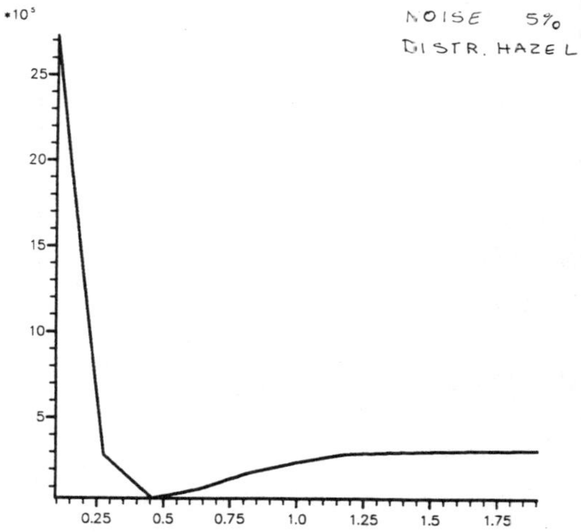

fig. 6

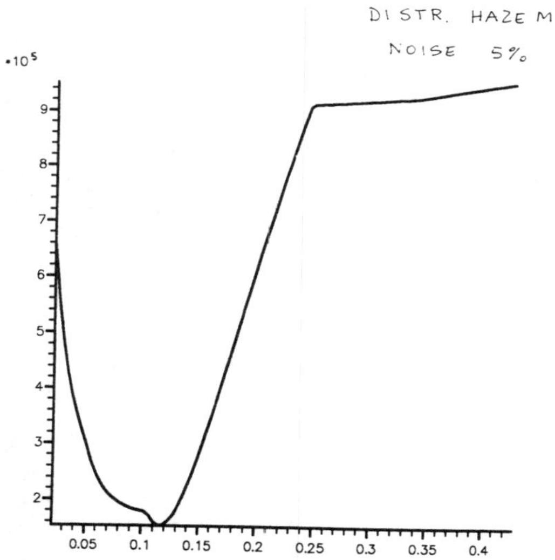

fig. 5

4. REFERENCES

[1] M.Born, E. Wolf: *Principles of Optics*, Pergamon Press, fourth edition, Oxford (1970).

[2] NAG Fortran Library Manual, Mark 14, Volume 8, H-X05, Oxford (1990).

[3] NAG Fortran Library Manual, Mark 14, Volume 1, Foreword-D01, Oxford (1990).

[4] G.W.Paltridge, C.M.R.Platt: *Radiative Processes in Meteorology and Climatology*, Elsevier Scientific Publishing Company, Amsterdam (1976).

[5] J.Lenoble (Ed.): *Radiative Transfer in Scattering and Absorbing Atmospheres: Standard and Computational Procedures*, A. DEEPAK Publishing, Hampton (1985).

[6] S. Twomey: *On Numerical Solution of Fredholm Integral Equations of First Kind by the Inversion of the Linear System Produced by Quadrature*, Jour. Asso. Comp. Mach., Vol. 10, 1963, pp. 97-101.

[7] G.I. Marcuk: Metodi del Calcolo Numerico, Editori Riuniti Edizioni Mir, I edizione, 1984.

Sensors and Environmental Applications of Remote Sensing, Askne (ed.) © *1995 Balkema, Rotterdam, ISBN 90 5410 524 0*

Analysis of cloud type and distribution using NOAA-AVHRR-APT data and ground observations in Central Europe

U. Heinzmann
Geographisches Institut der Universität Basel, Abteilung für Meteorologie/Klimaoekologie (GIB/AMK), Switzerland

ABSTRACT: A fuzzy logic classification technique (FLOP) is applied to a data set of more than 200 NOAA images in APT format. The satellite data was recorded at GIB/AMK between October 1990 and December 1991. Class definitions, calibration parameters and auxiliary data are stored in a related attribute data base and accessed during the classification procedure. For a test area in Central Europe, covering about 36.000 km2, maps of monthly cloud cover are derived. It is shown that cloud cover and weather type are correlated during the winter and summer months. Integrating surface based cloud observations, this correlation is used to extrapolate the classification results from 1991 to a climatological standard period of 30 years (1951-1980).

1 INTRODUCTION

Clouds play a crucial role in the climate system; they modulate the atmospherical radiation transfer as well as the latent heat flux. But the numerous processes involved in formation, transportation, modification and dissolution of clouds are so complex that our understanding of cloud-climate interactions is rather limited. Consequently, it is almost impossible to predict the role of clouds in man-made climate modifications. By now it is only accepted that clouds are likely to be important factors in climatic change, but neither the sign nor the magnitude of their effect is known. To fill this gap will be one of the most urgent tasks in the global change discussion. The meteorological satellites (NOAA and Meteosat series) have proven to be useful for monitoring of cloud distribution on large spatial scales at high temporal resolution. Satellite data with high spectral and spatial resolution has been applied for derivation of cloud physical parameters (e.g. optical depth, drop size, radiative properties) . Various efforts have been undertaken to derive a generally agreed on global cloud climatology; the two major projects are the ISCCP (Rossow 1985, 1991) and Nimbus-7 (Stowe 1988, 1989). But the fact that there is not even a consistent definition of cloudiness or a generally accepted method of deriving cloud cover from satellite data leads to large systematic differences between all cloud climatologies. This indicates a need for further investigation on the decision making process underlying the classification procedure. In this investigation, new decision making concepts are tested for cloud classification. The FLOP classification technique (Fuzzy Logic ClOud classification Procedure) has been developed for operational retrieval of cloud data over long time series. Here, FLOP is applied to a meso-scale test area, but in principle it can be applied on a global scale as well.

2 DATA

2.1 surface data

Visual observation is still the only operational method for estimation of cloud amount and type. The quality of ground-based cloud estimations is subject to large variations depending on the location of the station, cloud height, cloud type, illumination, etc. The observation is representative for a restricted zone around the observation point only. If these restrictions as well as the effect of different viewing geometries are considered, ground observations can be useful for the verification of the classification results. The FLOP results have been compared to daily 13:30 GMT ground observations of cloud cover at 53 stations in the German part of the Upper Rhine Valley and the Black Forest (provided by the German Weather Service). Mean monthly cloud cover data for the period 1951-1980 at 30 stations has been used for extrapolation to the 30- year period.

2.2 satellite data

The only sensors common to all operational geostationary and polar orbiting satellites with high temporal resolution are visible and thermal infrared. The method presented in this paper is based on NOAA APT data, a reduced data format processed from two of the NOAA-AVHRR HRPT bands (band 2 and band 4) at a spatial resolution of 3.3 by 3.3 km. From October 1990 to December 1991 daily NOAA-11 images were recorded at the Dartcom receiving station at GIB/AMK between 13:00 and 15:00 GMT.

3 DATA PROCESSING

The NOAA data sets are related to an attribute data base where the following information about each image is stored:
- Name, date, recording time, and data path of the original image (visible band)
- Name, date, recording time, and data path of the original image (thermal infrared band)
- Name and data path of the geocoded image (vis)
- Name and data path of the geocoded image (ir)
- mean square error of the geocoding
- area covered by the image (vis)
- area covered by the image (ir)
- Names and data paths of additional, auxiliary data layers used in the classification procedure
- remarks about noise, errors in the image
- cloud type used for normalization (vis)
- cloud type used for normalization (ir)
- weather type in Central Europe (Perret)
- cloud cover type over the test area
- cloud cover observed at ground stations
- number and names of classes to be derived
- physical definition of classes in each band

3.1 pre-processing

3.1.1 visible band

As it is intended to apply the FLOP classification method to full NOAA-APT images (which cover an area of more than 2500 * 2500 km), a correction for latitudinal insolation differences is undertaken. For this purpose, recording date and time are extracted from the data base and a correction coefficient matrix is calculated. The original image is then multiplied with that matrix and stored for further processing. The APT pixel values vary in scale and offset due to automatic contrast optimisation during recording of the signal. Consequently, the normalized images need to be calibrated before class definition. Calibration is performed by a linear interpolation between two surfaces defined in the attribute data base. These reference surfaces are usually ocean and Cumulonimbus. If no Cumulonimbus is present (especially in winter), the cloud with the highest albedo value is taken.

3.1.2 infrared band

Ocean and Cumulonimbus are taken as reference points for the interpolation. Where no Cumulonimbus is present, another cold cloud with high emissivity (no cirrus) is taken.

3.1.3 auxiliary data layers

To take advantage of typical patterns characterising certain cloud types, spatial variances of both calibrated input bands are computed using a convolution matrix size of 5*5 pixels. The results are stored as images and integrated in the classification.

3.2 class definition

The amount of radiation transmitted, absorbed or scattered by a cloud depends on parameters like particle phase and concentration, droplet size and size distribution, and cloud optical thickness. The fact that all these parameters show enormous variations over space and time make class definition a rather difficult task. Digitizing training data is probably not a good idea because of the lack of ground truth. Class definition in the FLOP procedure is therefore based on mean radiative properties calculated by radiative transfer models. Definition in the visible band is based on the albedo values published by Liou (1976). The values by Hunt (1973) are taken for class definition the infrared band.

4 UNCERTAINTY AND DECISION MAKING

Little attention has been paid to error propagation and handling of uncertainties in traditional image processing and GIS. Recently, the development of new decision making concepts has lead to a growing awareness in this field (Goodchild and Gopal, 1989; Eastman et al., 1993; Wang and Howarth, 1993, Van Gaans and Burrough, 1993). Now, it is accepted that a pixel value is only one of a whole range of possible values and that a mathematical model for dealing with these possible values is required. The traditional approach of describing their distribution and probabilities is probability theory. Probability theory, however, implies that the errors (or uncertainties) are random in nature, which - in many cases - they are not. The new approaches for characterising situations under uncertainty are based on a theory introduced in 1965 by Zadeh under the name fuzzy set theory. Within these approaches, uncertainty is a multidimensional concept. They are applied successfully to all kinds of problems where no precise mathematical model is known. These concepts are rather intuitive and robust to uncertainties and errors in the data. A number of uncertainties are inherent in the decision making process of classification. It seems straightforward to apply the terms "database uncertainty" and "decision rule uncertainty" as defined by Eastman et al. (1993): "Database uncertainty is that which resides in our assessments of the criteria that are enumerated in the decision rule." "Decision rule uncertainty arises from the manner in which criteria are combined and evaluated to reach a decision."
The following uncertainties can then be regarded as database uncertainty:

- the uncertainty inherent in the definition of "cloud"
- the systematic uncertainty due to different observation methods (or in this case wavelengths)
- stochastic measurement errors (e.g. noise in the satellite data)
- spatial uncertainty due to inaccuracy of geocoding
- temporal uncertainty due to the fact that we are dealing with nonstationary data.

The following uncertainties are decision rule uncertainty:

- the uncertainty in defining the class boundary
- the uncertainty in selecting the criteria (and their level of confidence) on which the decision is based
- the uncertainty in selecting the classification algorithm, parameters, fuzzy operators etc.

5 THE FLOP CLASSIFICATION PROCEDURE

The decision rule is a threshold technique with fuzzy class boundaries, i.e. the class boundary is expressed with some uncertainty or doubt. The mathematical means fuzzification is the Fermi-function. Position and fuzziness of the upper and lower class boundary in input band n are defined by the parameters xn and an of the Fermi function:

$$m(An\ i\ j) = \frac{1}{e^{(xn\ i\ j\ -\ x'n)/an} + 1}$$

m(An i j) : membership of pixel i,j to class A in input band n
xi j : pixel value of pixel i,j
x'n : pixel value where m(An) = 0.5
an : parameter describing the degree of fuzziness (For an=0 the class boundary is "hard" like in conventional threshold techniques)

For each class (A=1..z) defined in the attribute data base and each input band n the values m(An i j) are calculated. The membership values to class A in n bands are then related via fuzzy operators. After testing several operators representing different degrees of compensation, the minimum-operator, one of the basic operators suggested by Zadeh, was chosen.

This operation does not result in plane surfaces in multidimensional space like conventional threshold techniques. The Fermi function allows also a precise description of asymmetric pixel value distributions, as they are known from cirrus clouds for instance. Comparisons of FLOP to ML-classifications and ground observations are published in Heinzmann (1992 and 1993).

6 CLOUD CLIMATOLOGY

The methodology described in chap. 3 and 4 is applied to all geocoded images from 1991. The classification results are compared to ground observations for verification. The months January and August are of special climatological interest. For the test area, maps of mean cloud cover over a 30- year climatological standard period are derived integrating weather type statistics.

6.1 analysis of winter images

Some problems occur with the geocoding of very cloudy winter images, because of a lack of ground control points. These images are either skipped or classified as 100% cloudy after visual interpretation. No special attention had to be paid to the separation of low clouds from snow, as the year 1991 was characterized by very little snowfall in the test area. The classification results classified to 5 types of cloud spatial distribution:

1. cloud cover of 100%, or nearly 100%
2. cloud cover of 0%, or nearly 0%
3. high cloud cover either convective or advective
4. low density cumulus cells
5. low stratus in the Rhine Valley and/or the Swiss Mittelland

These classes are related to the following 6 weather types summarized from those defined by Perret (SMA, 1985) for Central Europe (shown with their relative frequencies 1951-1980):

1. Cyclonic weather type over Central Europe (32%)
2. Anticyclonic weather type over Central Europe (14%)
3. Anticyclone in the 500 hPa niveau over West- or Central Europe (23%)
4. Blocking Action (10%)
5. Cyclone at the 500 hPa niveau over West- or Central Europe (15%)
6. Low pressure trough over West Europe (6%)

Fig.1 shows the absolute frequency of three weather types for the cloud distribution classes for the December and January images. The dominating cloud distribution classes in the test area are class 1, 3, and 5. Class 1 is concurring with a Cyclonic weather type in 77% of the cases. Elevation depending cloud distribution (class 5) concurs with either an Upper Anticyclone (89%) or the Anticyclonic weather type (11%). This situation is characteristic for high pressure conditions in the test area and it is caused by temperature inversions. Only three images are cloud-free, all of them are concurring with an Upper Anticyclone.

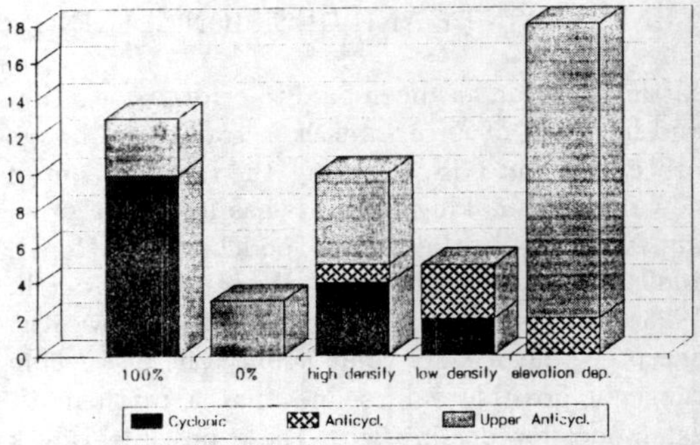

Fig.1 Cloud distribution and weather type in January and December 1991.

6.2 summer

For the summer images, cloud distribution class 5 is defined as "convective clouds related to topography". The other classes as well as the weather types are the same as for the winter images. The overall cloudiness in the area is significantly lower than in winter, while the spatial variance is higher. The dominating cloud distribution classes are classes 4 and 5. This indicates that the most important cloud formation process in summer is convection. The relative frequencies of the weather types for the period 1951-1980 are:

1. Cyclonic weather type over Central Europe (28%)
2. Anticyclonic weather type over Central Europe (13%)
3. Anticyclone at the 500 hPa niveau over West- or Central Europe (22%)
4. Blocking Action (6%)

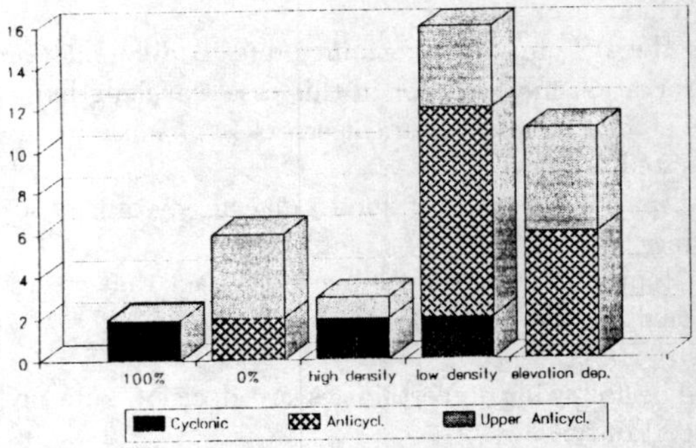

Fig.2 Cloud distribution and weather type in July and August 1991.

5. Cyclone in the 500 hPa niveau over West- or Central Europe (21%)
6. Low pressure trough over West Europe (10%)

7. RESULTS

Mean monthly cloud distributions for each weather type are calculated and then weighted according to the relative frequency of the weather type during the reference period. As a result, maps of mean cloud cover for January and August are published in the REKLIP Climatological Atlas. In addition, the frequency of situations with cloud cover less than 20% and more than 80% have been derived as well as the mean annual cover of cirrus, stratus and cumulus clouds for 1991.

8. CONCLUSIONS

It has been shown that cloud cover data on a regional scale can be derived from NOAA data in APT format. In combination with ground observations cloud cover for climatological standard normals can be derived relating cloud cover and type to weather types.

REFERENCES

Eastman, J.R., Kyem, P, Toledano, J. & W.Jin 1993. GIS and Decision Making. Explorations Geographic Information System Technology. UNITAR, Geneva.

Goodchild, M.F. & S. Gopal (Eds) 1989. Accuracy of Spatial Databases. London: Taylor and Francis.

Heinzmann, U. 1992. Cloud classification on the basis of NOAA-APT data using a fuzzy logic approach. International Archives of Photo grammetry and Remote Sensing. Vol. XXIX, part B7, Commission VII, XVIIth ISPRSS Congress 1992:908-913.

Hunt, G.E. 1973. Radiative Properties of terrestrial clouds at visible and infra-red thermal window wavelengths. Quart. J. R. Met. Soc. 99:346-369.

Liou, K.-N. 1976. On the Absorption, Reflection and Transmission of Solar Radiation in Cloudy Atmospheres. Journal of the Atmospheric Sciences. 33:798-805.

Rossow, W.B. et al. 1985. ISCCP Cloud Algorithm Intercomparison. Journal of Climate and Applied Meteorology 24/9:877-903.

Rossow, W.B. & R.A. Schiffer 1991. ISCCP Cloud Data Products. Bull. of the American Meteorological Society 72/1:2-20.

Stowe, L.L.; Wellemeyer, C.G.; Eck, T.F.; Yeh, H. and the Nimbus-7 Cloud Data Processing Team 1988. Nimbus-7 Global Cloud Climatology. Part 1: Algorithms and validation. Journal of Climate 1:445-470.

Stowe, L.L.; Yeh, H.Y.M.; Eck. T.F.; Wellemeyer, C.G.; Kyle, H.L. and the Nimbus-7 Cloud Data Processing Team 1989. Nimbus-7 Global Cloud Climatology. Part II: First year results. Journal of Climate 2:671-709.

Van Gaans, P.F.M. & P.A. Burrough 1993. The Use of Fuzzy Logic and Continuous Classification in GIS Applications: A Review. Proceedings of EGIS '93.

Wang, M. & P.J. Howarth 1993. Modelling Errors in Remote Sensing Image Classification. Remote Sensing of the Environment. 45: 261-271.

Sensors and Environmental Applications of Remote Sensing, Askne (ed.) © 1995 Balkema, Rotterdam, ISBN 90 5410 524 0

Synergy of Meteosat and NOAA/AVHRR data for the assessment of the role of clouds upon the air-sea sensible heat flux in the Northeastern Tropical Atlantic Ocean

O. Bauer & F. Cauneau
Centre d'Energétique, Groupe Télédétection & Modélisation, Ecole des Mines de Paris, France

ABSTRACT: Meteosat data is used to calculate both cloud cover and the downward shortwave flux over the Northeastern Tropical Atlantic Ocean. Using supplemental data coming from meteorological models, it is intended to used the Meteosat data to study the role of the cloud cover upon the sensible heat flux at the air-sea interface. The study shows a high variability for the shortwave flux for the period under concern.

1 INTRODUCTION

In order to study the interactions between oceans and atmosphere, it is necessary to know the order of magnitude of the various fluxes intervening at the ocean-atmosphere interface, fluxes which are estimated with relations set up between several geophysical parameters (in models) and the quantities measured by the satellites and/or *in situ*. The incoming solar flux is counterbalanced by the net exitance from the surface, the latent heat and the sensible heat fluxes. The shortwave radiation budget is always positive and depends upon the downward shortwave flux and the surface albedo; the longwave radiation budget requires the knowledge of the downward longwave flux, the sea surface temperature (SST) and the surface emissivity; it is generally negative because the surface exitance exceeds the downward longwave flux. The latent heat flux is directly linked to evaporation and the sensible heat flux is linked to the vertical temperature gradient. The purpose of this study is to investigate the potentials of multisatellite analysis to assess the influence of cloud characteristics upon the sensible heat flux. The geographical area under concern is part of the Northeastern Tropical Atlantic Ocean (0 - 40° N, 6° - 35° W) which is the subject of sea-truth campaigns made from August to October 1991 and 1992.

The global coverage provided by the geostationary Meteosat satellites, the high repetitivity of their measurements make these satellites an adequate means for the global estimation of all the terms in the total heat budget. Furthermore, the use of a special set of data called ISCCP-B2, *i.e.* one image produced every three hours permit an accurate description of the time-variability of the processes as well as a fast treatment and a simple handling of the data.

None of the relevant quantities from the budget is directly accessible by remote sensing because of the complex physical processes interfering within the atmosphere. Therefore, several retrieval methods for surface radiation budget from satellite data have been developed. The Heliosat method (Diabaté *et al.* 1988) is used for the mapping of the incident solar radiation at ground level. This method is based on the determination of a cloud index, related to a clearness index defined as the ratio of global solar radiation at ground on a horizontal surface to the horizontal irradiance outside the atmosphere (Cano *et al.* 1986). Its accuracy is about 10% (r.m.s), as demonstrated by numerous comparisons with ground measurements. The downward longwave flux is estimated by a method which makes use of two types of data, satellite data (Meteosat) and thermodynamic grid-point analyses data (Schmetz *et al.* 1986). Moreover this method utilises a clearness index, which is the same as in the Heliosat method, which is already computed.

The influence of clouds upon the sensible heat flux requires the knowledge of some cloud properties: optical like the optical depth and geometrical like the altitude and the fragmentation rate. the Avhrr instrument, on board the sun-synchronous polar Noaa satellites, offers a very large swath with a high radiometric accuracy and a high spectral resolution and a better spatial resolution as Meteosat. Some Noaa/Avhrr-11 data are used for the detection of clouds and the determination of their properties, and also for the determination of the SST which, according to the works of the Marine Environment Unit of the Institute for Remote Sensing Applications, can be related to the sensible heat flux via a sophisticated relationship.

2 THE HELIOSAT METHOD

The Heliosat method, developed at the Ecole Nationale Supérieure des Mines de Paris has two purposes: the estimation of incident solar radiation at ground level from images of the earth acquired by geostationary satellites, and the production of detailed maps of the spatial distribution of the global radiation in real time. It has been chosen for its accuracy as well as for its simplicity, for its computing-time, its possible level of automatization, for the kind of data required by the model and for real-time and operational constraints (Moussu *et al.* 1989 and Diabaté *et al.* 1989).

In preparation to the determination of the cloud index, a reference map of the planetary albedo for clear sky is constructed. Given a time-series of images, it is evaluated at each pixel in a recursive fashion by minimising the variance between the measured radiances and those resulting from the clear sky model, the cloudy cases being eliminated at each step.

2.1 Evaluation of the cloud index

The cloud index $n^t(i,j)$ at point (i,j) for given time t is defined as a function of the characteristic reference albedo $\rho(i,j)$, the apparent albedo at the same point measured by the satellite $\rho^t(i,j)$, and the average albedo of the cloud tops ρ_c (see figure 1).

The computation of the cloud albedo, ρ_c, is performed using the inverse of the algorithm -used for determining the reference albedo map- retaining only the cloudy areas. The histogram of this "only cloud" image provides an estimation of ρ_c using mode computation.

2.2 Statistical relationship between cloud index and clearness index

The clearness index K(i,j) is defined as the ratio of global radiation at ground on an horizontal surface G(i,j)

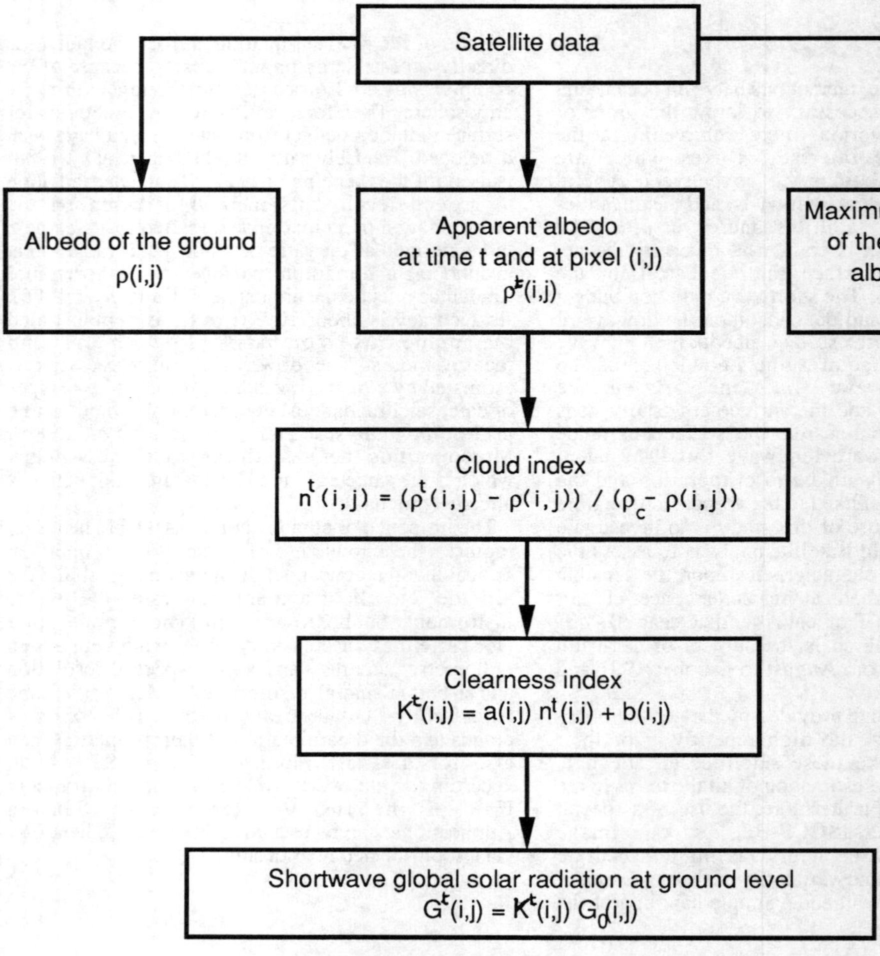

Figure 1: Synoptic scheme of the Heliosat method.

to the horizontal irradiance outside the atmosphere $G_0(i,j)$:

$$K(i,j) = G(i,j) / G_0(i,j)$$

Multiple previous studies did show a constant relationship between the cloud index and the clearness index:

$$K^t = -0.556\, n^t + 0.667$$

Then the hourly global radiation at ground level on a horizontal plane is determined according to the algorithm on figure 1.

2.3 Determination of monthly averages of global radiation at ground level

The monthly average of the hourly global radiation, for each hour interval, is obtained by averaging the images of hourly global radiation for that month.

The cloud index, already computed with the Heliosat method, is also used in the determination of the day-time downward longwave flux, estimated by a method described by Schmetz *et al.* (1986).

The detection (Wald 1991) and the classification of clouds can be done with Noaa/Avhrr images.

3 RESULTS AND DISCUSSION

This method is applied to the Northeastern Tropical Atlantic Ocean basin.

The Meteosat-4 image, taken at 11h30 UT the 09/03/92 (figure 2), displays the area under interest which is a part of the Northeastern Tropical Atlantic Ocean. The *in situ* measurements made in the region of the Canary Islands during the AMOR (Atlantic Measurement of Oceanic Radiation) experiment from 09/02 to 10/14/92 represents the contribution of the Institute for Remote Sensing Applications (IRSA), part of the JRC at Ispra, to studies on global change, especially to energy and mass exchanges at the ocean-atmosphere interface, which is of primary importance for the understanding of the great climate cycles on the earth. In particular, the various fluxes governing the ocean-atmosphere exchanges are well-known by their climatological mean. But, the advent of observational means from space has allowed to acquire a new dimension in these studies, particularly exhibiting the large variabilities, both spatial and temporal, of the various heat fluxes.

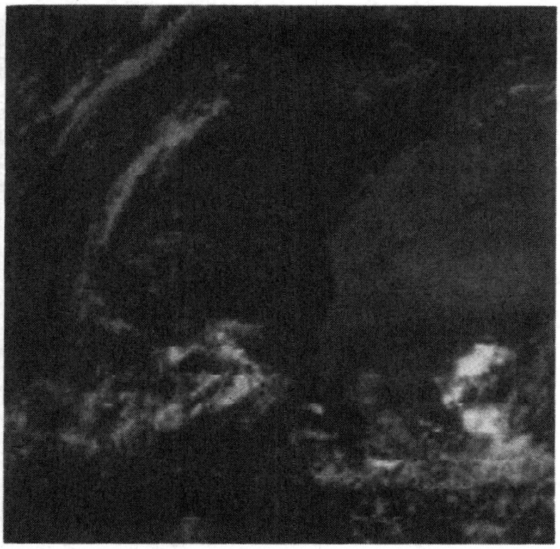

Figure 2: ISCCP-B2 raw image, 11h30 UT, 09/03/92.

The sensible heat flux, the smallest of all the terms contributing to the net heating at the ocean surface in global average, may vary substantially from one order of magnitude in the case of peculiar meteorological and/or oceanographical conditions, like over the western boundary currents for example (such zones are characterised by a strong temperature gradient between the sea and the air). The bulk formulation of the sensible heat flux depends linearly upon vertical temperature gradient, taken at the air-sea interface. Thus, the aim of this study is to assess the various effects of cloud occurrence over this term of the heat balance at the sea surface.

The two images on figure 3 represent monthly averages of the hourly global radiation for the month of September 1992 at two different slots (11h30 and 14h30 UT). The representation of the fluxes is made in grey levels. As expected, the solar radiation is decreasing northwards. These maps exhibit clearly basic structures of the general atmospheric circulation: the effects of both the middle northern latitudes storm track (top of image, at the Northwest of Morocco coast) and the ITCZ (bottom of image, west from Sierra Leone) over the radiation which appears to be lower there than in the other regions. One can also notice a zone of lower radiation located at the centre of the Northwest African upwelling area (along the coast of Morocco).

The two images on figure 4 display maps of hourly global solar radiation for two specific meteorological situations.

The left map (09/26/92 at 11h30 UT) shows up typical effect of depressions over the downward shortwave flux: near the Cape Verde Islands and in the Iberian area. A low radiation zone is also visible at the ITCZ. Thus, each cloudy system takes part in a large attenuation of the downward shortwave flux.

On the second map (10/04/92 at 11h30 UT), a great depression on the Morocco is visible; one must notice the large depression near the Cape Verde Islands: such phenomena are responsible for a local decrease of the solar radiation.

4 CONCLUSION

There are large variabilities, both temporal and spatial, in the shortwave radiation at ground level occurring in

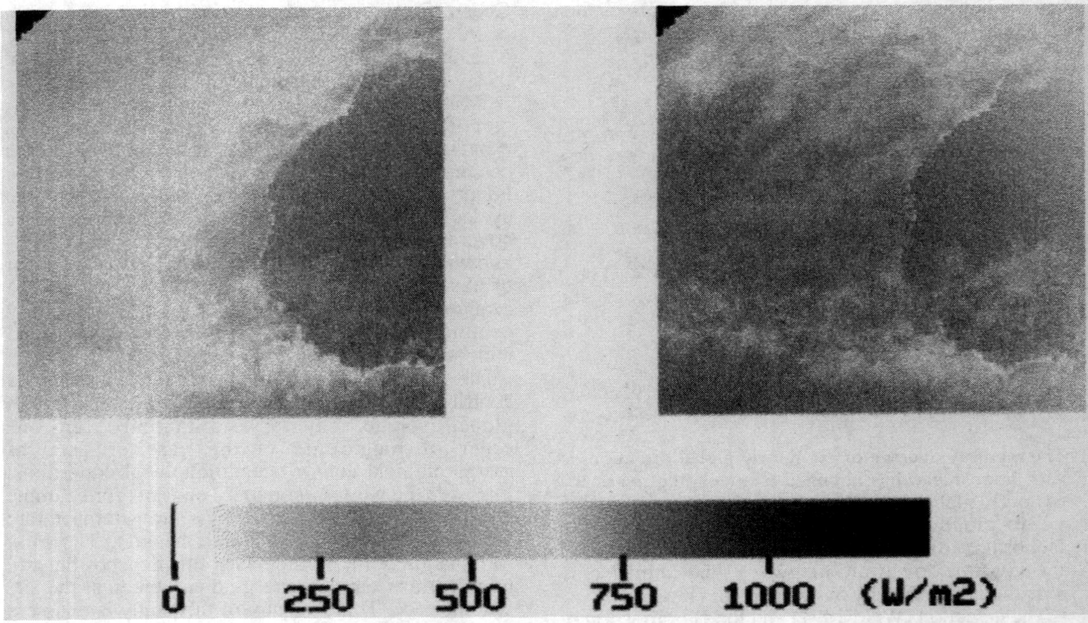

Figure 3: Monthly averages of hourly global solar radiation, September 1992, 11h30 and 14h30 UT.

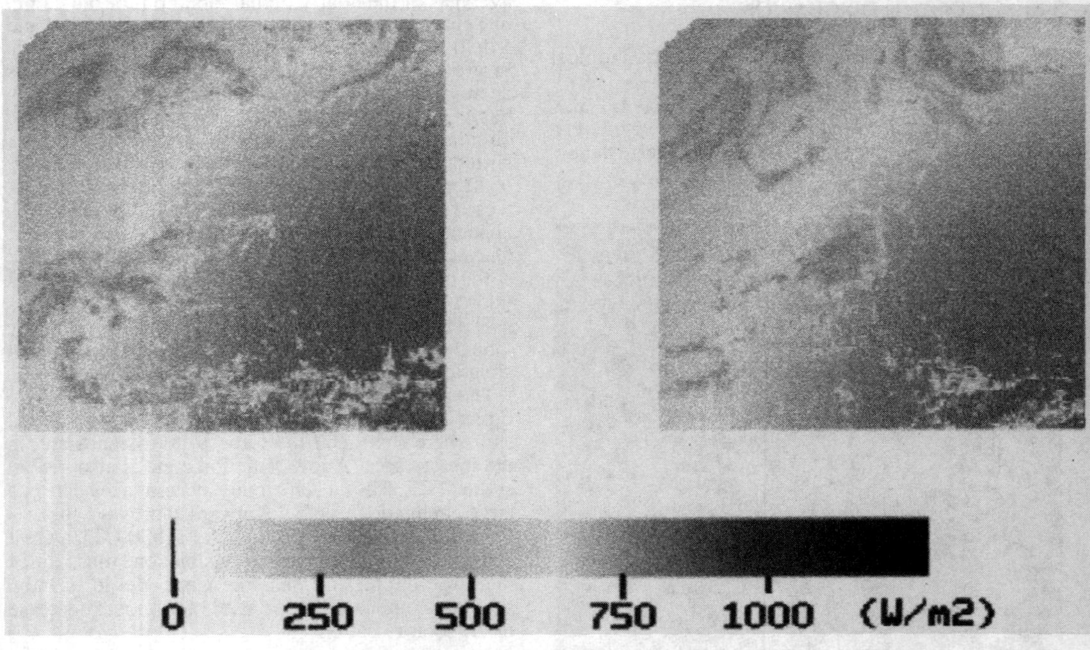

Figure 4: Hourly global solar radiation, 09/26/92 and 10/04/92, both at 11h30 UT.

the area of interest. Such variabilities should be taken into account when evaluating the roles played by the different terms of the air-sea energy balance. Furthermore these results enhance the use of geostationary satellites like Meteosat because of the high repetitivity of the images produced (every three hours).

Moreover the synergy of Meteosat and Noaa/Avhrr data should allow the mapping of the various fluxes and then will give a measure of their relative magnitudes at a local scale. They would also help to assess the influence of each flux on peculiar climatological and/or oceanographical well-marked processes as the upwellings, for example.

REFERENCES

Cano, D., J.M. Monget, M. Albuisson, H. Guillard, N. Regas & L. Wald 1986. A method for the determination of the global solar radiation from meteorological satellite data. *Solar energy*, 37 (1), 31-39.

Diabaté, L., H. Demarcq, N. Michaud-Regas & L. Wald 1988. Estimating incident solar radiation at the surface from images of the Earth transmitted by geostationary satellites: the Heliosat project. *Int. J. Solar Energy*, 5, 261-278.

Diabaté, L., G. Moussu & L. Wald 1989. Description of an operational tool for determining global solar radiation at ground using geostationary satellite images. *Solar Energy*, 42 (3), 201-207.

Moussu, G., L. Diabaté, D. Obrecht & L. Wald 1989. A method for the mapping of the apparent ground brightness using visible images from geostationary satellites. *Int. J. Remote Sensing,*, 10 (7), 1207-1225.

Schmetz, P., J. Schmetz & E. Raschke 1986. Estimation of daytime downward longwave radiation at the surface from satellite and grid point data. *Theor. Appl. Climatol.*, 37, 136-149.

Wald, L. 1991. Elaboration d'un algorithme de décontamination des nuages sur une série pendataire d'images de capteur à grand champ. Final report (CTAMN/91/R/02) 833/CNES/89/6078 of the Centre d'Energétique de l'Ecole Nationale Supérieure des Mines de Paris, France.

[illegible]

[illegible]

[illegible]

[illegible]

[illegible]

[illegible]

[illegible]

[illegible]

[illegible]

Hazards

Sensors and Environmental Applications of Remote Sensing, Askne (ed.) © 1995 Balkema, Rotterdam, ISBN 90 5410 524 0

Volcanic gas emission studies using optical remote sensing techniques

H. Edner, P. Ragnarson, S. Svanberg & E. Wallinder
Department of Physics, Lund Institute of Technology, Sweden

R. Ferrara
Istituto di Biofisica-CNR, Pisa, Italy

R. Cioni, B. Raco & G. Taddeucci
Istituto di Geocronologia e Geochimica Isotopica-CNR, Pisa, Italy

ABSTRACT: The total flux of sulphur dioxide from the Italian volcanoes Etna, Stromboli and Vulcano was determined using the differential absorption lidar (DIAL) technique. The measurements were performed from an oceanographic research ship making traverses under the volcanic plumes with the lidar system sounding vertically. Total fluxes of about 25, 180 and 1,300 tonnes/day for Vulcano, Stromboli and Mt. Etna, respectively were found. These data, collected with an active remote sensing technique, were compared with simultaneous recording with passive differential optical absorption spectroscopy (DOAS) using the sky radiation as the light source. Since the geometry of the light paths crossing the volcanic plume is not well defined in the passive measurements, a correction to the DOAS data is required. Lidar measurements on atomic mercury were also made for the plumes from Stromboli and Vulcano, but the system sensitivity and range only allowed estimates of upper limits for the Hg fluxes.

1. INTRODUCTION

Measurements of the flux of emitted gases from active volcanoes are of great importance both from an environmental point of view and for the monitoring of volcanic activity. Since the beginning of the 70's, a large number of measurements of the flux of sulphur dioxide have been performed from traverses under the plume using passive remote UV correlation spectroscopy (COSPEC) (Stoiber and Jepsen 1973; Caltabiano and Romano 1988; Hoff 1992). Volcanic effluents and aerosols injected into the stratosphere from cataclysmic eruptions have been studied by satellite instruments, e.g. AVHRR (Advanced Very High Resolution Radiometer) and SAGE I and II (Stratospheric Aerosol and Gas Experiment). The TOMS (Total Ozone Mapper Spectrometer) instrument on board Nimbus-7 has been monitoring SO_2 and ozone since 1978 by measuring the ratio of backscattered Earth radiation to incoming solar irradiance and the ultraviolet albedo. Events examined using the TOMS instrument include El Chichón in 1982, Nevado del Ruiz in 1985, Mt. Redoubt in 1989, Cerro Hudson in 1991, and Mt. Pinatubo in 1991 (Bluth et al. 1992). The lidar measurements of volcanic gas fluxes described in the present paper are the first, to our knowledge, in which an active remote sensing technique has been used. A previous attempt to measure the SO_2 flux at Vulcano using a lidar system was reported by Carapezza et al. (1989). However, wind conditions did not allow a reliable flux estimate to be made. For a measurement with a passive instrument, for example COSPEC (Millán and Hoff 1978), the zenith sky light is scattered within and under the volcanic plume, and the illumination source cannot be considered to be entirely above the plume. This increases the uncertainty in the integrated gas column evaluated with passive techniques. In the present experiments, simultaneous active and passive measurements were performed making it possible to address the evaluation problems associated with passive volcanic plume monitoring for the first time.

In the present paper we describe field experiments and discuss data on the sulphur dioxide emission from three Italian volcanoes: Mt. Etna (Sicily), Stromboli and Vulcano (Aeolian Islands). The data were obtained using optical remote-sensing techniques applied during underpasses of the volcanic plumes with the Italian research ship "Urania". All three volcanoes was studied during three consecutive days (September 3-5, 1992). A differential absorption lidar (DIAL) system was used for range-resolved measurements of concentration profiles, while a differential absorption spectroscopy (DOAS) system, employing the ambient sky radiation as the light source, was used for passive remote monitoring of the integrated gas column. Concentration measurements were combined with wind data to obtain the total flux of SO_2. Lidar measurements of the atomic mercury content in the volcanic plumes were also performed, but low concentrations only allowed estimates of upper limits of the mercury flux from Vulcano and Stromboli.

2. REMOTE GAS MONITORING EQUIPMENT

2.1 The lidar system

The lidar and DOAS remote sensing systems are illustrated in Fig. 1. The lidar system is an active sensor for the measurement of gaseous species. Differential absorption lidar uses two different wavelengths, on and off a resonance line of the species of interest. Time-resolved detection of the back-scattering from the short laser pulses enables range-resolved measurements to be made. The system is installed in a Volvo truck and forms a mobile laboratory system (Edner et al., 1987). Its specifications are summarised in Table 1a. Typical durations of measurement were 5 minutes for vertical scans, and 1 minute for horizontal traverses under the plumes.

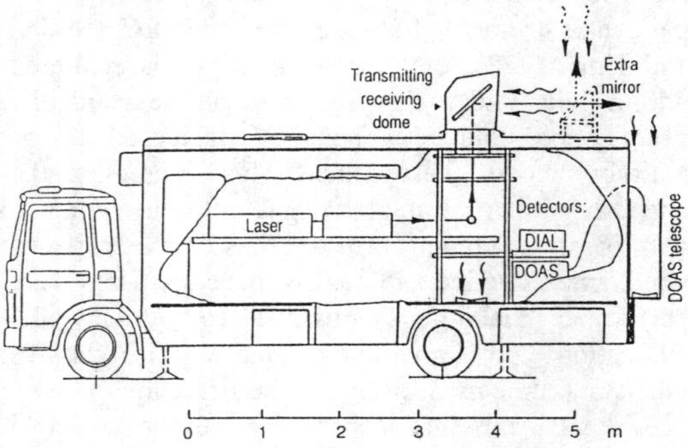

Figure 1. Principal arrangement of the lidar and the DOAS optical remote-sensing systems. The lidar system uses an extra mirror for vertical measurements. The passive DOAS system uses an external telescope fibre-optically connected to a spectrometer.

2.2 The DOAS system

The DOAS system was used as a passive remote sensor, using scattered sunlight collected with a telescope in the vertical position. The telescope was mounted at the back of the truck and the light was coupled to the analysing unit using an optical fibre. The basic lay-out of the system is similar to that described by Edner et al., (1993). The main specifications of the system are summarised in Table 1b. 15,400 spectral scans were added to give 3-minute measurements. Both DOAS and COSPEC data are from correlations with the atmospheric spectrum, but COSPEC does not allow spectral analysis of the data.

Table 1a. Data for the DIAL system

Laser:	Continuum YG682-20, Nd:YAG, rep. rate 20 Hz, 6-9 ns pulse length, 1200 mJ at 1064 nm, 600 mJ at 532 nm and 220 mJ at 355 nm
Wavelength conversion:	Frequency-doubled dye laser, Continuum, TDL60 Dye: R6G (SO_2) and Coumarine (Hg) Dual wavelength, alternate switching
Emitted wavelengths:	SO_2: 300.02 and 299.30 nm, 10 mJ Hg: 253.652 and 253.662 nm, 3 mJ
Telescope:	Newtonian, 40 cm diameter, *f*/2.5 computer-controlled steering mirror for beam direction
Detector:	Thorn-EMI 9816QA, S20 cathode 10^7 gain, ramped 2-10 μs
Digitizer:	LeCroy units 6102, 2xTR8818, 8013, 2xNM8103A, 8-bit resolution, 100 MHz sampling rate
Computer:	2xIBM-compatible PC 386´s

Table 1b. Data for the DOAS system

Receiving telescope:		Diameter 30 cm, *f*/3.3 optical fibre coupling
Spectrometer:		Spex 500 M, *f*/4
	grating	1200 grooves/mm, blaze at 300 nm
Detection system:		Rotating slotted disk
	slit width	100 μm
	scanning time	10 ms
	spectral width	40 nm
	spectral resolution	0.23 nm
Detector:		EMI 9558QA PMT S20 cathode
A/D converter:		12 bit
Multichannel analyser:		1024 channels
Computer:		IBM-compatible AT

3. MEASUREMENTS

3.1 Vulcano

Several vertical scans through the plume were made with the ship positioned in the bay east of La Fossa crater (Baia di Levante). The total flux of SO_2 was evaluated as the integrated concentration multiplied by the wind speed perpendicular to the scan. Wind data were obtained both on the ship and from the top of the volcano, with a vane anemometer. Scans through the plume were also obtained in traverses under the plume with the lidar system in a fixed optical direction. These traverses were performed during the evening and night and, therefore, no simultaneous measurements could be made with the passive DOAS system.

3.2 Stromboli

Five lidar measurements were made at Stromboli during the daytime and in the vertical mode, which enabled direct comparison with the vertical sounding DOAS system. However, due to clouds over the island, reliable DOAS data were only obtained from the second traverse under the plume. These clouds sometimes limited the evaluation range of the lidar measurements to the base of the cloud. The integrated SO_2 vertical column contents obtained from lidar data during the four last traverses are shown in Fig. 2 together with data from the passive DOAS system. Wind data for the flux determinations were obtained from the top of Stromboli.

3.3 Mt. Etna

A total of six traverses were made south of the Messina Straight along the east coast of Sicily. The north-westerly wind during the measurements enabled the plume to be captured at a typical distance of 23 km from the source. Wind data were received from observation stations at Etna at 2500 and 3000 metres above sea level. An example of a range-resolved lidar measurement of the SO_2 plume from Etna is shown in Fig. 3.

Results from the vertical measurements using lidar and passive DOAS, evaluated as the integrated vertical SO_2 column content, are shown in Fig. 4 for all six plume traverses. The data are evaluated up to 3.5 km which was the maximum range for the present set-up.

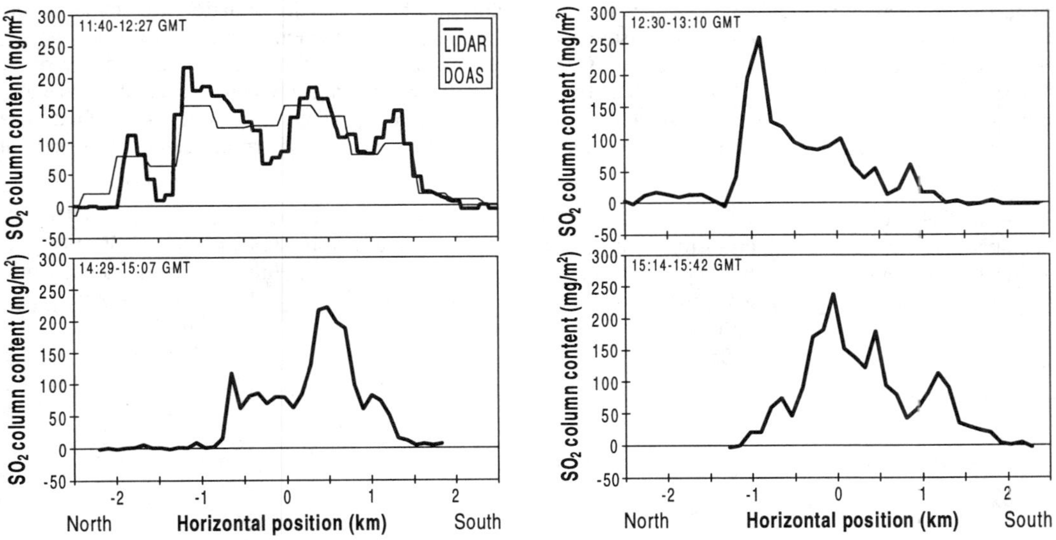

Figure 2. Diagrams of the integrated SO_2 vertical column as measured by lidar and DOAS for Stromboli as a function of ship position, Sept. 4 1992. Although traverses were performed in alternating directions, the scans are presented as if they were all north-south traverses.

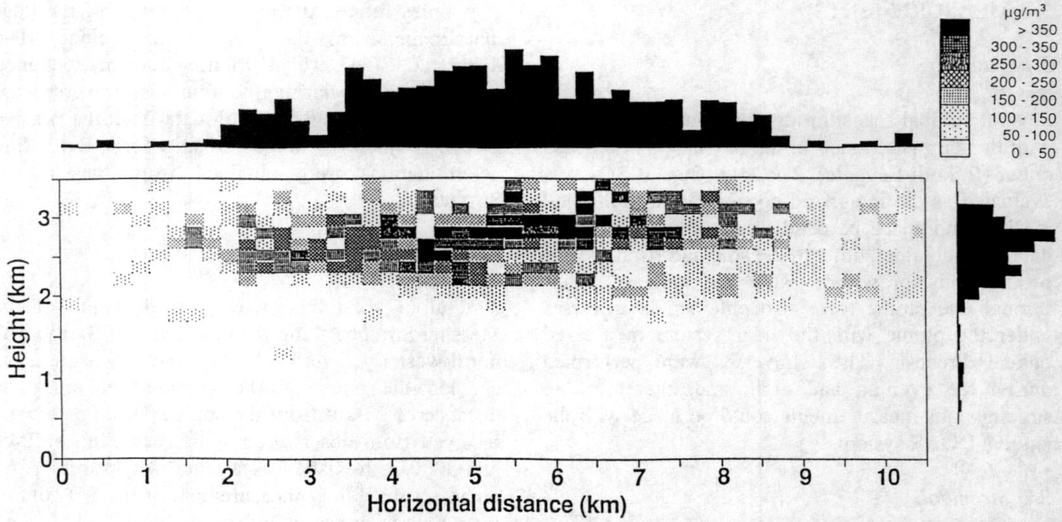

Figure 3. SO_2 distribution in the plume from Etna obtained from range-resolved lidar measurements. This measurement corresponds to traverse number 4 in Fig. 4, and the SO_2 flux was determined to be 1453 tonnes/day.

4. RESULTS

The results of all SO_2 flux determinations for the three volcanoes are given in Table 2. The values are from individual traverses of 40-80 min measurement time for the different volcanoes. Some of the Vulcano values are a mean of 3-6 vertical scans obtained from fixed position measurements. The results of these scans were very close to those obtained during traverses under the plume.

All the lidar data in Table 2 are raw values without any correction for limited range due to low clouds, which was the situation for some of the Stromboli measurements, or possible SO_2 => SO_4 conversion. Some of the Etna fluxes may also be slightly underestimated due to the fact that the entire plume could not be always caught. The DOAS data have not been corrected for the scattering of sunlight

Table 2. Data for SO_2 fluxes from Vulcano, Stromboli and Mt. Etna derived from DIAL and DOAS measurements. Fluxes are given in tonnes/day. The first 5 flux values from Vulcano were deduced from vertical lidar scans and the remaining 3 values are from traverses under the plume. The 4 last values from Stromboli and all the Etna values correspond to the traverses shown in Figs 2 and 4 in the same order.

	Vulcano (Sept. 3)	Stromboli (Sept. 4)		Mt. Etna (Sept. 5)	
	DIAL	DIAL	DOAS	DIAL	DOAS
	26	221		966	1622
	26	259	271	1645	2122
	24	132		1332	1756
	19	137		1453	2010
	17	161		1156	1237
	22			1217	1560
	24				
	26				
mean value	24	182	271	1295	1718
estimated uncertainty	±4	±30	±45	±250	±340

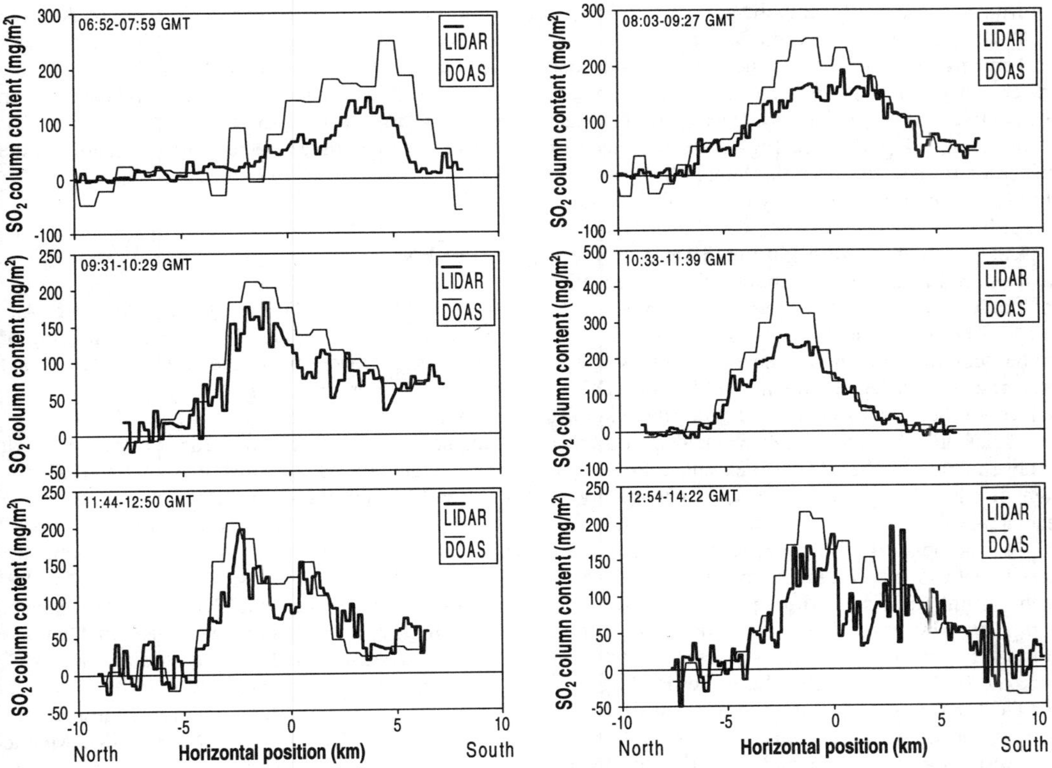

Figure 4. Diagram showing lidar and DOAS integrated SO_2 vertical column data for all six plume traverses performed under the Etna volcanic plume Sept. 5, 1992. Although traverses were performed in alternating directions, the scans are presented as if they were all north-south traverses.

within and under the plume. The uncertainty in the fluxes derived from both DIAL and DOAS measurements is normally determined by the uncertainty in the wind velocity determination. The mean values with the estimated uncertainties are also given in the table.

For some of the traverses at Vulcano and Stromboli, the lidar system was set for the detection of atomic mercury. However, the low concentrations at the distances from the craters necessary for the traverses, only allowed estimates of upper limits for the Hg flux. Taking into account the sensitivity and range for the Hg DIAL measurements, and assuming both Hg and SO_2 plumes to have the same extent, the atomic Hg flux was estimated to be less than 2.5 kg/day from Vulcano and less than 24 kg/day from Stromboli. No estimate of the Hg flux could be made for Etna due to the high altitude of the plume.

5. DISCUSSION

A distinct difficulty with passive optical measurements, compared with active ones, is establishing the effective path through the plume. In an ideal passive measurement all detected photons are scattered above the plume and penetrate the plume in vertical beams. The actual conditions are of course more complicated and a thorough investigation of this subject has been made by Millán, (1980). For a plume with the smallest horizontal extension exceeding the altitude of the plume base, an overestimation of 10-20% could be expected (Millán, 1980; Hoff, 1992). In our measurements, the horizontal extension of the plumes from Stromboli and Mt. Etna was about two and three times the height of the plume base, and the overestimation in the DOAS data should be greater for Mt. Etna than for Stromboli.

Due to clouds over Stromboli, probably formed on condensation nuclei from the volcano, only one traverse under this plume can be used to compare the active and passive systems. To facilitate comparison, all profiles in Fig. 2 are presented as measured from north to south, though the profiles in the right column were actually measured from south to north.

The Etna plume profiles in Fig. 4 are also shown as measured from north to south, (although profiles in the left column were measured from south to north). The first evident feature is the systematically higher column content measured by DOAS. The dramatic difference between DIAL and DOAS values for the first traverse could be attributed to the low solar angle of 29° which resulted in a greater overestimation of the SO_2 burden and also a lower light level which gave an increased uncertainty in the DOAS values.

Our comparative measurements indicate that passive DOAS measurements overestimate the flux from Stromboli by 5%, with a plume 2-2.5 km wide and 1 km above sea level. This value is inferred from only one traverse but is consistent with the estimate of Millán (1980). The corresponding data for the Etna plume, measured at a width of 10 km and a height of 2.7 km, suggest a DOAS overestimation of 33%, which is somewhat higher than predicted by Millán. This may partly result from a possible underestimation of integrated SO_2 concentrations by the DIAL system due to its limited measurement range.

6. CONCLUSIONS

Volcanic SO_2 flux measurements using DIAL and DOAS techniques have been demonstrated for the first time. DIAL, being an active technique, can measure the overhead burden more correctly than passive techniques, which often require correction for scattering within or below the plume. The DIAL technique also gives the height and vertical extension of the plume which are important parameters in the determination of the appropriate wind velocity. Our comparison between the two techniques indicates that correction of gas fluxes from passive optical remote sensors should be considered, although differences in meteorological conditions, horizontal extension of the plumes as well as the distances to them make it difficult to obtain a general correction factor. Our results agree with the estimates of Millán (1980) whose data can be used as guidance in estimating this factor. One advantage of DOAS compared with COSPEC is that in DOAS the complete spectrum is obtained, which makes it less sensitive to drifts and interference. It also makes it possible to determine the quality of the data.

REFERENCES

Bluth, G.J.S., S. D Doiron, C.C. Schnetzler, A.J. Krueger & L.S. Walter 1992. Global tracking of the SO_2 clouds from the June, 1991 Mount Pinatubo eruptions. *Geophys. Res. Lett.* 19: 151-154.

Caltabiano, T. & R. Romano 1988. Messa a punto di metodologie di misura con apparacchiatura COSPEC del flusso di SO_2 da vulcani attivi italiane. *Bollettino GNV, IV*: 135-145.

Carapezza, M., M. Dall´Aglio, G.A. Falchi, G. Elisei, A. Marzorati & E. Zanzottera 1989. Measurement of volcanic SO_2 in the atmosphere by DIAL remote sensing techniques, in *Man and his Ecosystem*, edited by L.J. Brasser and W.C. Mulder, 77-82, Amsterdam: Elsevier.

Edner, H., K. Fredriksson, A. Sunesson, S. Svanberg, L. Unéus & W. Wendt 1987. Mobile remote sensing system for atmospheric monitoring. *Appl. Opt.*, 26: 4330-4338.

Edner, H., P. Ragnarson, S. Spännare & S. Svanberg 1993. Differential optical absorption spectroscopy (DOAS) system for urban atmospheric pollution monitoring. *Appl. Opt.* 32: 327-333.

Hoff, R.M. 1992. Differential SO_2 column measurements of the Mt. Pinatubo volcanic plume. *Geophys. Res. Lett.* 19: 175-178.

Millán, M.M., & R.M. Hoff 1978. Remote sensing or air pollutants by correlation spectroscopy - instrumental response characteristics, *Atm. Environ.* 12: 853-864.

Millán, M.M. 1980. Remote sensing of air pollutants. A study of some atmospheric scattering effects. *Atm. Environ.*, 14: 1241-1253.

Stoiber, R.E. & A. Jepsen 1973. Sulfur dioxide contribution to the atmosphere by volcanoes. *Science* 182: 577-578.

Sensors and Environmental Applications of Remote Sensing, Askne (ed.) © 1995 Balkema, Rotterdam, ISBN 90 5410 524 0

The Near Ultraviolet Spectrometer (NUVS): A new instrument for volcanic sulphur dioxide measurements

S. Mattei, S. Panachia & V. Schena
COnsorzio per la RIcerca su Sistemi di Telesensori Avanzati (CORISTA), Naples, Italy
G. Hoover
Jet Propulsion Laboratory (JPL), Pasadena, Calif., USA

ABSTRACT: Monitoring the chemical composition and flux of volcanic eruptions is one of the prospective methods for eruptions prediction. Due to the world wide distributions of volcanoes in often inaccessible regions, and their unpredictable and hazardous natures, estimation of global volcanic SO2 production using ground-based methods is difficult. Aerospace observation can be a valuable tool for a global monitoring of volcanoes. This paper describes a miniaturized dual field of view fiber optic spectrometer system used for atmospheric and volcanological observations on board the NASA SR-71 high altitude high speed aircraft, during two preliminary engineering flights.

INTRODUCTION

The observation of the atmosphere in the ultraviolet (UV) region of the electromagnetic spectrum has been of increasing interest in the last decade due to many molecules of atmospheric interest that are active in this spectral region (O_3, SO_2, NO, CS_2, HNO_2, HNO_3, etc).
The improvement of atmospheric trasparency knowledge by means of UV remote sensing techniques plays an important role to understand the basic aspects of atmospheric and geophysical processes. In particular, remote sensing of chemical composition and flux of volcanic gas emissions provides a helpfull means to better understand volcanic phenomena. It can be considered as one of the prospective methods for eruption prediction since changes in gas chemistry or flux, such as fluctuations in SO_2 emission rate, are sensitive to the condition of magma (Menyailov 1975, Casadevall et al., 1983).

Sulfur and its compounds (e.g. S, SO_2, H_2S) are among the most environmentally active and common components of volcanic eruptions. Sulphur dioxide has been considered to be the principal precursor of stratospheric H_2SO_4 aerosols, which have been implicated in global cooling observed after past major eruptions. The SO_2 molecule has strong absorption bands in UV region, in particular at the wavelengths which are often used to measure atmospheric ozone (290÷320 nm) (figure 1).

While spacecraft observations have been able to track the emission of SO_2 from large volcanoes into the stratosphere (e.g. TOMS, MLS) (Krueger 1983, Read et al. 1993), these observations have inadequate spectral and spatial resolution to satisfy the needs of vulcanologists.

In the last two years CORISTA and JPL have focused their attention on the identification of the characteristics and performance of a high resolution imaging spectrometer to be flown on board an Earth Probe class polar orbit satellite dedicated to volcanic phenomena monitoring, including the detection and tracking of SO_2 emissions from volcanic eruptions world-wide.

As a first step, a new compact airborne fiber optic spectrometer was built, able to perform simultaneous upward and downward observations.

This paper is intended to describe its preliminary test-flights aboard the high-performance NASA SR-71 research aircraft.

2 INSTRUMENT DESCRIPTION

The Near Ultraviolet Spectrometer (NUVS) consists of two miniaturized fiber optic spectrometers. It is a modular, low cost, low power consumption instrument. The use of optical fibers to collect radiation into the spectrometers greatly simplifies the mechanical layout of the instruments and permits modular development of the sensor. This

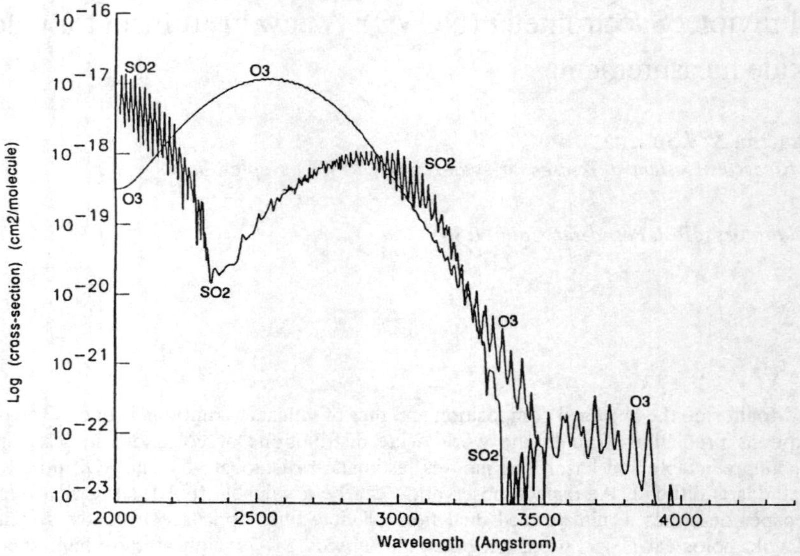

Figure 1: SO_2 and O_3 cross sections

allows the various subsystems (spectrometers, fore optics etc.) to be developed, aligned and tested independently of each other.

Figure 2 shows the instrument configuration adopted for the two engineering test flights on board the NASA SR71 aircraft. The system was assembled and hard-mounted in a little aluminium case (40×25×20 cm). Each small spectrometer (6.7×4.6×2 cm) accepts light from a single fiber, terminated with a standard SMA connector, and disperses it across a CCD linear array (NEC μPD3575) with 1024 photosensitive elements. Some of them have been obscured to give information on the dark current during the flights, since the instrument was not equipped with a shutter. The two analog video signals coming out from the focal plane detectors are digitized by a 12 bits analog to digital converter (ADC). The entire system is operated by a 80386/25MHz microprocessor. The acquired spectra, as well as the operating programs, are

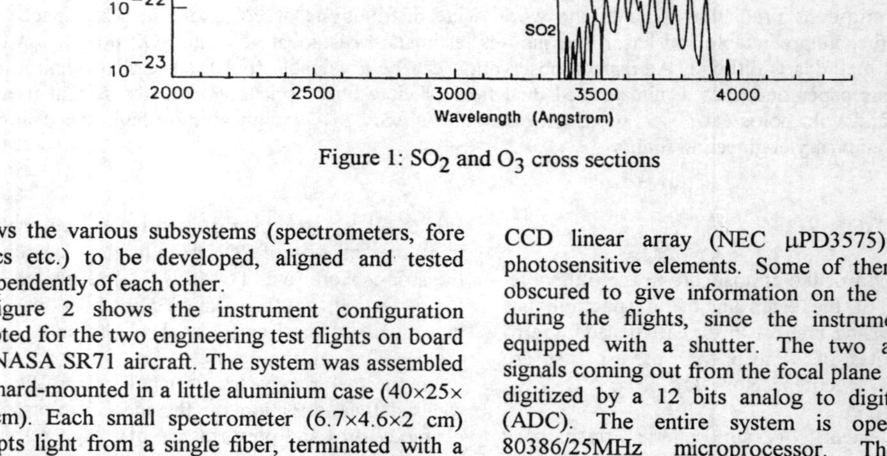

Figure2: high level block diagram of NUVS

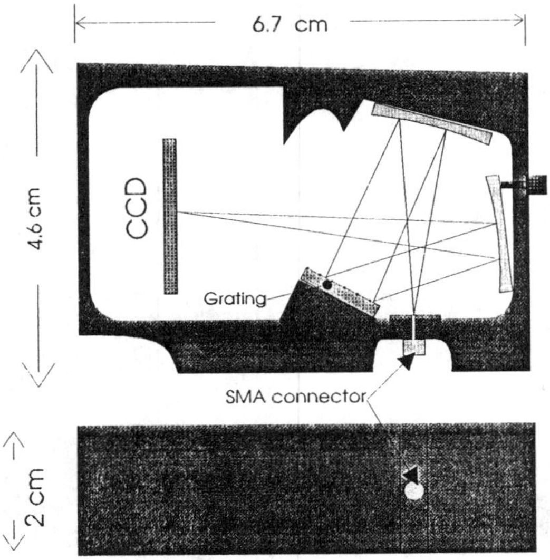

Figure 3: spectrometer optical layout

recorded in binary form on a solid state memory disk with a capacity of 8 Mbyte, that guarantees the system to operate in a non pressurized environment. A temperature sensor measures and registers the temperature of the system at regular interval of time during the flights in order to allow a reconstruction of "simulated" dark current by means of a set of temperature dependent multipliers determined by laboratory thermal characterization measurements of the system.

The spectrometers were each equipped with 2400 l/mm holographic gratings and UV grade silica fibers with 100 μm core diameter and polyimide buffer. The spectral bandpass was 250-365 nm, limited to 270-335 by interferential filters used to reduce stray light. The measured spectral resolution was ≈1.5 nm (Full Width Half Maximum, FWHM). The Instantaneous Field of View (IFOV) of each sensor was 25°, determined by the Numerical Aperture (NA=0.22) of the fibers. Therefore, for an aircraft flight altitude of 26 Km the GIFOV (Ground Instantaneous Field Of View) is a circle with a diameter of ≈11.7 km. It a cruise speed of 1 Km/sec, two subsequent GIFOV are about 50% superimposed.

3 SR71 DESCRIPTION AND ON-BOARD INSTRUMENT SET UP

The SR-71, built by Lockheed, is a long range, two places aircraft, powered by two Pratt and Whitney J58 turbojet engines. It is, for sustained flight, the highest flying, fastest aircraft in the world today. No other aircraft comes even remotely close to matching its performance for cruise flight at altitudes of 26 Km, at speeds of Mach 3 plus. The use of the SR-71 permits, among other things, a realistic simulation of the orbital observational environment with respect to SO_2 and O_3 (it flies above the zone, 20-25 Km, of highest ozone concentrations).

The SR-71 has a range of about 5550 km, without mid-air refuelling and virtually unlimited range with refuelling. Payloads can be carried internally in the mission equipment bays, located within the fuselage chines, in a small compartment (C-bay) forward of the nose wheel well, and in the detachable nose section.

Owing to its peculiar characteristics, small dimensions, fiber optic design and low power consumption, NUVS can be easily accommodated on the SR-71 airplane. Because of the operational requirements, the instrument was mounted in the C-bay, since from there it is not difficult for the fiber optical cables to reach the bottom and the top of the aircraft. Two little holes (about 5 mm diameter) were made in the skin of the aircraft to allow the mounting of two special titanium plates, with female SMA connector shape on one side, to house the blocking filters and fasten the terminal connectors (SMA connectors) of the fibers. The plates were tilted 6° to compensate for the angle of attack of the SR-71 in order to perform nadir observations. Figure 4 shows the geometry of the observation. An ON/OFF switch, to start and stop acquisition, was mounted in the aft cockpit.

The instrument flight configuration was characterized by an integration time of 2048 ms, an 6 sec time interval between two subsequent acquisitions, and an ADC (12 bits) dynamic range of 0-5 V.

The possibility of performing simultaneous upward and downward observations at altitudes above 20 Km, is very useful to analyze and separate the contributions of tropospheric and stratospheric gases (e.g. O_3).

4 LABORATORY CALIBRATIONS AND PRE AND POST-FLIGHT CHECKS

4.1 *Spectral calibration*

The dispersion relation, i.e. the wavelength λ(p) associated to each pixel p of the array, was

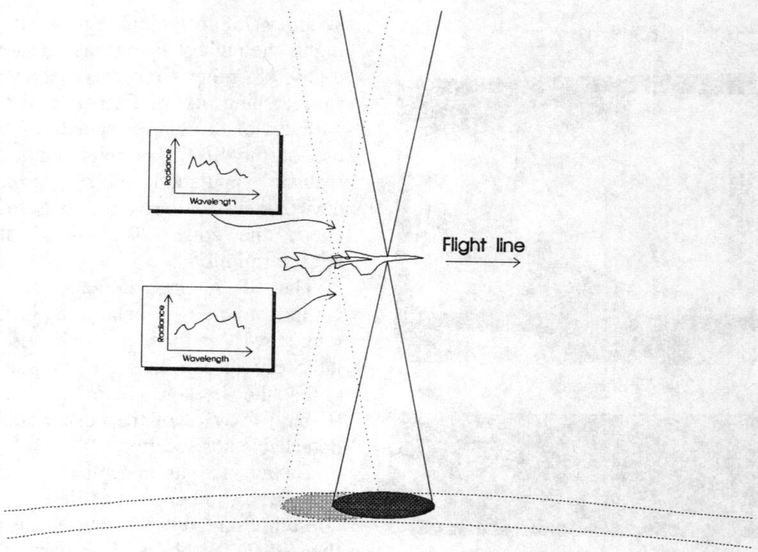

Figure 4: geometry of observation

determined using a Hg low pressure gas lamp as emitter of known spectral lines. Then a nonlinear least square fit procedure was performed to determine the coefficients of the polynomial

$$\lambda(p) = ap^2 + bp + c \quad (1)$$

that gives the appropriate dispersion law of the system. Table 1 reportes the a, b, c values calculated for the spectrometers.

Table 1: coefficients of the dispersion relation polynomial.

	a	b	c
uplooking spectrometer	-22×10^{-6} $\pm2\times10^{-6}$	0.143± 0.003	239.9±0.7
downlooking spectrometer	-23×10^{-6} $\pm3.\times10^{-6}$	0.147± 0.003	245.1±0.7

The spectral resolution (FWHM) was measured in ≈ 1.5 nm for the flight configuration.

The spectrometers were also characterized by measuring and calculating the Optical Modulation Transfer Function (OMTF) and the MTF associated with the focal plane array calculated from its geometrical characteristics (Young 1992). The optical MTF (Slater 1980, Young 1992) was found to be dominant, i.e.

$MTF_{optical}\times MTF_{detectors}\approx MTF_{optical}$
as shown by figure 5.

This means that the focal plane is "oversampled" by the detectors array as we can see from figure 6 where it is reported an Hg vapor lamp spectral line acquired by the downlooking spectrometer.

Due to the narrow width of the lines emitted by these type of lamps, the light intensity distribution can be considered the Impulse Spectral Response Function of the spectrometer and its FWHM is then an index of its spectral resolution. The continuos line in figure 6 is the plot of the function:

$$y(\lambda)=A\,|\,2*J_1(z)/z\,|^2+C \quad (2)$$

where J_1 is the first order Bessel function and $z=\pi/d(\lambda-\lambda_m)$.
This function is the one that best fits, in the sense of minimum squares, the distribution of the intensity in the focal plane produced by spectral line in input. The parameters d, λ_m, and C have been estimated by a non linear least squares fit. This fact suggests the possibility of grouping the pixels to form wider spectral channels to improve the SNR without reducing the spectral resolution of the systems as well as to use a different focal plane array reducing the data to be recorded during the flight.
The procedures above described were repeated after each flights, and produced within the experimental error, the same results. The sensor showed a very good stability.

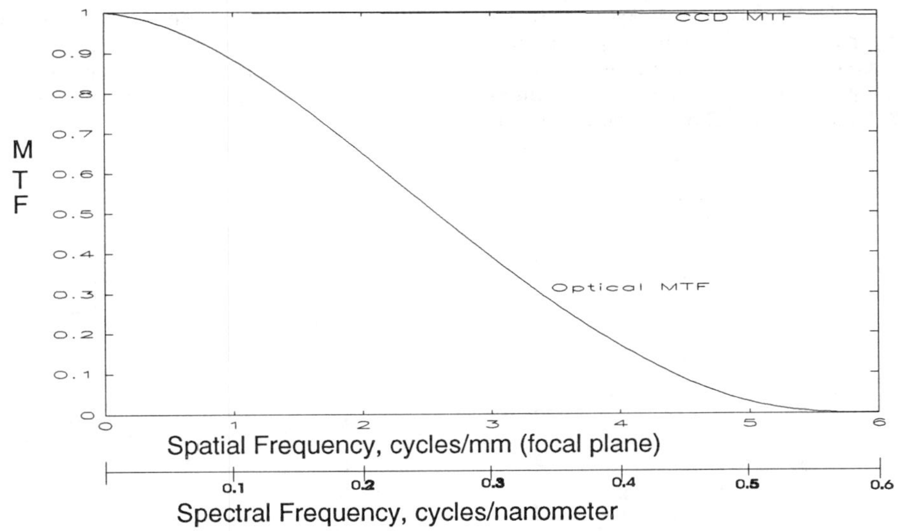

Figure 5: optical and array MTF

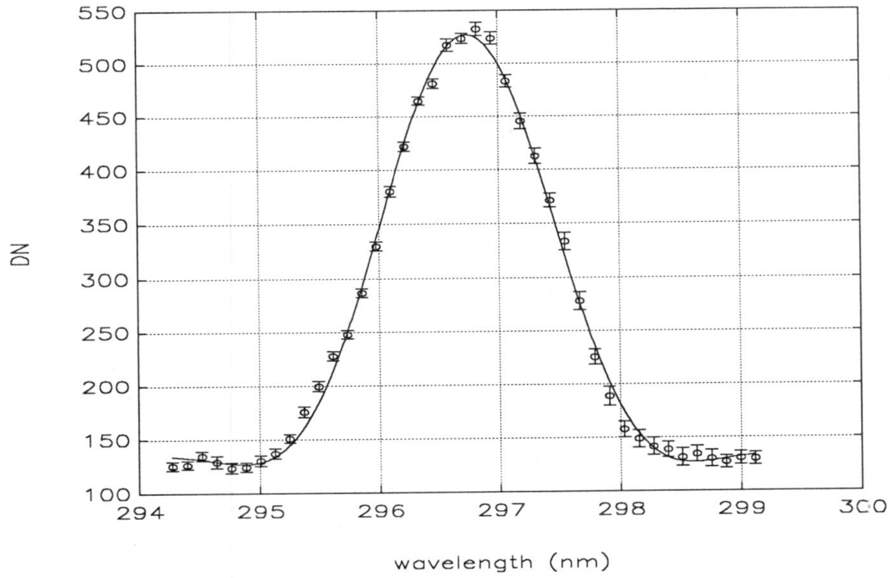

Figure 6: response of the spectrometer to a Hg spectral line

4.2 *Radiometric calibration*

The fundamental aim of the radiometric calibration is to determine a table of multipliers, MULT(λ), that, applied to the acquired data, equalizes the spectrum and converts the raw instrument data (digital numbers) to radiance values. If L(λ) is a reference known radiance in input at our sensor, then the multipliers are given by:

$$MULT(\lambda)=L(\lambda)/(DN(\lambda)-DARK(\lambda)) \quad (3)$$

where (DN(λ)-DARK(λ)) is the sensor response corrected by the dark current. In the above relation a linear response of the detectors is assumed. With our instrumentation it was not possible to perform an absolute radiometric calibration. However, it was possible to determine the instruments transfer function by positioning the fibers just in front of a stable UV lamp and varying the distance. It was found that the response is linear within 5%

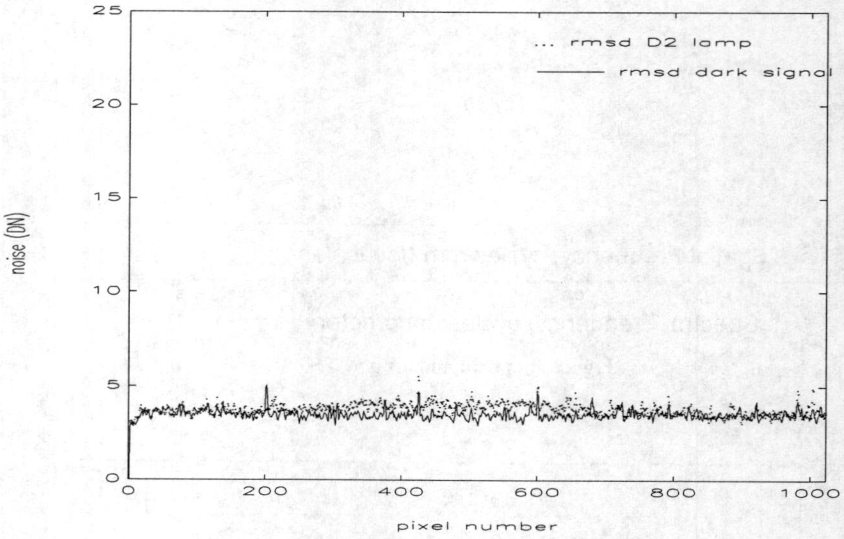

Figure 7: Measured noise (digital number)

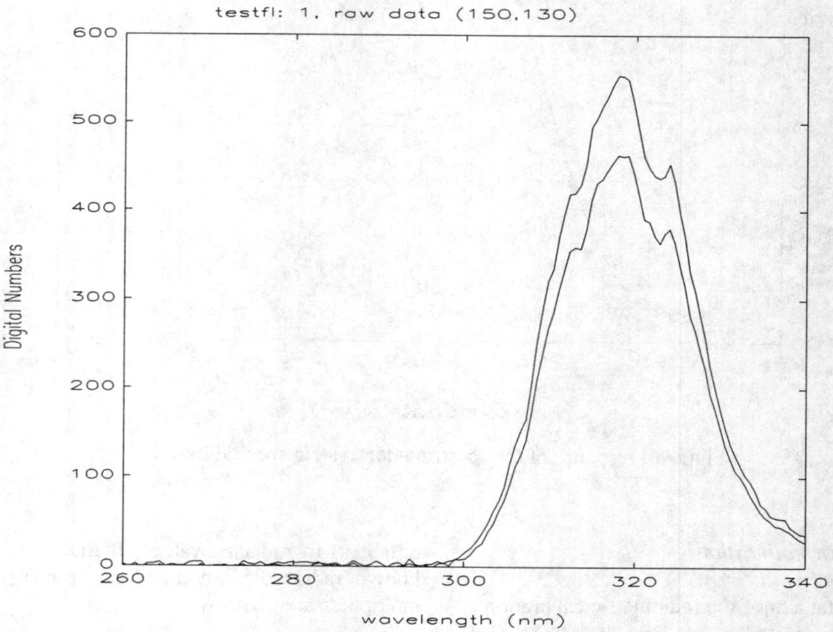

Figure 8: two spectra acquired (down-looking) from an altitude of ≈22 Km

4.3 *Thermal tests*

Thermal test of NUVS instrument was performed in the -10° to +40° C temperature using a thermal chamber made by BEMCO. A set of dark current measurements at different temperature were acquired. From them, we determined a set of temperature dependent multipliers, DM(T,i), to "reconstruct" the dark current during the flights using the count of a obscured pixel (e.g. #5) according to the relation

DARK(T,i)=c(5)*DM(T,i)

where c(5) is the count (Digital Number) of the obscured pixel in a scan during the flight, T is the temperature measured by the temperature sensor in the metallic box during the scan, i is the index (1-1024) of the pixel.

4.4 *Noise*

Noise has been calculated as the Root Mean Square Deviation (RMSD) of a sample of 50 spectra measured from a stabilized irradiance source. The conditions and operating parameters of the instrument during the measurements were set to be as close as possible to the flight ones. Figure 7 shows, for one of the spectrometers, the measured noise. In the figure is also reported the RMSD calculated with the lamp obscured.

As the arrays work far below their saturation levels, the two methods for noise calculations give similar results.

5 RESULTS OF TEST FLIGHTS

The principal objectives of our engineering test flights were essentially to verify the correct working of the sensor during the hypersonic and high altitude (≥22 Km) flights; to acquire a first set of flight data for a better definition of the operating parameters to be used in the next flights, such as integration time, dynamic range, etc; to check the resistance of the system to mechanical stress, especially during the take off and landing of the SR-71; to check the immunity of the system to the e.m. interference due to the other on-board electronics systems.

The test flights took place on 15/7/93 and on 28/7/93 with take off at 8 a.m. from Edwards Air Force Base.

During the flights the sensor worked as expected.

Moreover the pre and post flight checks have confirmed a good mechanical stability of the sensor. The spectral calibration and the radiometric sensitivity were conserved within the errors.

Figure 8 shows, as an example, two raw spectra acquired during the first flight at an altitude of about 22 km. They have been corrected for the dark current using the procedure explained in the previous paragraphs. Each five pixels were grouped to form a single wider spectral channel to improve ($\approx 5^{1/2}$) the SNR without reducing the spectral resolution.

The measured temperature inside the sensor was about 17° Celsius with little oscillation (±1°) during the whole period of the acquisition.

6 CONCLUSION

Main objective of this research program is to design an advanced high resolution very compact spectrometer system for ultraviolet atmospheric observations to be flown on board a free flying satellite.

In this framework the first step was the realization of a prototype of a miniaturized modular fiber optic, dual field of view, spectrometer system. This sensor had been tested with two engineering test flights on board the SR-71 high altitude hypersonic NASA aircraft. The engineering flights had been performed to check the correct working of the prototype system aboard the aircraft and explore the signal to noise ratio regimes for high altitude (>22Km) ultraviolet observations. The results have been satisfactory. The system worked as "expected". The miniature size and fibre optic design have proved to be fundamental for an easy installation on board the SR71. In particular only the use of optical fibres to bring the collected radiation to the spectrometers has permitted reaching the top of the aircraft from the C-bay, thus allowing simultaneous upward and downward acquisitions.

Further research is necessary to improve the system and perform reliable atmospheric high altitude UV observations. First of all a radiometric absolute radiance calibration of the system and the following hardware modifications:

- a shutter to acquire dark scans for a more accurate subtraction of the corresponding current during the flights;
- stabilization of the temperature and possibly cooling of the detectors;
- auxiliary visible cameras, synchronized with the spectrometers scans, to get imaging of the observed scenes;
- a spectral lines emitter to check the spectral calibration during the flight;

- a dedicated microcontroller to manage the system.

The use of additional fore-optics is under evaluation.

After these improvements, other high altitude flights will be performed in order to carry out a quantitative analysis on the acquired data and test atmospheric models and atmospheric inversion methods in the UV region.

At the same time it is being developed a first design of a high resolution spectrometer system to be flown aboard a free flying micro-satellite. The data acquired will integrate those of existing space-borne UV spectrometer systems, as well as those of future sensors such as Gome.

REFERENCES

Menyailov, I.A. 1975. Prediction of eruptions using changes in composition of volcanic gases. *JGR*. 39:112-125

Casadevall, T.J., Rose W., Gerlach T., Greenland L.P., Ewart J., Wundermann R. and Symonds R. 1983. Gas emissions and eruptions of Mount St. Helens through 1982. *Science*. 221:1383-1385

Krueger A.J. 1983. Sighting of El Chichon Sulphur Dioxide Clouds with the Nimbus 7 Total Ozone Mapping Spectrometer. *Science*. 220:1377-1379

Read W. G., Froidevaux L., Waters J. W. 1993. Microwave Limb Sounder (MLS) Measurements of SO_2 from Mt. Pinatubo Volcano". *sub. to Geophysical Research Letters*

Young M. 1992. *Optics and laser*. Berlin: Springer Verlag:

Slater N.S. 1980. *Remote sensing: Optics and Optical Systems*. London: Addison-Wesley Publishing Company

Sensors and Environmental Applications of Remote Sensing, Askne (ed.) © 1995 Balkema, Rotterdam, ISBN 90 5410 524 0

Estimating forest fire danger from AVHRR data

M. Pilar Martín, Luis Domínguez & Emilio Chuvieco
Department of Geography, Universidad de Alcalá de Henares, Spain

ABSTRACT: Estimation of fire danger is attempted from the analysis of AVHRR data. Vegetation indices were generated, that considered both absolute and relative variation, as well as surface temperatures. Because of a lack of weather data, fire danger was assumed from the estimation of potential evapotranspiration, using the Thornthwaite and Samani-Hargreaves methods. PET values were calculated whit data from a set of 18 weather stations in Central Spain and Portugal. The time span covers four biweekly periods of July and August, 1992. Correlations of PET and for the ratio of Maximum Value Composites over Surface Temperature was quite high for most of the periods and the accumulated summer value. Stepwise multiple regressions were calculated with R^2 values of 0.70 to 0.80 for single periods and 0.91 for the accumulated summer value.

1 REMOTE SENSING OF VEGETATION STRESS

Several studies have shown strong correlations between AVHRR spectral vegetation indexes and critical physiological variables. Some important physiological variables are: green biomass Leaf Area Index (LAI), evapotranspiration, primary productivity, chlorophyll activity and Active Photosynthetically Absorbed Radiation by the plant (Cihlar et al., 1991; Asrar et al., 1989; Sellers, 1989). Most of these studies found significant statistical correlations between the Normalized Difference Vegetation Index (NDVI) and vegetation trends, but the specific correlations for single periods were more complicated to establish (Deblonde and Cihlar, 1993).

Fire danger estimation demands frequent monitoring of vegetation stress. Vegetation moisture is a particularly difficult parameter to estimate as it accounts for little spectral variation with respect to other environmental factors (Cohen, 1991a). However, spectral characterization of vegetation stress is possible if temporal profiles are employed which emphasize contrast between the living and dead components (Cohen, 1991b).

The most common approach to the estimation of vegetation stress has been the analysis of NDVI multitemporal series. NDVI provides a good indication of vegetation healthiness. Therefore, decreases in NDVI are related to reduction of plant vigour and greenness which are connected with vegetation moisture content. Multitemporal accumulated NDVI composites have been successfully correlated to accumulated evapotranspiration, rainfall and crop moisture indices (Walsh, 1987; Kerr et al., 1989; Deblonde and Cihlar, 1993).

Several projects developed so far have addressed the operational application of NDVI multitemporal profiles in fire danger estimation (Chuvieco and Martín, 1994). To avoid atmospheric disturbances, most of these studies have used weekly or biweekly maximum value composites. Good correlations have been found for areas of homogeneous herbaceous vegetation such as the grasslands in Australia (Paltridge and Barber, 1988) and the United States (Sadowski and Westover, 1986; Eidenshink et al., 1989 and 1990; Burgan and Hartford, 1993).

An alternative to the use of NDVI series for vegetation moisture estimation is to follow the thermal dynamics of the vegetation cover. Vegetation moisture stress can be measured as the ratio of actual and potential evapotranspiration (ET/PET). ET has proven to be linearly related to net radiation and to the difference between air and surface temperature (T_s - T_a). Over long term periods and at regional scales, correlation coefficients of r=0.99 have been measured between accumulated values of T_s - T_a and Σ ET - R_n (Seguin et al., 1991). As T_a can be obtained from weather stations and T_s from AVHRR thermal channels, this relationship might successfully be applied to fuel moisture estimation, complementing the analysis of NDVI time series.

2 OBJECTIVES

This paper presents the results of a pilot project focused on the assessing the use of AVHRR data for fire danger estimation in mediterranean climate forest. The specific goals of this project may be summarized as follows:

1. Study the spatial relationship of fire danger indices with spectral vegetation indices and surface temperature
2. Estimate fire danger values from NDVI and surface temperature data.
3. Generate fire danger maps from AVHRR data.

The study area is located in the center-west part of the Iberian Peninsula (figure 1), covering approximately 250,000 sq. km. of Spain and Portugal. It includes a wide range of elevations and vegetation types, although mediterranean biomes are dominant. Grasslands are more frequent at lower elevations, while mountain areas are covered by evergreen oaks (*Quercus ilex*) and several types of pine trees (*P. pinaster, P. pinea, P. sylvestris*).

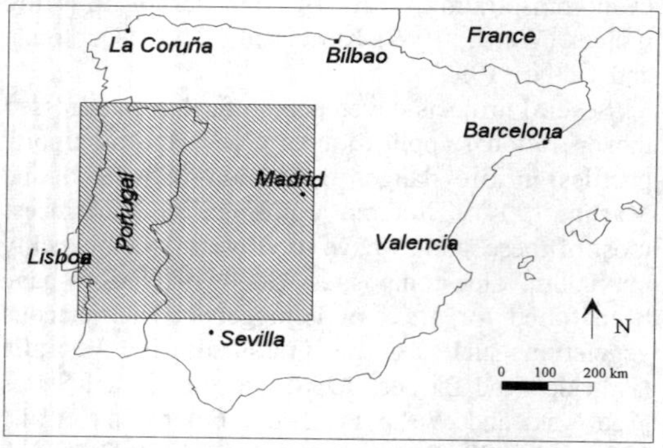

Figure 1. Location of the study area

3 METHODS

3.1 AVHRR image processing

A set of 37 NOAA-AVHRR images was acquired from the receiving station at Valladolid University, in Central Spain, covering the period between June 18 and August 31, 1992. The images were geometrically corrected by orbital models. After that, an image to image registration was applied to the whole series using control points. This last process is required for multitemporal analysis since orbital correction does not match images precisely among.

We hypothesized that several AVHRR-derived variables could be related to vegetation stress and, consequently, used to estimated forest fire danger. Vegetation indices should be related to greenness and vigour. Considering the landscape complexity of mediterranean areas, absolute as well as relative vegetation indices, were derived in order to better discriminate changes related to weather variability. On the other hand, surface temperature is also closely related to vegetation status, since it is directly associated to evapotranspiration.

NDVI images were generated after conversion from digital values to albedos, using standard calibration coefficients (Che and Price, 1992 and ESA, 1992). Multitemporal maximum value composites (MVC) were obtained for four biweekly periods from June to August. The number of images included in each period was related to the availability of cloud-free images, ranging from four (first half of July) to nine (first half of August).

Greenness images were generated from the formula proposed by Kogan (1990):

$$GRN_{rel} = \left(\frac{NDVI_o - NDVI_{min}}{NDVI_{max} - NDVI_{min}}\right) * 100 \qquad (1)$$

where GRN_{rel} is relative percent green, ND_o is the observed NDVI for an specific pixel, and ND_{min} and ND_{max} are the minimum and maximum NDVI values of that pixel for the whole study period.

MVC changes between periods were also calculated to improve the interpretation of water stress stages.

Surface temperatures were generated using the split window approach. This method is the most widely used for sea surface temperature retrieval (Sobrino et al., 1991). However, it is less reliable when applied to land surfaces because the emissivity effect has to be taken into account. An optimized split window algorithm developed for middle latitudes by Coll et al. (1994) was applied here:

$$T_s = T_4 + 2.13(T_4 - T_5) + 0.18 + 50(1-\epsilon_4) - 75\Delta\epsilon \qquad (2)$$

where T_s is the land surface temperature, T_4 and T_5 are respectively the brightness temperature in AVHRR channels 4 and 5, ϵ_4 is the channel 4 emissivity, and $\Delta\epsilon$ is the emissivity difference between channels 4 and 5. In a territory similar to our study area they found temperature errors of 0.7 Kelvin.

In order to obtain an estimation of ϵ_4 and $\Delta\epsilon$ for the different land covers found in our study area, we applied the method proposed by Kerr et al., (1992). which relies on estimating the fractional vegetation cover for each pixel and period based on the NDVI evolution. Their formula follows:

$$P_v = \left(\frac{NDVI_o - NDVI_{min}}{NDVI_{max} - NDVI_{min}} \right) * 100 \qquad (3)$$

where $NDVI_{min}$ and $NDVI_{max}$ are, respectively, the minimum and maximum NDVI values for the whole data set. The former corresponds to a bare soil pixel, and the latter to a fully vegetated pixel. In our case, limits were computed for the histograms of the four MVC periods, -0.2 was the minimum and 0.7 was the maximum. The fractional vegetation cover was computed for every date since cover varies with vegetation seasonal evolution. Cloud contamination of NDVI images was reduced by using a cloud mask generated from a ratio of visible and thermal channels.

Afterwards, channels 4 and 5 emisivities per pixel were calculated using standard emissivity values for bare soil and vegetation (Sobrino et al., 1993):

$$\epsilon_4 = 0.982\ P_v + 0.956\ (1 - P_v) \qquad (4)$$

$$\epsilon_5 = 0.986\ P_v + 0.967\ (1 - P_v) \qquad (5)$$

and finally,

$$\Delta\epsilon = \epsilon_4 - \epsilon_5 \qquad (6)$$

3.2 Weather data

Calculating the fire danger index currently used in Spain, was extremely difficult since most of the weather stations in the study area do not record air humidity and wind data. Consequently, fire danger was estimated from vegetation water stress as defined by potential evapotranspiration (PET) indices. These indices are used by other European countries as a basis for fire danger calculations (Sol, 1993).

Another problem in estimating fire danger from weather station data was related to their location. Frequently, weather stations are located in urban areas, and their data is difficult to extrapolate to nearby forested areas. Only data from weather stations with in a 20 km radius from forested areas, were used although this reduced the sample size. Finally, 18 weather stations were selected (15 in Spain and 3 in Portugal). They are primarily distributed the mountain areas, since the main basins of the Tajo and Duero rivers are covered by agricultural fields.

Since only temperature and rainfall measures were available for these areas, two simple water balance methods were made. The first one is the estimation of Potential Evapotranspiration (PET) proposed by Thornthwaite (TW). Although this method is based on monthly data, a correction formula may be applied to calculate daily data (Sánchez-Toribio, 1992):

$$PET = (135/30) * (t_m/26,5)^a \qquad (7)$$

where PET is the potential ET for an specific day; t_m is mean daily temperature and **a** is a constant related to the heat coefficient. In order to correlate PET measures with vegetation indices derived from AVHRR images, biweekly accumulation values were generated by summing up daily estimations over the same periods when satellite imagery were available.

Some authors have shown that TW method underestimates PET for semiarid climates (McKenney and Rosenberg, 1993). Since most of our study area falls within a mediterranean climate, we have also used the PET estimation method proposed by Samani-Hargreaves (SH). It offers reliable PET estimates in summer water shortage climates while only requiring temperature and latitude data (McKenney and Rosenberg, 1993):

$$PET = 0.00094 * R_a * T_m * (T_{max} - T_{min})\ ½ \qquad (8)$$

where R_a is daily extraterrestrial radiation in mm H_2O equivalent, and T_m, T_{max} and T_{min} are, respectively, mean, maximum and minimum daily temperatures. R_a values were calculated according to the method proposed by Allen et al. (1989). Fifteen days were also accumulated for the correlations with AVHRR-derived variables.

3.3 Estimation of PET from AVHRR data

To reduce problems caused by misregistration between images and/or by the coarse resolution of AVHRR images, the average from a 3 x 3 pixel window taken from a forested area close to each weather station was stored. Stepwise multiple regressions were computed between AVHRR variables (vegetation indices and surface temperatures) and PET estimations, both using TW and SH methods. For every biweekly period, PET was used as a dependent variable, the independent variables were derived from AVHRR data as follows: (i) maximum value composites (MVC), (ii) greenness values (GRE), (iii) accumulated difference of air and surface temperature (ST-AT), (iv) ratio of MVC over mean surface temperature (MVC/ST), and (v) decrement-increment of MVC values over the preceding period (ΔMVC). Table 1 includes the results of multiple regressions calculated for the time span of our study.

Table 1. For estimating multiple regression coefficients PET

Period	MVC	TSTA	MVC/TS	R^2
PET5	103	-1.13	-23.76	0.77
PET4	104.5	-	-23.43	0.72
PET3	125.5	-	-31.21	0.47
PET2	-	-	-16.31	0.51
PES5	70.37	-1.20	-25.56	0.73
PES4	64.55	-	-24.72	0.70
PES3	107.9	-1.42	-32.41	0.81
PES2	58.86	-	-22.29	0.72

PETn: Thornthwaite estimation of PET in period n
PESn: Samani-Hargreaves estimation of PET in period n (5 is the second half of August, 4 the first half of August, 3 the second half of July and 2 the first half of July).

Analysis of this table shows several conclusions:

(i) Assuming the SH estimation of PET is better adjusted to the water-short Mediterranean summer, AVHRR-derived variables can accurately estimate vegetation water stress. Determination coefficients (R^2) above 70 % have been obtained for all the biweekly periods. Less significant values were obtained with the TW method, although similar trends can be inferred.

(ii) The most significant AVHRR variables for estimating PET are MVC and the MVC to surface temperature ratio. The ratio accounts for 40 to 60% of the original variance.

(iii) Greenness variables and changes in MVC between periods were never significant for the multiple regressions, when $F < 2$.

(iv) The accumulated value of air versus surface temperature was usually not selected either.

Variables from previous time periods were also included in the multiple regressions, but most of the times they did not significantly improve determination coefficients. In the case of the SH value for the second half of August, R^2 only rose from 0.73 to 0.79 after the MVC/TS variable from the first half of August was included. The other variables were not significant.

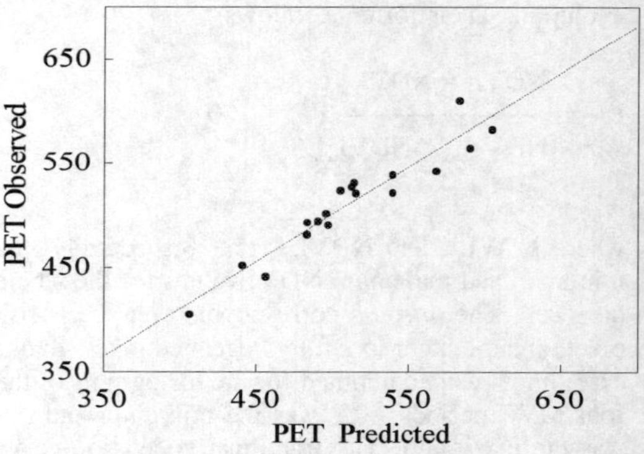

Figure 2. Estimation of the accumulated values of PET (Samani-Hargreaves) from AVHRR data

Finally, the accumulated PET values were correlated with the accumulated AVHRR variables, in order to study the seasonal relationships between weather and satellite data. The R^2 values obtained to estimate PET were 0.91 using the SH method (figure 2), and 0.737, using the TW method. The most significant variable in both cases was MVC/TS ($r=0.74$ for SH and $r=0.62$ for TW). The final equation for the accumulated value of SH was:

$$PET_{SH} = 57121.5 + 55.57\ MVC - 2.22\ TSTA - 30.21\ MVC/TS - 36.24\ \Delta MVC \quad (9)$$

This regression was significant at a 99.99 confidence level ($F=31.87$). This equation may be used to obtain maps of PET values by applying the coefficients to the previously generated AVHRR variables.

CONCLUSIONS

Using a set of 18 weather stations located in Central Spain, we have shown that PET estimation can be accurately derived from vegetation indices and surface temperature obtained from AVHRR images. The most significant variable is the ratio of MVC to surface temperature. Assuming that PET is closely related to vegetation water stress and, therefore, to probability of ignition and forest fire danger, this variable may be considered a good indicator of fire danger estimation. Further studies are required to correlate this variable with more precise estimations of PET, and will include wind and air humidity data.

ACKNOWLEDGMENTS

This research is funded by the MINERVE I project, supported by the European Commission Environmental program, under the supervision of DG-XII. Weather data was provided by the Instituto Nacional de Meteorología. Special thanks to José Ramón Marín, Anselmo Peinado and José Luis Albadalejo from the regional centers of Sevilla, Badajoz and Madrid and C.F. Warren of the I.C.E. at the U.Alcalá for linguistic assistance.

REFERENCES

Allen, R.G., M.E.Jensen, J.L.Wright & R.D. Burman 1989. Operational estimates of reference evapotranspiration. *Agron. J.* 81:650-662.

Asrar, G., R.M.Myneni & E.T.Kanemasu 1989. Estimation of plant canopy attributes from spectral reflectance measurements, In G.Asrar (ed.), *Theory and Applications of Optical Remote Sensing*: 252-296. New York: Wiley.

Burgan, R.E. & R.A.Hartford 1993. *Monitoring Vegetation Greenness with Satellite Data*. USDA Forest Service, INT-297, Ogden.

Che, N. & J.C.Price 1992. Survey of Radiometric Calibration Results and Methods for visible and Near Infrared Channels of NOAA-7, -9, and -11 AVHRRs. *Remote Sensing of Environment*. 25:19-27.

Chuvieco, E. & M.P. Martín 1994. Global fire mapping and fire danger estimation using AVHRR images. *Photogrammetric Engineering and Remote Sensing* (in press).

Cihlar, J., L.St.Laurent & J.A.Dyer 1991. Relation between the Normalized difference vegetation index and ecological variables. *Remote Sensing of Environment*. 35:279-298.

Cohen, W.B. 1991a. Response of vegetation indices to changes in three measures of leaf water stress. *Photogrammetric Engineering and Remote Sensing*. 57:195-202.

Cohen, W.B. 1991b. Chaparral vegetation reflectance and its potential utility for assessment of fire hazard. *Photogrammetric Engineering and Remote Sensing*. 57:203-207.

Coll, C., V.Caselles, J.A.Sobrino & E.Valor 1994. On the atmospheric dependence of the split-window equation for land surface temperature. *International Journal of Remote Sensing*.15:105-122.

Deblonde, G. & J.Cihlar 1993. A multiyear analysis of the relationship between surface environmental variables and NDVI over the Canadian landmass. *Remote Sensing Reviews*. 7: 151-177.

ESA 1992. *SHARP User's Guide*. Frascati: European Space Agency.

Eidenshink, J.C., R.H.Haas, R.M.Zokaites, D.O. Ohlen & K.P. Gallo 1989. *Integration of remote sensing and GIS technology to monitor fire danger in the Northern Great Plains*. U.S.Geological Survey, Final Report.

Eidenshink, J.C., R.E.Burgan & R.H.Haas 1990. Monitoring fire fuels condition by using time series composites of Advanced Very High Resolution Radiometer (AVHRR) data. *Proc. Resource Technology 90*. ASPRS, Washington, D.C.: 68-82.

Kerr, Y., J.Imbernon, G.De Dieu, O.Hautecoeur, J.P.Lagouarde & B.Seguin 1989. NOAA-AVHRR and its uses for rainfall and evapotranspiration monitoring. *International Journal of Remote Sensing*. 10: 847-854.

Kerr, Y.H., J.P.Lagouarde & J.Imbernon 1992. Accurate land surface temperature retrieval from AVHRR data with use of an improved split window algorithm. *Remote Sensing of Environment*. 41:197-209.

Kogan, F.N. 1990. Remote sensing of weather impacts on vegetation in non-homogeneous areas. *International Journal of Remote Sensing*. 11:1405-1419.

McKenney, M.S & N.J.Rosenberg 1993. Sensitivity of some potential evapotranspiration estimation methods to climate change. *Agricultural an Forest Meteorology*. 64:81-110.

Paltridge, G.W., & J.Barber 1988. Monitoring grassland dryness and fire potential in Australia with NOAA/AVHRR data. *Remote Sensing of Environment*. 25:381-394.

Sadowski, F.G., & D.E.Westover 1986. Monitoring the fire-danger hazard of Nebraska rangelands with AVHRR data. *Proc. 10th Canadian Symp. on Remote Sensing*, Edmonton: 355-363.

Sanchez-Toribio, M.I. 1992. Métodos para el estudio de la evaporación y evapotranspiración. *Cuadernos técnicos de la SEG*. 3:3-35. Logroño: Geoforma Ediciones.

Seguin, B., J.P.Lagouard & M.Savanne 1991. The assessment of regional water conditions from meteorological satellite thermal infrared data. *Remote Sensing of Environment*. 35:141-148.

Sellers, P.J. 1989. Vegetation-Canopy Spectral reflectance and Biophysical processes. In G.Asrar (ed.), *Theory and applications of Optical Remote Sensing*:297-335. New York: Wiley.

Sobrino, J.A., C.Coll & V.Caselles 1991. Atmospheric correction for land surface temperature using NOAA-11 AVHRR channels 4 and 5. *Remote Sensing of Environment*. 38:19-34.

Sobrino, J.A., C.Coll & V.Caselles 1993. La medida de la temperatura desde satélites. In *La teledetección en el seguimiento de los fenómenos naturales. climatología y desertifación*: 153-181. Valencia: Universitat de Valencia.

Sol, B. 1993. MINERVE project. Internal Report.

Walsh, S.J 1987. Comparison of NOAA-AVHRR data to meteorological drought indices. *Photogrammetric Engineering and Remote Sensing*. 53:1069-1074.

Sensors and Environmental Applications of Remote Sensing, Askne (ed.) © *1995 Balkema, Rotterdam, ISBN 90 5410 524 0*

Fire monitoring in Spain using NOAA thermal data and NDVI

P. Illera, A. Fernández & J. A. Delgado
Departamento Física Aplicada I, Universidad de Valladolid, Spain

ABSTRACT: Our aim is to design an operational system for fire monitoring using AVHRR images which we obtain by means of a HRPT receiver installed in our Department. It identifies forest fires using thermal data and albedo and provides an estimation of the affected area by means of the thermal channels and also comparing the NDVI values before and after the fire ocurrence. Finally, the possibility of generating a risk index based in the accumulation of vegetal combustible has been studied using the temporal evolution of the NDVI. The whole procedure has been applied to a set of fire events in Spain. The results have been compared to ground data generated by the Spanish Forest Service. The algorithm is suitable to be implemented in a fire monitoring network.

1 INTRODUCTION

Forest fires are a major environmental problem in Spain with large areas beeing affected every Summer. Remote Sensing allows us to obtain information about large wooded areas. It has been used over the last two decades in the study of forest fires following different approaches: fire identification, determining affected areas, analysis of the evolution of the vegetation after the fire and prediction of risk. The AVHRR radiometer, on board the NOAA satellites, is one of the best operational systems for fire monitoring. It provides multispectral data covering wide areas and offers excellent temporal sampling, as it passes over the same part of the earth several times a day.

A hihg temperatures, fire points emit high radiances in the atmospheric window of [3-5μm]. Channel 3 of the AVHRR is suitable for fire detection (Matson et al., 1984; Muirhead and Cracknell, 1985; Lee and Tag, 1990). In addition, the multispectral technique proposed by Dozier (Dozier, 1981; Matson and Dozier, 1981) allows the estimation of the fire temperature and affected areas, combining data from channels 3,4 and 5 (Matson et al., 1984; Flannigan and Vonder Haar, 1986; Robinson, 1991). As regards the prediction of areas at risk, the accumulation of vegetal fuel may be observed by studying the evolution of NDVI type vegetation indices, derived from AVHRR channels 1 and 2 (Paltridge and Barber, 1988; López et al., 1991).

Our aim is to desing an operational system for fire monitoring, using NOAA AVHRR images which we obtain by means of a HRPT receiver installed in our department. Firstly, an automatit algorithm for the identification of fires has been developed. It automatically detects the affected pixels, using a local threshold in the difference T_3-T_4 brightness temperature, gives UTM coordinates of the affected pixels and indicates whether they correspond to the same fire. Fire temperatures and affected areas are also calculated by means of Dozier´s Technique (Dozier, 1981). In a second step, this areas are compared to those derived from vegetation indices generated for days before and after the fire ocurrence and also to ground data provided by the Spanish Forest Service. Finally, the evolution of the NDVI over a period before the fire event has been studied in order to generate a risk index based in the accumulation of vegetal fuel.

The whole procedure has been applied to different situations in Spain. The results have been compared to ground data generated by the Spanish Forest Service. The comparison allows us to adapt the detection thresholds and to study the reliability of the method and any error in the calculations.

2 DATA SET

The study has been carried out using NOAA AVHRR images from April to October 1993. The five AVHRR channels have been employed. The images have been captured by the receiver in our department and georeferenced and calibrated in order to obtain brightness temperature and albedos by means of our own software (Illera et al., 1993).

Gound data provided by the Spanish Forest Service (ICONA) have also been used. They have allowed us to adapt the detection thresholds and to carry out the study about fire risk. Fire sizes and their locations in a ten by ten km grid are available. In figure 1, the data used in this work are shown. They correspond to 617 fire events in Valencia and East Andalucia that took place in the months of July, August and September, 1993.

3 FIRE IDENTIFICATION

An improvement of Dozier's multispectral technique has been aplied. Thermal channels brightness temperatures and channel 2 albedo are used. The mean value μ and standard deviation σ in 20*20 pixels areas are generated for every channel and for the difference in channels 3 and 4 temperatures. This is done at every pixel in the scene and the following tests are carried out:

1) Cloud detection using thresholds in channels 2, 4 and 5.

2) $T_3 - T_4 > \mu_{3\text{-}4} + 2\sigma_{3\text{-}4}$

3) $T_3 > \mu_3 + 2\sigma_3$

4) $\alpha_2 < \mu_2 + \sigma_2$ in daytime images to avoid saturation problems.

The algorithm automatically detects the affected pixels and indicates whether they correspond to the same fire. It also gives UTM coordinates of the pixels of possible fires which appear in the image. Finally, fires´ temperatures and affected areas are derived using Dozier´s multispectral technique. The possibility of carrying out a follow up of the fire´s temporal evolution is also considered, so that the final result is obtained by means of detection carried out over several images. Finally, it permits the treatment of large fires as well as those smaller than one pixel by means of a threshold based in Dozier's technique (Illera et al., 1993).

An example is presented in figure 2. It corresponds to a forest fire in Alfara, near the Ebro river delta. In table 1, some more examples are presented. The forest fires took place in different dates of July 1993. UTM coordinates correspond to the central pixel. In SAN MATEO´s fire Dozier´s algorithm does not work. This is due to that T_4 temperatures

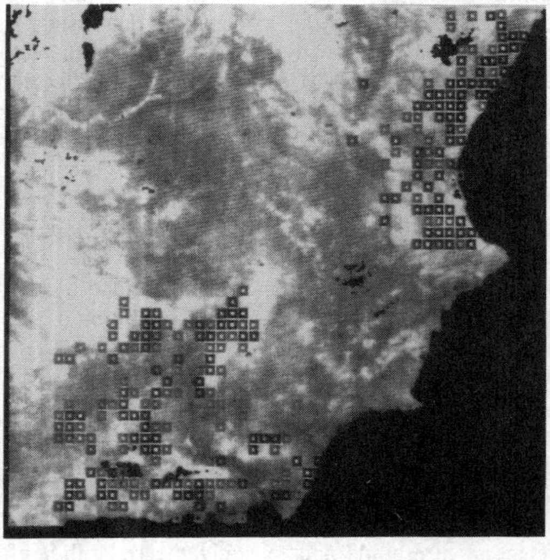

Figure 1: 617 fires in Valencia and east Andalucia located in a 10*10 km grid.

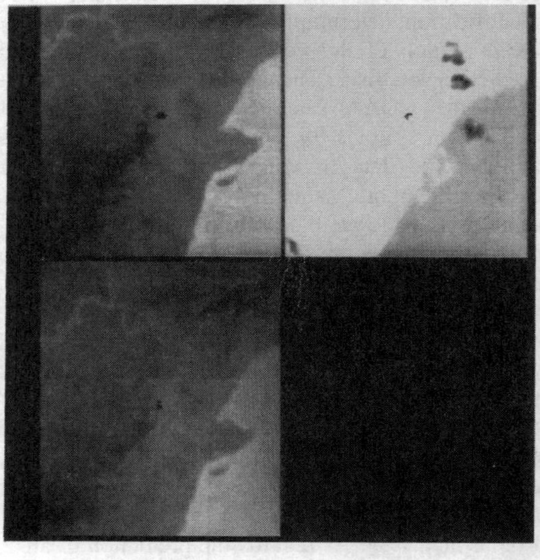

Figure 2: Forest fire in Alfara. It broke out on July 21 1993. The images correspond to days 21 (night) and 22 (day and night).

in some affected pixels are superior to those in pixels surounding the forest fire. This is a limitation of the method in terms of analysis, but it does not affect to the identification.

Table 1: fire identification and analysis using Dozier´s technique.

TOWNSHIP	XUTM	YUTM	ZONE	SIZE(ha)
ALFARA	285	4529	31	902
ARGELITA	725	4435	30	4400
SAN MATEO	265	4485	31	---
ENGUERA	705	4315	30	100

4 AREA ESTIMATION USING NDVI

The NDVI generated for days before and after the forest fire have been used to estimate affected areas by means of a threshold in the difference between both indices. An example, corresponding to the forest fire in Alfara, is shown in figure 3. Some problems of radiometric normalization can arise if the two images are not close enough in time. In table 2 we present the results for the forest fires considered before. The sizes provided by The Spanish Forest Service are also included. Comparing this results and the ones derived from Dozier´s technique (table 1), we observe that both are quite similar. The advantage of this procedure is that it can be always applied. In conclusion, both procedures are suitable to make a quick estimation of the affected areas.

Table 2: Area estimation using the difference between NDVI values before and after the fire. Ground data provided by the Spanish Forest Service (ICONA) are also included.

	SIZE NDVI(ha)	SIZE ICONA(ha)
ALFARA	1200	900
ARGELITA	4700	4896
SAN MATEO	2000	3520
ENGUERA	1100	300

5 FIRE RISK

The temporal evolution of the NDVI in the period before the fire has been studied in order to generate a risk index based in the amount of vegetal fuel. A total of 40 AVHRR images has been used to generate the NDVI MVC every ten day period. The possibility of evaluating the risk by means of the accumulated slope AS has been examined:

$$AS_n = \sum_{i=1}^{n} \frac{NDVI(t_i) - NDVI(t_{i-1})}{t_i - t_{i-1}} \qquad (1)$$

The AS has been calculated every ten days from

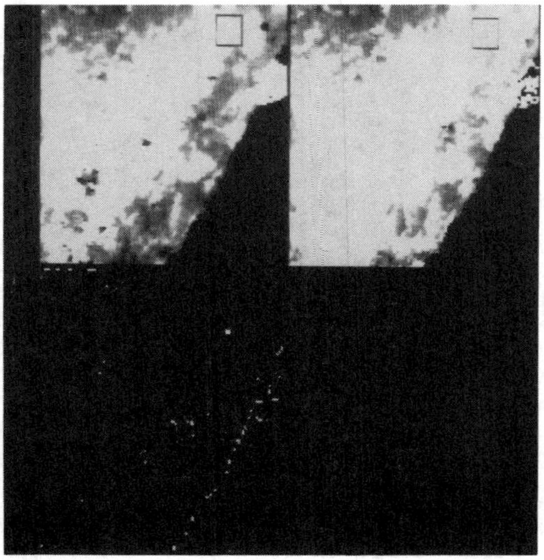

Figure 3: Estimation of the affected area. Forest fire in Alfara. The NDVI before and after the fire event have been used.

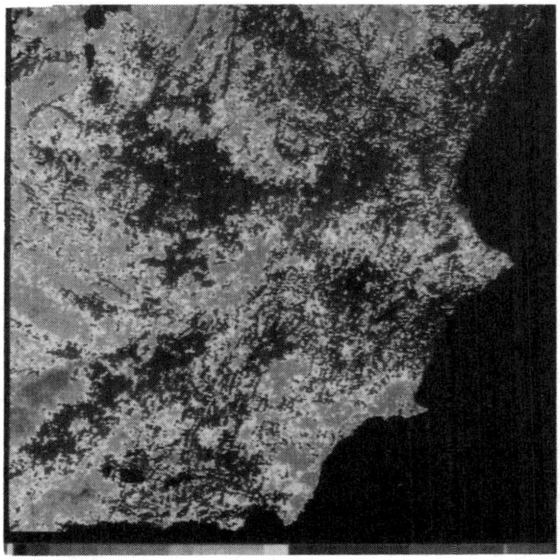

Figure 4: AS for the period from the 9th to the 18th of August.

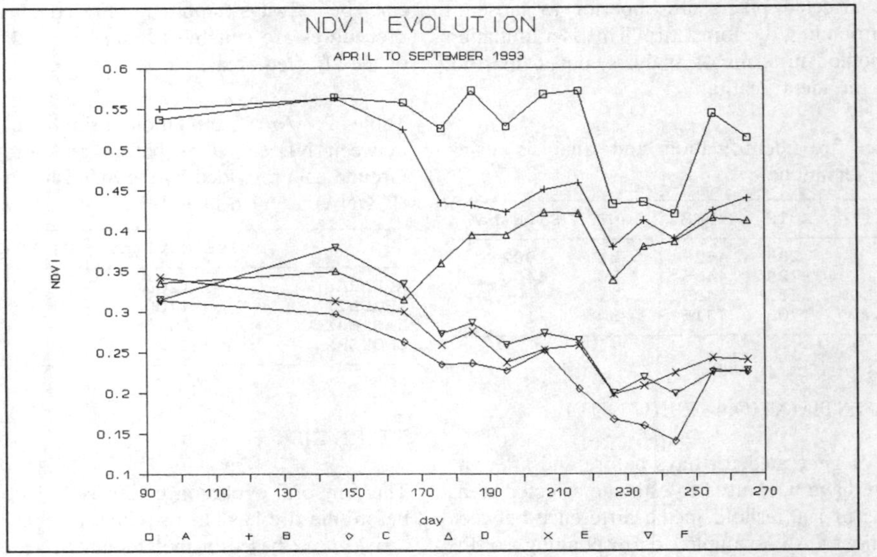

A: PINES
B: HOLM OAKS
C: SHRUBS
D: VINEYARD
E: OLIVE TREES
F: CROPS

Figure 5: NDVI evolution for different vegetal covers.

July to September, and the data of The Spanish National Forest Service have been superimposed to the images.

In figure 4, the AS for the period from the 9th to 18th of August is presented. There is some confusion between surfaces with similar slope values. The NDVI evolution for different vegetal covers is shown in figure 5. Most of them have negative AS values. Forest areas present intermediate values. The minimun ones can correspond to crops, shrubs or forest areas at risk.

In order to avoid possible mistakes between classes, the NDVI integral over the period has been calculated. Forest surfaces present high values, as can be seen in figure 6 that correspond to the same period as the AS shown above. In the bars diagrame (figure 7), ground and images data for the whole period have been superimposed. 58% of fires and 87% of burnt surface take place for integral values higher than the average over each period analysed. The frequencies are also higher than normal. Consequently, we propose to use the integral like a mask to determine which areas in the image can be at risk and to asses the risk grade by means of the AS value in a further step.

Consecuently, NDVI integral is used like a mask and only situations avove the average are considered at risk. The results of filtering the AS image (9th to 18th August) for Valencia and East Andalucia are presented in figure 8.

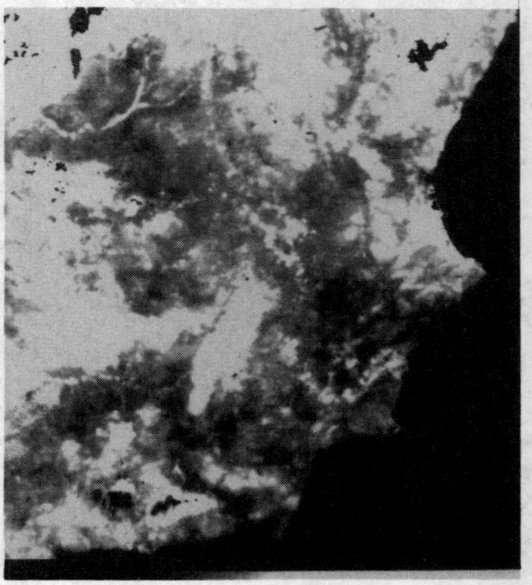

Figure 6: NDVI integral for the same period as AS in figure 4.

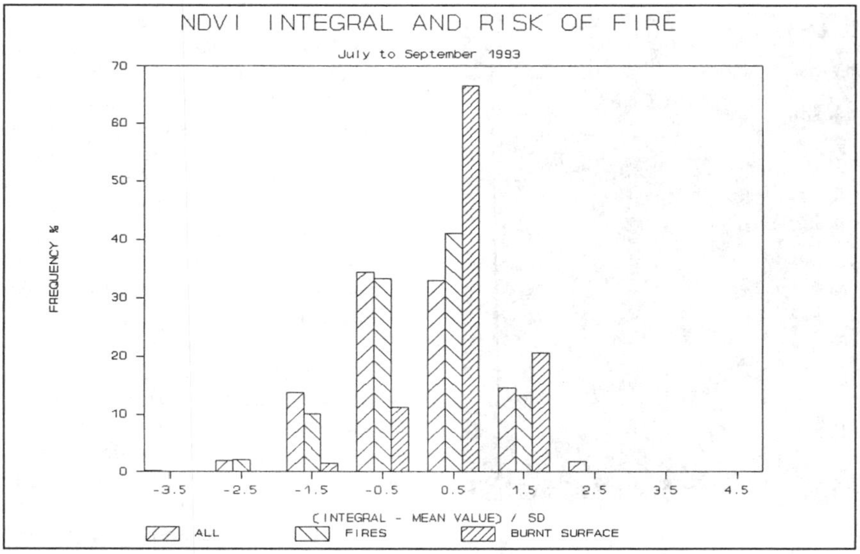

Figure 7: NDVI integral and fire occurrence. The classes correspond to typified values of the NDVI integral for the whole period analysed. Ground data have been superimposed. 58% of fires and 87% of burnt surface take place for integral values higher than the average over each period analysed.

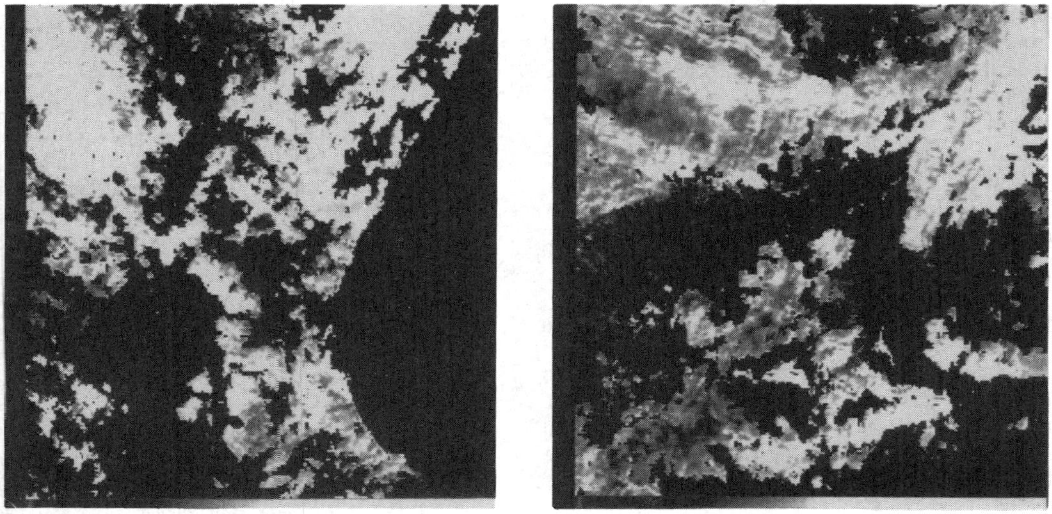

Figure 8: Result of filtering the AS values in figure 4 using NDVI integral values. The first image correspond to Valencia and the second one to East Andalucia.

Finally, the situations are classified in high, medium and low risk, using thresholds in the AS values. The index is generated in the 10*10 km grid used by the Spanish Forest Service. The results for the images in figure 8 are shown in figure 9. In table 3 the results for the whole set of risk situations are summarised. Image and ground data have been superimposed to asses the reliability of the method.

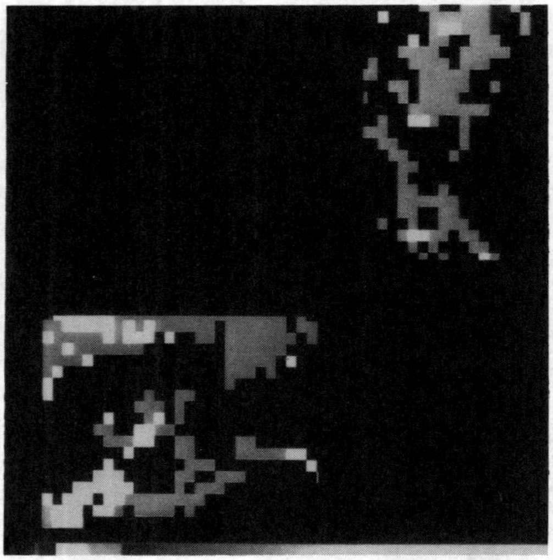

Figure 9: Risk in a 10*10 km grid. High risk is in bright grey, medium in intermediate and low in dark. It corresponds to the same period as the figures 4, 5 and 8.

Table 3: Risk for the whole period (July, August and September 1993). The risk index is derived from the AVHRR images by means of the proposed algorithm. Burnt surfaces are provided by the Spanish Forest Service.

	HIGH	MEDIUM	LOW
VALENCIA	9139 ha 46%	6323 ha 31%	4610 ha 23%
ANDALUCIA	1304 ha 15%	1168 ha 13%	6391 ha 72% *

*: 6223 ha in only one fire.

The index has a better behaviour in Valencia whith more homogeneous vegetal covers than in Andalucia where there are more variated situations. A local study of the normal values would be advisable before establishing any risk threshold in the AS.

6 CONCLUSIONS

An automatic algorithm for the detection and analysis of forest fires is presented. The identification of fires is based in determining local thresholds in the difference T_3-T_4 brightness temperatures. It automatically detects the affected pixels, gives its UTM coordinates and indicates whether they correspond to the same fire. The three thermal channels have also been used to estimate burnt areas using Dozier´s multispectral technique.

In a second step, those areas have been derived from the NDVI values in days before and after the fire event. Some problems of radiometric normalization can arise if the two images are not close enough in time. The results are compared to ground data provided by the Spanish Forest Service. In conclusion, both methods lead to similar estimations and can be applied in a quick assesment of affected areas (tables 1 and 2).

Finally, the evolution of the NDVI in a period before the fire ocurrence has been studied in order to generate a risk index based in the accumulation of vegetal fuel. The possibility of using the accumulated slope AS (equation 1) to asses the risk has been examined. The best results are reached by combining the AS and the NDVI integral over the period. The integral allows us to find areas with high amounts of vegetal fuel that can be at risk and the risk grade is derived from AS values. The results seem to be quite promising although a local study of the normal NDVI evolution would be advisable before establishing any risk threshold in the AS.

REFERENCES:

Dozier, J. 1981. A method for satellite identification of surface temperature fields of subpixel resolution. *Remote Sensing of Environment*. 11:221-229.

Flannigan, M. D. & L. Vonder Haar 1986. Forest fire monitoring using NOAA satellite AVHRR. *Canadian Journal of Forest Research*. 16:975-982.

Illera, P., J. A. Delgado & A. Calle 1993. Algoritmo de correccion geométrica de imágenes NOAA AVHRR. *Proc. V Reunión Científica de la Asociación Española de Teledetección*. Las Palmas de Gran Canaria, 10-12 noviembre 1993. (In press).

Illera, P., A. Fernández & J. L. Casanova 1993. Automatic algorithm for the detection and analysis of fires by means of NOAA AVHRR images. *Proc. EARSEL International Workshop on Satellite Technologie and GIS for Mediterranean Forest Mapping and Fire Management*. Thessalonikee, Greece. 4-6 November 1993. (In press).

Lee, F. & P. M. Tag 1990. Improved detection of hotspots using the AVHRR 3.7-μm channel. *American Meteorological Society*. 71(12):1722-1730.

López, S., F. González, R. Llop & J. M. Cuevas 1991. An evaluation of the utility of NOAA

AVHRR images for monitoring forest fire risk in Spain. *International Journal of Remote Sensing.* 12(9):1841-1851.

Matson, K. & J. Dozier 1981. Identification of subresolution high temperature sources using a thermal IR sensor. *Photogrammetric Engineering and Remote Sensing.* 47:1311-1318.

Matson, M., S. R. Schneider, B. Aldridge & B. Satchwell 1984. *Fire detection using the Noaa-series satellites.* US Department of Commerce, National Oceanic and Atmospheric Administration. NOAA Technical Report NESDIS 7, Washington DC.

Muirhead, K. & A. P. Cracknell 1985. Straw burning over Great Britain detected by AVHRR. *International Journal of Remote Sensing.* 6(5):827-833.

Paltridge, G. W. & J. Barber 1988. Monitoring grassland dryness and fire potential in Australia with NOAA-AVHRR data. *Remote Sensing of Environment.* 25:381-394.

Robinson, J. M. 1991. Fire from space: global fire evaluation using infrared remote sensing. *International Journal of Remote Sensing.* 12(1):3-24.

Sensors and Environmental Applications of Remote Sensing, Askne (ed.) © 1995 Balkema, Rotterdam, ISBN 90 5410 524 0

Emergency planning outside major hazard installations through territorial monitoring by means of satellite remote sensing

M. Ippoliti, L. Lepore & U. Poli
Italian Institute for Occupational Safety and Health (ISPESL), Environmental Department (DIPIA), Rome, Italy

M. Battista, G. Morrone & N. Napolitano
CSCT, Monteporzio Cantone, Rome, Italy

ABSTRACT: Emergency planning outside major installations (501/82 EEC Directive,the "Seveso Directive") becomes particularly important in relation to territorial vulnerability concerning population and environment.The methodological approach to this problem assumes different aspects in relation to the purposes of the public authority that is in charge of the emergency planning and to the aims to achieve as well.This paper presents an experimental method which makes use of satellite remote sensing to periodically survey areas potentially involved in accidents deriving from major hazard industrial activities.In particular the method is applied to an oil refinery.Risk areas are determined through application of specific mathematical models in order to exercise an effective control on land use and on urban and road expansion within a rational safety policy.

1 INTRODUCTION

This paper describes the first results of an application of satellite remote sensing to emergency planning outside a major hazard installation (oil refinery) included in the EEC Directive 501/82 (Seveso Directive).

Remote sensing has been considered to be a very useful and cost -effective tool at the disposal of the public authorities with the following in application:

- territorial vulnerability concerning population safety and health;
- pollution of the environment;
- surveillance and monitoring of anthropic developments and environmental modifications in the areas potentially at risk;
- bring up to date and optimize the emergency plan in relation to territorial and industrial modifications.

The purpose of this application is to investigate the environmental and sanitary impact of accidents from a major hazard installation by means of satellite remote sensing techniques.

This work was performed within the TELSAT project.

2 DESCRIPTION OF THE INSTALLATION AND TEST SITE

The major hazard installation under consideration (oil refinery) is located in a coastal zone of about 600.000 m^2 and employs 400 workers.The oil refinery includes the following plant units:

- oil distillation,refining and other oil product and subproduct trasformations;
- storage of flammable liquids and LPG(refrigerated and under pressure);
- sea equipment and fittings for loading/unloading ships anchored to a dock (small ships),to an artificial isle located at about 4 km (ships up to 90.00 DWT [Dead Weight Tons]),to a sea platform located to about 16 km from the coast(ships up to 300.000 DWT).

In the neighbourhood of the refinery,within a radius of 2 km,there are residencial buildings (urban centre of about 2.000 people)to SE,barracks to S,a river to W and the sea to NW,N and NE.The area of the plant is crossed by a railway,an arterial road and a civil airport.A motorway is also located at about 5 km from the plant.According to the meteorological situation on the site,where the typical sea and land breeze is present,the stability category of Pasquill D associated with 4m/s wind speed had been assumed as input values into mathematical models utilized in the evaluation of consequences from accidents.

The following items were considered for emergency purposes:

- sea pollution from ship tanker oil release;
- urban street development;

- land-use modification;

in Figure 1 is reported the flow chart of the procedure adopted to implement information related to emergency planning.

3 RISK ASSESSMENT

A hazard operability (HAZOP) analysis has been performed to determine the major accidents and the critical point of each unit of the plant.Risks were identified and assessed using mathematical models in order to define areas involved in:

- heat radiation (pool-fire of flammable liquids storage tanks, BLEVE/Fireball of LGP storage tanks);
- explosion (UVCE,projection of fragments);

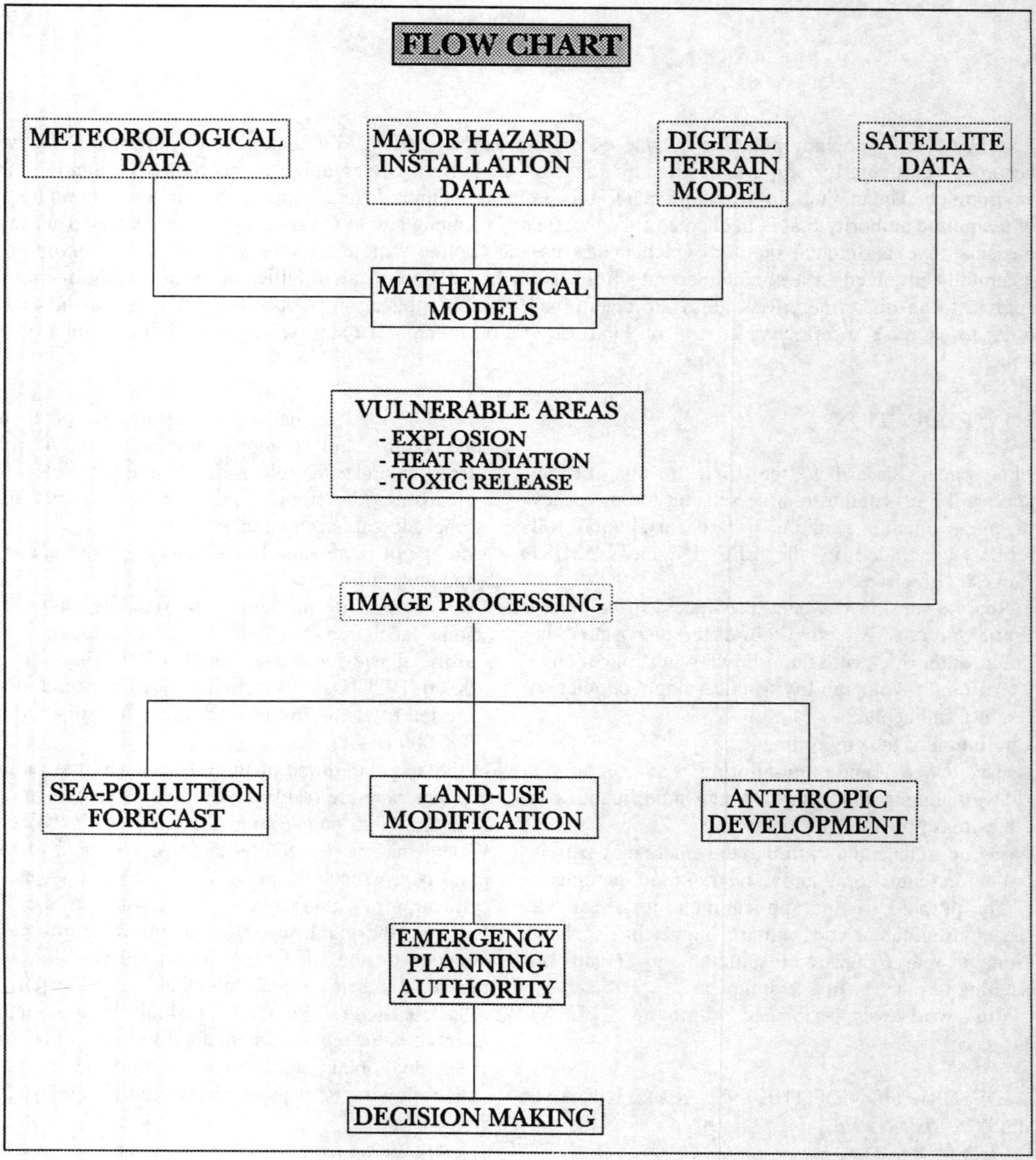

Fig. 1 Flow Chart

- release of toxic substance (tetramethyl lead TML and tetraethyl lead TEL).

The scenarios under consideration for emergency planning and the distances related to lethality and injuries are reported in Table 1 together with the relative probability of occurrence.Figure 2 shows the territorial areas involved in heat radiation .The image is obtained by merging LANDSAT TM (Bands 752) and SPOT (Panchromatic).The vulnerable areas were investigated and a surveillance and monitoring experimental programme has been set up.

Table 1.Accidents scenarios related to the oil refinery

Plant Unit	Risk typology	Damage distances(m) in D/4 Lethality	Injury	Probability (occ./y)
LPG spher.tan	Bleve/Fireball	1060	1410	$9.3*10^{-8}$
LPG cylindr.tank	Bleve/Fireball	280	410	$9.9*10^{-6}$
LPG refrig.tan	Bleve/Fireball	574	766	$1.0*10^{-5}$
Gasoline tank	Heat Radiatio	negl.	45	$1.1*10^{-4}$
Crude-oil tank	Heat Radiatio	negl.	120	$1.1*10^{-4}$
Process unit	UVCE	120	250	$2.0*10^{-3}$
Antiknock lead unit	Fragments Ejection	negl.	600	-
Antiknock lead unit	Toxic Release	negl.	750	-

4 IMAGE PROCESSING AND DATA SET

4.1 *Image Data Set*

For this application LANDSAT TM (sept '86 and sept '91) and SPOT (may '87 and sept 92) satellites were utilized together with the ground level data.A software (EASI-PACE) on HP Apollo 9000 Series 700 computer was used.

4.2 *Geometric and radiometric correction*

A nearest neighbour algorythm has been used to

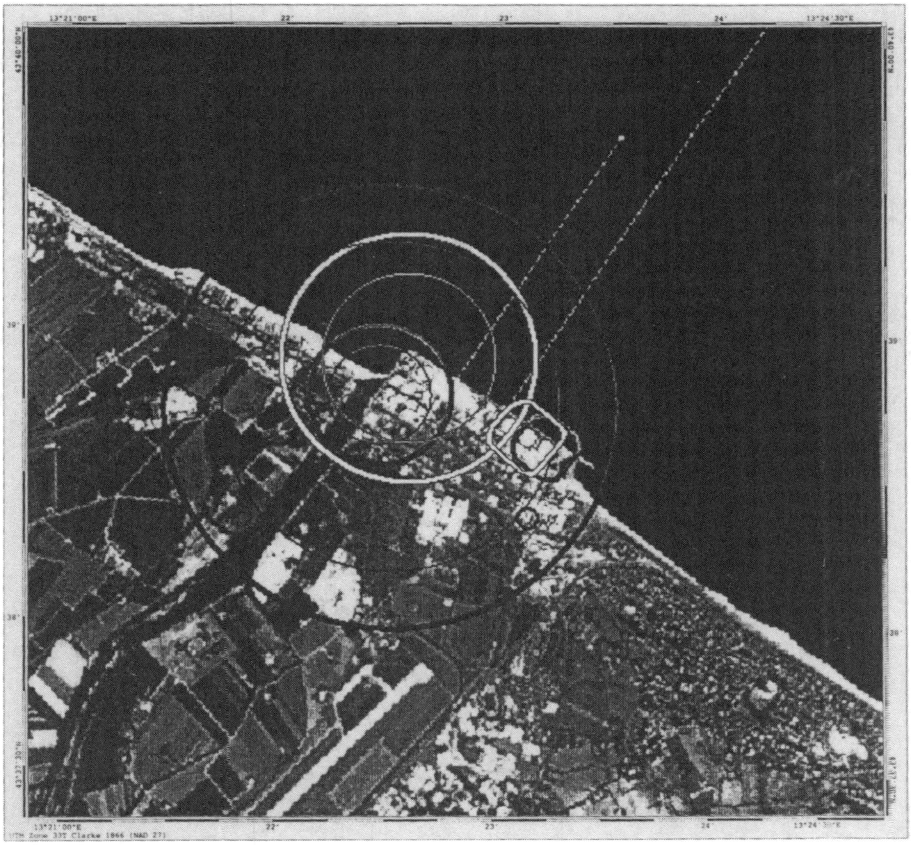

Fig. 2 Territorial areas involved in heat radiation

reference the image to an UTM set of coordinates.

4.3 *Image processing*

For image processing,a multitemporal approach has been performed,in particular:

- Landsat TM band 6,on whic atmospheric correction and a linear stretch were applied to define the sea surface streams,the oil-slicks spots and its time-varying.A frequency analysis for texture study has been utilized to distinguish the sea surface current lines through the temperature gradients. A Stream Index(SI)will been derived in order to forecast the possible direction of the oil slick.
- Landsat TM bands 3 and 4 were utilized and corrected for atmospheric disturbance to determine the land-use modification related to vegetated soils included in the at risk areas.A Vegetation Index (NDVI) has been determined to detect the modifications in the cultivated areas.
- Spot panchromatic band was used to assess the urban and street modification.An Anthropic Index(AI) will be calculated to recognize the territorial modifications related to urban developments and street modifications.

5.RESULTS AND CONCLUSIONS

Figure 3 shows the results of the application concerning urban and street development.

The results encourage authors to continue the investigation of this application in the field of emergency planning.First of all,more images and different satellites are going to be examined in order to increase the statistical significance of the computed parameters(indexes).Landsat TM images overlapped with Spot panchromatic images have given good geometric resolution for land-use scopes,but for sea surface streams the SAR sensor will be assessed.For surveillance and monitoring purposes it will be more useful to extend the application to other operative

Fig. 3 Urban and street development

satellites with a higher frequence of passage such as NOAA.

REFERENCES

R.Graziani, L.L, L.O & U.P "Studio di rilevamento da Satellite LANDSAT-TM di gradienti termici della superficie del mare, *Scienza Diritto e Economia dell'ambiente n. 1/89*

R.Graziani,L.L.,U.P. "Metodo per l'analisi e la valutazione delle conseguenze di eventi incidentali connessi a determinate attività industriali, *Fogli di informazione ISPESL 1°suppl.monografico anno IV 1993*

L.Bretherick, "Handbook of Reactive Chemical Hazards", *2nd ed.Butterworths,1979*

C.H.Huang, "A theory of dispersion in Turbulent Shear Flow", *Atmospheric Environment vol.13,(1979)*

Paolo Bragatto,A.B.,R.G.e tal., "Il Progetto Telsat:il Telerilevamento per il controllo delle aree ad alto rischio", *in Atti VI Convegno AIT*

Sensors – Optical technique

Sensors and Environmental Applications of Remote Sensing, Askne (ed.) © 1995 Balkema, Rotterdam, ISBN 90 5410 524 0

Reduction of high resolution spectra: Application to characterization of salinity status of soils

László Pásztor
Research Institute for Soil Science and Agricultural Chemistry of the Hungarian Academy of Sciences, Hungary

Ferenc Csillag
Centre for Surveying Science, University of Toronto, Ont., Canada

ABSTRACT: In this paper we present a technique for reduction of spectra based on the methods of multivariate statistical analysis. The procedure was developed in the context of a pedological problem, namely for spectral characterization of salinity status of soils, but, according to the resulted characteristics of the method (efficiency, optimality, convergence, robustness), it is proposed for general use in processing of digital, high resolution spectra. The recursive band selection method can be applied in studies for weighting (original) spectral bands according their sensitivity to a predefined classification scheme. Additionally, since the role of original bands is provided, definition of synthetic medium and broad band systems can be possible, which can efficiently substitute the original spectrum.

1 INTRODUCTION

Remote sensing is one of the key tools in monitoring local, regional and global environmental processes (NASA, 1988). Commercially available Earth-observing optical satellite systems have been dominated by sensors with broad bands, not always appropriate for specific remote sensing applications. A recent and important area in research and development of remote sensing is imaging spectrometry (Goetz et al., 1987). Such observations lead to very large dimensional data sets having tens and even hundreds of narrow spectral bands. Efficient processing and interpretation of these kind of data sets require different approaches in data processing (Chen and Landgrebe, 1989). In a general sense, efficiency of information processing primarily means reduction of data dimensionality with constraints on keeping high proportion of information, while in particular cases it requires specific, calibration-type preprocessing for well-defined tasks of specific elements of the Earth system (Henderson et al., 1989).

A general approach to the problem is the introduction of stochastic models and the application of the various methods of multivariate statistical analysis. There are some evidences in recent works that the automatizable multivariate descriptive data processing methods can be efficiently applied to the analysis and reduction of high resolution spectra not depending on the spectral ranges.

Beside the most noteworthy feature of multivariate statistical methodology, namely the flexible ability to describe and analyze large data matrices, two additional advantages are emphasized by Murtagh and Heck (1987) in the context of their use in spectrum analysis; the exemption of any a priori bias arising from existing schemes, and the possibility of work in a multidimensional parameter space opposed to general two dimensions of classical morphological classifications.

An important area of environmental research, where the application of (imaging) spectrometry can play an important role, is soil science, namely detection of different soil degradation processes. Soils are really essential components of the terrestrial ecosystem. Soil degradation related to salinization and alkalization represents an increasing environmental hazard to extensive natural and agricultural ecosystems, and consequently to the human environment (Scharpenseel et al., 1990). Timely detection and diagnosis of these processes have an especially significant role in the prevention and reclamation of the salt-affected areas.

The progenitor of present method was developed and applied in this context (Csillag et al. 1993). The modified and improved version of the method (first presented in details in Pásztor 1993) was also applied to the same problem (data set) for the sake of possibility of comparing efficiency of the various versions.

2 THE PROBLEM

1.1 Mathematical approach:

In many cases the variables characterizing objects of a study may be ranked into two categories (the change-over from elder feature variables to newer ones is a frequent example). The relation of these variable groups may be analyzed by various multivariate techniques (Multiple Regression Analysis, Canonical Correlation Analysis etc.). An important case is when one of the two variable groups is of qualitative nature, either because of inherent or technical reasons. In these cases the observations can be categorized according to the qualitative data. Thus the relationship between the variable groups can be described in a way when the multivariate methods are focussed on the recognition of a predefined classification scheme.

Another important characteristic of multivariate data sets is their inclination to contain redundant information. Generally, the inherent information content does not require the usage of all the descriptive variables since they are greatly correlated among themselves. In this case a few number of certain, artificial variables (factors) may provide (almost) the same informativity on the complex data sets.

A very good example of data sets with the above described features is provided by observations with very high resolution spectra which take over the role from previously used variables in the characterization of various physical objects. The qualitative ex-variables are generally based on morphological, locational etc. features or on spectral characteristics registered in different spectral (most often in optical) ranges.

1.2 Pedological aspect:

Reflectance spectra of salt-affected surface soil samples have been measured at 10 nm spectral resolution between 495 nm and 2395 nm. The samples were collected in California and Hungary to compare the effect of different salinization-alkalization processes on reflectance properties. Sampling sites were carefully selected to represent the significantly different quality and quantity of salts in the soil. In California salinization is primarily related to irrigation and inadequate drainage, where large quantities of soluble salts can accumulate in irrigated soils in a relatively short period of time. In Hungary, however there is a wide variety of factors contributing to salinization. Most of them are related to the hydrologically closed characteristics of the Carpathian Basin, where the major source of sodium salts is subsurface water and the dominant accumulation forms are Na_2CO_3 and $NaHCO_3$.

There were 41 samples collected in the United States and 49 in Hungary from the surface layer of soil. Field samples were split into two subsamples: one was analyzed in detail to characterize soil chemical and physical properties; the other was prepared for laboratory spectral reflectance measurements. The final data set consisted of 272 spectra, additionally a category value was attributed to each sample, since based on three key soil chemical properties (electrical conductivity [EC], pH and exchangeable sodium percentage [ESP]) the samples were ranked into 16 classes. Then the reflectance data set was statistically analyzed using the following band selection method.

3 THE METHOD

The presented procedure is a modified and improved version of the technique detailed in Csillag at al. (1993). Essentially, the introduced method assigns weights to the original spectral bands, which weights characterize their sensitivity to a predefined classification scheme, defined independently of the spectral features. Our recursive band selection procedure eliminates, step by step, the least informative band of the original high resolution spectrum. The procedure is based on the methods of multivariate data analysis which have become popular and widely used in various remote sensing applications (Swain and Davis, 1978; Richards, 1986; Mather, 1987). The output of the method consists of a subset of the original bands providing almost the same classification accuracy as the whole spectrum; 'goodness' values of the final bands characterizing their sensitivity; and a matrix for the evaluation of subsequent measurements.

Performance of our procedure is as good as that of the methods which select optimum bands as linear combination of the entire data set (Chen and Landgrebe, 1989), although the resulted number of final bands in our study is slightly higher then in earlier reports (Price, 1990). This is due to keeping original bands rather than their linear combinations, and to the complex nature of soil salinity.

Our recursive band selection procedure eliminates, step by step, the less informative ('noisiest') band of the original high resolution spectrum. Every step is combined as follows (the scheme of the method is outlined in Fig.1):

(i) Principal Component Analysis on the (remaining) spectral bands.

Principal Component Analysis is quite widely applied in multiwavelength image processing for reducing dimensionality. It has been successfully used to identify a linear combination of spectral measurements, leading to a smaller number of uncorrelated axes (or dimensions) in the measurement (or feature) space. This approach determines the importance of each original band in each resultant feature and finds optimum

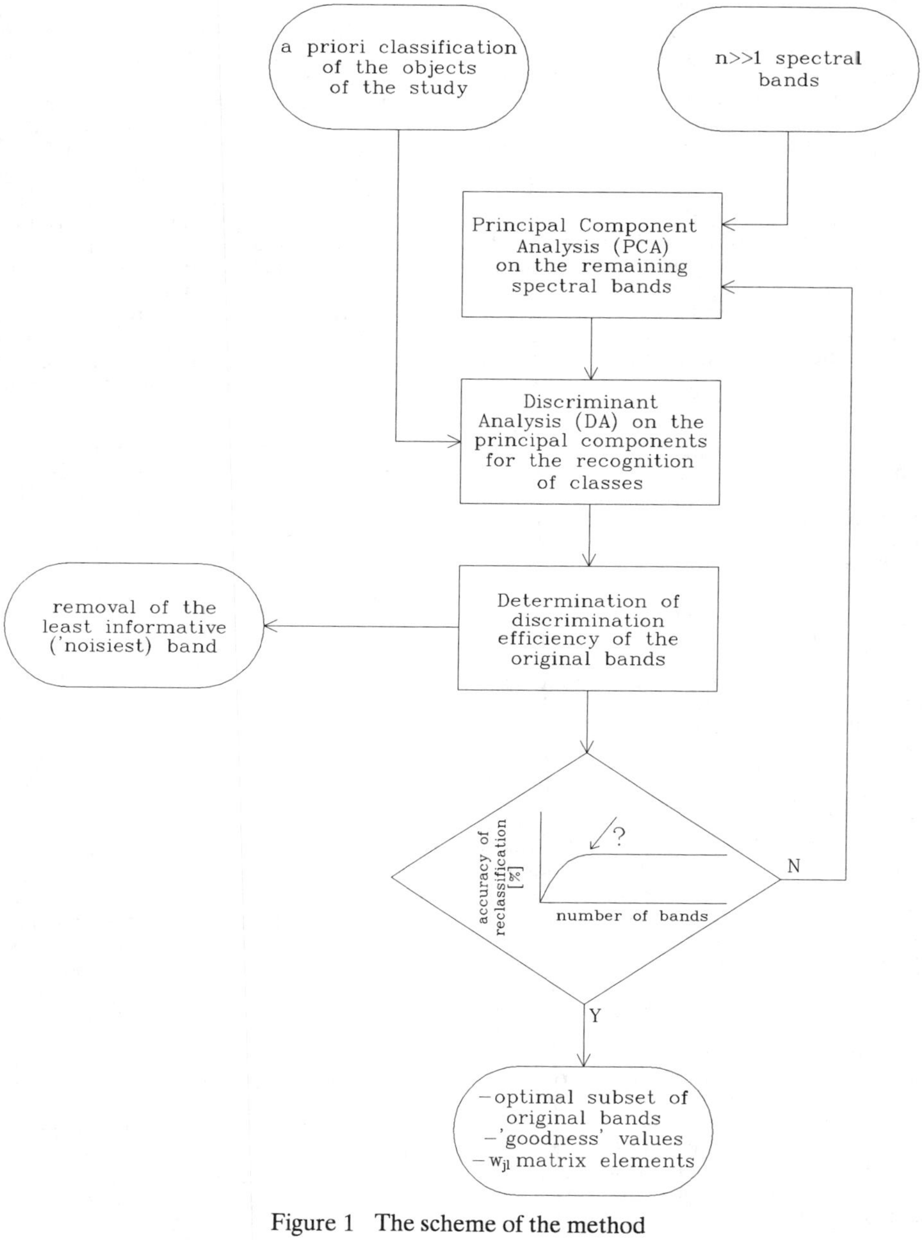

Figure 1 The scheme of the method

features in the sense that the number of features can be much less than the number of original bands without significant loss of information in terms of proportion of variance.

Let us recall the mathematical formalization of the problem. We find that such linear combinations of input variables which are not correlated have maximum variance and the sum-of-square of coefficients is 1. If X is the nxm data matrix (m is the number of variables, n is the number of measurements) and the variables are standardized, the correlation matrix is $R=X'X/n$. A principal component can be written in the following form: $y=Xb$, where b is the m-element vector of coefficients. The expected value of a principal component is 0, and its variance is: $y'y/n=(Xb)'(Xb)/n=b'Rb$. Since the linear combination should be found which has the highest variance under the constraint of $b'b=1$, a Lagrange multiplier can be introduced, which leads to $(R-\mu E)b=0$ which is an eigenvalue/ eigenvector problem (μ is the eigenvalue and b is the eigenvector of R).

The correlation matrix is said to be totally explained by the m principal components. In many cases however $k<m$ principal components already account for the vast majority of the variance. The efficiency of this substitution can be easily estimated by the ordered eigenvectors.

Our band selection procedure is based on the principle that it may not be necessary to keep all the original bands to obtain data with a large amount of spectral detail, if we can identify those bands which are most sensitive to a given set of classes, or variables, and if they provide sufficient recognition probability or separability power without combination. This may also be useful for given tasks if processing capacity or sampling for reasonable statistics are limited.

(ii) Discriminant Analysis on the formerly computed principal components for the recognition of classes.
Once principal components are resulted discriminant analysis can be performed to test the performance of 'conventional' PCA-derived features on the predefined classification scheme. Formally, it is quite similar to the equations used in PCA, because it also leads to an eigenvalue/eigenvector problem.

Let matrix T be the covariance matrix which can be decomposed into two components $T=B+W$, where W and B correspond to the within-classes and between-classes variances. Then we construct linear discriminant functions d with coefficients c, where $c'd=1$. Then $d'd=c'Tc=c'Bc+c'Wc$. According to our goal, $L=[c'Wc]/[c'Bc']$ need to be maximized. Taking the logarithm and partial derivative, we obtain $(W^{-1}B-LE)c=0$, which is the already known form of an eigenvalue/eigenvector problem.

Here the eigenvectors provide the discriminant factors (functions), while the eigenvalues characterize their merits. The number of discriminant factors is determined by the number of groups (g) and variables (m). If $g-1 < m$, the space of discriminant factors is of $d=g-1$ dimension, otherwise of $d=m$.

(iii) Determination of discrimination efficiency of the original bands, and removal of the least informative ('noisiest') band.
Thus the discriminant functions are obtained with another set of weights applied on features, which are themselves linear combinations of the original bands. Therefore, the final role of the lth original band in the jth discriminant function can be summarized in a matrix element $w_{jl}=\Sigma^{m}_{k=1}c_{kj}b_{lk}$.

Here c_{kj} is the weight of the kth principal component in the jth discriminant factor and b_{lk} is the weight of the lth original variable in the jth principal component.

For selection of bands a final weight assignment is proposed. The resulted weights for the original bands in a discriminant function are normalized and then multiplied by the total between-groups variability attributable to the given function. These values are then summarized for each original band. The resulted values (called 'goodness') well characterize the overall role of an original band in the recognition of classification.

The original band having the lowest 'goodness' value will be considered as noisiest and removed from further processing. When this band is dropped, it is one iterative step in our procedure.

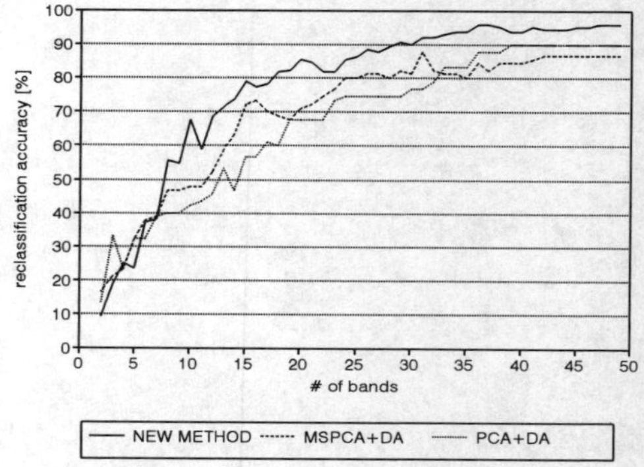

Figure 2
Comparison of various methods. MSPCA+DA refers to former version of the present method. PCA+DA had been used as a reference for testing the efficiency of the original procedure. The overall improvement in the reclassification accuracy of the new version is evident.

(iv) Suspension or continuation of the procedure.
The effectiveness of a set of discriminant functions can be measured by the reclassification accuracy resulted from test data. According to the general behaviour of multivariate data sets and based on our experiences, the efficiency of reclassification shows saturation versus the number of bands (Fig.2). That is beyond a threshold, the increase in the number of bands does not provide improvement in the goodness of reclassification. The recursion need to be stopped reaching this threshold value.

The output of the method consist of a subset of the original bands providing almost the same classification accuracy as the whole spectrum; the 'goodness' values of the bands in the subset and the $w_{jl}=\Sigma^{m}_{k=1}c_{kj}b_{lk}$ matrix elements for the evaluation of subsequent measurements.

4 CONCLUSION

The presented band selection procedure provides an optimum subset of original spectral bands in a sense that it is composed of the least (generally much less than originally) number of spectral bands, which, however, contains all the relevant information as for recognition of the predefined classification scheme. If, because of any reason, the resulted number of bands is even more than what is acceptable, the usage of a smaller subset is proposed, which consists of the bands with the best 'goodness' values (although such truncation of data sets may already result in much more significant loss of information).

Careful study of the algorithm shows that our procedure is stable and robust (Pásztor 1993), however its convergence is not well established yet.

In our pedological application the resulted bands clustered along the spectrum (Csillag et al., 1993), which fact made the definition of broader bands meaningful. Usage of these synthetic bands resulted in much better recognition.

Compared to recent reports on optimum band selection suggesting combination on 'purely' statistical grounds, our procedure leads to an application-sensitive measure of the potential of the data set in terms of classification accuracy.

The present improved version proved to be significantly more efficient than it was developed in its original form (Fig.2), since its application-sensitive aspect had been improved.

Summarizing, our procedure proved to be an efficient tool for the characterization of salt affected soils, and it is proposed for application in other fields with similar conditions (i.e. multiwavelength data set versus predefined classification scheme).

ACKNOWLEDGEMENTS

The financial support of the Hungarian Academy of Sciences and the National Science Foundation (Grant No. INT-8721949) are gratefully acknowledged. The authors are indebted to S. Kabos for valuable discussions.

REFERENCES

Chen, C.T. and Landgrebe, D.A., 1989, *IEEE Trans. Geosci. Rem. Sen.* 27, 891.

Csillag, F., Pásztor L., Biehl L.L., 1993, *Remote Sens. Environ.* 43, 231.

Goetz, A.F.H., Vane, G., Solomon, J.E. and Rock, B.N., 1987, *Science* 228, 1147.

Henderson, T.L., Szilágyi, A., Baumgardner, M.F., Chen, C.T. and Landgrebe, D.A., 1989, *Soil. Sci. Soc. Am. J.* 53, 1778.

Mather, P.M., 1987, *Computer Processing of Remotely-Sensed Images.* Wiley.

Murtagh F., Heck A., 1987, *Multivariate data analysis.* Reidel, Dordrecht

NASA, 1988, *Earth System Science: A Closer View.* Earth System Science Committee, NASA, Washington, DC.

Pásztor, L, 1993, *Ph.D. thesis* (in Hungarian) Eötvös University, Budapest

Price, J.C., 1990, *Remote. Sens. Environ.* 33, 113.

Richards, J.A., 1986, *Remote Sensing Digital Image Analysis.* Springer-Verlag.

Scharpenseel, H.W., Schomaker, M. and Ayoub, A., 1990, *Soils on a Warmer Earth, Developments in Soil Science 20.* Elsevier.

Swain, P.H. and Davis, S.M., Eds., 1978, *Remote Sensing; The Quantitative Approach.* McGraw-Hill.

Sensors and Environmental Applications of Remote Sensing, Askne (ed.) © 1995 Balkema, Rotterdam, ISBN 90 5410 524 0

An improved method of estimating measurement error in remotely sensed images

Peter M. Atkinson
Department of Geography, University of Southampton, Highfield, UK

Andrew R. Harrison
Department of Geography, University of Bristol, UK

ABSTRACT. Previously a method has been developed to estimate the signal-to-noise ratio of airborne visible infrared imaging spectrometer (AVIRIS) imagery. The intercept on the ordinate of the modelled experimental variogram, known as the nugget variance, is used to estimate noise. However, the nugget variance is due not only to measurement error, but also to sampling effects, fitting an inappropriate model, and short-range spatial variation that has not been measured. Consequently, the method may be unreliable. An improved method is presented which involves a check on the accuracy of the estimate of measurement error provided by the nugget variance. The improved method is demonstrated with an example using simulated Landsat TM imagery of Belmonte in central Spain.

1. INTRODUCTION

A common approach to estimating measurement error in spatially distributed variables involves measuring repeatedly at the same place, thereby building an estimate of the distribution function of the error. Generally, this is not possible for remotely sensed images which provide a single observation at each pixel location. To overcome this limitation Curran and Dungan (1989) developed a method of estimating the signal-to-noise ratio in remotely sensed images based on spatial autocorrelation. This method may be unreliable where the model fitted to the experimental variogram is unrepresentative. The objective of this paper is to present an improved method of estimating measurement error in remotely sensed images.

It has been realized for some time that the identification and proper treatment of measurement error is important for the successful application of remotely sensed and other digital spatial data, particularly where such data are to be brought together in geographical information systems (Goodchild and Gopal 1989, Piwowar and LeDrew 1990, Davis *et al.* 1991, Lunetta *et al.* 1991). The method in this paper is presented in the context of these long standing concerns.

Spatial dependence, also referred to as spatial autocorrelation, is the likelihood that observations close in space are more alike than those further apart. It has been recognized for some time that most geographical phenomena are to some extent spatially dependent (Cliff and Ord 1973, Rossi *et al.* 1992). Further, spatial dependence can exist simultaneously at a variety of different scales (Lam and Quottrochi 1992).

The variogram is a function that represents spatial dependence. The semivariance is half the expected squared difference between the random functions (RF) $Z(\mathbf{x})$ and $Z(\mathbf{x}+\mathbf{h})$ at a particular distance and direction of separation or lag $\mathbf{h}$ (Matheron 1965, 1971, Journel and Huijbregts 1978). The variogram is then the function that relates the semivariance at a particular lag to that lag (Equation 1).

$$\gamma(\mathbf{h}) = 1/2\ \mathrm{E}[\{Z(\mathbf{x})-Z(\mathbf{x}+\mathbf{h})\}^2] \qquad (1)$$

The experimental variogram $\gamma_v(\mathbf{h})$ can be estimated for $p(\mathbf{h})$ pairs of observations, $z_v(\mathbf{x}_l)$ and $z_v(\mathbf{x}_l+\mathbf{h})$, $l=1,2,\ldots,p(\mathbf{h})$, by Equation 2.

$$\gamma_v(\mathbf{h}) = 1/2p(\mathbf{h}) \sum_{l=1}^{p(\mathbf{h})} \{z_v(\mathbf{x}_l) - z_v(\mathbf{x}_l+\mathbf{h})\}^2 \qquad (2)$$

The experimental variogram has been computed and applied widely in remote sensing in recent years (for example, Curran 1988, Ramstein and Raffey 1989,

Clark 1990, Cohen *et al.* 1990, Chavez 1992, Gohin and Langlois 1993, McGwire *et al.* 1993).

Measurements of spatially continuous properties have three important characteristics that affect the experimental variogram. First, all such measurements are regularized, that is, are made over a support v of positive size (where the support is the size, geometry and orientation of the space over which an observation is defined).

Second, each measurement is expected to contain a component of variation due to measurement error. It is possible to measure without error, but this can never be known for certain. Measurement error can arise from many sources, but a basic distinction may be drawn between internal and external errors. Internal measurement error arises when the correct underlying variation is measured incorrectly. External measurement error amounts to adding some external source of variation, for example, instrument noise to the true values.

Third, all measurements constitute samples. Even where the region of interest is covered completely, for example, as with remotely sensed imagery the integration of the underlying variation over the support constitutes a summary of information.

Let us examine how measurement error appears in the experimental variogram. Realizations $z_v(\mathbf{x}_0)$ of a RF $Z_v(\mathbf{x})$ (defined for a single time t_0) at the same place $\mathbf{x}_0$ and over the same support v must take the same value. Therefore, the experimental variogram of the underlying variation is continuous through the origin.

Unlike the underlying variation measurement error is not spatially continuous. No matter how close two realizations of measurement error they will be unrelated. Even realizations at the same place $\mathbf{x}_0$ will be unrelated. Therefore, measurement error appears in the experimental variogram as some finite positive value at a lag of zero. The underlying variation contributes nothing at a lag of zero because the true value is always the same at a given place.

Internal measurement error results in an increase in semivariance at a lag of zero, and no increase at very large lags, with increases of intermediary magnitude between these extremes. External measurement error contributes the same amount of semivariance at all lags and, therefore, amounts to shifting the variogram upwards.

Certain types of error do not affect the experimental variogram at all. For example, if all the measurements constituting the image are shifted in space by an equal vector amount each observation will be in error relative to the truth at the ground, but the variogram will be unaffected. This is because the variogram deals only with relative differences and not absolute values. This limits the nature of the measurement errors that are detectable from the variogram.

Because the experimental variogram is computed from a spatial sample of the property it represents it comprises a set of values of semivariance at an ordered set of discrete lags. For remotely sensed images the closest observations are at a lag equal to or less than the one-dimensional size of the experimental support. It follows that this is the smallest lag represented in the experimental variogram. The experimental semivariance at a lag of zero $\hat{\gamma}_v(0)$ is not measurable because satellite and airborne sensors generally are not designed to make repeated measurements over the same support v, at the same location $\mathbf{x}_0$.

To make the variogram continuous for further analysis it is necessary to fit a mathematical model to the experimental values. The model fitted must be chosen from a select few that are known to be conditional negative semi definite (CNSD). This criterion ensures that if the model is used with techniques for estimation such as kriging (for example, Gohin and Langlois 1993), negative estimation variances cannot result. Webster and Oliver (1990) list most of the CNSD models in common use.

Once fitted, the model will usually have a positive intercept on the ordinate known as the nugget variance c_0. The nugget variance is due to measurement error, but it is also due to the effects of estimating the variogram from a sample, fitting an inappropriate mathematical model, and short-range spatial variation that has not been measured. For remotely sensed and other types of imagery where the supports of the observations are adjacent or overlapping, the latter effect is small.

In Curran and Dungan's (1989) method for estimating the signal-to-noise ratio the nugget variance is treated as noise and the spatially dependent component of the variogram is treated as signal. They applied their method to experimental Airborne Visible InfraRed Imaging Spectrometer (AVIRIS) data, and found that AVIRIS data are inherently noisy. While this concept is of fundamental importance, Curran and Dungan's method is unreliable because it does not account for the contribution to the nugget variance of sampling effects, fitting an inappropriate model, or short-range variation.

It is not possible to account for sampling effects without further measurement. However, it is possible to account for an inappropriate choice of model and the effect of short-range variation. The method presented in this paper is an extension of Curran and Dungan's method. This improved method is applied to a Landsat Thematic Mapper (TM) image of the region surrounding Belmonte in central Spain.

2. REGULARIZING THE VARIOGRAM

The variogram defined on one size of support is related to and predictable from the variogram defined on another size of support through the process of regularization (Matheron 1971, Clark 1977, Journel and Huijbregts 1978, Jupp *et al.* 1988, Isaaks and Srivastava 1989, Webster 1991).

Here, a method for regularizing the variogram is used to improve the precision with which measurement error may be estimated.

2.1 *Regularizing the variogram*

The regularized variogram, $\gamma_v(\mathbf{h})$, on some support v is readily obtained from the punctual or point variogram by Equation 3.

$$\gamma_v(\mathbf{h}) = \bar{\gamma}(v,v_\mathbf{h}) - \bar{\gamma}(v,v) \qquad (3)$$

Here $\bar{\gamma}(v,v_\mathbf{h})$ is the integral of the punctual semivariance between two supports of size $|v|$ whose centroids are separated by $\mathbf{h}$ and $\bar{\gamma}(v,v)$ is the average punctual semivariance within an observation of size $|v|$.

Given the punctual variogram, the regularized semivariance at a given lag may be computed from Equation 3 by numerical integration, that is, by approximating the support as a regular grid of points. Supports with different geometries can be approximated by varying the spatial positions of the grid nodes. If the regularized semivariance is computed at enough lags the regularized variogram can be estimated and a model fitted.

Figure 1a shows an example of regularizing a spherical model, given by Equation 4.

$$\gamma(h) = c_1 \{1.5h/a_1 - (0.5h/a_1)^3\} \qquad (4)$$

Here, c_1 is the spatially dependent component of variation and a_1 is the limit to spatial dependence or range. The upper curve in Figure 1a shows a spherical model with c_1 equal to 10 units and a_1 equal to 10 m. The lower curve (Figure 1a) is its regularization over square supports of side 1 m.

In practice, one does not have the punctual variogram. The experimental variogram of remotely sensed imagery is always regularized. Therefore, the first task is to estimate the punctual variogram by deregularizing the experimental variogram. One should then be able to predict the regularized semivariance for any given size of support using Equation 3.

2.2 *Deregularizing the variogram*

This section describes a method for deregularizing the experimental variogram based on Equation 3.

Given that a standard authorized (spherical) model is fitted to the experimental variogram, a standard model will not represent satisfactorily the corresponding punctual variogram. The larger gradient at the origin and the greater convex upwards curvature at short lags may be obtained by representing the punctual model as the sum of the model fitted to the experimental variogram plus a term that represents the short-range variation encompassed within the experimental support ($\bar{\gamma}(v,v)$).

An exponential term, $c_e\{1 - \exp(-h/a_e)\}$ (where a_e is the range and is set between 0.1 and 0.5 times the one-dimensional size of the experimental support), is added to the model of the experimental variogram to estimate the punctual variogram. By iteratively regularizing the estimated punctual model, the sill of the short-range term c_e is adjusted until the sill of the regularized variogram equals that of the model fitted to the experimental variogram within an acceptable tolerance. This combination of the original function and an exponential term has been found to provide a close approximation to the original experimental variogram when regularized over supports that are square or similar to square.

If a spherical model with c_1 equal to 10 units and a_1 equal to 10 m is fitted to the experimental variogram then the model of the punctual variogram achieved using the method described above is given by Equation 5.

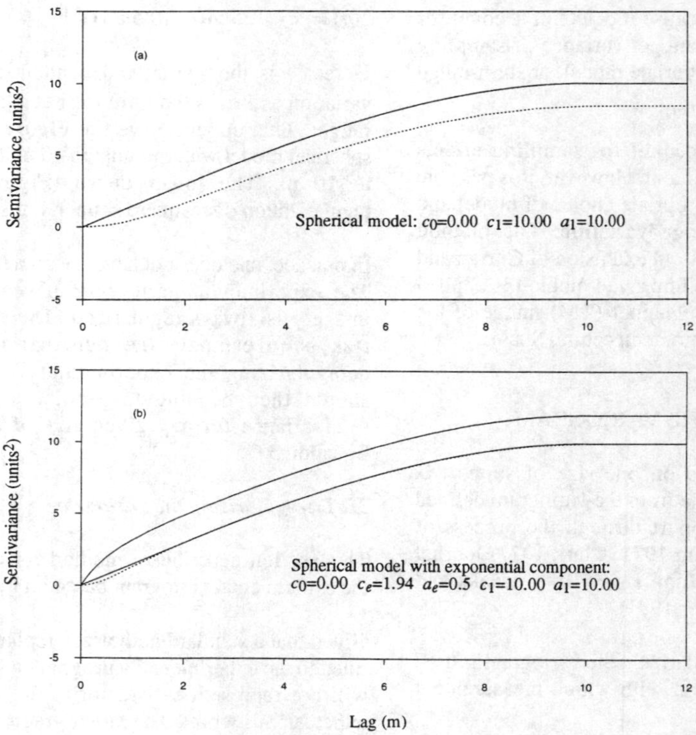

Figure 1. Spherical model (a) regularized over a support of 1 m by 1 m (lower curve) and (b) deregularized to estimate the punctual variogram (upper curve) and subsequently regularized over the experimental support (lower dashed curve).

$$\gamma(h) = c_e\{1\text{-exp}(-h/a_e)\} + c_1\{1.5h/a_1-(0.5h/a_1)^3\} \qquad (5)$$

where c_e is 1.937, c_1 is 10 units, a_e is 0.5 m, and a_1 is 10 m. Figure 1b shows the estimated punctual model, the variogram regularized over the experimental support (dashed line) and the spherical model (solid line) that it matches.

The punctual variogram obtained in this way provides the basic function from which to regularize the variogram.

2.3 *Short-range and long-range variation*

The method of deregularizing and regularizing the variogram assumes that the model fitted to the experimental estimates of semivariance is accurate. The processes of deregularization and regularization are sensitive to values of semivariance at very short lags, and in particular, the semivariance at a lag of zero. Therefore, it is important to estimate accurately the variogram at the shortest lags. The following discussion shows how a model fitted to the variogram is often in error at the ordinate.

Underlying variation may, for the present purposes be split into two classes; long-range and short-range variation. Long-range variation refers to spatially dependent variation with a range larger than the one-dimensional size of the support. An example is provided by the upper curve in Figure 2a where the range is approximately 10 m. This represents the underlying punctual variogram. If the variation is regularized on supports of 1 m by 1 m, the regularized variogram will look like that in the lower curve in Figure 2a with a concave upwards form near the origin. Clearly, if a model is fitted to the experimental estimates of semivariance at lags 1,2,3..., the experimental semivariance at a lag of zero will be underestimated.

Short-range variation may be defined as spatially dependent variation with a range shorter than the one-

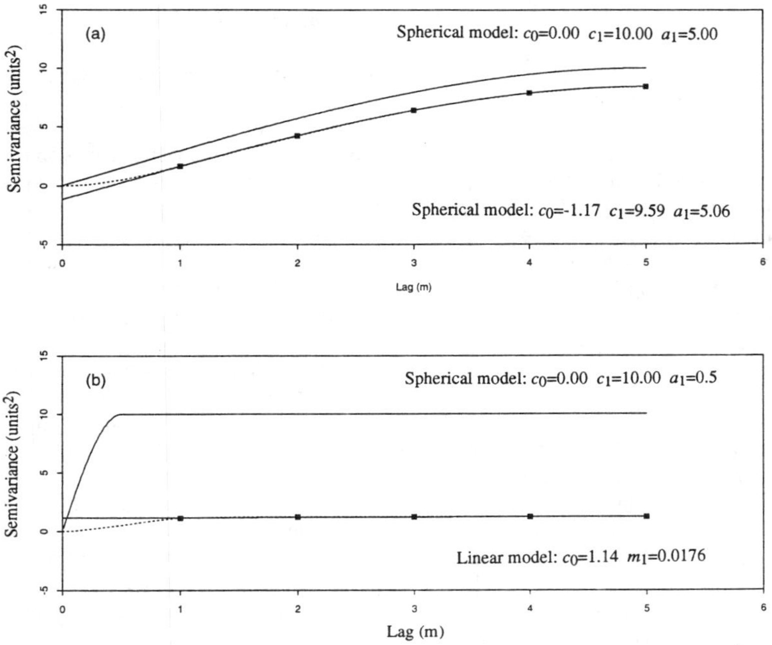

Figure 2. Spatial variation that is (a) long-range and (b) short-range regularized over the experimental support.

dimensional size of the support. A variogram of underlying variation with range equal to 0.1 m is shown in the upper curve in Figure 2b. The lower curve shows this variogram regularized on a support of 1 m by 1 m. In this case, the regularized variogram is convex upwards near the origin. A mathematical model fitted to the experimental semivariance at positive lags would overestimate the experimental semivariance at a lag of zero.

In summary, the experimental semivariance at a lag of zero may be under or overestimated by the nugget variance of the model fitted. It is for this reason that Curran and Dungan's (1989) method may be unreliable.

2.4 *Regularizing measurement error*

Equation 3 above applies to the spatially dependent underlying variation only. It does not apply to measurement error. Therefore, methods of regularizing the variogram depend on being able to estimate accurately measurement error. Measurement error must be removed from the variogram prior to deregularization and regularization, regularized separately and added back in to estimate the regularized variogram.

Possible sources of measurement error are many and varied (for example, Epema 1991) and the way that each source of error behaves with regularization may be different. For example, variation due to sensor noise may be the same regardless of spatial resolution. However, most measurements of properties at the ground have a larger signal-to-noise ratio, the larger they are.

Regularizing measurement error is a matter of, first, identifying the sources of measurement error and judging their contribution to the total and, second, judging how each varies with size of support.

Since the ability to regularize the variogram depends on an accurate estimate of measurement error, it follows that if one can assess the accuracy of the regularized variograms one can also assess the accuracy of the estimate of measurement error. Here, the spatial resolution of the remotely sensed imagery is coarsened, experimental variograms are computed for each new spatial resolution and these variograms are used to check the accuracy of the variograms regularized using Equation 3.

2.5 *Regularizing the imagery*

Remotely sensed imagery usually provide complete coverage. The spatial resolution of the imagery can, therefore, be coarsened simply by averaging the original pixel values to estimate values for the larger cells. Then experimental variograms may be computed for each new spatial resolution directly from the coarsened imagery. The experimental variograms computed in this way do not depend on the model fitted to the experimental variogram. Therefore, the experimental variograms so computed provide a means of checking the regularized estimates obtained using Equation 3.

Measurement error cannot be removed from the values of the image and is, therefore, convolved with the spatially dependent variation. Consequently, measurement error varies with size of support in inversely in proportion to the number of pixels in each new larger cell.

Equation 3 may be used to estimate regularized variograms for sizes of support comparable to those to which the imagery has been coarsened. Measurement error (estimated by the nugget variance) must be removed from the experimental variogram, regularized according to the inverse relation above and added back in.

The two sets of variograms may then be compared and any discrepancies will highlight an error in the estimate of the punctual variogram and, most importantly, an error in the nugget variance as an estimate of measurement error.

3. EXAMPLES

The following example illustrates the method described above.

3.1 *Field site and experimental data*

A Landsat TM image recorded on 12 August 1991 was obtained in the red waveband (0.63 μm - 0.69 μm). The scene is of semi-natural vegetation surrounding Belmonte in central Spain. The Landsat imagery is of orbit 200, frame 33, and comprises 1024 by 1024 pixels each representing a support of 30 m by 30 m at the ground. Homogeneous sub-scenes of 500 lines by 400 pixels were selected for analysis.

3.2 *Experimental variograms*

An experimental variogram was computed for the sub-scene and was fitted with several CNSD variogram models using a weighted least squares procedure where the weights are chosen in proportion to the number of paired comparisons at each lag (Ross 1987). The exponential model was found to provide the best fit in terms of Akaike's Information Criterion (AIC) (Akaike 1973). The nugget variance of this model was approximately zero, implying that measurement error was negligible in this case.

To provide more control over the experimental data the spatial variation was conditionally simulated using a sample of the original image and the modelled variogram (Journel and Huijbregts 1978, Isaaks and Srivastava 1989, Rubin and Journel 1991). Sequential Gaussian simulation was applied to the sample values using the GSLIB software (Deutsch and Journel 1992). With conditional simulation it is possible to provide images with chosen amounts of measurement error simply by increasing the nugget variance of the variogram model. An image was simulated from the original modelled variogram with a nugget variance of 94.4 DN^2 and a sample of 1/25th of the original number of pixels. An experimental variogram was computed from the simulated image (Figure 3) and fitted with several CNSD models as above. The exponential model with coefficients as shown in Figure 3 provided the best fit in terms of the AIC.

3.3 *Coarsening the spatial resolution of the image*

The simulated image was degraded to successively

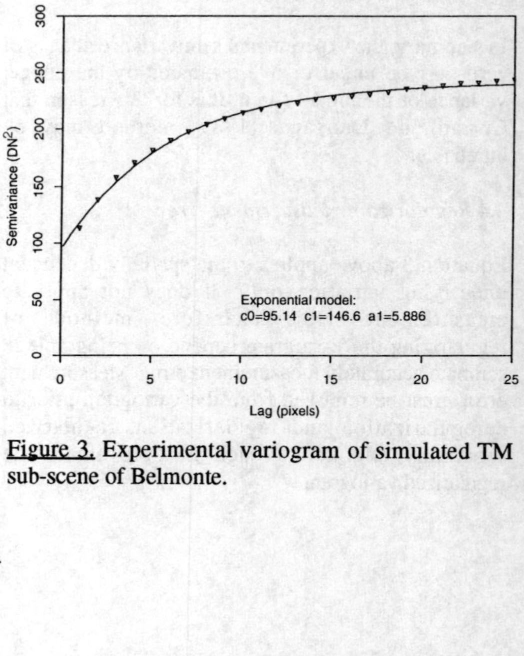

Figure 3. Experimental variogram of simulated TM sub-scene of Belmonte.

coarser spatial resolutions simply by summing the original 30 m by 30 m pixels to estimate values for the larger cells. Experimental variograms were computed for four sizes of support, namely 5 pixels, 10 pixels, 15 pixels and 20 pixels (squared) (the symbols in Figure 4).

3.4 *Regularizing the experimental variograms*

The experimental variogram representing the original size of pixel was regularized to sizes of support of 5 pixels, 10 pixels, 15 pixels and 20 pixels (squared) by deregularizing and regularizing the variogram as described above. The support was assumed to be square in shape and to have a uniform distribution over its extent (square wave response) to match the variograms obtained directly from the coarsened images. The regularized estimates are shown in Figure 4 (lower curves).

The correspondence between the experimental variograms computed from the coarsened imagery (symbols) and the variograms regularized using Equation 3 (lines) may be checked from Figure 4. In this case there is no discrepancy between the two sets of regularized variograms and, therefore, one may assume that the estimate of measurement error provided by the nugget variance (95.1 DN^2) is accurate. In this case, the nugget variance and,

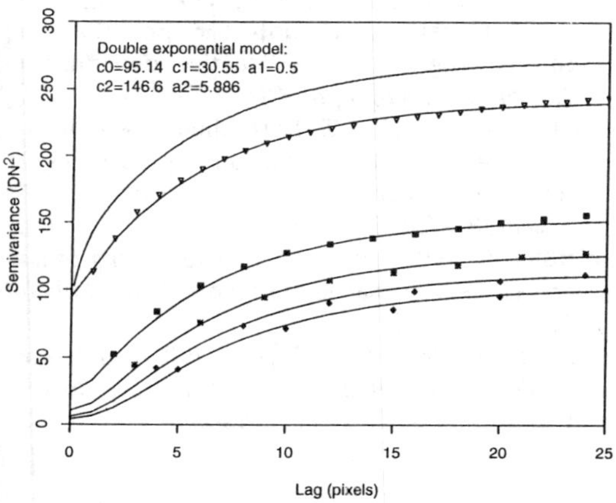

Figure 4. Experimental variogram of simulated TM sub-scene deregularized to estimate the punctual variogram (upper curve) and subsequently regularized to supports of 5 pixels, 10 pixels, 15 pixels and 20 pixels (squared) (lower curves). The symbols on the lower curves represent the corresponding experimental variograms computed from coarsened imagery.

therefore, measurement error, was set at 94.4 DN^2 confirming that the prediction is accurate.

In other examples, a discrepancy between the two sets of regularized variograms may point to an inaccurate estimate of measurement error. Then the investigator should go back and search for a more representative model for the experimental variogram.

4. SUMMARY

An improved method of estimating measurement error in remotely sensed images has been presented. The variogram of underlying spatially continuous variation is continuous through the origin. Therefore, the experimental semivariance at a lag of zero $\hat{\gamma}_v(0)$ is due solely to measurement error. For remotely sensed images $\hat{\gamma}_v(0)$ cannot be computed directly. Rather it must be estimated by fitting a model to the experimental values of semivariance at larger lags and extrapolating the model to the ordinate. If the model fitted is inappropriate, the nugget variance of the model may be inaccurate as an estimate of $\hat{\gamma}_v(0)$.

Methods for regularizing the variogram depend on an accurate estimate of measurement error. Therefore, by checking the accuracy of regularized variograms obtained using a simple equation it is possible to assess the accuracy of the model fitted to the experimental variogram and, in particular, the accuracy of the nugget variance as an estimate of measurement error.

The method proposed in this paper is as follows. The spatial resolution of the remotely sensed imagery are coarsened and experimental variograms are computed directly for each new spatial resolution. These variograms may then be used to check the accuracy of the regularized variograms obtained using Equation 3.

ACKNOWLEDGEMENTS

The authors acknowledge the financial support of the Leverhulme Trust (Grant F182T) at the University of Bristol. Prof. P.J. Curran, Dr R. Dunn and Prof. R. Webster are thanked for help at various stages of the research.

REFERENCES

Akaike, H. 1973. Information theory and the extension of maximum likelihood principle. in *2nd*

International Symposium on Information Theory (edited by B.N. Petrov & F. Csaki). Budapest: Akademia Kiado. pp. 267-281.

Chavez, P.S. 1992. Comparison of spatial variability in visible and near-infrared spectral images. *Photogrammetric Engineering and Remote Sensing*. 58: 957-964.

Clark, C.D. 1990. Remote sensing scales related to the frequency of natural variation: an example from paleo-ice-flow in Canada. *IEEE Transactions on Geoscience and Remote Sensing*. 28: 503-515.

Clark, I. 1977. Regularization of a semivariogram. *Computers and Geosciences*. 3: 341-346.

Cliff, A.D. & Ord, J.K. 1973. *Spatial Autocorrelation*. London: Pion.

Cohen, W.B., Spies, T.A. & Bradshaw, G.A. 1990. Semivariograms of digital imagery for analysis of conifer canopy structure. *Remote Sensing of Environment*. 34: 167-178.

Curran, P.J. 1988. The semi-variogram in remote sensing: an introduction. *Remote Sensing of Environment*. 3: 493-507.

Curran, P.J. & Dungan, J.L. 1989. Estimation of signal to noise: a new procedure applied to AVIRIS data. *IEEE Transactions on Geoscience and Remote Sensing*. 27: 620-628.

Davis, F.W., Quottrochi, D.A., Ridd, M.K., Lam, N. S-M., Walsh, S.J., Michaelson, J.C., Franklin, J., Stow, D.A., Johannsen, C.J. & Johnston, C.A. 1991. Environmental analysis using integrated GIS and remotely sensed data: some research needs and priorities. *Photogrammetric Engineering and Remote Sensing*. 57: 689-697.

Deutsch, C.V. & Journel, A.G., 1992. *GSLIB Geostatistical Software Library*. Oxford: Oxford University Press.

Epema, G.F. 1991. Studies of errors in field measurements of the bidirectional reflectance factor. *Remote Sensing of Environment*. 35: 37-49.

Gohin, F. & Langlois, G. 1993. Using geostatistics to merge in situ measurements and remotely sensed observations of sea surface temperature. *International Journal of Remote Sensing*. 14: 9-19.

Goodchild, M. & Gopal, S., (ed.s) 1989. *Accuracy of Spatial Databases*. London: Taylor and Francis.

Isaaks, E.H. & Srivastava, R.M. 1989. *Applied Geostatistics*. Oxford: Oxford University Press.

Journel, A.G. & Huijbregts, C.J. 1978. *Mining Geostatistics*. London: Academic Press.

Jupp, D.L.B., Strahler, A.H. & Woodcock, C.E. 1988. Autocorrelation and regularization in digital images I. Basic theory. *IEEE Transactions on Geoscience and Remote Sensing*. 26: 463-473.

Lam, N.S.-N. & Quottrochi, D.A. 1992. On the issues of scale, resolution and fractal analysis in the mapping sciences. *Professional Geographer*. 44: 88-98.

Lunetta, R.S., Congalton, R.G., Fenstermaker, L.K., Jensen, J.R., McGwire, K.C. & Tinney, L.R. 1991. Remote sensing and geographic information system data integration: error sources and research issues. *Photogrammetric Engineering and Remote Sensing*. 57: 677-687.

McGwire, K., Friedl, M. & Estes, J.E. 1993. Spatial structure, sampling design and scale in remotely sensed imagery of a California savanna woodland. *International Journal of Remote Sensing*. 14: 2137-2164.

Matheron, G. 1965. *Les Variables Régionalisées et Leur Estimation*. Paris: Masson.

Matheron, G. 1971. *The Theory of Regionalized Variables and its Applications*. Fontainebleau: Centre de Morphologie Mathématique de Fontainebleau.

Piwowar, J.M. & LeDrew, E.F. 1990. Integrating spatial data: a user's perspective. *Photogrammetric Engineering and Remote Sensing*. 56: 1497-1502.

Ramstein, G. & Raffey, M. 1989. Analysis of the structure of radiometric remotely sensed images. *International Journal of Remote Sensing*. 10: 1049-1074.

Ross, G.J.S. 1987. *Maximum Likelihood Program*. Oxford: Numerical Algorithms Group.

Rossi, R.E., Mulla, D.J., Journel, A.G. & Franz, E.H. 1992. Geostatistical tools for modeling and interpreting ecological spatial dependence. *Ecological Monographs*. 62: 277-314.

Rubin, Y. & Journel, A.G. 1991. Simulation of non-gaussian space random functions for modelling transport in ground water. *Water Resources Research*. 27: 1711-1721.

Webster, R. 1991. Local disjunctive kriging of soil properties with change of support. *Journal of Soil Science*. 42: 301-318.

Webster, R. & Oliver, M.A. 1990. *Statistical Methods for Soil and Land Resources Survey*. Oxford: Oxford University Press.

Sensors and Environmental Applications of Remote Sensing, Askne (ed.) © 1995 Balkema, Rotterdam, ISBN 90 5410 524 0

Specially cultivated fields as test targets for satellite imagery

Risto Kuittinen & Juhani Laaksonen
Finnish Geodetic Institute, Helsinki, Finland

ABSTRACT: In the summer 1992 and 1993 the Finnish Geodetic Institute carried out experiments to examine how useful spacially cultivated fields would be as test figures. The reason for these experiments was the requirement of cheap targets for testing the spatial resolution of satellite images. The results showed that it is possible to cultivate test figures and the spatial resolutions obtained were comparable with the results determined using other targets in images.

1 INTRODUCTION

Test fields have been used for a long time to evaluate the quality, especially the spatial resolution of aerial imagery. These fields are mostly made by painting figures on asphalt or plates. For earth resources satellite imagery, these kinds of test fields have not been made because of the huge costs of construction. Other kinds of solutions for testing satellite imagery have been studied and these are some reports on the use of agricultural fields e.g. Kaufmann and Fraser (1984).

In 1992 and 1993 the Finnish Geodetic Institute organized in 1992 and 1993 experiments for testing how useful were fields in the evaluation of the spatial resolution of Landsat TM and SPOT XS images. The reasons were the following.

a) The contrasts between different terrain types are in general much lower than presented in the literature when reporting the characteristics of the sensors. It is important to test the sensors by using low contrast targets.

b) Atmospheric conditions can significantly effect on the quality, especially the contrasts, of the images. To obtain practical results on the usefulness of satellite images targets should be imaged under different atmospheric conditions.

c) The Finnish Geodetic Institute is interested, together with other Finnish institutes, to develop a crop growth estimation method for Finnish agricultural statistics. Due to the relatively small average size of single fields in Finland it is important to know the effect of the practical spatial resolution on the measured spectral signmatures of the fields.

The aim of these experiments was to see if it was technically possible to construct test figures by using fields of different sizes and how good these figures would be in determining the spatial resolution of SPOT and Landsat images.

2 METHODS AND DATA

2.1 Calculation of contrasts

Modulation (M) and contrast ratio (C) were used in the data analysis. The modulations in the image and on the field were calculated as follows.

$$M = \frac{f_t(x,y) - f_o(x,y)}{f_t(x,y) + f_o(x,y)} \quad \text{or} \quad M = \frac{L_t(\lambda) - L_o(\lambda)}{L_t(\lambda) + L_o(\lambda)} \tag{1}$$

$$C = \frac{f_t(x,y)}{f_o(x,y)} \quad \text{or} \quad C = \frac{L_t(\lambda)}{L_o(\lambda)} \tag{2}$$

where

$f_t(x,y)$ is the mean of the digital numbers of the target.

$f_o(x,y)$ is the mean of the digital numbers of the background.

$L_t(\lambda)$ is the measured ground radiance of the target.

$L_0(\lambda)$ is the measured ground radiance of the background.

These equations were used for edges and test-bars. To obtain average values of modulation for whole images, mean, maximum and minimum modulations were computed by using 3x3 subimages Kuittinen and Laaksonen (1993).

2.2 Edges

In the edge gradient analysis (EGA) straight edges are used for determining the spatial resolution of images. This method was reported by Scott (1964) and the method has been used since to determine the spatial resolution of aerial photos. For digital images methods have been developed by Kamiya (1988) and Wong (1991).

When the EGA-method is used for digital images the length of the edge in the image has to be at least fifteen times the side of the pixel to obtain enough measurements for determining the line spread function of the edge. This means that for testing Landsat TM and Spot images, the length of edges must thus vary from 150 m to 450 m. This is possible also in practice when fields are used.The second requirement for the edge is that it is almost parallel with the rows and/or columns of the image. The third important characteristic of the edge is the width of the homogenous areas which form the edge. The width should be 5 - 7 times the pixel size to obtain stable maximum and minimum digital number values for determining the edge gradient.

The MTF of the satellite image was computed by using edge function E(x), its maximum and minimum values and the Fourier transformation of the first derivate of the E(x):

$$MTF(\upsilon) = [absFT(E'(x))]\frac{1}{\int E'(x)} \qquad (3)$$

Edges between fores and field were used to determine the MTF-values.

2.3 Test-bar figures

Test-bar figures are widely used and obtain good results used in image quality analysis. In these targets different bars are used in groups of 3-4 bars and the ability to separate bars from each other in the image is used as the measure of the spatial resolution. The size of the bars varies from one group to another. When the modulations of these test-bars figures are known in the field and in the image, one can determine the modulation transfer function values. Satellite image figures were made in the Metsähovi Research Station in the summers 1992 and 1993 to examine how useful the test-bars were. Table 1 shows the results.

Table 1. Test-bar figures for satellite imagery in the Metsähovi Research Station.

Year	Bar sizes (mxm)	Number of bars	Vegetation
1992	40x150	2/2	Oilseed rape/bare soil
1992	20x150	3/2	Oilseed rape/bare soil
1993	30x100	2/2	Phacelia&clover/Bare soil
1993	18x180	2/2	Phacelia&clover/Bare soil
1993	15x90	3/2	Phacelia&clover/Bare soil
1993	10x90	2/2	Phacelia&clover/Bare soil

The figures were made in the spring of each year by sowing the bar areas and leaving the areas between the bars unsown. Weeds were distroyed when the weather permitted.

The spatial resolution was determined on the basis of the visual observations of the satellite images as well as the modulation values in the images.

2.4 Data

Landsat TM data and SPOT XS&Pan data were used for the testing how useful the figures were. In summer 1993 the spectral reflectances of the vegetation in bars and the soil between the bars were measured by using a spectrometer. The dates of imaging and field measurements are presented in Table 2.

3 RESULTS

3.1 Spatial resolution determined by using the EGA

Figure 1a and 1b shows the spatial resolution of the studied SPOT and Landsat images as a function of target and image modulations determined by the EGA-method. The image modulations are 0.05, 0.1, 0.2 and 0.5 and target modulations vary from 0 to 1. The average modulations in the images were 0.075

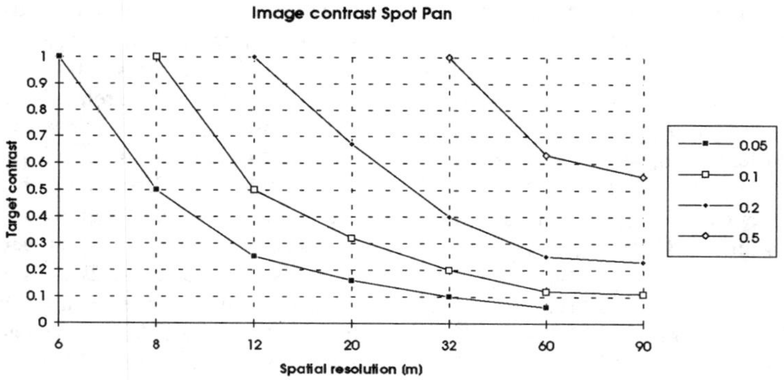

Figure 1a. The spatial resolution of the image as function of target and image contrast. (Here contrast=modulation). SPOT/Pan.

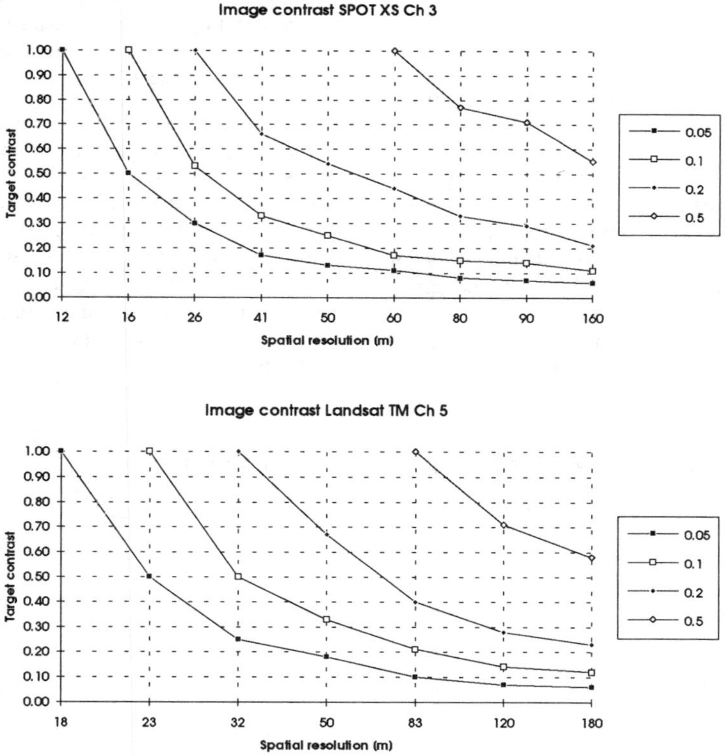

Figure 1b. The spatial resolution of the image as function of target and image contrast.(Here contrast=modulation). SPOT/ XS and Landsat/TM images.

and in general more than 50% of the minimum modulations in the images were below 0.035, Kuittinen and Laaksonen (1993).

These results show that when satisfied with very low image modulations one can obtain reasonable spatial resolutions. The IR channels gave the best spatial resolutions because the modulations in these channels were the highest.

Table 2. Satellite images and field measurements used in the study for spatial resolution determination.

Satellite measurement	Date of measurement	Method
Landsat TM	26 June 1992	Test-bar
Landsat TM	31 July 1993	Test-bar
Landsat TM	8 Sept 1993	Test-bar
SPOT XS	14 June 1986	EGA
SPOT XS	19 June 1992	Test-bar
SPOT XS	10 July 1992	Test-bar
SPOT XS	10 July 1993	Test-bar
SPOT XS	14 Sept 1993	Test-bar
SPOT Pan	20 July 1987	EGA
Field measurements		
Spectrometer	13 July 1993	
Spectrometer	8 Sept 1993	

3.2 Spatial resolution determined by using test-bar targets

The construction of bars succeeded well and they could be used as test figures. As Table 2 shows the widths of the bars were close to the pixel size in terrain because the aim was to obtain information about the maximal spatial resolution of images in practical conditons.

Let the pixel value of the bar be $f_b(x,y)$ and the pixel value of the bare soil $f_s(x,y)$. If the width of the bar is smaller than the area of the pixel, then the measured radiance $f_t(x,y)$ is smaller than $f_b(x,y)$. If the modulation of $f_t(x,y)$ and $f_s(x,y)$ is drawn as a function of the pixel area which covers the bar, Figure 3 for Landsat TM images and Figure 4 for SPOT XS images are obtained. In these figures we see that when the area of the bar covered by the pixel area in field increases, the modulation increases, too. If the contrast between $f_t(x,y)$ and $f_s(x,y)$ is large enough. For 30 m bars only TM channels 4 and 5 gave results where as the other channels could not detect the bars. 18-20 m bars were imposible to detect. Figure 2 shows the field reflections on 13 July 1993. For Spot XS images only channel 3 can detect 18-20 m bars whereas all channels of Spot XS can detect 30 m bars.

Besides of the modulation and MTF calculations the spatial resolution of the available images were evaluated by visual interpretation too. These results are presented in Table 3.

The spatial resolution of Landsat TM images determined by visual interpretation of the test bars corresponds with those published by Schowengert (1985), if MTF values of 0.4 are used. If an image modulation of 0.03 is accepted for the minimum detectable modulation, spatial resolutions of 25 m for Landsat TM/4 images and 18 m for SPOT XS/3 images are obtained. Spatial resolutions are poorer for other channels which result from lower modulation values.

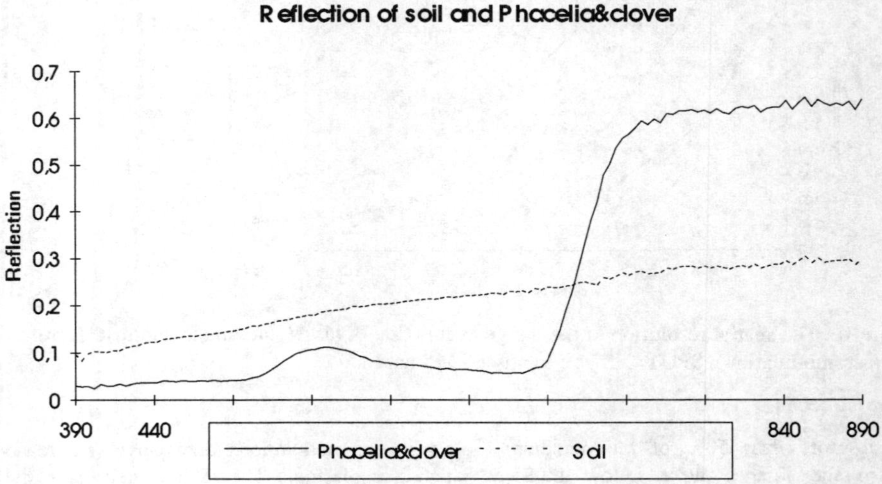

Figure 2. The reflection of the test-bar figures as a function of the wavelenght on 13 July 1993. Phacelia & clover and bare soil.

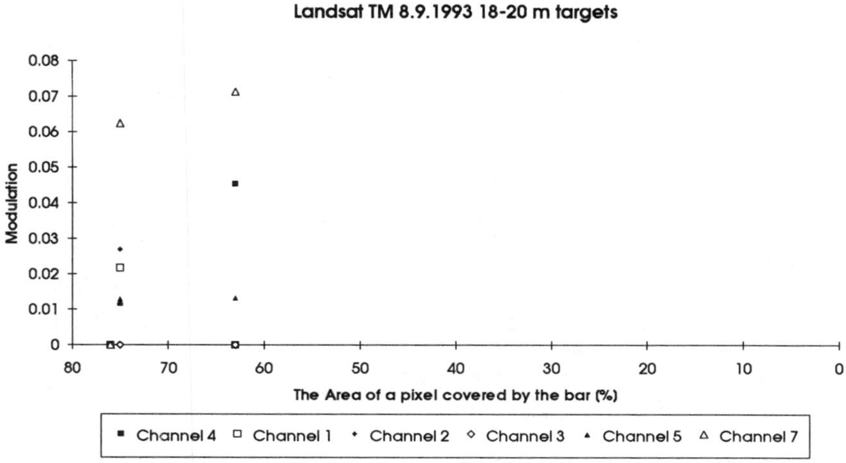

Figure 3a. The Landsat TM image (8 September 1993) modulations as a function of the bar width and the area of a pixel covered by the bar.

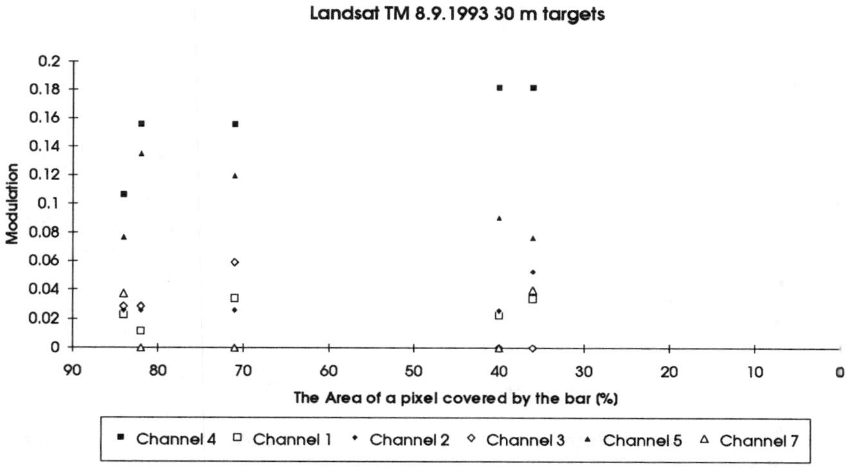

Figure 3b. The Landsat TM image (8 September 1993) modulations as a function of the bar width and the area of a pixel covered by the bar.

4 DISCUSSION

These results show that cultivated test-bar figures can be used. This method is however limited to be used for images having a pixel size in terrain less than 50x50 m^2, because elsewhere the figures become too large. The following things have to be taken into account when using these test-bars or edges:

-Soil type must be similiar and the test area flat, which produces targets

-If a bare soil areas between the bars are used these need the careful removal of weeds, otherwise results will not be achieved.

-The vegetation in figures must be dense, to give spatially uniform reflectances at part of the figures.

-Field measurements of radiances very much help the evaluation of the results because the contrasts and modulations of the figures change during the summer.

-In general it is not possible to obtain in field modulations larger than 0.4. Thus the results

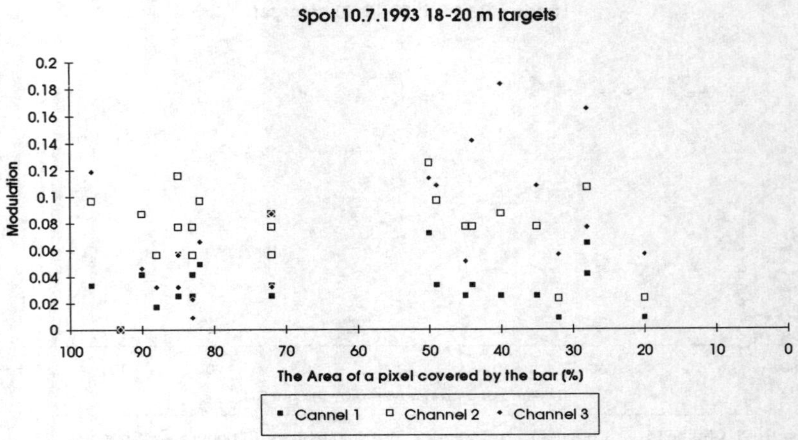

Figure 4a. The SPOT XS image (10 July 1993) modulations as a function of the bar width and the area of a pixel covered by the bar

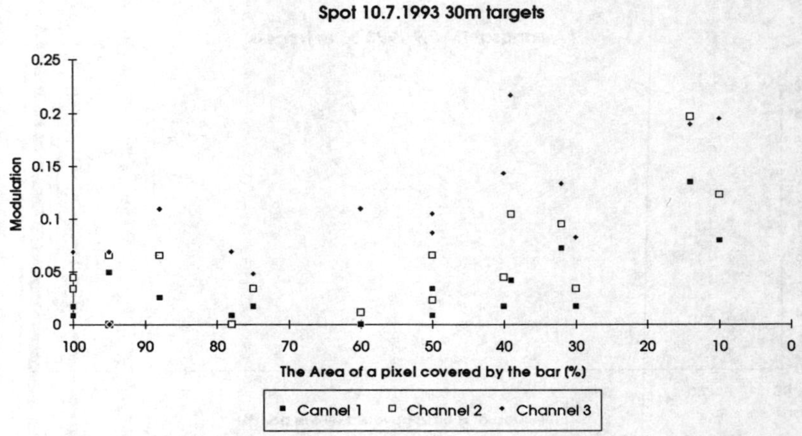

Figure 4b. The SPOT XS image (10 July 1993) modulations as a function of the bar width and the area of a pixel covered by the bar.

obtained present average conditions in satellite images.

-The determined spatial resolutions are in general lower than those determined for the sensors. This is due to the effect of the atmosphere and the sampling of measurements.

Correct spectral signatures cannot be obtained with maximal spatial resolution. As correct signatures are needed for good interpertation, only spatial frequencies giving MTF values close to one can be used. This means that the field sizes must perhaps be in the case of Landsat and SPOT images between 60 - 120 m.

In future research these test-bar figures will be used when developing image enchancement methods to obtain a better spatial resolution for images.

Table 3. The spatial resolution of Landsat TM and SPOT XS satellite images evaluated by visual interpretation of cultivated test-bar figures.

Image	Date	Chs	MI	MF	SR (m)
TM	26.6.1992	1-3,5,7	0.02-0.21		>40
		4			25
	31.7.1993	1-3,5,7			<30
		4			20
	8.9.1993	1-7	0.05		<30
XS	19.6.1992	2-3	0.07-0.18		20
		1			30
	10.7.1992	2-3	0.03-0.20		20
		1			30
	10.7.1993	1,2	0.18-0.33		30
	10.7.1993	3	0.50		25
	14.9.1993	1-3			>30

MI=Image modulation, MF=Field modulation
SR=Spatial resolution

REFERENCES

KAUFMANN, Y. J., FRASER, R. S., 1984. Atmospheric Effect on Classification of Finite Fields. *Remote Sensing of Environment*, 15, 95-116.

KAMIYA, I., 1988. A Practical Method of Remotely Sensed Digital Image's Resolving Power. In: Int. Arch. of Photogrammetry and Remote Sensing, Commission I, part B1, Vol. 27, Kyoto, 64-72.

KUITTINEN, R., LAAKSONEN, J., 1993. The effect of the spatial resolution of the satellite image on the object interpretation accuracy. Reports of the Finnish Geodetic institute,40.

SCHOWENGERDT, R. A., ARCHWAMETY, C., WRIGLEY, R. C., 1985. Landsat Thematic Mapper Image-Derived MTF. *Photogrammetric Engineering & Remote Sensing*, 9, Vol. 51, 1395-1406.

SLATER , P. N., 1980. Remote sensing, optics and optical systems. Addison-Wesley, Massachusets, 575.

WONG, H., 1991. Effect of Knife-Edge Skew on Modulation Transfer Function Measurements of Charge-Coupled Device Imagers Employing a Scanning Knife Edge. *Optical Engineering*, 9, Vol. 9. 1394-1398.

Sensors – Microwave-technique

Sensors and Environmental Applications of Remote Sensing, Askne (ed.) © 1995 Balkema, Rotterdam, ISBN 90 5410 524 0

The ENVISAT Advanced Synthetic Aperture Radar (ASAR)

S. Karnevi
ESA Directorate for Observation of the Earth and its Environment, ESTEC, Noordwijk, Netherlands

ABSTRACT: An Advanced Sythetic Aperture Radar (ASAR) has been selected as the imaging radar sensor for the ENVISAT-1 payload. In addition to providing continuity of data with the single-swath, single polarisation Actiev Microwave Instrument (AMI) currently flying on ERS-1 (and ERS-2), the ASAR has the additional advantage of offering: greater coverage, with more flexibility in the choice of swath positions, a wider range of incidence angles, and a wide swath capability; dual polarisation; and improved wave mode capabilities.

1 INTRODUCTION

In Munich (1991) and in Granada (1992) the European Ministers endorsed the Agency's Earth-Observation Programme, where one aspect is the first of the Polar Platform missions namely Envisat-1 .

When a possible orientation for this first Polarplatform mission was considered, particular note was taken of the increased concern over the Earth's atmosphere and its environment.
The objectives of the ENVISAT-1 mission were therefore formulated as:

(a) to provide for continuity of the observations started with the ERS satellites, including those obtained from radar-based observations;

(b) to provide for enhancement of the ERS-1 mission, notably the ocean and ice mission;

(c) to extend the range of parameters observed to meet the need to increase knowledge of factors determining the environment;

(d) to make a significant contribution to environmental studies, notably the areas of atmospheric chemistry and ocean studies (including marine biology).

Coupled with two linked secondary objectives:

- to allow more effective monitoring and management of the Earth's resources
- to better understand solid Earth processes.

This article discusses one of the instruments onboard the Envisat-1 satellite, namely the Advanced Synthetic Aperture Radar (ASAR).

2 THE OBJECTIVES OF ASAR

The objectives of the Advanced SAR are very similar to those of the Active Microwave Instrument (AMI) on ERS-1 and ERS-2. The main task will again be monitoring of the environment in order to collect more precise information on global changes. The key point of interest is the global climate and all possible factors which may have an influence on it.

Several SAR applications are at present being studied with ERS data by a great number of scientific groups and pilot projects. The outcome of the these studies will certainly make SAR data more attractive for a wide range of users.

It therefore is essential to offer as part of the ENVISAT-1 mission on one hand an instrument which offers data continuity with ERS-1 and ERS-2 but at the same time is a step forward with respect to system flexibility and scientific and operational value of the data sets.

To satisfy this need the ASAR Design incorporates:

- a flexible swath position capability which will offer the choice between several image swath positions at various distances from the subsatellite track with different incidence angles
- a second polarisation that in addition to the vertical polarisation of ERS, will offer horizontal or vertical polarisation for imaging

These two features are used and combined such that, in addition to the single swath image mode of ERS, a host of new modes are created.

The capability of ASAR to switch between different swaths offers very fast multiplexing between swaths so that an overall swath of more than 400 km width can be imaged in one go.

This mode of operation is called "Wide Swath Mode" and will improve the capability of monitoring large areas.

In addition to this wide swath mode ASAR can be set to operate at a coarse spatial resolution of only 1000 m. The data rate will then be low enough to store on the tape recorders on board the space craft. This mode is called "Global Monitoring Mode". Due to its independence of ground station coverage, brought about by its lower data rate, this mode offers a tool for the monitoring of features such as ice coverage, snow coverage, deforestation, desertification or humidity without interruption on a global scale.

The capability to switch between two polarisation allows a special "Alternating Polarisation Mode" to be implemented which permits half of the looks of an image to be acquired in horizontal and the other half in vertical polarisation, so that in a single pass an image of horizontal and vertical polarisation will be taken from the same scene. This Alternating Polarisation Mode will considerably increase the ASAR capability for target classification. This will be especially so if it is used in conjunction with multitemporal imaging of the same scene.

ASAR will further - like ERS - have a Wave Mode for taking images of 5km x 5km but more frequently in 100km distance and selectable in two different swaths over the oceans to supply samples of ocean wave spectra on a global basis.

3 REQUIREMENTS

The requirements for the most important operational modes of ASAR are summarised in Table I.

In addition to these requirements an absolute radiometric accuracy of 1.75dB is required in all modes over orbit and mission. On one hand this sets very narrow tolerances for on-board and overall calibration but on the other hand allows the comparison of images taken at different times during the lifetime of the satellite. It also makes possible quantitative comparisons of ASAR images with

Table I ASAR Performance summary

Mode	Polarisation	Spatial Resolution	Radiometric Resolution	Swath Width	Incidence Angles
Image	HH or VV	30 m	2.5 dB	7 Swaths 100 to 56 km	20 - 45^{0}
Alternating Polarisation	HH + VV	30 m	3.5 dB	7 Swaths 100 to 56 km	20 - 45^{0}
Wide Swath	HH or VV	150 m	2.0 dB	> 400 km	17 - 42^{0}
Global Monitoring	HH or VV	1000 m	1.5 dB	> 400 km	17 - 42^{0}
Wave	HH or VV	30 m	2.0 dB	Two 5km in any swath	20 - 45^{0}

images taken by other satellites.

Ambiguity and side lobe levels in the images are specified to be better than -25dB in all modes.

4 INSTRUMENT DESCRIPTION

A functional block diagram of the ASAR instrument is shown in Figure 1. It divides functionally into two sub-assemblies:

- Antenna Sub-Assembly (ASA) - containing the phased array antenna electronics distributed over 20 Tiles. Each Tile containes 16 transmit/receive modules controlled by a Tile Control and Interface Unit. The T/R module performs the final amplification of the signal and gives the signal an amplitude and/or phase weight so that the total result of the 320 T/R modules creates the desired 3 dimensional elevation antenna pattern. Each T/R module is connected to a radiator with 24 annular slots fed from behind by a stripline feed system. The strip can be driven such that they produce a vertically or horizontally polarised radiation pattern.
- Central Electronics Sub-Assembly (CESA) - providing rf signal generation and reception, control signals, power conditioning and data formatting.

The physical configuration of ENVISAT-1 is shown in Figure 2. From this can be seen that the ASAR antenna consists of five hinged panels. Each panel contains four Tiles.

The CESA equipments are mounted on three of the Payload Equipment Bay panels inside the spacecraft. There is then a short connection to the ASA. This connection involves power lines to 80 Tile power supply units, two waveguide feeds and the serial control bus for data, commands and telemetry.

5 MODES OF OPERATION

The operational modes of ASAR can be divided into three different categories.

5.1 *High spatial resolution imaging modes*

These are the Image mode and the Alternating polarisation mode. In both modes the ASAR operates as an imaging radar with high spatial resolution (typically a few tens of meters), and relatively narrow swaths (up to 120 km). The radar geometry

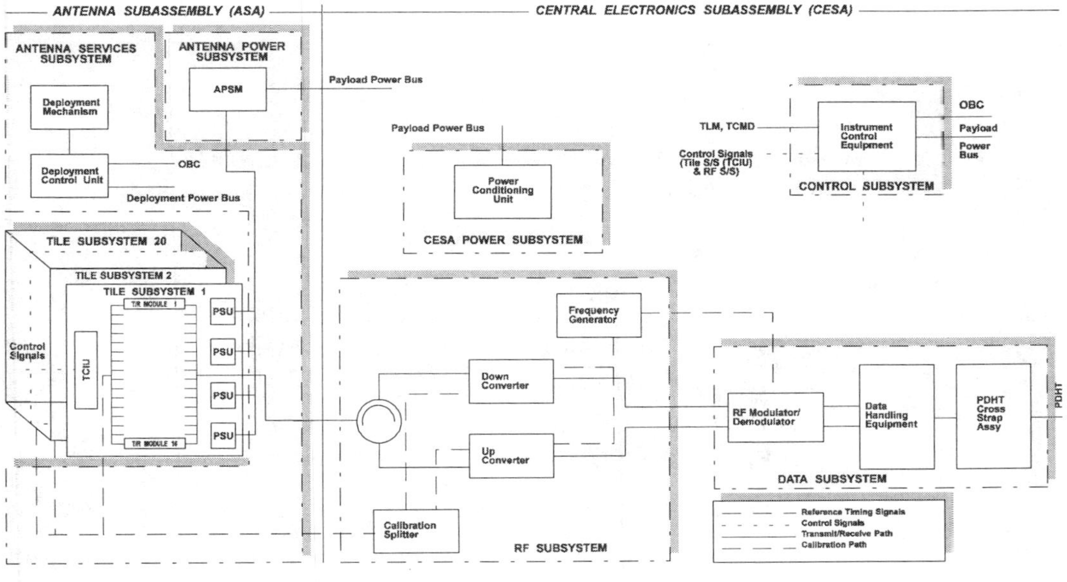

Figure 1 ASAR Blockdiagram

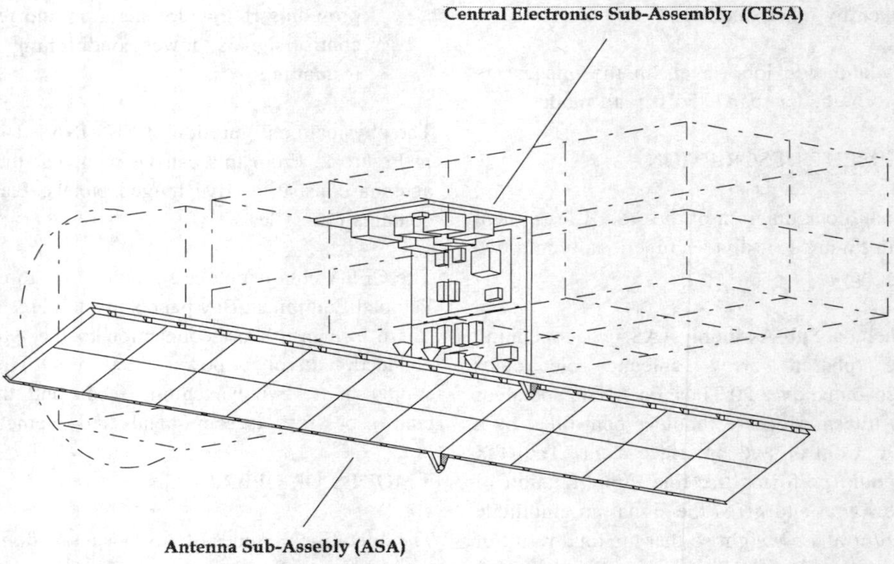

Figure 2 ENVISAT-1 Configuration

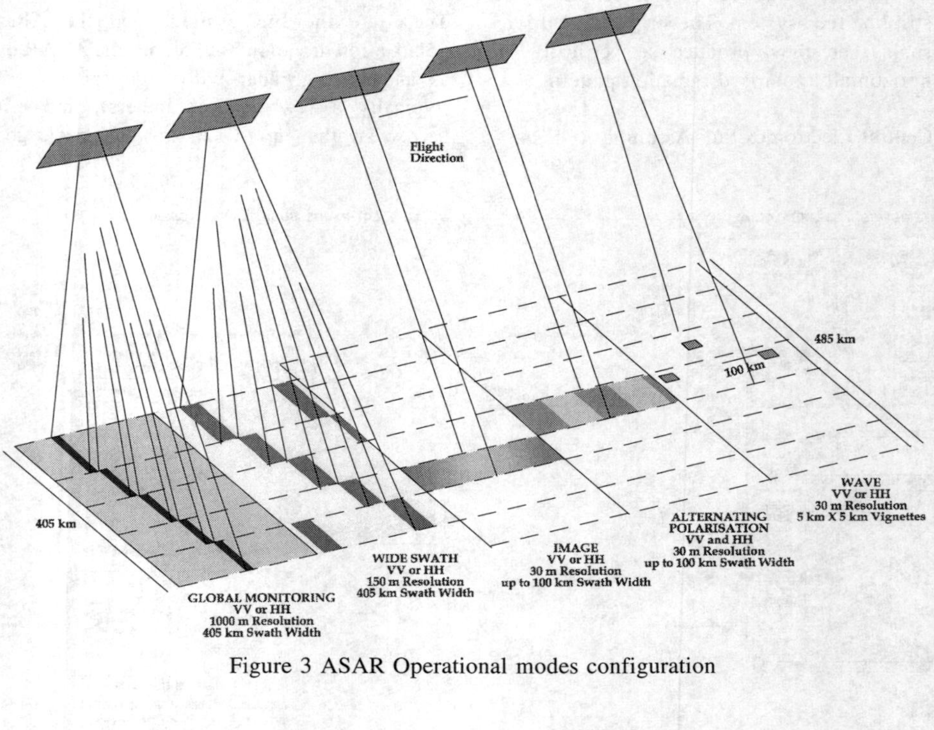

Figure 3 ASAR Operational modes configuration

when operating in this manner can be seen in fig. 3.

Image Mode

The Image mode provides continuous coverage over a single swath nominally 100km wide. The swath can be selected from anywhere within a 500km region (incidence angle range 15° - 45°). For nominal operation seve byn swaths, IS1 to IS7, are defined over this region, see figure 4. The imaging is performed transmitting a continuous series of pulses and acquiring the required echo information after the appropriate return trip delay.

Alternating Polarisation Mode

The Alternating Polarisation mode provides imaging in VV and HH polarisations of the same image by interleaving looks of each polarisation along track within the synthetic aperture. Effectively, a ScanSAR technique is used but without varying the subswath. The echo measurement is made within repetition cycles containing two bursts of transmissions on each of the polarisations. The only interruption in the sequence is when the internal calibration requires updating (as for Image mode). This update is made on alternative VV or HH cycles.

5.2 Low Spatial resolution imaging modes

These are the wide Swath mode and the Global Monitoring mode. In these modes the ASAR operates as an imaging radar with low spatial resolution (between 100m to 1 km), and relatively wide swath (more than 400 km). To obtain global coverage, for the global monitoring mode, the data is stored on-board the satellite and transferred to ground when the satellite pass by a ground station.

Wide swath mode

The Wide Swath mode provides continuous coverage over a swath nominally 400km wide. This swath is divided into 5 subswaths ranging from 60 to 100km, see figure 4. Using the ScanSAR technique ASAR transmits bursts of pulses to each of the subswaths in turn such that on return to any subswath a continuous along track image is formed.

Global Monitoring Mode

The Global Monitoring mode provides continuous along track sampling across a 400km swath. This is achieved in a similar manner to Wide Swath mode, using ScanSAR techniques. The mode has a low data

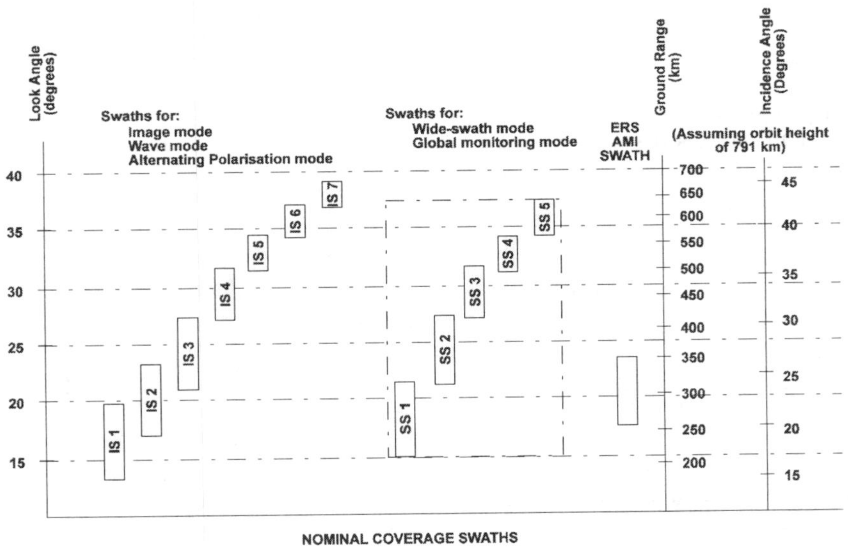

Figure 4 ASAR Swath coverage

rate due to a slightly reduced along track duty ratio and the use of digital filtering for reduction in the across track direction. The same subswaths as defined for the Wide Swath Mode are used.

5.3 Wave mode

In the wave mode, the ASAR utilises the facilities of the image mode to image small areas of the ocean surface. The ASAR measures the change in radar backscatter from the sea surface due to ocean waves. The imaging is carried out to a high spatial resolution over one or two areas across a swath of approximately 5km X 5km at intervals of nominally 100 or 200 km in the along track direction.

6 PRODUCT PROCESSING

The images collected by ASAR will be processed on ground, and distributed mainly from ESOC in Frascati. The products are divide into four main categories each described below.

6.1 *High Resolution Image*

Generated within 24 hours of sensing and is normally distributed by non-electronic media. The product is similar to the Precision Image (PRI) in ERS-1.

6.2 *Fast Delivery*

Delivered within 3 hours of sensing and is delivered via electronic link. Product has a lower geometric resolution (~100m pixels).

6.3 *Browse*

Delivered within 3 hours, collated from measurements, and has very low resolution. Used for browse catalogues.

6.4 *Wave imagette & spectra*

This is a product similar to ERS-1 wave spectra with the improvements introduced in ASAR compared to ERS-1.

7 BIBLIOGRAPHY

1. S. 1. Austin D. J. Q. Carter and J. D. McCaughey, ERS- I Spacecraft - Active Microwave Instrumentation: A Major Achievement, *The GEC Journal of Research*, 09/01, 2-12

2. F. G. Sawyer, S. S. Hartley and M. A. Brown, Versatile SAR for the First Polar Platform, *The GEC Journal of Research* 09/01, 13-22.

3. D. J. Q. Carter, S. S. Hartley and G. Graf, The Advanced Synthetic Aperture Radar (ASAR) for POEM-1, *Proceedings ISY-92, ESA SP-341.*

4. C.J. Readings and P.A. Dubock, Envisat-1: Europe's Major Contribution to Earth Observation for the Late Nineties, *ESA Bulletin* 76/93, 15-28.

5. S. Karnevi, E. Dean, D.J.Q. Carter and S.S Hartley, Envisat's Advanced Synthetic Aperture Radar: ASAR, *ESA Bulletin* 76/93, 30-35.

Sensors and Environmental Applications of Remote Sensing, Askne (ed.) © 1995 Balkema, Rotterdam, ISBN 90 5410 524 0

The calibration of the ERS-1 synthetic aperture radar using UK-PAF imagery

P.J. Meadows
GEC, Marconi Research Centre, Chelmsford, Essex, UK

ABSTRACT: Calibration of the ERS-1 synthetic aperture radar has shown that users are required to make corrections for replica pulse power variations and for the non-linear response of the on-board analogue to digital convertor (ADC). The replica pulse variations can be as much as 1.8 dB while the ADC can lead to underestimates of up to 3.3 dB in radar cross-sections measurements. The precision of the corrections is estimated to be better than 0.5 dB. A full expression to enable distributed and point target radar cross-sections to be calculated is given.

1 INTRODUCTION

The calibration of any synthetic aperture radar (SAR) is an essential requirement to ensure that the user community receives high quality imagery. For the ERS-1 SAR, calibration is carried out via the imaging of external calibration targets on the Earth's surface, instrument generated internal calibration data, and quality information derived from SAR raw data. The calibration exercise not only enables monitoring of the performance of the SAR instrument but also verification of the SAR products.

The principal aim of radiometric calibration is to enable SAR product pixel values to be converted into absolute radar cross-section measurements via a calibration constant. Once this calibration constant has been derived, the secondary aim of radiometric calibration is to determine any variation or drift in the radiometric performance of the SAR. This is achieved by the derivation of radiometric stability (the standard deviation of the transponder radar cross-section measurements) and radiometric accuracy of the SAR (the difference between the actual and measured transponder radar cross-sections). These enable the precision of any radar cross-section measurement to be determined and if necessary re-calculation of the calibration constant. The calibration results can also be used to determine image quality parameters, such as spatial resolution.

Geometric calibration determines the spatial accuracy of an image. The determination of the image localisation accuracy can be achieved via the known location and measured position of features within an image.

This paper begins with an investigation into radiometric calibration where an expression is derived for SAR users to determine accurate radar cross-section measurements. The geometric calibration results are then presented. All these results have been derived using products processed at the UK Processing and Archive Facility (UK-PAF).

2 RADIOMETRIC CALIBRATION

Radiometric calibration of ERS-1 SAR imagery is achieved by carrying out three separate activities: external calibration, internal calibration and raw data quality assessment. External calibration involves the analysis of the impulse response function of bright man-made point targets of known radar cross-section. These are used to determine the relationship between image pixel values and radar cross-section for both point and distributed targets. The results of external calibration can also be used to assess the quality of an image. Internal calibration examines the parameters that are generated on-board the satellite to aid in the calibration of the SAR. Finally, raw data quality assessment determines the properties of the SAR data prior to processing into imagery. This activity involves comparing the raw data with its expected properties. Each of these three activities are discussed further below together with their implications for the determination of an expression for point and distributed radar cross-sections. The derivation of radiometric stability and accuracy will also be addressed.

2.1 *External calibration*

At the UK-PAF, the three ESA transponders (Woode et al. 1992) have been used to perform external calibration. These have been deployed since the start of the ERS-1 mission at either Flevoland or Zeeland in the Netherlands. The selection of the deployment site has been determined by the phase of the mission, as Table 1 indicates (based on Eurimage 1993). The transponders were built specifically to perform ERS-1 SAR calibration and have a quoted stability of better than ± 0.1 dB (Woode et al. 1992). For each mission phase the transponders were deployed at three different locations such that they appeared at different positions in the across-track (range) direction, while being approximately at the same along-track (azimuth) position. During the 3 day commissioning and ice phase repeat periods the transponders were always imaged during ascending passes, with each transponder being at approximately the same range position. For the 35 day multi-disciplinary phase the transponders were imaged during ascending and descending passes - some passes imaged all three transponders while others imaged only two.

Table 1. ERS-1 Mission Phases and Transponder Locations.

Phase	Location	Orbits
Commissioning	Flevoland	126 to 2103
Ice	Zeeland	2354 to 3695
Multidisciplinary	Flevoland	3901 to 12725

ERS-1.SAR.PRI images containing the transponders have been analysed using a Point Target Quality Assessment tool at the UK-PAF Calibration & Validation facility (Scott et al. 1994). The purpose of this tool is to derive various properties of the transponder impulse response function (IRF) and hence to derive the absolute radar cross-sections for the transponders and image quality parameters. Briefly, the tool extracts a region centred on the point target, derives the mean background amplitude which is subtracted from the extracted region, interpolates by a factor of 8, converts to intensity and then derives the IRF parameters. The image quality parameters include the spatial resolution (the half intensity width of the IRF), the total power in peak (energy within the IRF mainlobe), integrated sidelobe ratio (ratio of the energy in the sidelobes to the energy in the mainlobe) and peak sidelobe ratio (ratio of the intensity of the most intense sidelobe of the IRF to the peak intensity in the mainlobe of the IRF).

Using these IRF parameters, the radar cross-section of the transponders, σ, can be calculated thus:-

$$\sigma = \frac{I_p}{C_f}\frac{A}{K}\frac{\mathrm{Sin}\,\alpha_p}{\mathrm{Sin}\,\alpha_{ref}} \qquad (1)$$

where I_p is the total power in the IRF peak, C_f is the relative power in the sidelobes, A is the area of a pixel, K is the calibration constant, α_p is the transponder point target incidence angle and α_{ref} is the reference incidence angle (23°).

Table 2 shows the mean quality assessment parameters for transponder 2 during the commissioning phase. The azimuth spatial resolution compares well with the theoretical azimuth resolution of 20.76m (Brown 1992). In range, the

Table 2. ERS-1 Transponder Impulse Response Function Parameters.

Parameter	Transponder 2
Azimuth Resolution (m)	20.07 ± 1.25
Range Resolution (m)	25.48 ± 0.75
Integrated Sidelobe Ratio (dB)	-13.90 ± 1.26
Peak Sidelobe Ratio (dB)	-21.21 ± 1.36
Number of Measurements	13

spatial resolution varies across the swath with the largest resolution occurring at near range. The theoretical range resolution at the mid-swath position is 25.33m (Brown 1992) which corresponds approximately to the position of the transponder in Table 2 - a comparison with the actual value is quite good. Note that the variation in the spatial resolution can be accounted for, in part, by the precision to which this parameter is measured (1/8th of a pixel or 1.56m). The integrated and peak sidelobe ratios are all consistent with theoretical results (Brown 1992) which give low sidelobe ratios and show that the transponder IRFs are not contaminated by nearby targets. The other transponders give similar results for this and the other mission phases.

Figure 1 shows the relative radar cross-section for the three transponders from the commissioning phase through to the end of the multidisciplinary phase (i.e. up to the end of 1993). The relative radar cross-section is the transponder radar cross-section relative to its nominal radar cross-section as determined by on-ground calibration prior to each mission phase. Note that the calibration constant, K, has been derived using the transponder 2 IRF measurements from an image acquired on 13th October 1991 (Orbit 1273) (Meadows 1993). Thus, the relative radar cross-section for this transponder on this date is

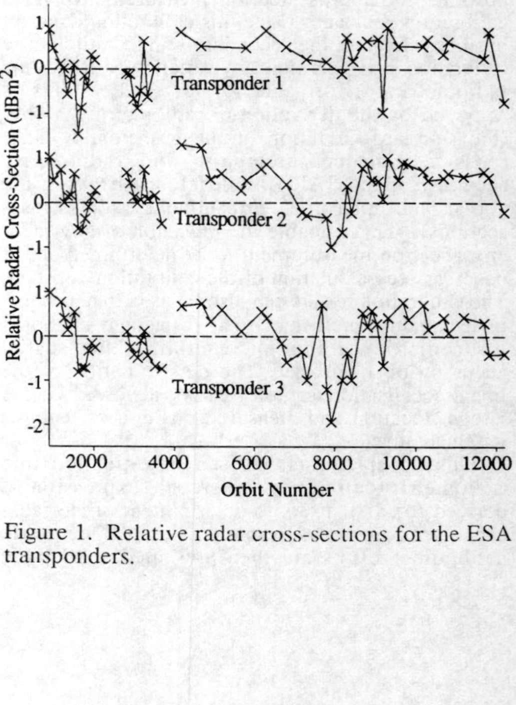

Figure 1. Relative radar cross-sections for the ESA transponders.

defined as being zero. Figure 1 also shows quite clearly that the variation in relative radar cross-section is much larger than the quoted stability of the transponders. For example, transponder 3 in the multidisciplinary phase has a peak to peak variation of some 2.69 dB while the corresponding radiometric stability is 0.67 dB. The other transponders during this and the other phases also show large peak to peak and stability values. Clearly the source of these large variations needs to be identified and corrections made.

2.2 *Internal calibration*

At the start and end of each imaging sequence (the operation period of the SAR), a set of 8 noise measurements and 4 calibration pulse measurements are made. During the imaging sequence, copies of the transmitted pulses (replicas) are generated and appended to the raw data. These quantities are required to ensure that the SAR image products are internally calibrated, especially as the ERS-1 SAR does not have an automatic gain control system. The system gain can drift due to temperature changes and ageing. The gain changes are monitored via the calibration and replica pulse powers as they are passed through the SAR system. The calibration pulse measures the majority of the gain drift from image sequence to image sequence with the replica pulse being used to monitor the gain drift during the imaging sequence when the more representative calibration pulse is not available (Lancashire 1987).

Figures 2 and 3 show the replica and calibration pulse powers during November 1992 for imaging sequences acquired from the Salmijaervi ground station (Kiruna, Sweden) and archived at the UK-PAF (Wright 1993). Figure 2 shows that the replica pulse power between the start and end of an imaging sequence differ by typically 0.1 dB, with the higher power being at the end of an imaging sequence. However, from imaging sequence to imaging sequence the replica pulse power can be seen to vary

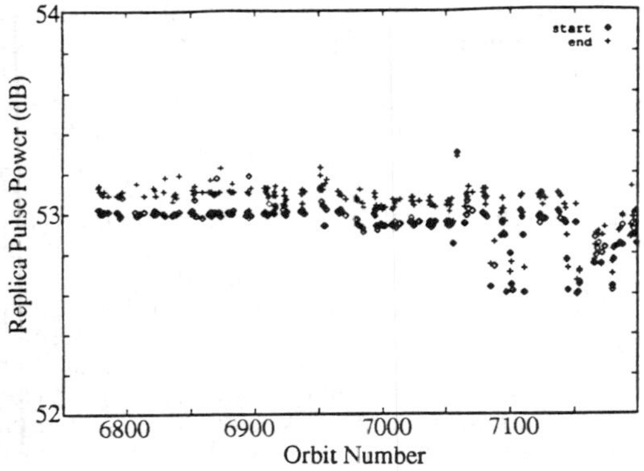

Figure 2. Replica pulse powers for November 1992.

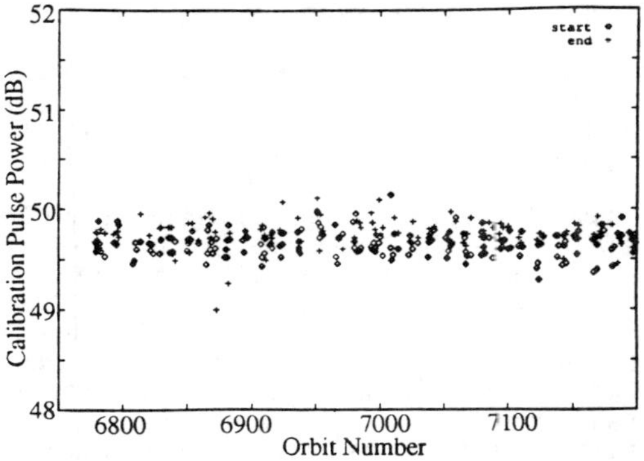

Figure 3. Calibration pulse powers for November 1992.

by as much as 0.8 dB. Between the period September 1991 and December 1992, variations of up to 1.8 dB have been identified (Wright 1993). The calibration pulse power variation is much smaller compared to the replica pulse: as Figure 3 shows, the variation is typically less than 0.5 dB. However allowances should be made for the calibration pulse power measurement accuracy (a 1σ value of 0.17 dB compared to only 0.018 dB for the replica pulse power (Lancashire 1987)). Over longer time periods, the calibration pulse power variation is also less than 0.5 dB (Laur et al. 1993). It should be noted that both the replica pulse and calibration powers are within their expected variation range of ±1.0 dB (Lancashire 1987).

The UK-PAF SAR processor uses a complete replica pulse from the raw data corresponding to the required SAR image for gain normalisation and pulse compression. The gain normalisation scales the raw data prior to processing. No use is made of the calibration pulses (as these are only available at the start and end of an imaging sequence and may not even be available due to the possibility of being generated outside the reception of the ground station). Use of this scheme to perform automatic internal calibration successfully requires that the replica pulse power is directly proportional to the transmitter pulse power. If this were the case any transmitter pulse power variations would be compensated for in the resultant SAR image. Unfortunately this is not so (Meadows 1993) and consequently the replica pulse power variations introduced by the SAR processor need to be removed to ensure fully calibrated SAR imagery. This is achieved by the comparison of the product replica pulse power (contained within the product CEOS header) and the replica pulse power for the reference calibration image of 13th October 1991 (Orbit 1273).

Any variation in transmitter pulse power is monitored via the external calibration results and if necessary via the introduction of a new calibration constant, K.

2.3 *Raw data quality assessment*

SAR raw data obeys a certain statistical distribution for each of the in-phase (I) and quadrature (Q) channels. This distribution is a zero mean, Gaussian amplitude and uniform phase. For ERS-1 SAR image mode the I and Q channels are quantised to 5 bits each within the Analogue to Digital Convertor (ADC). This allows the quantised data to take integer values in the range -16 to 15, with each channel having a Gaussian distribution. This distribution should have a maximum at a value of 0 or -1 and should fall to zero at -16 and 15. If the distribution does not fall to 0 at the upper and lower quantisation levels, one of two situations occur: saturation or bit redundancy. Saturation occurs when the input signal level to the ADC is high. The condition is characterised by a large number of data values occupying the lowest and highest quantisation levels, as revealed by a 'tail' in a histogram of the ADC distribution. Regions of high radar cross-section from distributed targets can lead to saturation. Bit redundancy occurs when the ADC input is low, resulting in only some of the available quantisation levels being used. Distributed targets of low radar cross-section can lead to bit redundancy.

The 'health' of the ADC can be assessed by measuring the standard deviation of the ADC output. If the standard deviations are significantly higher than approximately 5.3 (Brown et al. 1988) then saturation occurs, while if the standard deviation is less than this ideal value bit redundancy is present. If either of these situations persists for all the raw data, this indicates that the gain prior to the ADC is either too high or too low.

To produce perfectly calibrated SAR imagery, the ADC should preserve the power of the raw data (i.e. the input power to the ADC should equal the output power). The ERS-1 ADC I and Q channels can be simulated via the use of a zero mean Gaussian distribution (Meadows 1993). Figure 4 shows the ADC power loss (the difference between input and output power as a function of output ADC standard deviation). This shows that for ADC standard deviations greater than 6 there are significant SAR image power losses, which would, if present, have severe implications for the calibration of ERS-1 SAR imagery.

Figure 4. The ADC power loss verses ADC standard deviation. The dashed line refers to a linear ADC.

The effect of either ADC bit redundancy or saturation is illustrated here via the examination of one Zeeland scene from the Ice Phase (Orbit 3165, 22nd February 1992). As Figure 5 shows, this scene contains approximately 40% land and 60% ocean and has a high ocean radar cross-section.

Figure 5. ERS-1.SAR.PRI image of Zeeland for Orbit 3165 (22nd February 1992).

For this scene an ERS-1.SAR.RAW product has been examined and a series of raw data blocks extracted. For each block the I channel standard deviation and ADC power loss have been calculated. The block size is determined, in azimuth, by the along-track length equivalent to the synthetic aperture integration period, while in range, by the across-track length equivalent to the chirp length. This corresponds to approximately 5 km in azimuth and 5 km in slant range or 15 km in ground range.

Figure 6 shows the I channel standard deviation contour plot where values up to 11.5 are present. Note that in the range direction, the contour plot shows slant range rather than ground range. The drop at extreme near range is due to the influence of the reduced gain of the SAR elevation antenna pattern. The corresponding power losses amount to 3.3 dB at near range, reducing to 0.5 dB at far range.

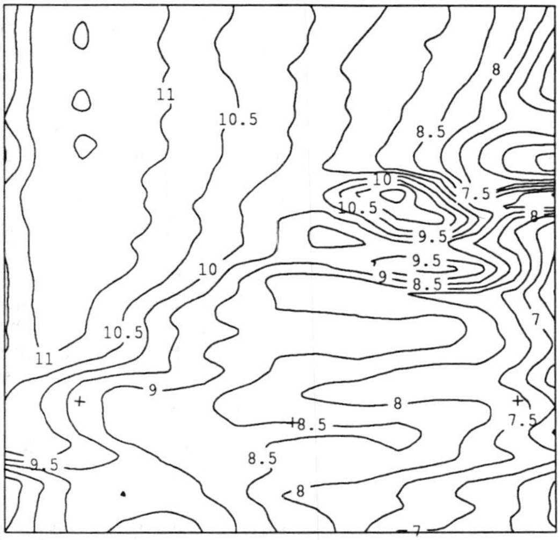

Figure 6. ERS-1.SAR.RAW Zeeland I channel standard deviation variation for Orbit 3165. The crosses indicate the location of the three ESA transponders.

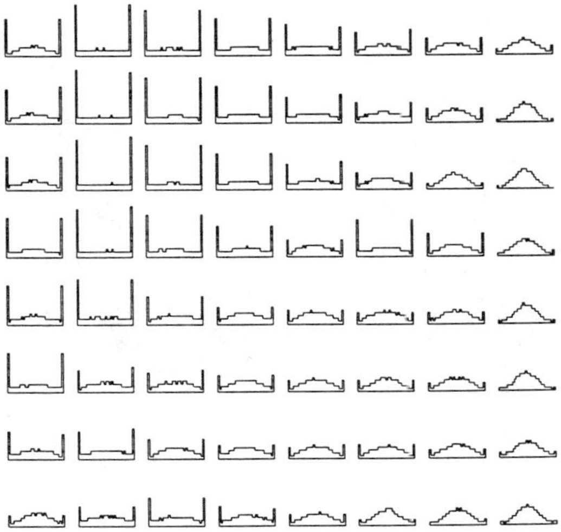

Figure 7. ERS-1.SAR.RAW histograms of ADC quantisation levels for Zeeland (Orbit 3165, 22nd February 1992). Maximum vertical scale = 19%.

To determine the cause of the high standard deviations, a histogram for a selection of raw data blocks is shown in Figure 7. The vertical scale of each histogram has been kept the same to enable the relative distribution of ADC output within the product to be shown. The ADC histograms for the high standard deviation region show severe saturation in the higher and lower quantisation levels. In fact up to 19% of the samples can occur in the upper and lower quantisation levels. This clearly illustrates that saturation is indeed the cause of the high standard deviations. Note that the water between the individual islands and peninsulas is still giving high backscatter values, and because of the large size of the raw data blocks this influences the standard deviation of the neighbouring land regions. Thus even coastal regions are influenced by the ocean conditions. It is only for land regions well away from the ocean that its influence can be neglected. The area near Rotterdam and The Hague shows high standard deviations. This region is dominated by a large number of bright point targets (which have significantly higher radar cross-sections than the ESA transponders). For regions like this, where the high standard deviation values are caused by a large number of bright point targets, any power loss derived would not be applicable to the underlying large scale distributed radar cross-section values.

These results clearly have implications for any radar cross-section measurements made using ERS-1.SAR.PRI imagery. For any region for which there is a corresponding power loss, the measured radar cross-section values will be underestimated by the power loss. For this Zeeland scene, ocean backscatter measurements could be underestimated by as much as 3.3 dB. Other studies have also shown that high I channel standard deviations are predominately from the ocean (Laur et al. 1993). However, rather curiously, standard deviation of up to 12.2 have been found for Greenland (Meadows & Stapleton 1994) with corresponding power losses of 4.5 dB.

ADC saturation is also a problem for calibration using the ESA transponders. At both Flevoland and Zeeland, the transponders are located near to the coast and are thus under its influence. Figure 6 also shows the location of the transponders at Zeeland. To enable the ERS-1 SAR radiometric stability and accuracy to be derived, corrections need to be made for each transponder for the effects of ADC saturation.

The relationship between ERS-1.SAR.PRI amplitude and ERS-1.SAR.RAW ADC power loss is shown in Figure 8. This has been derived by the comparison of corresponding ERS-1.SAR.PRI and ERS-1.SAR.RAW products for the same Zeeland image used above (Orbit 3165). The ERS-1.SAR.PRI product has been sufficiently smoothed to mimic the raw data block size (5km in azimuth and 15 km in ground range). ERS-1.SAR.PRI products have the elevation antenna pattern and range spreading loss applied, while ERS-1.SAR.RAW products do not. Thus, these need to be removed from the ERS-1.SAR.PRI product prior to smoothing. To enable direct comparison of the ERS-1.SAR.PRI and ERS-1.SAR.RAW products, the latter needs to be converted to ground range and the coordinates of each raw data block converted into the ERS-1.SAR.PRI coordinates. The ERS-1.SAR.RAW I channel standard deviation values are

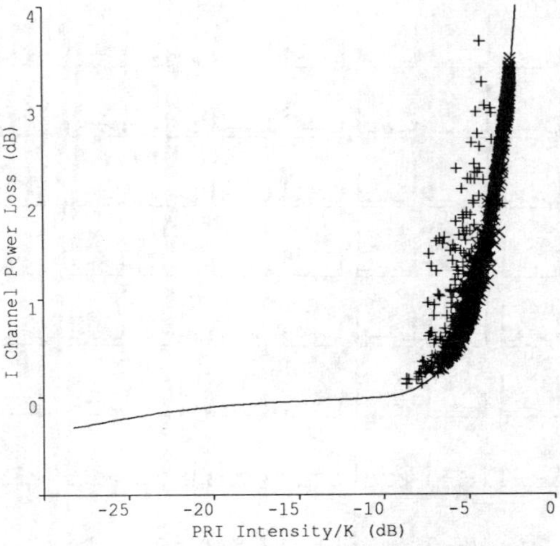

Figure 8. Relationship between ERS-1.SAR.PRI intensity/K and ADC power loss for the Zeeland scene (Orbits 3165).

converted to ADC power loss via the relationship illustrated in Figure 4.

Within Figure 8 two different symbols have been used. The 'plus' symbol is for the region surrounding Rotterdam/The Hague (where there are a large number of bright point targets present leading to the high I channel standard deviations and no corresponding increase in smoothed amplitude values). The remainder of the scene is indicated by the 'cross' symbol. The solid curve also presented in Figure 8 is the expected relationship between ADC power loss and PRI smoothed amplitude, based on the expected linear relationship between ERS-1.SAR.PRI amplitude and I Channel Standard deviation (Wilson, 1986) together with the inclusion of an appropriate noise level. As can be seen there is a good fit with the Zeeland data.

Differences in ADC power loss derived from the ERS-1.SAR.RAW product and the corresponding ERS-1.SAR.PRI product (via the solid curve shown in Figure 8) show that ADC power loss can, in general, be estimated from an ERS-1.SAR.PRI product to a precision of better than 0.5 dB (Meadows & Stapleton 1994). This shows that the ERS-1.SAR.PRI image can be used quite successfully to determine the ADC power loss. Exceptions to this are regions consisting of numerous point targets (such as cities).

2.4 *Implications for external calibration results and for ERS-1 SAR users*

Applying the replica pulse power variations and ADC power losses to the expression for point target radar cross-sections (equation 1) gives:-

$$\sigma = \frac{I_p}{C_f} \frac{A}{K} \frac{\sin \alpha_p}{\sin \alpha_{ref}} \frac{\text{Image Replica Power}}{\text{Reference Replica Power}} \text{Power Loss} \quad (2)$$

Figure 9 shows the replica pulse power correction and ADC power loss correction for the three ESA transponders derived directly from the ERS-1.SAR.PRI products. For these Flevoland and Zeeland images, the replica pulse power is particularly small, with at most a correction of 0.4 dB and typically less than 0.1 dB. The ADC power correction is far more significant with power losses up to 2.6 dB being found. Notice that for the transponders at Flevoland (commissioning and multidisciplinary phases) all three transponders experience similar power losses. However, for the transponders at Zeeland, the power loss decreases from transponder 1 to transponder 3. This is due to the fact that in Flevoland, all three transponders are near to the open water region of Ijsselmeer and Marker Meer, while for Zeeland, transponder 1 is on the coast and transponder 3 is furthest inland. As mentioned previously, it is open water/ocean regions that predominantly introduce the high power losses.

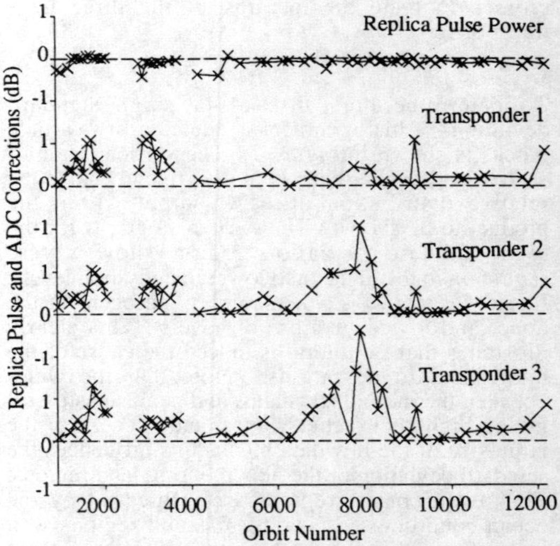

Figure 9. Replica pulse and ADC power loss corrections for the ESA transponders.

One consequence of applying the power loss correction to the transponder radar cross-section measurements is that the reference transponder relative radar cross-section is no longer zero. The ADC correction for this transponder is 0.39 dB. This implies that the calibration constant, K, needs to be updated thus:-

$$K \text{ (updated)} = K + 0.39 \text{ dB} \quad (3)$$

Figure 10 presents the transponder radar cross-sections with the corrections applied for each mission

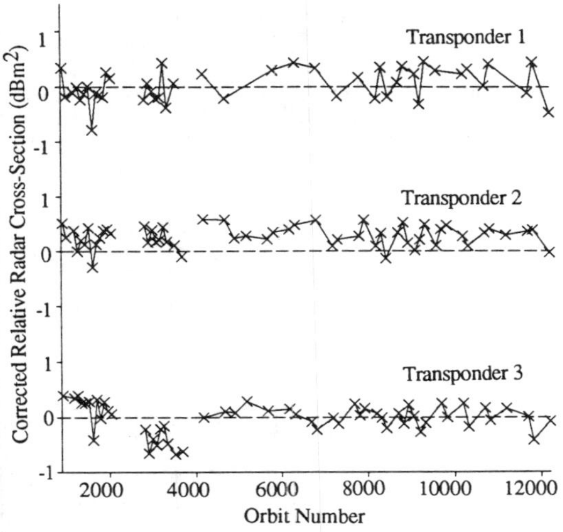

Figure 10. Corrected relative radar cross-sections for the ESA transponders.

phase and using the updated calibration constant. This shows a dramatic improvement in the radiometric stability: from an average uncorrected value of 0.50 dB for all transponder and phases to an average corrected value of 0.22 dB. Also the peak to peak radar cross-section variation has been similarly reduced - from an average value of 1.81 dB to 0.77 dB. The radiometric accuracy has an average value of 0.06 dB.

It should be stressed that this updated calibration constant is consistent with the previous K value. This is because the peak to peak variation of the original K was estimated to be ± 0.75 dB (Laur et al. 1993) whereas the derived peak to peak value established above is much smaller at ± 0.39 dB. Thus the updated K with its reduced peak to peak variation lies within the original K and its larger peak to peak variation.

For users wishing to determine distributed target radar cross-section measurements, the full expression is:-

$$\sigma^o = \frac{<I>}{K} \frac{\sin \alpha_D}{\sin \alpha_{ref}} \frac{\text{Image Replica Power}}{\text{Reference Replica Power}} \text{Power Loss} \quad (4)$$

where < I > is the average pixel intensity of the distributed target, α_D is the distributed target incidence angle while the other parameters are as per equation 2 and are derived in exactly the same manner. Note that even though a distributed target radar cross-section may be low, an ADC power loss correction may still be required if neighbouring distributed targets (within a region of 5km in azimuth and 15 km in range) are high enough to cause ADC saturation.

3 GEOMETRIC CALIBRATION

ERS-1 SAR image localisation accuracy has been determined by using the known location of the three ESA transponders, calculating their expected positions and comparing these with their actual positions within the image. This requires the conversion of the transponder latitude, longitude and elevation to azimuth time and slant range of closest approach to the satellite. The transponder pixel positions are then calculated using the zero Doppler azimuth time (for the azimuth pixel position) and the zero Doppler range time (for the range pixel position) given in the product CEOS header. Included in the calculation of the range pixel position is the transponder time delay which can displace its IRF by some 600 m.

The accuracy of the localisation depends on several factors: the accuracy of the transponder latitude, longitude, elevation and time delay, the orbit data and the zero Doppler times. The quoted latitude and longitude accuracy of the transponders is ± 50 m (Guignard et al. 1991). An attempt to remove any orbit errors has been performed by deriving the pixel coordinates of the image corners and image centre using their latitude and longitude values. The difference in pixel coordinates gives an offset to be applied to the derived transponder coordinates. The accuracy of the image corners and centre has been found to be approximately ± 30 m (Meadows 1993).

Based on a selection of 14 images spread throughout the period from the commissioning to multidisciplinary phase gives an average localisation accuracy of 97.0 ± 15.9 m, which is quite satisfactory considering the accuracies involved.

4 CONCLUSIONS

The calibration of the ERS-1 SAR using the three ESA transponders, examination of the instrument internal calibration data and analysis of raw data has enabled a complete expression to be derived for users to calculate radar cross-section values. The expression includes corrections for replica pulse power variations and ADC power losses. The replica pulse correction is applied by extracting the relevant replica pulse power from the ERS-1.SAR.PRI CEOS header. A technique has been presented whereby ADC power losses can be determined directly from ERS-1.SAR.PRI imagery. Applying both of these corrections to the ESA transponder measurements leads to significant improvements to the ERS-1 SAR radiometric stability measurement. Image quality results indicate excellent spatial resolutions and low sidelobe ratios. Geometric calibration analysis gives a localisation accuracy of approximately 100m.

It should be stressed that even though UK-PAF imagery has been used throughout, products from other ERS-1 PAFs will require similar replica pulse and ADC corrections.

ACKNOWLEDGEMENT

This work has been carried out under contract to the National Remote Sensing Centre Ltd funded by the European Space Agency.

REFERENCES

Woode, A.D., Desnos, Y-L, & Jackson, H., 1992. The Development and First Results from the ESTEC ERS-1 Active Radar Calibration Unit. IEEE Transactions on Geoscience and Remote Sensing, vol 30, pp. 1122-1130.

Eurimage & ERS User Services Section ESRIN, 1993. ERS User Handbook. ESA Publication SP-1148, Revision 1

Scott, R.F., Elkington, M.D. & McIntyre, N.F, 1994. ERS-1 processing chain development and product support by the EODC science team, the algorithm development facility and the calibration and validation facility. Int. J. Remote Sensing, Vol 15, pp 761-784.

Brown, M.A., 1992. Comparison of Actual and Theoretical Azimuth IRF Shapes. GEC-Marconi Technical Report, DC-TN-MRC-CV-0003, Issue 1.

Meadows, P.J., 1993. ERS-1.SAR.PRI Status Report. GEC-Marconi Technical Report, PF-TN-MRC-0006, Issue 1.

Lancashire, D.C., 1987. ERS-1 AMI Calibration and Characterisation Plan. ER-PL-MSS-AM-0008, Issue 13.

Wright, P.A., 1993. ERS-1 SAR Continued Assessment and Trends Analysis of the Raw Data Quality Parameters and the Internal Calibration Parameters for September 1991 to December 1992. GEC-Marconi Technical Report, PF-TN-MRC-0010, Issue 1.

Laur, H., Meadows, P.J., Sanchez, J.I., & Dwyer, E., 1993. ERS-1 SAR Radiometric Calibration. Proceeding of the CEOS SAR Calibration Workshop, ESA Publication WPP-048.

Brown, L.M.J., Smith, D.J., Bestwick, D.J., Blacknell, D., Deane, J.M., Freeman, A. & Meadows, P.J., 1988. SAR Data Quality Assessment and Rectification. GEC-Marconi Technical Report, MTR 87/107.

Meadows, P.J. & Stapleton, C.M., 1994. ERS-1 SAR Analogue to Digital Convertor Saturation. GEC-Marconi Technical Report, PF-TN-MRC-0012, Issue 2.

Wilson, J.J. W., 1986. The Effect of a Non-Linear AMI Receiver System Gain Characteristic on ERS-1 SAR Imagery. GEC-Marconi Technical Report, ER-TN-MRC-AM-0225, Issue 2.

Guignard, J-P., Desnos, Y-L., Brooker, G. & Caspar, C., 1991. ERS-1 SAR Fast Delivery Reference Image. ESA, ER-TN-JP-ESA-0009, V1.2.

Sensors and Environmental Applications of Remote Sensing, Askne (ed.) © 1995 Balkema, Rotterdam, ISBN 90 5410 524 0

ERS-1 Scatterometer data processing over the Mediterranean sea: Algorithm development and validation activity at I-PAF

A. Bartoloni & C. D'Amelio
TELESPAZIO SpA, Rome, Italy

G. Milillo & F. Nirchio
ASI, Centro di Geodesia Spaziale, Contrada Terlecchia, Matera, Italy

ABSTRACT: As a result of the activity developed at the Italian PAF in order to process the ERS-1 scatterometer data, the off-line product ERS-1.WSC.DWP (Scatterometer Wind and Pressure Field) has been defined.

In this paper we describe the validation activity that has been carried out in order to check the scientific quality of the ERS-1.WSC.DWP product over a set of case studies representing different geographic and climatic conditions: the Oceanic case study relative to the Haltenbanken Validation Campaign data set and the Mediterranean case study relative to the data sets provided by METEOFRANCE and by ECMWF.

1. PRODUCT DESCRIPTION

The ERS-1.WSC.DWP product contains the off-line processed wind vector fields (referred at a height of 10 meters in a neutral stable atmosphere) retrieved from the σ_0 backscattering coefficients measured by the ERS-1 scatterometer for each node of the illuminated grid constituted by 19x19 cells covering an area of 475x475 km^2.

The product contains both the two wind fields (rank1 and rank2 solutions), that is possible to retrieve from the scatterometer data, in order to allow further analysis through visual inspection or comparison with meteorological maps.

Moreover the product contains the surface pressure field generated from the computed wind field on the same grid of measure.

The measurements of the σ_0 values are obtained by three antennas pointing in different directions: the mid beam antenna, which illuminates the sea surface in the across satellite track direction and the for and aft beam antennas which look respectively at an angle of 45° degrees before and after the mid beam. Hence for each point on the sea surface three echo power values obtained from three different angles are used to reconstruct the surface wind vectors (wind speed and direction).

However, due to the blinding of one antenna and by the scatterometer switching on/off, the grid points on the sea surface may have only two or one sigma nought measurements (incomplete data). The procedure used to generate the wind field contained in this product is applicable also to nodes provided with only two sigma nought measurements but only if at least one node with three sigma nought measurements is available.

The capability to process nodes with incomplete data is very important in the Mediterranean area, to which the I-PAF is particularly devoted, where the occurrence of such kind of data is very high.

2. ALGORITHMS DESCRIPTION

The wind vector field reconstruction procedure is essentially based on the minimization for each node of the scatterometer measure grid of the following function :

$$f(w,\phi) = \sum_{k\in\{A\}} p^k\left[\sigma_0^k - \sigma_0(w,\phi+\alpha^k)\right]^2 \qquad (2.1)$$

where:

$\{A\}$ = set of antennas giving valid data at the node

w = wind speed at the node

ϕ = upwind direction at the node with respect to the mid-beam direction

α^k = aspect angle of the k-th antenna
p^k = weight assigned to the k-th antenna
$\sigma_0{}^k$ = backscattering coefficients measured by the k-th antenna
$\sigma_0(w,\phi)$= C-band model used for the wind field reconstruction

The basic idea is to solve the minimization problem (2.1) by using a suitable combination of local and global minimization algorithms.

From the wind vector field obtained by processing the scatterometer data it is possible to compute the pressure field on the sea surface.

Our computation is based on a discretized version of the stationary Navier-Stokes equations that connect the pressure gradient to the wind vector field:

$$(\underline{u},\nabla)\underline{u}+2\underline{\Omega}x\underline{u}=-\frac{1}{\rho_0}\nabla p+\mu\Delta\underline{u} \qquad (2.2)$$

where:

∇ : gradient operator;
Δ : Laplace operator;
$2\underline{\Omega}x\underline{u}$: Coriolis force acting at the latitude θ;
ρ_0 : mass density per unit volume;
μ : viscosity coefficient.

The main steps of the reconstruction procedure are:

- to build an approximation $\underline{w}$ to ∇p;
- to project $\underline{w}$ on the curl free subspace;
- to compute the line integral of the curl free projected vector field to obtain the approximate pressure field $p(\underline{x})$.

3. VALIDATION ACTIVITY

The methodology we applied in order to validate the product ERS-1.WSC.DWP is based on the definition of the following case studies:

1. Oceanic data set
2. Mediterranean data sets

The Oceanic data set is constituted by the data provided by the Haltenbanken Validation Campaign (September-December 1991): 77 collocation files obtained collecting ERS-1 data together with data from the platforms involved in the campaign and surface wind fields from numerical models provided by meteorological centres.

The Mediterranean data sets are constituted by:

- the forecasted wind fields provided by the PERIDOT model (METEOFRANCE) for the first week of December 1992;
- the analyzed wind fields provided by the ECMWF for the following periods:

PERIOD A :	March 15 - April 15 1992
PERIOD B :	September 20 - October 15 1992
PERIOD C :	December 1 - 20 1992
PERIOD D :	December 21 - January 10 1993

These particular periods of time were selected in order to verify the capability of the scatterometer data processing chain in recovering particular situations in the Mediterranean area such as Mistral winds or cyclogenesis.

Our validation data sets (66 and 1212 collocation files respectively for PERIDOT and ECMWF data) have been realized through a collocation process between the meteorological wind data and the ERS-1 scatterometer measurements, using the following steps:

– selection, among the four wind fields provided by the Meteorological data (0:00 - 6:00 - 12:00 - 18:00), of the two wind fields closest, in time, to the satellite UTC time;
– generation of two dimensional surfaces interpolating the (u,v) components of the two wind fields and evaluation of the interpolating function at the scatterometer measurements data points;
– temporal linear interpolation of the two wind fields;
– generation of the collocation files containing the Meteorological derived wind field corresponding to the scatterometer data.

The analysis of the data sets has been organized in the following steps :

– σ_0 data analysis

Estimation of the quantity $\Delta\sigma = \frac{|\sigma_0 - \sigma_0^M(u,\phi)|}{\sigma_0}$ representing the percentual difference between the measured σ_0 and the $\sigma_0{}^M$ obtained by the CMOD model on the analyzed wind field contained in the collocation files.

– Processing with method A)

It is an algorithm based on a priori information on the wind field to be reconstructed; the basic idea is to search for a minimizer of the objective function in the vicinity of a first guess wind field (as close as possible in space and time to the scatterometer data) externally supplied to the algorithm.

– Processing with method B)

It is a self consistent algorithm, based only on the data measured by the scatterometer: aimed at the reconstruction of the wind field without using a priori information on the wind field to be reconstructed. The basic idea is to solve the minimization problem of the objective function combining local and global minimization methods.

– Processing with method C)

It is essentially the method B) aimed at reconstructing the wind field without using a priori information; it differs from the method B) because it uses the a priori information in ranking the two solutions determined by the algorithm B). The basic idea is to choose as correct solution the wind field corresponding to the smaller wind direction root mean square difference computed with respect to the a priori wind field.

- Processing results assessment

In order to assess the processing results obtained by the methods A), B) and C) the following quantities have been estimated:

– Wind speed and direction root mean square differences

$$RMS\_SPEED = \sqrt{\frac{\sum_{i=1}^{N_p}\left[(v_A^i - v_P^i)^2\right]}{N_p}}$$

$$RMS\_ANGLE = \sqrt{\frac{\sum_{i=1}^{N_p}\left[(\phi_A^i - \phi_P^i)^2\right]}{N_p}}$$

– Mean value of difference vector

$$MV\_DIFFVECT = \frac{\sum_{i=1}^{N_p}\sqrt{(w_{Ax}^i - w_{Px}^i)^2 + (w_{Ay}^i - w_{Py}^i)^2}}{N_p}$$

where:

v_A, ϕ_A : analyzed wind speed and direction;

v_P, ϕ_P : processed wind speed and direction;

w_{Ax}, w_{Ay} : (x,y)-components of analyzed wind field;

w_{Px}, w_{Py} : (x,y)-components of processed wind field;

N_P : number of processed nodes.

4. RESULTS OVER THE OCEANIC DATA SET

The results obtained by the three methods considered are shown in Table1 by estimating the values assumed by the root mean square differences for the wind speed and wind direction and by the mean value of difference vector .

It is important to note that although the area interested by the Campaign (the Norwegian sea) is different from the area of interest for the I-PAF (the Mediterranean sea), this data set has been used in order to demostrate the capacity of the I-PAF scatterometer processing chain on a data set considered as the reference data set for the ERS-1 scatterometer. In fact, in case of processing method A), if we select points according to the recommended threshold value IQ≥5 [6], we obtain a good agreement between the processed and analyzed wind fields (RMS_ANGLE=27.7degrees) that becomes highly comparable (RMS_ANGLE=21.8degrees) to the value RMS_ANGLE=21degrees obtained on the same data set by D.Offiler [6].

5. RESULTS OVER THE MEDITERRANEAN DATA SETS

With regard to the first week of December 1992 the wind fields provided by METEOFRANCE and ECMWF have been simultaneously used.

The results assessed by the estimation of the predefined quantities are shown in Table 2 and Table 3 respectively for the METEOFRANCE and ECMWF data sets.

The results obtained by the processing methods A), B) and C) for the four different periods considered are respectively shown in Table 4, Table 5 and Table 6.

6. ANALYSIS OF THE RESULTS

The results of the method A) show that the use

Table 1: HALTENBANKEN data

METHOD	RMS_SPEED (m/s)	RMS_ANGLE (degrees)	MV_DIFFVECT (m/s)
A)	2.59	27.7	3.5
B)	2.6	102.0	9.9
C)	2.59	51.8	4.68

Table 2: METEOFRANCE data

METHOD	RMS_SPEED (m/s)	RMS_ANGLE (degrees)	MV_DIFFVECT (m/s)
A)	1.9	28.5	3.4
B)	1.9	92.9	6.3
C)	1.9	34.5	3.5

Table 3: ECMWF data

METHOD	RMS_SPEED (m/s)	RMS_ANGLE (degrees)	MV_DIFFVECT (m/s)
A)	1.8	22.0	2.6
B)	1.9	95.1	6.0
C)	1.9	23.0	2.7

Table 4: Method A) results

PERIOD	RMS_SPEED (m/s)	RMS_ANGLE (degrees)	MV_DIFFVECT (m/s)
A)	2.0	32.0	2.6
B)	1.9	34.0	2.5
C)	2.0	31.0	2.7
D)	2.0	31.0	2.7

Table 5: Method B) results

PERIOD	RMS_SPEED (m/s)	RMS_ANGLE (degrees)	MV_DIFFVECT (m/s)
A)	2.2	100.0	5.7
B)	2.0	89.0	4.6
C)	2.1	96.0	6.2
D)	2.0	88.0	5.0

Table 6: Method C) results

PERIOD	RMS_SPEED (m/s)	RMS_ANGLE (degrees)	MV_DIFFVECT (m/s)
A)	2.2	45.0	3.0
B)	2.0	55.0	3.0
C)	2.0	52.0	3.2
D)	2.0	52.0	3.2

of the a priori information before the data processing gives wind fields in good agreement with the considered ground truth.

Comparing the results obtained by the method A) to the ones obtained by the method C) we can conclude that the use of the a priori information after the processing is less effective by far.

Of course, the results obtained on the Haltenbanken data set by the method A) significantly improve (wind direction root mean square difference = 21.8 degrees) if we filter the data according to the quality index IQ≥5 recommended in [6] to ensure the contribution of independent in situ measurements.

Comparing this last result to the results obtained on the Mediterranean data sets we have to take into account two important aspects:

1) the intrinsic better quality of the Haltenbanken data set which include in situ measurements that are not included by the METEOFRANCE and ECMWF data derived from meteorological models;
2) the different climatic conditions of the Norwegian sea with respect to the Mediterranean sea.

The fact that for the Haltenbanken data set the wind conditions are within the range of validity of the C-band model function (wind speed greater than 4m/s and less than 24m/s) surely influence positively the processing results; in fact if we analyze the other data sets by considering only points with wind speed values greater than 4m/s we obtain the results shown in Table 7 which show an improvement of about 8 degrees.

The same considerations can be applied to the results obtained by the processing method A) for the ECMWF data of the first week of December (Table 3) where the better agreement between the analyzed and processed wind field (RMS_ANGLE = 22 degrees) is due to the occurrence of a Mistral wind condition for the most part of the first week of December 1992.

In order to better understand the influence of the a priori information on the processing results and to find out its best implementation in the ERS-1.WSC.DWP generation processing chain, we have to consider the histograms of Fig.1 and Fig.2. They represent, for the periods A) and B), the distribution of frames number according to the wind direction root mean square difference (RMS_ANGLE) values obtained by the processing method A).

Looking at these histograms we see that about the 60-70% of the frames are in good agreement with the corresponding a priori information (RMS_ANGLE values less than 40 degrees).

The first action that has been undertaken in order to investigate the presence of the other frames (with RMS_ANGLE values higher than 40 degrees) has been the exclusion, in the computation of the wind direction root mean square errors, of the points with wind speed values out of the validity range (that is points with wind speed values less than 4m/s); the results are shown in the histograms of Fig.3 and Fig.4 where we can observe the reduction of about 15% of frames with RMS_ANGLE values higher than 40 degrees.

From these considerations we deduce that the problems of the processing method A) are partially due to the low wind conditions; the remaining

Table 7: Method A) results
Points with wind speed > 4m/s

DATA SET	RMS ANGLE	RMS ANGLE
HALTENBANKEN	27.7 21.8 (IQ≥5)	21.8
METEOFRANCE (92/12/01 - 92/12/07)	28.5	26.5
ECMWF (92/12/01 - 92/12/07)	22.0	19.7
ECMWF PERIOD A	32.0	22.3
ECMWF PERIOD B	34.0	26.7
ECMWF PERIOD C	31.0	24.0
ECMWF PERIOD D	31.0	25.0

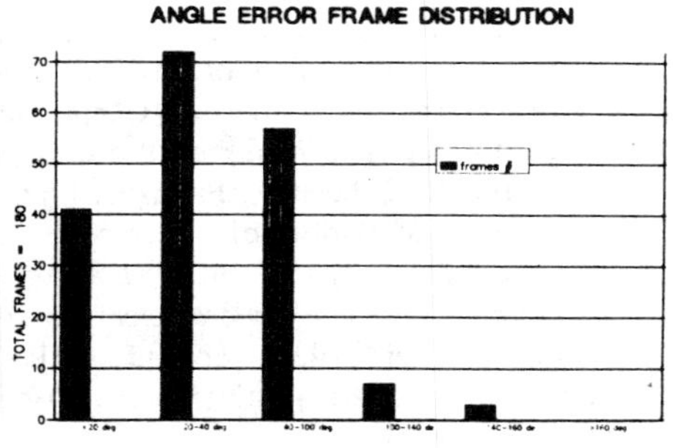

Fig.1 : Distribution of frames w.r.t. RMS_ANGLE Period A

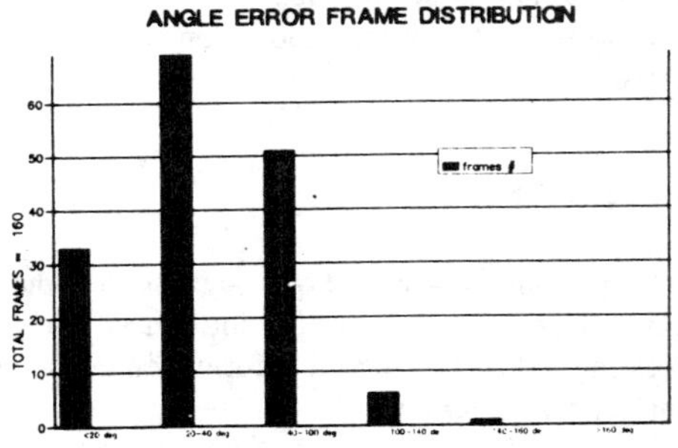

Fig.2 : Distribution of frames w.r.t. RMS_ANGLE Period B

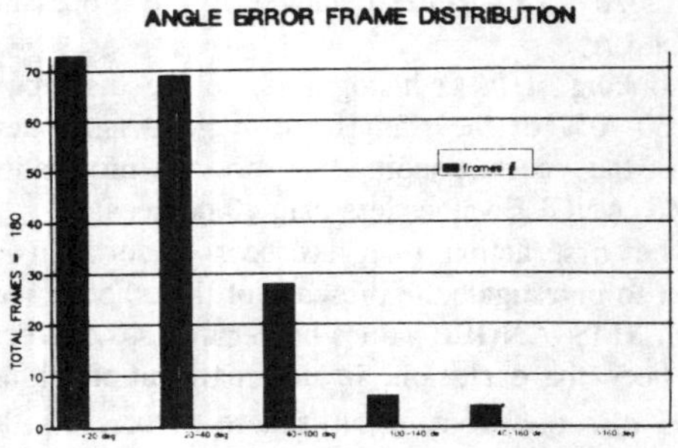

Fig.3 : Distribution of frames w.r.t. RMS_ANGLE
Points with wind speed > 4m/s
Period A

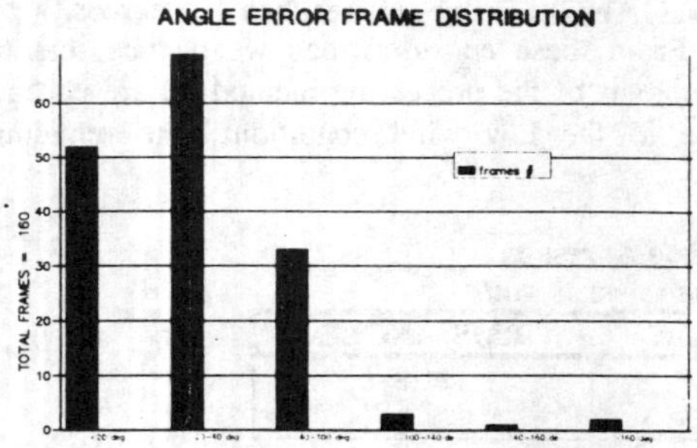

Fig.4 : Distribution of frames w.r.t. RMS_ANGLE
Points with wind speed > 4m/s
Period B

problems (about 15% of the total frames number) are caused by a real disagreement between the a priori wind field and the wind field derived by the scatterometer data. In such cases it is better to rely on the scatterometer data only avoiding the use of the a priori information at all (method B).

As a consequence, when there is no agreement with the meteorological data, the a priori information has to be introduced in the processing not destroying the information contained in the scatterometer data.

7. CONCLUSIONS

The previous considerations suggest that the a priori information has to be introduced in the processing not destroying the information contained in the scatterometer data.

This can be obtained by introducing some criteria for the assessment of compatibility between the processed and analyzed data during the processing; that is, if a first run of the processing method A) determines RMS_ANGLE values higher than a threshold value (about 40 degrees) then, the a priori information is considered incompatible with the scatterometer data and the two solutions ranked automatically by the method B) are provided.

This fact have suggested the opportunity, already stressed by P.I.s involved in the validation activity [7], [8], to provide the ERS-1.WSC.DWP product with two ranked solutions together with the indication of the method used in ranking the two wind fields (Autonomous / Meteorological).

REFERENCES

[1] A.Bartoloni, C.D'Amelio, F.Zirilli: *"Wind reconstruction from Ku-band or C-band scatterometer data"*, IEEE Transactions of Geoscience and Remote Sensing, Vol.31, No.5, September 1993.

[2] A.Bartoloni, C.D'Amelio, F.Zirilli: *"An algorithm for the reconstruction of pressure fields from scatterometer data"*, Proceedings of IGARSS '92 Symposium, 1008-1010.

[3] *"ERS-1 Geophysical Validation Campaign"*, Workshop Proceedings, 27-30 April 1992, Penhors, Bretagne, France.

[4] H.L.Pham: *"Historique du Modelè d'adaptation et de preèvision locale:le modèle PERIDOT"*, La Météorologie VIIè série, 28, 11-27, 1989.

[5] A.Bartoloni, C.D'Amelio, G.Milillo, F.Nirchio: *"ERS-1 scatterometer data processing over the Mediterranean sea: algorithm development and validation activity at I-PAF"*, Proceedings of the Second ERS-1 Symposium, 1149-1154, October 1993, Hamburg, Germany.

[6] D.Offiler: *"Wind measurements from the MRF C-130 and the derivation of analyses from RENE-91 winds"*, ERS-1 Geophysical Validation Campaign, Workshop Proceedings, 11-16, April 1992, Penhors, Bretagne, France

[7] S.Zecchetto, M.Bortoletto: *"Examples of Mesoscale meteorology in the Mediterranean Sea using ERS-1 scatterometer data"*, Proocedings of the Second ERS-1 Symposium, 815-818, October 1993, Hamburg, Germany

[8] G.Borzelli, A.Cappelletti, G.Manzella, S.Marullo, R.Santoleri: *"Use of ERS-1*

scatterometer data in numerical models", Proocedings of the Second ERS-1 Symposium, 571-575, October 1993, Hamburg, Germany

Sensors and Environmental Applications of Remote Sensing, Askne (ed.) © 1995 Balkema, Rotterdam, ISBN 90 5410 524 0

Use of C-band active radar calibrators in spaceborne and airborne SAR missions

G. De Carolis
GNCB-CNR, Bari, Italy

S. Ponte
Faculty of Engineering, University of Naples 'Federico II', Italy

F. Posa
Physics Department, Polytechnic of Bari, Italy

V. D. Schena
Consortium CORISTA, Napoli, Italy

ABSTRACT: This paper is addressed to the description and performance analysis of the C-band ARCs developed under a collaboration between the Consortium for Research on Advanced Remote Sensors (CO.RI.S.T.A., Naples, Italy) and the University of Bari (Bari, Italy). The main design considerations and the hardware performance of the active devices are presented. Furthermore, the laboratory measurements of radar cross-section (RCS) and propagation delay time are compared to the ones obtained from SAR data acquired during spaceborne (ERS-1) and airborne (AIRSAR-TOPSAR) campaigns. The radiometric calibration of the SAR images, performed with the intensity integration method, allowed us to recover the RCS of the ARCs within ±0.8 dB: the RCS values obtained from the calibrated images are in good agreement with the experimental measurements, within the laboratory calibration error (±0.6 dB).

1 INTRODUCTION

With respect to their passive counterpart, i. e. the trihedral corner reflectors (CR), Active Radar Calibrators (ARC) offer to the field of external calibration of synthetic-aperture radar (SAR) systems unique advantages such as design flexibility, reduced dimensions, relatively small sensitivity to positioning errors, large and adjustable radar cross-section (RCS). This paper deals with the C-band ARCs developed since 1991 by the consortium CO.RI.S.T.A. (Naples, Italy) and the University of Bari, Italy. Currently, three C-band ARCs are operational, and have been widely used during past external SAR calibration campaigns.

The main laboratory measurements performed to characterise the hardware performance are presented, together with the procedures aimed at the characterisation with respect to the variation of parameters critically related to the RCS. In addition, ARC responses collected from spaceborne (ERS-1, ESA) and airborne (AIRSAR-TOPSAR, NASA/JPL) SAR imagery are considered.

Results from laboratory analysis and calibrated SAR images provide a comprehensive understanding of the ARCs' performance, allowing one to gain insight into the use of active devices in ground calibration missions. Moreover, with a full characterisation of an ARC, the experimenters can exploit the great potentialities of such devices in different applications, such as the evaluation of the SAR's end-to-end system response function, cross- and co-polarised, the assessment of in-flight azimuth antenna patterns, the acquisition of amplitude and phase characteristics of multi-channel and multi-polarisation SARs.

2 DESCRIPTION OF THE ARC

Figure 1 shows a block diagram of the C-band ARC.

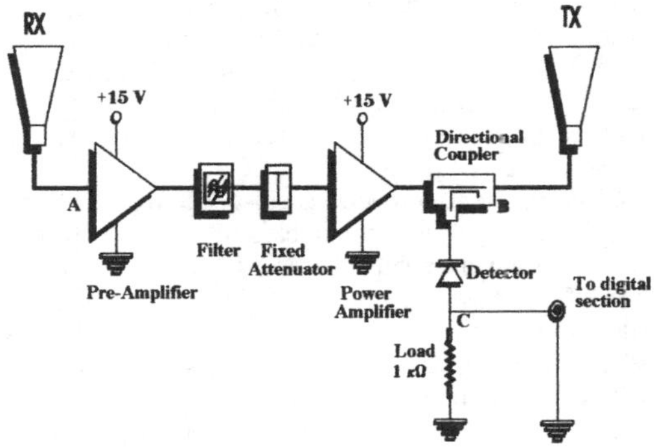

Figure 1. C-band ARC: block diagram of the RF section.

The active device basically consists of a receiving (RX) and a transmitting (TX) antenna coupled with a radio-frequency (RF) amplifier section. Signal detection circuitry has been designed with the aim of feeding a portion of the received power to a digital data acquisition section, capable of indicating with a

LED the duration of the illumination corresponding to the overpass of the microwave sensor. It is possible to store the received signal, digitally converted by an A/D section and routed to a parallel interface, by means of an acquisition card (Ground Receiver), in order to evaluate the azimuth antenna pattern.

The receiving horn antenna is connected to a preamplifier, built in hybrid technology and cascaded with a band-pass filter, centred on the desired frequency (5.3 GHz) and with a maximum available bandwidth of ≅400 MHz. The chosen bandwidth is a compromise between the need of filtering out spurious signals and noise, and the possibility of using the same ARC for future or different SAR systems. A second amplification stage is provided by a power amplifier connected after a fixed attenuator, which arranges the dynamic range of the received signal between the two stages. During the prototypal realisation, a variable attenuator was used to have a suitable range of RCS values, so as to avoid both saturation of the SAR impulse response function and positive feedback in the RF section.

The directional coupler spills a fraction of the input power (10 dB below the power amplifier output) to a detector diode, connected to ground by a 1-kΩ resistor which adjusts the sensitivity. The result of the detection is the envelope of the radar signal in a negative pulse train at the pulse repetition frequency (PRF). Finally, the signal is re-radiated to the passing-by sensor through a horn with the same characteristics of the receiving one.

In the following we summarise the main aspects and characteristics of the ARC.

1: packaging. Electronics and antennas are weatherproof and the structure is able to withstand normal winds without readjustment. Deployment and orientation are easily accomplished, and the adjusting/locking mechanisms are manual, independent for each antenna which can be oriented at 45°, 90° and 180° with respect to the other. The device weighs about 20 kg.

2: power requirements. Lead-gel airtight batteries supply 12 V, 10 Ah, and an electronic power step-up DC-DC converter supplies 15 VDC to the RF section. The power supply allows 10-hours full operation, with a current absorption of 650 mA. The external power source is easy to connect and replace, with circuitry to protect against wrong insertions.

3: amplifier section. A LED turns on when the ARC is receiving a power greater than or equal to the peak expected power. Both the preamplifier and the power amplifier have gain flatness of ±0.05 dB over the available bandwidth (Table 2), with noise figure of 3.6 dB at the centre frequency.

4: antennas. Both RX and TX antennas, labelled for easy identification, are C-band pyramidal horns with rectangular aperture (13.2x9.4 cm). The -3 dB beamwidths in the E and H-planes are 25° and 34° respectively, values which overcome orientation problems. To achieve good cross-coupling, the distance between the centre of symmetry of the two horns has been set to 62 cm. At 5.3 GHz the antennas have a bandwidth of 300 MHz. The peak gain declared by the manufacturer is 15 dB (Table 3), with typical sidelobe levels of -20 dB which reduce multipath effects and background noise. In addition to the orientation of the antennas with respect to the flight line, a different polarisation state is achieved by rotating each antenna about its boresight at 45° steps, with a system of hexagonal-head bolts on a frame structure. Echo-absorbing material surrounds the horns, in order to avoid resonance effects and to minimise re-radiation effects of the structure (Fig. 2).

Figure 2. C-band ARC.

3 MEASUREMENTS AND VALIDATION

In order to obtain a full characterisation of the active device, the following measurements were carried out:

1. Output power as a function of the input power (i.e. transfer function of the device).
2. Gain over the frequency bandwidth.
3. Thermal characterisation.
4. Antenna characterisation.

3.1 Transfer function measurement

The measurement set-up for the evaluation of the RF gain of the ARC as a function of the input power is shown in Figure 3. A synthesised sweeper generator is connected to the input of the device under test (DUT) with a step attenuator whose output is set 30 dB below the input level. The RF output of the DUT (point B in the figure) is connected to a spectrum analyser. A digital multimeter was used for voltage measurements. During the measurements, the temperatures of the room and the device were monitored by a K-type probe (NiCr-NiAl thermocouple) and a digital multimeter acting as a thermometer. A chirp signal with 5.3 GHz carrier frequency and 600 MHz

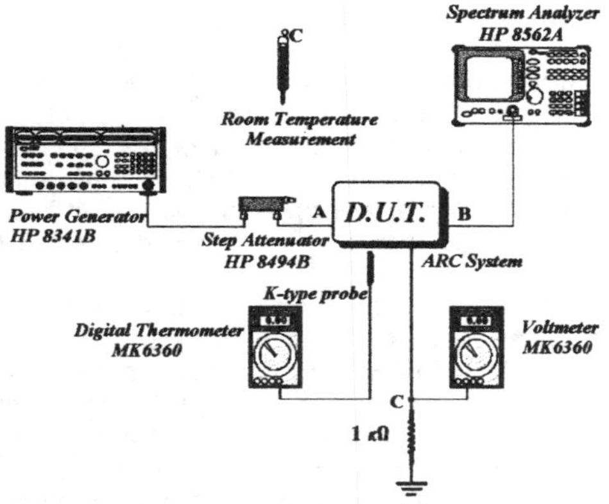

Figure 3. Experimental set-up for gain measurement.

bandwidth, produced by the signal generator, represented the input to the DUT.

The step attenuator allowed us to scan power levels from -82.00 dBm to -31.85 dBm using a high and constant value of the generated signal. In this way the signal -to-noise ratio is improved. However, given an input signal level, the effect of the noise was taken into account by averaging a set of measurements and considering the root mean square (rms) value. An error equal to ± 0.05 dB was found, but ± 0.5 dB was measured for the two lowest input values. Great stability on the output results was observed, so that we assumed an error of ≅0 dB. This behaviour is due to the so-called *ballast effect* introduced by the resistors compounding the device and the double stage of the ARC amplifier section (Figure 1) (Cheung and Owens 1981, Carlson 1992).

We used two custom amplifiers produced by JCA, a pre-amplifier (model JCA56-353) which includes into the package a band-pass filter centred on 5.3 GHz with a bandwidth of 300 MHz, and the JCA56-519 power amplifier. Both amplifiers are made in hybrid technology. Table 1 reports the factory technical characteristics. By means of Table 1 and the Friis' formula for cascaded amplification stages (Carlson 1992), the maximum value of the noise figure of the ARC amplifier was estimated as 2.5 dB.

Table 1. Technical data of the JCA amplifiers.

Device name	JCA56-353	JCA56-519
Frequency Range	5.1 to 5.5 GHz	5.0 to 5.6 GHz
Gain (dB)	27 (min.)	43 (min.)
Gain Flatness (dB)	±1.0 max.	±1.5 max.
Noise Figure (dB)	2.5 max.	4.5 max.
Power Output (dBm)	10 min. (-1dB)	25 min. (-1 dB)
Input VSWR	2:1 max.	2:1 max.
Output VSWR	2:1 max.	2:1 max.
DC Power	+15 V @ 150 mA	15 V @ 550 mA

The room temperature was constant (20.5 °C), whereas the device took 15 minutes to reach the temperature of 36.6 °C, which did not change during the measurement procedure. In Figure 4 the transfer function of the amplifier section is shown.

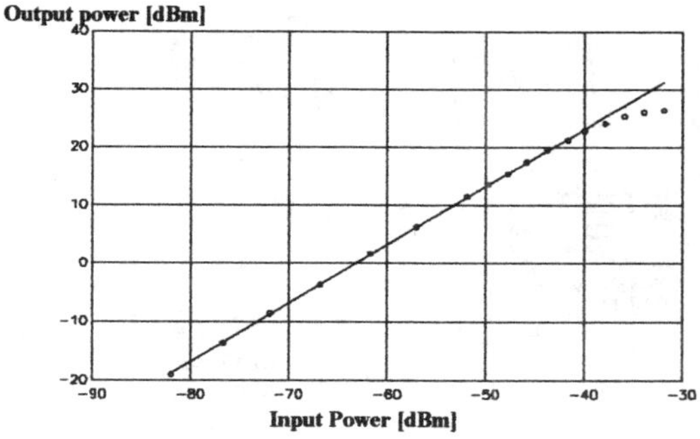

Figure 4. Measured transfer function of the C-band ARC.

The transfer function exhibits linearity between -95 dBm and about -40 dBm. In this range a linear least-square fit (Bevington and Robinson 1992) was performed to evaluate the line slope and intercept with their uncertainty values. Table 2 reports the numerical values of the parameters of the linear fitting. The value 2.262 assures that the true value of the sample is bounded within the given uncertainty with 97.5% probability (Capiluppi *et al.* 1978), whereas the rms errors on the slope (m) and on the intercept (n) are indicated with Δm and Δn, respectively.

Table 2. Parameters of the straight-line fitting.

	Theoretical formula	Numerical value [dBm]
Slope	$m \pm 2.262\Delta m$	1.005 ± 0.006
Intercept	$n \pm 2.262\Delta n$	-63.310 ± 0.078

Above the input value of -40 dBm, saturation of the amplifier section occurs. The 1-dB compression point, i.e. the input value where the amplifier's output power deviates from the extrapolated value by 1 dB, is located at -37.85 dBm. The absolute value of the intercept represents the mean gain of the amplifier section calculated after integration on the bandwidth. This value has been found to be 63.31 dB. The same measurements were carried out for the available three ARCs. In Table 3 the corresponding gains are reported, together with the standard deviations (the contribution from the antennas is not considered).

Table 3. C band ARC data.

Device	Gain [dB]	Standard deviation [dB]
ARC #1	60.55	±0.06
ARC #2	60.76	±0.08
ARC #3	63.31	±0.08

3.2 *Gain over the frequency bandwidth*

The measurement set-up used to characterise the amplifier section of the ARC with respect to the frequency is shown in Figure 5.

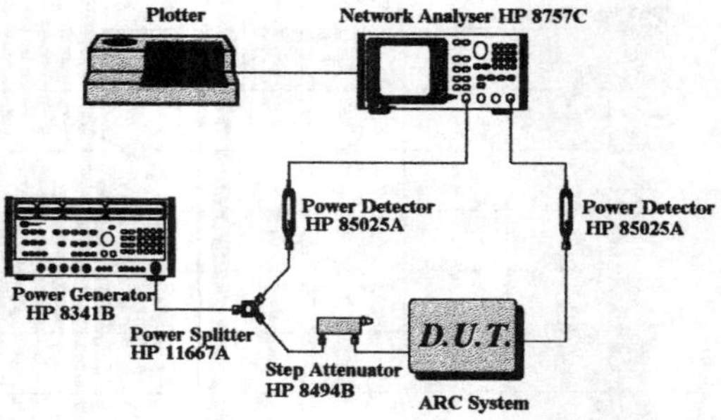

Figure 5. Measurement set-up for the frequency-domain analysis.

This kind of measurement allowed us to evaluate the performance of the filter and the overall gain flatness. The signal from the synthesised sweeper generator is divided by a power splitter. One of the outputs feeds the DUT and a power detector, and the other one enters another power detector. The outputs from the power detectors are sent to a scalar network analyser connected to a plotter. In Figure 6 the output diagram provided by the scalar network analyser is shown.

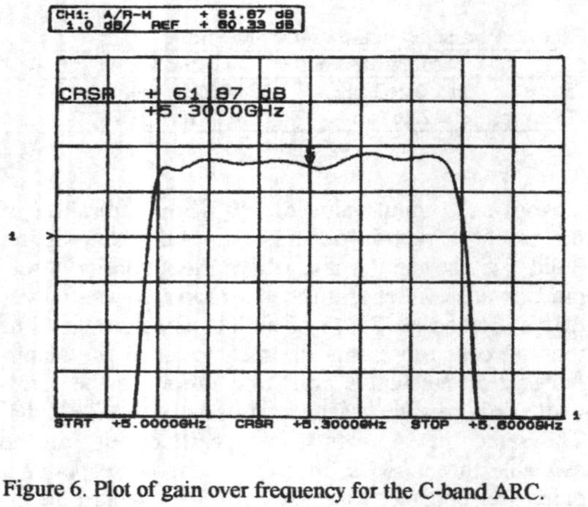

Figure 6. Plot of gain over frequency for the C-band ARC.

The measured bandwidth of the filter was in good agreement with the nominal one. The gain flatness is about ± 0.2 dB, which is better than the values reported on the data sheet of the single components of the amplifier section (Table 1). This improvement can be again explained as a ballast effect.

3.3 *Thermal characterisation*

The purpose of thermal measurements was to evaluate changes of the ARC's RF gain with respect to variations of room temperature. The "thermal" assessment is very important for ARCs because they are expected to work in open-air sites at different temperature and humidity conditions during SAR campaigns.

The DUT was placed into a thermal room and the device input was connected to a synthesised CW generator (Model HP8671B) set up on a power level of -60 dBm, a typical value of the received power during *in situ* operation. The DUT output was connected to a spectrum analyser (Model TK2792), and the temperature was detected by a K-type probe, connected to a digital thermometer. The temperature was changed with 10 °C steps from - 5 °C to + 55 °C every 30 minutes, in order to get stabilised values. The uncertainty on the temperature measurements is about 0.5 °C. In this way a wide range of air conditions were simulated.

In Figure 7 the experimental RCS values as a function of the temperature and both the linear and fourth-order polynomial fittings are reported. The rms errors of the fit are also given. It can be noted that the rms values of the fits are quite similar.

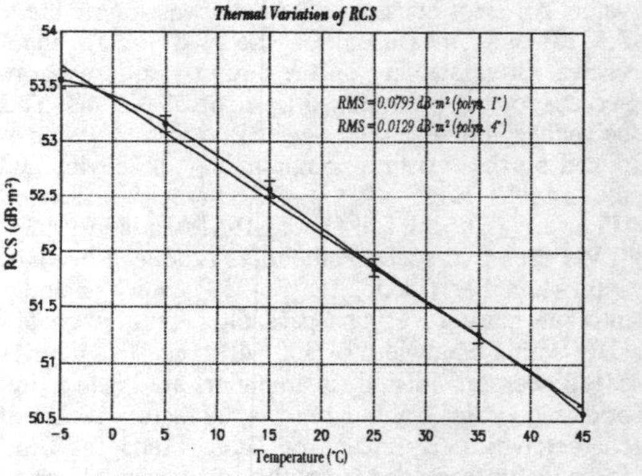

Figure 7. Comparison between 4-th order polynomial and linear fitting on the thermal data.

In order to evaluate the radar cross-section of the ARCs, the following formula was used (Brunfeldt and Ulaby 1984):

$$\sigma_{ARC} = \frac{\lambda^2}{4\pi} G_{ARC} G^2_{Ant} \tag{3.1}$$

where σ_{ARC} is the calculated RCS, λ is the wavelength at the centre frequency (0.05656 m), G_{ARC} is the measured gain of the RF section with mismatch and insertion losses, and G^2_{Ant} is the product of the receiving and transmitting antenna gains. The linear fit allowed us to calculate ΔRCS/ΔT, i.e. the change rate of the RCS with respect to the temperature, given by the slope of the straight line. We found:

$$\frac{\Delta RCS}{\Delta T} \cong 0.06\,\mathrm{dBm^2\,{}^\circ C^{-1}} \tag{3.2}$$

which indicates good thermal stability.

3.4 *Antennas characterisation*

The two pyramidal horn antennas are built with a short waveguide containing a radiator cut on λ/4, resonating at 5.3 GHz. The radiator can be connected to an external coaxial cable by means of a SMA connector. In Table 4 the factory technical data are reported.

Table 4. Antenna factory data.

Specification	Value
Frequency [GHz]	5.3
Bandwidth [MHz]	300
Gain at -3 dB [dB]	15 (min.)
Beamwidth at boresight [deg]	15
Return loss [dB]	17 (min.)
Cross polarisation isolation [dB]	32 (min.)
Sidelobe level* [dB]	-17 (min.), -20 (min.)

*For horizontal (H) and vertical (V) polarisation.

For fine characterisation of the antennas, some measurements were made into an anechoic chamber: gain, in order to evaluate return loss (RL), voltage standing wave ratio (VSWR), mismatch loss (ML) and decoupling with respect to polarisation.

The well-known standard-gain antenna method has been used for gain measurement. The measurement set-up consisted of a standard antenna (NARDA Model 643) with known gain G_{st}, connected by a fixed attenuator (-3 dB) to a CW generator (Model HP8671B) working at 5.3 GHz. The antenna under test was connected by a fixed attenuator to a spectrum analyser and placed at a distance R from the standard antenna. The unknown gain, G_{Test}, was evaluated as (Laverghetta 1981):

$$G_{Test} = \frac{4\pi R^2}{\lambda}\frac{P_r}{P_t}\frac{1}{G_{st}} \tag{3.3}$$

where P_r and P_t are respectively the received and transmitted power. The measured gain ranges from 15.50 dB to 16.06 dB for all C-band antennas. The other measurements were made by a double-output directional coupler connected to the antennas under test and to a CW generator. Table 5 reports the measured values: VSWR is the average voltage standing wave ratio. As a final remark, it is worth saying that the measurement accuracy was 0.6 dB, including the error contributions of the gain-measuring system (up to 0.3 dB), instabilities of the antenna gain over temperature (0.1 dB), thermal noise and multipath errors (of the order of 0.2 dB).

Table 5. Antenna parameters: measured values.

Description	Value	
Gain [dB]	15.5 - 16.06	
Return loss [dB]	20.00	
Mismatch loss [dB]	0.05	
VSWR	1.22	
Decoupling*	$V_1 V_2 > 60$ dB	$H_1 H_2 > 54$ dB
	$V_1 H_2 > 82$ dB	$H_1 V_2 > 82$ dB

*Pedices indicate TX antenna (1) and RX antenna (2).

4 USE OF C-BAND ARCs IN SAR EXTERNAL CALIBRATION CAMPAIGNS

The validation of the laboratory measurements and the assessment of correct ARC performance during ground calibration campaigns were performed by analysing C-band calibrated data, gathered from different SAR systems. The main goals of the calibration experiments, namely image quality assessment and radiometric calibration, have been accomplished by deploying on both the test-sites two ARCs and several trihedral corner reflectors with different RCS, used as reference targets. In the following, after a brief description of the strategy adopted for radiometric calibration of SAR images, the main results of our analysis on radar data will be presented.

4.1 *Radiometric calibration*

The images were calibrated by using the intensity integration method (Gray *et al.* 1990), insensitive to focusing errors and processor coherence. It is based on the estimates of E_{gp}, the energy associated to a reference reflector deployed on the ground and imaged in the data set, E_{gu}, the background clutter energy, and the image noise power, whereas the accuracy of the estimates derives from the standard deviation of E_{gp}, evaluated on different reference targets. The average backscattering coefficient in correspondence of each point target, $\langle\sigma^0\rangle$, is given by:

$$\langle\sigma^0\rangle = \sigma_{ref}\frac{E_{gu}}{A_u E_{gp}}\sin\vartheta_u \tag{4.1}$$

where A_u is the area swept by the slant range and azimuth spacings, and viewed under a look angle of ϑ_u. The reference RCS, σ_{ref}, and the estimates of the energy term E_{gp} were derived from the CRs deployed on the test site, as it will be described later. The amplitude-calibrated image, proportional to $\langle\sigma^0\rangle$, was used to recover the RCS of the ARCs, by evaluating the product $E_{gp}A_p$ on the area A_p, containing the response of the active calibrator. The accuracy of the method, which depends on the speckle autocovariance function and on the background clutter (Ulander 1991) has been found to be of the order of 0.7 dB for the TOPSAR data set, and 0.8 dB for the ERS-1 images.

4.2 MAC-Europe 1991 Campaign (TOPSAR)

The multisensor airborne campaign (MAC) organised in Europe in 1991 gave to our research group the possibility of setting up a calibration experiment on the Matera test-site, gathering data from the NASA DC-8 Airborne SAR (AIRSAR), augmented with a pair of C-band antennas displaced across-track (Topographic Synthetic-Aperture Radar, TOPSAR, Zebker *et al.* 1992). Three C-band ARCs and 26 trihedral CRs were deployed on homogeneous background, and interferometric data were acquired during five passes (Alberti *et al.* 1994). One of the images of the interferometric couple, processed by CO.RI.S.T.A., is shown in Figure 8: the brightest ARC, No. 1, is visible on the centre of the image, just above the upper diagonal line of CRs with different leg lengths (the four brightest CRs, on the left, have 180 cm leg length).

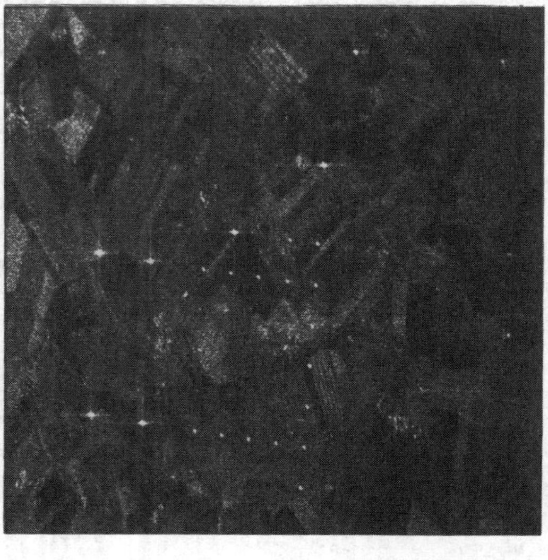

Figure 8. TOPSAR image of the test-site (June 25, 1991 pass).

A measurement of the propagation delay time of the ARC was carried out and compared with the laboratory data. The calibrated images allowed us to recover the ARCs' RCS and delay times in good agreement with the anechoic chamber tests, as reported in Table 6.

Table 6. Comparison between laboratory measurements and estimates from TOPSAR data (ARC No. 1).

Parameter	Laboratory	SAR image
RCS [dBm^2]	43.28 ± 0.6	43.6 ± 0.8
Delay time [ns]	30 ± 0.5	44 ± 11

The uncertainty on the delay time, measured on the SAR images, is due to the ratio between the velocity of light and the slant range spacing, and is ± 11 ns. The values are referred to ARC No. 1: similar results were obtained from ARC No. 3, whereas ARC No. 2 did not work during the overpass of the sensor, due to hardware malfunctioning.

4.3 Experiment I-7, 1991-92 (ERS-1)

In the framework of the Experiment I-7, ERS-1 SAR data were acquired during the October 23, 1991 night pass over the Salento Peninsula and June 21 and July 7, 1992 day passes over Matera (Southern Italy). Two ARCs and various corner reflectors were deployed on both sites. The analysed data sets are extracted from Single-Look Complex (SLC) products provided by the SAR Verification Mode Processor (VMP) installed at the Earthnet Program Office of the European Space Agency (ESA-ESRIN, Frascati, Italy) and from Precision Image (PRI) products processed at the Italian Processing and Archiving Facility (I-PAF), Matera. A detailed report on the first results of this experiment can be found in Cafforio *et al.* 1993.

The two PRI images provided by the I-PAF processor were calibrated by using the ESA transponder responses of the Oct. 13, 1991 Flevoland acquisition, according to ESA's criteria. The energy terms were estimated on a region of 4-times interpolated 51x51 pixels, by evaluating E_{gp} on a subset of 31x31 pixel centred on the mainlobe peak, and the background clutter from the remaining side pixels. The computed calibration constant was applied to the data sets, in order to estimate the RCS of the active devices deployed on the test- sites. The calibration results are summarised in Table 7, where σ_{exp} is the expected RCS. The differences from the expected (laboratory) values are mainly due to the lack of thermal stabilisation, a problem that was solved later in 1993. Moreover, a temperature variation of about 25 °C was recorded, due to different seasons and acquisition times. The mean look angle was 22 degrees.

Table 7. Radiometric analysis of the PRI 1991/92 products.

Date	Oct. 23, 1991		July 7, 1992	
ARC #	1	2	1	2
σ_{exp} [dBm^2]	54.80	56.80	52.00	52.00
σ [dBm^2]	56.25	59.51	55.45	54.70

4.4 ERS-1, 1994

The thermal stabilisation of the ARC, performed via a compensation look-up table in the prototypal version, was improved in late 1993 with cooling fins. During the first months of 1994, two "revised" C-band ARCs were deployed on the salt-works of Margherita di Savoia (Puglia, Southern Italy), a test-site with very low clutter echo return, due to the presence of large evaporite ponds (average depth: 80 ÷ 100 cm) surrounding the ARCs' locations. The images analysed (PRI product, delivered by the I-PAF, in slant range format, with 3 azimuth looks) refer to two night passes of the satellite, namely, January 28th, 1994 and February 3rd, 1994 (Fig. 9).

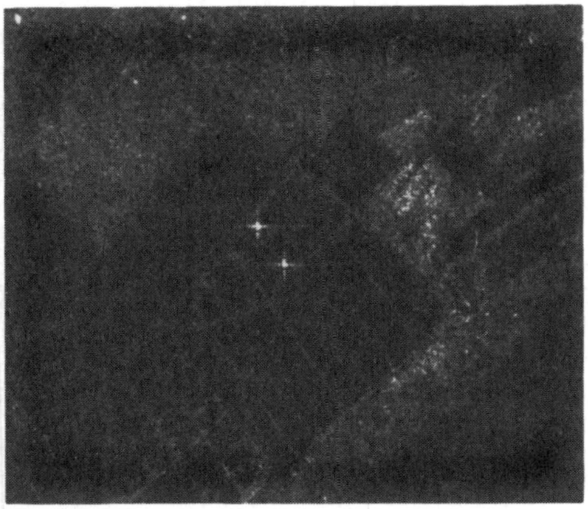

Figure 9. PRI image of the ARCs (Feb. 3, 1994 pass).

The mean temperature during the acquisitions was 9 °C, and the wind speed was 40 knots during the first pass (Jan. 28th), about 10 knots during the second pass. The active devices were deployed on the embankments in central position, at a distance of about 200 m, to minimise coupling effects. The results of the radiometric calibration, performed with the intensity integration method, are shown on Table 8, whereas a point target analysis confirmed the good geometric and radiometric processor quality, with the resolutions, Integrated Sidelobe Ratio (ISLR) and Peak Sidelobe Ratio (PSLR) well within the ESA specifications. The spatial resolution is 25x25 m^2. The accuracy on the radiometric calibration is rather good (about 0.5 dB), even if a small number of estimates of E_{gp} were available, because of the low background clutter from the ponds.

Table 8. Radiometric analysis on 1994 PRI products.

Date	Jan. 28, 1994		Feb. 3, 1994	
ARC #	1	2	1	2
σ_{exn} [dBm²]	52.6	52.6	52.6	52.6
σ [dBm²]	51.5	53.0	53.5	51.7

5 CONCLUSIONS AND FURTHER WORK

This paper has dealt with the development and use of C-band ARCs in SAR external calibration campaign. Laboratory measurements and results from radiometrically calibrated SAR images have shown good stability and performance of the active devices, which are the natural substitutes of conventional CRs, overcoming the stability and size problems of passive devices with the same radar cross-section. In addition, unique characteristics could be exploited, such as the possibility of tuning the propagation delay time with suitable devices (for example, microwave fibre-optic delay lines). This characteristic could be useful for calibration applications: forcing the point target response in a more convenient location on the image, the energy terms of Eq. (4.1) could be estimated very accurately by minimisation of the backscattering from pixels surrounding the ARC response Moreover, a tunable delay time means a controlled phase shift on the complex echo return, and this could be useful for interferometrical applications, allowing one the estimation of processor-induced phase errors.

On the basis of the experience gained by our research group during the last few years, and starting from the satisfactory results obtained from laboratory and on-the-spot tests, future developments on our active devices could be summarised in the following points:

1. Development of a ground acquisition section, in order to monitor the in-flight azimuth antenna pattern by detecting the strength of the signal received during the synthetic-aperture formation time. Successful laboratory tests on a digital section interfaced to a PC strongly encouraged us to test the ARC by attempting to record the SIR-C/X-SAR azimuth pattern, last April. As soon as the official data from this sensor are available, we will verify and validate our results. The possibility of collecting the range antenna pattern is obviously related to the number of ARCs necessary to sample the range swath: at the moment, such a sampling is far from adequate, due to the little number of ARCs developed.

2. Accurate laboratory measurement of the phase stability of the ARC. By characterising the phase shift induced by the active device up to fractions of radians, it should be possible to trace our ARCs as a reliable phase reference. Work is currently underway on the implementation of such measurements.

3. Development of a set of multi-frequency ARCs. With suitable X-band and L-band ARCs it will be possible to calibrate data from new operational multi-frequency SAR systems (such as SIR-C/X-SAR), extracting information on the electromagnetic behaviour of the imaged area at different wavelengths, and analysing the antenna patterns with appropriate ground receiver sections.

ACKNOWLEDGEMENTS

This work is being carried out under an Italian Space Agency (ASI) contract. We would also like to thank Eng. L. Garamone, responsible of the Microwave Laboratory of the ASI Spatial Geodesy Center (Matera, Italy).

REFERENCES

Alberti, G., Vetrella, S., Moccia, A., Ponte, S. 1994. Analysis and results of the TOPSAR experiments in Southern Italy. *EARSeL Advances on Remote Sensing*, Vol. III, No. 1, in press.

Bevington, P. R., Robinson, D. K. 1992. *Data reduction and error analysis for the physical sciences.* New York: McGraw-Hill.

Brunfeldt, D. R., Ulaby, F. T. 1984. Active Reflector for Radar Calibration. *IEEE Trans. on Geoscience and Remote Sensing*, GE-22: 165-169.

Cafforio, C., De Carolis, G., Mattia, F., Moccia, A., Ponte, S., Posa, F., Schena, V. D., Sergi, R., Smacchia, P., Sozzi, V., Vetrella, S. 1993. Preliminary results of the experiment I-7. *Proc. 1st ERS-1 Symposium, Cannes 4-6 Nov. 1992.* ESA SP-359: 161-166.

Capiluppi, A., Postpischl, D., Randi, P. 1978. *Introduzione alla elaborazione dei dati sperimentali.* Roma: CLUEB.

Carlson, A. B. 1992. *Communication systems.* Singapore: McGraw-Hill.

Cheung, W. S., Owens, R. 1981. *Microwaves made simple*. Norwood: Artech House.

Gray, A. L., Vachon, P. W., Livingstone, C. E., Lukowski, T. I. 1990. Synthetic aperture radar calibration using reference reflectors. *IEEE Trans. on Geosc. Rem. Sens.*, GE-28: 374-383.

Laverghetta, T. S. 1981. *Handbook of microwave testing.* Norwood: Artech House.

Ulander, L. M. H. 1991. Accuracy of using point targets for SAR calibration. *IEEE Trans. Aerosp. Electronic Syst.*, 27: 139.148.

Zebker, H. A., Madsen, S. N., Martin, J., Wheeler, K. B., Miller, T., Lou, Y., Alberti, G., Vetrella, S., Cucci, A. 1992. The TOPSAR interferometric radar topographic mapping instrument. *IEEE Trans. on Geosc. Rem. Sens.*, 30: 933-940.

Sensors and Environmental Applications of Remote Sensing, Askne (ed.) © 1995 Balkema, Rotterdam, ISBN 90 5410 524 0

CARABAS VHF synthetic aperture radar: Image quality and radiometric calibration

L.M.H.Ulander
Chalmers University of Technology, Göteborg, Sweden

H.Hellsten
National Defense Research Establishment, Linköping, Sweden

ABSTRACT: The performance of the CARABAS SAR system is evaluated in terms of spatial resolution and radiometric properties. Analysis of imagery from October 1992 indicates a resolution of about 8 m x 5 m in azimuth and slant range, respectively. The azimuth resolution is limited by the truncated raw data and the lack of motion compensation. Radiometric calibration using near-resonant horizontal dipoles is verified and applied to data from 1992 and 1993 with an estimated accuracy of better than ± 2 dB. The importance of ground reflection interference is verified and an example of sub-surface detection is given.

1 INTRODUCTION

CARABAS ("Coherent All Radio Band Sensing") is an airborne synthetic-aperture radar (SAR) operating in the 20 - 90 MHz frequency range, i.e. in the lowest part of the VHF band. It has been designed and constructed by the National Defence Research Establishment (FOA) in Linköping, Sweden. The original objective of the system design was to achieve a large relative bandwidth and to improve object detection in severe clutter environments (Hellsten and Andersson 1987, Hellsten 1990, Gustavsson *et al.* 1993, Hellsten *et al.* 1994). The lower VHF band was chosen to achieve a usable swath width and to provide significant foliage penetration.

System construction started in 1987 and the first radar test trials were conducted in 1992. Several flight campaigns have thereafter been completed for evaluating system performance over different terrain types (Hellsten *et al.* 1994, Mirkin *et al.* 1994). In this paper, we discuss results from image quality and radiometric calibration experiments. Complementary results on this topic can be found in (Ulander and Hellsten 1994).

2 SYSTEM DESCRIPTION

The main system parameters for CARABAS are summarised in Table 1.

The antenna system consists of two wide-band dipoles which are mounted close to each other at the rear of the aircraft. The tubular antennas have an overall length of 5.5 m and are made of canvas. The tubes are closed in the back end but have an opening in the front mounted on an air inlet. With this design, the antennas are stabilised in flight by ram air flowing into the tube. A single dipole antenna provides essentially an omni-directional gain pattern and two antennas are required to separate the left-right ambiguity.

A major limitation to system performance is radio source interference which results in large (mainly narrow band) variations in the received power. The number of bits in the analogue-to-digital converter (ADC) was therefore chosen to 12 to achieve a dynamic range approaching 70 dB. However, current high-speed devices with 12 bits/sample are not capable of providing Nyquist sampling of the total signal bandwidth. The transmitted waveform was therefore designed as a stepped frequency-modulated signal with a small instantaneous bandwidth. Each frequency step of 1.25 MHz is 0.4 μs long and the received signal is digitised at an offset frequency of 2.5 MHz using a sampling rate of 10 MHz. The latter can be achieved with 12 bits/sample. The peak transmitted power is 1 kW but the low duty factor results in an average power of only 1 W when all frequency steps and both dipole antennas are used.

The phase of the transmitted signal varies randomly from each frequency step (pulse) to the next. A sample of each pulse is therefore routed through an optical delay line and digitised in front of the nadir echo. The system is thus coherent-on-receive which is a prerequisite for coherent signal integration.

Table 1. CARABAS SAR system parameters

Frequency	20 - 90 MHz
Peak power	1 kW
Polarisation	Horizontal
Antenna	2 Wide-band dipoles
Transmitted waveform	Stepped FM
Frequency step	1.25 MHz
Pulse (step) duration	0.4 µs
Number of steps, *n*	≤ 57
System PRF	10 kHz
Effective PRF[1]	10/2/*n* kHz
ADC sampling rate	10 MHz
ADC quantisation	12 bits
Maximum slant range	7.5 km
Nom. altitude	1500 m
Nom. speed	100 m/s
Data rate	80 Mbits/s

Aircraft platform is a Rockwell Sabreliner normally operating at 1500 m altitude and with a speed of 100 m/s. The imaged swath is determined by a fixed digitising window of 50 µs from transmission which is equivalent to 7.5 km in range.

The processing of the received raw data into a focused image is performed using a Fourier-Hankel inversion (Hellsten and Andersson 1987). It is an exact solution to the problem of reconstructing a reflectivity function from its circular averages.

3 SPATIAL RESOLUTION: THEORY

3.1 Fourier relations

The minimum spatial resolution which can be achieved by a linear imaging system may be determined using fundamental relations from Fourier analysis. The width Δs of the impulse response which contains a range of wavenumbers in an interval Δk is constrained by

$$\Delta k \cdot \Delta s \geq 2\pi \quad (1)$$

The slant range (across track) coordinate R is the measured two-way distance to an object divided by two. This means that the resolution is half of which a one-way distance measurement could achieve. The wavenumber corresponding to the received slant range signal is given by

$$k_R = 2k = \frac{4\pi f}{c} \quad (2)$$

[1]The effective PRF is determined by the PRF, the number of frequency steps and the number of transmitting antennas.

where k and f are the wavenumber and frequency of the transmitted signal and c is the velocity of light. The wavenumber interval is thus proportional to the transmitted frequency bandwidth B, i.e. the slant range resolution is constrained by

$$\Delta R \geq \frac{2\pi}{\Delta k_R} = \frac{c}{2B} \quad (3)$$

The azimuth resolution may be estimated by determining the wavenumber content of the received signal when the aspect angle ϑ changes (see Fig. 1). For one frequency component f of the transmitted signal, the received azimuth signal is given by

$$a(x) \sim \exp\left(-j2k\sqrt{x^2 + y^2}\right) \quad (4)$$

where we have ignored any power scaling factors. A Fourier expansion of (4) shows that the azimuth wavenumber at the stationary phase is given by

$$k_x = -2k \sin \vartheta \quad (5)$$

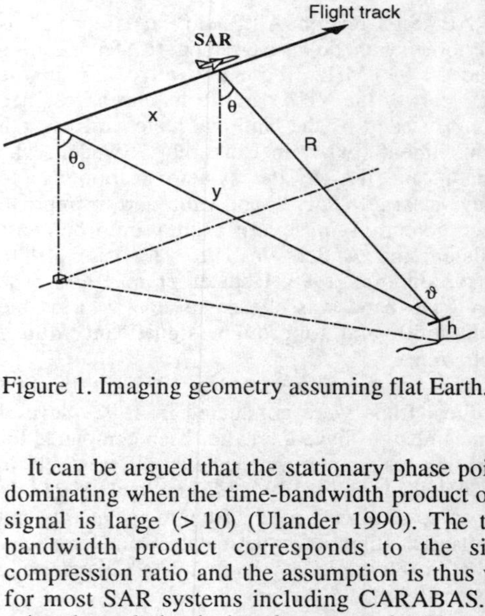

Figure 1. Imaging geometry assuming flat Earth.

It can be argued that the stationary phase point is dominating when the time-bandwidth product of the signal is large (> 10) (Ulander 1990). The time-bandwidth product corresponds to the signal compression ratio and the assumption is thus valid for most SAR systems including CARABAS. The azimuth resolution is therefore constrained by

$$\Delta x \geq \frac{\pi}{\Delta(k \sin \vartheta)} = \frac{\lambda_{min}}{2(\sin \vartheta_{max} - \sin \vartheta_{min})} \quad (6)$$

where λ_{min} is the minimum transmitted wavelength. We note in passing that (6) reduces to the conventional (narrow band) azimuth resolution $D/2$ when $D >> \lambda$, where D is the antenna aperture length and $\vartheta_{max} = -\vartheta_{min}$. For wide-band systems, spatial resolution is not a clear-cut quantity since it depends on frequency-selective characteristics of the object. A more reliable estimate of the achievable resolution is therefore obtained by evaluating the combined system and object transfer functions.

3.2 System transfer function

The two-dimensional wavenumber spectrum of the received signal is illustrated in Fig. 2a. The signal spans a range of wavenumbers according to (2) and (5). The processed bandwidth of the transmitted signal spectrum results in an interval of slant range wavenumber k_R. The processed "beamwidth" of the illuminated aspect angles results in an interval of azimuth wavenumbers k_x which increases linearly with k_R. The rectangular box enclosing the wavenumber spectrum symbolises the finite sampling of the signal. After processing of the raw data by the Fourier-Hankel inversion algorithm, the wavenumber spectrum changes to that in Fig. 2b. The inversion algorithm includes a variable transformation according to

$$k_y = \sqrt{k_R^2 - k_x^2} \qquad (7)$$

which causes the distortion of the spectrum. The final SAR image is obtained by a two-dimensional inverse Fourier transform of the spectrum in Fig. 2b, which thus also outlines the system transfer function.

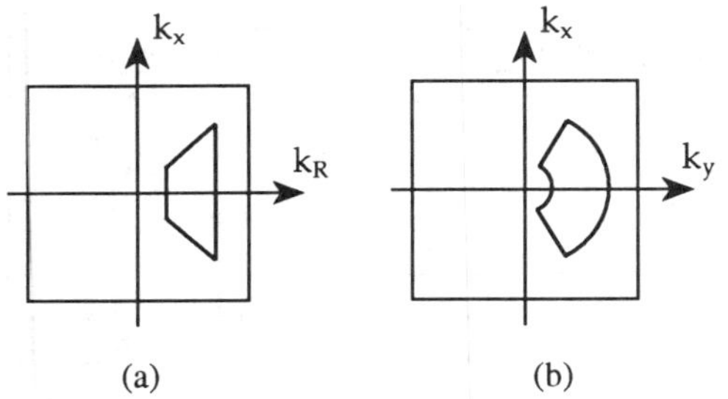

Figure 2. Wavenumber spectrum of (a) raw data and (b) final image. Wavenumbers in azimuth and slant range are denoted k_x and k_y, respectively. The radial wavenumber corresponding to the received signal in slant range is k_R.

A detailed evaluation of the spatial resolution according to (1) can now be performed using both system and object characteristics. In this paper, however, we adopt a first-order approximation and use (3) and (6) but with the centre wavelength (λ_c) instead of the minimum wavelength (λ_{min}) in the latter, i.e.

$$\Delta x \gtrsim \frac{\lambda_c}{2(\sin\vartheta_{max} - \sin\vartheta_{min})} \qquad (8)$$

where λ_c is defined as the wavelength corresponding to the centre frequency. This approximation is justified if the processed beamwidth is relatively small.

4 SPATIAL RESOLUTION: DATA ANALYSIS

In this section, we investigate the achieved spatial resolution for the CARABAS SAR system in relation to the theoretical limits. Raw data have been processed using the Fourier-Hankel inversion algorithm and a clipping procedure to reduce radio interference noise. There has been no attempt to perform compensation for the deviation of the flight track from an ideal straight line, i.e. no motion compensation. To improve azimuth focus, however, it was found necessary to tune the average velocity of the aircraft.

4.1 Data set

The analysed images are from an experiment carried out close to Ottenbylund on the island Öland in the Baltic Sea during October 1992. Several different reference objects were deployed for system calibration studies, i.e. horizontal near-resonant dipoles, trihedrals and log-periodic antennas. We will mainly focus on the dipoles, since their (complex) scattering amplitude may be accurately calculated including reflection interference from the ground (Ulander and Hellsten 1994). The dipoles (length 3.3 m, diameter 2.5 cm) were oriented with their axis' parallel to the flight track and positioned at different heights above the ground surface. Heights of 2.0, 3.0, 3.5, 4.0, and 5.0 m have been used in the present investigation. Additional dipoles were also deployed at 1.0 m and 4.5 m but have not been used. The signal-to-noise ratio of the 1.0 m dipole was too low, and the 4.5 m dipole had fallen down prior to the SAR flights. Details of the dipole scattering are found in Sections 5.2 and 5.3.

The characteristics of two analysed images are summarised in Table 2, together with the minimum resolution computed from (3) and (8). The two images were acquired on 10 October 1992 with different incidence angles and the processed frequency bands are also slightly different. The resulting minimum resolutions are therefore different, but both are of the order 2 m x 4 m in azimuth and slant range, respectively.

Table 2. Characteristics of the analysed images at the dipoles. Incidence angle is θ, $\vartheta_{min} = -\vartheta_{max}$, and the minimum azimuth and slant range resolutions are Δx_m and Δy_m, respectively.

Image	f (MHz)	θ	ϑ_{max}	Δx_m (m)	Δy_m (m)
2:3:1	30 - 66	59°	39°	2.5	4.1
2:4:4	30 - 80	67°	36°	2.3	3.0

4.2 Methodology

The impulse response function corresponds to the image response of an ideal point target, i.e. which has a constant scattering amplitude for all wavenumbers. A near-resonant horizontal dipole does not have a constant scattering amplitude, and the frequency response varies with the slant range wavenumber due to ground reflection interference. These effects result in broadening of the image response. For sufficiently large dipole heights, the reflection interference results in splitting of the response into three separated peaks. This means that the dipoles are mainly useful for estimating the azimuth resolution.

An examination of the dipole responses showed that (1) the main peak was distorted by noise ripple and (2) high sidelobes were present. The reason for the first observation is probably to be found in the implementation of the Fourier-Hankel inversion algorithm. It presently assumes ideal raw data using all possible wavenumbers and does not consider raw data of finite extent. For example, the algorithm is based on the maximum azimuth wavenumber interval ($\vartheta_{max} = -\vartheta_{min} = 90°$) although the raw data is limited by the input data matrix. In the presence of noise, we therefore expect unnecessary distortions of the impulse response. The reason for the second observation is probably that no motion compensation is performed. A small change in along-track velocity will therefore cause a spatial shift of the image response with multiple peaks appearing in azimuth. Based on these observations, we choose to smooth the image responses and define spatial resolution as the maximum 3 dB width in slant range or azimuth of the smoothed image.

The dipole responses also show broadening due to ground reflection interference. We therefore decided to exclude dipoles where the reflection interference phase is close to π radians, since the latter would bias the resolution estimate to an unphysical small value.

4.3 Results

The results from estimating the spatial resolution is shown in Tables 3 and 4. We also include the average and standard deviation for each image. For the slant range resolution, these values only include the three dipoles closest to an in-phase condition for the reflection interference.

The azimuth resolution is considerably larger than the theoretical values given in Table 2. On average, the measured azimuth resolution is 3 times the theoretical value. The reason for this discrepancy is most likely the lack of motion compensation. Flight track data of sufficient precision is not available for the Ottenbylund data, but missions after 1992 have acquired carrier-phase GPS data which enables motion compensation. Results from the latter will be given in the future.

The slant range resolution is also larger than the theoretical values in Table 2. The broadening is about 1.5 and 2.3 for images 2:3:1 and 2:4:4, respectively. This is not unexpected due to the effect of reflection interference and variable scattering amplitude. Calculations show that we expect a width of about 5 m due to the latter effect. It is clear that the dipoles do not accurately estimate the slant range resolution of the system, and only indicate that it is not worse than 6-7 m. As a comparison, the 3 dB width of the response from a trihedral (4.9 m short side length) was measured to 4.7 m in the image 2:3:1, which more likely reflects the true slant range resolution.

Table 3. Resolution measurements of image 2:3:1. The values for the 5.0 m dipole is missing due to low signal-to-noise ratio. Azimuth resolution is Δx and slant range resolution is Δy.

Dipole height (m)	Δx (m)	Δy (m)
2.0	7.3	6.4
3.0	8.5	7.5
3.5	6.1	5.0
4.0	9.8	(8.0)
5.0	(-)	(-)
Average ± St.dev.	7.9 ± 1.6	6.3 ± 1.3
Minimum (Table 2)	2.5	4.1

Table 4. Resolution measurements of image 2:4:4. Azimuth resolution is Δx and slant range resolution is Δy.

Dipole height (m)	Δx (m)	Δy (m)
2.0	9.6	(4.9)
3.0	5.5	5.0
3.5	6.8	8.5
4.0	6.8	7.2
5.0	6.3	(3.8)
Average ± St.dev.	7.0 ± 1.5	6.9 ± 1.8
Minimum (Table 2)	2.3	3.0

5 RADIOMETRIC CALIBRATION: THEORY

5.1 Radiometric calibration strategy

Radiometric calibration involves establishing the relation between the SAR image output and the radar cross section (RCS) σ for a point scatterer or the backscattering coefficient σ° for a distributed scatterer. A robust procedure is to determine the "energy" E from the response of a known point scatterer. The energy is defined as the area integral of the power response with the background and noise subtracted. The backscattering coefficient in a nearby flat area may then be determined from the following equation (Gray *et al.* 1990, Ulander 1991)

$$\sigma^\circ = \left(\frac{\sigma \sin\theta_i}{E}\right) P_{\sigma^\circ} = \frac{\sin\theta_i}{K} P_{\sigma^\circ} \tag{9}$$

where θ_i is the local incidence angle, P_{σ° is the average power of the distributed scatterer, and K is the calibration constant. Equation (9) assumes azimuth and slant range image coordinates, that signal ambiguities are negligible, and that the point and distributed scatterers are ideal, i.e. σ and σ° are constant during image formation. The latter is clearly not a valid assumption for a broad-band SAR system, but we may still use (9) in an average sense. In the present formulation, we assume that σ varies with both slant range and azimuth frequency. It can then be shown that (9) remains valid if we substitute σ for σ_e (Ulander and Hellsten 1994), the latter defined by

$$\sigma_e = \frac{\iint \sigma \left|F^2 G\right|^2 \frac{k_c^2}{k^2}\, dk_x dk_y}{\iint |G|^2\, dk_x dk_y} \tag{10}$$

where $G(k_x,k_y)$ is the system transfer function and F is the propagation factor. The calibration strategy thus reduces to using point scatterers with accurately known σ and F. The near-resonant horizontal dipoles have a large RCS in relation to their size and weight but the effect of surface reflection interference is significant and must be taken into account. An alternative point scatterer which is often used is the trihedral retro-reflector. The trihedrals are normally much larger than the radar wavelength which clearly makes life difficult in the VHF band. The RCS of trihedrals close to resonance is also not well known, in particular when surface interaction must be considered. We therefore choose to base the radiometric calibration on the dipoles.

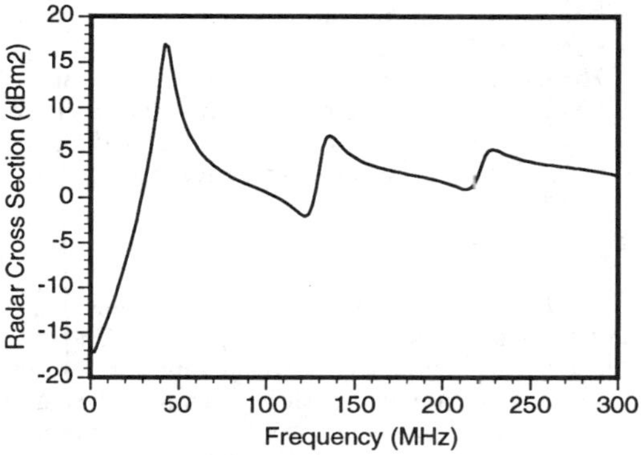

Figure 3. RCS of near-resonant and perfectly conducting dipole. Incident wave polarisation is assumed along dipole axis. The dipole has a length of 3.3 m and a diameter of 2.5 cm.

5.2 Dipole scattering amplitude

The near-resonant dipoles are thin cylinders which have a length of the same order of magnitude as the wavelength. For particular lengths, which are nearly a multiple of half a wavelength, the dipole becomes resonant. This means that its RCS is much larger at these wavelengths. The RCS of these objects can only approximately be calculated using closed-form expressions. Convenient expressions for the perfectly conducting case are given in (Van Vleck *et al.* 1947). The resulting RCS is accurate to about 0.6 dB at the resonance peak (Ruck 1970). An example of the RCS versus frequency for a 3.3 m long dipole is shown in Fig. 3. A frequency of 45.5 MHz makes the dipole length equal to half a wavelength and the resonance peaks are observed to be close to an odd multiple of this frequency.

The RCS of the dipole also varies with azimuth position along the flight track, which may be translated to azimuth frequency using (5). In general, the angular response of the RCS has multiple peaks related to the resonance order, but is a smooth symmetrical curve with a peak at broad-side for frequencies close to the first resonance. The incident polarisation at the dipole also varies along the synthetic aperture and must be taken into account.

5.3 Surface reflection interference

Consider a scattering object situated at a height h above a smooth surface as illustrated in Fig. 4. The total incident field on the object is the sum of the direct and surface-reflected plane wave, which takes multi-path into account but neglects mutual coupling between the surface and the object. A virtual image of the object can be introduced which simplifies derivation of the geometrical factors.

The propagation factor F is defined as the ratio of the electric field at a point in space with the surface present to that which would have existed without the surface. For H polarisation, this reduces to

$$F = 1 + \rho \exp(-j2kh\cos\theta) \qquad (11)$$

where k is the wavenumber, θ is the incidence angle and ρ is the Fresnel reflection coefficient. The latter depends on the constitutive parameters of the ground, frequency and incidence angle. The phase factor in (11) is conveniently derived as the phase difference between the object and its image in the incident wave field.

The interference with the surface-reflected wave has a profound effect on the backscattered field, the latter being proportional to the two-way propagation factor F^2. The backscattered power is thus proportional to

$$|F|^4 = \left[1 + |\rho|^2 + 2|\rho|\cos(2kh\cos\theta - \varphi)\right]^2 \qquad (12)$$

where $|\rho|$ and φ is the magnitude and phase of ρ, respectively.

The phase angle φ controls the phase of the interference pattern, and, conversely, the location of maxima and minima may be used to determine the phase angle. For H polarisation and a non-magnetic and non-conducting medium, the Fresnel reflection coefficient is given by

$$\rho = -\frac{\sqrt{\varepsilon - \sin^2\theta} - \cos\theta}{\sqrt{\varepsilon - \sin^2\theta} + \cos\theta} \qquad (13)$$

where ε is the (effective) complex dielectric constant of the ground including conductivity. It is defined by

$$\varepsilon = \varepsilon_r - j\frac{\sigma_c}{\omega\varepsilon_o} \qquad (14)$$

where ε_r is the relative dielectric constant, σ_c is the conductivity (S/m) and ω is the angular frequency. Equation (13) shows that the phase angle is close to 180° when the imaginary part of ε is much smaller than the real part. This is true for many practical situations, and significant deviations from 180° is in fact only found when the real and imaginary parts are of the same order of magnitude and $|\varepsilon| < 10$.

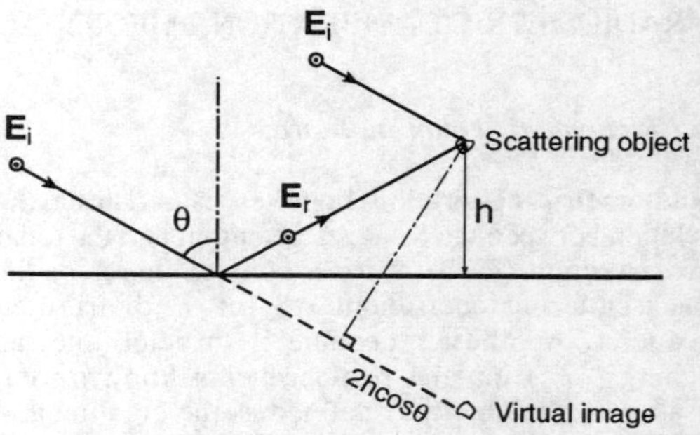

Figure 4. Geometry of scattering object above a smooth surface.

6 RADIOMETRIC CALIBRATION: ANALYSIS

6.1 Data set

Two radiometric calibration experiments have been analysed, which were jointly performed by FOA and MIT Lincoln Laboratory under ARPA sponsorship (Hellsten *et al.* 1994, Mirkin *et al.* 1994). The first data set was acquired on 10 October 1992 during the Ottenbylund experiment, while the second was acquired on 5 June 1993 close to Yuma in Arizona. Dipoles (length 3.3 m and diameter 2.5 cm) were deployed at different heights. Data have been processed using transmitted frequencies between 36 and 48 MHz and aspect angles between ± 10°. This corresponds to a resolution of 12 m by 10 m in slant range and azimuth, respectively.

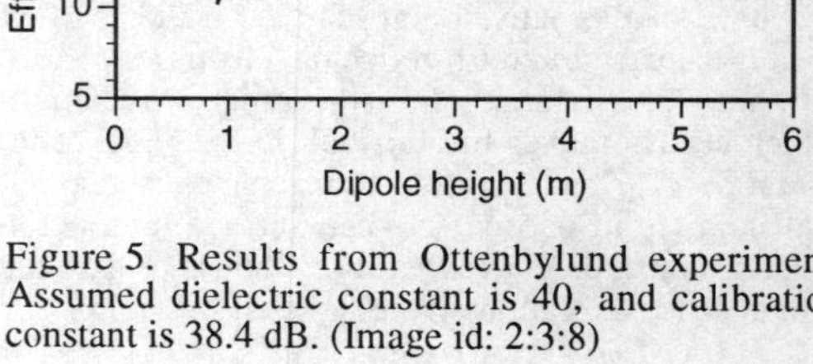

Figure 5. Results from Ottenbylund experiment. Assumed dielectric constant is 40, and calibration constant is 38.4 dB. (Image id: 2:3:8)

6.2 Results

The effective RCS was calculated using (10) and the RCS model including surface reflection interference. The "energy" of the dipole responses was measured from complex image data, and the calibration constant in (9) was computed as the weighted average difference between the model (σ_e) and measurement (E) in dB units.

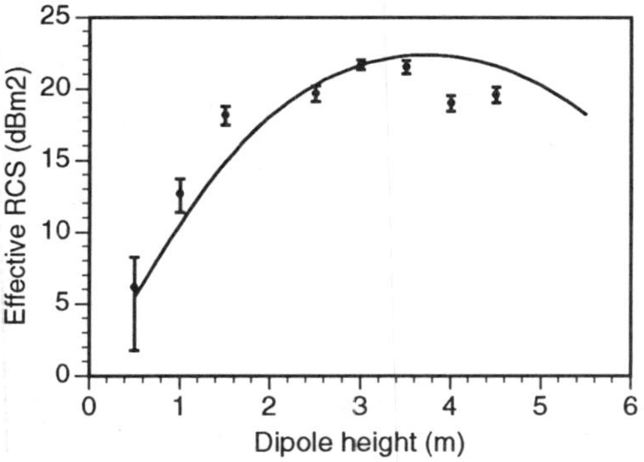

Figure 6. Results from Yuma experiment. Assumed dielectric constant is 3. (Image id: 2:6:10)

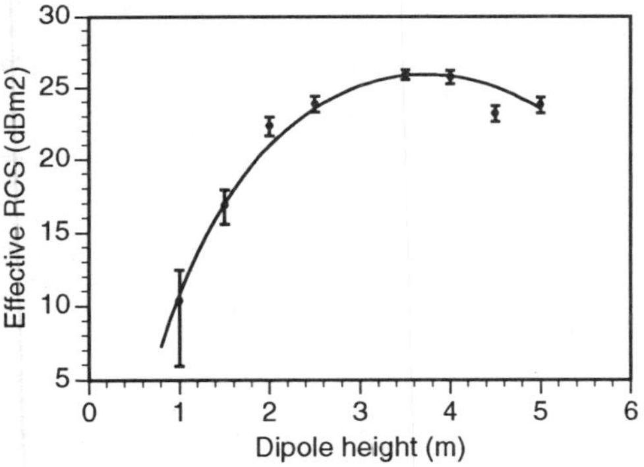

Figure 7. Results from Yuma experiment. Assumed dielectric constant is 40, heights are offset by 0.5 m, and calibration constant is 46.6 dB. (Image: 2:6:10)

The result for the Ottenbylund data is shown in Fig. 5. The reflection interference was computed assuming a dielectric constant of 40 (humid soil). A good agreement is observed which supports the model assumptions. The result for the Yuma data is shown in Fig. 6 with less good agreement. In this case, reflection interference was computed assuming a dielectric constant of 3 (dry soil) since the surface was dry sand. However, ground truth suggests that a conductive sand layer was present only a few dm below the surface. A new model calculation was therefore made with a dielectric constant of 40 and the dipole heights offset by 0.5 m. This improved the agreement as seen in Fig. 7. A calculation shows that the offset corresponds to a 0.2 m thick dry sand layer assuming a dielectric constant of 3.

The sensitivity of the effective RCS to the dielectric constant of the ground is illustrated in Fig. 8. The four curves correspond to a (real) dielectric constant of 5, 10, 20 and 40, which show a maximum difference of about 2.5 dB at the peak. We note that the different dipole heights may be used to estimate the dielectric constant. The calibration uncertainty is thus of the order ± 1-2 dB without knowledge of the ground dielectric constant.

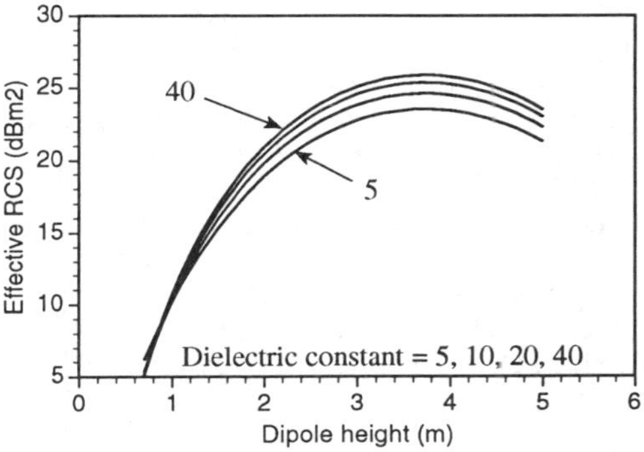

Figure 8. Effective RCS for Yuma experiment with different ground dielectric constants. Image parameters as in Figs. 6 and 7.

7 CONCLUSIONS

CARABAS is an airborne SAR operating in the lowest part of the VHF band. The system design uses novel techniques to achieve its objectives, e.g. two wide-band dipole antennas, a stepped frequency-modulated waveform, and a receiver with a very large dynamic range. Processing of the received raw data is performed using a Fourier-Hankel inversion algorithm.

The spatial resolution of imagery produced from the Ottenbylund experiment in October 1992 is estimated to 8 m x 5 m in azimuth and slant range, respectively. Corresponding theoretical resolutions are 2 m x 4 m with truncated raw data. The large discrepancy in azimuth is most likely due to the lack of motion compensation.

Radiometric calibration has been performed based on the response of horizontal and near-resonant dipoles. Their frequency-dependent RCS is computed using models for the dipole RCS and the surface reflection interference. Images from the Ottenbylund experiment in October 1992 and the Yuma experiment in June 1993 have been processed to a resolution of 10 m x 12 m in azimuth and slant range. A comparison between measurements and models shows good agreement after shifting the Yuma dipole heights by 0.5 m. The latter indicates that reflection from a dielectric interface located at 0.2 m below the surface is dominating, which is in agreement with *in situ* observations. The uncertainty of the calibration procedure is dominated by the ground dielectric constant and is estimated to be better than ± 2 dB.

REFERENCES

Gray, A.L., P.W. Vachon, C.E. Livingstone and T.I. Lukowski 1990. Synthetic Aperture Radar Calibration using Reference Reflectors. *IEEE Transactions on Geoscience and Remote Sensing*, 28:374-383.

Gustavsson, A., P.O. Frölind, H. Hellsten, T. Jonsson, B. Larsson and G. Stenström 1993. The Airborne VHF SAR System CARABAS. Proc. IGARSS'93. Tokyo, Japan: p. 558-562.

Hellsten, H. 1990. Inverse Scattering Analysis of Diffraction Limited SAR. *IEEE Transactions on Antennas and Propagation*, 38:1517-1522.

Hellsten, H. and L.E. Andersson 1987. An Inverse Method for the Processing of Synthetic Aperture Radar Data. *Inverse Problems*, 3:111-124.

Hellsten, H., P.O. Frölind, A. Gustavsson, T. Jonsson, B. Larsson, G. Stenström, B.T. Binder, M. Mirkin and S. Ayasli 1994. Ultra-Wideband VHF SAR - Design and Measurements. Proc. SPIE. Orlando, Fl, 4-8 April 1994.

Mirkin, M., T.O. Grosch, T.J. Murphy, S. Ayasli, H. Hellsten, R. Vickers and J.M. Ralston 1994. Results of the June 1993 Yuma Ground Penetration Experiment. Proc. SPIE. Orlando, Fl, 4-8 April 1994.

Ruck, G.T. 1970. *Radar Cross Section Handbook*. New York: Plenum Press.

Ulander, L.M.H. 1990. Theory of SAR Radiometric Calibration. Research Report 167, Department of Radio and Space Science, Chalmers University of Technology, Gothenburg, Sweden

Ulander, L.M.H. 1991. Accuracy of Using Point Targets for SAR Calibration. *IEEE Transactions on Aerospace and Electronic Systems*, 27:139-148.

Ulander, L.M.H. and H. Hellsten 1994. Calibration of the CARABAS VHF SAR System. Proc. IGARSS'94. Pasadena, CA, 8-12 August 1994.

Van Vleck, J.H., F. Bloch and M. Hamermesh 1947. Theory of Radar Reflection from Wires or Thin Metallic Strips. *Journal of Applied Physics*, 18:274-294.

Remote sensing methodology

Sensors and Environmental Applications of Remote Sensing, Askne (ed.) © 1995 Balkema, Rotterdam, ISBN 90 5410 524 0

Neural networks in remote sensing multisensor data processing

V.Cappellini & A.Chiuderi[1]
University of Florence, Department of Electronic Engineering, Italy
S.Fini
Fondazione Scienza per l'Ambiente, Florence, Italy

ABSTRACT: In this paper neural networks are investigated as one of the tools that can be implemented for remote sensing multisensor data processing. An example of land use inventory by means of multisensor data processed with neural networks is reported and discussed.

Three are the experiments reported in the following: as a first case study, the TM image is classified by means of the Counterpropagation neural network, successively, the same image is first pre-processed by a decorrelation algorithm and then classified by the same network, leading to remarkable improvements in the classification accuracies, finally the TM image is integrated with a second data set acquired by a Synthetic Aperture Radar and, again, processed by the same type of network. The results of the three experiments are discussed and compared.

The capabilities of this kind of neural networks are also investigated, the problems connected with the use of remotely sensed data are outlined and some possible solutions are proposed. In particular the potentiality of the Counterpropagation network to implement data fusion will be analyzed as one of the possible ways to integrate and merge multisensor in order to obtain better classification accuracies.

1. INTRODUCTION

Remote sensing activities generate a very large amount of data necessary for environmental phenomena monitoring. Multisensor data offer the possibility to process information on the same area collected by sensors operating at different altitudes with different spatial resolutions, and at different times. An integrated processing of data (data fusion) can produce a greater knowledge of the phenomena which are to be studied than an interpretation of each channel regardless of other collected data. Many investigators are nowadays involved in the analysis of data fusion possibilities and are developing innovative techniques for multisensor data management.

The development of new techniques is aimed at overcoming the difficulties related to the intrinsic nature of the data to be processed which are not collected by one single sensor. For instance when processing is performed on data collected by one sensor, it is possible to make hypotheses on data distribution so that the application of specific algorithms (i.e. maximum likelihood classification for land-use) is made possible.

When processing is performed on multidimensional data it becomes more difficult to make hypotheses on data distribution so as to create theoretical fundamentals for the application of processing algorithms. The use of neural networks allows us to overcome these theoretical problems and to approach the task of merging different sources of information in a totally different way.

In the last years neural networks have received a great interest in all fields of applied sciences and in particular in pattern recognition and classification; as far as remote sensing is concerned, neural networks have been successfully employed for landcover classification purposes as in (Benediktsson et al. 1990), (Bishof et al. 1992), (Kanellopoulos et al. 1993) and for physical parameter inversion. See, as an example (Chen et al. 1992), (Zurk et al. 1992).

In all these papers the authors employ the well-known Back Propagation (BP) algorithm (Rumel-

[1] The author is supported by a grant from the Italian Space Agency (A.S.I)

hart et al. 1986), a totally supervised method. Even though this training algorithm represents the most accurate and studied one, it must be pointed out that, in remote sensing, the collection of the training data (ground truth data) is a very difficult task, with respect both to time and to money.

The network proposed in this paper, on the other hand, is composed by two layers which can be trained separately: the first layer, called the Kohonen layer trained by an unsupervised procedure, and the second layer, the Grossberg layer, which can be trained in a second time, by means of a supervised algorithm.

The great advantage of this type of network is given by the fact that at the end of the first (unsupervised) stage, a first result is obtained: the area to be classified is clustered according to the similarity of the patterns. If the ground truth data are available the results obtained so far can be refined and a classification of the area under examination can be performed.

2. MATERIALS AND METHODS

In the following two sections a brief description of both the data and the network employed is given. In the former the technical characteristics of the sensors employed is given, while in the latter the Counterpropagation network is described and some of the drawbacks connected with the use, as input channels, of remotely sensed data are reported.

2.1 The Data

For this study two different sets of data were available: data provided by the Daedalus airflown Thematic Mapper Simulator (TMS), operating in the visible and infra-red region of the electromagnetic spectrum and data acquired by the multifrequency, multipolarized JPL AIR-SAR, operating in the L, P, and C bands (Held et al. 1988).

Both data sets were acquired during the MAC-Europe'91 campaign, a co-operation of NASA-JPL and the Italian Space Agency (A.S.I.), in an agricultural area situated in Montespertoli, in the surroundings of Florence (Italy) (Bechini et al. 1992). An on site campaign was also carried out in order to collect the ground truth data for the experiments (Coppo et al. 1992).

The two data sets have different spatial resolution: 25m by 25m for the TMS data while 6m in range and 12m in azimuth for the SAR data. In or-

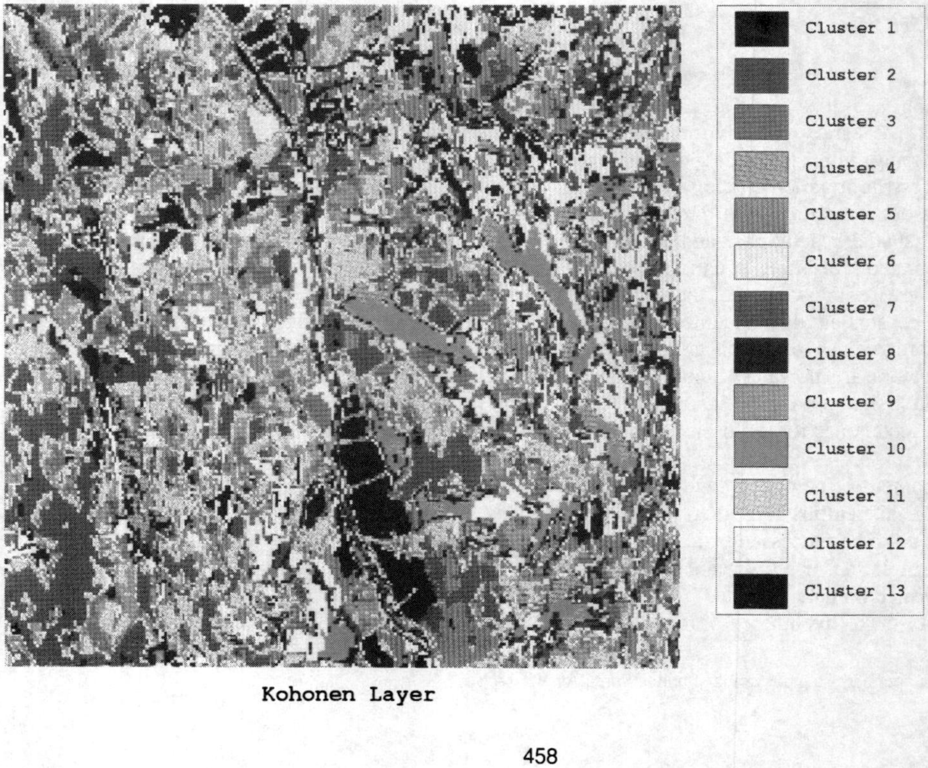

Kohonen Layer

der to co-register the two images, SAR data has been first resampled, by averaging, and successively coregistered by means of a 5th order polynomial, leading to residuals of 0.284 and 0.330 in the x and y directions respectively.

2.2 The Counterpropagation network

The network consists of three layers: the input layer which acts just as a fan-out, a hidden layer trained by means of an unsupervised procedure, and the output layer, called the Grossberg layer, trained by the supervised Widrow-Horf rule. Figure 1 illustrates the general topology of the CP network.

The input of the network is given by the multidimensional vector associated to each pixel of the image under consideration.

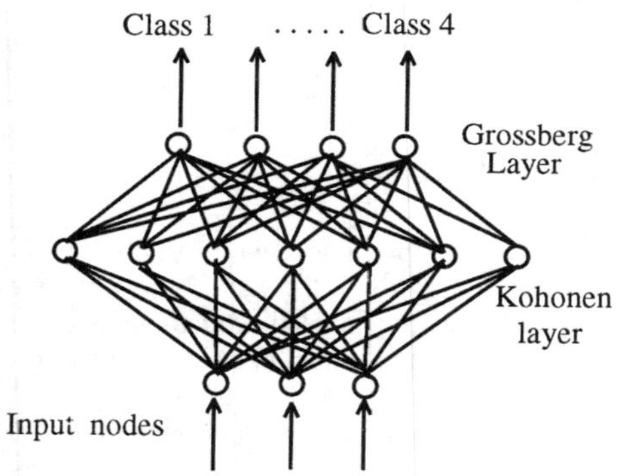

Fig. 1 The Counterpropagation network

The first layer of the network is trained by the competitive learning strategy, the metrics being the scalar product: at every pattern presentation only the unit which achieves the maximum value of the scalar product with the current input is updated, while the remaining ones are left unchanged.

At the end of this first phase, the pixels of the input image are subdivided into several clusters according to their spectral similarities, or equivalently, according to their position in the feature space. This preliminary result, especially when dealing with remotely sensed data, can be extremely interesting and useful: if a training set is not available, or not particularly representative of the area under consideration, the clustered image can be however considered as a classification, which can be refined in a second time either by a human expert or by a supervised algorithm.

If, as it is the case in the present study, a training set is available, the second layer of the network is trained in order to refine automatically the results obtained so far. The second (output) layer contains as many nodes as the classes to be discriminated on the scene, every node being connected to all the units of the Kohonen layer. Every pattern of the training set is presented to the network together with the class it belongs to.

At every step of the supervised procedure, a pattern of the training set is presented to the network, the unit which achieves the maximum scalar product is identified, and only the weights connecting this unit to the output layer are updated. More details can be found in (Wasserman 1989) and (Chiuderi 1994).

The main problem connected with the use of this type of network when dealing with remotely sensed data is given by the fact that in order to employ the scalar product as a measure of closeness, both the input vector and the weights vector must be normalized. In this way the scalar product is given by the cosine of the angle formed by the two vectors: the closer the vectors the higher the value of the scalar product.

Unfortunately, when remotely sensed data is concerned, the normalization might cause some loss of information, especially if the data are highly correlated. As a matter of fact the normalization identifies all the points which, in the feature space, fall on the same line passing through the origin, as shown in figure 2. If the patterns are highly correlated, the values which are fed into the network are, after the normalization, quite similar one to another, making the task of separating them even more difficult.

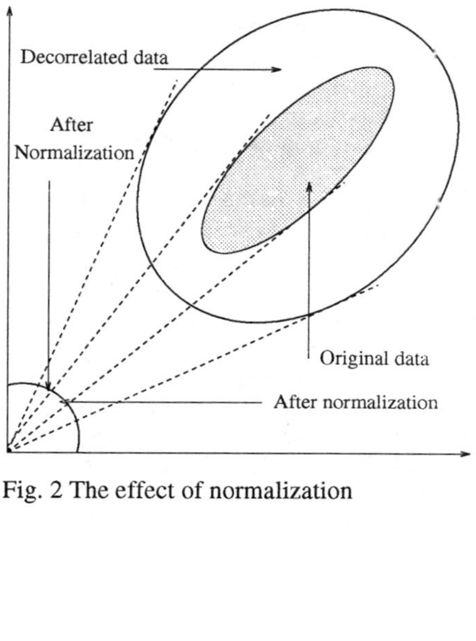

Fig. 2 The effect of normalization

In the first two experiments reported in this paper, we show that, when dealing with just one source of data, the introduction of a decorrelation procedure as a pre-processing step to the input data for the network is of great importance in order to improve the classification accuracies.

3. THE EXPERIMENTS

The results of three different experiments are reported and discussed: as a first case we applied the Counterpropagation network to the TMS image alone, successively the TMS data were first decorrelated by means of the procedure described in (Gillespie 1986) and then input to the network, while in the last case the whole data set available, TMS and SAR data, was employed.

For the first two experiments, the ones dealing with the TMS data alone, the input was given by the spectral values associated with each image pixel in the 6 bands.

In the SAR + TMS experiments 9 were the input channels employed: the 6 as in the previous cases plus the three polarizations of the L band provided by the microwave sensor.

In all three the experiments, the unsupervised layer was constituted by 18 neurons, while 6 were the nodes of the output layer corresponding to the following ground truth classes:

1. Wheat.
2. Forest
3. Alfalfa
4. Vineyards
5. Baresoil
6. Grassland

The unsupervised layer of the network was trained for 5000 on the whole image (256 x 256 pixels), while the supervised layer was trained for 2000 iteration over a training set constituted of 992 pixels. The results of the training were validated employing a test set composed of 236 pixels.

In tables 1 and 2 the results of the classification of the raw TMS and decorrelated TMS data respectively are reported. In the following tables Cli indicates the ground classes while NNi indicates the neural network output (i = 1, ..., 6).

The improvement introduced by the decorrelation of the TMS data is very impressive: not only all the classes are identified, but also the single per class classification accuracies are quite good, especially if compared with those obtained on the TMS raw data.

Tab 1: Classification accuracies - TMS raw data

	Cl1	Cl2	Cl3	Cl4	Cl5	Cl6
NN1	51	0	0	0	0	0
NN2	0	25	35	0	0	1
NN3	0	0	0	0	0	0
NN4	5	1	0	83	17	0
NN5	0	0	0	0	0	0
NN6	0	5	2	0	0	11

Tab 2: Classification accuracies - Decorrelated data

	Cl1	Cl2	Cl3	Cl4	Cl5	Cl6
NN1	51	0	0	0	0	0
NN2	0	26	0	8	1	3
NN3	0	4	30	0	0	1
NN4	2	1	0	71	0	0
NN5	3	0	0	4	16	0
NN6	0	0	7	0	0	8

As stated in the previous section this improvement is certainly due to the fact that the decorrelation stretching spreads the input patterns all over the feature space, making the loss of information introduced by normalization less dramatic. The results of the experiment employing both the optical-infrared and the microwave data are reported in table 3. In this case, as can be seen, the introduction of radar data, has diminished the overall correlation of the multidimensional data, allowing us to obtain some good results without the need of pre-processing the TMS data.

Moreover, for all classes except the vineyard, the classification accuracy is higher than the one obtained on the decorrelated data, and this highlights the importance of the use of multisensor data for

Tab 3: Classification accuracies - TMS + SAR data

	Cl1	Cl2	Cl3	Cl4	Cl5	Cl6
NN1	51	0	0	0	0	0
NN2	0	30	0	0	0	0
NN3	0	0	35	0	0	1
NN4	5	1	0	83	13	0
NN5	0	0	0	0	4	0
NN6	0	0	2	0	0	11

landcover classification purposes.

As far as vineyard is concerned, the errors in the classification by the neural network trained on both the SAR and TMS can also be thought as a more precise distinction between two different categories: the network might have learned, due to the presence of the microwave data, a stronger characterization of the two classes "vineyard" and "baresoil" classifying the pixels of the test set generally labeled "vineyard" in the most appropriate of the two classes.

4. CONCLUSIONS

In this paper the use of neural networks for remote sensed image classification was investigated. In particular the use of a combined unsupervised-supervised training procedure was presented.

The network was employed first on the data provided by a single sensor operating in the optical-infrared region of the electromagnetic spectrum.

The results obtained on the raw data were quite disappointing, due both to the high correlation of the data and on the nature of the training procedure. After pre-processing the input data by means of the decorrelation stretching algorithm, all the landcover classes were well identified and the classification accuracies obtained by the same kind of network were much higher.

Successively the same type of neural network, was trained on the merged TMS (as in the previous case) and SAR data set.

As a matter of fact, optical data provide infor mation on the reflectance of the observed area and allow to estimate the chemical properties of the target, infrared channels provide information on the surface temperature, while microwave data are sensitive both to the dielectric properties of the ground and to the soil moisture content providing some useful information for the estimation of surface roughness.

The results obtained in this last case are a confirmation of the fact that the integration of different types of sensors and the processing of the obtained data set as a whole is very important for landcover classification purposes.

Moreover it must be pointed out that when dealing with multisensor data the statistical algorithms, such as the maximum likelihood, should not be employed because the hypothesis on which they are based are not always satisfied and thus neural networks become, rather than an alternative to a more classical approach, a very useful tool for these kind of tasks.

ACKNOWLEDGMENTS

The authors would like to thank the Joint Researc-Center of the European Community situated in Ispra (Varese) Italy: most of the processing reported in the present paper were carried out during a stage of A. Chiuderi at the Environmental Modeling and Mapping Group Laboratories.

REFERENCES

Bechini, C., Chiarantini, L., Ciotti, P., Moretti, S., Pettinelli, E., Pierdicca, N., (1992): MAC91 on Montespertoli: Preliminary assessment of Land Polarimetric Features - Proc. IGARSS, Vol. I, pg. 395 - 397

Benediktsson, J. A., Swain, P. H., Ersoy, O. K. (1990): Neural Network approaches versus statistical methods in classification of multisource remote sensing data - *IEEE Transactions on Geoscience and Remote Sensing*, Vol 28, N. 4, pg. 540 - 552

Bishof, H. Schneider, W. and Pinz, A. J. (1992): Multispectral classification of Landsat-images using neural networks - *IEEE Transactions on Geoscience and Remote Sensing*, Vol 30, N. 3, pg. 482 - 490

Chen Z., Davis D. T., Tsang L., Hwang J. N., Chang A. T. C. (1992): Inversion of Snow Parameters by a Neural Network with Iterative Inversion - *Proc. IGARSS92* Houston, Vol. II, IEEE Press, pg. 1061 - 1063

Chiuderi, A. (1994): Reti Neurali: uno strumento per la fusione di dati telerilevati - Ph. D Thesis University of Florence (Italy)

Coppo, P., Ferrazzoli, P., Luzi, G., Paloscia, S., Schiavon, G., Susini, C.,(1992): MAC91 on Montespertoli: Preliminary analysis of multifrequency SAR sensitivity to Soil and Vegetation Parameters - *Proc. IGARSS'92*, Vol. I, pg. 489 - 491

Gillespie, A. R., Kahle, A. B., Walker, R. E. (1986): Color Enhancement of Highly Correlated Images. I. Decorrelation stretching and HSI Contrast Stretches - *Remote Sensing of Environment,* Vol. 20, pg. 209 - 235

Held, D. N., Brown, W. E., Freeman, A. Klein, J. D., Zebker, H., Sato, T., Miller, T., Nguyen, Q., Lou, Y. (1988): The NASA/JPL Multifrequency, Multipolarization Airborne SAR System - *Proc. IGARSS*'88, Vol. I, pg. 345 - 348

Kanellopoulos, I., Wilkinson, G. G., Megier, J. (1993): Integration of Neural Network and statistical Image Classification for Land Cover Classification - *Proc. IGARSS93*, Tokio, Vol. II, pg. 511 - 513

Rumelhart, D. E., Hinton, G. E., Williams, R. J. (1986): Learning internal representations by error propagation - In *Parallel Distributed Processing: Explorations of the Microstructure of Cognition*, J. L. McClelland & D. E. Rumelhart ed. MIT Press, pg. 318 - 362

Wasserman, P. D., (1989): *Neural Computing, Theory and Practice* - Van Nostrand Reinhold, New York

Zurk, L. M., Davis, D., Njoku, Tsang, L., Hwang, J. N. (1992): Inversion of Parameters for Semi-arid Regions by a Neural Network - *Proc. IGARSS92* Houston, Vol. II, IEEE Press, pg. 1075 - 1077

Sensors and Environmental Applications of Remote Sensing, Askne (ed.) © 1995 Balkema, Rotterdam, ISBN 90 5410 524 0

Comparison of neural network and maximum likelihood classifications in an urban area

Lena Halounová
Czech Technical University, Prague, Czech Republic

ABSTRACT: One branch of sanitary engineering - rainfall runoff from urban areas - seems to have gained great importance, and is attracting much attention. The input data set into rainfall runoff models is of various types. One type consists of data describing the surface of all drained areas from the point of view of their perviousness. Description of a surface was and still is performed either manually from maps or from aerial photos, with one exception when aerial scanner data was used in England. Six bands of Thematic Mapper satellite image from 16 August 1991 were processed. The study area in Prague was divided into 32 subcatchments. The boundaries of all subcatchments were digitized and the image window of the part of Prague was geocoded into them. Then all the roads and houses were digitized to get "ground truth data". Twelve classes of built-up areas were chosen (asphalt, industrial roofs, two types of family house area, old built-up area, four types of vegetation, two types of bare soil, and water). The training set was then applied in maximum likelihood and neural network classifications. The results of both classifications were compared to data obtained from digitization.

1 INTRODUCTION

Rainfall runoff modelling and the modelling of water quality in receiving waters are necessary for the study of sewer network systems. Both depend on a more or less exact description of surface types and their area size. The determination of both these surface characteristics has mostly been done by using maps, aerial photos or detection *in situ*. An aerial scanner data set was used in 1989 in England, which was the first attempt to computerize this quite time-consuming part of the preparation phase of work. In order to obtain an aerial scanner data set it is necessary, in the conditions of my country, to order the scanning abroad, and thus it is not an inexpensive matter. Digitization of maps is far cheaper, but is reliable only in the case of updated maps. The best way could be to use satellite images, thanks to their low cost, objective source of information and quite easy updating. The possibility of application of this kind of data has been investigated in the south-eastern part of Prague.

2. DATA AND METHODS

One Thematic Mapper scanner satellite image[1)] from 16 August 1991 was processed. A window 256 x 256 pixels was chosen, which covers the south eastern area of Prague.

A map in the scale of 1:5000 was used as ground truth data supplying the layout of all subcatchments. The area size of the subcatchments varied from 5.771 sq. m to 186.670 sq. m. The image window was geocoded into this map and resampled. The new pixel size was 5x5 sq. m. The work can be divided into the following steps:

1. choice of bands
2. enhancement of bands
3. choice of training sites
4. classification of enhanced bands
5. verification of results

[1)]Satellite data used courtesy of CSRC University of New Hampshire

2.1 Choice and enhancement of bands

All bands except the sixth were chosen for image processing. The normalized vegetation index (NDVI) was calculated according to the following formula, in order to make a better separation of the vegetation (pervious surface)

$$NDVI = (R_4 - R_3)/(R_4 + R_3)$$

where R_4, R_3 are reflectances in the third and fourth bands.

Principal component analysis (PCA) was applied as an additional source of data. The first three eigenchannels were chosen as input channels into Martin-Taylor Enhancement (MTE). The final set of bands consisted of the six original TM bands, NDVI and red, green, and blue bands of MTE.

2.2 Choice of training sites

According the map and knowledge of Prague, eight classes of built-up areas were chosen at first:
1. asphalt roads
2. industrial roofs
3. family house area type 1
4. family house area type 2
5. vegetation type 1
6. vegetation type 2
7. vegetation type 3
8. vegetation type 4.

During the verification step the number of classes was found to be insufficient and some extra classes were added:
9. old built-up area
10. bare soil type 1
11. bare soil type 2
12. water.

These classes were chosen for the needs of rainfall runoff models, even though that may not appear to be the case at first sight.

Pervious surfaces may include all types, with the exception of paved surfaces, asphalt roads and roofs. However, it was not possible to make them into a single category. Several types of pervious surface were chosen and the sum of them will be the total pervious surface.

The family house classes represent areas where various types of surface can occur: roofs, pavements, and gardens with a pervious surface. The final result of area classes encompasses all the information. The solution to the problem was undertaken as the next step after classification, when the sample areas were explored and the real parts of the 'subclasses' were determined: what part of the class consists of gardens, roofs, and other impervious surfaces?

Table 1

class	total area sq m	area of roofs	% of total area	area of roads	% of total area
family houses 1	6350	490	8	640	10
family houses 2	10725	450	4	3100	29
old built-up area	16250	3600	22	2253	14

The layout of the chosen class 'old built-up area' is in Fig. 3, together with digitized roofs and roads.

2.3 Classification of Enhanced Bands

Two methods of classification were applied - maximum likelihood and neural network classification. Neural networks are computational systems, either hardware or software, which mimic the computational abilities of biological systems by using simple, interconnected artificial neurons. A neuron is the fundamental building block of the nervous system. Artificial neurons are simple emulations of biological neurons; they take input information, perform very simple operations on this data, and selectively pass the results on to other artificial neurons. Neural network classifiers process imagery by pattern recognition. The programs use a back-propagation network that is probably the most well-known and widely used of neural network systems. The term 'back-propagation' refers to the training method by which the connection weights of the network are adjusted. This method was used together with the classic maximum likelihood method in order to show its possibilities as a new, modern classification method in remote sensing.

The following group of programs was used to perform the classification using back-propagation neural network processing: NNCREAT which creates a back-propagation neural network segment given the set of the eight channels (in this case) and training site bitmaps (the same bitmaps for the maximum likelihood classification), NNTRAIN which trains a back-propagation neural network segment created by NNCREAT, and NNCLASS which performs classification using a back-propagation neural network segment trained by NNTRAIN. One input unit was used in NNCREAT and 1 and 2 hidden layers were tested. The result with 1 hidden layer showed to be better. Momentum rate (MOMEN) was 0.9, learning rate (LEARN) was 0.01, maximum normalized total error (MAXTERR) was set up 0.03, maximum individual error (MAXIERR) 0.003, and maximum number of iterations (MAXIT) 1000 iterations.

3. RESULTS AND DISCUSSION

The results of both classifications were compared and are shown in Fig 1. and 2. The percentage of areas of roofs and roads in each subcatchment were evaluated in neural network results, in maximum likelihood results and digitized data ("ground truth data"). The differences between both classifications and digitized data were obtained by subtracting the percentage of digitized data for roofs and roads in all subcatchments from percentage evaluated by both classifications. The review of differences is in Table 2.

Table 2. Differences of maximum likelihood and neural network classifications in comparison with digitized data

class		<11%	<30%	<50%	>50%
roofs	neur.n	30	1	1	0
	max.l.	18	10	4	0
roads	neur.n.	18	10	3	1
	max.l	21	9	2	0

The results of neural network classification are undoubtedly better than the results of maximum likelihood classification especially in "roofs"class.

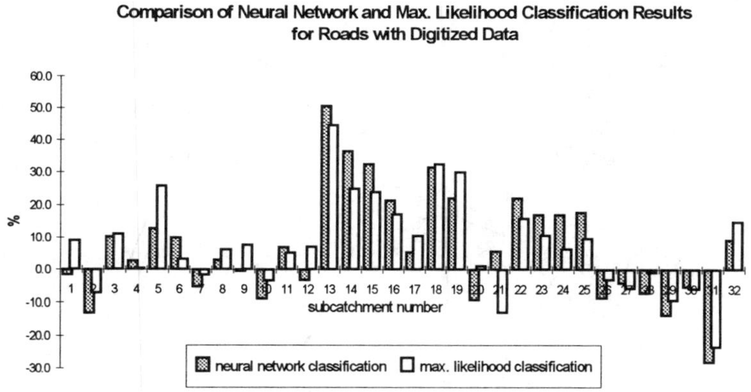

Fig. 1

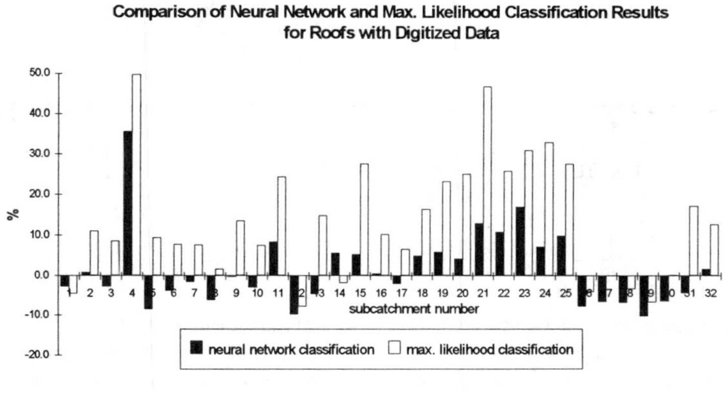

Fig. 2

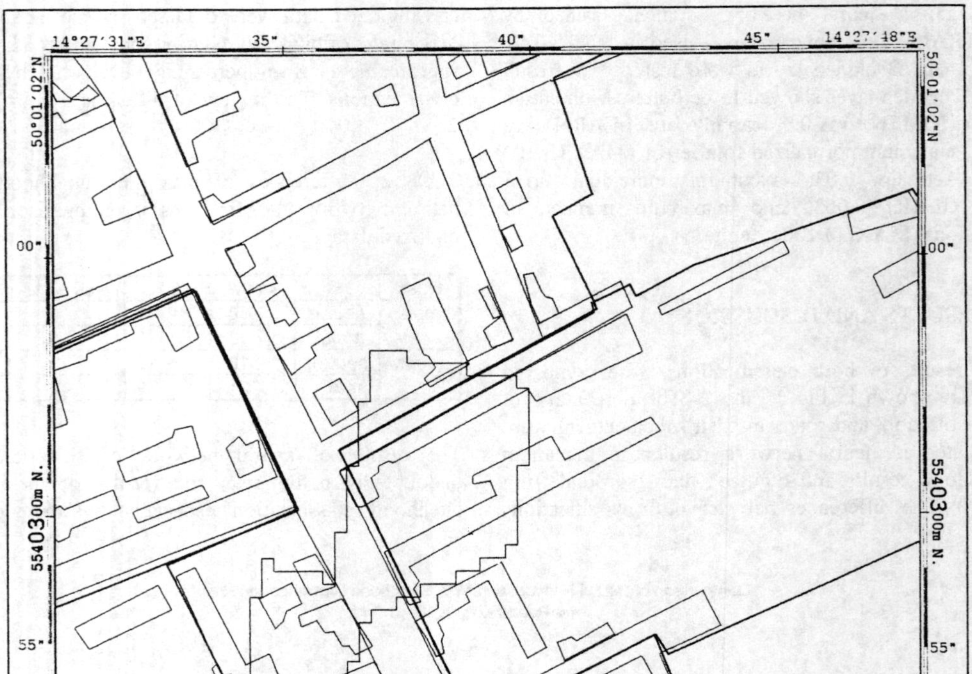

CHOICE OF DENSELY BUILT-UP AREA CLASS

subcatchments, roads, houses

Fig. 3

The places with the biggest differences between this method and digitized data are industrial areas (subcatchments 13 - 19, 22, 23, 25). The actual state of roads (paved areas) is not correctly mapped and the classification shows a more real situation in these subcatchments than the map does. Subcatchment 24 is a new spacious residential area with with separate blocks of houses connected by asphalt paths which

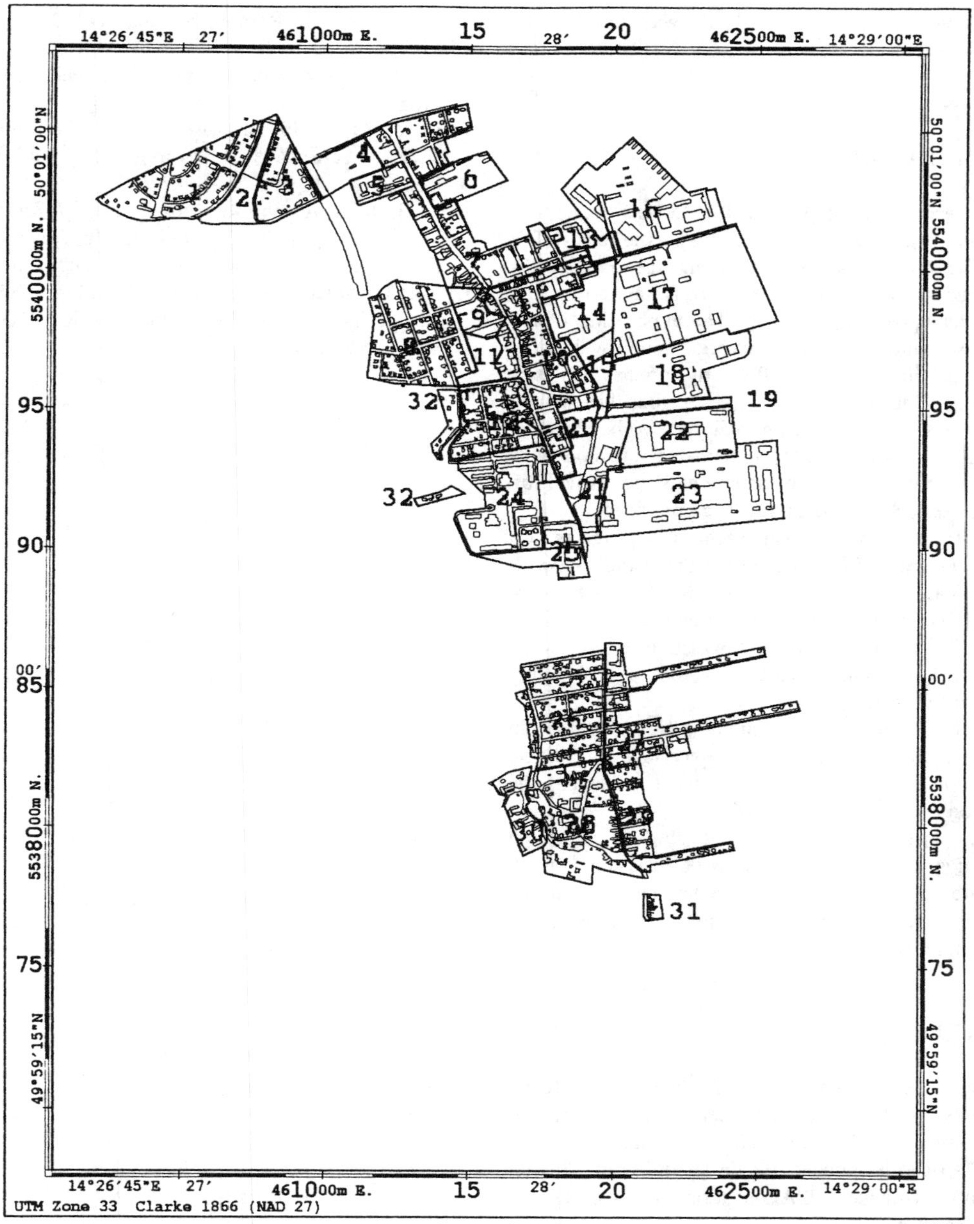

Map of study area

South-eastern part of Prague - Libus, Pisnice

Fig.4

are not on the map which was used. The classifications shows reality better than the map does.

4. CONCLUSIONS

The precision of the proposed method is better than 90 per cent. Taking into account the cost of a satellite image and of processing the image, this method is cheaper than the use of aerial photos, and is better up-dated than maps.The application of the data set into a sewer network model will show whether the data is sufficiently precise for this purpose. The final decision on applying this method to sewer network modelling will have to be based on comparison with results of mathematical models of sewer networks and measurements performed on these networks *in situ* and their comparison..

Thorough *in situ* measurement is in progress in a different area of Prague, where the neural network classification will be used for surface data determination into the sewer network model of the area. The results will then show the effectiveness of the proposed remote sensing method for obtaining data of this type.

Careful choice of the training set is necessary in order to gain a relatively good overview of possible built-up urban area types. If this is achieved, the method can be recommended.

The paper is only the first step in thorough investigation of neural network classification in remote sensing. This first step has shown that the neural network classification is better than the classic maximum likelihood.

REFERENCES:

Bernier, M., Bonn, F., and Gagnon, P. (eds.) 1989. *Télédétection et Gestion des Resources*. Volume VI: Compte rendu du sixieme congres de l'Association québecoise de télédetection, Sainte-Foy, Québec: L'Association québecoise de télédetection. 388p.

Benediktsson, J.A., Swain, P.H., and Ersoy, O.K. 1990. *Neural Network Approaches Versus Statistical Methods in Classification of Multisource Remote Sensing Data*. IEEE Transactions on Geoscience and Remote Sensing, Vol. 28, No. 4, July 1990.

Bischof, H., Schneider, W., and Pinz, A.J. 1992. *Multispectral Classification of Landsat-Images Using Neural Networks*. IEEE Transactions on Geoscience and Remote Sensing, Vol. 30, No. 3. May 1992.

Kolář, J. 1990. *Remote Sensing* (in Czech). SNTL, Prague.

Pao, Y. H. 1989.*Adaptive Pattern Recognition Neural Networks*. Addison-Wesley Publishing Company.

Finch, J., Reid, A., and Roberts, G. *Use of Remotely Sensed Imagery in Determining the Impervious Cover in Urban Catchments for Input to the WASSP Model* , Institute of Hydrology, NERC Computing Services, Wallingford, OXON OX10 8BB, UK, 1989.

Sensors and Environmental Applications of Remote Sensing, Askne (ed.) © 1995 Balkema, Rotterdam, ISBN 90 5410 524 0

Scaling in the fractal spatial structure of environment and implementation for imagery interpretation

Leonid Vasiliev
Institute of Geography, Russian Academy of Sciences, Moscow, Russia

ABSTRACT: The goal of this work is to better understand the scaling property of fractal spatial structure of geosystems. Emphasis is placed on the determination of the limit intervals of self-similarity and self-affinity of 2D structures and 3D radiation fields. The frequency distribution of sizes must be hyperbolic with an exponent equal to the fractal dimension. Hence, it is possible to determine interpolation and extrapolation interval to low and high scale levels and their limits. Using MSS imagery from "COSMOS 1939" MSU-E with 40 m spatial resolution it has been demonstrated that the urban green areas in Moscow form a self-similar fractal structure with fractal dimension $D = 1.86 \pm 0.02$ and that vegetation areas obey hyperbolic distribution. Using the hyperbolic distribution function a number of vegetation areas has been extrapolated from a pixel size to 1/16 of its size (2.5 x 2.5 m) on a scaling range of self-similarity. The self-affine field of spectral radiance in Moscow is closely related to small areas of the greening with crossover scale 2 km.From practical point of view the study shows that the use of fractal geometry for mathematical description of natural and man-made spatial structures may be useful for measurements on mixed- and sub-pixel levels. There are some difficulties and limitations in a geographical interpretation of scaling interval if relevant geosystems' models do not exist.

1 INTRODUCTION

In the technogenic urban geosystem vegetation strongly influences the climate and improves sanitary and hygienic conditions in the city (Landsberg 1981). The International Geographical Union Study Group on Urban and Local Climatology showed that the radiation and thermal conditions as well as the air pollution bioclimatic complex are the three principal problems to investigate in urban climatology (Paszynski 1986). The urban microclimate does not depend much on large territorial units, of greater importance are small housing developments, various types of surface. Within surrounding streets and buildings vegetation may be one of the main climate-generating factors comparable in its role to urban park areas. As the role of vegetation in the horizontal heat exchange is not so essential, special attention is given to vegetation structures surrounded by buildings and in the streets - from the viewpoint of microclimatic differentiation (Katzschner 1986).

Vegetation is also an important factor in the formation of a thermal field structure from the viewpoint of spatial distribution of gaseous and dust substances in the atmosphere. Numerous investigations made in large cities proved that there indeed exists a relationship between the thermal field and atmospheric pollution.

The estimation of the spatial structure of trees and shrub vegetation in the city implies the assessment of the area they cover, the areals' distribution and the relationship between the biomass and population. Topoclimatologic simulation needs information about the species composition of trees and shrubs and the bulk of their biomass. Of course, it is extremely difficult to make estimates of the interaction of numerous factors - as to how they affect population in a city which is a heat island. And it is for this reason that the problems of architecture and planning of the urban development should rely on actual ecological conditions rather than on norms and standards.Hence the assessment of the pattern of vegetation covering about half the territory of any city is the attribute sine-qua-non in the urban geoinformation system.

The spatial structure of a geosystem is most often related to the nature of the land use / land cover. It may be described as follows: $\bigcup_{i=1}^{n} C_i$ is the union of all classes of land, $\sum_{i=1}^{n} S_i$ - is the total area comprising the areas occupied by each class, $\Pr(S)$ is the size distribution within each class and $A_r(C_i, C_j)$ is the adjacency matrix for elements of the classes C_i and C_j, where r denotes the relationship between C_i and C_j. The elements of a spatial structure are most accurately identified on multiband imagery. The determination of the size distribution $\Pr(S)$ and size-interval depend on imagery spatial resolution. In practice there is always some discrepancy in practice between the spatial resolution and the geometrical accuracy. Recently, interest in the sub-pixel support has re-arisen through the fractals in spatial structures of environment.

Fundamentally, however, this problem appears to be nontrivial. Indeed, what should be considered the boundary and the area of a space covered with trees among buildings or around a single house? The area and boundaries could be related with the percentage cover of vegetation and the smaller the measurement unit the more detailed indentedness would be identified on images. The difficulty with mapping, bringing about some uncertainty, distinctly manifests itself on 1:2000 topographic plans of Moscow, where the boundaries of vegetated areas are somewhat conventional whereas in a number of cases plantings along the streets, individual trees, allies are shown by symbols without observing the scale. That is why even large-scale maps could not help measuring exactly the overall area under vegetation in the city. Thus green areas in the city form a spatial structure of "islands" of very different size, and with wriggling boundaries that suggests its possible fractality (Vasiliev, Kachalin, Tyuflin 1994).

We would like to note in passing that the fractal character of an urban structure component, e.g. subway network in Paris or growth pattern of certain British cities according to the model of diffusion limited aggregation, implies that anthropogeneous activity has, though not necessarily deliberately, a fractal character (Benguigui, Daoud 1991; Batty 1991).

Most complete and reliable mapping of urban vegetation can be made using air or space images but in that case their spatial resolution would affect the accuracy of area estimates. The present study shows that fractality of the spatial structure of vegetation planting in Moscow permitted the area they cover to be estimated from satellite imagery to an accuracy of fractions of a pixel.

2 EXPERIMENT: THEORETICAL CONCEPT

Let us cover a set formed by vegetation areals by a grid of squares with an r-side and denote as $N(r)$ the number of cells that overlap with this set. If a square with a different side is assumed, r_1, we get a new number of cells $N(r_1)$. At least one picture element classified as vegetation would fall into a cell. In case this ensemble is fractal the $N(r)$ dependence behavior would be described by:

$$N(r) \propto r^{-D_G} \tag{1}$$

where D_G is the grid dimension.

The mass dimension D_M may be a practical alternative to the calculation of a fractal dimension and is given by

$$M(L) \propto L^{D_M} \tag{2}$$

where $M(L)$ is a number of pixels in a cell of the L-size.

When natural fractals are studied, of greatest interest is the extrapolation of asymptotic behavior, described by Eqs. (1) or (2). Adequate values of $N(r)$ or $M(L)$ may be derived using air or space imagery. However, the spatial resolution typical of a chosen remote sensing system would limit the lower value of the scaling range and, therefore, the accuracy with which the area $N(r)r^2$ covered has been estimated. Of course, an attempt can be made to solve this problem with the help of large-scale photos which would require processing about 250 images of 1:10000 scale.

To interpret green areas within the Moscow circular high-way region, space photos were used from the COSMOS 1939 MSU-E scanner (Fig. 1) as well as a SPOT system image. To get a classified image for the analysis of the spatial pattern of vegetated areas, MSU-E images taken in three spectral bands were used, their resolution being 40 x 40 m. Clusters of trees and shrubs within a 3D spectral signature space are outlined with the help of reference plots in Forest and Park Belts and urban parks. To estimate the accuracy of classification from summer space photos the measurement data for the areas covered with these vegetation species were compared with the data derived from 1:2000 maps for various parts of Moscow. The reliability of ground-truth measurements was verified by field observations. The plots with full percentage cover yielded a relative error of about 2 percent. Naturally if we are to estimate the classification accuracy, compact vegetation plots should be selected so as to avoid as much as possible the inclusion of mixed picture elements (mixels). It is such mixels that restrict the accuracy of area measurements, thus forcing specialists to find how to overcome this limit by means of fractal geometry.

Fir. 2 shows the results of grid dimension calculation from Eq. (1) in log-log scale. Hence it

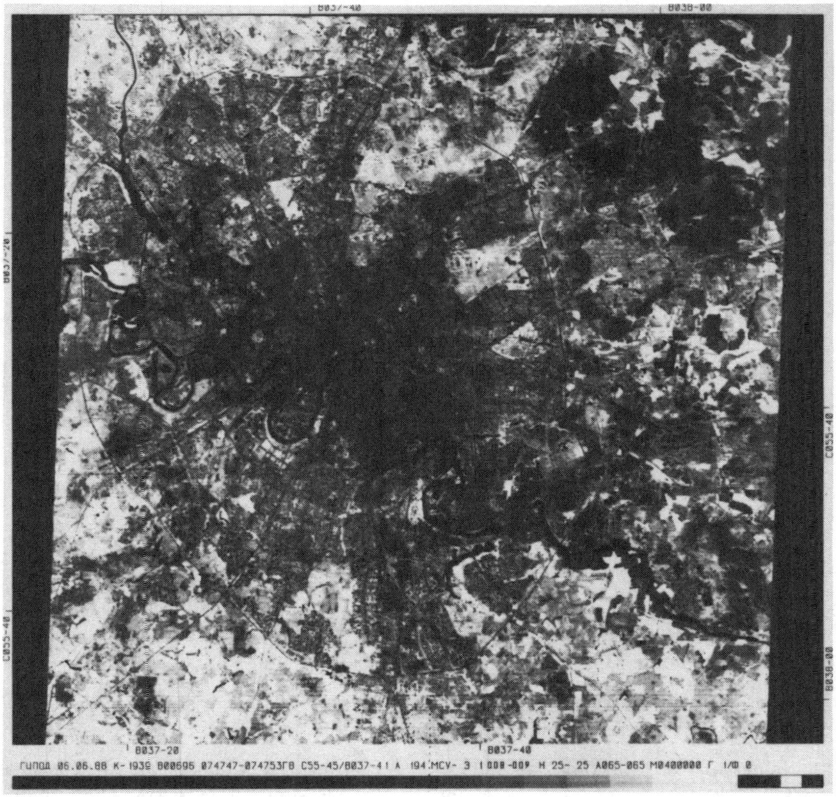

Figure 1 Cosmos 1939 MSU-E scanner image of Moscow, located on the Moskva River, Moscow Circular Highway. Shades of white denote vegetation. June 6, 1988. Spectral band 0.8-0.9 μm

follows that green areals in Moscow have fractal structure with $D = 1.86 \pm 0.02$. Note that the estimation of fractal dimensionality made separately for the southern and the northern half of the city (partitioned through the center of Moscow), does not lead to more than 0.03 differences in D. The estimate of the scaling range for vegetated areas depends on the results of their classification from space photos with limited spatial resolution. Assume that trees and/or shrubs falling into a pixel form there different projective covers. Obviously only a fully covered pixel enters a given class of vegetation. If the projective cover is 0.5 the pixel may be referred to a given class only with a 0.5 probability. Because of this, while counting the grid dimension D using $N(r)$ values $r_{\min} = 1$, all pixels with the coverage less than 0.5 will not be taken into account. Therefore $N(r=1)$ would evidently be less than its actual value. Similarly reduction of $N(r=2)$, $N(r=3)$ should be expected as is seen in Fig. 2. Hence it follows that the estimation of vegetated areas in the city from space imagery with a spatial resolution of 20 to 40 m would not be correct.

Let us emphasize again that the computed D_G and D_M fractal dimensions and parameters of the irreversible growth model only determine scaling properties and self-similarity interval $[L_{\min}, L_{\max}]$ or $[r_{\min}, r_{\max}]$. It is absolutely obvious that in case the imagery is used, the lower limit of the self-similarity interval, i.e. $\inf r_{\min}$ depends on the pixel size whereas $\sup r_{\max}$ is governed by the area of fractal homogeneity. Theoretically self-similar fractal sets, however, allow infinite extrapolation and seem attractive since extrapolation of the asymptotic behaviour of Eqs. $N(r) \propto r^{-D_G}$ or $M(L) \propto L^{D_M}$

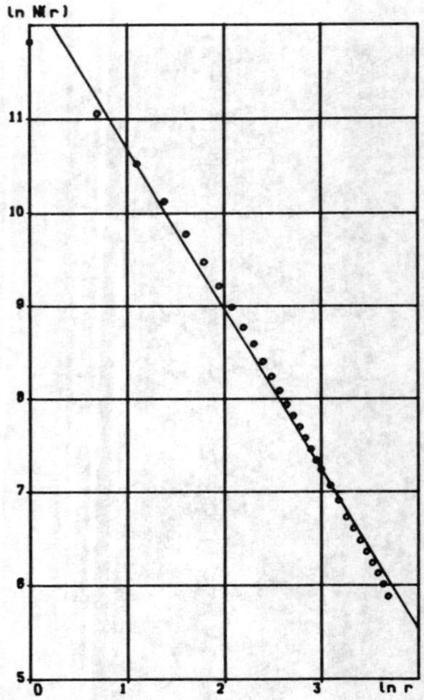

Figure 2 The greening of Moscow. The log-log plot of number of boxes ln *N(r)* versus side length of square boxes ln*(r)*. (Cosmos 1939 MSU-E). Grid dimension D = 1.86 ± 0.01

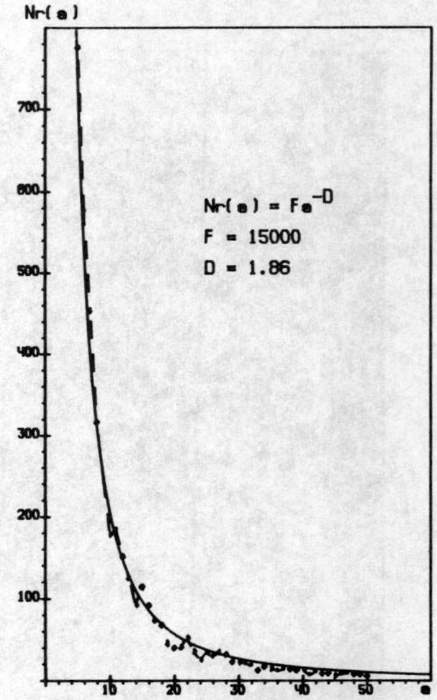

Figure 3 Comparision between theoretical and experimental results of the hyperbolic distribution of the greening of Moscow

toward smaller values of r or L becomes possible as well as the curve straightening in the log-log plot and calculations of the limit $D = \lim_{r \to 0} \ln N(r) / \ln r$. The arguments rely on the fact that the power law relationships for a self-similar fractal set is valid at all scales (Mandelbrot 1983; Vasiliev, Tyuflin 1991; Bittner, Sernetz 1991).

Self-similar fractal structures are most closely related with the hyperbolic distribution

$$P(a) = P(A > a) = Fa^{-D} \tag{3}$$

where an exponent is equal to the fractal dimension D, and a is the "green island" area of elements making a fractal set. If we are going to be careful in setting $a_{\min}$, then $\mathrm{Nr}(A > a)$ will be finite whereas the integral taken of $P(a) = Fa^{-D}$ determines the area covered.

3 RESULT OF EXPERIMENT

Relying upon this idea, measurement data acquired from the imagery had been extrapolated beyond the spatial resolution specified by the pixel size. The experiment was conducted to determine the pattern of vegetation in Moscow from multiband photos made with the COSMOS-1939 MSU-E system whose spatial resolution was 40 m.

The log-log plot (Fig. 2) shows the determination of the fractal behaviour of urban green areas in Moscow leading to the fractal dimension $D = 1.86 \pm 0.02$. Note that the grid and mass dimensions do not differ by more than 0.02 whereas the self-similarity range derived from the Equation $M(L) \propto L^D$ is 5 to 127 pixels or 210 to 5000 sq.m. The $M(L)$ values for $L < 5$ pixels, drop from the line since they are strongly affected by mixed pixels which cause errors in pixel classification. Extrapolation to small scale of green islands is based on the hyperbolic distribution $P(a) = P(A > a) = Fa^{-D}$, where the fitting constant F is derived from the a - distribution over the interpreted image

$$F = \sum_{i=1}^{n} a_i^{-D} N_i(a_i) \Big/ \sum_{i=1}^{n} \left(a_i^{-D}\right)^2 .$$

The result of extrapolation to a = 1/16 of a pixel size, which corresponds to the physical interpretation

of one tree crown, with the size of the square 2.5 x 2.5 m is shown in Fig. 3.
The probability distribution calculated using Eq. (3) and the experimental data illustrated by Fig. 3 demonstrate their fairly good agreement which permits moving further, to the goal of our study, that is, to the determination of the area of vegetation by integrating Eq. (3). Obviously the amount of vegetated "boxes" $\mathrm{Nr}(A > a)$ for areals covering pixels corresponding to full projective cover, may be derived from a classified image. When hyperbolic distribution parameters are known the $\mathrm{Nr}(A > a)$ estimate is extrapolated to the preset limit $a_{\min}$ which amount to only fractions of a resolution cell. The minimal fraction of pixels $a_{\min}$ = 1/16 pixel corresponds , in physical interpretation, to the crown of a single tree, 2.5 x 2.5 m - size; the equation $Nr(A > a) = 15000a^{-1.86}$ yields the following estimates

Total area, percent	45.5
Number of green areals	2, 829, 200
Tree plantation area and parks, percent	11.2
Vegetation among buildings, squares, alleys, percent	34.3

It is interesting that among the blocks of houses vegetated areals covering up to 40 x 40 m amount to 91 percent.

According to the official data in 1969 among the 87,500 ha of Moscow within the Moscow Circular Highway vegetated areas covered 28,300 ha, or 32,3 percent. The annual average increment of this vegetation used to be 600 ha from 1958 to 1968, the same figure was envisaged for the years to come. This implied that by 1991 the total vegetated area in Moscow would grow up to 40,300 ha and cover 46 percent of the entire territory of the city. The results summarized in Table show good agreement with the scheduled planting though the area of park-forest belt and city parks decreased from 11.9 to 11.2 percent. Note that Master Plans of urban development usually suggest that about 50 to 65 percent of the overall residential territory should be allocated to green areas. Thus, the way the green areas are arranged in Moscow is characterized by fractal properties with the scaling ratio $M(\lambda L) = \lambda^{-D} M(L)$. It means that the structure of a sample area with a larger step proportional to λ, is statistically self-similar to the entire area with the coefficient $\lambda^{-1.86}$, that is, for instance, the spatial structure of greening in a city or its part is, on the average, self-similar to the situation in an individual block of houses.

To conclude this part of the experiment, we will see what practical results are achieved by stating that vegetated areas in Moscow have fractal properties.

1. A property of self-similarity meaning that the organization of a spatial structure of vegetation in a certain part of the city is similar to that of the entire city. This implies the scaling of vegetation distribution for all kinds of territorial units.

2. The fractal structure unequivocally suggests the hyperbolic distribution of areals and thus well substantiates the rule of extrapolation to small dimensions within a self-similarity scaling range beyond the limits of the imagery's spatial resolution to be used. Owing to this, the estimate of the vegetated area involves the areals from 4 to 1/16 pixel in size whereas the relative error of area determination decreases from 7 to 1 percent.

4 FRACTALS IN URBAN RADIATION FIELD

Naturally, green areas of Moscow amounting to about half of its territory considerably affect the structure of surface temperature and albedo fields. To better understand the interrelationship of convective heat flows at micro- and macroscopic levels, it is useful to determine the properties of the 3D - radiation field. Radiance (or albedo) fields were analyzed using the same space images of

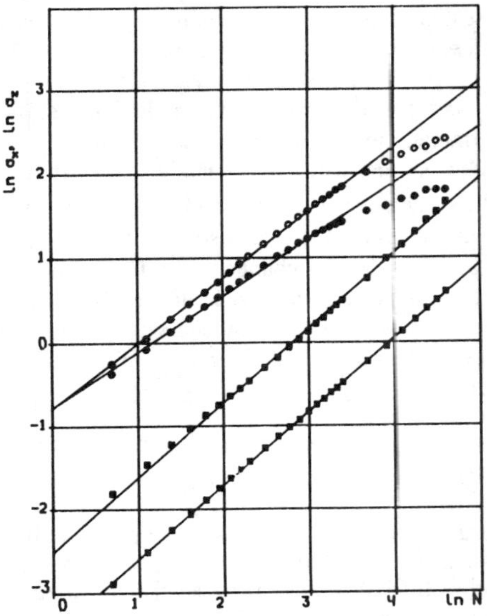

Figure 4 Transect profile Moscow. St.d. σ_x , σ_z of x, z coordinates vs curve length N in sections of the 3D spectral albedo.

Channel 2 o - $v_z = 0.775$, ⊗ - $v_x = 0.883$
Channel 3 • - $v_z = 0.666$, ⊙ - $v_x = 0.892$

the MSU-E in its 2nd and 3rd channels with 0.6 to 0.7 and 0.8 to 0.9 µm - bands, over several transect profiles. It is demonstrated in (Vasiliev 1993) that remotely sensed geophysical fields are self-affine fractals. For a self-affine fractal curve which is an albedo field cross-section, the relationship between the length of the curve within the Na_0 section and the standard deviations σ_x, σ_z of a_0 unit length projections onto x and z axes is described by Eq. (4) (Matsushita, Ouchi 1989) :

$$\sigma_x \propto N^{\nu_x},\ \sigma_z \propto N^{\nu_z},\ \text{and}\ H = \nu_z / \nu_x \qquad (4)$$

Since the albedo field of Moscow is governed by the fine structure of buildings, roads and green areas, the unit length a_0 smaller than the crossover x_c $(a_0 < x_c)$ should be used for measuring the lengths of fractal curves. In this case the fine structure of the albedo profile is determined by the fractal dimension $D = 1/H$.

It follows from the log-log plots that $z(x)$ albedo profile scaling over x,z axes is made with different coefficients ν_x, ν_z (Fig. 4). Eqs. (4) are used to derive local fractal dimensions of profiles $D = 1.13$ in the spectral band 0.6 to 0.7 µm and $D = 1.36$ in the spectral band 0.8 to 0.9 µm. From $\sigma_z \propto N^{\nu_z}$ dependences over straightened portions of a curve the scaling interval is determined - which is the length equal to 40 to 50 - pixels, that is 1600 to 2000 m.

It was already proved that radiation fields are multifractal, hence, if the spatial resolution,or the value of the unit length a_0 is made larger, up to 160 m, then fractal dimensions would be similar to $D = 1+H$ and characterize larger structures. Hence, D is 1.54 and 1.36 on the channels 2 and 3, respectively, whereas the scaling interval is 480 to 6400 m. The result obtained shows convincingly how scaling properties of fractal fields change for different spatial resolution.

5 CONCLUSION

A large city is a combination of fine-structure elements which determine the scaling properties of radiation fields as well as the spatial distribution of gaseous impurities in the near-surface layer. Following the commonly accepted view that the thermal field and impurity concentrations are interrelated an approach may be developed for determining their spatial distribution in the city. Relying upon the identified fractal properties of the radiation field an assumption can be made that the spatial structure of the air pollution field would also be fractal. Thus interpolation and extrapolation intervals would be the basic relationship between various ground measurements and remotely sensed signatures. Thus estimated properties of the spatial distribution of green areas in Moscow are to be used in the Geoinformation System for planning and for topoclimatological modelling.

REFERENCES

Batty, M. 1991. Sities as fractals: simulating growth and form. In A.Crilly, R.Earnshaw, H.Jones (eds), *Fractals and chaos*: 43-69. N. Y.: Springer.

Benguigui, L. & M.Daoud 1991. Is the suburban railway system a fractal ? *Geographycal Analysis.* 23: 362-268.

Bittner, H.R. & M.Sernetz 1991. Selfsimilarity within limits: description with the log-logistic function.*fractals in the fundamental and applied sciences.* In H.-O.Peitgen, J.Henriques & L.Penedo (eds), *Fractals in the fundamental and applied sciences:* 47-58. North-Holland: Elsevier.

Katzschner, L. 1986. The urban structure and its effect on climate. *Int. Symp. on Urban and Local Climatology. Freiburg (Br.), Feb. 20-21 1986. 26: 42-56.* Freiburg: Selbstverlag des Institutes für Physische Geography der Universität Freiburg.

Landsberg, H.E. 1981. *The Urban Climate.* N.Y.: Academic Press.

Mandelbrot, B.B. 1982. *The Fractal Geometry of Nature.* San Francisco: W. H. Freeman.

Matsushita, M. & S.Ouchi 1989. On the self-affinity of various curves. *Physica D*, 38: 246-251.

Paszynski, J. 1986. Topoclimatology - Progress. Methods. Results. *Int. Symp. on Urban and Local Climatology. Freiburg (Br.), Feb. 20-21 1986. 26: 7-10.* Freiburg: Selbstverlag des Institutes für Physische Geography der Universität Freiburg.

Vasiliev, L.N. & A.S.Tyuflin 1992. The fractal nature of the spatial structure of geosystem. *Sov. J. Remote Sensing* 9: 630-642.

Vasiliev, L.N. 1993. Multifractal nature of geophysical fields measured by remote sensing techniques. *Proc. 12th Symposium EARSeL* : 279-287. Rotterdam: Balkema.

Vasiliev, L.N., A.B.Kachalin & A.S.Tyuflin 1994. The spatial structure of greening in Moscow and estimation by using satellite imagery. *Earth Research from Space* 1: 86-92.

Sensors and Environmental Applications of Remote Sensing, Askne (ed.) © 1995 Balkema, Rotterdam, ISBN 90 5410 524 0

The temporal observation of natural phenomenon – Preliminary report

Bruce Bullen & Robin Vaughan
University of Dundee, UK
Gary Wright & David Miller
Macaulay Land Use Research Institute, Aberdeen, Scotland

Abstract: The progression of photogrammetry from photographic and map products to digital imagery and vector products has enabled the integration of remote sensing and aerial photography. We report upon work in progress that looks at the issues involved in the development and validation of a methodology to measure coastal erosion via the high resolution capabilities of ERS-1 AMI, SPOT HRV and Landsat TM in conjunction with aerial photography.

1 Objective

The original intent of the study was to perform a thorough multi-sensor evaluation of the coastal zone, with emphasis on the development of methodologies for coastal mapping. The mapping definition included delineation of coastal geomorphology and image classification. The inherent difficulty in working within the zone is that it is not well resolved by the traditional space-borne sensors. In addition, the structures of primary interest in coastal geomorphology are composed of beach material, they are difficult to identify from a distance. Consequently, the focus of the study was shifted to a measurement of coastal erosion, an issue of concern to the proponents.

1.1 Study Area

The study area is the North Head of Richard's Island in the Mackenzie Delta region of the Canadian Beaufort Sea (figure 1). The Beau-fort Sea lowland region is a thermokarst terrain extensively underlain by permafrost. Ground ice is a major component of the permafrost occurring as pore ice, layers and

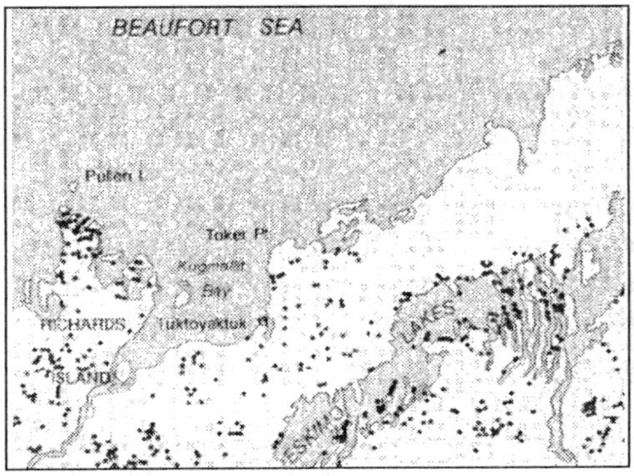

Figure 1 The study area is located to the south of Pullen Island on the north head of Richard's Island (Harry 1989).

lenses of segregated and intrusive ice, and ice wedges (Harry 1988). The surficial geology
of the study area is discontinuous hummocky and rolling moraine mixed with exposures of massive deltaic sands overlain by lacustrine sediments, (Rampton 1988). The topographic relief, 30 metres above sea level, is believed to be due to excess ground ice (French 1974). The topographic variation is the result of active geomorphic processes of retrogres-

sive thaw failure and permafrost aggradation.

The vegetative overburden is a tussock tundra vegetative community, dominated by sedge tussocks (Mackay 1963). The other constituents of the community are primarily mosses and lichens (Hope 1993).

2 Coastal Erosion

Coastal retreat results from retrogressive thaw failures along the cliff face which are removed by wave action (figure 2). The wave action in this region is limited by shore fast ice remaining at the foot of the cliff well into the summer season and by the damping effect of offshore ice on wave magnitude. Significant erosional events are limited to a frequency of one to two storms per annum (French 1976).

A storm event accelerates the frequency of retrogressive thaw failures along the cliff face by removing the slumped material from the base of the cliff and by cutting into the cliff base at the high tide mark. This process accelerates thaw slumping and may induce a collapse of the cliff face. The net effect is that the process of erosion is accelerated at the land water interface such that the annual rate of erosion is on the order of 1 metre per annum.

2.1 Study Description

This study assesses the utility of remote sensing to observe coastal erosion in the Beaufort Sea lowlands. As the land / water interface is a narrow linear feature, special considerations are required to locate the interface within the image data sets. One must account for the effective field of view of the various sensors and devise strategies to optimize the spatial resolution that might be applied to the measurements.

The measurement of erosion by the space - borne sensors will be compared with a photogrammetrically derived measurement using aerial photography spanning the period 1947 to 1985. The amount of erosion is derived by subtracting the coordinates comprising the 1947 coastline from the coastline coordinates observed by each sensor. The satellite measurements will then be compared to the photogrammetric results.

Figure 2 An illustration of wave action in the Beaufort Sea lowlands from Harry, (1989).

The study will conclude with an analysis of the difficulties inherent in the application of a space borne approach to a terrestrial photogrammetric problem.

3 Photogrammetric Aspects

Work to date has concentrated upon photogrammetric issues inherent when working with photographic products acquired in 1947. Lens distortion, camera calibration, and the resolving power of the photographic system differ somewhat from modern day photogrammetric systems (figure 3). The most problematic aspect however, has been the lack of signalised ground control points.

Given the remoteness of the study area, we have opted to base the absolute orientation of the 1947 model on air triangulation points from a 1985 photographic model constructed by the mapping division of Energy, Mines and Resources Canada, and to which all data sets in the study will be georeferenced.

The photogrammetric control points have been generated by a statistical selection process where each control point has been recorded a number of times over a number of models to estimate the mean absolute coordinates of each control point from the 1985 model.

In a similar manner the relative orientation of the 1947 model has been based upon the average model coordinates for each control point. This statistical approach to parallax removal over poorly defined control points is possible on an analytical stereo-plotter but would be greatly facilitated by the addition of a few statistical procedures to the relative orientation algorithm.

Despite the statistical approach to the acquisition and entry of control points on the analytical stereo-plotter the results to date have not been satisfactory. This may be due to the lack of an adequate lens calibration, which the National Air Photo Library of Canada are unable to locate, and which may be on the order of 150 mm. peak for a 6" Ross lens (Hothmer 1957).

Recently, we have been able to construct an accurate model on an analog stereo-plotter with a residual error vector about the control points of 1.5 metres after absolute orientation. This confirms the adequacy of the control points and also shows that regardless of the degree of lens distortion present in the stereo pair, the distortion is not so bad with respect to the film resolution that it may not be removed during the relative orientation process. This is, of course, not the place where lens distortion should be removed but it is the only feasible approach at present.

The relative accuracy between the 1947 and the 1985 stereo models has been verified by a comparison of the thermokarst lakes within the area of interest. Work is underway on a statistical measure of the goodness of fit between the models which will be extended to

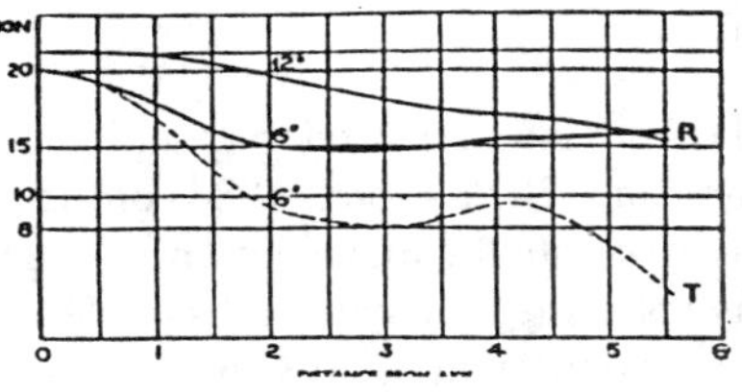

Figure 3 Resolution of the 1947 photography, (Fish 1953).

assess the relative accuracy of the satellite imagery when it is georeferenced.

3.1 Photogrammetric Data Capture

The process of capturing data for the coastline several times has been instructive. Cartographic data capture from maps has been the subject of considerable interest in the GIS literature, (Burrough 1986, Bolstad 1990, Dunn 1990, Gong 1992, and Walsby 1994). We attest that the errors observed during map digitization also occur in digital data capture from a stereo-plotter (figures 4 and 5).

A standard photogrammetric approach to the classification of error is to ensure that the map adheres to a set of criteria describing the accuracy of the product, (CCM, 1989, NATO 1983). This approach is not feasible with aerial photography acquired 47 years ago and for which adequate control points are unavailable. We have therefore adopted an empirical approach in the present work.

4 Meta Data

In a like manner to the capture of control points for the 1947 model we have opted to capture the coastline from each photogrammetric model several times and have employed a simple algorithm to detect and discard outliers while generating a mean coastline from the remaining data (figure 4). The output is a single coastline representing the mean datum.

Standard deviation is used to quantify the epsilon band (Blakemore, 1984) of the points used to generate the mean datum, though strictly speaking, we have not sampled an adequate population of possible coastlines.

Never the less, the term is useful for the discrimination of outliers and has more utility in the present context than distance to mean. The standard deviation for each point comprising the mean coastline is attached as an attribute, (Dutton 1992), in the subsequent GIS coverage.

The present meta data is an incomplete record of the precision of measurement. One has to consider the residual errors about the ground control points incurred during absolute orientation and couple those residuals to the precision of measurement (Chrisman, 1982). We should like to use the statistical measure of the goodness of fit between the thermokarst shorelines in each model to effect this coupling,

A measure of the goodness of fit between the shorelines is a preferred indicator of the accuracy of the orientation error between the two models. This is because the measure is derived from a large distribution of points spread throughout the models. This is in contrast to a measure based upon the control point residuals, which have been targeted by a least squares algorithm, and are neither numerous nor are they widely distributed. We would point out that the photogrammetric process is mathematically rigorous and that normally we would accept the control point residuals as adequate indicators of absolute orientation. In the present case we have introduced uncertainties related to the lens distortion of the 1947 photography and there is a certain fuzziness of location associated with

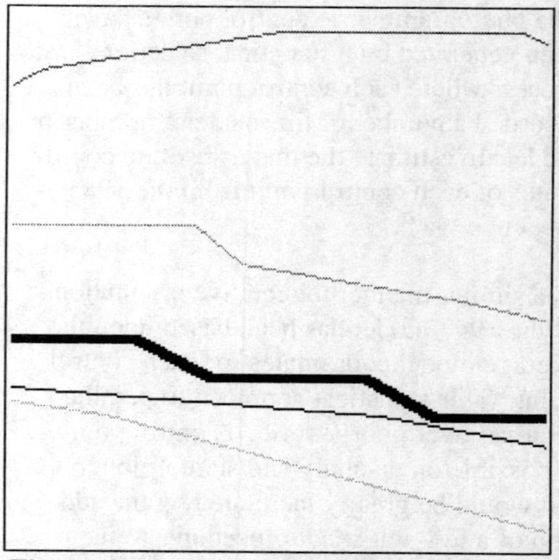

Figure 4 Showing the position of various possible coastlines. the heavy line is the mean datum, the top line is an excluded outlier.

Figure 5 A bi-modal distribution attributable to poor feature contrast.

the ground control points. The fuzziness is due to:

1. poor film resolution and feature contrast.
2. 38 years of geomorphic activity.
3. the absence of signalised control points.

5 Concluding Remarks

We have begun a study of coastal erosion with the intent of assessing the ability of space borne sensors to perform the measurement. Work to date has concentrated on the photogrammetric issues required to derive the ground truth. We have considered the issue of the accuracy of the photogrammetric data capture process and have opted for an empirical approach to categorize the epsilon band of the data.

6 References

Blakemore M., 1984, Generalisation and error in spatial databases. *Cartographica,* 21, pp. 131 - 139.

Bolstad P., Gessler P. and Lillesand T., 1990. Positional uncertainty in manually digitized map data. *Int. J. Geographical Information Systems,* Vol. 4, No. 4, pp. 399 - 412.

Canada Centre for Mapping, 1989 - Map Evaluation Report, Topographical Survey Division, Canada Centre for Mapping, June 6, 1989.

Chrisman, N., 1982. A Theory of Cartographic Error and It's Measurement in Digital Data Bases. *Auto Carto* 5, pp. 159 - 168.

Dunn R., Harrison A.R., White J.C., 1990. Positional accuracy and measurement error in digital databases of land use: an empirical study. *Int. J. Geographical Information Systems,* 1990, Vol. 4, No. 4, pp. 385 - 398.

Dutton G., 1992. Handling Positional Uncertainty in Spatial Databases. *Proceedings of the 5th Int. Symposium on Spatial Data Handling,* Charleston South Carolina, pp. 460 - 469.

Fish R.W., 1953. Lens and Film Resolution *The Photogrammetric Record,* Vol. 1 No.2

French H.M. 1976. the Periglacial Environment. Longman, London.

Gong P., Chen J., 1992. Boundary Uncertainties in Digitized Maps I: Some Possible Determination Methods. *Proceedings GIS/LIS,* pp 274 - 281.

Harry D.G., Dallimore S.R., 1989. Permafrost, Ground Ice and Global Change in the Beaufort Sea Coastlands. *GEOS,* Vol. 18, No. 3. Summer pp. 48 - 53.

Hothmer J., 1959. Possibilities and Limitations for Elimination of Distortion in Aerial Photographs. *The Photogrammetric Record,* Vol. II, No. 12, pp. 426 - 445, Vol. III, No. 13, April 1959, pp. 61 - 78.

Lasserre M., 1989. Errors Derived From Uncertainty in the Accuracy of the Canadian National Topographic Series 1:50000 Maps. *Canadian Journal of Remote Sensing,* Vol. 15, No. 3, pp. 152 - 159.

NATO, 1983, STANAG 2214. Standardization Agreement, Evaluation of Land Maps Edition 4, December 13, 1983.

Walsby J., 1993. The Causes and Effects of Manual Digitising On Error Creation in Data Input to GIS. *Proceedings of the GIS Research UK 1994 Conference,* pp. 321 - 326.

Sensors and Environmental Applications of Remote Sensing, Askne (ed.) © 1995 Balkema, Rotterdam, ISBN 90 5410 524 0

Training in remote sensing of the environment, from research to application: The TELSAT project

M. Ippoliti, L. Lepore & U. Poli
Italian Institute for Occupational Safety and Health (ISPESL), Environmental Department (DIPIA), Rome, Italy

P. Bragatto, N.A. Minciotti & S. Mugnoli
CSCT, Monteporzio Catone, Rome, Italy

ABSTRACT:Industrial hazardous areas must be continuously surveyed to avoid major injuries to the inhabitants and to the natural resources. Scientific research has shown that remote sensing is particularly useful in the study of such areas. At present a successful application is hindered only by the lack of trained personnel. The TELSAT project intended by I.S.P.E.S.L., the scientific department of the Italian Ministry of Health, is now training experts from various scientific background in the control of hazardous areas. Cartography updating, land use and dumping location are studied. Water quality is evaluated on the basis of thermal gradients, plumes and suspended matter. Moreover the state of vegetation is studied to gather indirect information about air quality. A mobile laboratory allows the verificaton "in situ" of pollution parameters.

1. ENVIRONMENTAL PROBLEMS AND HAZARDS FOR INDUSTRIAL DISTRICTS.

Many industrial plants release pollutants into the atmosphere, the sea, the rivers and underground waters. At the same time within these plants major hazards exist and they concern not only employees, but also the health, property and life of the local inhabitants. Moreover industrial activities are often associated with the production of huge quantities of harmful and toxic wastes which are difficult to dispose of and which create further environmental risks.

The concentration of industrial plants in small areas exacerbates these problems, which together create particularly critical situations. These areas are classified as "*hazardous areas*" and often coincide with dense urbanization. They are given particular attention as a greater understanding of the situation may help the proper authorities to avoid further environmental damage and at the same time to mitigate the most dangerous aspects of the problem.

2. REMOTE SENSING AND MONITORING OF INDUSTRIAL AREAS

The Telsat project began with the idea that analysis of aerial and satellite data may be useful in the monitoring of highly hazardous areas. Although image processing cannot substitute data received from direct analysis of the air and water of these zones, it can provide information which is continually updated and which can improve the efficacy of ground level analyses.

2.1 *Environment study by Remote sensing*

The state-of-the-art-technology of remote sensing allows the extraction from aerial and satellite images of exhaustive and up-to-date information about the condition of the territory together with partial or indirect information relating to pollution.

The most suitable parameters for study by remote sensing are coastal waters, soil, vegetation and urban expansion.

As regards *water*, thermal gradients and plumes are detected in portions of sea in front of industrial plants. Furthermore the study of suspended matter is useful in determining the dispersion of pollutants.

As regards *soil,* in highly industrialised areas a systematic control of the evolution of land use is carried out near major hazard plants. Remote sensing particularly helps in the detection of uncontrolled dumping areas and in the study of fires in wood zones near industrial plants.

The survey of the *vegetation* health, carried out by the multitemporal comparison of green indexes, provides indirect information about environmental quality, which is very useful for planning air and water measurements by means of mobile laboratories.

Finally continous *cartographic update* of hazardous areas allows the developments of urban plans according to environmental safety standards.

2.2 *An information system for hazardous areas*

The information collected by remote sensing is always indirect and partial, but very useful both for more effective pollution controls and for better land planning. To this end measurements and processing must be combined in a *Land Information System (LIS)*. This database should also include compulsory security reports supplied by firms to the Ministry. The architecture of the system is shown in fig.1.

The data consist of information about toxic or dangerous substances which are used in the production cycle of almost three thousand Italian plants which are considered to be hazardous. The combination of industrial data, cartographic material, information extracted by remote sensing and direct survey provides the updating of a detailed report on the environment.

Historical evolution of pollution data recorded in the LIS is also analyzed. Adequate modelling is able to forecast the evolution of phenomena through temporal analysis.

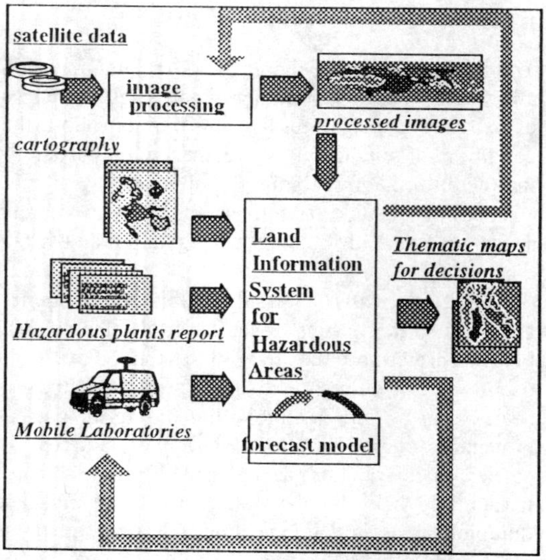

fig.1 Methodology for Hazardous Area Control

3. TELSAT PROJECT

The validity of a systematic monitoring of hazard areas through remote sensing is well known, but the realization of such monitoring is difficult for lack of appropriate personnel. Infact this monitoring must be carried out by a qualified team of experts skilled in very different fields such as hydrogeology, geomorphology, pedology, agronomy, cartography, vegetal biology, chemistry of pollution, hydrodynamics, geoinformatics and land planning.

3.1 *Personnel training*

I.S.P.E.S.L., the scientific organ of the Ministry of Health which is qualified for monitoring of hazard areas, is remeding the lack of personnel by preparing about one hundred young graduate and high school graduate technicians in a 3 year long professional training programme.

Training activity is carried out in the Monteporzio research center near Rome and in the branch office in Manfredonia (Puglia). The Telsat project is training personnel who will be employeed at the facilities of the Ministry of Health and other public administration bodies.

3.2 *Hazardous area monitoring center.*

Together with the professional training a fully equipped monitoring center has been created; at the end of the experimental phase it will go into an effective service using its trained personnel.

The center is equipped with more than 30 work stations dedicated to the processing of remote sensing data. A mobile laboratory for direct analysis has been prepared to verify the information that has been extracted by remote sensing. A fixed chemical laboratory, equipped with two gas cromatographs, a ion cromatograph, an atomic absorption spectrometer and a spectrophotometer, allows a more detailed analysis of samples of fresh and salt water, soil, dust and air collected during the ground surveys.

Thus training in image processing and cartography is completed by the direct ground control of environmental quality, which is the essential step of "*ground truth*" in the adopted procedure.

REFERENCES

R.Graziani, L.L, L.O & U.P "Studio di rilevamento da Satellite LANDSAT-TM di gradienti termici

della superficie del mare, *Scienza Diritto e Economia dell'ambiente n. 1/89*
E. Barnaba, W.R.P&A.W.,1991, "The Use of Aerial Photographs in County Inventories of Waste-Disposal Sites", *Photogrammetric Engineering & Remote Sensing*, vol. 57,n.10, p. 1289-1296
P.A.Brivio,I..D&E.Z,1993, "Aspects of Spatial Autocorrelation of Landsat Data for the Inventory of Waste Disposal Sites *Photogrammetric Engineering & Remote Sensing,* vol. 59, n. 9.
L. Palla e G.C, 1988, "Il Telerilevamento nello studio delle situazioni ambientali a rischio", *Inquinamento* ,n. 12, Dicembre 1988
P.Bragatto, A.B., R.G., E.M., A.N. & E.Z., 1994 "Il Progetto Telsat: Il Telerilevamento per il controllo delle aree a rischio", *Atti Sesto Cenvegno Nazionale A.I.T"*
J.G.Lyon, 1987, "Use of maps, aerial photographs and remote Sensed Data for evaluation of hazardous sites", Photogrammetric Engineering and Remote Sensing. vol 53,n.5 p.515-519.

Author index

Ahlcrona, E. 19
Aiken, J. 289
Albuisson, M. 259
Allewijn, R. 273
Alpers, W. 177
Arbelo, M. 329, 333
Arst, H. 281
Askne, J. 97, 203, 211, 243
Atkinson, P.M. 399

Bartoloni, A. 337, 431
Battista, M. 385
Bauer, O. 349
Bijaoui, J. 311
Borg, C.-G. 15
Boxall, S. 289
Bragatto, P. 481
Brüsch, W. 253
Bullen, B. 475
Büttner, G. 31

Calle, A. 87
Cappellini, V. 457
Carlström, A. 235
Casanova, J.L. 87
Caselles, V. 329
Castagnas, L. 259
Cauneau, F. 259, 311, 349
Cavalieri, D.J. 211
Chan-Chuan Wen 319
Chiuderi, A. 457
Chuvieco, E. 371
Cioni, R. 357
Csillag, F. 393
Cuevas, J.M. 87

Dabrowska-Zielinska, K. 111
D'Amelio, C. 431
De Carolis, G. 439
de Groof, H. 45
Dekker, A.G. 273
Delgado, J.A. 377
Díaz, A. 333
Díaz, J.P. 333
Dierking, W. 235
Domínguez, L. 371
Dowdeswell, J.A. 263

Edner, H. 357
Emiliani, L. 3
Expósito, F.J. 329, 333

Fernández, A. 377
Ferrara, R. 357
Fini, S. 457
Fog, B. 37
Folving, S. 131, 135
Förster, B. 91

González Alonso, F. 87
Gruszczynska, M. 111
Gudmandsen, P. 103, 197

Halounová, L. 463
Harrison, A.R. 399
Hayling, K. 189
Heinzmann, U. 343
Hellsten, H. 447
Herrera, F. 329, 333
Hill, J. 53, 143
Hoover, G. 363
Hunner, G. 91

Igualada, F.J. 23
Illera, P. 87, 377
Ippoliti, M. 385, 481
Israelsson, H. 97

Janowska, M. 111
Jiankang Ji 103
Jory, I.S. 183
Jullien, S. 311

Kahru, M. 315
Karnevi, S. 417
Kennedy, P. 135
Koch, B. 91
Kohl, H.-G. 45
Kokke, J.M.M. 273
Kostka, R. 155
Krämer, W. 155
Kuittinen, R. 407
Kutser, T. 281

Laaksonen, J. 407
Larsson, H. 69
Lepore, L. 385, 481

Mäekivi, S. 281
Mansell, W. 319
Marshall, G.J. 263
Martín, M.P. 371
Martinec, J. 253
Mattei, S. 363
Maucha, G. 31
McCormick, N. 135
Meadows, P.J. 423
Mégier, J. 9, 53, 131
Mehl, W. 53, 143
Michelson, D.B. 123
Milillo, G. 337, 431
Miller, D. 475
Minciotti, N.A. 481
Mochi, M. 337
Morrone, G. 385
Mróz, M. 45
Mugnoli, S. 481

Napolitano, N. 385
Nezry, E. 45
Nirchio, F. 431
Noréus, J. 189
Nyborg, M. 189

Olaizola, M. 305
Olsson, K.B.-E. 79

Panachia, S. 363
Pásztor, L. 393
Pedersen, L.T. 227
Pettersson, M.I. 211
Poli, U. 385, 481
Pons, I. 259
Ponte, S. 439
Posa, F. 439
Poulsen, J.N. 37

Raco, B. 357
Radeloff, V. 143
Ragnarson, P. 357
Rees, W.G. 263
Rosenbaum, M.S. 117

Rud, O. 315

Sandholt, I. 37
Schena, V.D. 363, 439
Schlittenhardt, P.M. 305
Seaquist, J.W. 123
Seidel, K. 253
Skriver, H. 37, 103, 219, 227
Sloggett, D.R. 183, 289
Sommer, S. 53
Somogyi, P. 63
Srokosz, M. 289
Stankiewicz, K. 111
Steinmeier, C. 253
Stjernholm, M. 37
Sun, Y. 243
Svanberg, S. 357
Syrén, P. 75

Taddeucci, G. 357
Thomsen, B.B. 227
Toutin, T. 163

Ulander, L.M.H. 235, 447

van der Linde, D.W. 305
Vasiliev, L. 469
Vaughan, R.A. 319, 475
Vos, R.J. 273

Wallinder, E. 357
Wiedemeier, J. 253
Winkler, P. 63
Wladis, D. 117
Wright, G. 475

Zelei, Gy. 63

Colour plates

Plate 1. Example of photomap sheets produced by FÖMI RSC for CORINE Land Cover Hungary (Büttner & Maucha, page 34).

Figure 1. Classification results (Kohl et al., page 50).

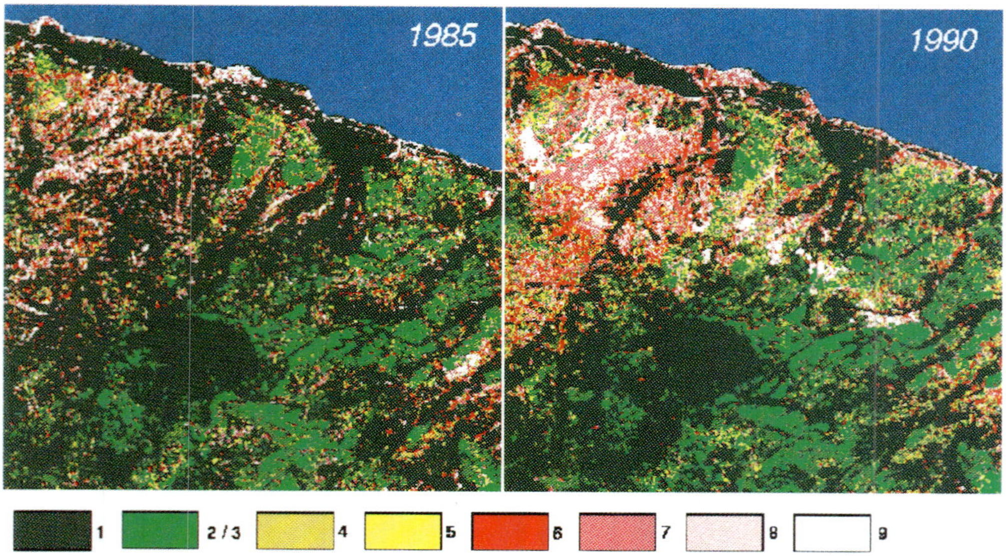

Figure 3. Peloponnese Greece: Comparison of land surface properties based upon a synoptic evaluation of Landsat-derived information layers on soil conditions and vegetation abundance. The numbers refer to 'desertification risk levels' in the sense of varying susceptibility to further degradation, ranking from 'stable' (1: vegetation cover more than 50% and soil condition index I) to 'irreversible degradation' (9: vegetation cover less than 20% and soil condition index IV) (Hill et al., page 57).

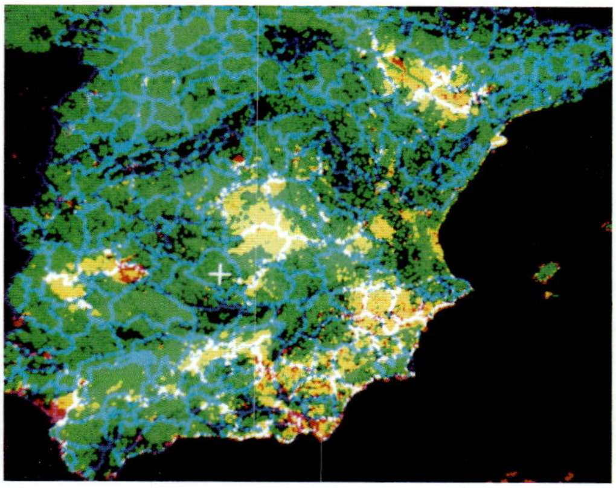

Photograph 1. 1993 drought zones (in yellow) (González Alonso et al., page 89).

Figure 4. Classified radar image overlaped by the border of the forest land out of a forest map. Forest = green; agricultural land = sand; housing areas = black (Koch et al., page 94).

Figure 1. Digital elevation model of the Heiderscheid study area, illuminated according to the conditions during the Landsat satellite overpass from 21 September 1989 (top of page is north) (Hill et al., page 144).

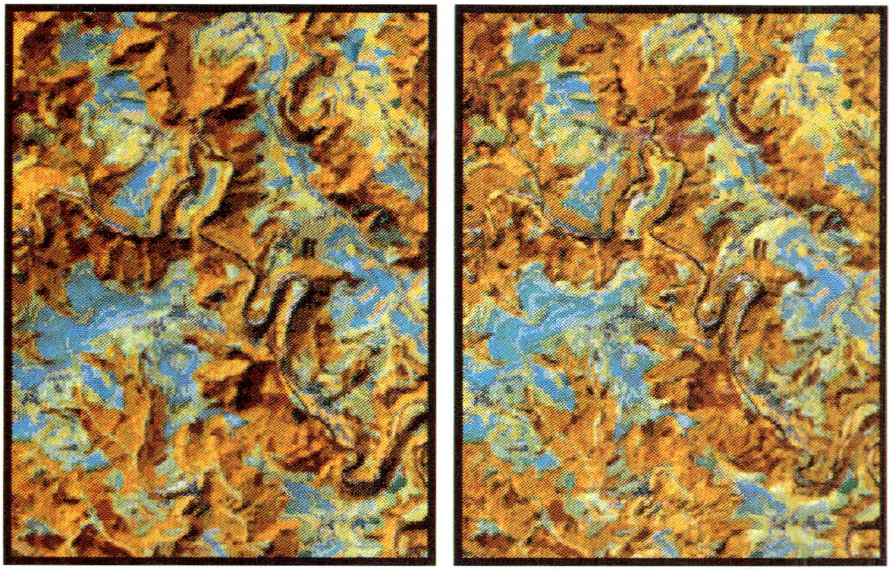

Figure 4. Colour composite of the Landsat image from 20 August 1989 (band combination is 4-5-3 in red-green-blue): atmospheric correction only (left) and atmospheric/topographic correction (right) (Hill et al., page 148).

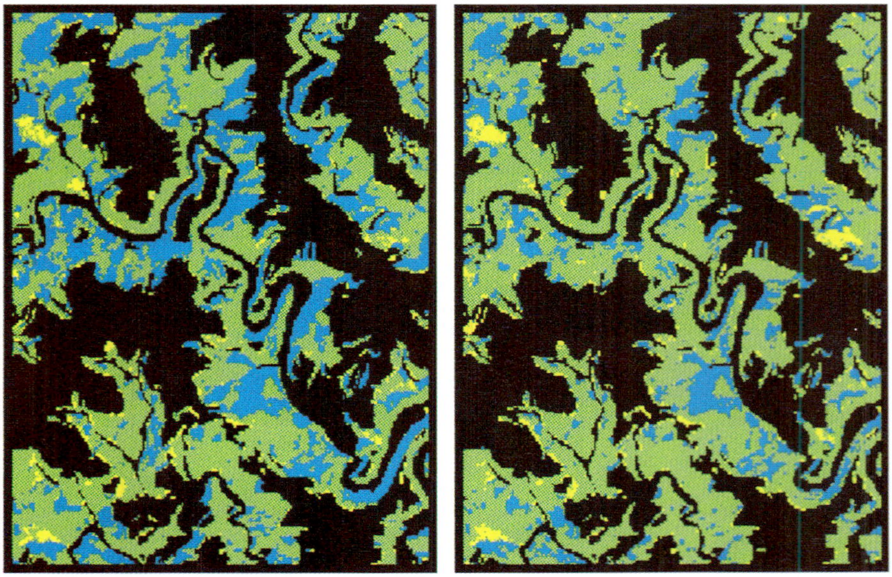

Figure 7. Comparison of forest mapping results for the Landsat TM scene from 20 August 1989, obtained with the neural network classifier before and after topographic corrections (blue – conifers, green – deciduous, yellow – storm damage) (Hill et al., page 150).

Figure 2. Enlargement of the ortho-mosaic (17 × 15 km; 10 m pixel size) (Toutin, page 169).

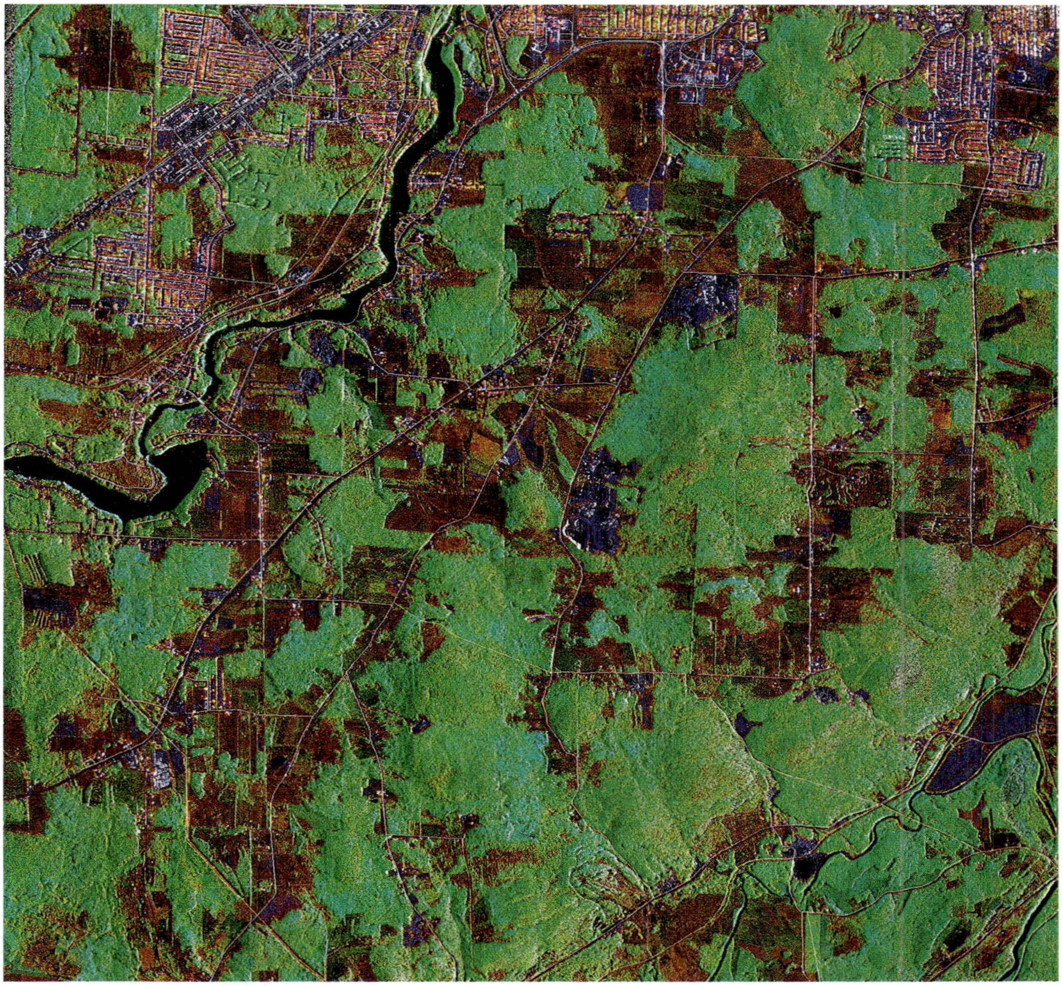

Figure 3. Enlargement (10 × 10 km; 5 m pixel size) of an IHS colour composite SPOT-P and airborne SAR (Toutin, page 170).

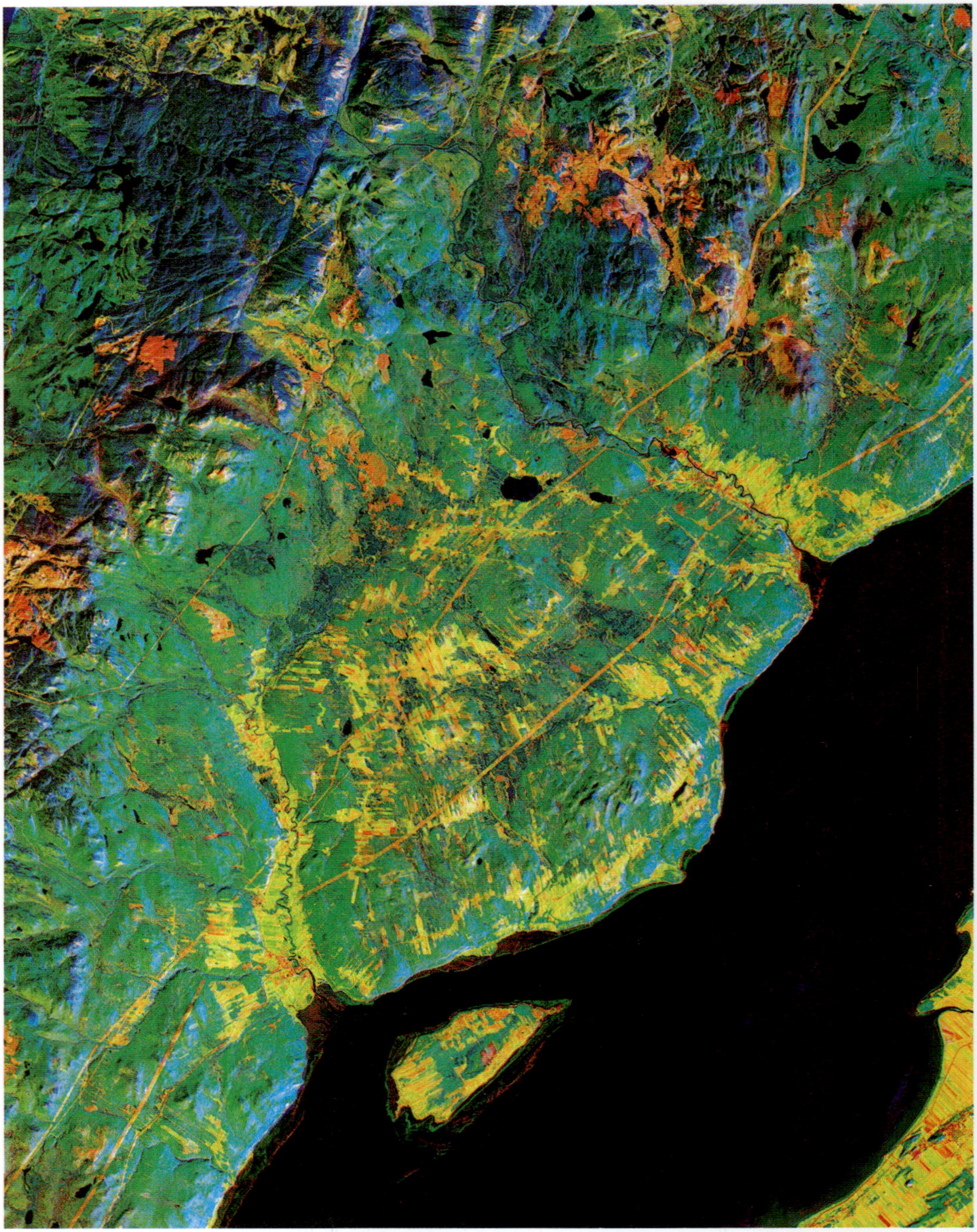

Figure 4. Integration of a RGB colour composite Landsat-TM (7,4) and ERS-1-SAR (67 × 52 km; 30 m pixel size) (Toutin, page 171).

Figure 2.5. CZCS Image of Coccolithophores in the Skagerak on 6 July 1983. Land and cloud boundaries are masked with a white line. In the water green corresponds to a high, yellow to a medium, blue to a low pigment concentration (Courtesy of DLR) (Sloggett et al., page 298).

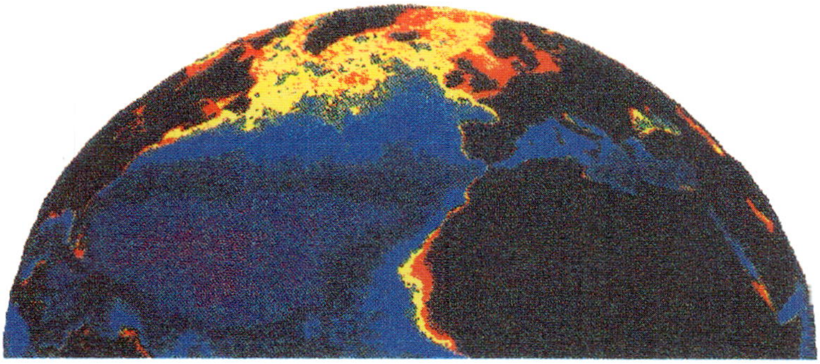

Figure 1. Composite false colour image representing typical relative chlorophyll concentrations for the North Atlantic ocean based on remotely sensed ocean colour. Black = land masses; red = high; yellow = medium; blue = low; and purple = lowest chlorophyll concentrations (Olaizola et al., page 310).

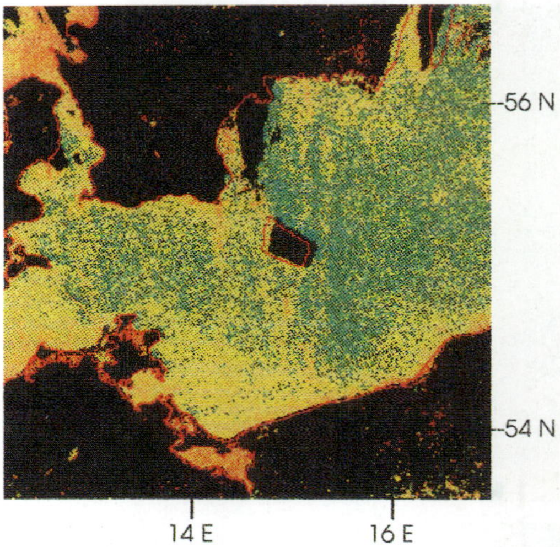

Figure 6. False colour image representing relative chlorophyll concentrations for the Southern Baltic on March 13, 1983, based on remotely sensed ocean colour. Black = land masses; orange = high; yellow = medium; and green = low chlorophyll concentrations (Olaizola et al., page 310).

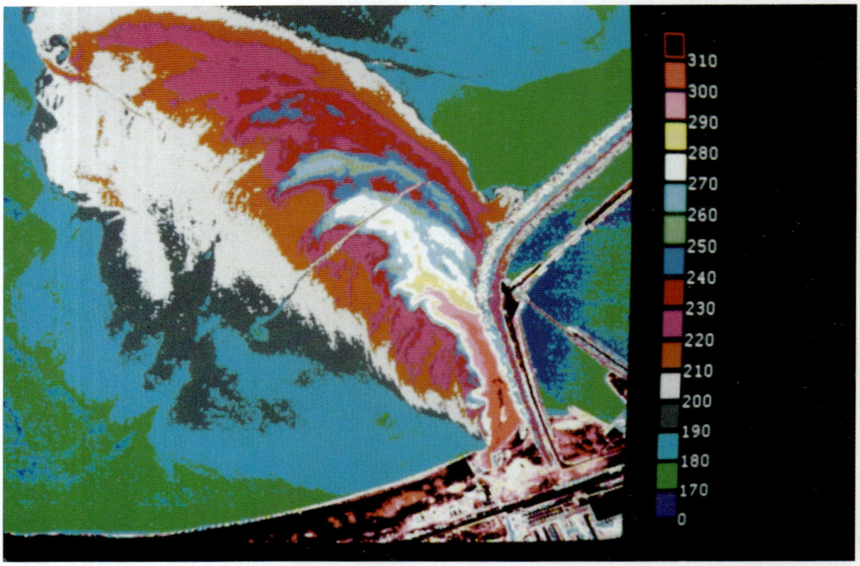

Plate 1. Density sliced image of the thermal plume at 0900h (1° per slice). The plume originates at the bottom right and clings to the wall before dispersing off shore. The sea was calm at this time of day and the prevailing alongshore current was small. Note the track of a boat diagonally across the centre of the image demonstrating poor vertical mixing (Vaughan et al., page 321).

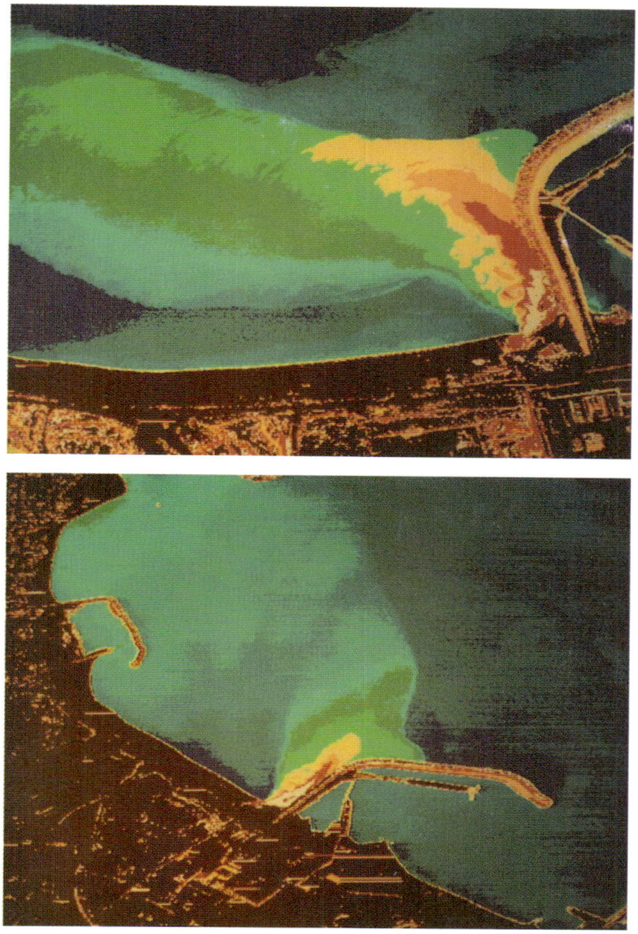

Figure 2. Plate 1 shows the form of the thermal plume at 0900h when the sea was calm. As the day wore on, an increasing on-shore breeze turned the plume northwards along the coast as seen above (upper image, 1030h, lower image, 1330h) (Vaughan et al., page 322).

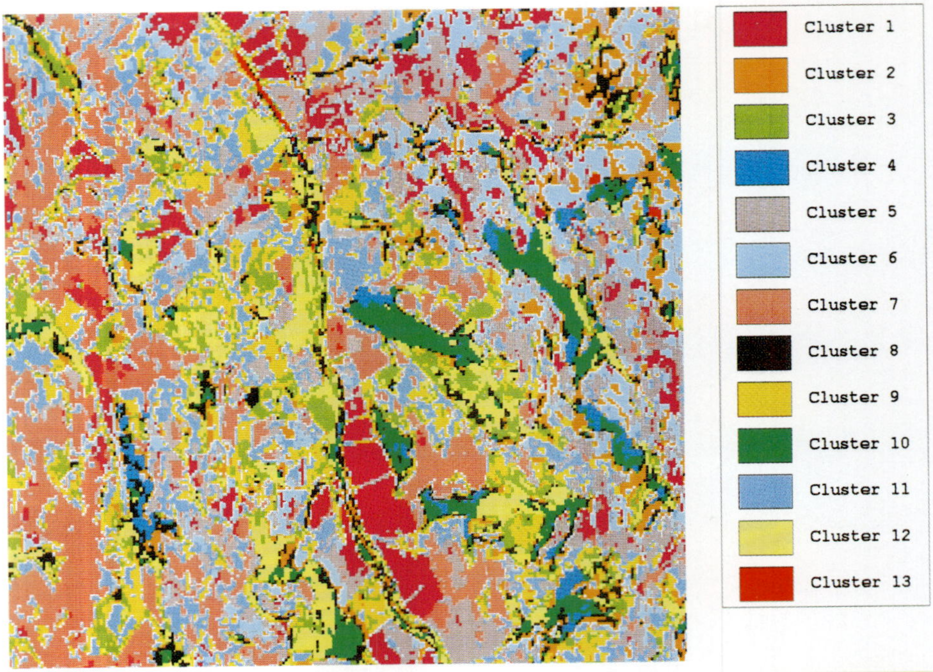

Kohonen Layer
(Cappellini et al., page 458).